T0400335

Underwater Acoustic Modeling and Simulation

Underwater Acoustic Modeling and Simulation

Fifth Edition

Paul C. Etter

CRC Press is an imprint of the
Taylor & Francis Group, an **informa** business

CRC Press
Taylor & Francis Group
6000 Broken Sound Parkway NW, Suite 300
Boca Raton, FL 33487-2742

Printed on acid-free paper

International Standard Book Number-13: 978-1-1380-5492-9 (Hardback)

Library of Congress Cataloging-in-Publication Data

Names: Etter, Paul C., author.
Title: Underwater acoustic modeling and simulation / Paul C. Etter.
Description: Fifth edition. | Boca Raton : CRC Press, Taylor & Francis Group, 2018. | Includes bibliographical references and indexes.
Identifiers: LCCN 2017049571| ISBN 9781138054929 (hardback : alk. paper) | ISBN 9781315166346 (ebook)
Subjects: LCSH: Underwater acoustics–Mathematical models.
Classification: LCC QC242.2 .E88 2018 | DDC 534/.23015118–dc23
LC record available at https://lccn.loc.gov/2017049571

Visit the Taylor & Francis Web site at
http://www.taylorandfrancis.com

and the CRC Press Web site at
http://www.crcpress.com

To my wife Alice
and to my two sons and their families:
Gregory, Sophia, Elaina, and Renae
and
Andrew, Michelle, and Kale

Contents

Preface

This fifth edition of *Underwater Acoustic Modeling and Simulation* addresses advances in the development and utilization of underwater acoustic models since 2013. The organization of material into the 13 chapters of the fourth edition has served well and therefore remains unchanged. Major new developments are described in newly created subsections of the existing chapters. Other existing sections have been enhanced to further elaborate on notable developments. The order of presentation of the first 10 chapters follows the structure suggested by a hierarchical method of sonar model construction. Chapter 1 introduces the types of underwater acoustic models, provides a framework for the consistent classification of modeling techniques, and defines the terminology common to modeling and simulation. Aspects of oceanography essential to an understanding of acoustic phenomena in the ocean are presented in Chapter 2. Chapters 3 through 10 address the observations and models dealing with propagation, noise, and reverberation in the sea. In Chapter 11, the information from Chapters 3 to 10 is integrated into sonar performance models. Chapter 12 describes the process of model evaluation. Chapter 13 discusses the application of simulation in underwater acoustics. In addition, there are four appendices: A—Abbreviations and Acronyms; B—Glossary of Terms; C—Websites; and D–Problem Sets. Finally, References, Author Index, and Subject Index complete the book contents.

Many emerging trends and challenges in applied underwater acoustic modeling have been motivated by marine-mammal protection research focused on the mitigation of naval-sonar, seismic-source, and pile-driving noise, among others. In particular, channel modeling, underwater acoustic networks, and communications technologies have evolved to support the increased bandwidths needed for undersea data collection. Energy-flux models, not traditionally used in naval sonar applications, have proved useful for assessing marine-mammal impacts. Collectively, these trends have added new analytical tools to the existing inventory of propagation, noise, reverberation, and sonar performance models.

Models originally developed for traditional sonar applications have matured and rapidly evolved over the past several years to support a much more diverse community of users. This broadened user community embraces a diverse group of ocean scientists, marine biologists, and engineers who require high-fidelity prognostic and diagnostic tools. Notable developments in inverse sensing include seismic oceanography, which employs low-frequency marine seismic reflection data to image ocean dynamics. Interest in the polar regions has increased due to the well-publicized effects of global warming. Moreover, underwater acoustic models now serve as enabling tools for assessing noise impacts associated with the installation and operation of marine-hydrokinetic energy devices in coastal regions.

As broadly defined in the previous editions, modeling is a method for organizing knowledge accumulated through observations or deduced from underlying principles, while simulation refers to a method for implementing a model over time. The field of underwater acoustic modeling and simulation translates our physical understanding of sound in the sea into mathematical models that can simulate the performance of complex acoustic systems operating in the undersea environment. The core summary tables have been updated with the latest underwater acoustic propagation, noise, reverberation, and sonar performance models. Subsequent to the fourth edition, the inventory of underwater acoustic models has increased by approximately 10%, thus demonstrating a continued expansion of R&D efforts. Underwater acoustic models thus appear as proxy indicators of these R&D investments.

This book discusses the fundamental processes involved in simulating the performance of underwater acoustic systems and emphasizes the importance of applying the proper modeling resources to simulate the behavior of sound in virtual ocean environments. Guidelines for selecting and using these various models are highlighted. Specific examples of each type of model are discussed to illustrate model formulations, assumptions, and algorithm efficiency. Instructive case studies in simulation are included to demonstrate practical applications. The problem sets in Appendix D are intended to reinforce an understanding of these applications.

The term *soundscape* continues to appear frequently in the scientific and technical literature relevant to underwater acoustic modeling. A soundscape is a combination of sounds that characterize, or arise from, an ocean environment. The study of a soundscape is sometimes referred to as acoustic ecology. Recent observations have indicated that the ocean soundscape has been changing due to anthropogenic activity (e.g. naval-sonar systems, seismic-exploration activity, maritime shipping, and windfarm development) as well as natural factors (e.g., climate change and ocean acidification). Disruption of the natural acoustic environment results in noise pollution. In response to these developments, new regulatory initiatives have placed additional restrictions on uses of sound in the ocean; mitigation of marine-mammal endangerment is now an integral consideration in acoustic-system design and operation. Modeling tools traditionally used in underwater acoustics have undergone a necessary transformation to respond to the rapidly changing requirements imposed by this new ocean soundscape.

As stated in the preface to the first edition, this book is intended for those who have a fundamental understanding of underwater acoustics but who are not familiar with the various aspects of modeling. Sufficient mathematical derivations are included to demonstrate model formulations, and guidelines are provided to assist in the selection and proper application of these models. Comprehensive summaries identify the available models and associated documentation. Where available, links to appropriate websites have been provided. The level of technical detail presented in this book is appropriate for a broad spectrum of practitioners and students in sonar technology, acoustical oceanography, marine engineering, naval operations analysis, systems engineering, and applied mathematics.

As in the second, third, and fourth editions, I have retained descriptions of earlier developments (including the older models) to provide a historical account of the progress that has been achieved over the cumulative period of record covered by these various editions. I trust that this new edition will continue to serve as a useful source of information for all those engaged in modeling and simulation in underwater acoustics.

I have continued to teach short courses with the Applied Technology Institute (Riva, Maryland, USA) on the topic of underwater acoustics using this book as the principal text. The critical feedback from my students has always been encouraging and enlightening. Recent book reviews have provided additional feedback useful in the improvement of this latest edition.

As a way to explain the motivation for my long-term involvement in underwater acoustic modeling, I often relate the following story to the students who attend my short courses. My interest in this field had its genesis during my period of active duty in the US Navy (1969–1973). During that time, I served as anti-submarine warfare (ASW) officer aboard frigates (then called destroyer escorts) where I was responsible for the tactical operation of the ships' sonar systems. In that capacity, I managed well-educated teams of sonar technicians who maintained and operated the sonar systems. Since sonar performance varies with ocean location, time of day, and time of year, all sonar-equipped ships received daily forecast messages from shore-based prediction facilities. (At that time, there were no modern equivalents of on-board prediction capabilities.) These forecast messages contained model-generated sonar ranges based on the expected environmental-acoustic conditions for that particular location and time. However, the sonar ranges predicted in these messages did not always agree with those obtained at sea. When there were divergences, the sonar ranges predicted by the models appeared to be more optimistic. Since the ships' commanding officers were also aware of these messages, any divergence between our sonar performance and the model prediction was often interpreted as a deficiency in operator performance. I thought that there must be an explanation for such divergences; however, not being well versed in the art of acoustic modeling at that time, I was ill prepared to offer convincing explanations to the commanding officers. Upon release from active duty, I completed my long-planned (but temporarily deferred) graduate work in physical oceanography. In my first position in private industry, I was able to revisit this issue of underwater acoustic modeling. My original intention was to compare model predictions against observed sonar performance and thus vindicate the performance of my former sonar technicians. My sponsor in the US Navy, however, felt that I should first identify and review the available modeling assets before conducting any such comparisons, particularly since there was no real understanding of what models already existed. The resulting review (first published in 1978) was both widely promulgated and well received within the naval sonar modeling community. This review provided the community with a comprehensive accounting of the modeling efforts being conducted by government, academic, and industrial research laboratories. My subsequent work has built upon that original review of modeling efforts, but with an expanded scope now including international efforts. Comparisons of sonar performance against model predictions have since been undertaken by a number of evaluation groups, some of which are discussed in Chapter 12 (Model Evaluation).

Paul C. Etter
Odenton, Maryland

Preface to the Fourth Edition

Ideas and observations concerning underwater sound have continued to accumulate over the past decade and have found increasing relevance in the marine sciences. This trend has encouraged the further expansion of this book into a fourth edition to embrace these new developments, especially where they were found to stimulate advancements in the field of underwater acoustic modeling and simulation.

Broadly defined, modeling is a method for organizing knowledge accumulated through observation or deduced from underlying principles while simulation refers to a method for implementing a model over time. The field of underwater acoustic modeling and simulation translates our physical understanding of sound in the sea into mathematical models that can simulate the performance of complex acoustic systems operating in the undersea environment.

This book discusses the fundamental processes involved in simulating the performance of underwater acoustic systems and emphasizes the importance of applying the proper modeling resources to simulate the behavior of sound in virtual ocean environments. Guidelines for selecting and using the various models are highlighted. Specific examples of each type of model are discussed to illustrate model formulations, assumptions, and algorithm efficiency. Instructive case studies in simulation are included to demonstrate practical applications.

The material in this fourth edition is organized into 13 chapters. The order of presentation of the first 10 chapters follows the structure suggested by a hierarchical method of sonar model construction. Chapter 1 introduces the types of underwater acoustic models, provides a framework for the consistent classification of modeling techniques, and defines the terminology common to modeling and simulation. Aspects of oceanography essential to an understanding of acoustic phenomena are presented in Chapter 2. Chapters 3 through 10 address the observations and models dealing with propagation, noise, and reverberation in the sea. A new Chapter 6 on *Special Applications and Inverse Techniques* has been added; this material had originally been part of Chapter 5 in previous editions, but as the volume of material increased, it became necessary to create a stand-alone chapter. In Chapter 11, the information from Chapters 3 to 10 is integrated into sonar performance models. Chapter 12 describes the process of model evaluation. Chapter 13 discusses the application of simulation in underwater acoustics. Since simulation is a method for implementing a model over time, it is fitting that this topic is addressed only after a firm foundation of modeling and evaluation has been established. There are four appendices: A—Abbreviations and Acronyms; B—Glossary of Terms; C—Websites; and D—Problem Sets. Appendix D, which is new to this fourth edition, is intended to assist students and instructors in assimilating the information contained in this fourth edition. Finally, References, Author Index, and Subject Index complete the book contents.

The term *soundscape* has appeared frequently in the recent scientific and technical literature. A soundscape is a combination of sounds that characterize, or arise from, an ocean environment. The study of a soundscape is sometimes referred to as acoustic ecology. Recent observations have indicated that the ocean soundscape has been changing due to anthropogenic activity (e.g. naval-sonar systems, seismic-exploration activity, maritime shipping, and windfarm development) and natural factors (e.g., climate change and ocean acidification). Disruption of the natural acoustic environment results in noise pollution. In response to these developments, new regulatory initiatives have placed additional restrictions on uses of sound in the ocean; mitigation of marine-mammal endangerment is now an integral consideration in acoustic-system design and operation. Modeling tools traditionally used in underwater acoustics have undergone a necessary transformation to respond to the rapidly changing requirements imposed by this new ocean soundscape.

As stated in the preface to the first edition, this book is intended for those who have a fundamental understanding of underwater acoustics but who are not familiar with the various aspects of modeling. Sufficient mathematical derivations are included to demonstrate model formulations, and guidelines are provided to assist in the selection and proper application of these models. Comprehensive summaries identify the available models and associated documentation. The level of technical detail presented in this book is appropriate for a broad spectrum of practitioners and students in sonar technology, acoustical oceanography, naval operations analysis, systems engineering, and applied mathematics.

As in the second and third editions, I have retained descriptions of earlier developments (including the older models) to provide a historical account of the progress that has been achieved over the cumulative period of record covered by these four editions. I trust that this new edition will continue to serve as a useful source of information for all those engaged in modeling and simulation in underwater acoustics.

I have continued to teach short courses with the Applied Technology Institute (Riva, Maryland, USA) on the topic of underwater acoustics using this book as the principal text. The critical feedback from my students has always been encouraging and enlightening. In addition, numerous book reviews have provided constructive feedback useful in the improvement of this latest edition.

Paul C. Etter
Rockville, Maryland

Preface to the Third Edition

Broadly defined, modeling is a method for organizing knowledge accumulated through observation or deduced from underlying principles while simulation refers to a method for implementing a model over time. The field of underwater acoustic modeling and simulation translates our physical understanding of sound in the sea into mathematical models that can simulate the performance of complex acoustic systems operating in the undersea environment.

This book discusses the fundamental processes involved in simulating underwater acoustic systems and emphasizes the importance of applying the proper modeling resources to simulate the behavior of sound in virtual ocean environments. Summary tables identify available propagation, noise, reverberation, and sonar performance models. Guidelines for selecting and using these various models are highlighted. Specific examples of each type of model are discussed to illustrate model formulations, assumptions, and algorithm efficiency. Instructive case studies in simulation are reviewed to demonstrate practical applications.

Over the past decade, rapid changes in the world situation have opened new avenues for international collaboration in modeling and simulation. Concurrent advances in electronic communications have greatly facilitated the transfer of modeling and simulation technologies among members of the international community. The Internet now provides unprecedented access to models and databases around the world. Where appropriate, references to pertinent websites are incorporated in this edition.

The level of technical detail presented in this book is appropriate for a broad spectrum of practitioners and students in sonar technology, acoustical oceanography, naval operations analysis, systems engineering, and applied mathematics. The material is organized into 12 chapters. The order of presentation of the first 10 chapters follows the structure suggested by a hierarchical method of sonar model construction. Chapter 1 introduces the types of underwater acoustic models, provides a framework for the consistent classification of modeling techniques, and defines the terminology common to modeling and simulation. Aspects of oceanography essential to an understanding of acoustic phenomena are presented in Chapter 2. Chapters 3 through 9 address the observations and models dealing with propagation, noise, and reverberation in the sea. In Chapter 10, the information from Chapters 3 to 9 is integrated into sonar performance models. Chapter 11 describes the process of model evaluation. Chapter 12, which is new to this edition, discusses the application of simulation in underwater acoustics. Since simulation is a method for implementing a model over time, it is fitting that this topic is addressed only after a firm foundation of modeling and evaluation has been established. The title of this edition has been changed to *Underwater Acoustic Modeling and Simulation* to reflect the inclusion of material on simulation.

Rather than purging older material from the third edition, I have intentionally retained descriptions of earlier developments (including the older models) to provide a historical account of the progress that has been achieved over the cumulative period of record covered by these three editions. I trust that this new edition will continue to serve as a useful source of information for all those engaged in modeling and simulation in underwater acoustics.

Paul C. Etter
Rockville, Maryland

Preface to the Second Edition

The subject of underwater acoustic modeling deals with the translation of our physical understanding of sound in the sea into mathematical formulas solvable by computers. This book divides the subject of underwater acoustic modeling into three fundamental aspects: the physical principles used to formulate underwater acoustic models; the mathematical techniques used to translate these principles into computer models; and modeling applications in sonar technology and oceanographic research.

The material presented here emphasizes aspects of the ocean as an acoustic medium. It shows mathematicians and physical scientists how to use this information to model the behavior of sound in a spatially complex and temporally variable ocean. This approach diminishes the need for discussions of engineering issues such as transducers, arrays, and targets. Aspects of hardware design and modeling in underwater acoustics are discussed in other excellent texts.

Recent developments in underwater acoustic modeling have been influenced by changes in global geopolitics. These changes are evidenced by strategic shifts in military priorities as well as by efforts to transfer defense technologies to nondefense applications.

The strategic shift in emphasis from deep-water to shallow-water naval operations has focused attention on improving sonar performance in coastal regions. These near-shore regions, which are sometimes referred to as the littoral zone, are characterized by complicated and highly variable acoustic environments. Such difficult environments challenge the abilities of those sonar models intended for use in deep-water scenarios. This situation has prompted further development of underwater acoustic models suitable for forecasting and analyzing sonar performance in shallow-water areas.

The policy of defense conversion has encouraged the transfer of sonar modeling technology to nondefense applications. Much of this transfer has benefited the growing field of environmental acoustics, which seeks to expand exploration of the oceans through acoustic sensing. Such technology conversion is exemplified by the utilization of naval underwater acoustic models as both prognostic and diagnostic tools in sophisticated experiments employing inverse acoustic sensing of the oceans.

These rapid developments in modeling have created a need for a second edition. The intent is to update recent advances in underwater acoustic modeling and to emphasize new applications in oceanographic research. This edition also reflects a broader international interest in the development and application of underwater acoustic models. The coming years promise to be challenging in terms of defining research directions, whether for defense or industry, and this edition should provide technology planners with a useful baseline.

The original organization of material into 11 chapters has served well and therefore remains unchanged. When required, new material has been arranged into additional subsections.

Comments from users of the first edition have evidenced appeal from acousticians, as well as oceanographers, who have enthusiastically endorsed this book as both a practical tool and an instructional aid. In this latter regard, several academic institutions have utilized this book as an adjunct text for graduate-level courses in applied mathematics and ocean sciences.

This edition has benefited from a continuation of my short courses which, since 1993, have been offered through the Applied Technology Institute (Clarksville, Maryland, USA). Continued exposure to the insightful questions posed by my students has provided me with the opportunity to further refine my presentation.

Despite the appearance of several new books in the field of ocean acoustics, this book remains unique in its treatment and coverage of underwater acoustic modeling. It is a pleasure to note that the first edition has been recognized as an authoritative compendium of state-of-the-art models and is often cited as the standard reference.

Paul C. Etter
Rockville, Maryland

Preface to the First Edition

The subject of underwater acoustic modeling deals with the translation of our physical understanding of sound in the sea into mathematical formulas solvable by computers. These models are useful in a variety of research and operational applications including undersea defense and marine seismology. There has been a phenomenal growth in both the number and types of models developed over the past several decades. This growth reflects the widespread use of models for the solution of practical problems as well as the considerable advances made in our computational abilities.

The primary motivation for the development of underwater acoustic models is defense related. Researchers involved in anti-submarine warfare (ASW) and associated undersea defense disciplines use models to interpret and forecast acoustic conditions in the sea in support of sonar design and sonar operation. Consequently, the emphasis in this book is placed on those models that are particularly useful in solving sonar performance problems.

Users and potential users of models are commonly ill acquainted with model formulations. As a result, the capabilities and limitations of the models are poorly understood and the models are often improperly used. Moreover, the sheer number of available models complicates the process of model selection.

This book is intended for those who have a fundamental understanding of underwater acoustics but who are not familiar with the various aspects of modeling. Sufficient mathematical derivations are included to demonstrate model formulations, and guidelines are provided to assist in the selection and proper application of these models. Comprehensive summaries identify the available models and associated documentation.

The material is organized into 11 chapters. The order of presentation follows the structure suggested by a hierarchical method of sonar model construction. Chapter 1 introduces the types of underwater acoustic models, provides a framework for the consistent classification of modeling techniques, and defines the terminology common to modeling work. Aspects of oceanography essential to an understanding of acoustic phenomena are presented in Chapter 2. Chapters 3 through 9 address the observations and models dealing with propagation, noise, and reverberation in the sea. In Chapter 10, the information from Chapters 3–9 is integrated into sonar performance models. Finally, Chapter 11 describes the process of model evaluation.

Since 1982, I have developed and taught a series of intensive short courses for the Technology Service Corporation (Silver Spring, Maryland, USA). Earlier versions of this course were taught in collaboration with Professor Robert J. Urick of the Catholic University of America. Professor Urick would discuss underwater acoustic measurements while I would review the related modeling techniques. As the course evolved into one in which I became the sole instructor, I borrowed heavily from Professor Urick's several books (with permission) in order to preserve the continuity of the course material. The success of this course encouraged me to publish my class notes as a book.

Many notable books have been published in the field of underwater acoustics. None, however, has dealt exclusively with modern developments in modeling, although some have addressed aspects of propagation modeling. This book is unique in that it treats the entire spectrum of underwater acoustic modeling including environmental, propagation, noise, reverberation, and sonar performance models.

I have intentionally preserved the notation, terminology, and formalism used by those researchers whose work I have cited. I have also intentionally emphasized aspects of oceanography since my experience has indicated that many acousticians have little appreciation for the complex role played by the ocean as an acoustic medium. Conversely, oceanographers frequently fail to appreciate the great potential of underwater acoustics as a remote sensing technique.

Paul C. Etter
Rockville, Maryland

Acknowledgments

The students who have attended my short courses over the past 35 years have provided both a receptive and critical audience for much of the material contained in this book. Many of my colleagues have provided useful insights and suggestions. In particular, I want to recognize Dr. Michael A. Ainslie, Dr. Aubrey L. Anderson,* Dr. Stanley A. Chin-Bing, Dr. Richard B. Evans, Dr. Robert W. Farwell, Dr. Richard P. Flanagan, Dr. Charles W. Holland, Dr. Robert L. Martin,* Dr. Peter M. Ogden, Dr. Frederick D. Tappert,* and Dr. Henry Weinberg. Robert S. Winokur provided administrative guidance in the early stages of my work in underwater acoustic modeling.

Professor Robert J. Urick* provided much encouragement and graciously allowed me to liberally borrow material from his several books. Professor John D. Cochrane* of Texas A&M University inspired the scholarly discipline that facilitated creation of this book.

* Deceased.

Author

Paul C. Etter has worked in the fields of ocean-atmosphere physics and environmental acoustics for the past 45 years while employed by federal and state agencies, academia, and private industry. He received his BS degree in physics and his MS degree in oceanography at Texas A&M University. Etter served on active duty in the US Navy as an anti-submarine warfare officer aboard frigates, where he was responsible for the tactical operation of the ships' sonar systems. He is the author or coauthor of more than 200 technical reports, professional papers, and books addressing environmental measurement technology, underwater acoustics, and physical oceanography.

1 Introduction

1.1 BACKGROUND

1.1.1 Setting

Underwater acoustics entails the development and employment of acoustical methods to image underwater features, to communicate information via the oceanic waveguide, or to measure oceanic properties. In its most fundamental sense, *modeling* is a method for organizing knowledge accumulated through observation or deduced from underlying principles. *Simulation* refers to a method for implementing a model over time.

Historically, sonar technologists initiated the development of underwater acoustic modeling to improve sonar system design and evaluation efforts, principally in support of naval operations. Moreover, these models were used to train sonar operators, assess fleet requirements, predict sonar performance, and develop new tactics. Despite the restrictiveness of military security, an extensive body of relevant research accumulated in the open literature, and much of this literature addressed the development and refinement of numerical codes that modeled the ocean as an acoustic medium. This situation stimulated the formation of a new subdiscipline known as *computational ocean acoustics*. Representative developments in computational ocean acoustics have been documented by Merklinger (1987), Lee et al. (1990a,b,c, 1993), Lau et al. (1993), Murphy and Chin-Bing (2002), and Jensen (2008).

As these modeling technologies matured and migrated into the public domain, private industry was able to apply many aspects of this pioneering work. Subsequently, there has been much cross-fertilization between the geophysical-exploration and the sonar-technology fields as the operating frequencies of both fields began to converge. Recently, acoustical oceanographers have employed underwater-acoustic models as adjunct tools for inverse-sensing techniques (see Section 1.6) that can be used to obtain synoptic portraitures of large ocean areas or to monitor long-term variations in the ocean.

Underwater acoustic models are now routinely used to forecast acoustic conditions for planning at-sea experiments, designing optimized sonar systems, and predicting sonar performance at sea. Modeling has become the chief mechanism by which researchers and analysts can simulate sonar performance under laboratory conditions. Modeling provides an efficient means by which to parametrically investigate the performance of hypothetical sonar designs under varied environmental conditions as well as to estimate the performance of existing sonars in different ocean areas and seasons.

1.1.2 Framework

A distinction is made between physical (or "physics-based") models and mathematical models, both of which are addressed in this book. Physical models pertain to theoretical or conceptual representations of the physical processes occurring within the ocean; the term "analytical model" is sometimes used synonymously. Mathematical models include both empirical models (those based on observations) and numerical models (those based on mathematical representations of the governing physics). The subject of analog modeling, which is defined here as controlled acoustic experimentation in water tanks employing appropriate oceanic scaling factors, is not specifically addressed in this book. Barkhatov (1968) and Zornig (1979) have presented detailed reviews of acoustic analog modeling.

The physical models underlying the numerical models have been well known for some time. Nevertheless, the transition to operational computer models has been hampered by several factors: limitations in computer capabilities, inefficient mathematical methods, and inadequate oceanographic and acoustic data with which to initialize and evaluate models. Despite continuing advances in computational power, together with the development of more efficient mathematical methods and the dramatic growth in databases, the emergence of increasingly complex and sophisticated models continues to challenge available resources.

This book addresses three broad types of underwater acoustic models: *environmental models*, *basic acoustic models*, and *sonar performance models*.

The first category—environmental models—includes empirical algorithms that are used to quantify the boundary conditions (surface and bottom) and volumetric effects of the ocean environment. Such models include, for example, sound speed, absorption coefficients, surface and bottom reflection losses, and surface, bottom, and volume backscattering strengths.

The second category—basic acoustic models—comprises propagation (transmission loss), noise, and reverberation models. This category is the primary focus of attention in this book.

The third category—sonar performance models—is composed of environmental models, basic acoustic models, and appropriate signal-processing models. Sonar performance models are organized to solve specific sonar-applications problems such as submarine detection, mine hunting, torpedo homing, and bathymetric sounding.

Figure 1.1 illustrates the relationships among these three broad categories of models. As the applications become more and more system specific (i.e., as one progresses from environmental models toward sonar performance models), the respective models become less universal in application. This is a consequence of the fact that system-specific characteristics embedded in the higher-level models (for example, signal-processing models) restrict their utility to particular sonar systems. Thus, while one propagation model may enjoy a wide variety of applications, any particular sonar performance model is, by design, limited to a relatively small class of well-defined sonar problems.

At the base of the pyramid in Figure 1.1 are the *environmental models*. These are largely empirical algorithms that describe the boundaries of the ocean (surface and bottom) and the water column. The *surface* description quantifies the state of the sea

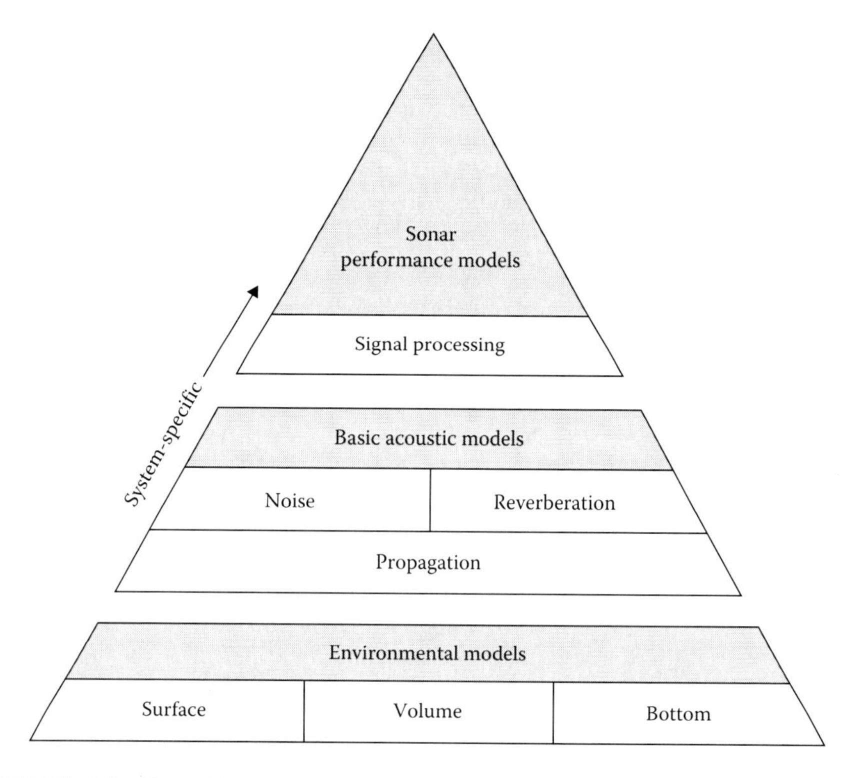

FIGURE 1.1 Generalized relationships among environmental models, basic acoustic models, and sonar performance models.

surface including wind speed, wave height, and bubble content of the near-surface waters. If ice covered, descriptions of the ice thickness and roughness would also be required. Surface-reflection coefficients and surface-scattering strengths are needed to model propagation and reverberation. The *bottom* description includes the composition, roughness, and sediment-layering structure of importance to acoustic interactions at the sea floor. Bottom-reflection coefficients and bottom-scattering strengths are needed to model propagation and reverberation. The *volume* description entails the distribution of temperature, salinity and sound speed, absorption, and relevant biological activity. Volume scattering strengths are needed to model reverberation. Once the marine environment is adequately described in terms of location, time and frequency (spatial, temporal and spectral dependencies), the *basic acoustic models* can be initialized. In order to proceed higher up in the pyramid, it is first necessary to generate estimates of acoustic *propagation*. If a passive sonar system is being modeled, it is necessary to understand the propagation of sound as it is radiated from the target toward the sonar receiver. Furthermore, any interfering noises must be propagated from their source to the sonar receiver. The behavior of the noise sources can be quantified using *noise* models, which must include a propagation component. If an active sonar system is being modeled, the contribution of reverberation must be considered in addition to those factors already mentioned above with regard to noise sources. Moreover, since the active sonar transmits a signal, sound must be propagated out to the target and back. Also, the transmitted pulse will be scattered and returned by particulate matter in the ocean volume. *Reverberation* models quantify the effects of scatterers on the incident pulse and its subsequent scattering and propagation back to the sonar receiver and, therefore, must include a propagation component. Finally, the outputs of the environmental models and basic acoustic models must be combined with *sonar performance models*, in conjunction with appropriate passive or active sonar *signal processing* models. In concert, these models generate metrics useful in predicting and assessing the performance of passive or active sonars in different ocean environments and seasons. The inclusion of sonar system and target parameters is not explicitly discussed at this level but will be treated in Section 11.4.2.

The wide breadth of material covered in this book precludes exhaustive discussions of all existing underwater acoustic models. Accordingly, only selected models considered to be representative of each of the three broad categories will be explored in more detail. However, comprehensive summary tables identify all known basic acoustic models and sonar performance models. These tables also contain brief technical descriptions of each model together with pertinent references to the literature. Notable environmental models are identified and discussed in appropriate sections throughout this book.

Modeling applications will generally fall into one of two basic areas: research or operational. Research-oriented applications are conducted in laboratory environments where accuracy is important and computer time is not a critical factor. Examples of research applications include sonar-system design and field-experiment planning. Operationally oriented applications are conducted as field activities, including fleet operations at sea and sonar-system training ashore. Operational applications generally require rapid execution, often under demanding conditions; moreover, modeling accuracy may be subordinate to processing speed.

1.2 MEASUREMENTS AND PREDICTION

The scientific discipline of underwater acoustics has been undergoing a long transition from an observation phase to a phase of understanding and prediction. This transition has not always been smooth: direct observations have traditionally been limited, the resulting prediction tools (models) were not always perfected, and much refinement remains to be completed.

Experimental measurements in the physical sciences are generally expensive due to instrumentation and facility-operation costs. In the case of oceanographic and underwater acoustic data collection, this is particularly true because of the high costs of platform operation (ships, aircraft, submarines, and satellites). Acoustic datasets obtained at sea are limited by their inherent spatial, temporal and spectral dimensions. Consequently, in the field of underwater acoustics, much use is made of what field measurements already exist. Notable large-scale field programs that have been conducted successfully in the past include AMOS (Acoustic, Meteorological, and Oceanographic Survey) and LRAPP (Long Range Acoustic Propagation Project). More recent examples include ATOC (Acoustic Thermometry of Ocean Climate) and other basin-scale tomographic experiments.

The National Science Foundation's ocean observatories initiative (OOI) is a new program designed to implement global, regional and coastal-scale observatory networks (along with cyberinfrastructure) as research tools to provide long-term and real-time access to the ocean. Three acoustic-measurement technologies were proposed for incorporation (Duda et al., 2007): long-range positioning and navigation, thermometry, and passive listening. Passive-listening techniques can be used for marine-mammal and fishery studies, for monitoring seismic activity, for quantifying wind and rain, and for monitoring anthropogenic noise sources.

According to the Office of Naval Research (US Department of the Navy, 2007), ocean acoustics is considered to be a US national naval responsibility in science and technology. The three prime areas of interest are shallow-water acoustics, high-frequency acoustics, and deep-water acoustics.

- Shallow-water acoustics is concerned with the propagation and scattering of low-frequency (10 Hz to ~3 kHz) acoustic energy in shallow-water ocean environments. Specific components of interest include: scattering mechanisms related to reverberation and clutter; seabed acoustic measurements supporting geoacoustic inversion; acoustic propagation through internal waves and coastal ocean processes; and the development of unified ocean-seabed–acoustic models.
- High-frequency acoustics is concerned with the interaction of high-frequency (~3 kHz to 1000 kHz) sound with the ocean environment, with a view toward mitigation or exploitation of the interactions in acoustic detection, classification, and communication systems. Specific components of interest include: propagation of sound through an intervening turbulent or stochastic medium; scattering from rough surfaces, biologics, and bubbles; and penetration and propagation within the porous seafloor.
- Deep-water acoustics is concerned with low-frequency acoustic propagation, scattering and communication over distances from tens to thousands of kilometers in the deep ocean where the sound channel may, or may not, be bottom limited. Specific components of interest include effects of environmental variability induced by oceanic internal waves, internal tides, and mesoscale processes; effects of bathymetric features such as seamounts and ridges on the stability, statistics, spatial distribution, and predictability of broadband acoustic signals; and the coherence and depth dependence of deep-water ambient noise.

Modeling has been used extensively to advance scientific understanding without expending scarce resources on additional field observations. The balance between observations and modeling, however, is very delicate. Experimenters agree that modeling may help to build intuition or refine calculations, but they argue further that only field observations give rise to genuine discovery. Accordingly, many researchers find mathematical models most useful once the available observations have been analyzed on the basis of simple physical models.

The relationship between experimentation and modeling (in the furtherance of understanding and prediction) is depicted schematically in Figure 1.2. Here, physical

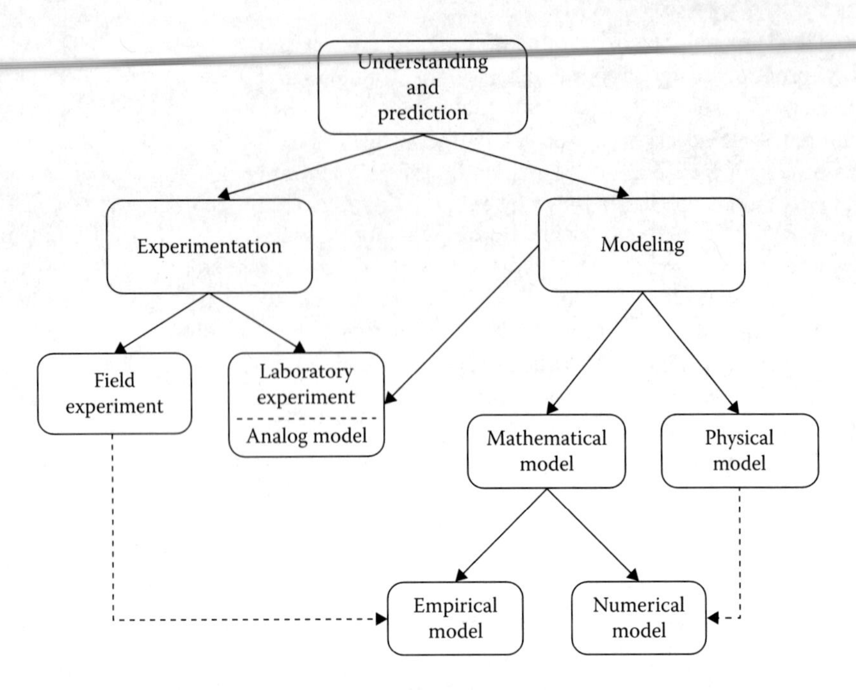

FIGURE 1.2 Schematic relationship between experimentation and modeling.

models form the basis for numerical models while experimental observations form the basis for empirical models. Moreover, analog modeling is represented as a form of laboratory (versus field) experimentation.

Scientists are becoming more aware of the connection between physical processes and computation, and many now find it useful to view the world in computational terms. Consequently, computer simulation is sometimes viewed as a third form of science midway between theory and experiment. Furthermore, understanding can be enhanced through the use of advanced computer graphics (visualization) to convert large volumes of data into vivid and comprehensible patterns.

The term *e-Science* refers to those technical activities that are performed through distributed global collaborations enabled by the Internet in concert with very large data collections, tera-scale computing resources and high-performance visualization. Thus, *e-Science* is an enabling technology that provides the marine and energy industries with meteorological and oceanographic (METOC) data in formats that facilitate easy computation of future states of the environments in which they operate. Applications for operational services concerned with maritime surveillance and security require the further integration of remotely sensed Earth-observation data. The drivers for *e-Science* are grounded in advances in computing, communications, remote sensing, and modeling. The term *marine informatics* is sometimes used to denote those activities in data services, visualization services, grid-computing facilities and the supporting infrastructure needed to generate the necessary ocean products (Graff, 2004).

Because of national security concerns, some existing datasets are limited in accessibility. Also, because of the wide range of acoustic frequencies, ocean areas and geometries of interest to researchers, it is virtually impossible to accommodate all potential observational requirements within normal fiscal constraints. To make matters worse, acoustic data are sometimes collected at sea without the supporting oceanographic data. Thus, models cannot always replicate the observed acoustic results because they lack the necessary input parameters for initialization. This situation has been improving with the advent of modern, multidisciplinary research that necessitates the inclusion of oceanographers in the planning and execution of complex field experiments.

Satellites, together with other remote-sensing techniques, provide a useful adjunct to the prediction of underwater-acoustic conditions. Specifically, many dynamic features of the ocean affect the behavior of sound in the sea. Knowledge of the location and size of such dynamic features can improve the prediction of sonar performance. Although satellite-borne sensors detect only surface (or near-surface) features of the ocean such as thermal contrast, color, or surface roughness, these "surface expressions" can generally be associated with dynamic oceanographic features below the surface, particularly when comprehensive climatological databases already exist with which to establish such associations. Thus, for example, satellite imagery can be used to provide timely and accurate position information on variable ocean features such as fronts and eddies—features that are known to have a significant impact on the propagation of acoustic signals in the sea. In a recent development, Jain and Ali (2006) demonstrated the capability of estimating sound-speed profiles using surface observations obtained from satellites in conjunction with an artificial neural-network method to infer subsurface temperature profiles from the satellite observations. By invoking assumptions regarding the subsurface salinity profiles, the desired sound-speed profiles could be computed.

Wainman (2012) investigated a naval operational concept using available sea-surface temperature (SST) information obtained from an infrared sensor on an Earth-orbiting satellite to match with vertical temperature values of a suitably representative (temporal and spatial) synthetic profile. This temperature profile could then be converted into a sound-speed profile for use with an existing sound-propagation model. The proposed method could potentially improve ship routing, strategy, and tactics as a force multiplier. This technique could also contribute to maritime security and protection of sovereign rights in a state's exclusive economic zone (EEZ). The hypothesis explored by Wainman (2012) was that thermal characteristics of the water column in the vicinity of the Benguela Current in the southeastern Atlantic Ocean (refer to Figures 2.17 and 2.20) can be numerically modeled and deduced from a single SST value, if provided with sufficient historic temperature-depth profiles for that region. For operational use, the SST would ideally be provided from near-real-time remotely sensed satellite data. The methodology would use suitably pre-processed historical temperature-depth measurements in an artificial neural network, self-organizing map (SOM) analysis as a basis for selecting a representative temperature profile. The proposed method combined static (climatological) temperature profiles and dynamic (near-real-time) surface temperatures to form a quasi-dynamic solution. Finally, for naval applications, the predicted temperature profile would be converted to a sound-speed profile and incorporated into existing sound modeling and ranging software (SMOD). A wide range of SMOD setup options (source frequency, beam width, water depth, and sea floor composition, among others) were kept constant and used repeatedly to show the final outcome in terms of ray trace and probability of detection plots. The SMOD software package uses a suite of empirical sonar equations to determine sound-propagation paths, transmission loss, and probability of detection of signals emanating from a predefined underwater sound source. SMOD was developed by the Institute for Maritime Technology, a Division of Armscor Defence Institutes (Pty) Ltd., for use by the South African Navy.

Tactical oceanographic data collection in support of naval operations has been augmented by drifting buoys, which use satellite relays to transmit data to mobile or stationary receiving stations, and by autonomous underwater vehicles (AUVs) to access remote ocean areas such as shallow-water and under-ice regions (Brutzman et al., 1992; Selsor, 1993; Dantzler et al., 1993; Etter, 2001b).

The problem of operational sonar prediction embraces many disciplines, one of which is modeling. Such modern operational applications involve not only underwater acoustic models but also oceanographic models (Etter, 1989). The coupling of these two types of models provides a valuable set of prediction tools to naval force commanders by enabling them to respond to the changing environmental conditions that affect their sonar performance. The remote-sensing data now available to naval forces afloat can be used in conjunction with oceanographic models to accurately forecast the locations and characteristics of dynamic ocean features (Robinson, 1992, 1999). This information can then be input to the appropriate acoustic models to assess the resultant impacts on sonar performance. These sonar systems can then be optimized for performance in each region of operation at any given time of the year.

Advances in sonar technologies have rendered modern sonar systems useful for *in situ* measurements of the ambient marine environment. For example, through-the-sensor measurements of the ocean impulse response (Smith, 1997) have enabled modern sonars to perform collateral functions as "tactical environmental processors." This aspect is addressed in more detail in Section 6.17.

1.3 DEVELOPMENTS IN MODELING

A goal of science is to develop the means for reliable prediction to guide decision and action (Ziman, 1978). This is accomplished by finding algorithmic compressions of observations and physical laws. Physical laws are statements about classes of phenomena, and initial conditions are statements about particular systems. Thus, it is the solutions to the equations, and not the equations themselves, that provide a mathematical description of the physical phenomena. In constructing and refining mathematical theories, we rely heavily on models. At its conception, a model provides the framework for a mathematical interpretation of new phenomena.

In its most elemental form, a model is intended to generalize and to abstract. A perfect model is one that perfectly represents reality. In practice, however, such a perfect model would defeat its purpose: it would be as complex as the problem it is attempting to represent. Thus, modeling in the physical sciences is normally reduced to many, more easily managed, components. Oreskes et al. (1994) argued that the primary value of models in the earth sciences is heuristic (i.e., an aid to learning, as through trial-and-error methods) and that the demonstration of agreement between observation and prediction is inherently partial since natural systems are never closed. The ocean is a natural system and, as an acoustic medium, it is not a closed (or deterministic) system. As will be demonstrated, most underwater acoustic models treat the ocean as a deterministic system. This can create problems when evaluating models against field data that are, by nature, nondeterministic (i.e., stochastic or chaotic). Thus, evaluation is an important aspect of any discussion of modeling. Frequently, models become data limited. This means that observational data are lacking in sufficient quantity or quality with which to support model initialization and model evaluation.

Models, which are virtual repositories of accumulated knowledge, can be utilized as pedagogical tools. Specifically, those individuals new to a field can familiarize themselves with phenomenological aspects by self-navigating these models. Bartel (2010) argued that writing an ocean-acoustic model could provide a further pedagogical benefit. The theory of computational ocean-acoustics is complex, and available treatments are often heavily mathematical. One way to learn is to create one's own model and explore its outputs. By writing a model, one gains insight into what other computer codes are trying to do in addition to an understanding of some of the pitfalls to be avoided in such codes. This experience can prove beneficial when interpreting the output of established computer models, particularly since any given model can generate erroneous results under pathological conditions.

With the advent of digital computers, modeling in the physical sciences advanced dramatically. Improvements in computer capabilities over the past several decades have permitted researchers to incorporate more complexity into their models, sometimes with little or no penalty in run time or computer costs (e.g., Hodges, 1987; Runyan, 1991). Although computational capabilities have increased dramatically over the past several decades, so too have the expectations placed on software performance. Consequently, software efficiency still remains a very critical issue—we cannot look to unlimited computing power as a panacea for inefficient software. Furthermore, with the dramatic increase in autonomous, self-guided systems such as AUVs and unmanned undersea vehicles (UUVs) (National Research Council, 1996, 1997), many of which use self-contained modeling and simulation (M&S) technologies, issues of verification, validation, and accreditation (VV&A) will assume even greater importance in maintaining and improving system reliability.

As modeling techniques continue to proliferate within the underwater acoustics community, it becomes increasingly difficult to take stock of the various models already in existence before launching a new effort to develop yet more models. Moreover, analysts confronted with sonar performance problems have difficulty in determining what models exist and, of those, which are best for their particular situation. This book had its genesis in just such a dilemma. The US Navy sponsored a small study in 1978 (Etter and Flum, 1978) to review the availability of numerical models of underwater acoustic propagation, noise and reverberation as well as the availability of databases with which to support model development and operation. Results of this work, and extensions thereto (Etter and Flum, 1980; Etter et al., 1984), have subsequently been presented at meetings of the Acoustical Society of America. Since 1979, the inventory of basic acoustic models and sonar performance models has been updated at 8-year intervals (Etter and Flum, 1979; Etter,

1987b, 1995, 2003b, 2011). Since 1979, approximately five models have been added to the total inventory each year (refer to Section 1.8 for more details). Moreover, progress has been documented in periodic literature review articles and related presentations (Etter, 1981, 1984, 1987a, 1990, 1993, 1999, 2001a, 2009, 2012a,b, 2014; Etter et al., 2014, 2015). An enhanced version of the first review article (Etter, 1981) was included as Chapter 3 in a book by Urick (1982). Collectively, this work later evolved into a series of lectures and culminated in the first edition of this book (Etter, 1991). Second, third and fourth editions (Etter, 1996, 2003a, 2013) were prepared to address the rapid advances unfolding in this area. The present edition continues and expands this review work.

As new models have been developed and older models have fallen into disuse, it is fair to ask why the older material has not been purged from the newer editions of this book. Simply stated, this book serves two purposes. First, the book introduces a complicated topic to people of varied backgrounds, including those who do not routinely work in the field of underwater acoustic M&S. In this sense, the older material provides an historical perspective and identifies the pioneering names that are taken for granted by the seasoned professionals in the field. Second, for those who do routinely work in this field, retention of the older material provides an inverted roadmap of past exploration.

The technical literature cited in this book includes many unpublished reports (so-called "gray" literature) since no other sources of documented technical information were available. Unpublished reports comprised nearly 40% of the literature cited in the first edition. In this fifth edition, reliance on unpublished reports decreased to about 24%. This trend is attributed, in part, to the continued maturing of underwater acoustic M&S technologies and their subsequent migration into the academic literature. Approximately 32% of the literature cited in the all five editions was drawn from the *Journal of the Acoustical Society of America*, evidence of this journal's role in communicating progress in the field of underwater acoustic modeling. As work has advanced in simulation, progress has been reported in related academic journals as well, accounting for approximately 34% of all references. Published books accounted for approximately 10% of all references.

References to Internet websites are now included as sources of information and specific sites of interest are indicated in Appendix C. While websites are useful sources of information, they are problematic as references since the addresses for these websites sometimes change or disappear entirely.

Other researchers have conducted reviews of modeling that provide useful sources of information. These reviews have tended to be more in-depth but more narrowly focused than the work presented in this book. Weston and Rowlands (1979) reviewed the development of models with application to underwater acoustic propagation over the period 1963–1978. DiNapoli and Deavenport (1979) provided a highly mathematical examination of a select number of propagation models. Brekhovskikh and Lysanov (1982, 2003) presented a comprehensive Russian perspective on underwater acoustics with a limited treatment of modeling. Piskarev (1992) and Godin and Palmer (2008) provided an account of state-of-the-art Soviet research in underwater acoustic propagation modeling up to 1989. Jensen et al. (2011) provided a lucid and comprehensive review of recent theoretical developments in ocean acoustic propagation modeling. Lurton (2002) introduced a systems-oriented view of underwater acoustics while Ainslie (2010) addressed broader issues in sonar performance modeling.

1.4 ADVANCES IN SIMULATION

Broadly defined, simulation refers to a method for implementing a model over time. The term "modeling and simulation" will refer collectively to those techniques that can predict or diagnose the performance of complex systems operating in the undersea environment.

The functions of M&S can be categorized as either prognostic or diagnostic. Prognostic functions include prediction and forecasting, where future oceanic conditions or acoustic sensor performance must be anticipated. Diagnostic functions include systems design and analyses, which are typically encountered in engineering tradeoff studies.

In the context of naval operations, simulations can be decomposed into four fundamental levels: engineering, engagement, mission, and theater (National Research Council, 1997). Engineering-level simulation comprises environmental, propagation, noise, reverberation and sonar performance models. Engagement-level simulation executes engineering-level models to generate estimates of system performance in a particular spatial and temporal ocean environment when operating against (engaging) a particular target. Mission-level simulation aggregates multiple engagements to generate statistics useful in evaluating system concepts within the context of well-defined mission scenarios. Finally, theater-level simulation aggregates mission-level components to analyze alternative system-employment strategies. Figure 1.3 illustrates the hierarchical relationship between engineering-level simulations and underwater-acoustic models. Aspects of simulation will be discussed in greater detail in Chapter 13.

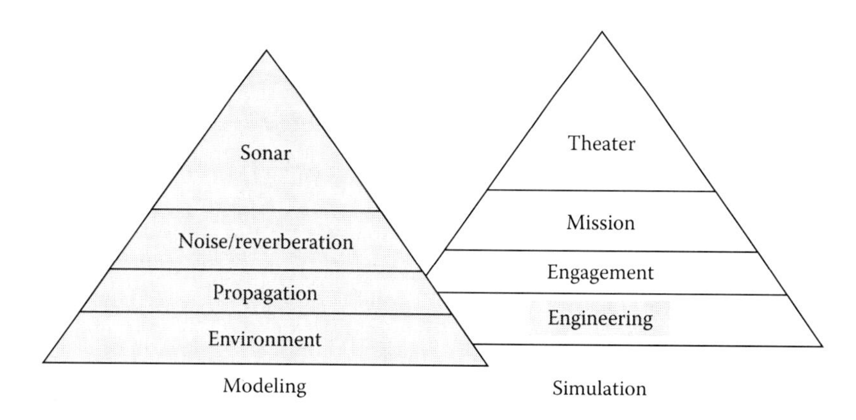

FIGURE 1.3 Modeling and simulation hierarchies illustrating the relationship between underwater acoustic models (left) and simulations (right). In this context, engineering-level simulations comprise environmental, propagation, noise, reverberation, and sonar performance models. (Adapted from Etter, P.C., *J. Sound Vib.*, 240, 351–383, 2001a.)

A relatively new term used in the context of M&S is a *community of interest* (COI), which refers to a collaborative group of users who exchange information in pursuit of their shared goals, interests and missions. These users benefit from having a shared vocabulary for the information they exchange. To be effective, these COIs try to gain semantic and structural agreement on their shared information and also try to keep their sphere of shared agreements as narrow as possible. Typically, a COI is a functional entity that crosses organizational boundaries and includes producers and consumers of data as well as developers of systems and applications. As a COI matures, it may evolve into a more formal organization such as a professional society and sponsor technical-exchange symposia.

1.5 OPERATIONAL CHALLENGES

M&S products aid the development and employment of acoustical techniques used to image underwater features, communicate information via the oceanic waveguide, or measure oceanic properties (Etter, 2000). Representative applications of these techniques are summarized in Table 1.1.

Technology-investment strategies driven by the geopolitical realities of the past several decades have greatly influenced the direction of research and development (R&D) in general, and of M&S in particular. These investment strategies have adversely impacted government and academia by diminishing budgets for undersea research, reducing the number of field experiments, reducing at-sea-training time and limiting asset modernization. Other economic factors have adversely affected the offshore industries. This situation creates new opportunities and challenges for M&S. As will be discussed below, M&S can be leveraged as enabling technologies to meet the technical and programmatic challenges in naval operations, offshore industries and oceanographic research.

TABLE 1.1
Summary of Acoustical Techniques and Representative Applications

Acoustical Techniques	Applications
Image underwater features	Detection, classification and localization of objects in the water column and in the sediments using monostatic or bistatic sonars.
	Obstacle avoidance using forward-looking sonars.
	Navigation using echo sounders or sidescan sonars to recognize seafloor topographic reference features.
Communicate information via the oceanic waveguide	Acoustic transmission and reception of voice or data signals in the oceanic waveguide.
	Navigation and docking guided by acoustic transponders.
	Release of moored instrumentation packages using acoustically activated mechanisms.
Measure oceanic properties	Measurement of ocean volume and boundaries using either direct or indirect acoustical methods.
	Acoustical monitoring of the marine environment for regulatory compliance.
	Acoustical surveying of organic and inorganic marine resources.

Underwater acoustics also plays a role in the international monitoring system (IMS), which comprises a network of stations that monitor Earth for evidence of nuclear explosions in all environments to ensure compliance with the comprehensive nuclear-test-ban treaty (CTBT). The system employs seismic, hydroacoustic and infrasound stations to monitor the underground, underwater and atmosphere environments, respectively. Bedard and Georges (2000) reviewed the application of atmospheric infrasound for such monitoring. Newton and Galindo (2001) described aspects of the hydroacoustic-monitoring network. Moreover, Farrell and LePage (1996) described the development of a model to predict the detection and localization performance of this hydroacoustic-monitoring network.

Whitman (1994) reviewed defense conversion opportunities in marine technology and made a distinction between dual use and conversion. Dual use suggests the deliberate pursuit of new research, development or economic activity that is applicable within both military and civilian domains. Conversion implies seeking new uses for existing defense resources. An example of conversion is the utilization of existing undersea surveillance assets as a National Acoustic Observatory. Research uses include stock and migration monitoring of large marine mammals, remote ocean observations, seismic and volcanic monitoring, acoustic telemetry, and fisheries enforcement (Amato, 1993; Carlson, 1994).

1.5.1 Naval Operations

Over the past several decades, naval mission requirements have shifted from open-ocean operations to littoral (or shallow-water) scenarios. This has not been an easy transition for sonar technologists since sonar systems that were originally designed for operation in deep water seldom work optimally in coastal regions. This has also held true for M&S technologies, which have undergone a redefinition and refocusing to support a new generation of multistatic naval systems that are intended to operate efficiently in littoral regions while still retaining a deep-water capability. A corresponding shift has been reflected in the research directions of the supporting scientific community as technical priorities have been realigned.

Shallow-water geometries increase the importance of boundary interactions, which diminish acoustic energy through scattering and also complicate the detection and localization of submerged objects due to multipath propagation. Moreover, the higher levels of interfering noises encountered in coastal regions combine with higher levels of boundary reverberation to mask signals of interest.

Naval operations in littoral regions often rely on multistatic acoustic sensors, thus increasing the technical challenges associated with the field-intensive experiments necessary to test multistatic geometries. Acoustical oceanographers have conducted supporting research using traditional direct-sensing methods in addition to more sophisticated inverse-sensing techniques such as acoustic tomography, full-field processing and ambient-noise imaging (see Section 1.6). Due to an increased awareness of the potential technological impacts on marine life, naval commanders and acoustical oceanographers must now comply with new environmental regulations governing the acoustic emissions of their sonar systems.

M&S can mitigate these technical and programmatic challenges in four ways. First, reduced at-sea training opportunities can be offset through the use of computer-based training (CBT). Second, simulation testbeds can facilitate system-design efforts aimed at maximizing returns on diminished asset-modernization expenditures. Third, the operation of existing systems can be optimized through the application of high-fidelity M&S products. Fourth, system-design tradeoffs can be evaluated using M&S products as metrics. Such efforts are important components of so-called simulation-based acquisition (SBA). Simulation-based acquisition comprises product representations, analysis tools (for design optimization and cost estimation), and an infrastructure that allows the product representations and analysis tools to interact with one another. These issues will be addressed in Chapter 13.

In advance of naval operations, it is necessary to collect METOC data from remote or hostile coastal environments in order to forecast acoustic sensor performance. Coupled atmosphere-ocean-acoustic models could reduce the need for hazardous *in situ* data collection by numerically computing initial states for the embedded acoustic models.

Specific solutions may include integration of M&S technologies in autonomous or unmanned undersea vehicles (AUV/UUV) to create an advanced generation of environmentally adaptive acoustic-sensor systems for naval operations and for oceanographic research. This environmental adaptation is accomplished by making *in situ*, through-the-sensor measurements of environmental conditions in conjunction with a sonar controller using an environmental feedback loop.

The U.S. Navy has identified nine missions for UUVs (Martin, 2012): (1) intelligence, surveillance and reconnaissance; (2) mine countermeasures; (3) anti-submarine warfare; (4) inspection and identification; (5) oceanography; (6) communication and navigation network nodes; (7) payload delivery; (8) information operations; and (9) time-critical strike. One of these missions, network nodes, is discussed in more detail in Section 6.15.3.

It may also be possible to leverage M&S technologies to enhance evolving network-centric data fusion and sensor-integration functions (Morgan, 1998), as has already been demonstrated in model-based signal processing approaches (Candy and Sullivan, 1992). A network-centric operation derives its power from the strong networking of well-informed but geographically dispersed entities. The networked platforms can include AUVs, surface ships, submarines, aircraft or satellites. The elements that enable network centricity include distributed sensor networks, a high-performance information grid, access to pertinent information sources (*in situ* and archival), precision and speed of response, and command and control processes. Network centricity fuses common tactical and environmental pictures, thus reducing uncertainties in measurements and modeling. Simulated volumetric (3D) visualizations of the undersea battlespace derived from M&S technologies could further enhance the efficient management and deployment of critical resources. The National Research Council (2003b) examined the use of environmental information by naval forces. The study recommended the utilization of network-centric principles in the collection and dissemination of METOC information.

High-fidelity, multistatic sonar-performance models can also be used to gauge compliance with environmental noise regulations concerning marine-mammal protection. Moreover, controlling underwater-radiated noise and sonar self-noise on naval vessels is critically important, and simulation can be used to predict the noise environments on surface ships and submarines.

For coastal-defense scenarios, sensor performance is an important input parameter for the judicious allocation of radar and sonar assets. The unification of radar and sonar sensor-performance modeling could reduce the amount of information that multi-sensor operators have to process. In fact, unified modeling and fused visualization of sensor coverage has been demonstrated for radar and sonar in a simple scenario (van Leijen et al., 2009). The radar-range equation and the active-sonar equations have strong similarities (Collins, 1970).

Mansour and Leblond (2013) proposed the design of a novel system to perform sustainable and long-term monitoring of coastal marine ecosystems and to enhance port surveillance capability. The proposed system was evaluated based on the analysis, classification and the fusion of a variety of heterogeneous data collected using different sensors including hydrophones, sonars, and various camera types.

For optimal sensor utilization, van Valkenburg-Haarst et al. (2010) advocated incorporation of electro-optical sensor systems within automated sensor-management systems. Moreover, automatic information extraction techniques should be available to support or relieve the human operator.

1.5.2 Offshore Industries

In the commercial sector, acoustic sensing methods have found numerous applications including acoustic Doppler current profilers (ADCPs) for measuring currents, compact sonars for obstacle location and avoidance by AUVs (e.g., Brutzman et al., 1992), fish-finding devices, underwater communications systems for divers, fathometers for bathymetric sounding and navigation, and side-scanning sonars for topographic mapping of the sea-floor relief. Some point-source and nonpoint-source pollution studies now use acoustic backscatter measurements to monitor the marine environment.

Offshore industries, particularly oil and gas, have undergone profound changes over the past several decades in response to global economic factors. Specifically, the contribution of offshore oil production to the total non-OPEC supply increased from about 25% in 1990 to about 30% in 1995. Approximately 80% of the significant growth in non-OPEC supply up to 2000 was offshore. An appreciable contribution to the growth in offshore production was made by new technologies such as 3D seismic evaluations, horizontal drilling, sub-sea completions, multi-phase pipelines, and floating production storage and off-loading (FPSO) vessels (International Energy Agency, 1996).

As the exploitation of offshore oil reserves has increased, exploration and production (E&P) operations have expanded into deeper waters. For example, recent Angolan oil-exploration concession areas comprised three water-depth bands: shallow blocks (<500 m), deep blocks (500–1500 m), and ultra-deep blocks (1500–2500 m). By comparison, significant sub-sea oil production in the Gulf of Mexico has typically occurred in water depths approaching 1200 m. At such depths, it is not possible to build fixed oil rigs. Instead, floating platforms are anchored to the seabed. Since the equipment needed to operate each well is too heavy to install on the floating platforms, the equipment is placed on the sea floor where it is maintained by remotely operated vehicles (ROVs) deployed from the floating platforms. Because these formerly topside systems were designed for direct (human) intervention rather than remote intervention, the tasks necessary to install and maintain these systems are difficult (if not impossible) to perform with traditional ROV-based tools and techniques. Automation of remote-intervention tasks can make use of commercial-off-the-shelf (COTS) technologies such as acoustic positioning, acoustic imaging and enhanced user interfaces integrated into a single system (Schilling, 1998). These technologies can also be used for inspection, maintenance, and repair (IMR) operations.

Pipeline routes are planned to be as short as possible to reduce costs. Moreover, bottom slopes that could put stress on unsupported pipes are avoided, seabed sediments are mapped to identify unstable areas, and pipe-burial options are assessed. Surveys of potential pipeline routes commonly utilize data derived from sidescan sonars.

METOC data-collection efforts in support of offshore operations often employ acoustic Doppler current profiling (ADCP) sensors to measure surface and subsurface ocean currents. These data are required to determine the vertical and horizontal current shears that can impact the siting and placement of offshore structures.

Employing volumetric (acoustic) plume detection to identify hydrocarbon seepages of natural or man-made origins can fulfill environmental monitoring mandates. Oil-spill tracking, prediction, and containment operations, as well as disposal monitoring, can also employ volumetric acoustic methods.

Offshore work in marginal ice zones (MIZs) requires knowledge of ice thickness and under-ice features (especially keels). Information on under-ice features can be obtained from AUVs or ROVs equipped with upward-looking (acoustic) echo sounders. This information can complement independent surface (altimeter) measurements of ice ridges to obtain estimates of total ice thickness.

Acoustic systems are used widely in the offshore industry for ROV tracking, seismic-towfish tracking, and drilling operations. These systems must perform in noisy, shallow-water environments. Acoustic transponders function as navigational beacons and as remote-control release mechanisms in the deployment and recovery of instrumentation packages. Moreover, subsea drilling-rig supply operations employ acoustic beacons for navigation and docking evolutions. Similarly, divers often rely on portable acoustic devices for communication and navigation.

Noise-control design of planned facilities and noise-control retrofit of existing plants entail environmental-noise monitoring, M&S, and development of noise-control procurement specifications.

Offshore industries can benefit most from recent advances in M&S by integrating such technologies directly into ROV/AUV control software in order to improve responsiveness to changing environmental conditions. Furthermore, increasing the use of M&S in system-design and operator-training functions may derive additional technical and economic benefits.

1.5.3 Operational Oceanography

The term "operational oceanography" has become a topic of frequent discussion in the contemporary trade literature, although the activities normally associated with this term have been in existence for some time. The three principal attributes that characterize operational oceanography are (1) routine and systematic measurements of the oceans and atmosphere; (2) modeling, simulation, analysis, and interpretation of these measurements to generate useful information products; and (3) rapid dissemination of these products to the user communities. The user communities typically comprise government, industry, regulatory authorities, research institutions, and the general public.

In data assimilation centers, numerical forecasting models process the data and generate information products. The utility of these simulated products is further enhanced (value engineered) by subject-matter experts in such disciplines as marine transportation, marine construction, public health, and seawater quality. Different applications require different products. This implies that an array of information products must be tailored to satisfy the needs of specific user communities, who have been identified in advance through socio-economic or cost-benefit analyses.

Civil applications of operational oceanography in coastal regions are the most visible and include warnings against such hazards as coastal floods, waves, coastal erosion, and effluent contamination. Commercial applications in the open ocean include guidance on optimal ship routing. Defense applications of operational oceanography, as defined by the US Department of the Navy (2000b), include the development of oceanic and atmospheric observations and models to provide on-scene commanders with predictive capabilities, especially in the littoral zone. The US Navy's Geophysics Fleet Mission Program Library (GFMPL) contains meteorological, oceanographic, electromagnetic, and acoustic software for use as aids in planning naval operations in the open ocean as well as in the littoral zone. Clancy (1999) and Clancy and Johnson (1997) provided useful overviews of naval operational ocean modeling products and applications.

1.6 INVERSE ACOUSTIC SENSING OF THE OCEANS

Useful information about the ocean can be derived from both forward and inverse applications of underwater sound. Direct (or forward) methods include traditional sonar applications. Inverse methods extract information from direct measurements of the physical properties of the ocean (Buchanan et al., 2004). These inverse methods combine direct physical measurements with theoretical models of underwater acoustics. The objective is to estimate detailed underwater-acoustic fields from sparse physical measurements using the theoretical models as guides.

Inverse sensing techniques that employ acoustics have been used in several subdisciplines of geophysics including seismology, meteorology, and oceanography. Seismologists have used tomographic techniques to infer the bulk properties of the lithosphere (e.g., Menke, 1989). Atmospheric scientists have employed naturally generated, low-frequency sound (microbaroms) to probe the upper layers of the atmosphere in an inverse fashion (Donn and Rind, 1971); Coulter and Kallistratova (1999) discussed acoustic remote sensing of the lower atmosphere. Drob et al. (2010) presented an up-to-date review of the utility of inverted infrasound signals for passive remote sensing of the atmosphere.

In oceanography, inverse acoustic data provide estimates of spatially integrated and temporally averaged oceanic conditions that are not readily available from a traditional constellation of point sensors (e.g., Bennett, 1992). Collins and Kuperman (1994b) presented a broad discussion of inverse problems in ocean acoustics and methods for solving them. Parameters of interest included sound speed in the water column, sediment properties, and boundary roughness. The importance of forward models in solving inverse problems was stressed.

Diachok et al. (1995) documented the proceedings of a conference on full-field inversion methods in ocean and seismo-acoustics, which was sponsored by the NATO SACLANT Undersea Research Centre in Italy in June 1994. At this conference, it was demonstrated that inversion methods could exploit the amplitude and phase information detected on hydrophone arrays or geophone arrays to infer environmental information about the ocean. Furthermore, proceedings of an international workshop on *Tomography and acoustics: Recent developments and methods* (University of Leipzig, March 2001) were documented in a special issue of the journal of the European Acoustics Association (*ACUSTICA · acta acustica*, Vol. 87, No. 6, 2001). This two-day workshop addressed (1) tomography, (2) acoustics, (3) atmosphere applications, and (4) ocean applications.

Inverse acoustic sensing of the oceans utilizes one of three natural phenomena: propagation, noise, or reverberation. Table 1.2 summarizes selected inverse ocean acoustic sensing techniques according to the natural phenomenon utilized. The specific techniques identified in Table 1.2 will be discussed below and in appropriate sections throughout this book.

Acoustic propagation characteristics in the deep oceans are determined largely by the refractive properties of the water column and, to a lesser extent, by the surface and bottom boundary conditions. Propagation measurements can be used to infer bulk properties of the water column such as temperature, sound speed, density and currents. In shallow-ocean areas, where propagation characteristics can be strongly affected by the bottom boundary, propagation measurements can be used to infer properties of the

TABLE 1.2
Summary of Inverse Ocean-Acoustic Sensing Techniques

Propagation	Noise	Reverberation
Matched field processing	Field inversion	Field inversion
Source localization	Wind speeds	Sea-floor imaging
Marine environment characterization	Rainfall rates	
Ocean acoustic tomography	Acoustic daylight	Time reversal mirror (TRM) nulling
Density field (eddies, currents)	Object imaging	Reverberation attenuation
Temperature (climate monitoring)		
Deductive geoacoustic inversion	Geoacoustic inversion	
Sediment parameters	Seabed acoustics	
Sea-floor scattering characteristics		
TRM		
Signal refocusing		

sea floor such as composition and scattering characteristics. Caiti et al. (2000) reviewed recent progress in experimental acoustic inversion methods for use in shallow water environments based on papers presented at a workshop sponsored by the Portuguese Foundation for Science and Technology in March 1999. Taroudakis and Makrakis (2001) edited a collection of papers that addressed a wide spectrum of inverse problems in underwater acoustics including estimation of geoacoustic parameters, acoustic thermometry, and shallow-water characterization. Inverse acoustic sensing methods utilizing the propagation (controlled-source) characteristics of the oceans include matched field processing, ocean acoustic tomography and deductive geoacoustic inversion. A new technique known as a time reversal mirror (TRM) uses inverse methods to refocus received signals back to the source (see Section 6.11).

The ambient noise field in the oceans is described by the spectral, spatial, and temporal characteristics of sound generated by both natural and industrial sources. Measurements of these characteristics can provide useful information regarding the nature of the noise sources themselves as well as physical features within the oceans. Examples of inverse applications of the noise field include wind speed determination, rainfall measurements, object imaging ("acoustic daylight"), and geoacoustic inversion (see Chapter 7).

The reverberation field in the oceans is the product of acoustic scattering by the surface and bottom boundaries, and by inhomogeneities within the oceans. The utility of the reverberation field as an inverse sensing technique is analogous to that of the ambient noise field. For example, the reverberation field can be inverted to image the sea floor. A new development uses TRM methods to attenuate reverberant returns (see Section 9.4).

Inverse acoustic sensing techniques presently constitute adjuncts to direct measurement methods. However, the application of inverse acoustic sensing techniques to dynamical studies of the oceans' boundaries and interior show great promise for three reasons. First, such data can be used to establish comparative baselines for other remote sensors, such as satellites, by providing synoptic portraitures of the interior oceans together with concurrent groundtruth data at the sea surface. Second, inverse acoustic techniques often afford useful insights into a broad class of oceanic phenomena since their successful employment relies heavily on the use of numerical models first to understand the role of the oceans as an acoustic medium. Third, inverse data provide estimates of spatially integrated and temporally averaged oceanic conditions that are not readily available from traditional oceanographic sensors.

Wille (2005) provided an overview of acoustic imaging applications in an atlas that facilitated an interdisciplinary understanding of underwater acoustics and its diagnostic capabilities.

1.7 STANDARD DEFINITIONS

A consistent vocabulary and standard system of units is essential for work in any scientific discipline. Such a system facilitates efficient and unambiguous communication among members of the community. The underwater acoustics community has struggled with a common vocabulary and standard system of units for quite some time. This situation derives, in part, from the fact that many of the participants in this community have been trained in other disciplines and later migrated into this field.

Many investigators have introduced suggestions for a standard system of units to satisfy the requirements that are unique to the underwater acoustics community. Recently, for example, Carey (1995) clarified the use of SI metric units for measurements and calculations used in underwater acoustics and bioacoustics while Hall (1995) re-examined the dimensions of units for source strength, transmission loss, target strength, surface-scattering strength and volume-scattering strength. (Hall [1996] presented important corrections to his 1995 paper.) In addition, technical dictionaries (e.g., Morfey, 2000) provide useful guidance on proper terminology and usage.

Work in M&S also requires a consistent vocabulary. In this regard, the Institute of Electrical and Electronics Engineers (1989) published a glossary of definitions for M&S terms. The US Defense Modeling and Simulation Coordination Office (DoDM&SCO) assembled the *DOD M&S Glossary*, which prescribes a uniform M&S terminology, particularly for use throughout the Department of Defense. In addition to the main glossary of terms, this highly useful manual includes a list of M&S-related abbreviations, acronyms, and initials commonly used within the Department of Defense (refer to Appendix C for the DoDM&SCO website).

Recent communications in the forum of the *Journal of the Acoustical Society of America* (Clay, 1999; Hickling, 1999; Chapman, 2000) highlight confusion over use of the decibel (dB). In Chapter 3, the use of the decibel in underwater acoustics will be explained in more detail. Confusion over the decibel has created significant technical and legal problems for those acoustical oceanographers and sonar technologists who must interact with their counterparts in aeroacoustics and bioacoustics. In air acoustics, a reference sound pressure of 20 micropascal (μPa) is used while the current choice in underwater acoustics is 1 μPa. Part of the confusion arises from the use of decibels to represent ratios of dimensionless quantities as well as ratios of absolute quantities having physical dimensions. When reporting ratios of dimensionless quantities (such as reflection losses), decibels can be used without further qualification. However, when decibels are used to represent ratios of absolute quantities having physical dimensions (such as radiated noise levels), the reference quantities (pressure and distance) must be clearly stated. Since acoustical measurements are first made in SI units (sound pressure in pascals and radiated source power in watts) and then converted to decibels, some researchers have argued that confusion caused by the decibel could be removed by reporting acoustical measurements in SI units.

The need for standardized definitions continues to motivate research. In an investigation of the definition of propagation loss, EPWI (equivalent plane wave intensity) was contrasted with MSP (mean square pressure) formulations of propagation loss (Ainslie, 2004), with a preference expressed for an MSP ratio (Ainslie and Morfey, 2005).

Ambient sound levels in air and underwater were compared using units of intensity spectral density (W m^{-2} Hz^{-1}) over the frequency range 10 Hz to 100 kHz (Dahl et al., 2007). The intensity spectral density ranged from 10^{-16} to 10^{-4}. It was found that the intensity spectral density of a quiet residential environment (with distant traffic influence) exceeded that of nominal high-level underwater ambient noise conditions.

Finally, Carey (2006) reviewed definitions from the *American National Standards on Acoustics* that are applied to common sources used in both underwater acoustics research and naval sonar system operation. Metrics were recommended for both continuous and transient sound sources.

1.8 HISTORICAL GROWTH CURVES

As noted earlier in Section 1.3 (developments in modeling), approximately five models have been added to the total inventory each year since 1979. This count includes propagation, noise, reverberation, and sonar performance models. With the perspective afforded by a 40-year period of study, it is now possible to provide reliable growth curves for each of the four types of models as well as the combined total (Figure 1.4).

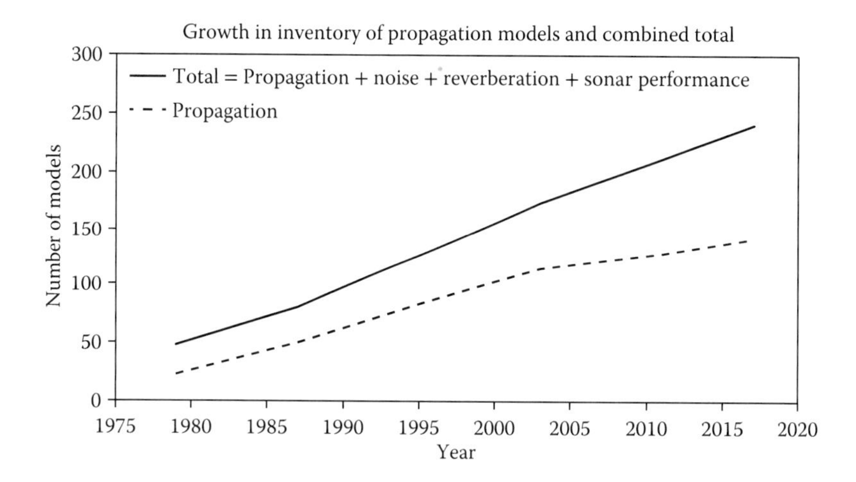

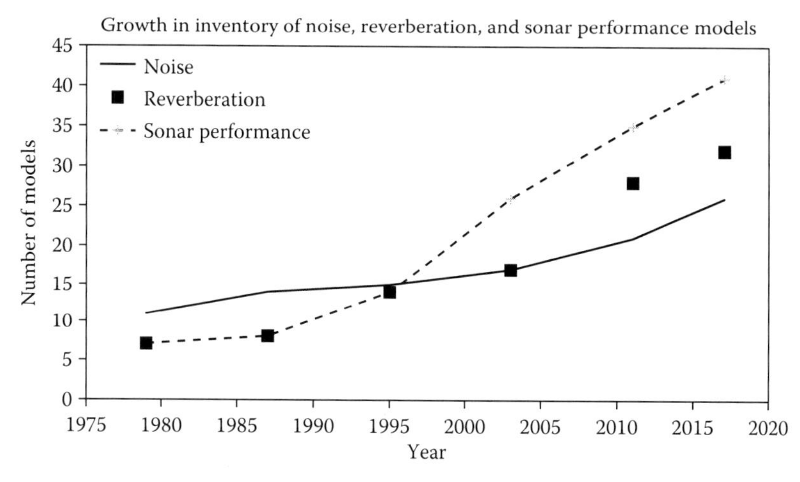

FIGURE 1.4 Growth in inventories of underwater acoustic models: Upper panel—combined totals and propagation models; Lower panel—noise, reverberation, and sonar performance models.

These curves serve to illustrate the sustained growth of underwater acoustic modeling even as the motivations and applications have evolved over the past four decades. The combined curve exhibits a nearly linear gradient of five models per year over the period of study. However, the growth of propagation models has saturated in the last decade while the growth of noise, reverberation, and sonar performance models has accelerated. This pattern suggests that, as propagation modeling technology has matured, resources have been redirected to improving the technology behind noise, reverberation, and sonar performance models.

2 Acoustical Oceanography

2.1 BACKGROUND

Acoustical oceanography describes the role of the ocean as an acoustic medium by relating oceanic properties to the behavior of underwater acoustic propagation, noise, and reverberation. Consequently, acoustical oceanography includes both the study of acoustics in the ocean and the use of acoustics to study the ocean. Acoustical oceanography crosses four other branches of oceanography: physical, chemical, geological, and biological oceanography.

Sound propagation is profoundly affected by the conditions of the surface and bottom boundaries of the ocean as well as by the variation of sound speed within the ocean volume. The single most important acoustical variable in the ocean is sound speed. The distribution of sound speed in the ocean influences all other acoustic phenomena. The sound speed field, in turn, is determined by the density (or temperature and salinity) distribution in the ocean. Advection of the underwater sound field by water currents is also important. Refraction of sound by fronts, eddies and other dynamic features can distort the propagation of acoustic signals. Knowledge of the state of the sea surface as well as the composition and topography of the seafloor is important for specification of boundary conditions. Bathymetric features can block the propagation of sound. Biological organisms contribute to the noise field and also scatter underwater sound signals. The balance of this chapter will address (1) physical and chemical properties, (2) sound speed, (3) boundaries, (4) dynamic features, and (5) biologics.

A number of books and published papers already exist on these subjects and appropriate citations will be made to them. Notable text and classic reference books of a general nature include those by Apel (1987), Broecker (2010), Gill (1982), Medwin and Clay (1998), Neumann and Pierson (1966), Peixoto and Oort (1992), Pickard and Emery (1990), Roll (1965), and Sverdrup et al. (1942).

2.2 PHYSICAL AND CHEMICAL PROPERTIES

Temperature is basic to any physical description of the oceans. It is the easiest and therefore the most common type of oceanographic measurement made. The exchange of heat between the ocean and the atmosphere depends strongly on temperature in the marine boundary layer. The density field and resulting stratification of the ocean depend largely on temperature. The speed of sound in the upper layers of the ocean is most strongly dependent on temperature. Temperature further influences the kinds and rates of chemical reactions occurring in the ocean. The distribution of nutrients and other biologically important substances depends on temperature and the resulting density stratification.

Sea water is a binary fluid in that it consists of various salts in water. The presence of salts affects a number of oceanic parameters including compressibility, sound speed, refractive index, thermal expansion, freezing point and temperature of maximum density.

Colligative properties are those properties of sea water that depend on the number of dissolved particles in solution, but not on the identities of the solutes. The four commonly studied colligative properties are freezing-point depression, boiling-point elevation, vapor-pressure lowering, and osmotic pressure. Since these properties yield information on the number of solute particles in solution, they can be used to obtain the molecular weight of the solute.

Salinity is a term used to measure the quantity of salts dissolved in sea water and is expressed in units of parts per thousand (‰ or ppt). The precise definition of salinity is complicated. Fofonoff (1985) reviewed the development of the modern salinity scale and the equation of state for sea water. The *Practical Salinity Scale 1978* was introduced to rectify shortcomings associated with the traditional chlorinity-conductivity relationship used to establish salinity (Lewis, 1980; Perkin and Lewis, 1980; Culkin and Ridout, 1989). In the new scale, the existing link between chlorinity and salinity was broken in favor of a definitive salinity-conductivity relationship. The new practical standard is IAPSO (International Association for the Physical Sciences of the Ocean) Standard Seawater, produced and calibrated by the IAPSO Standard Seawater Service. Salinity is now a dimensionless quantity (psu, or practical salinity unit) because the algorithms in the new scale were adjusted to eliminate the ‰ (or ppt) used in previous scales.

The density of sea water is related to temperature, salinity and pressure (which is nearly proportional to depth) through the equation of state (e.g., Fofonoff, 1985). Density provides a measure of the hydrostatic stability in the ocean. Specifically, a stable water column is one in which density increases monotonically with increasing depth.

Sea water is compressible, although less so than pure water. The compressibility of sea water can be expressed by the coefficient of compressibility, which relates fractional changes in water volume to the corresponding changes in pressure (e.g., Apel, 1987).

Compressibility of sea water is an important factor in several applications: the precise determination of the density of sea water, particularly at great depths; the computation of adiabatic temperature changes in the ocean (in an adiabatic process, compression results in warming, while expansion results in cooling); and most importantly, the computation of sound speed in sea water.

The speed of sound (c) in sea water is related to the isothermal compressibility (K) as

$$c = \sqrt{\frac{\gamma}{K\rho}} \tag{2.1}$$

where γ is the ratio of specific heats of sea water at constant pressure and constant volume, and ρ is the density of sea water. The isothermal compressibility is easier to measure experimentally than is the adiabatic compressibility.

2.2.1 Temperature Distribution

The distribution of temperature at the surface of the oceans is zonal in nature, with isotherms (lines of constant temperature) oriented in an east-west pattern. The annual mean temperature distribution shown in Figure 2.1 illustrates this general zonal gradation. This pattern is due largely to the zonal distribution of the solar energy received at the sea surface. Specific exceptions to this pattern occur in regions of upwelling (where colder water from below is brought to the surface through the action of the winds), and in the vicinity of major (baroclinic) current systems such as the Gulf Stream (where the temperature field is distorted). The relatively low equatorial and tropical sea-surface temperatures in the eastern Pacific and Atlantic Oceans are generally ascribed to the effects of upwelling. The more meridional trend of the isotherms off the northeast coast of the United States, for example, is evidence of the Gulf Stream current system. Examining only annual averages, however, can sometimes be misleading. The monsoon circulation in the Indian Ocean, for example, makes interpretation of an annual mean temperature field questionable.

The temperature field in the ocean exhibits a high degree of stratification with depth. Since the isotherms are nearly parallel to the horizontal plane, this type of structure is referred to as horizontal stratification. This is evidenced in Figure 2.2, which presents zonal (annual) averages of temperature in the Atlantic Ocean by one-degree latitude belts. These zonal averages do not include the Mediterranean Sea, Baltic Sea, or Hudson Bay.

2.2.2 Salinity Distribution

The distribution of salinity at the surface of the ocean is shown in Figure 2.3, and notable features have been summarized by Levitus (1982). Specifically, subtropical

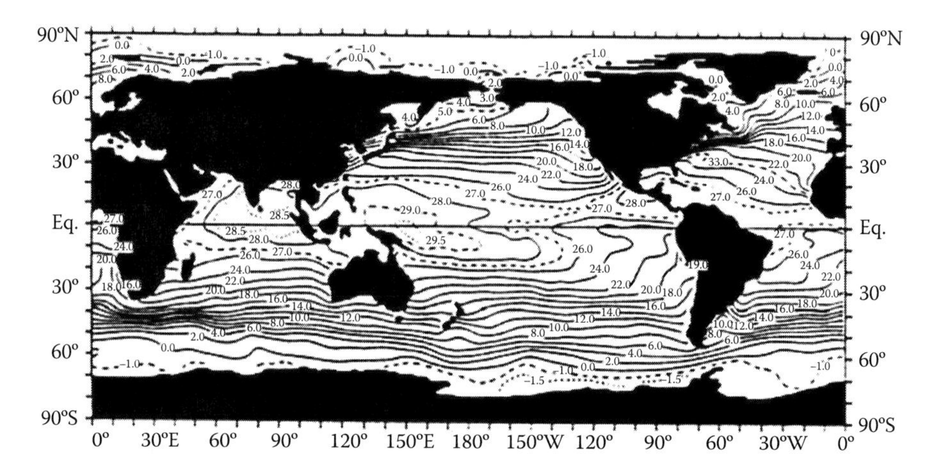

FIGURE 2.1 Annual mean temperature (°C) at the sea surface. The distribution of surface temperatures shows a strong latitudinal dependence due largely to the zonal distribution of solar energy received at the sea surface. (Adapted from Levitus, S., Climatological atlas of the world ocean. NOAA Professional Paper 13, 1982.)

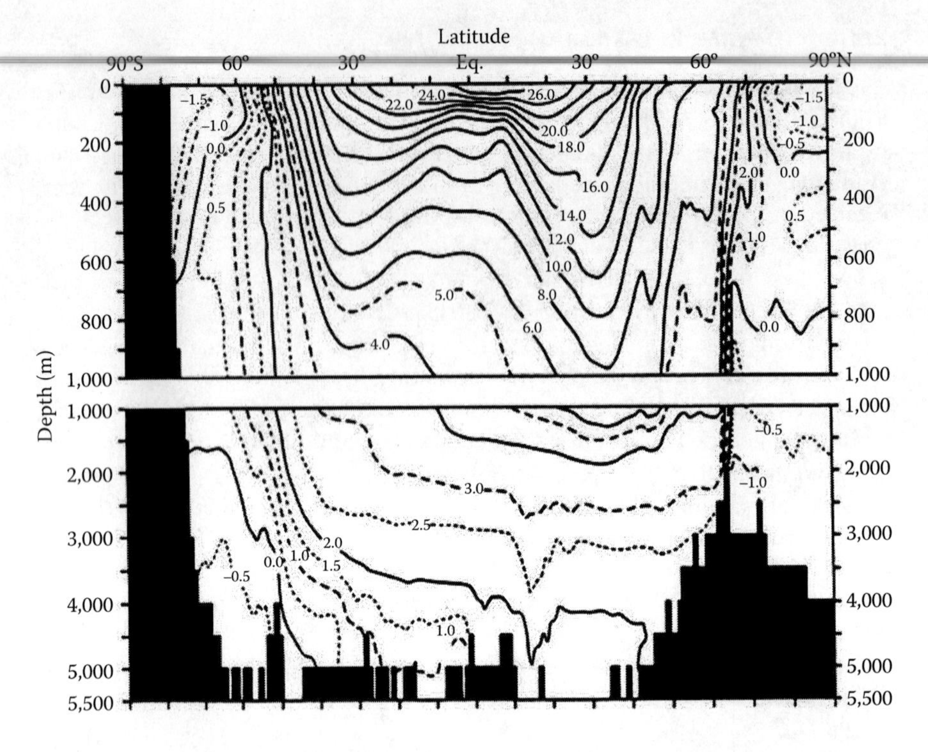

FIGURE 2.2 Annual mean Atlantic Ocean zonal average (by one-degree squares) of temperature (°C) as a function of depth. Note the break in the depth scale at 1000 m. (Adapted from Levitus, S., Climatological atlas of the world ocean. NOAA Professional Paper 13, 1982.)

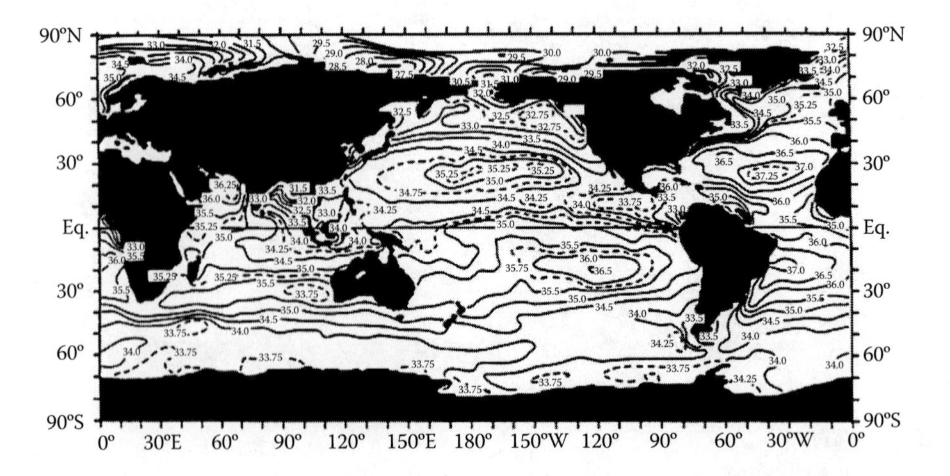

FIGURE 2.3 Annual mean salinity (ppt) at the sea surface. (Adapted from Levitus, S., Climatological atlas of the world ocean. NOAA Professional Paper 13, 1982.)

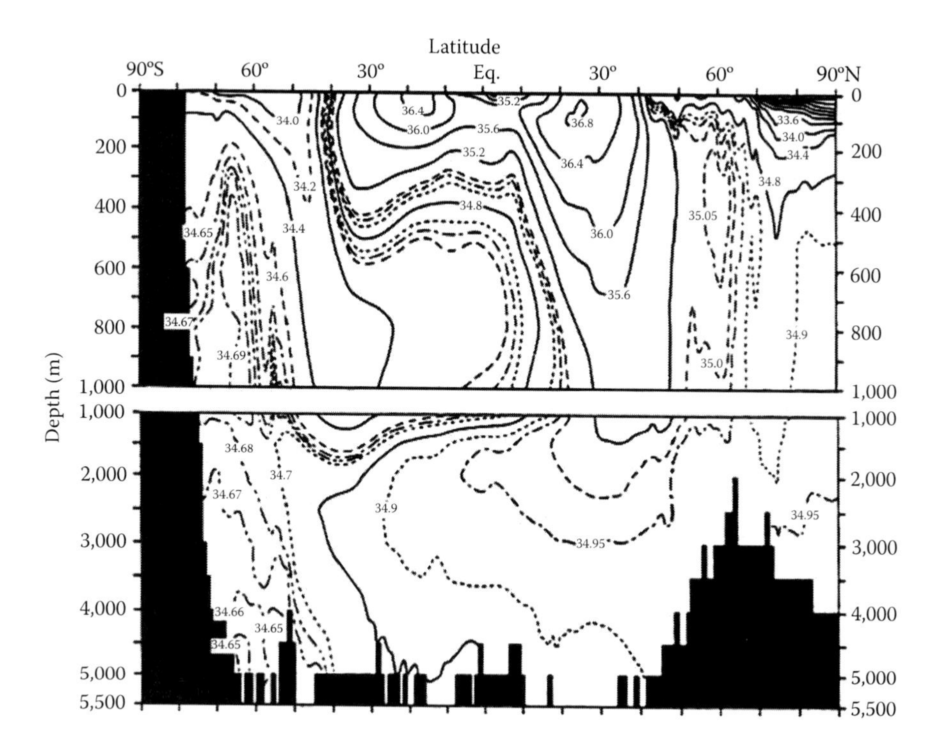

FIGURE 2.4 Annual mean Atlantic Ocean zonal average (by one-degree squares) of salinity (ppt) as a function of depth. Note the break in the depth scale at 1000 m. (Adapted from Levitus, S., Climatological atlas of the world ocean. NOAA Professional Paper 13, 1982.)

salinity maxima associated with the excess of evaporation (E) over precipitation (P) appear in all the individual oceans (regions where $E - P > 0$). Subpolar regions exhibit low salinities associated with the excess of precipitation over evaporation (regions where $E - P < 0$). Low-salinity tongues associated with runoff from major river systems, such as the Amazon, are also apparent.

Unlike the temperature fields, salinity does not exhibit a consistent stratification with depth, as for example in the Atlantic Ocean (Figure 2.4). These patterns reflect the complex movements of water throughout the oceans.

2.2.3 Water Masses

Using the concept of water masses facilitates descriptions of seawater characteristics and motions. This concept is analogous to that employed by meteorologists to describe air masses in weather patterns. Air masses are identified by characteristic combinations of air temperature and moisture content (humidity). These characteristics allow meteorologists to identify the history (or source regions) of the various air masses. Examples of common air masses include continental polar (cold, dry air formed over high-latitude land areas) and maritime tropical (warm, moist air formed over equatorial ocean areas).

Oceanographers (e.g., Sverdrup et al., 1942) have convincingly demonstrated that certain characteristic combinations of water temperature and salinity are associated with water masses formed in particular regions of the world's oceans. After formation, these water masses spread both vertically and laterally to occupy depth ranges of the water column consistent with their density. Moreover, these water masses are distinguishable from one another when plotted on a graph of temperature-versus-salinity, referred to as a *T–S* diagram. The *T–S* relations of the principal water masses of the Atlantic Ocean are presented in Figure 2.5 (Naval Oceanographic Office, 1972). The names of these water masses suggest their relative positions in the water column (e.g., central, intermediate, deep, and bottom). From Figure 2.5, for example, one can conclude that the waters of the South Atlantic Ocean are fresher (i.e., less saline) than those of the North Atlantic. Emery and Meincke (1986) provided an updated review and summary of the global water masses.

The movement of Antarctic intermediate water (AAIW), for example, can now be identified in the distribution of salinity illustrated previously in Figure 2.4. Specifically, the movement of AAIW is evidenced by a low-salinity tongue (~ 34.0–34.6 ppt) extending downward from the surface in the latitude belt 60°S to 70°S, then northward at a depth of about 700–800 m, and finally upward near the equator.

The distribution of sound speed in the ocean can be related to the local water-mass structure. Knowledge of the water-mass structure, then, can greatly enhance understanding of the large-scale spatial and temporal variability of the sound-speed field in the ocean.

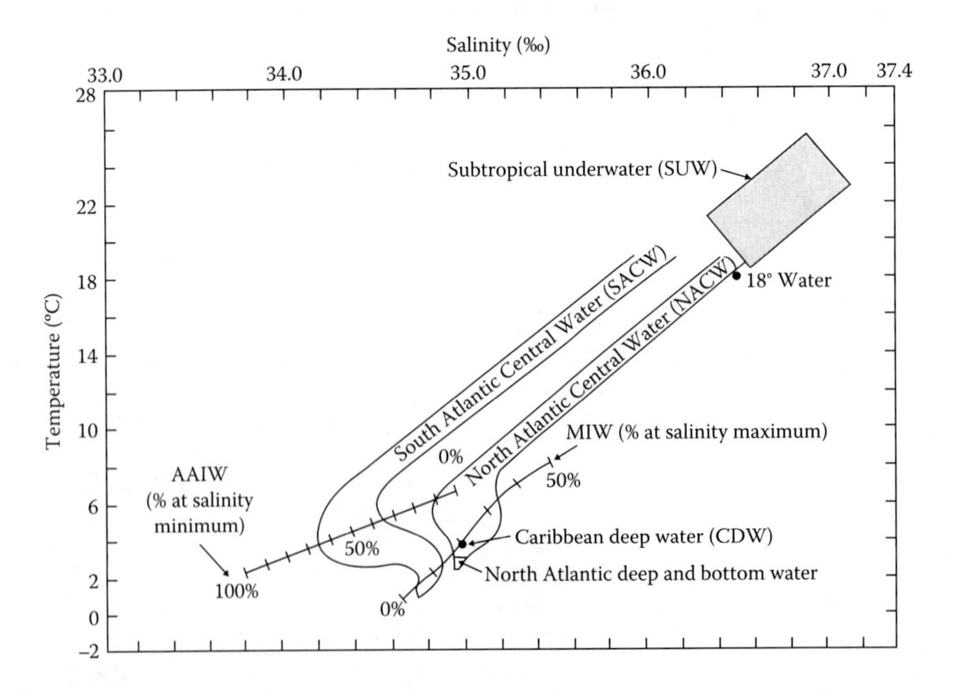

FIGURE 2.5 Temperature-salinity (*T–S*) diagrams of the major water masses of the Atlantic Ocean. (Adapted from Naval Oceanographic Office, *Environmental-Acoustics Atlas of the Caribbean Sea and Gulf of Mexico. Volume II—Marine Environment,* SP-189II, US Dept. of the Navy, Washington, DC, 1972.)

2.3 SOUND SPEED

The speed of sound in sea water is a fundamental oceanographic variable that determines the behavior of sound propagation in the ocean. Equation 2.1 defined the speed of sound in sea water as a function of the isothermal compressibility, the ratio of specific heats of sea water at constant pressure and constant volume, and the density of sea water.

The term "sound velocity" is sometimes used synonymously with "sound speed." The symbol c, which is commonly used to indicate sound speed, is derived from the word "celerity."

Since Equation 2.1 is difficult to compute in practice, considerable work has been devoted to expressing the speed of sound in sea water in terms of more commonly observed oceanographic parameters. Specifically, it is known empirically that sound speed varies as a function of water temperature, salinity, and pressure (or depth). Moreover, the speed of sound in sea water increases with an increase in any of these three parameters.

2.3.1 Calculation and Measurements

Many empirical relationships have been developed over the years for calculating sound speed using values of water temperature, salinity and pressure (or depth). Frequently used formulas include those of Wilson (1960), Leroy (1969), Frye and Pugh (1971), Del Grosso (1974), Medwin (1975), Chen and Millero (1977), Lovett (1978), Coppens (1981), Mackenzie (1981), and Leroy et al. (2008, 2009). As summarized in Table 2.1, each formula has its own ranges of temperature, salinity, and pressure (or depth). Collectively, these ranges are referred to as "domains of applicability." Calculations outside of the specified domains may lead to errors. The individual equations are not reported in Table 2.1. However, the number of terms comprising each equation is indicated in the far-right column. The simplest equation (Medwin, 1975) contains six terms while the most complicated equation (Wilson, 1960) contains 23 terms.

For convenience, the nine-term algorithm developed by Mackenzie (1981) is presented here:

$$\begin{aligned} c = {} & 1448.96 + 4.591T - 5.304\times10^{-2}T^2 + 2.374\times10^{-4}T^3 \\ & + 1.340(S-35) + 1.630\times10^{-2}D + 1.675\times10^{-7}D^2 \\ & - 1.025\times10^{-2}T(S-35) - 7.139\times10^{-13}TD^3 \end{aligned} \quad (2.2)$$

where c is the speed of sound in sea water (m s^{-1}), T is the water temperature (°C), S is the salinity (‰) and D is the depth (m). Mackenzie (1981) also discussed the mathematical relationship between pressure and depth in the ocean. In related work, Leroy and Parthiot (1998) developed convenient equations for converting pressure to depth and *vice versa* (also see Leroy, 2007).

Leroy et al. (2008, 2009) proposed a new 14-term equation for the calculation of sound speed in sea water as a function of temperature, salinity, depth, and latitude in

TABLE 2.1
Summary of Sound Speed Algorithm Parameter Ranges

Reference	Temperature Range	Salinity Range	Pressure or Depth Range[a,b]	Standard Error	Number of Terms (Comments)
Wilson (1960)	–4–30 °C	0–37 ppt	1–1,000 kg cm^{-2}	0.30 m s^{-1}	23 terms
Leroy (1969)	–2–34 °C	20–42 ppt	0–8,000 m	0.2 m s^{-1}	13 terms (complete [1st] formula)
Frye and Pugh (1971)	–3–30 °C	33.1–36.6 ppt	1.033–984.3 kg cm^{-2}	0.10 m s^{-1}	12 terms
Del Grosso (1974)	0–35 °C	29–43 ppt	0–1,000 kg cm^{-2}	0.05 m s^{-1}	19 terms
Medwin (1975)	0–35 °C	0–45 ppt	0–1,000 m	~ 0.2 m s^{-1}	6 terms (simple formula)
Chen and Millero (1977)	0–40 °C	5–40 ppt	0–1,000 bars	0.19 m s^{-1}	15 terms (requires correction at low temperature and high pressure [Millero and Li, 1994])
Lovett (1978)	0–30 °C	30–37 ppt	0–10,000 dbars	0.063 m s^{-1}	13 terms (3rd equation)
Coppens (1981)	–2–35 °C	0–42 ppt	0–4,000 m	0.1 m s^{-1}	8 terms (Lovett's 3rd equation simplified)
Mackenzie (1981)	–2–30 °C	25–40 ppt	0–8,000 m	0.07 m s^{-1}	9 terms
Leroy et al. (2008)	–1–30 °C	0–42 ppt	0–12,000 m	0.2 m s^{-1}	14 terms

[a] For each 10-m increase in water depth, pressure increases by approximately 10 dbars, 1 bar or 1 kg cm^{-2}.
[b] Refer to Mackenzie (1981) or Leroy and Parthiot (1998) (and Leroy [2007]) for algorithms useful in converting pressure to depth.

all oceans and open seas, including the Baltic Sea and the Black Sea. The proposed equation agrees to better than ±0.2 m s^{-1} with two reference equations. The only exceptions are isolated hot brine spots that may be found at the bottom of some seas. The equation is of polynomial form, with 14 terms and coefficients of between one and three significant figures. This is a substantial reduction in complexity compared to the more complex equations using pressures that need to be calculated according to depth and location. The equation uses the 1990 universal temperature scale, where an elementary transformation is given for data based on the 1968 temperature scale:

$$\begin{aligned} c = {} & 1402.5 + 5T - 5.44 \times 10^{-2} T^2 + 2.1 \times 10^{-4} T^3 \\ & + 1.33S - 1.23 \times 10^{-2} ST + 8.7 \times 10^{-5} ST^2 \\ & + 1.56 \times 10^{-2} Z + 2.55 \times 10^{-7} Z^2 - 7.3 \times 10^{-12} Z^3 \\ & + 1.2 \times 10^{-6} Z(\Phi - 45) - 9.5 \times 10^{-13} TZ^3 \\ & + 3 \times 10^{-7} T^2 Z + 1.43 \times 10^{-5} SZ \end{aligned} \quad (2.3)$$

where c = speed of sound (m s^{-1}), T = water temperature (°C), S = water salinity (‰), Z = water depth (m) and Φ = latitude (degrees). Refer to Table 2.1 for the domains of application for this and other equations.

Recent tomographic measurements of sound speed in the oceans (see Section 6.10) have implications for very-long-range ducted propagation (Spiesberger and Metzger, 1991b). Specifically, previous algorithms derived from laboratory measurements have been found to overpredict sound speeds due to pressure effects at great depths (Dushaw et al., 1993). This matter is still the subject of investigation. Millero and Li (1994) corrected the earlier Chen and Millero (1977) sound speed formula to improve its applicability to low temperatures and high pressures. This correction is especially important for tomographic applications.

Two practical devices that are commonly used to measure sound speed as a function of depth in the ocean are the bathythermograph (BT) and the velocimeter. Expendable versions of the BT, designated XBT, actually measure temperature (via a thermistor) as a function of depth (via a known fall rate). Sound speed can then be calculated using one of the algorithms identified in Table 2.1, often on the assumption that salinity is constant, or nearly so. This assumption is justified by the observation that the typical range of salinities in the open ocean is usually small and that the corresponding impact on sound speed is negligible from a practical standpoint. In coastal areas, and near rivers or ice, this assumption is not generally valid and a velocimeter, which measures sound speed directly in terms of the travel time of sound over a fixed (constant-length) path, would be preferable. An expendable version of this instrument is also available (often designated XSV).

The velocimeter would seem to be the preferred instrument for obtaining measurements of sound speed in support of naval operations. From a broader scientific perspective, however, information on the temperature distribution in the ocean directly supports many other naval applications including ocean-dynamics modeling, air-sea interaction studies, and various marine-biological investigations, to name a few. The demonstrated accuracy of the various algorithms (when used within the specified domains of applicability) is sufficient for most operational applications. In terms of expendable sensors, measurements of temperature represent a better scientific investment than do measurements of sound speed. Nevertheless, naval operations in coastal areas or in ice-covered regions may be better supported by the velocimeter. Increasing emphasis on naval operations in the littoral zone has placed greater importance on measuring, in real time, coastal processes that are characterized by extreme temporal and spatial variability.

Precise scientific measurements are normally accomplished using expensive, but recoverable, instrumentation. Such instruments are typically deployed from surface ships that maintain station while the instrument package is lowered and then raised in what is termed a hydrographic cast. The instrument package can contain an assortment of sensors for measuring water temperature, salinity (or conductivity), pressure, sound speed, currents, and dissolved oxygen, among others.

2.3.2 Sound Speed Distribution

Typical North Atlantic winter and summer profiles of sound speed versus depth are shown in Figure 2.6. These profiles represent a region of the North Atlantic Ocean located near 23°N and 70°W (Naval Oceanographic Office, 1972). Temperature-salinity (*T–S*) diagrams for winter and summer seasons, based on actual measurements, are also presented to show their relationships to the sound speed profiles. Since the *T–S* diagrams indicate the ocean depths corresponding to the measured temperature-salinity pairs, the individual temperature and salinity profiles for both winter and summer can be reconstructed. Principal water masses are also noted on the sound speed profiles. The so-called 18-degree water (refer back to Figure 2.5) marks a change in the sound-speed gradient at a depth of about 300 m, and Mediterranean intermediate water (MIW) occupies the region of the water column near the sound-speed minimum (at about 1200 m).

The sound speed profiles in Figure 2.6 are representative of those encountered in many tropical and subtropical deep-ocean areas. Such profiles may be divided into arbitrary layers, each having different characteristics and occurrence (Figure 2.7). Just below the sea surface is the sonic layer where the speed of sound is influenced by local changes in heating, cooling and wind action. The base of the sonic layer is defined as the sonic layer depth (SLD), which is associated with the near-surface maximum in sound speed. This surface layer is usually associated with a well-mixed layer of near-isothermal water. Oceanographers refer to this well-mixed region as the mixed layer. The base of this layer is then termed the mixed layer depth (MLD).

Below the mixed layer lies the thermocline, a region of the water column where the temperature decreases rapidly with increasing depth. This region is characterized by a negative sound speed gradient (i.e., sound speed decreases with increasing depth).

Below the thermocline, and extending to the sea floor, is the deep isothermal layer. This layer has a nearly constant temperature in which the speed of sound increases with depth due to the effects of pressure (pressure increases with increasing depth). In this region of the water column, the sound speed profile becomes nearly linear with a positive gradient of about 0.017 (m s^{-1}) m^{-1} or 0.017 s^{-1}. Between the negative sound speed gradient of the thermocline and the positive sound speed gradient of the deep isothermal layer is a sound-speed minimum. The depth corresponding to this sound-speed minimum is referred to as the sound channel axis. At high latitudes, the deep isothermal layer extends nearly to the sea surface. That is, the sound channel axis shoals as one approaches the polar regions. This behavior is vividly

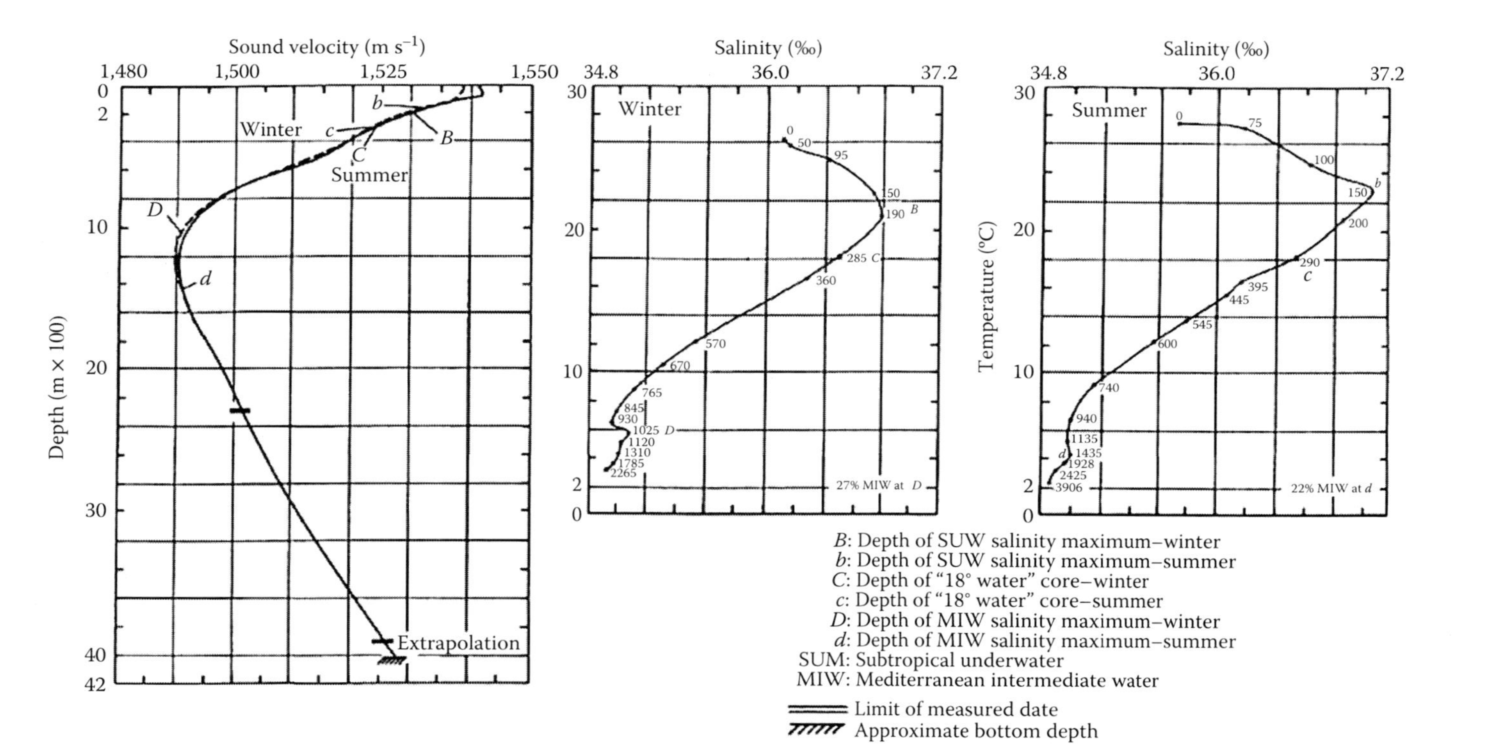

FIGURE 2.6 Sound speed profiles (winter and summer) and T–S comparisons for the North Atlantic Ocean near 23°N, 70°W. (Adapted from Naval Oceanographic Office, *Environmental-Acoustics Atlas of the Caribbean Sea and Gulf of Mexico. Volume II—Marine Environment,* SP-189II, US Dept. of the Navy, Washington, DC, 1972.)

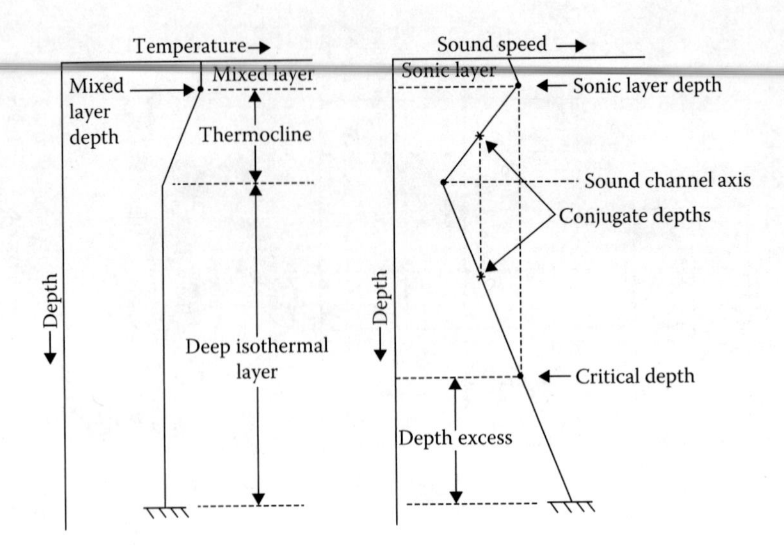

FIGURE 2.7 Schematic relationship between temperature and sound speed profiles in the deep ocean.

demonstrated in Figure 2.8. The top panel presents contours of the depth of the minimum sound speed (m) and the bottom panel presents the sound speed (ms^{-1}) on this axial surface (Munk and Forbes, 1989). At low latitudes, the depth of the sound channel axis is typically near 1000 m. At high latitudes, the axis is located near the sea surface. The associated sound speeds on this axial surface generally decrease away from the equatorial regions.

In profiles containing a sound channel axis, a critical depth can be defined as that depth below the axis at which the sound speed equals the near-surface maximum value. (The near-surface maximum value of sound speed is usually located at the SLD.) The vertical distance between the critical depth and the sea floor is referred to as the depth excess. Other pairs of points can be identified on the sound speed profile that have the same value of sound speed but which lie on opposite sides of the sound channel axis. Such pairs are referred to as conjugate depths. In Figure 2.7, the critical depth is actually a conjugate of the SLD.

It is convenient at this juncture to discuss additional points of morphology relating to the sound speed profile illustrated in Figure 2.7. These points will be useful in later discussions concerning full channels, half-channels, and ducts. A full channel is formed around the sound channel axis and, for purposes of discussion, is limited above the axis by the SLD and below the axis by the sea floor. Either the upper portion or the lower portion of the full channel represents a half-channel. The upper half-channel is characterized by a negative sound speed gradient (as is found in some shallow-water regions) while the lower half-channel is characterized by a positive sound speed gradient (as is found in the Arctic or in some shallow-water regions). In Figure 2.7, the sea surface and the SLD define the upper and lower boundaries, respectively, of a duct. The properties of acoustic propagation in a full channel, half-channel, and duct will be discussed in Chapter 3.

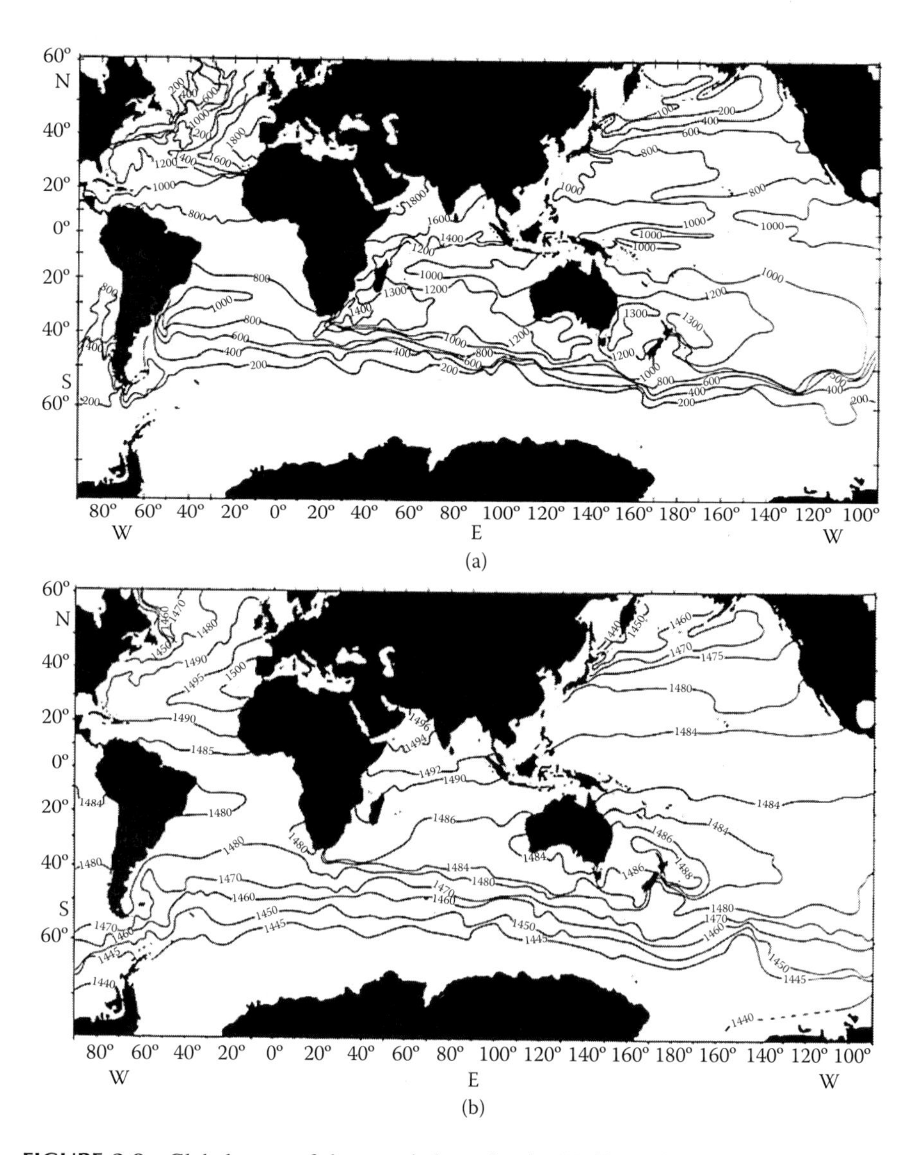

FIGURE 2.8 Global maps of the sound channel axis. (a) Channel depth (m). (b) Channel speed (m s^{-1}). (From Munk, W.H. and Forbes, A.M.G., *J. Phys. Oceanogr.*, 19, 1765–78, 1989. Copyright by the American Meteorological Society. With permission.)

Additional sound speed profiles typical of the winter season in different ocean areas of the world are presented in Figure 2.9. These additional profiles demonstrate that the simple model described above (specifically in Figure 2.7), which is representative of many tropical and subtropical deep-ocean areas, is not applicable to high-latitude ocean areas or to some smaller water bodies.

The depth dependence of sound speed in the ocean poses a particular problem for echo sounders, which use near-vertical acoustic paths to measure the depth of the sea

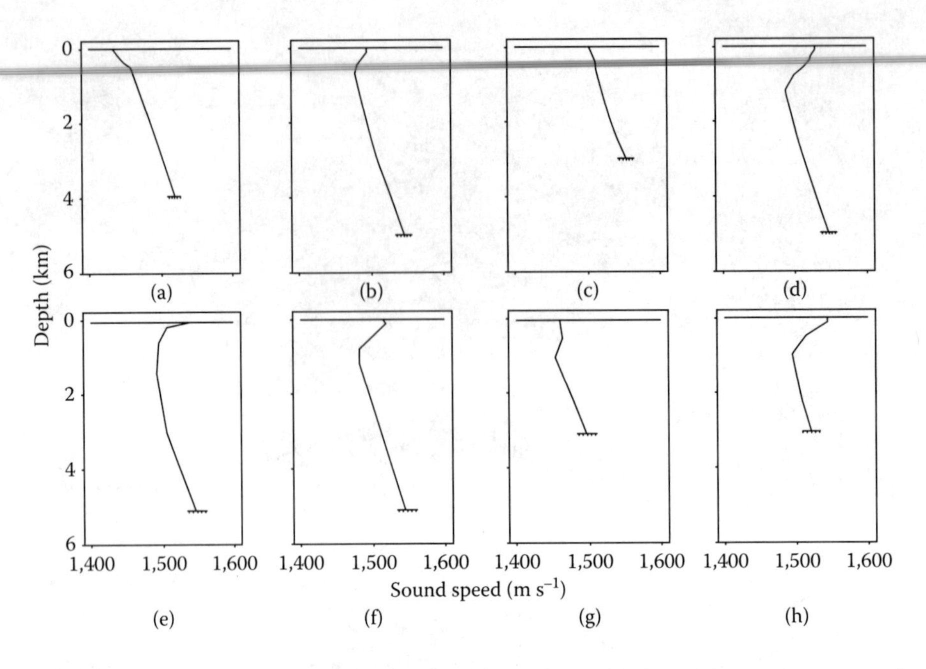

FIGURE 2.9 Characteristic winter sound-speed profiles for selected deep-ocean areas of the world. (a) Canada Deep; (b) North Pacific; (c) Western Mediterranean; (d) North Atlantic; (e) North Indian; (f) South Atlantic; (g) Norwegian Sea; and (h) Eastern Gulf of Mexico.

floor based on the two-way travel time of the signal. Echo sounders are set to read the depth directly by assuming a constant speed of sound in the water column, usually 1463 m s^{-1} or 1500 m s^{-1}. When the actual depth-integrated (or mean) sound speed departs from the assumed value, a correction must be applied to the observed readings. Bialek (1966: 63), for example, tabulated such corrections according to ocean area. Depending on the particular ocean area and water depth, these corrections can be on the order of several percent of the true water depth. It is important to check navigational charts to see if any such correction has been applied to the soundings. If no corrections have been made, then care should be exercised in ascertaining the true bottom depth before undertaking any deep-water operations in proximity to the sea floor.

Underwater acoustic propagation problems involving long ranges may not be able to ignore horizontal variations in either the sound speed or bathymetry. Modeling developments, therefore, generally distinguish between range-independent (1D, where the ocean varies only as a function of depth) and range-dependent (2D, where the ocean varies as a function of both depth and range) problems. Parameters other than sound speed and water depth may also be considered in range-dependent problems such as surface losses, bottom losses, and absorption. If, in addition to the 2D problem, there are azimuthal variations, then the problem is considered to be 3D.

An east-west cross section of the North Atlantic Ocean between 23°N and 24°N is presented in Figure 2.10 to illustrate the variability of both sound speed and bathymetry over moderate range scales. Also noteworthy are changes in the depth of the sound channel axis and the absence of a critical (or limiting) depth in some basins.

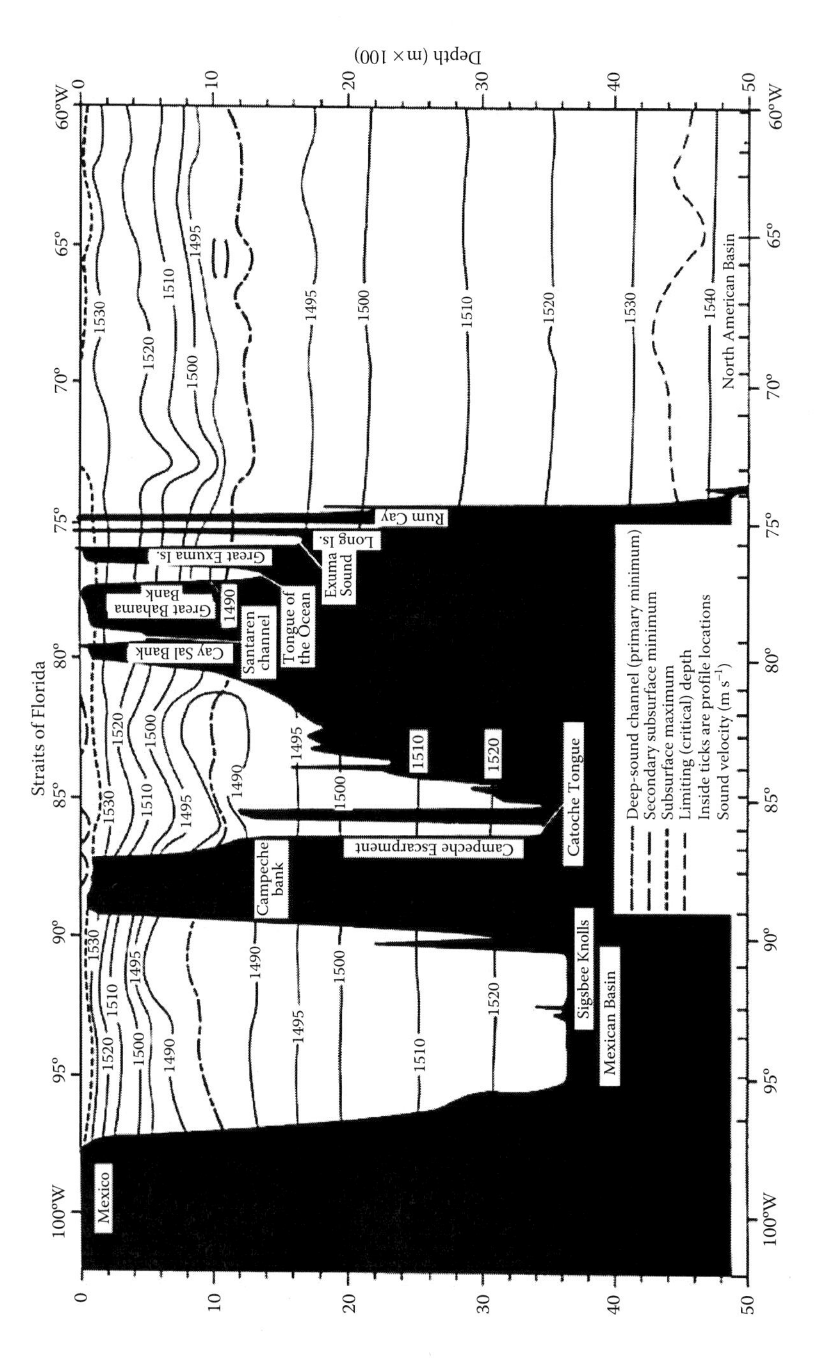

FIGURE 2.10 East-west sound speed cross section between 23°N and 24°N and between 60°W and 100°W for the period February–April. (Adapted from Naval Oceanographic Office, *Environmental-Acoustics Atlas of the Caribbean Sea and Gulf of Mexico. Volume II—Marine Environment*, SP-189II, U.S. Dept. of the Navy, Washington, DC, 1972.)

The sound speed profiles presented previously in Figure 2.6 are consistent with the data of Figure 2.10 at a longitude of about 70°W.

2.4 BOUNDARIES

The boundaries of the water column—the sea surface and the sea floor—can exert a profound influence on the propagation of acoustic energy through the actions of reflection, scattering and absorption.

2.4.1 Sea Surface

The surface of the sea is both a reflector and a scatterer of sound. If the sea surface were perfectly smooth, it would form an almost perfect reflector of sound due to the acoustic impedance mismatch at the air-water interface. As the sea surface becomes rough, as it does under the influence of wind (refer to Table 2.2), reflection losses are no longer near zero.

Sea-surface roughness is typically specified in terms of wave height. At sea, however, weather observers generally record wind speed and not wave height as a description of sea state (Bowditch, 1977). A number of statistical relationships exist with which to quantitatively associate these two parameters (e.g., Earle and Bishop, 1984). These relationships are very precise as to the height above sea level at which the wind speed is measured and the type of statistical wave height that is considered. Also factored in are the duration of the wind and the fetch (i.e., the distance along open water over which the wind acts from the same direction). Based on the Pierson–Moskowitz spectrum (Moskowitz, 1964; Pierson and Moskowitz, 1964; Pierson, 1964, 1991), a fully developed significant wave height can be calculated from the observed wind speed as

$$H_{1/3} = 0.566 \times 10^{-2} V^2 \tag{2.4}$$

where V is the wind speed (knots), measured at a height of 19.5 m, and $H_{1/3}$ is the average height (m) of the one-third highest waves.

Two other commonly used wave-height descriptors are the rms wave height (H_{rms}) and the one-tenth significant wave height ($H_{1/10}$). These are related to the one-third significant wave height as (Earle and Bishop, 1984)

$$H_{rms} = 0.704 H_{1/3} \tag{2.5}$$

and

$$H_{1/10} = 1.80 H_{rms} \tag{2.6}$$

The average wave height (H_{avg}) can be related to the rms wave height (H_{rms}) as

$$H_{avg} = \frac{\sqrt{\pi}}{2} H_{rms} = 0.886 H_{rms} \tag{2.7}$$

TABLE 2.2
Beaufort Wind Scale with Corresponding Sea State Codes

	Wind Speed					Estimating Wind Speed			Sea State	
Beaufort Number or Force	Knots	Miles per Hour (mph)	Meters per Second	km per Hour	World Meteorological Organization (1964)	Effects Observed Far from Land	Effects Observed Near Coast	Effects Observed on Land	Term and Height of Waves (m)	Code
0	Under 1	Under 1	0.0–0.2	Under 1	Calm	Sea like mirror.	Calm.	Calm; smoke rises vertically.	Calm, glassy (0)	0
1	1–3	1–3	0.3–1.5	1–5	Light air	Ripples with appearance of scales; no foam crests.	Fishing smack just has steerage way.	Smoke drift indicates wind direction; vanes do not move.		
2	4–6	4–7	1.6–3.3	6–11	Light breeze	Small wavelets; crests of glassy appearance, not breaking.	Wind fills the sails of smacks, which then travel at about 1–2 miles per hour.	Wind felt on face; leaves rustle; vanes begin to move.	Calm, rippled (0–0.1)	1
3	7–10	8–12	3.4–5.4	12–19	Gentle breeze	Large wavelets; crests begin to break; scattered whitecaps.	Smacks begin to careen and travel about 3–4 miles per hour.	Leaves, small twigs in constant motion; light flags extended.	Smooth, wavelets (0.1–0.5)	2
4	11–16	13–18	5.5–7.9	20–28	Moderate breeze	Small waves, becoming longer; numerous whitecaps.	Good working breeze, smacks carry all canvas with good list.	Dust, leaves, and loose paper raised up; small branches move.	Slight (0.5–1.25)	3
5	17–21	19–24	8.0–10.7	29–38	Fresh breeze	Moderate waves, taking longer form; many whitecaps; some spray.	Smacks shorten sail.	Small trees in leaf begin to sway.	Moderate (1.25–2.5)	4
6	22–27	25–31	10.8–13.8	39–49	Strong breeze	Larger waves forming; whitecaps everywhere; more spray.	Smacks have doubled reef in mainsail; care required when fishing.	Larger branches of trees in motion; whistling heard in wires.	Rough (2.5–4)	5
7	28–33	32–38	13.9–17.1	50–61	Near gale	Sea heaps up; white foam from breaking waves begins to be blown in streaks.	Smacks remain in harbor and those at sea lie-to.	Whole trees in motion; resistance felt in walking against wind.	Very rough (4–6)	6

(Continued)

TABLE 2.2 (Continued)
Beaufort Wind Scale with Corresponding Sea State Codes

	Wind Speed					Estimating Wind Speed			Sea State	
Beaufort Number or Force	Knots	Miles per Hour (mph)	Meters per Second	km per Hour	World Meteorological Organization (1964)	Effects Observed Far from Land	Effects Observed Near Coast	Effects Observed on Land	Term and Height of Waves (m)	Code
8	34–40	39–46	17.2–20.7	62–74	Gale	Moderately high waves of greater length; edges of crests begin to break into spindrift; foam is blown in well-marked streaks.	All smacks make for harbor, if near.	Twigs and small branches broken off trees; progress generally impeded.		
9	41–47	47–54	20.8–24.4	75–88	Strong gale	High waves; sea begins to roll; dense streaks of foam; spray may reduce visibility		Slight structural damage occurs; slate blown from roofs.		
10	48–55	55–63	24.5–28.4	89–102	Storm	Very high waves with overhanging crests; sea takes white appearance as foam is blown in very dense streaks; rolling is heavy and visibility is reduced.		Seldom experienced on land; trees broken or uprooted; considerable structural damage occurs.	High (6–9)	7
11	56–63	64–72	28.5–32.6	103–117	Violent storm	Exceptionally high waves; sea covered with white foam patches; visibility still more reduced.		Very rarely experienced on land; usually accompanied by widespread damage.	Very high (9–14)	8
12	64 and over	73 and over	32.7 and over	118 and over	Hurricane	Air filled with foam; sea completely white with driving spray; visibility greatly reduced.			Phenomenal (over 14)	9

Source: Adapted from Bowditch, N, *American Practical Navigator*, Vol. I. Defense Mapping Agency Hydrographic Center, Pub. No. 9, Washington, DC, 1977. (Continuously maintained since first published in 1802.)

These relationships can be compared with those presented in Table 2.2, which closely follow the values of H_{avg}.

This aspect is important since many underwater acoustic propagation models require an input of wind speed or wave height (and sometimes both) for specification of sea-surface roughness. This input is used to initialize an internal submodel that generates surface losses. If the wrong statistical relationship is used for the wave heights, then errors can be introduced into the computed solution due to an improper characterization of surface roughness. It is therefore important to understand what type of statistical wave height is expected as an input by the propagation model being used.

Wind-wave generation in coastal regions may be limited by the geometry of the water body, which is often irregular. Therefore, it may be necessary to consider the effects of fetch shape (both distance and width) in order to estimate wave spectra in coastal environments, especially when sea conditions are not fully developed. The Pierson–Moskowitz spectrum, which was discussed earlier for open-ocean applications, assumed that sea conditions were fully developed.

Air bubbles are produced by the breaking of waves and are carried beneath the surface by turbulence. They are also generated in the wakes of ships where they can persist for long periods of time. Free air bubbles in the sea are quite small since the larger bubbles tend to rise quickly to the surface. Bubbles only form a very small volumetric percentage of the sea. However, because air has a markedly different density and compressibility from that of sea water and because of the resonant characteristics of bubbles (e.g., Leighton, 1994), the suspended air content of sea water has a profound effect upon underwater sound. Urick (1983: 24954) summarized these effects, which include resonance and changes in the effective sound speed. Norton et al. (1998) developed a numerical procedure to parameterize bubble clouds in terms of an effective complex index of refraction for use in high-fidelity models of forward propagation.

Aside from reflection losses, there are other acoustic effects associated with interactions with the sea surface. A moving sea surface produces frequency-smearing and shifting effects on constant-frequency signals. Large and rapid fluctuations in amplitude or intensity are also produced by reflection at the sea surface. Furthermore, Lloyd mirror (or image-interference) effects produce a pattern of constructive and destructive interference between direct and surface-reflected signals. This effect is diminished when the sea surface is roughened by wind.

2.4.2 Ice Cover

When the sea surface is covered by an ice canopy, as in the polar regions (see Figures 2.11 and 2.12), acoustic interaction with the surface is further complicated by an irregular under-ice surface. The Arctic environment can be segregated into three distinct regions according to the type of ice cover: (1) pack ice, (2) marginal ice zone (MIZ), and (3) open ocean. Field measurements have shown that forward scatter from a rough anisotropic ice canopy is a function of acoustic frequency, geometry and the statistical (spatial correlation) properties of the under-ice surface.

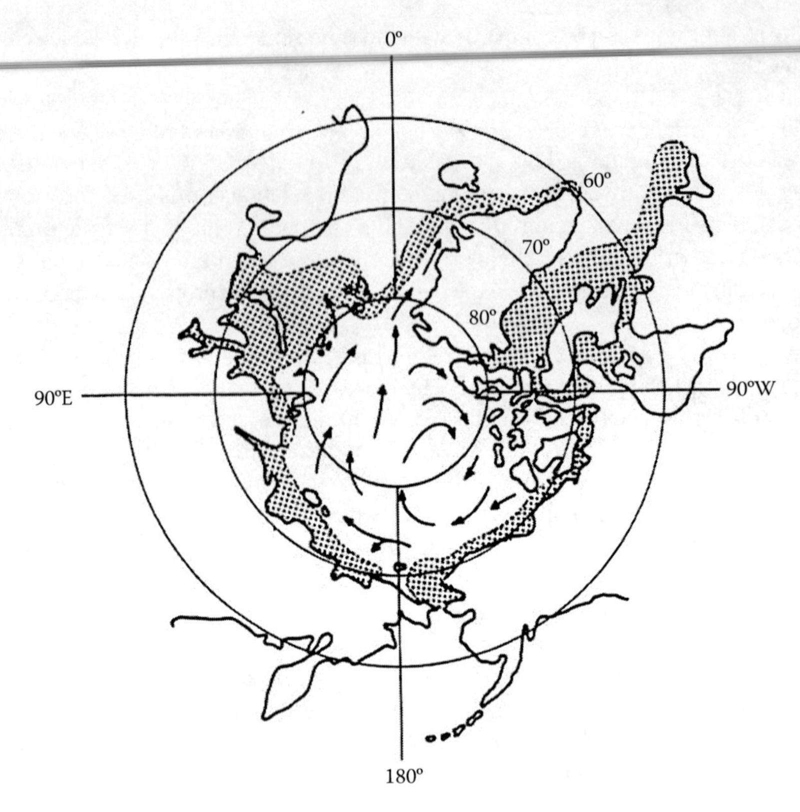

FIGURE 2.11 Average boundaries of sea ice (coverage at least 5-8 tenths) in autumn and spring in the Arctic. Arrows indicate the general drift pattern. The width of the stippled area indicates the range of ice limits between autumn and spring. (From Untersteiner, N., In *The Encyclopedia of Oceanography*, Van Nostrand Reinhold, New York, pp. 777–781, 1966. With permission.)

2.4.3 Sea Floor

The sea floor is a reflecting and scattering boundary having a number of characteristics similar in nature to those of the sea surface. Its effects, however, are more complicated than those of the sea surface because of its diverse and multilayered composition. Specifically, the sea floor is often layered, with a density and sound speed that may change gradually or abruptly with depth or even over short ranges. Furthermore, the sea floor is more variable in its acoustic properties since its composition may vary from hard rock to soft mud. One feature that is distinct from the sea surface is that the bottom characteristics can be considered to be nearly constant over time (except in regions of high bioturbation or in areas characterized by high rates of sediment deposition or erosion), whereas the configuration of the sea surface is statistically in a state of change as the wind velocity changes. McCall and Tevesz (2013) discussed the effects of marine benthos on the physical properties of sediments as well as mathematical models of bioturbation.

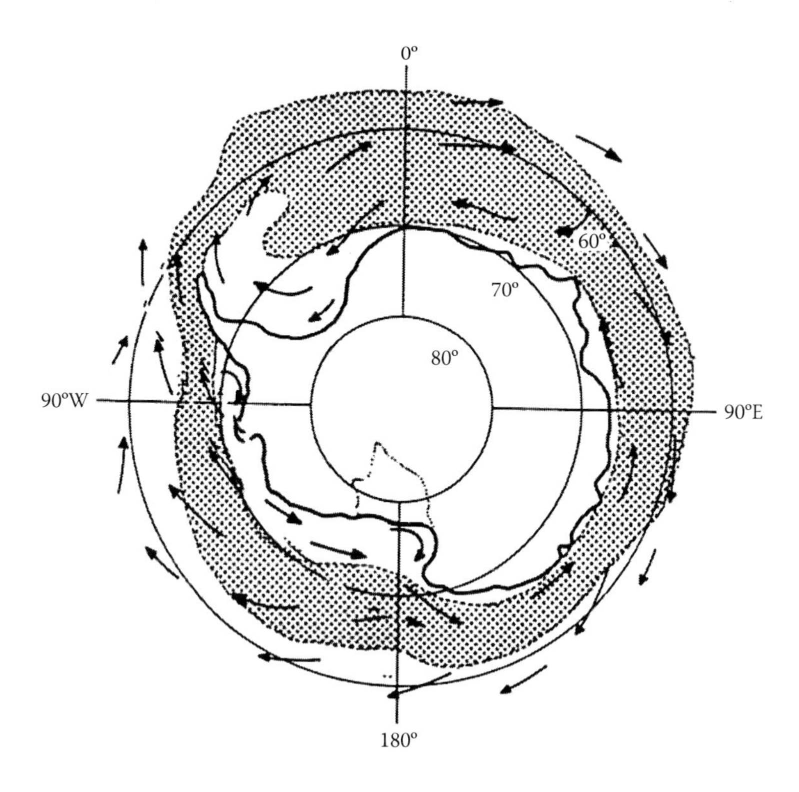

FIGURE 2.12 Average boundaries of sea ice (coverage at least 5-8 tenths) in autumn and spring in the Antarctic. Arrows indicate the general drift pattern. The width of the stippled area indicates the range of ice limits between autumn and spring. (From Untersteiner, N., In *The Encyclopedia of Oceanography*, Van Nostrand Reinhold, New York, pp. 777–781, 1966. With permission.)

Because of the variable stratification of the bottom sediments in many areas, sound is often transmitted into the bottom where it is refracted or internally reflected. Thus, the bottom often becomes a complicated propagating medium that is characterized by both shear and compressional sound speeds.

The topography of the sea floor exhibits a diversity of features not unlike those of the continental landmasses. Figure 2.13 presents an artist's conception of common ocean basin features. The undersea features noted in Figure 2.13 are defined in the Glossary (Appendix B). Stagpoole et al. (2016) described a new web resource that enables scientists to standardize the naming of seamounts, trenches, and other undersea features, thus reducing ambiguity in identification and communication. Newly discovered undersea features are often given informal names that, with repeated use over time, become incorporated into maps and scientific papers. Historically, this has sometimes led to confusion and misidentification. The general bathymetric chart of the oceans (GEBCO) subcommittee on undersea feature names (SCUFN) is developing new web applications to facilitate collaboration and coordination for naming undersea features within the world's oceans. These future

FIGURE 2.13 Common seafloor features. (Adapted from Bowditch, N., *American Practical Navigator*, Vol. I. Defense Mapping Agency Hydrographic Center, Pub. No. 9, Washington, DC, 1977. Continuously maintained since first published in 1802.)

applications will assist users in completing name proposals and also facilitate the review and approval process.

Underwater ridges and seamounts can effectively block the propagation of sound, an occurrence that is referred to as bathymetric blockage. Moreover, when actively ensonified, seamounts can mask targets of interest by either providing false targets or by shadowing targets of interest.

Marine seismic studies have greatly improved our understanding of the crustal structure of that part of Earth covered by the oceans (Bryan, 1967). A typical bottom-structure section is presented in Figure 2.14. This typical section consists of 5–6 km of water, about 0.5 km of unconsolidated sediments, 1–2 km of basement rock and 4–6 km of crustal rock overlying the upper mantle.

The global pattern of ocean-sediment thickness can be characterized by (1) a thick apron around the continents; (2) an equatorial band originating from high biogenic productivity due to upwelling; and (3) vast areas underlain by sediments that are very thin, reflecting the young age of many of the ocean basins, with the thinnest sediments found along the mid-ocean ridge systems. Major portions of the sea floor are covered with unconsolidated sediments with an average thickness of approximately 500 m. Sediments can be classified according to their origin as either terrigenous or pelagic, although no single classification scheme has universal approval. The general distribution of sediments according to Arrhenius (1963) is presented in Figure 2.15.

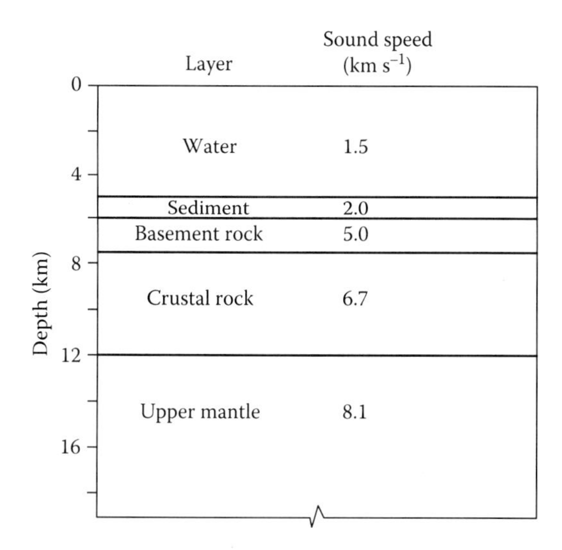

FIGURE 2.14 Typical oceanic section with associated depths and sound speeds in the water column and subbottom layers. (From Bryan, G.M., In *Underwater Acoustics*, Vol. 2, Plenum Press, New York, pp. 351–361, 1967. With permission.)

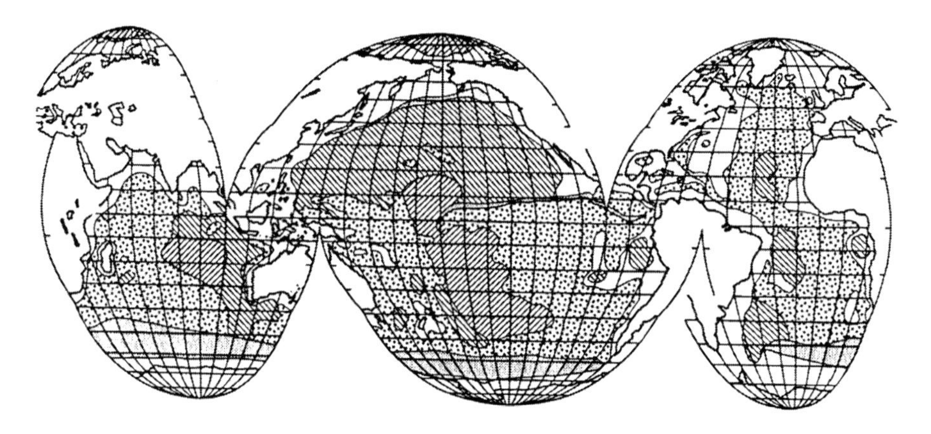

FIGURE 2.15 Idealized distribution of marine sediments. Hatched areas indicate inorganic pelagic and terrigenous sediments. Stippled areas denote organic pelagic sediments comprising calcareous and silicious oozes. (Adapted from Arrhenius, G. *The Sea*, 3, 655–727, 1963. Copyright Wiley-VCH Verlag GmbH & Co. KGaA. Reprinted with permission.)

Terrigenous sediments are derived from land and are particularly prominent near the mouths of large rivers. These sediments are generally classified as silt, sand, and mud. Pelagic sediments are derived from either organic or inorganic sources. Organic pelagic sediments comprise the remains of dead organisms and are further classified as either calcareous or siliceous oozes. Inorganic pelagic sediments are derived from materials suspended in the atmosphere and are generally classified as clay.

2.5 DYNAMIC FEATURES

It is convenient to categorize dynamic features of the ocean according to characteristic time and space scales. While no precise terminology is universally accepted, it is common to recognize space scales as large (> 100 km), meso (100 m–100 km), and fine (< 100 m). Time scales are less precise, but generally distinguish among seasonal, monthly, inertial, tidal, and other feature-specific time scales appropriate to currents, eddies, waves (both surface and internal), and turbulence, among others.

2.5.1 Large-Scale Features

The large-scale circulation of the ocean can be classified either as wind-driven or thermohaline. The former is due to wind stress acting on the sea surface, while the latter is due, in part, to density changes arising from variations in temperature and salinity (e.g., Pickard, 1963).

The wind-driven component of circulation is usually horizontal in nature and is restricted primarily to the upper few hundred meters of the ocean. In the case of upwelling or downwelling near coasts, the original horizontal flow is forced to become vertical due to the basin geometry. In the open ocean, bands of upwelling and downwelling can be created by divergence and convergence, respectively, of wind-generated surface currents. Such currents are commonly referred to as Ekman drift currents.

The thermohaline component normally originates as a vertical flow arising from imbalances in the heat or freshwater (salt) fluxes near the sea surface. These vertical flows eventually become horizontal at a depth that is consistent with the density of the newly formed water. See van Aken (2006) for additional details on the oceanic thermohaline circulation.

Broecker (1997) coined the term thermohaline "conveyor" circulation to refer to a globally interconnected system of thermohaline ocean currents that govern Earth's climate by transporting heat and salt around the planet (salt is an inverse measure of the freshwater content). At present, the driving force of the "conveyor" is the cold, salty water of the North Atlantic Ocean. This water is denser than the surrounding warm, fresh water and thus sinks to the ocean bottom. This deep current runs southward to the southern tip of Africa, where it joins an undersea stream that encircles Antarctica. Here, the "conveyor" is recharged by cold, salty water created by the formation of sea ice, which leaves salt behind when it freezes. This renewed sinking propels water back northward, where it gradually warms again and rises to the surface in the Pacific and Indian Oceans. In the Indian Ocean, surface waters are too warm to sink. Moreover, northern Pacific waters are cold, but not salty enough to sink into the deep. This is primarily due to the input of fresh water from coastal rainfall along the western United States and Canada. The "conveyor" comes full circle, eventually propelling warm surface waters back into the North Atlantic. It is estimated that a roundtrip on the "conveyor" would take on the order of 1000 years to complete. In winter, these warm waters transfer their heat to the overlying cold air masses that advect off ice-covered Canada, Greenland, and Iceland. These eastward-moving air masses temper the climates of northern Europe, making them warmer in winter than at comparable latitudes in North America. Broecker (2010) further elaborated on this concept in his recent book.

The circulation of the surface waters corresponds closely to the prevailing wind patterns. The water currents set in motion by the winds would encircle the globe in the absence of land masses. The land masses, however, obstruct the flow and force the water along the coasts, thus forming completed loops (or gyres) in the Pacific (Figure 2.16) and Atlantic Oceans (Figure 2.17). The effect of changing wind patterns due to monsoons is evident in the Indian Ocean (Figure 2.18). Only in the Antarctic Ocean does the absence of land masses permit true circumpolar flow at a latitude of about 60°S (Figures 2.16 through 2.18).

2.5.2 Mesoscale Features

Mesoscale oceanic features of importance to underwater acoustics include fronts, eddies (or rings) and internal waves.

2.5.2.1 Fronts and Eddies

Ocean frontal features are frequently associated with major ocean currents, or with vertical circulation patterns in areas of upwelling or downwelling. An example of

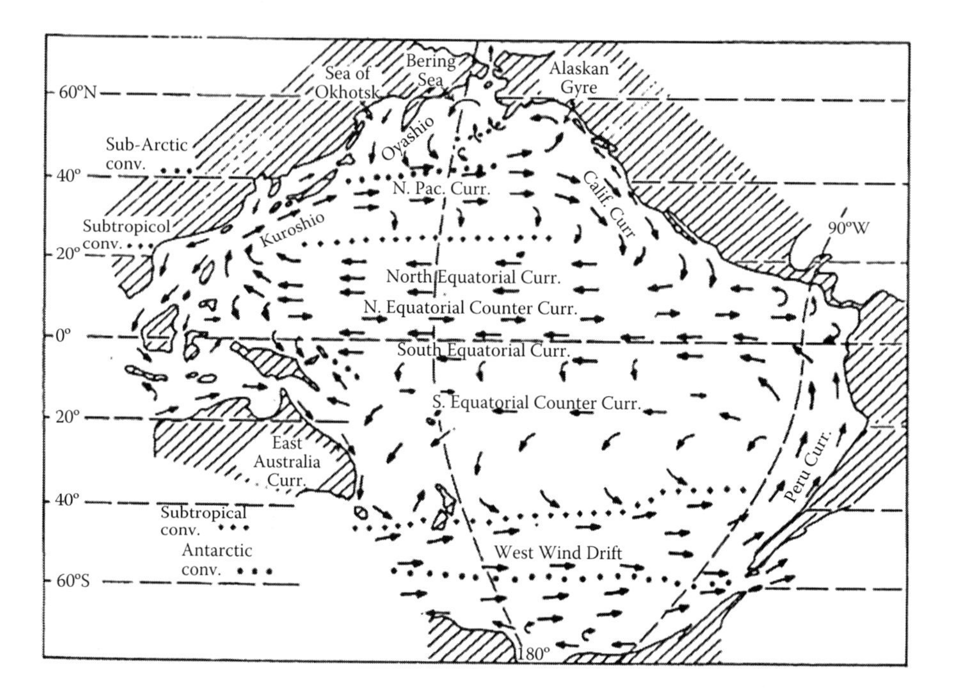

FIGURE 2.16 Surface circulation of the Pacific Ocean. (Reprinted with permission from Pickard, G.L., *Descriptive Physical Oceanography: An Introduction*. Pergamon Press, New York. Copyright by Pergamon Press Plc, 1963.)

FIGURE 2.17 Surface circulation of the Atlantic Ocean. (Reprinted from Pickard, G.L., *Descriptive Physical Oceanography: An Introduction*. Pergamon Press, New York. Copyright by Pergamon Press Plc, 1963. With permission.)

ocean fronts associated with major ocean currents is shown in Figure 2.19 (Naval Oceanographic Office, 1967). This figure represents a range-depth section crossing the North Atlantic Ocean between Newfoundland (left) and Senegal (right). The hydrographic station numbers used in creating this section are shown at the top. Between station 6 and station 23, the isotherms are relatively horizontal. Two adjacent frontal features are evidenced by the vertically oriented isotherms between station 3 and station 6. These features correspond to the cold, southward-flowing Labrador Current (centered at station 4) and the warm, northward-flowing Gulf Stream (centered at station 5). The effects of the Gulf Stream frontal system are evident at depths exceeding 1000 m.

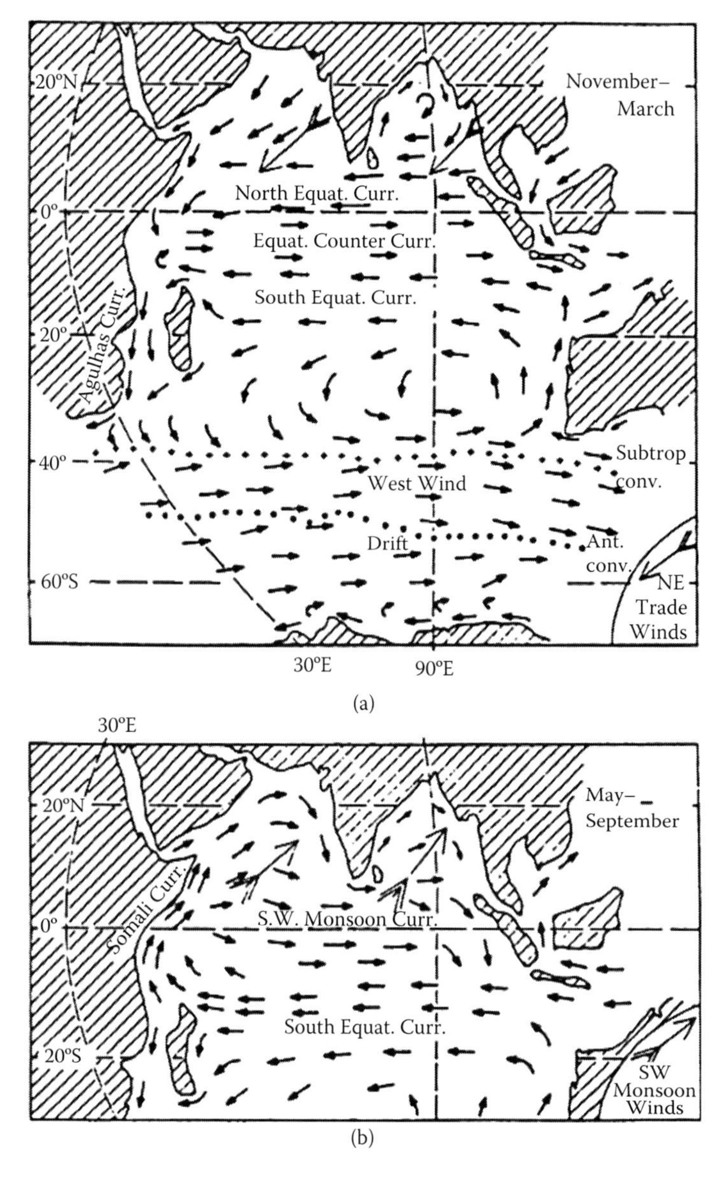

FIGURE 2.18 Surface circulation of the Indian Ocean showing effects of the monsoon seasons: (a) NE trade winds; (b) SW monsoon winds. (Reprinted from Pickard, G.L., *Descriptive Physical Oceanography: An Introduction*. Pergamon Press, New York. Copyright by Pergamon Press Plc, 1963. With permission.)

Different water masses are commonly separated by a transition zone referred to as an ocean front. The degree of abruptness in the change of water-mass characteristics (particularly temperature and salinity, and thus sound speed) determines whether the front is classified as a strong or weak front. Ocean fronts are similar in concept to the more familiar fronts encountered in meteorology that separate different air masses.

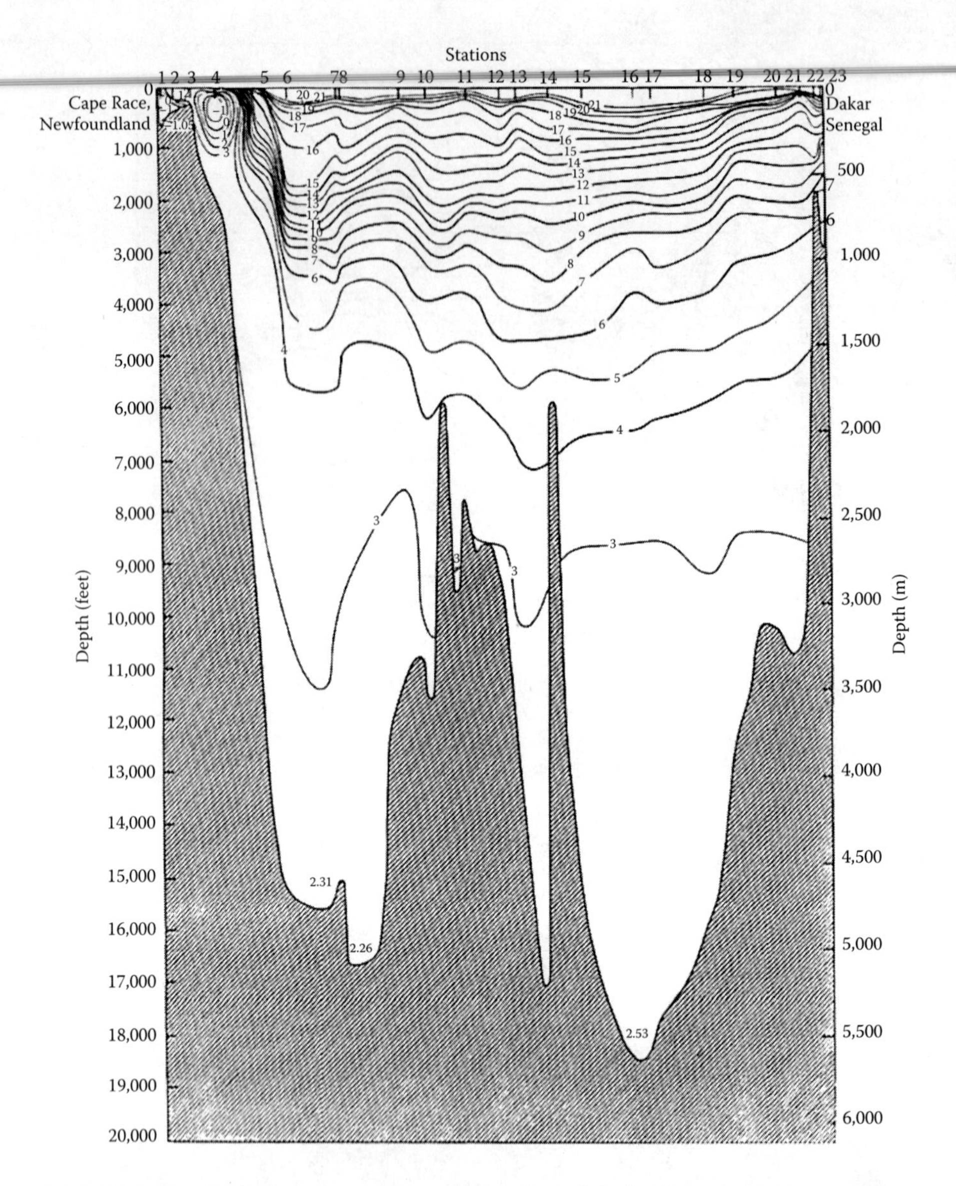

FIGURE 2.19 Vertical section of water temperature (°C) for the transect between Cape Race, Newfoundland (left), and Dakar, Senegal (right). (Adapted from Naval Oceanographic Office, *Oceanographic Atlas of the North Atlantic Ocean. Section II—Physical Properties,* Pub. No. 700, US Dept. of the Navy, Washington, DC, 1967.)

Cheney and Winfrey (1976) summarized the classification and distribution of ocean fronts. For underwater acoustic applications, they recommended the following definition of an ocean front: A front is any discontinuity in the ocean that significantly alters the pattern of sound propagation and transmission loss. Thus, a rapid change in the depth of the sound channel, a difference in SLD, or a temperature inversion would denote the presence of a front. In terms of acoustics, Cheney and Winfrey (1976) identified the following significant effects:

1. Surface sound speed can change by as much as 30 m s^{-1}. Although this is due to the combined effect of changing temperature and salinity, temperature is usually the dominant factor.
2. Differences in SLD on the order of 300 m can exist on opposite sides of a front during certain seasons.
3. A change in in-layer and below-layer gradient usually accompanies a change in surface sound speed and SLD.
4. Depth of the sound channel axis can change by 750 m when crossing from one water mass to the next.
5. Increased biological activity generally found along a front will increase ambient noise and reverberation levels.
6. Enhanced air-sea interaction along a frontal zone can cause a dramatic change in sea state and thus increase ambient noise levels and surface roughness.
7. Refraction of sound rays as they pass through a front at an oblique angle can cause bearing errors in sonar systems.

A summary of the positions of prominent fronts is presented in Figure 2.20 (accompanied by Table 2.3). Each front is characterized as strong, moderate, or

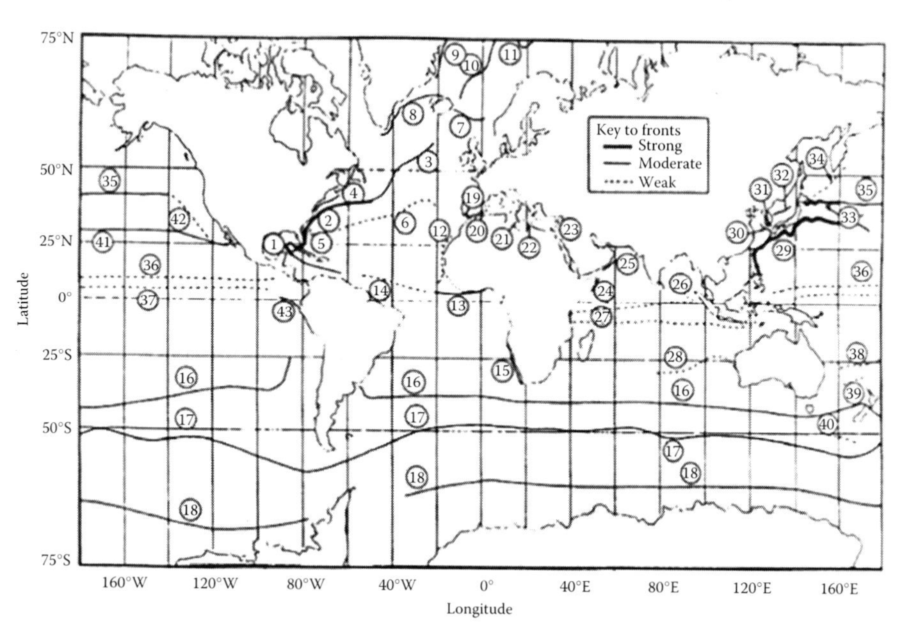

FIGURE 2.20 Global distribution of ocean fronts. Numbers correspond to the generally accepted names, as listed in Table 2.3. Only the mean positions are presented. Due to seasonal variability, frontal locations can shift by as much as 200 km. (Adapted from Cheney, R.E. and Winfrey, D.E., Distribution and classification of ocean fronts. Nav. Oceanogr. Off., Tech. Note 3700-56-76, 1976.)

TABLE 2.3
Names of Ocean Fronts Shown in Figure 2.20

Atlantic Ocean Fronts

1. Loop Current (Gulf of Mexico)
2. Gulf Stream
3. North Atlantic Current (north polar front)
4. Slope front
5. Sargasso Sea front
6. Subtropical convergence
7. Iceland-Faeroe Islands front
8. Denmark Strait front
9. East Greenland polar front
10. Greenland-Norwegian Sea front
11. Bear Island front
12. Northwest African upwelling
13. Gulf of Guinea front
14. Guiana Current
15. Benguela upwelling
16. Subtropical convergence
17. Antarctic convergence (south polar front)
18. Antarctic divergence

Mediterranean Sea Fronts

19. Huelva front
20. Alboran Sea front
21. Maltese front
22. Ionian Sea front
23. Levantine Basin front

Indian Ocean Fronts

24. Somali upwelling
25. Arabian upwelling
26. Indian Ocean salinity front
27. Equatorial Countercurrent fronts
28. West Australian front

Pacific Ocean Fronts

29. Kuroshio front
30. Yellow Sea Warm Current
31. Korean coastal front
32. Tsushima Current
33. Oyashio front
34. Kuril front
35. Subarctic front
36. North Doldrum salinity front
37. South Doldrum salinity front
38. Tropical convergence
39. Mid Tasman convergence
40. Australian Subarctic front
41. Subtropical front
42. California front
43. East Pacific equatorial front

weak depending on representative values for (1) the maximum change in sound speed across a front; (2) change of SLD; (3) depth to which the front extends; and (4) persistence.

Detached eddies, such as are found in the Atlantic Ocean near the Gulf Stream and in the Pacific Ocean near the Kuroshio Current, represent another class of ocean fronts since they are separate water-mass entities contained within their own enclosed circulation and surrounded by water having different characteristics (Kerr, 1977).

Cold-core Gulf Stream rings form on the south side of the Gulf Stream. These rings are typically 100–300 km in diameter and rotate cyclonically (i.e., counterclockwise in the northern hemisphere) at speeds up to 3 knots at the surface (The Ring Group, 1981). These rings generally drift in a southwestward direction in the Sargasso Sea and can persist for up to a year before losing their identity or being reabsorbed into the Gulf Stream (Figure 2.21). It has been estimated that five to eight cold-core rings form each year (Lai and Richardson, 1977). These rings have cold cores of slope water encircled by warm Sargasso Sea waters, as illustrated in the temperature section of Figure 2.22. Therefore, these rings represent strong thermal anomalies.

Warm-core Gulf Stream rings form on the north side of the Gulf Stream, trapping pockets of warm Sargasso Sea water in the cold slope water (Figure 2.21). The

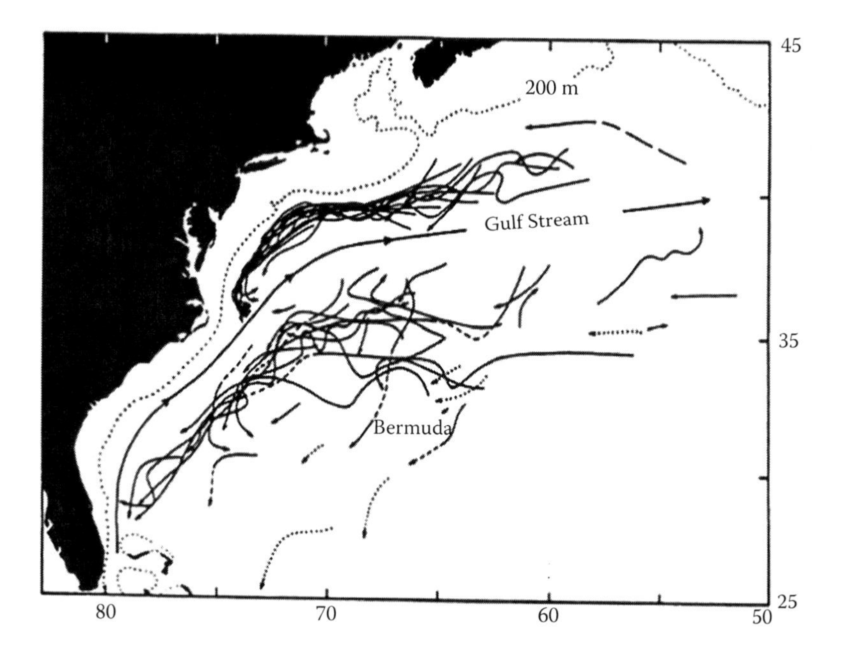

FIGURE 2.21 Trajectories of Gulf Stream rings. Cold rings south of the stream: unbroken lines represent inferred ring time series; dashed and dotted lines indicate estimated trajectories of rings based on incomplete data. Warm rings north of the stream: trajectories are confined between the continental slope and the Gulf Stream. (From Lai, D.Y. and Richardson, P.L., *J. Phys. Oceanogr.*, 7, 670–683, 1977. Copyright by the American Meteorological Society. With permission.)

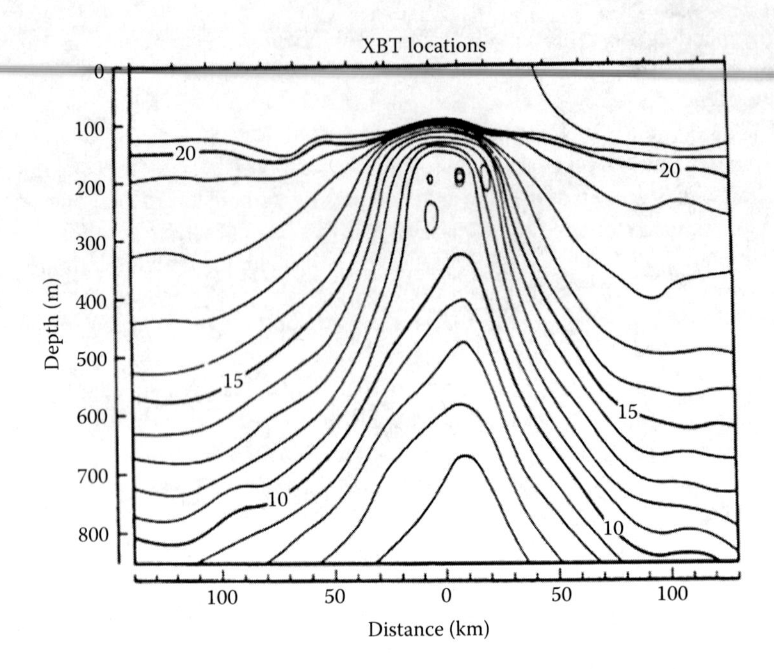

FIGURE 2.22 Temperature (°C) section through a cyclonic (cold-core) Gulf Stream ring as observed during December 1975 near 36°N, 58°W. This cross section, which is oriented southwest to northeast, shows that the central core of the ring contains cold, low-saline slope water with isotherms rising about 500 m above the normal depth in the surrounding Sargasso Sea water. The overall size of the ring is approximately 200 km. (From Lai, D.Y. and Richardson, P.L., *J. Phys. Oceanogr.*, 7, 670–683, 1977. Copyright by the American Meteorological Society. With permission.)

resulting circulation is anticyclonic (i.e., clockwise in the northern hemisphere). An average of 22 warm-core rings are formed annually between 75°W and 44°W (Auer, 1987).

Another example of rings is provided by Goni and Johns (2001), who conducted a census of anticyclonic rings shed by the North Brazil Current in the western tropical Atlantic off the northeast coast of South America near 8° N latitude. (The Guiana Current, which is noted on the ocean frontal chart of Figure 2.20, is fed by the North Brazil Current and is a major source of water for the Caribbean Sea.) These rings were tracked over the period October 1992 through December 1998 using sea-height anomaly data derived from TOPEX/Poseidon satellite radar altimeters. On average, five rings were formed each year with estimated translation speeds of 14 km per day. One in six of these rings penetrated into the Caribbean Sea through the southern Lesser Antilles while the rest followed a northern trajectory past Barbados. Available data suggest that the vertical structures of these rings can vary widely. Some rings may have intense subsurface structure but weak surface signatures, which would make them difficult to detect from satellites.

The acoustic impacts of ocean fronts and eddies will be further explored in Section 6.8.1.

2.5.2.2 Internal Waves

Internal waves are subsurface waves that propagate along interfaces separating fluid layers of different densities. They can also exist within fluid layers where vertical density gradients are present. These waves can be generated by a number of mechanisms including surface waves, wind forcing, submarine earthquakes, submarine landslides, air pressure changes, and current shears, among others.

In the open ocean, internal waves appear to take the form of progressive waves. In partially closed water bodies, standing internal waves are generally found. Internal waves are commonly observed over continental shelves. Such waves are probably generated by the scattering of the barotropic tide into baroclinic modes at the edge of the shelf, and then propagate shoreward on the shelf where they are absorbed and reflected as they break on the sloping bottom. Evidence suggests that such internal waves may have lifetimes of several days on the shelf. Over the shelf, internal waves may be manifested as solitary wave packets (or solitons). A helpful review paper regarding the occurrence of solitons in the ocean was prepared by Apel et al. (2007). Also see the comprehensive treatment of nonlinear ocean waves in a book by Osborne (2010).

The distribution of internal wave amplitudes as a function of depth is influenced by the vertical density distribution. Specifically, as the density boundary weakens, the amplitudes become larger. Internal waves will normally have amplitudes several times greater than surface waves. Crest-to-trough wave heights for internal waves can be on the order of 10 m. Wavelengths can range from a few hundred meters to many kilometers. At the long-wavelength end of the internal wave spectrum are internal tides.

In theory, free internal waves can only exist in the frequency range bounded at the lower limit by the inertial frequency (which is a function of latitude) and at the upper limit by the buoyancy frequency (which is a function of depth). The inertial frequency (ω_i) is equivalent to the Coriolis parameter (f). Specifically, $\omega_i = f = (2\pi \sin L)/12$ rad h^{-1}, where L is the latitude (deg). The associated period (T) is 12 h/sin L. At latitude 30°N, the inertial period is 24 h. The buoyancy frequency (N), also referred to as the Brunt-Väisälä frequency, is related to water density (ρ), depth (z), gravitational acceleration (g), and sound speed (c) by

$$N = \sqrt{-g\left(\frac{1}{\rho}\frac{\mathrm{d}\rho}{\mathrm{d}z} + \frac{g}{c^2}\right)} \text{ rad s}^{-1} \tag{2.8}$$

where water depth (z) is measured in the negative (downward) direction (Apel, 1987: 169–70). The buoyancy frequency measures the stability of the water column against small vertical perturbations: when N is real (a stable water column), buoyant oscillations are initiated; when N is imaginary (an unstable water column), rising or sinking motions are initiated. In a typical stable water column, the buoyancy frequency decreases approximately exponentially with depth with a maximum value ($N/2\pi$) of about 10 cycles h^{-1} near the surface and an e-folding depth of about 1.5 km (Spindel, 1985).

When present, internal waves can be detected by making continuous temperature measurements at one location (as with a fixed vertical array of thermistors) over a given period of time. Their presence is evidenced by oscillations in the temperature record with periods consistent with those of internal waves. Figure 2.23 shows the passage of internal waves as recorded in the changed depths of the isotherms.

One problem in testing theories of acoustic propagation in inhomogeneous media is the inability to determine accurately the spatial and temporal scales associated with fluctuations in the index of refraction. These fluctuations can be caused by tides, internal waves and fine-scale features. Ewart and Reynolds (1984) reported results from the mid-ocean acoustic transmission experiment (MATE). This experiment was designed to measure phase and intensity fluctuations in sound pulses transmitted at 2, 4, 8, and 13 kHz over a wholly refracted path (i.e., no boundary interactions). Two receiver towers were placed on Cobb Seamount (in the northeastern portion of the North Pacific Ocean). A sister tower located 20 km to the southwest was used for placement of the acoustic transmitter. This geometry minimized transducer motion and also assured the presence of wholly refracted paths between source and receiver. The environmental program conducted in support of MATE was specifically designed to oversample the internal-wave variability within the context of the Garrett-Munk model (Garrett and Munk, 1979). This archive of environmental measurements contains sufficient data with which to test existing internal-wave models as well as test future models of internal-wave variability. In related work, Macaskill and Ewart (1996) refined numerical solutions of the fourth-moment equation for acoustic intensity correlations, particularly the temporal cross correlation between acoustic signals of different frequencies propagating through the same medium.

Internal waves are considered to be a limiting factor in the propagation of acoustic energy, particularly in the frequency range 50 Hz to 20 kHz. The effects are manifested as amplitude and phase variations. Internal waves may also limit both the temporal and spatial stability of acoustic paths (Flatté et al., 1979). Below 50 Hz, the relatively long acoustic wavelengths (> 30 m) are less likely to be affected by internal waves. Above 20 kHz (with acoustic wavelengths less than a few centimeters), the effects of fine-scale features are probably more important. The acoustic impacts of internal waves will be further explored in Section 6.8.2.

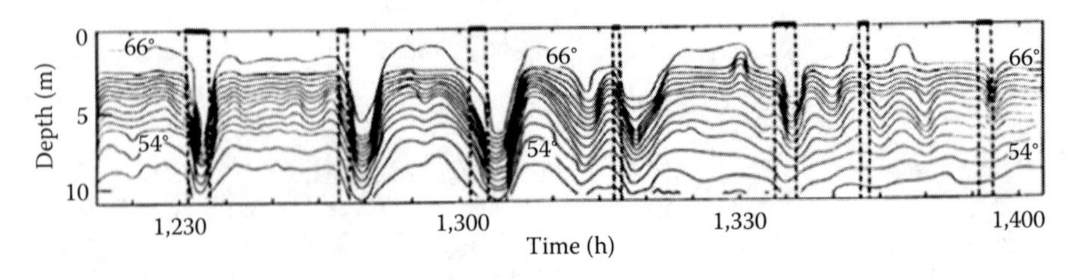

FIGURE 2.23 Passage of internal waves observed at a stationary observation point. The internal waves are evidenced by temporal fluctuations in the isotherm patterns (°F). Vertical dashed lines indicate the relationship between the internal wave structure and the location of sea-surface slicks. (From LaFond, E.C., *The Sea*, 1, 731–751, 1962. Copyright Wiley-VCH Verlag GmbH & Co. KGaA. Reprinted with permission.)

The effects of upper-ocean stirring were added to the effects of internal waves to account for the major part of acoustic variability in the ocean (Dzieciuch et al., 2004). The critical dependence on mixed-layer processes suggested a scheme for acoustically monitoring the upper ocean from sensors deployed at a depth of 3–5 km.

2.5.3 Fine-Scale Features

This section will discuss selected fine-scale features including thermohaline staircases and Langmuir circulation.

2.5.3.1 Thermohaline Staircases

One type of fine-scale oceanic feature is the "thermohaline staircase." These staircases are generally found in the main thermocline and are evidenced by layers of uniform temperature and salinity on the order of 10-m thick separated by thin, high-gradient interfaces on the order of a few meters in thickness (Figure 2.24). The incidence of well-developed staircases appears to be limited to less than 10% of the available high-resolution profiles taken in the North Atlantic Ocean (Schmitt, 1987; Schmitt et al., 1987).

Staircase structures are most frequently associated with a strongly destabilizing vertical gradient of salinity (caused by the confluence of fresh and saline water masses). These features have been observed northeast of South America in the tropical Atlantic Ocean just outside the Caribbean Sea, and also in the eastern Atlantic Ocean outside the Mediterranean Sea. In the tropical Atlantic near Barbados at a depth of approximately 400 meters, layers of saline ocean water intermix with layers of fresher ocean water, thus creating a staircase (or layering) driven by a small-scale convection known as "salt fingers." In this region, warm, high-salinity subtropical water lies over cooler, fresher water flowing northward from Antarctica, thus creating

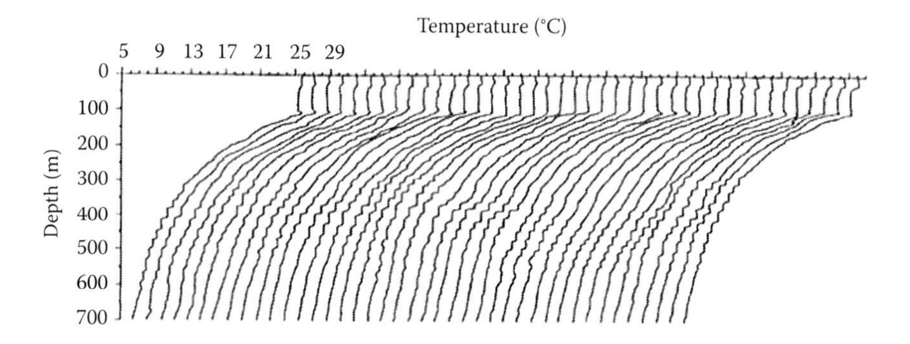

FIGURE 2.24 Series of expendable bathythermograph (XBT) temperature profiles taken from an east–west section east of Barbados in the Atlantic Ocean. The profiles are separated by a distance of 5.5 km; the total distance covered is 220 km. The temperature scale is correct for the profile at the extreme left (west), and each subsequent profile is offset by 1.6°C. Undulations of the thermocline caused by internal waves and mesoscale eddies can also be seen. (From Schmitt, R.W., *EOS, Trans. Am. Geophys. Union*, 68, 57–60, 1987. Copyright by the American Geophysical Union. With permission.)

a unique horizontal stratification with distinct layers of water. The thermohaline staircases of the western tropical Atlantic are characterized by as many as 10–15 mixed layers, each 5–40 m thick, separated by thinner, high-gradient interfaces 0.5–5 m thick. Interfacial temperature jumps can reach 1.0°C. The mixed layers are maintained by convective overturning driven by "salt fingers" on the interfaces. These fingers release energy from the unstable salt distribution by the greater diffusion of heat over salt between adjacent fingers, thus leading to the term *double-diffusive convection.*

Historically, double-diffusive staircases have been observed since the advent of continuous profiling systems of the late 1960s through to the present time. Though continuously modified by eddies, these layers appear to persist like geologic strata for decades and cover an area on the order of 10^6 km^2. Before the dynamics of thermohaline staircases were well understood, these features were sometimes dismissed as malfunctions in the oceanographic sensors that recorded them.

Significant (2–3 m s^{-1}) sound-speed changes are associated with these interfaces, which makes it likely that they will be good reflectors of sound. They have also been shown to have large horizontal-coherence scales, making then ideally suited to detection by seismic reflection profiling (Schmitt et al., 2005). Recent work by Kormann et al. (2010) showed that multichannel seismic (MCS) systems could provide detailed information on oceanic fine structure. Specifically, they analyzed the suitability of high-order numerical algorithms for modeling the weak wavefield scattered by oceanic fine structure. Synthetic shot records were generated along a coincident seismic and oceanographic profile acquired across a Mediterranean salt lens in the Gulf of Cadiz. A 2D finite-difference time-domain propagation model was applied together with second-order complex frequency-shifted perfectly matched layers (PMLs) at the numerical boundaries. (A PML is an artificial absorbing layer for wave equations that is commonly used to truncate computational regions in numerical methods simulating problems with open boundaries, especially in the finite-difference time-domain and finite-element methods.) A realistic sound-speed map with the lateral resolution of the seismic data was used as a reference. It was shown that this numerical propagator created an acoustical image of the oceanic fine structure (including the salt lens) that accurately reproduced the field measurements (also see related discussions in Section 6.19).

Chin-Bing et al. (1994) studied the effects of thermohaline staircases on low-frequency (50 Hz) sound propagation. Several propagation models were used to generate transmission loss as a function of range from source to receiver based on a sound speed profile containing staircase features. A source was placed (in depth) at the center of the staircase features while receivers were placed above, below and at the center of the features. These results were then compared to baseline (control) simulations based on a profile in which the effects of the staircase features were effectively averaged out. The greatest effects were observed when both the source and receiver were placed at the center of the features. These effects were attributed to a redistribution of intensity caused by the staircase features. Chin-Bing et al. (1994) also noted that backscatter can occur when the step-structured discontinuities of the thermohaline staircase are on the order of an acoustic wavelength. Thus, at frequencies greater than about 3 kHz, backscatter from the thermohaline steps could become significant.

Ross and Lavery (2010) explored the feasibility of using high-frequency acoustic scattering techniques to map the extent and evolution of double-diffusive convection in the ocean. First, they developed a model to describe acoustic scattering from double-diffusive interfaces in the laboratory, which accounted for much of the measured scattering in the frequency range 200–600 kHz. Next, this model was used (in conjunction with published *in situ* observations of diffusive-convection interfaces) to make predictions of the attendant acoustic-interface scattering for a range of locations throughout the global ocean. These results were then corroborated with measurements from a multifrequency acoustic backscattering experiment conducted near the western Antarctic Peninsula.

2.5.3.2 Langmuir Circulation

Another example of fine-scale features is Langmuir circulation. On a trip across the Atlantic Ocean, Langmuir (1938) noticed patterns of floating seaweed organized into long, parallel strips called windrows. He subsequently studied the dynamics of this circulation in a lake. Updated reviews of Langmir circulation have been provided by Leibovich (1983) and Thorpe (2004).

Representative spatial dimensions are 4–6 m in depth and 10–50 m between windrows, although these dimensions vary widely. Cell lengths have been observed to vary from a few meters to many kilometers in the ocean. The cell axes are typically aligned with the wind direction, but may vary by as much as 20° degrees. When the wind direction changes, the cells gradually shift to realign with the new direction, typically within about 20 minutes. Cells have also been observed to coalesce at Y-junctions, typically at angles of 30°, with the base of the Y pointing downwind (Figure 2.25).

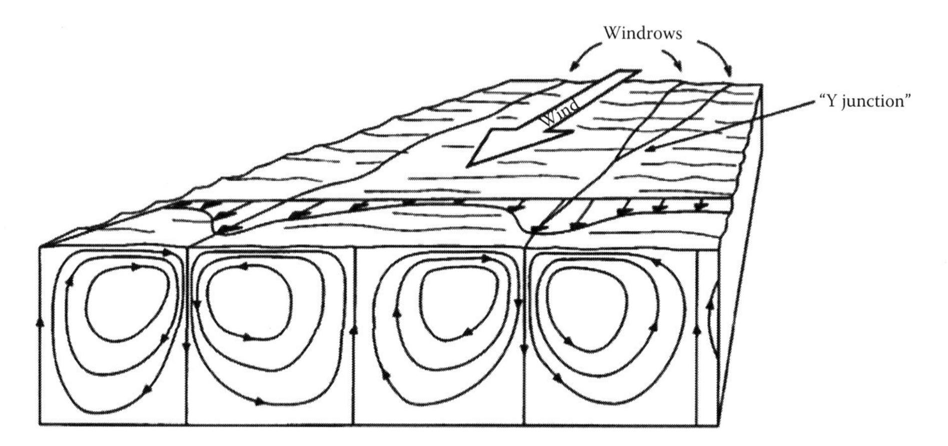

FIGURE 2.25 Pattern of mean flow in an idealized Langmuir circulation. The windrows may be 2–300 m apart, and the cell form is roughly square. Typically, the flow is turbulent, especially near the water surface, and the windrows amalgamate and meander in space and time. Bands of bubbles or buoyant algae may form within the downward-going (or downwelling) flow. (From Thorpe, S.A. *Ann. Rev. Fluid Mech.*, 36, 55–79, 2004. With permission.)

Wind speeds must typically reach 3 m s^{-1} to generate Langmuir cells, although speeds within the cells are only a fraction of the winds that create them. If water conditions are turbulent, cells may not develop at all. At higher wind speeds, instabilities start to dominate and the cells may disintegrate, amalgamate or regenerate. Evolving theories suggest that Langmuir cells are generated by a distortion of the near-surface vorticity flow by the Stokes drift associated with wind-driven surface gravity waves. Heat convection and heat exchange may also play roles in forming and damping circulation cells.

Langmuir circulation may pose implications for mixed-layer dynamics, plankton distributions, and contaminant dispersal. Although it is not entirely clear how Langmuir circulation impacts the mixed layer, Li et al. (1995) found that that the mixed layer can be deepened to a depth of 200 m in the presence of these cells. Langmuir circulation patterns can also aggregate upward-swimming fish and zooplankton into near-surface patches in otherwise well-mixed regions of the ocean (Pershing et al., 2001). Moreover, the ability to predict and track the distribution of surface contaminants may be affected by Langmuir circulation patterns, which cause oil-buoyant material to concentrate in long windrows.

The strength and scales of Langmuir circulation in the ocean can be determined from acoustic Doppler current profilers (ADCPs) that measure the near-surface velocity structure (Pershing et al., 2001). Patterns of Langmuir circulation may also be observed using side-scan sonars, which detect bands of subsurface bubbles (acting as scatterers of high-frequency sound) in the downwelling region beneath the surface convergence of the circulation cells (Thorpe, 1992).

Farmer and Li (1995) used sonar observations to examine the temporal evolution of bubble distributions in relation to Langmuir circulation (also see Thorpe et al. [1994] for a discussion of particle dispersion associated with Langmuir circulation). Li and Garrett (1997) examined the interaction between wind-driven Langmuir circulation and existing stratification to explain the role of Langmuir circulation in deepening the ocean surface mixed layer. The influences of Langmuir circulation on upper-ocean diurnal and seasonal changes in stratification were further investigated by Li et al. (1995).

2.6 BIOLOGICS

Marine organisms can be segregated into four major categories: plankton, nekton, benthos and algae. Plankton (or floaters) include both plants (phytoplankton) and animals (zooplankton). The zooplankton have little or no swimming ability and thus drift with the currents. Phytoplankton are typically smaller than 0.5 mm while zooplankton are smaller than 1 cm.

Nekton (or free swimmers) are animals that are capable of swimming purposefully. Nekton include fish and mammals, and occur over the entire depth range of the ocean. Benthos are dwellers on, in, or near the bottom of the ocean. Fouling organisms such as barnacles would also be included in this category. Algae include marine plant life, such as seaweed.

Biological organisms can affect underwater sound through noise production, attenuation and scattering of signals, presentation of false targets, and fouling of sonar transducers. Certain marine animals, many of which are found over the continental shelves, produce sounds that increase the background noise levels. These include snapping shrimp, whales, porpoises, and various fish such as croakers and drum fish. Organisms that may cause attenuation are schools of fish, dense populations of plankton, and floating kelp, for example. False targets are commonly presented to active sonars by whales or large schools of fish or porpoises.

While fouling organisms such as barnacles do not directly affect sound, indirectly they can degrade sonar performance by fouling sonar domes and transducer faces. Furthermore, such organisms can contribute to an increase in hull noise of ships and submarines through the generation of turbulence as the vessels move through the water. This effect is also referred to as self noise, as distinguished from ambient noise.

Perhaps the most notable impact of marine organisms on active sonars (particularly those operating at frequencies near 10 kHz) is known as the deep scattering layer (DSL). The DSL is a dense accumulation of marine organisms at depth below the sea surface. The strong scattering nature of the DSL is attributed primarily to fish and other marine animals with swim bladders and gas floats, although plankton and nekton are also present. The DSL is typically encountered in temperate regions. Moreover, the DSL exhibits a diurnal migration in depth, being shallower at night and deeper during the day. The DSL will be further discussed in Section 9.2.1.

3 Propagation I: Observations and Physical Models

3.1 BACKGROUND

The propagation of sound in the sea has been studied intensely since the beginning of World War II when it was recognized that an understanding of this phenomenon was essential to the successful conduct of anti-submarine warfare (ASW) operations. These early measurements were quickly transformed into effective, albeit primitive, prediction tools. Naval requirements continue to motivate advances in all aspects of underwater acoustic modeling, particularly propagation modeling.

The study of sound propagation in the sea is fundamental to the understanding and prediction of all other underwater acoustic phenomena. The essentiality of propagation models is inherent in the hierarchy of acoustic models illustrated previously in Figure 1.1.

Advances in propagation modeling have been achieved by both marine seismologists and underwater acousticians, although the motivating factors have been quite different. Marine seismologists have traditionally used earthborne propagation of elastic waves to study the solid earth beneath the oceans. Underwater acousticians have concentrated on the study of waterborne, compressional-wave propagation phenomena in the ocean as well as in the shallow subbottom layers (Akal and Berkson, 1986). As research in underwater acoustics has extended to frequencies below several hundred hertz, it has overlapped with the spectral domain of marine seismologists. Moreover, marine seismologists have become more interested in exploring the velocity-depth structure of the uppermost layers of the sea floor using higher frequencies. This area of overlapping interests has been recognized as a subdiscipline of both communities and is referred to as "ocean seismo-acoustics."

The emphasis in this chapter is focused on applications in underwater acoustics. Developments in marine seismology will be discussed when the applications to sonar modeling are clearly evident. Much research has been performed in the marine seismology community that is theoretically and conceptually applicable to underwater acoustics. Such practical research includes the development of sophisticated, yet robust, mathematical methods.

Propagation models have continued to be used for the prediction of sonar performance. They have also found great utility in analyzing field measurements, in designing improved sonar systems, and in designing complicated inverse-acoustic field experiments.

As modeling has continued to grow in prominence in many aspects of underwater acoustics, it is prudent to reassess the state of the art in modeling techniques and the relationship to available measurements. Ideally, such an assessment should identify those areas requiring further measurement support as well as those that are firmly understood and hence properly modeled.

This chapter addresses the observations that have been made in the field and the physical (i.e., physics-based) models that have been developed. Aspects of propagation phenomena including ducts and channels, boundary interactions, volumetric effects, and coherence are described. Chapter 4 addresses the mathematical models that have been developed for underwater acoustic propagation. Specialized aspects of surface ducts, shallow water areas, and Arctic regions are discussed in Chapter 5.

3.2 NATURE OF MEASUREMENTS

Field measurement programs are usually quite complex and typically involve multiple platforms (e.g., ships, buoys, towers, aircraft, submarines, or satellites).

A wide variety of experimental field techniques have been used in underwater acoustic propagation studies. Some of the more typical types of measurement platforms and experimental geometries that have been utilized include (Urick, 1982: Chapter 1):

1. Two ships—one a source ship and the other a receiving ship. The range between them is changed as transmission runs are made in order to yield level versus range.
2. Single ship—using a suspended transmitter and either sonobuoys or a hydrophone array for reception.
3. Ship and aircraft—where the aircraft drops explosive sound sources while flying toward or away from the ship.
4. Single aircraft—using sonobuoys for reception and recording on board the aircraft.
5. Bottomed hydrophone array—with a cable connected to shore, receives signals transmitted from a ship or the explosive shots dropped by an aircraft.
6. Two bottomed transducers—one acting as a source and the other as a receiver. This geometry is typically used in studies of the fluctuation of sound transmission between two fixed points in the sea.

A simple experimental geometry illustrating method (2) above is presented in Figure 3.1. Here, the transmitting ship is receiving signals via radio directly from the array. Fully integrated oceanographic and acoustic field experiments are required in order to obtain a comprehensive portraiture of the temporal, spatial, and spectral scales necessary to characterize the marine environment for a full understanding of the governing acoustic phenomena.

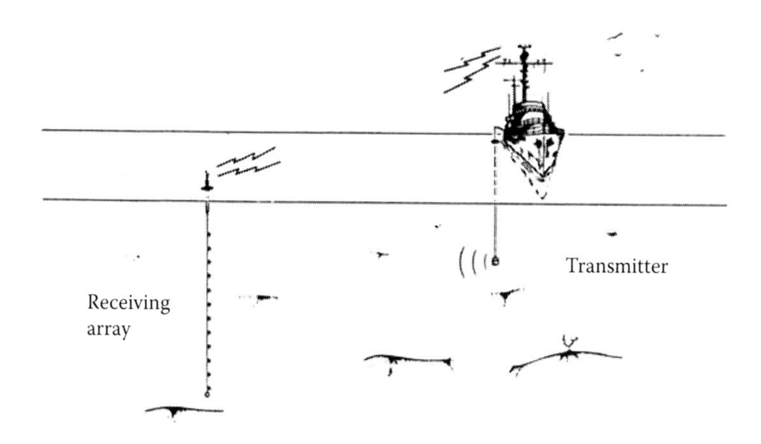

FIGURE 3.1 Example of a simple experimental geometry. (Adapted from Ingenito, F. et al., 1978. Shallow water acoustics summary report (first phase). Nav. Res. Lab., Rept 8179.)

3.3 BASIC CONCEPTS

The standard unit of measure of underwater acoustic propagation is acoustic intensity (I), which is sound pressure flow (power) per unit area (reported in units of watts per square meter):

$$I = \frac{p^2}{\rho c} \tag{3.1}$$

where p is the instantaneous pressure amplitude of a plane wave, ρ is the density of sea water and c is the speed of sound in sea water. Sound intensity is actually a vector quantity, but in the far-field approximation it is represented as a scalar quantity based on sound pressure squared. The product ρc is commonly referred to as the characteristic acoustic impedance.

Transmission loss (TL) is defined as 10 times the log (base 10) of the ratio of the reference intensity (I_{ref}), measured at a point 1 m from the source, to the intensity (I), measured at a distant point, and is expressed in units of decibels (dB):

$$\text{TL} = 10\log_{10}\frac{I_{\text{ref}}}{I} \tag{3.2}$$

The standard metric unit for pressure (force per unit area) is 1 micopascal, which is equivalent to 10^{-6} N m^{-2}, and is abbreviated 1 μPa.

TL has conventionally been plotted for each frequency, source depth and receiver depth as a function of range, as illustrated in Figure 3.2. This type of display is easily generated by all propagation models. Certain types of propagation models can also generate and display acoustic TL in the entire range-depth plane for all receiver depths and ranges, given a fixed source depth (Figure 3.3).

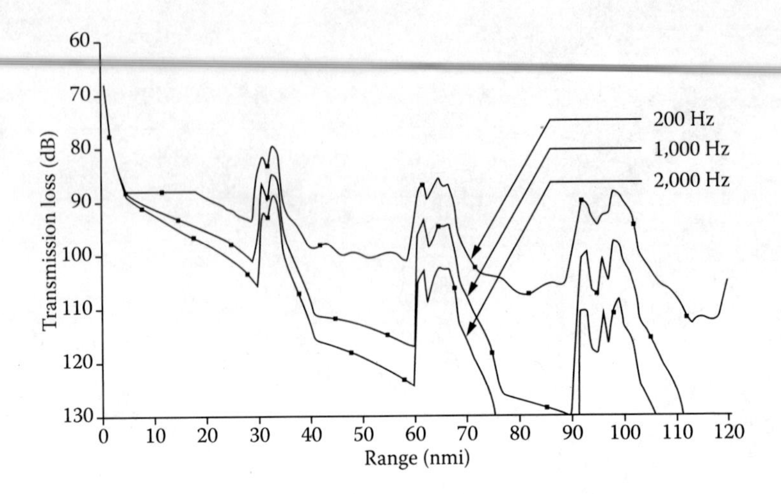

FIGURE 3.2 Example of standard transmission loss curves generated by the FACT model for each combination of frequency, source depth, and receiver depth. Here, the source and receiver depths are fixed at 150 and 90 m, respectively. The peaks (minimum TL values) correspond to convergence zones. Note the increase in transmission loss (TL) with increasing frequency due to absorption.

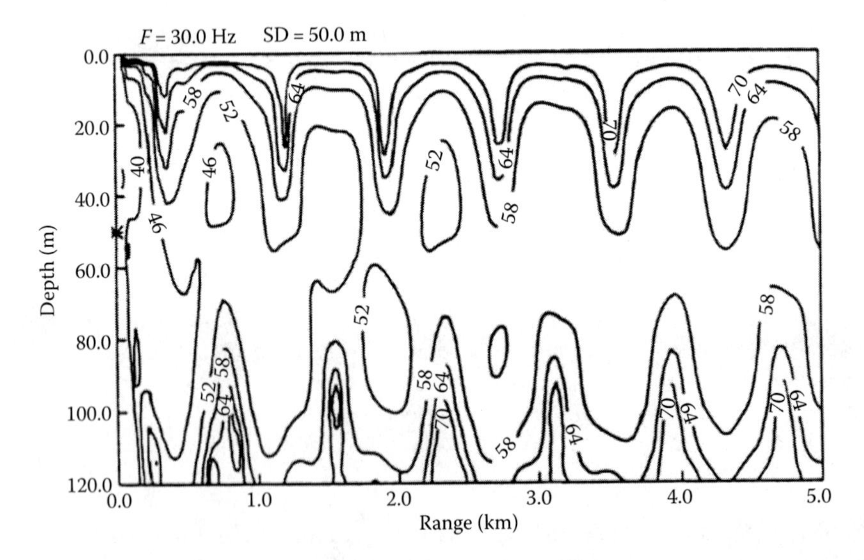

FIGURE 3.3 Example showing contours of transmission loss plotted in the range-depth plane. This plot is valid for one frequency (30 Hz) and one source depth (50 m), but can be used to determine the transmission loss at any receiver location in the range-depth plane. The contour interval is 6 dB. (Adapted from Schmidt, H. 1988. SAFARI: Seismo-acoustic fast field algorithm for range-independent environments. User's guide. SACLANT Undersea Res. Ctr, Rept SR-113.)

Sonar performance is commonly described in terms of a figure of merit (FOM). The FOM is a quantitative measure of sonar performance. Specifically, the larger the FOM value, the greater the performance potential of the sonar. Numerically, the FOM is equal to the allowable one-way TL in passive sonars. The FOM is further described in Chapter 11 within the context of the sonar equations.

The display method illustrated in Figure 3.2 is very useful in evaluating passive sonar performance. Specifically, once an FOM has been calculated for a particular sonar operating in a particular ocean environment against a particular target, a horizontal line can be drawn on the plot equating the numerical value of the FOM to TL. Then, any area below the TL curve, but above the FOM line, represents a sonar detection area. Figure 3.4 shows a hypothetical relationship between the FOM and the sonar detection areas (upper panel), and the correspondence between the

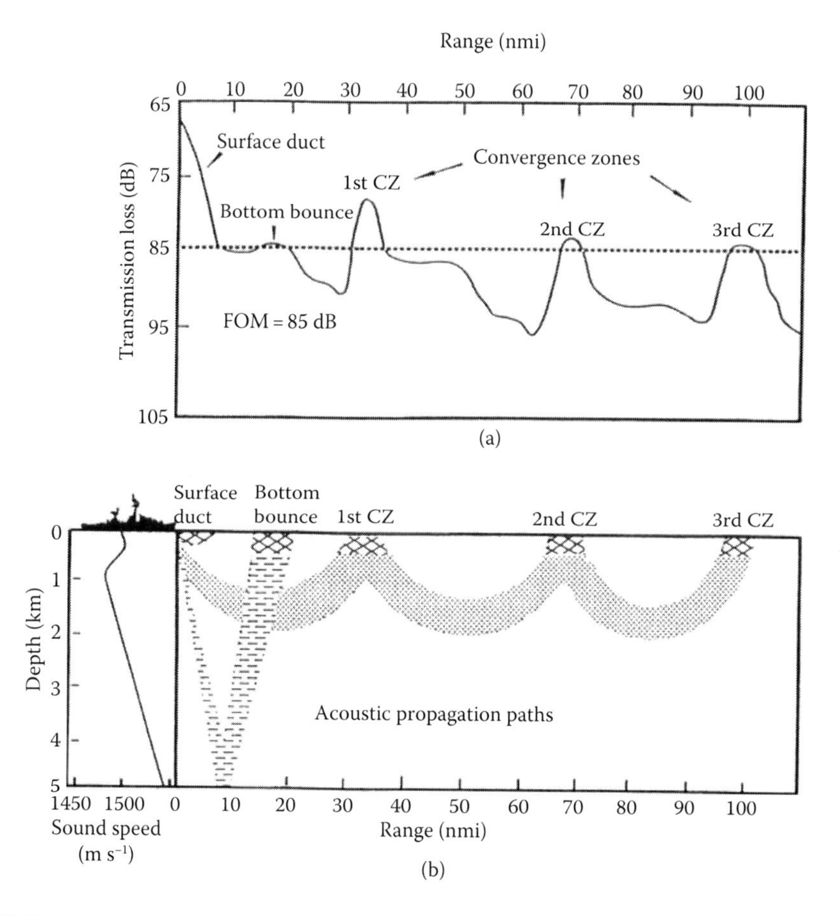

FIGURE 3.4 Hypothetical relationship between (a) transmission loss (TL) curve and (b) the corresponding propagation paths and detection zones (cross-hatched areas near the sea surface) associated with a figure of merit (FOM) of 85 dB. A plausible sound speed profile is shown at the left side of panel (b). Both the source (target) and receiver (ship's sonar) are positioned near the surface.

TL curve and the ray paths as propagated in the water column (lower panel). These ray paths are consistent with those resulting from a shallow source (target) and shallow receiver (sonar) positioned in a water column characterized by the sound speed profile shown on the left side of the lower panel of Figure 3.4.

Sound propagates in the sea by way of a variety of paths. The particular paths traveled depend upon the sound speed structure in the water column and the source-receiver geometry. The six basic paths include direct path, surface duct, bottom bounce, convergence zone, deep sound channel, and reliable acoustic path (RAP). These six paths are illustrated in Figure 3.5. Depending upon the ocean environment, propagation over combinations of paths may be possible for any give source-receiver geometry; this situation is referred to as multipath propagation. Four of these paths (surface duct, deep sound channel, convergence zone, and RAP) are strongly affected by the sound speed structure in the water column and will be discussed in detail in Sections 3.7, 3.8, 3.9, and 3.10, respectively. The remaining two paths (direct path and bottom bounce) are relatively unaffected by the refractive properties of the sound speed structure: direct paths span relatively short distances, and bottom-bounce paths penetrate the refractive layers at steep angles.

In basic ray tracing, Snell's law is used in one form or another. This law describes the refraction of sound rays in a medium in which sound speed varies as a function of depth, but is constant within discrete horizontal layers of the water column. Consider Figure 3.6 where a ray (which is normal to the acoustic wavefronts) is traveling from medium 1 (with sound speed c_1) into medium 2 (with sound speed c_2), where $c_1 \neq c_2$.

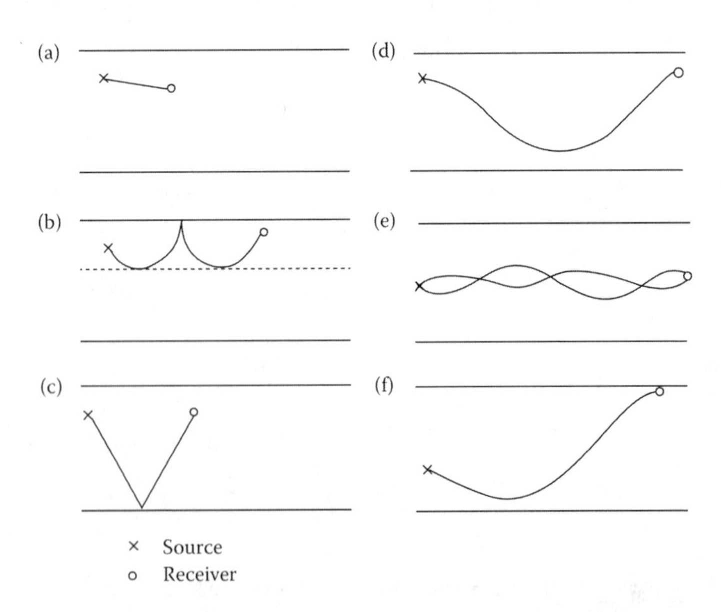

FIGURE 3.5 Six basic propagation paths in the sea: (a) direct path (DP); (b) surface duct (SD); (c) bottom bounce (BB); (d) convergence zone (CZ); (e) deep sound channel (DSC); and (f) reliable acoustic path (RAP).

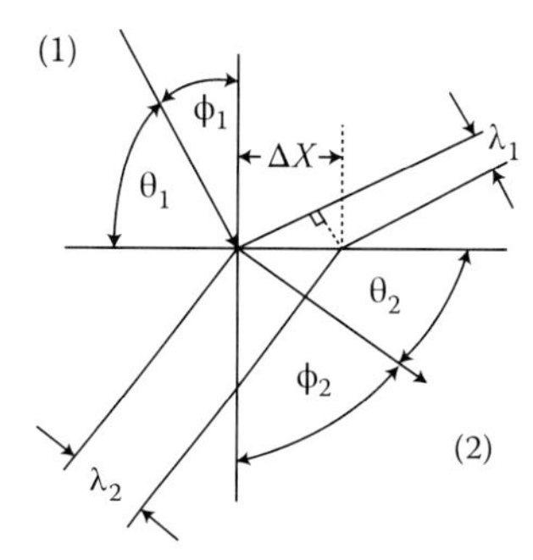

FIGURE 3.6 Geometry for Snell's law.

Let λ_1 be the distance between successive wavefronts (i.e., the wavelength) in medium 1 and λ_2 the corresponding value in medium 2. Then, as defined in Figure 3.6

$$\lambda_1 = \Delta x \sin\phi_1 = c_1 \Delta t \qquad \text{and} \qquad \lambda_2 = \Delta x \sin\phi_2 = c_2 \Delta t$$

where Δt is an increment of time. Rearranging terms, we obtain the familiar relationship

$$\frac{\sin\phi_1}{c_1} = \frac{\sin\phi_2}{c_2} \tag{3.3a}$$

or, equivalently from Figure 3.6

$$\frac{\cos\theta_1}{c_1} = \frac{\cos\theta_2}{c_2} \tag{3.3b}$$

As a matter of convention, ϕ is referred to as the incidence angle while θ is referred to as the grazing angle.

In a homogeneous medium, acoustic TL varies as the inverse of the range squared. This relationship is easily derived, as demonstrated below.

Let I = intensity, P = power, and A = area. Then

$$I = \frac{P}{A}$$

For spherical spreading (see Figure 3.7a)

$$I_1 A_1 = I_1\left(4\pi \cdot r_1^2\right) \qquad I_2 A_2 = I_2\left(4\pi \cdot r_2^2\right)$$

where r_1 and r_2 are the radii of concentric spherical sections.

Power (P) is conserved; therefore, $P_1 = I_1 A_1 = I_2 A_2$ and

$$I_2 = I_1\left(\frac{r_1^2}{r_2^2}\right)$$

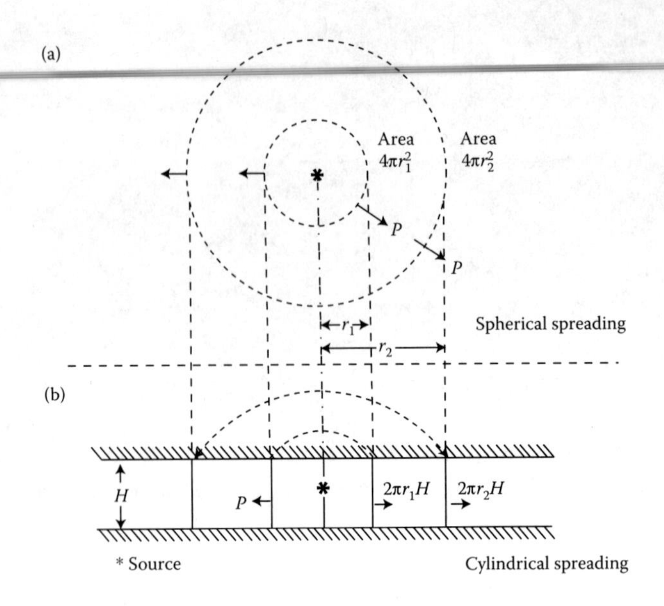

FIGURE 3.7 Geometry for (a) spherical spreading and (b) cylindrical spreading. (Reproduced from Urick, R.J., *Principles of Underwater Sound*, 3rd ed. McGraw-Hill, New York, 1983. With permission.)

but r_1 represents a unit reference distance and thus

$$I_2 = \frac{I_1}{r_2^2}$$

Since intensity is power per unit area, and since the area of a sphere increases as the square of its radius, the intensity falls off as the inverse square of the radius (or range) in order that power remains constant.

The corresponding TL is defined as

$$\mathrm{TL} = 10\log_{10}\frac{I_1}{I_2} = 10\log_{10} r_2^2 = 20\log_{10} r_2 \tag{3.4}$$

This relationship is valid for an isotropic deep ocean with no absorption effects.

An analogous expression can be derived for cylindrical spreading. This spreading law would be appropriate for a duct, or in shallow water, where the water is homogeneous and the boundaries are perfect reflectors. Then, referring to Figure 3.7(b)

$$I_1 A_1 = I_1(2\pi \cdot r_1 H) \qquad I_2 A_2 = I_2(2\pi \cdot r_2 H)$$

where H is the depth of the duct or of the water column. Since power is conserved, $P_1 = I_1 A_1 = I_2 A_2$ and

$$I_2 = I_1 \frac{r_1}{r_2}$$

For unit radius r_1

$$I_2 = \frac{I_1}{r_2}$$

Thus, the intensity falls off as the inverse of the radius (or range). The corresponding TL is defined as

$$\text{TL} = 10 \log_{10} \frac{I_1}{I_2} = 10 \log_{10} r_2 \tag{3.5}$$

This relationship is valid in an isotropic ocean with no absorption effects.

3.4 SEA-SURFACE BOUNDARY

The sea surface affects underwater sound by providing a mechanism for

1. Forward scattering and reflection loss
2. Image interference and frequency effects
3. Attenuation by turbidity and bubbles
4. Noise generation at higher frequencies due to surface weather
5. Backscattering and surface reverberation

Urick (1982: Chapter 10) provided a comprehensive summary of sound reflection and scattering by the sea surface. Items (1) through (3) will be discussed below. Item (4) will be addressed in Chapter 7 and item (5) will be discussed in Chapter 9.

The mechanisms operating at the sea surface can be incorporated into mathematical models through the specification of appropriate "boundary conditions." These boundary conditions can range from simplistic to complex depending upon the sophistication of the model and the availability of information concerning the state of the sea surface.

3.4.1 Forward Scattering and Reflection Loss

When a plane sound wave in water strikes a perfectly smooth surface, nearly all of the energy is reflected at the boundary in the forward (or specular) direction as a coherent plane wave. As the sea surface roughens under the influence of wind, sound is also scattered in the backward and out-of-plane directions, and the intensity of sound reflected in the forward direction is accordingly reduced. The backward-directed (backscattered) energy gives rise to surface reverberation (see Chapter 9). Eckart (1953) developed a theoretical treatment of scattering by a sinusoidal boundary as a way to approximate reflection from a wind-roughened sea surface. Marsh et al. (1961) developed simple formulas to express scattering losses at the sea surface. Eller (1984a) reviewed the availability of simple surface loss algorithms appropriate for incorporation into propagation models.

The sea surface is most commonly modeled as a pressure release surface (see Kinsler et al., 1982: 126–7). This is a condition in which the acoustic pressure at the air-water interface is nearly zero, the amplitude of the reflected wave (in water) is almost equal to that of the incident wave, and there is a 180°-phase shift. This is also known as the Dirichlet boundary condition (Frisk, 1994: 32).

It is also common practice to use the term "reflection coefficient" to express the amount of acoustic energy reflected from a surface or from a boundary between two media. This coefficient depends upon the grazing angle and the difference in the acoustic impedance between the two media. A reflection loss is then defined as $10\log_{10}$ (reflection coefficient). This reflection loss is referred to as "surface loss" when describing the reflection of sound from the sea surface, or "bottom loss" when describing the reflection of sound from the sea floor.

A measure of the acoustic roughness of the sea surface is provided by the Rayleigh parameter R through the relationship

$$R = 2ka\sin\theta \tag{3.6}$$

where $k = 2\pi/\lambda$ is the acoustic wavenumber, λ is the acoustic wavelength, a is the root-mean-square (rms) amplitude of the surface waves ($2a$ is the crest-to-trough rms wave height, or H_{rms}—refer back to Equation 2.5 in Chapter 2) and θ is the grazing angle (measured relative to the horizontal plane). When $R \ll 1$, the sea surface is considered to be acoustically smooth; when $R \gg 1$, the sea surface is acoustically rough.

Sea-surface wave spectra can be generated numerically by executing available spectral ocean wave models in the hindcast mode. Hindcasting is usually the only means available for obtaining sufficiently long record lengths from which to generate reliable statistics. A statistical analysis of these hindcast data produces probability distributions of critical parameters for use in estimating future sea surface conditions.

Kuo (1988) reviewed and clarified earlier formulations of sea-surface scattering losses based on perturbation methods and also presented new predictions based on numerical integration in a complex domain.

Godin (2006) studied the energy characteristics of sound emitted into air by an underwater point source. The energy transfer due to inhomogeneous waves was shown to cause the phenomenon of anomalous transparency of the interface for low-frequency sound. The anomalous transparency is evidenced by an energy flux through the interface that increases with decreasing frequency of sound; moreover, at sufficiently low frequencies, almost all of the acoustic energy produced by the underwater source is emitted into the air. Conversely, at high frequencies, when the contribution of the inhomogeneous waves becomes negligible, the water-to-air interface is similar to a perfectly reflecting surface and almost all of the acoustic energy produced by the source is emitted into water. The anomalous transparency phenomenon changes the conventional opinion on the possibility of acoustic coupling between points in water and air and on the role played by physical processes evolving in the water column in generating atmospheric acoustic noise.

Ghadimi et al. (2015b) investigated the enhanced sound transmission and anomalous transparency of the air-water interface by calculating the sound generated by a submerged shallow-depth monopole point source situated at depths (D) less than one-tenth of the acoustic wavelength (λ_w) in water ($D < 0.1\ \lambda_w$). They solved the coupled Helmholtz wave equations in a two-phase air-water medium using finite-element methods incorporating cylindrical-radiation conditions prescribed for the air and water boundaries. Continuity of the acoustic pressure at the air-water interface was considered based on the anomalous transparency of air-water interface theory. This enhanced transmission of sound into the air occurred because of evanescent waves that caused higher rates of the emitted sound passing the air-water interface. The transmitted acoustic pressure through the interface into the air versus the nondimensional ratio D/λ_w was examined. It was concluded that, as the ratio D/λ_w increased, the ratio of on-axis microphone and out-of-the-cone angle microphone amplitudes decreased. This trend of reduction was also observed for the ratio of on-axis microphone to the hydrophone amplitudes. This numerical approach displayed good agreement with both experimental and theoretical data.

3.4.2 Image Interference and Frequency Effects

When the surface is smooth, an interference pattern is produced between direct-path sound and sound reflected from the sea surface. The sound reflected from the sea surface may be considered to originate from an image source located on the opposite (mirror image) side of the surface (Figure 3.8). This image signal will have an amplitude nearly equal to that of the incident signal, but will be out of phase. The resulting sound field can be divided into three parts (Figure 3.9): (1) the near field

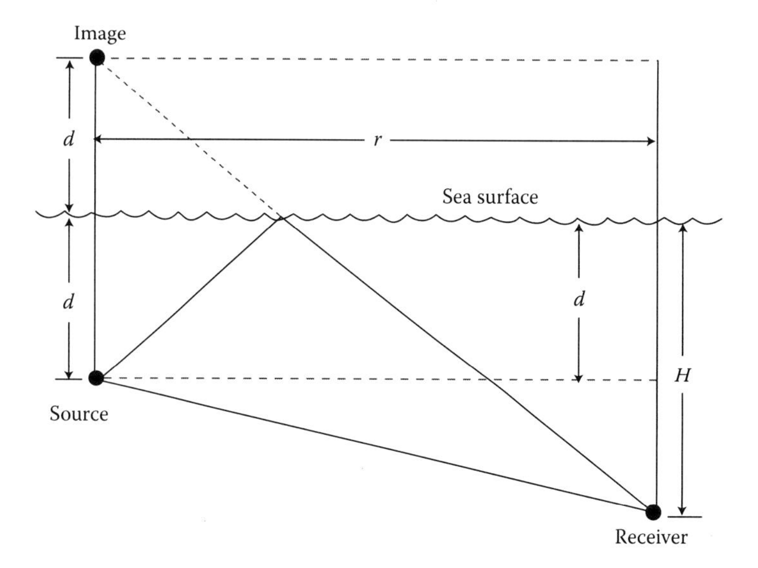

FIGURE 3.8 Geometry for image interference effect.

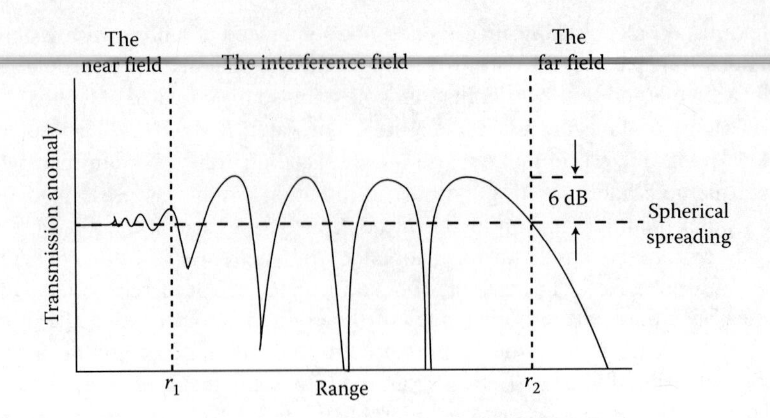

FIGURE 3.9 Illustration of image interference effects. The transmission anomaly (TA) represents the difference between the observed transmission loss (TL) and losses due to the effects of spherical spreading (TA = $20\log_{10}(r)$ – TL), where the range (r) is measured in meters. (Adapted from Urick, R.J., *Sound Propagation in the Sea*, US Government Printing Office, Washington, 1979. With permission.)

close to the source in which the image source is too far away, and the reflected sound is too weak to produce appreciable interference; (2) an interference field in which there are strong peaks and nulls in the received signal as range increases; and (3) the far field in which there is an increasingly out-of-phase condition between the source and image, and the intensity falls off as the inverse fourth power of range. This phenomenon, also known as the Lloyd Mirror effect, diminishes with increasing surface roughness.

Assuming an acoustically smooth sea surface ($R \ll 1$) and a shallow source depth (d) in deep water, the ranges r_1 and r_2 in Figure 3.9 can be approximated as (Urick, 1982, 1983):

$$r_1 \approx 2\sqrt{dH} \qquad r_2 \approx \frac{4\pi \cdot dH}{\lambda}$$

where λ is the acoustic wavelength and H is the receiver depth.

The image effect can be used to estimate the depth of a submerged object at short ranges. In Figure 3.8, if the source is replaced by a submerged object that has been ensonified by a single, short pulse, then the depth (d) of the object is approximated by

$$d \approx \frac{rc\Delta t}{2H} \tag{3.7}$$

where c is the speed of sound and Δt is the difference in time between receipt of the direct and the surface-reflected pulses (Albers, 1965: 50–1).

The concept of surface interference can also be used to solve relatively simple propagation problems. The approach is called the "method of images" and is valid for all frequencies. The solution is normally expressed as a sum of contributions from all images within a multilayered space. Although this method is usually cumbersome, it is commonly employed as a physical model against which to check the results of more elaborate mathematical models (to be discussed in Chapter 4). Kinsler et al. (1982: 427–30) provided a detailed discussion of this method together with several examples of its application. Tolstoy and Clay (1966: 33–6) discussed solutions in waveguides.

When the sea surface is rough, the vertical motion of the surface modulates the amplitude of the incident wave and superposes its own spectrum as upper and lower sidebands on the spectrum of the incident sound. Moreover, when there is a surface current, the horizontal motion will appear in the scattered sound and cause a Doppler-shifted and Doppler-smeared spectrum.

3.4.3 Turbidity and Bubbles

3.4.3.1 Open Ocean

The presence of bubble layers near the sea surface further complicates the reflection and scattering of sound as a result of the change in sound speed, the resonant characteristics of bubbles and the scattering by bubbly layers (e.g., Leighton, 1994).

Hall (1989) developed a comprehensive model of wind-generated bubbles in the ocean. The effects on the transmission of short pulses in the frequency range 1.25–40 kHz were also examined. For long-range propagation, Hall concluded that the decrease in the near-surface sound speed due to bubbles does not significantly affect the intensity of the surface-reflected rays.

3.4.3.2 Coastal Ocean

Coastal waters are often characterized by suspensions of solid-mineral particles that are agitated by waves, currents or river outflows, in addition to microbubbles that are generated at the sea surface by wind and wave action or at the sea floor by biochemical processes (Richards and Leighton, 2001a,b). Suspended solid particles and microbubbles jointly modify the complex acoustic wavenumber, thus influencing the acoustic properties of the medium and thereby affecting the performance of acoustic sensors operating in such turbid and bubbly environments.

Consequently, the acoustic attenuation coefficient (Section 3.6) in shallow coastal waters is of interest to designers and operators of Doppler-current profilers, sidescan-surveying sonars, and naval mine-hunting sonars operating in the frequency range from tens of kHz to several hundred kHz and possibly up to 1 MHz. At these frequencies, attenuation due to suspended particulate matter is an important contribution to the total attenuation coefficient (Richards, 1998). Typical suspensions contain particles in the size range 1–100 μm, where a variety of shapes and concentrations from 0.1–4 kg m^{-3} are possible (Brown et al., 1998). Microbubbles with radii in the range 10–60 μm will be resonant in the frequency interval 50–300 kHz. Preliminary calculations using viscous-damping theory suggest that particulate concentrations on the order of 0.1 kg m^{-3} may be important, even possibly reducing the detection range of sonars by a factor of two relative to clear water at a frequency of 100 kHz. Observations have shown that concentrations of this level, or greater, often occur in coastal waters and have been detected several tens of kilometers offshore of the

Amazon River, in the Yellow Sea, and in the East China Sea off shore of the Yangtze and Yellow Rivers (Richards et al., 1996).

The presence of microbubbles increases acoustic attenuation through the effects of thermal and viscous absorption and scattering. Unlike particles, however, the resonant scattering of bubbles can be important—the scattering cross-section of a bubble near resonance can be much larger than its geometric cross-section. Moreover, bubbles cause the compressibility of the medium to be complex, thereby resulting in dispersion. The effect of bubbles on the phase speed should be used to modify the sound speed profile when computing ray paths in bubbly layers. A numerical procedure was developed by Norton et al. (1998) to parameterize bubble clouds in terms of an effective complex index of refraction for use in high-fidelity models of forward propagation.

The effective attenuation coefficient in turbid and bubbly environments can be expressed as (Richards and Leighton, 2001a)

$$\alpha = \alpha_w + \alpha_p + \alpha_b$$

and

$$\alpha_p = \alpha_v + \alpha_s$$

where α is the total volume attenuation coefficient of sea water containing suspended particles and microbubbles, α_w is the physico-chemical absorption by clear sea water (see Section 3.6), α_p is the plane-wave attenuation coefficient due to a suspension of solid particles (neglecting thermal absorption), and α_b is the attenuation coefficient for a bubbly liquid. Furthermore, α_p is composed of two terms: α_v is the attenuation coefficient associated with the visco-inertial absorption by suspended particles, and α_s is the attenuation coefficient associated with scattering by suspended particles.

Wall et al. (2006) designed a sampling protocol and computational procedure to process acoustic Doppler current profiler (ADCP) data to determine suspended-sediment discharge, which was computed using measures of echo intensity (EI) and velocity from an ADCP at a site in the freshwater-tidal Hudson River. Adjustments to EI data included (1) an instrument-specific and beam-specific EI conversion to decibels; (2) normalizations for temporal and instrument variations in transmit power and length; (3) beam-to-beam variability; and (4) range-dependent corrections (including acoustic absorption by water, computed according to Schulkin and Marsh [1962, 1963]). Cross-sectional SSC estimates, based on boat-mounted ADCP measurements, were used to adjust data collected by the fixed-position, upward-looking ADCP to conditions in the river cross-section. Water discharge was estimated through multiple-regression relations between boat-mounted ADCP-measured discharge and wind stress, river stage, and upward-looking ADCP measures of depth-averaged velocity. Net suspended-sediment discharge was computed by filtering time-series data of instantaneous suspended-sediment discharge with a low-pass digital filter that used a fast-Fourier transform to remove the semidiurnal tidal signal in the data.

3.4.4 Ice Interaction

Acoustic interaction with an ice canopy is governed by the shape of the under-ice surface and by the compressional wavespeeds (typically 1300–3900 m s^{-1}) and shear wavespeeds (typically 1400–1900 m s^{-1}) (see Untersteiner, 1966; Medwin et al., 1988).

McCammon and McDaniel (1985) examined the reflectivity of ice due to the absorption of shear and compressional waves. They found that shear wave attenuation is the most important loss mechanism from 20° to 60° incidence for smooth ice at low frequencies (≤2k Hz).

In Arctic regions, the presence of a positive-gradient sound speed profile and a rough under-ice surface (with a distribution of large keels) may lead to significant out-of-plane scattering. The acoustic impacts of this scattering are twofold. First, significant beam widening may result from the multiple interactions with the randomly rough under-ice surface. Second, the presence of ice keels in the vicinity of the receiver leads to multiple source images or beam-steering errors arising from interactions of the acoustic signal with the facets of the local under-ice surface.

Because of the overwhelming effect of ice on the propagation of sound in the Arctic, the magnitude of the excess attenuation observed under the ice should be determined by the statistics of the under-ice surface. Available ice-ridge models can be used to generate such statistics. These models can be categorized according to two classes: discrete models and continuous statistical models. These two classes of ice-ridge models are briefly described below.

Discrete ice-ridge models prescribe a representative ridge shape, or an ensemble of ridge shapes, to calculate the statistics of the surface from the discrete statistics of the known ice structure. Continuous statistical ice-ridge models treat the under-ice surface as a stochastic process. This process is then analyzed using the techniques of time-series analysis in which the under-ice surface can be characterized by its autocorrelation function. Continuous statistical models can give a more complete description of the under-ice roughness than can the discrete models; however, they are limited in application to those surfaces that can be completely specified by a Gaussian depth distribution.

The model developed by Diachok (1976) will be described since it is considered to be representative of the class of discrete ice-ridge models known to exist and because of its intuitive appeal. The discrete models are also more robust (i.e., require less knowledge of the under-ice surface) than the continuous statistical models. Furthermore, Diachok's model has been incorporated into existing propagation models with some success.

According to Diachok's model, sea ice may be described as consisting of floating plates, or floes, about 3 m thick, occasionally interrupted by ridges, which are rubble piles formed by collisions and shear interactions between adjacent floes. Ridge dimensions vary widely, but are nominally about 1-m high, 4-m deep, and 12-m wide, with the ridge lengths generally being much greater than the depths or widths. A representative average spacing between ridges (the spacing is random) is about 100 m. Ice-ridge orientation is commonly assumed to be directionally isotropic, although limited empirical data suggest that, at least locally, there may be a preferred orientation. The physical model of reflection developed by Twersky (1957) was used.

A comparison between measured contours and simple geometrical shapes suggests that ridge keel contours may reasonably be represented by a half-ellipse (as in Figure 3.10), and that ridge sail contours may be described using a Gaussian distribution function. The relative dimensions of this geometrical model are indicated in Figure 3.10. The exact solution of under-ice scattering off a flat surface with a single semi-elliptical cylindrical boss of infinite extent was developed by Rubenstein and Greene (1991).

Goff (1995) developed a robust set of tools for estimating the stochastic properties of sea-ice drafts. Five parameters were estimated along with their formal uncertainties: mean draft, rms topographic variation, characteristic length, fractal dimension, and normalized skewness. These five parameters provided an objective basis for terrain classification. Data profiles obtained from an upward-looking submarine sonar survey were compared with synthetic profiles generated from the statistical parameters estimated from the data. These provided a means of qualitatively assessing the success of the stochastic model at characterizing the principal morphological properties. In general, the comparisons were good, thus providing confidence in the applicability and versatility of the stochastic model.

LePage and Schmidt (1994) extended the applicability of perturbation theory to under-ice scattering at low frequencies (10–100 Hz) by including the scattering of incident acoustic energy into elastic modes, which then propagate through the ice. Kapoor and Schmidt (1997) developed a canonical model in which the under-ice scattering surface was represented as an infinite elastic plate with protuberances.

Alexander et al. (2013) provided an overview of the acoustic properties of sea ice and assessed the influence of ice-canopy and water-column properties on acoustic TL for propagation within 20 km of a sound source situated at a depth of 20 m. The influence of the ice canopy was first assessed as a perfectly flat surface, and then as a statistically rough surface. A Monte Carlo method was used for the inclusion of ice deformation and roughness. This involved the creation of sets of synthetic ice profiles based on a given sea-ice thickness distribution followed by statistical methods for combining the output of individually evaluated ice realizations. The experimental situation considered was that of an AUV operating within 50 m of

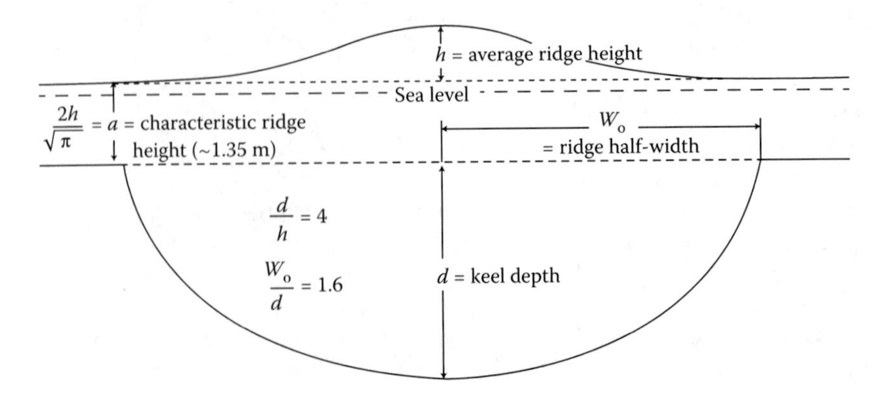

FIGURE 3.10 Geometrical model of sea-ice ridges. (From Diachok, O.I., *J. Acoust. Soc. Am.*, 59, 1110–1120, 1976. With permission.)

the surface. This scenario was associated with the frequency band 9–12 kHz and a horizontal range up to 20 km. The situation was evaluated for a set of typical ice statistics using ray and beam acoustic propagation techniques. The observed sound speed profile resulted in a strong defocusing of direct-path signals at ranges from 9–20 km and for depths shallower than 50 m. The reduction in signal strength of the direct path created areas where the influence of surface-reflected paths became significant. At ranges of 9–20 km, the inclusion of a perfectly flat ice layer reduced the TL by 15–50 dB. When the ice layer was included as a rough surface layer, the results showed an increase of signal strength of up to 8 dB in the small regions of maximum defocusing.

3.4.5 Measurements

Three basic experimental techniques have been employed to measure forward reflection losses at the sea surface:

1. Comparing the amplitude or energy of pulses returned from the surface with that of the direct arrival
2. Using the Lloyd Mirror effect and observing the depth of the minima as the frequency is varied
3. Measuring the attenuation in the surface duct

Based on a compilation of results in the literature by Urick (1982: Chapter 10), it appears that surface losses are less than 1 dB (per bounce) at frequencies below 1 kHz, and rise to about 3 dB (per bounce) at frequencies above 25 kHz.

3.5 SEA-FLOOR BOUNDARY

The sea floor affects underwater sound by providing a mechanism for

1. Forward scattering and reflection loss (but is complicated by refraction in the bottom)
2. Interference and frequency effects
3. Attenuation by sediments
4. Noise generation at lower frequencies due to seismic activity
5. Backscattering and bottom reverberation

Items (1) through (3) will be discussed below. Item (4) will be discussed in Chapter 7 and item (5) in Chapter 9. Urick (1982: Chapter 11) provided a comprehensive summary of sound reflection and scattering by the sea floor. The single most important physical property that determines the acoustic characteristics of sediments is their porosity.

The return of sound from the sea floor is more complex than from the sea surface for several reasons: (1) the bottom is more variable in composition; (2) the bottom is often stratified (layered) with density and sound speeds (both shear and compressional) varying gradually or abruptly with depth; (3) bottom characteristics (composition and roughness) can vary over relatively short horizontal distances; and (4) sound can propagate through a sedimentary layer and either be reflected back into the water by subbottom layers or be refracted back by the large sound-speed gradients in the sediments.

These mechanisms can be incorporated into mathematical models through the specification of appropriate "boundary conditions." The complexity of these boundary conditions will depend on the level of known detail concerning the composition and structure of the sea floor, and also to some degree on the sophistication of the mathematical model being used.

The specification of boundary conditions at the sea floor has assumed greater importance due to increased interest in the modeling of sound propagation in shallow-water areas. Such propagation, by definition, is characterized by repeated interactions with the bottom boundary. Acoustic interactions with highly variable sea-floor topographies and bottom compositions often necessitate the inclusion of both compressional and shear wave effects, particularly at lower frequencies. A fluid, by definition, cannot support shear stresses. Therefore, in modeling acoustic propagation in an ideal (boundless) fluid layer, only compressional wave effects need be considered. As an approximation, saturated sediments are sometimes modeled as a fluid layer in which the sound speed is slightly higher than that of the overlying water column. The basement, however, can support both compressional and shear waves, and rigorous modeling of acoustic waves that interact with and propagate through such media must consider both types of wave effects. As an approximation, shear-wave effects are sometimes included in the form of modified attenuation coefficients.

3.5.1 Forward Scattering and Reflection Loss

3.5.1.1 Acoustic Interaction with the Sea Floor

Westwood and Vidmar (1987) summarized pertinent developments in the modeling of acoustic interaction with the sea floor. It is convenient to partition the discussion according to low-frequency and high-frequency bottom interaction. The transition between low and high frequencies is imprecise but can be considered to occur near 200 Hz.

At low frequencies and low grazing angles, acoustic interaction with the sea floor in deep ocean basins is simple and well understood. The relatively long acoustic wavelengths are insensitive to details of small-scale layering in the sediments. Moreover, for low grazing angles, there is little interaction with the potentially rough substrate interface. Accordingly, the sea floor can be accurately approximated as a horizontally stratified and depth-dependent fluid medium. The major acoustical processes affecting interaction with the sea floor are (1) reflection and transmission of energy at the water-sediment interface, (2) refraction of energy by the positive sound speed gradient in the sediments, and (3) attenuation within the sediments. Modeling of this interaction is further enhanced by the availability of established methods for estimating the geoacoustic profile (i.e., sound speed, density and attenuation as functions of depth) of deep-sea sediments, given the sediment type and physiographic province.

In contrast, bottom interaction at high frequencies is not well understood. The relatively short wavelengths are more sensitive to the small-scale sediment layering. These layers are reported to have an important effect on the magnitude and phase of the plane-wave reflection coefficient. Stochastic techniques with which to analyze the effects of the near-surface sediment layering are being developed, but they do not yet incorporate potentially important acoustical processes such as refraction and shear wave generation. Modeling at high frequencies is further frustrated by the high spatial variability of sediment layering.

The concept of "hidden depths" (Williams, 1976) states that the deep ocean sediment structure well below the ray turning point has no acoustical effect. This concept is important because it focuses attention on those low-frequency processes occurring in the upper regions of the sediments (see Knobles and Vidmar, 1986).

3.5.1.2 Boundary Conditions and Modeling

The ideal forward reflection loss of sound incident on a plane boundary separating two fluids characterized only by sound speed and density was originally developed by Rayleigh (1945, Vol. II: 78). This model is commonly referred to as Rayleigh's law. In the simplest model incorporating absorption, the bottom can be taken to be a homogeneous absorptive fluid with a plane interface characterized by its density, sound speed, and attenuation coefficient. In the case of sedimentary materials, all three of these parameters are affected by the porosity of the sediments. (Appendix D contains a study question involving Rayleigh's law.)

In underwater acoustics, a common idealized model for the interaction of a point-source field with the sea floor is the so-called Sommerfeld model (after A.N. Sommerfeld). This model consists of an isospeed half-space water column overlying an isospeed half-space bottom. The bottom has a higher sound speed than the water. Thus, a critical angle exists in the plane-wave reflection coefficient. For large grazing angles, energy is partially reflected and partially transmitted at the water-bottom interface. For small grazing angles, energy is totally reflected back into the water column. Energy incident near the critical angle produces a complex phenomenon known as the lateral, or head, wave (Chin-Bing, et al., 1982, 1986; Westwood, 1989a). See also the discussions by Clay and Medwin (1977: 262–3) and Frisk (1994: 32). This boundary condition is referred to as an "impedance (or Cauchy) boundary."

Another commonly assumed boundary condition for the sea floor is the homogeneous Neumann bottom boundary condition. Here, the derivative of the pressure normal to the boundary vanishes (Frisk, 1994: 32–3). There is no phase shift in the reflected wave. For harmonic time dependence and constant density, this condition is also termed a "rigid boundary."

Frisk (1994) conveniently summarized four ideal boundary-reflection cases, two each for impedance boundaries and penetrable boundaries. Here, the sediment and water densities are represented by ρ_s and ρ_w while the sediment (compressional) and water sound speeds are represented by c_s and c_w.

1. For an impedance boundary:
   a. Pressure release (Dirichlet) at the sea surface
   b. Neumann (rigid) at the sea floor

2. For a penetrable boundary (impedance or Cauchy boundary condition):
   a. An angle of intromission exists for water-silty-clay interfaces ($\rho_s > \rho_w$ and $c_s < c_w$)
   b. A critical angle exists for water-sand interfaces ($\rho_s > \rho_w$ and $c_s > c_w$).

Hall and Watson (1967) developed an empirical bottom reflection loss expression based largely on the results of Acoustic, Meteorological and Oceanographic Survey (AMOS) (Marsh and Schulkin, 1955). Ainslie (1999) demonstrated that much of the complexity of bottom interaction could be represented in simple equations for the reflection coefficient when expressed in the form of a geometric series. Such simplifications can be useful in modeling acoustic propagation in shallow water where repeated interactions with the seabed are expected. Moreover, Ainslie et al. (1998b) (and erratum in Ainslie and Robins [2003]) presented benchmarks for bottom reflection loss versus angle at 1.5 Hz, 15 Hz, and 150 Hz for four different bottom types, each comprising a layered fluid sediment (representing sand or mud) overlying a uniform solid substrate (representing limestone or basalt). These benchmarks provide ground-truth reference solutions against which the accuracy of other models can be assessed. The benchmarks are calculated using exact analytical solutions where available (primary benchmarks) or they are calculated using a numerical model (secondary benchmark). While the secondary benchmarks are approximate, they provide useful diagnostic information. Robins (1991) developed a FORTRAN program called PARSIFAL to compute plane-wave reflection coefficients from a sediment layer modeled as an inhomogeneous fluid overlying a uniform substrate.

Tindle and Zhang (1992) demonstrated that the acoustic-reflection coefficient for a homogeneous fluid overlying a homogeneous solid with a low shear speed could be approximated by replacing the solid with a fluid having different parameters. Zhang and Tindle (1995) subsequently simplified these expressions by approximating the acoustic-reflection coefficients of solid layers with a fluid described by suitably chosen (proxy) parameters.

Westwood and Vidmar (1987) developed a ray-theoretical approach called CAPARAY for simulating the propagation of broadband signals interacting with a layered ocean bottom. CAPARAY can simulate a time series at a receiver due to an arbitrary source waveform by constructing a frequency domain transfer function from the eigenray characteristics.

Jones et al. (2013) found that the coherent reflection loss from a rough seafloor may be estimated accurately by summing the dB loss obtained for a flat surface of the candidate seafloor material together with the roughness loss obtained for a lossless surface of the same surface profile.

3.5.1.3 Geoacoustic Models

Geoacoustic models of the sea floor more properly account for the propagation of sound in sediments (Anderson and Hampton, 1980a,b; Hamilton, 1980). As summarized by Holland and Brunson (1988), geoacoustic models of marine sediments can be formulated in one of three ways: (1) by empirically relating geoacoustic and geophysical properties of the sediments (e.g., Hamilton, 1980); (2) by using the Biot–Stoll model to relate sediment geoacoustic properties to geophysical properties on the basis of physical principles (Biot, 1956a,b; Stoll, 1974, 1980, 1989); and (3) by using an inversion technique to generate sediment geophysical parameters from bottom loss measurements (e.g., McCammon, 1991; Hovem et al., 1991: see especially Section 3, Modelling and inversion techniques; Rajan, 1992; Dosso et al., 1993; Hovem, 1993; Frisk, 1994).

Chotiros (2017) presented a model of seabed acoustics with input parameters that allow the model to cover a wide range of sediment types. The seabed is a fluid-saturated porous material that obeys the wave equations of a poroelastic medium, which are significantly more complicated than the equations of either a liquid or a solid. Guigné and Blondel (2016) provided support to the offshore industries (oil and gas, renewables, harbour building) by drawing on innovative acoustic approaches highlighting the design, refinement, and commercial adoption of new approaches to acoustic interrogation of the seabed and sub-seabed, linking research and application.

The Biot–Stoll model (Biot, 1956a,b; Stoll, 1974, 1980, 1989) provides a comprehensive description of the acoustic response of linear, porous materials containing a compressible pore fluid. The model predicts two types of compressional waves and one shear wave. Recent applications in underwater acoustics with references to the key historical literature were provided by Beebe et al. (1982) and by Holland and Brunson (1988). Routine operational employment of this model is complicated by the input of more than a dozen geophysical parameters, some of which are difficult to obtain even in laboratory environments.

Williams (2001) presented an acoustic propagation model that approximated a porous medium as a fluid with a bulk modulus and effective density derived from Biot theory. Within the framework of Biot theory, it was assumed that the porous medium had low values of frame bulk and shear moduli relative to the other moduli of the medium, and these low values were approximated as zero. This led to an effective density fluid model (EDFM). It was further shown that for saturated sand sediments, the dispersion, transmission, reflection, and in-water backscattering predicted with this EDFM were in close agreement with the predictions of Biot theory. It was demonstrated that the frame bulk and shear moduli played only a minor role in determining several aspects of sand acoustics. Thus, for many applications the EDFM is an accurate alternative to full Biot theory and is much simpler to implement. For typical environmental parameters, the accuracy of the EDFM relative to Biot theory confirmed that the frequency dependence of sound speed and attenuation (over the range of 10 Hz–100 kHz) seen in the Biot model for sand is almost entirely due to changes in the motion of the water relative to the sand grains. The agreement between Biot theory and EDFM thus demonstrated that the bulk frame and shear moduli play a minor role in reflection, transmission, and in-water backscattering. Thus, for many applications the EDFM is an accurate alternative to full Biot theory and is much simpler to implement. The EDFM requires eight input parameters: the density and bulk modulus of sediment grains; density, bulk modulus, and viscosity of water; and the porosity, permeability, and tortuosity of the sediment. Measurement uncertainties of grain bulk modulus, and sediment permeability and tortuosity, produce the greatest EDFM uncertainty.

An EDFM was developed for unconsolidated granular sediments and was applied to sand (Williams, 2013). This model was a simplification of the full Biot porous media model. Two other effects were added to the EDFM model that led to additional dispersion and attenuation: heat transfer between liquid and solid at low frequencies, and granularity of the medium at high frequencies. The frequency range studied was 100 Hz to 1 MHz. The analytical sound speed and attenuation expressions obtained had no free parameters. The resulting model included only environmental parameters that could be measured separately from any acoustic measurements. At low frequencies, heat transfer reduced the sand sound speed; at high frequencies, granularity had the same effect. These results were in qualitative agreement with available experimental data.

The EDFM was developed by Williams (2001) to approximate the behavior of sediments governed by Biot's theory of poroelasticity. Previously, it had been shown that the EDFM predicted reflection coefficients and backscattering strengths that are in close agreement with those of the full Biot model for the case of a homogeneous poroelastic half-space. However, it has not yet been established to what extent the EDFM can be used in place of the full Biot–Stoll model for other cases. Bonomo et al. (2015, 2016) used the finite element method together with the flat-interface reflection and rough-interface backscattering predictions of the Biot–Stoll model to compare against the EDFM for the case of a poroelastic layer overlying an elastic substrate. It was shown that considerable differences between the predictions of the two models can exist when the layer is very thin and has a thickness comparable to the wavelength of the shear wave supported by the layer, with a particularly strong disparity under the conditions of a shear wave resonance. For thicker layers, the predictions of the two models were found to be in closer agreement, approaching nearly exact agreement as the layer thickness increases. In order to ascertain how accurately the EDFM approximated the full Biot–Stoll model for the case of a poroelastic layer overlying an elastic substrate, a series of numerical experiments were performed using the commercial FEM code COMSOL Multiphysics, which was used for all meshing and solving. Frequencies of 100 Hz and 10 kHz were considered. At each frequency, both the bottom loss for a flat water-layer interface and the backscattering strength for a rough water-layer interface as predicted by the two models were compared for layer thicknesses of 0.3, 3, and 30 m. This disparity between the EDFM and the full Biot–Stoll model was strong (1) when the poroelastic layer was both very thin compared to an acoustic wavelength and on the order of the wavelength of the shear wave supported by the layer (as in the 100 Hz cases); and (2) when the layer had a thickness equal to an odd multiple of quarter wavelengths of the shear wave, as in the cases with thickness of 0.3 m. The mechanism causing this disparity appeared to be the shear-wave resonances that occurred under these conditions. From the 10 kHz cases, it appeared that the backscattering strength predictions of the EDFM were closer to those of the full Biot–Stoll model than the bottom-loss predictions. From this result, it could be inferred that the presence of the rough surface obstructed the substrate. In general, however, both the bottom loss and backscattering strength predictions of the EDFM were in very close agreement with those made by the Biot–Stoll model when the layer thickness was on the order of the acoustic wavelength or larger.

McCammon (1988) described the development of a geoacoustic approach to bottom interaction called the thin layer model. This model, which is based on an inversion technique, contains a thin surficial layer, a fluid sediment layer and a reflecting subbottom half-space. There are 10 input parameters to this model: sediment density, thickness, sound-speed gradient and curvature, attenuation, and attenuation gradient;

thin-layer density and thickness; basement reflectivity; and water-sediment velocity ratio (Figure 3.11). The model generates bottom-loss curves as a function of grazing angle over the frequency range 50–1500 Hz. The model makes several assumptions: it relies upon the "hidden depths" concept of Williams (1976); the sediments are isotropic; the roughness of the sediment and basement interfaces, as well as multiple scattering within the layers, is neglected; and shear wave propagation is ignored.

Sample outputs from this thin layer model are presented in Figure 3.12. A ratio $(c_s/c_w) > 1$ (where c_s is the sound speed in the upper sediment and c_w is the sound speed at the base of the water column) predicts a critical angle $\theta_c = \cos^{-1}(c_w/c_s)$, below which most of the incident energy is reflected; that is, the bottom loss is nearly zero. By comparison, a ratio $(c_s/c_w) < 1$ would refract the incident energy into the sediments and result in greater losses at small angles.

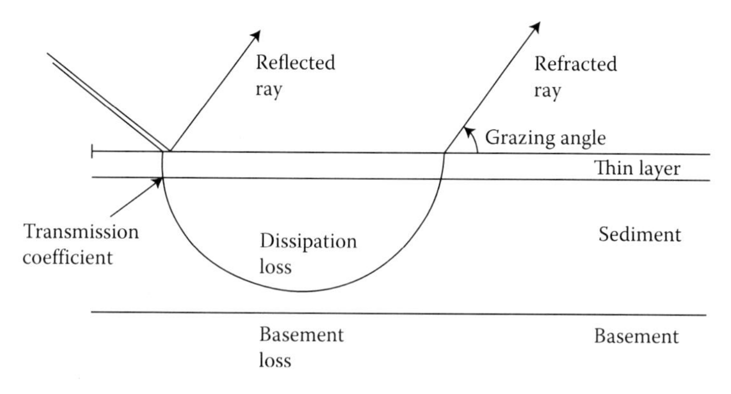

FIGURE 3.11 Thin layer model for sediment reflected and refracted paths. (Adapted from McCammon, D.F., *J. Geophys. Res.*, 93, 2363–1369, 1988. Published by the American Geophysical Union.)

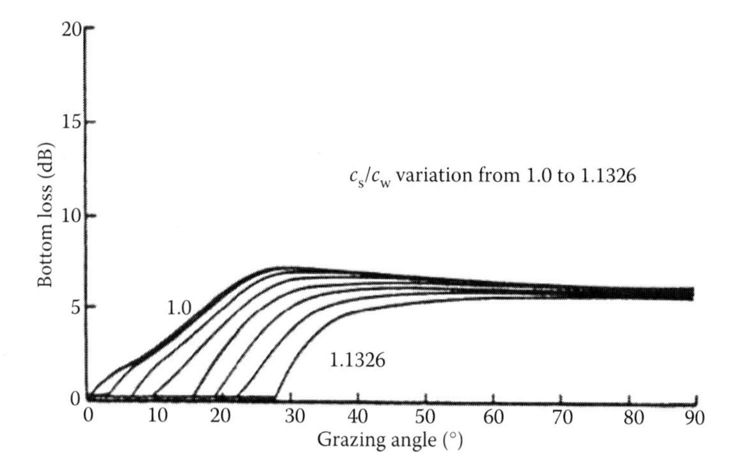

FIGURE 3.12 Variation of bottom loss (dB) as a function of grazing angle for a frequency of 1000 Hz. Curves are presented for various values of the ratio of the upper sediment sound speed (c_s) to the sound speed at the base of the water column (c_w). (Adapted from McCammon, D.F., *J. Geophys. Res.*, 93, 2363–2369, 1988. Published by the American Geophysical Union.)

A qualitative comparison of bottom loss versus grazing angle for $(c_s/c_w) \geq 1$ and $(c_s/c_w) < 1$ is presented in Figure 3.13. It has been demonstrated that low-porosity sediments (e.g., hard sands) have compressional wave speeds greater than that of the overlying water while high-porosity sediments (e.g., mud and silt) have sound speeds less than that of the overlying water (Urick, 1983: 138–9; Apel, 1987: 386). Qualitatively, then, the comparison presented in Figure 3.13 contrasts the effects of high-porosity $[(c_s/c_w) < 1]$ and low-porosity sediments $[(c_s/c_w) \geq 1]$.

The bottom loss upgrade (BLUG) model, which was a modular upgrade designed for incorporation into existing propagation models to treat bottom loss, was based on a geoacoustic (or inverse) approach. The low-frequency bottom-loss (LFBL) model has subsequently replaced the BLUG model.

3.5.2 Interference and Frequency Effects

Stratification and attendant scattering within the bottom produce pulse distortion, as does reflection at grazing angles less than the critical angle. Zabal et al. (1986) developed a simple geometric-acoustic model to predict frequency and angle spreads as well as coherence losses to sonar systems. The sea floor was modeled by homogeneous and isotropic slope statistics. The facets are planar and reflect specularly, thus giving rise to the name "broken mirror" model.

When close to, or on the seabed, a hydrophone receives the direct and bottom reflected signals. The resulting multipath interference leads to an effective receive beam pattern that depends not only on source and receiver position and water depth, but also on the seabed characteristics, which affect the phase and magnitude of the reflection coefficient (Parsons and Duncan, 2011).

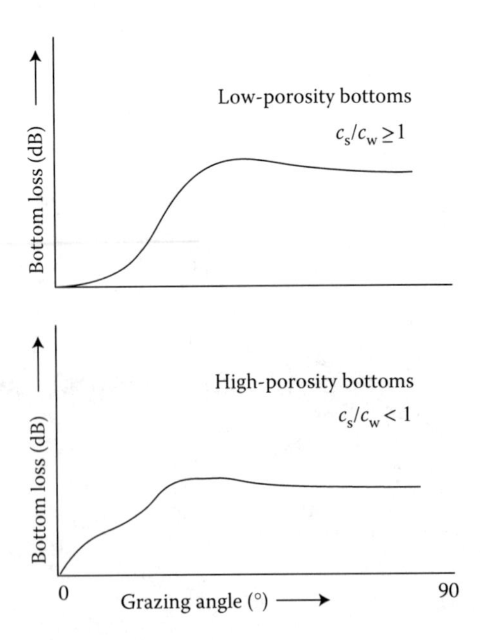

FIGURE 3.13 Qualitative illustration of bottom loss versus grazing angle for low-porosity and high-porosity bottoms.

3.5.3 Attenuation by Sediments

Sound that propagates within sediment layers is subject to the effects of attenuation. A variety of sediment attenuation units are commonly used in the underwater acoustic and marine seismology communities. The relationships among these units can become very confusing when attempting to enter values into propagation models.

Mathematically, acoustic attenuation (α) is expressed in the exponential form as $e^{-\alpha r}$ using the units of nepers (Np) per unit distance for α. The acoustic attenuation can be converted to the units of dB m^{-1} using the relationship

$$\begin{aligned}\alpha\left(\text{dBm}^{-1}\right) &= 20\left(\log_{10}\text{e}\right)\alpha\left(\text{Npm}^{-1}\right)\\ &= 8.6859\alpha\left(\text{Npm}^{-1}\right)\end{aligned}$$

Some propagation models require that the attenuation be specified in units of dB per wavelength (λ):

$$\alpha(\text{dB}\,\lambda^{-1}) = \lambda(\text{m})\alpha\left(\text{dB m}^{-1}\right)$$

Another term commonly encountered in underwater acoustic modeling is the attenuation coefficient (k), which is based on the concept that attenuation (α) and frequency (f) are related by a power law:

$$\alpha\left(\text{dB m}^{-1}\right) = kf^{n}$$

where f is measured in kilohertz and n is typically assumed to be unity. Over the frequency range of interest to underwater acousticians, attenuation is approximately linearly proportional to frequency.

A compilation of sediment attenuation measurements made by Hamilton (1980) over a wide frequency range showed that the attenuation in natural, saturated sediments is approximately equal to $0.25\,f$ (dB m^{-1}) when f is in kilohertz. There was a tendency for the more dense sediments (such as sand) to have a higher attenuation than the less dense, higher-porosity sediments (such as mud). The attenuation in sediments is several orders of magnitude higher than in pure water.

In the Arctic, acoustical parameters of the sea floor and subbottom are poorly known. Difficulties in obtaining direct core samples to great depths limit the database from which to extract the parameters needed to determine many of the major acoustical processes in bottom interaction. Estimating these geoacoustical parameters based on data from contiguous areas may not be meaningful since the basic processes of sedimentation at work under the pack ice are unique to that environment. Sedimentation rates are very low, being dominated by material carried by the ice rather than by material of biologic origin, as is the case in more temperate areas. The ice pack may also carry large boulders of glacial origin and deposit them in the Arctic Ocean. The low sedimentation rate leaves the boulders exposed as potential scatterers for acoustic energy over a wide range of frequencies.

3.5.4 Measurements

The standard method for measuring bottom loss is to use pings or explosive pulses and to compare the amplitude, intensity or energy density (integrated intensity) of the bottom pulse with that of the observed or computed pulse traveling via a direct path.

Bottom loss data typically show a loss increasing with angle at low angles, followed by a nearly constant loss extending over a wide range of higher angles (refer back to Figure 3.13). High-porosity bottoms (having a sound speed less than that of the overlying water) tend to have a maximum loss at an angle between 10° and 20° where an angle of intromission (i.e., no reflection, but complete transmission into the bottom) would be expected to occur in the absence of attenuation in the bottom. Holland (2002) used data from field measurements to show this intromission phenomenon with remarkable clarity. When narrowband pulses are used, measured losses are often irregular and variable, showing peaks and troughs due to the interference effects of layering in the bottom. Measured data rarely show a sharp critical angle (as would be inferred from the Rayleigh reflection model) because of the existence of attenuation in the bottom (Urick, 1983: Chapter 5).

Analytical formulas describing the ray properties of a family of sediment sound-speed profiles were compiled by Ainslie et al. (2004). These profiles were described as "generalized-power law" profiles since they were generalizations of much simpler power-law profiles. The specific properties considered included ray paths, caustic envelopes, and cusp coordinates due to a point source in isovelocity water. To demonstrate the utility of these analytical formulas, a few caustic shapes were generated for comparison by computing the pressure field using a normal-mode program.

Harrison (2004a) applied spectral factorization to the modulus of the seabed's reflection coefficient so that a subsequent Fourier transformation of the new complex reflection coefficient produced a minimum-phase impulse response for each angle. This method required the reflection coefficient to be known over a range of frequencies, and the grazing angles had to be above the critical angle. This technique was developed to support a method for extracting the seabed's plane-wave reflection coefficient from ambient-noise data measured on a moored or drifting vertical line array. This method thus offers the possibility of conducting subbottom profiling from a single platform using only ambient noise.

Jensen (1998) established stair-step discretization criteria for an accurate representation of smoothly varying bathymetry in numerical models. The strictest criteria were found to apply to backscatter calculations, where the horizontal stair-step size (Δx) must be a small fraction of an acoustic wavelength (λ): $\Delta x \leq \lambda/4$. It was found that the forward-scatter problem (assuming weak backscatter) could be solved accurately with order-of-magnitude larger step sizes.

3.6 ATTENUATION AND ABSORPTION IN SEA WATER

Sound losses in the ocean can be categorized according to spreading loss and attenuation loss. Spreading loss includes spherical and cylindrical spreading losses in addition to focusing effects. Attenuation loss includes losses due to absorption, leakage out of ducts, scattering and diffraction. Urick (1982: Chapter 5) summarized the relevant literature pertaining to this subject.

Absorption describes those effects in the ocean in which a portion of the sound intensity is lost through conversion to heat. Field measurements of the absorption coefficient (α), typically expressed in units of dB km^{-1}, span the frequency range 20 Hz–60 kHz. In practice, absorption loss (in dB) is computed as the product of α and range (r) using self-consistent units for range.

The dependence of α on frequency is complicated, reflecting the effects of different processes or mechanisms operating over different frequency ranges. The equation developed by Thorp (1967) is probably the best known and is valid at frequencies below 50 kHz:

$$\alpha = 1.0936\left[\frac{0.1f^2}{1+f^2} + \frac{40f^2}{4100+f^2}\right] \tag{3.8}$$

where α is the absorption coefficient (dB km^{-1}) and f is the frequency (kHz). The factor 1.0936 converts the original formula from units of dB kyd^{-1} to dB km^{-1}. More recent formulas for the absorption coefficient have been described by Fisher and Simmons (1977) and by Francois and Garrison (1982a,b). Ainslie and McColm (1998) simplified a version of the Francois–Garrison equations for viscous and chemical absorption in sea water by making explicit the relationships among acoustic frequency, depth, sea-water absorption, pH, temperature and salinity. An older dataset that has received renewed attention is that reported by Skretting and Leroy (1971), which included measurements of sound attenuation in the western Mediterranean Sea.

In practice, the effects of absorption and attenuation are considered jointly. Then, the frequency dependence can conveniently be segregated into four distinct frequency regions over which the controlling mechanisms can be readily identified. These regions are (in order of ascending frequency) (1) large-scale scattering or leakage; (2) boric acid relaxation; (3) magnesium sulfate relaxation; and (4) viscosity. Fisher and Simmons (1977) summarized these effects graphically (Figure 3.14). Research conducted by other investigators provides regional formulas for absorption and attenuation (Skretting and Leroy, 1971; Kibblewhite et al., 1976; Mellen et al., 1987a,b,c; Richards, 1998). Absorption is regionally dependent due mainly to the pH-dependence of the boric acid relaxation. Attenuation due to turbidity and bubbles was discussed in Section 3.4.3.

The attenuation of low-frequency sound in the sea is pH-dependent; specifically, the higher the pH, the greater the attenuation (Browning et al., 1988). Thus, as the ocean becomes more acidic (lower pH) due to increasing CO_2 emissions, the attenuation will diminish and low-frequency sounds will propagate farther, effectively making the ocean noisier.

The Applied Physics Laboratory, University of Washington (1994) documented high-frequency (approximately 10–100 kHz) acoustic models with potential application to sonar simulation and sonar system design efforts. These models treat: volumetric sound speed, absorption and backscattering; boundary backscatter and forward loss for the sea surface and the sea floor; ambient-noise sources and levels; and Arctic attenuation and under-ice losses.

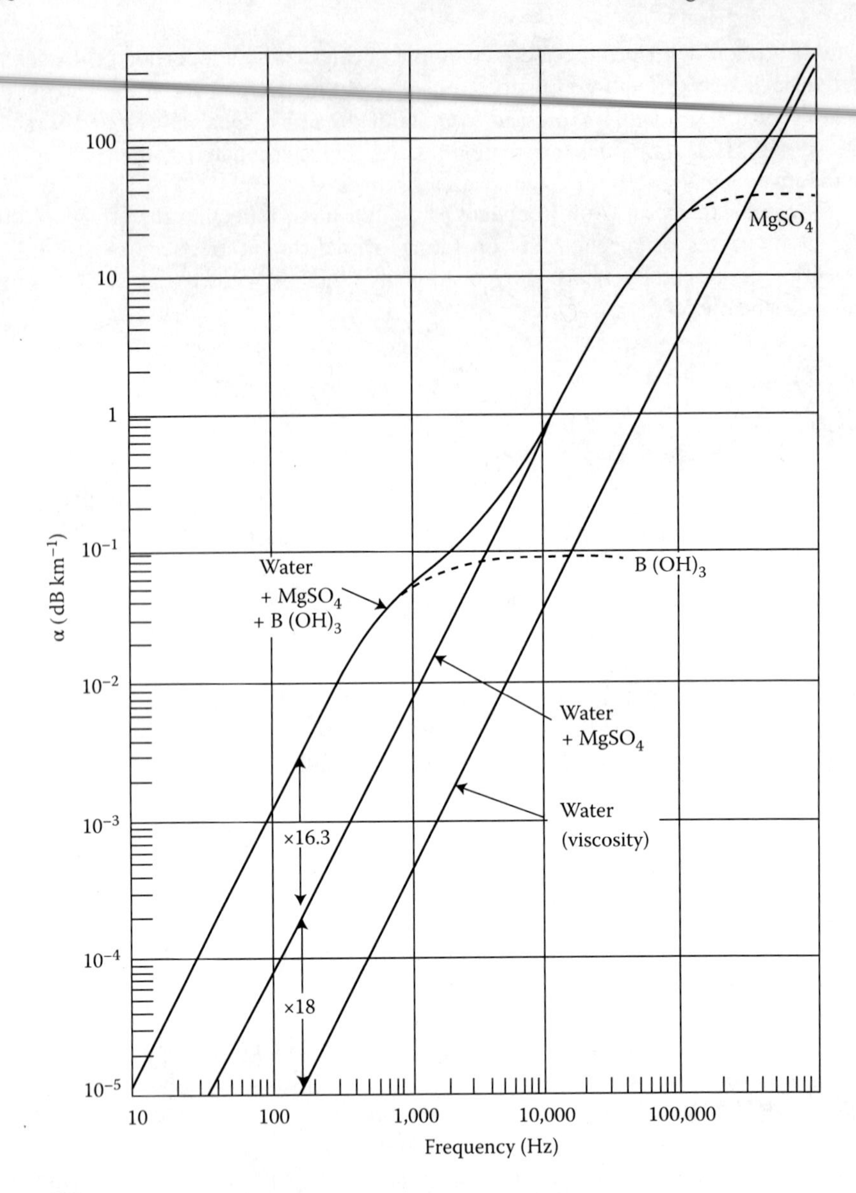

FIGURE 3.14 Absorption coefficients for sea water at a temperature of 4°C at the sea surface. Dashed lines indicate contributing absorption rates due to relaxation processes. (From Fisher, F.H. and Simmons, V.P., *J. Acoust. Soc. Am.*, 62, 558–564, 1977. With permission.)

A new empirical formula for the absorption of sound in seawater (dB km^{-1}) was derived in an inverse fashion by van Moll et al. (2009). This nonlinear inverse problem was solved using a global search to find the parameter setting that yielded the best fit to *in situ* absorption measurements. The new formula contains three terms: (1) boric acid contribution, with explicit dependence on frequency, pH, temperature, and salinity; (2) magnesium sulfate contribution, with explicit dependence on frequency, depth, temperature, and salinity; and (3) freshwater absorption, from Francois and Garrison (1982a,b), with explicit dependence on frequency, temperature, and depth. The domains of applicability for this formula were specified as follows:

- frequency 0.16–650 kHz
- pH 7.69–8.18 (NBS scale)
- salinity 8–40.5 g kg^{-1}
- temperature −1.75°C–22°C
- depth 0.013–3.35 km

The formula is complicated and is not reproduced here. Those interested in using the equations should consult the cited reference.

In the case of monochromatic waves propagating in a quiet medium, the wave amplitude is a constant, and the attenuation is proportional to $\exp[-\omega^2]$, where $\omega = 2\pi f$. When the medium becomes turbulent, the wave amplitude shows fluctuations. Lacaze (2007) demonstrated that turbulent and motionless propagation can be approached by a conditional Gaussian process that gives a random character to the attenuation.

3.7 SURFACE DUCTS

Sound travels to long distances in the ocean by various forms of ducted propagation. When sound travels in a duct, it is prevented from spreading in depth and remains confined between the boundaries of the duct. The surface duct is a zone bounded above by the sea surface and below by the sonic layer depth (SLD). Within the surface duct, sound rays are alternatively refracted and reflected. A surface duct exists when the negative temperature gradient within it does not exceed a value determined by the effect of pressure on sound speed (refer to Chapter 2). Specifically, the surface duct is characterized by a positive sound speed gradient. For example, in isothermal water (and ignoring the effects of salinity), the pressure effect will produce a positive sound speed gradient of 0.017 s^{-1}.

The surface duct is the acoustical equivalent of the oceanographic mixed layer, although they are defined differently. While the SLD is normally defined in terms of the sound speed gradient, the mixed layer depth (MLD) is defined in terms of temperature, or more precisely, in terms of density (which is a function of temperature, salinity and pressure). The mixed layer is a quasi-isothermal layer of water created by wind-wave action and thermohaline convection. Algorithms for the prediction of surface duct propagation will be discussed in Chapter 5. These algorithms use the depth of the mixed layer as an input variable.

3.7.1 Mixed-Layer Distribution

Oceanographers have extensively studied the dynamics of the mixed layer. Variations in the temperature and depth of the mixed layer are closely related to the exchange of heat and mass across the air-sea interface and are thus of interest to scientists engaged in studies of the global climate. Lamb (1984) presented bimonthly charts of the mean MLDs for the North and tropical Atlantic Oceans. Bathen (1972) presented monthly charts of MLDs for the North Pacific Ocean.

Levitus (1982) presented charts of MLDs on a global basis. Distributions of MLDs were calculated using both a temperature criterion and a density criterion. The temperature criterion was based on a temperature difference of 0.5°C between the surface and the depth referred to as the MLD. The density criterion was based on a difference of 0.125×10^{-3} g cm^{-3} between the surface and the depth of the mixed layer. The use of the density criterion recognized the importance of salinity in determining the stability of the mixed layer and hence, from an acoustics viewpoint, the true depth of the sonic layer. For example, in subarctic regions, isothermal conditions (or even temperature profiles with inversions) combine with a salinity profile that stabilizes the water column to control the depth of mixing. MLDs for the months of March and September are presented in Figures 3.15 and 3.16, respectively, comparing the global distributions resulting from both the temperature and the density criteria.

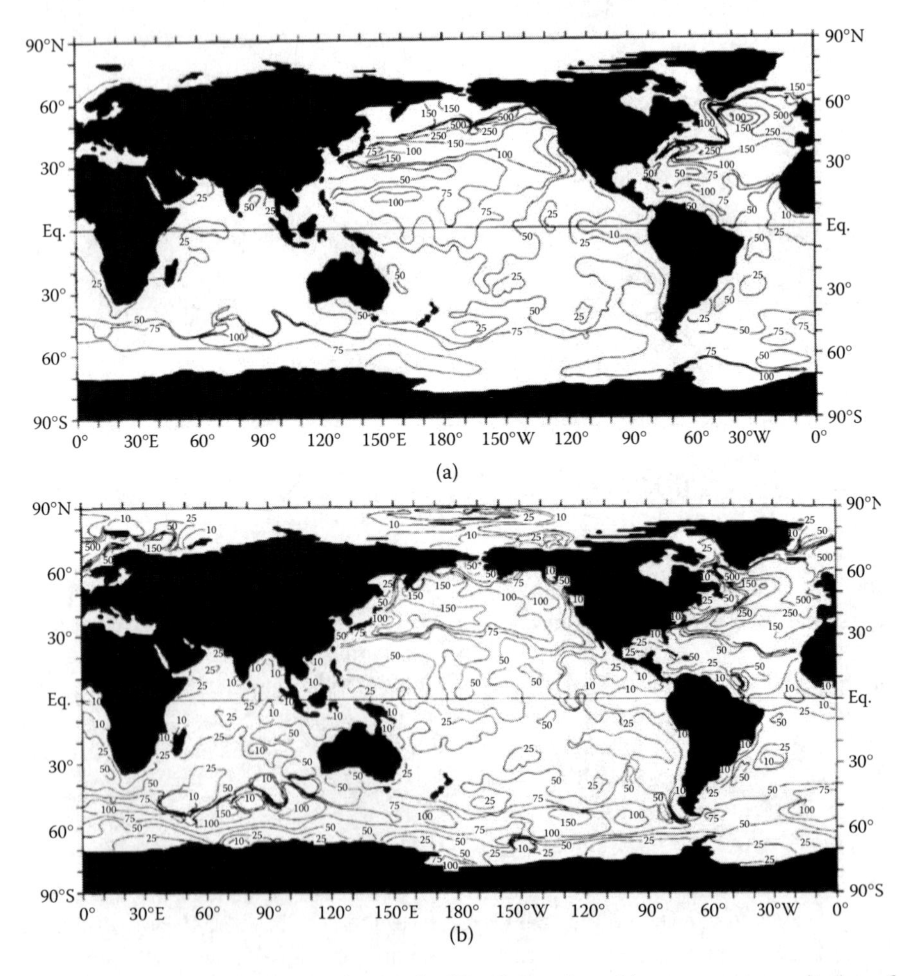

FIGURE 3.15 Mixed layer depths (m) for March based on (a) a temperature criterion of 0.5°C and (b) a density criterion of 0.125×10^{-3} g cm^{-3}. (Adapted from Levitus, S. 1982. Climatological atlas of the world ocean. NOAA Professional Paper 13.)

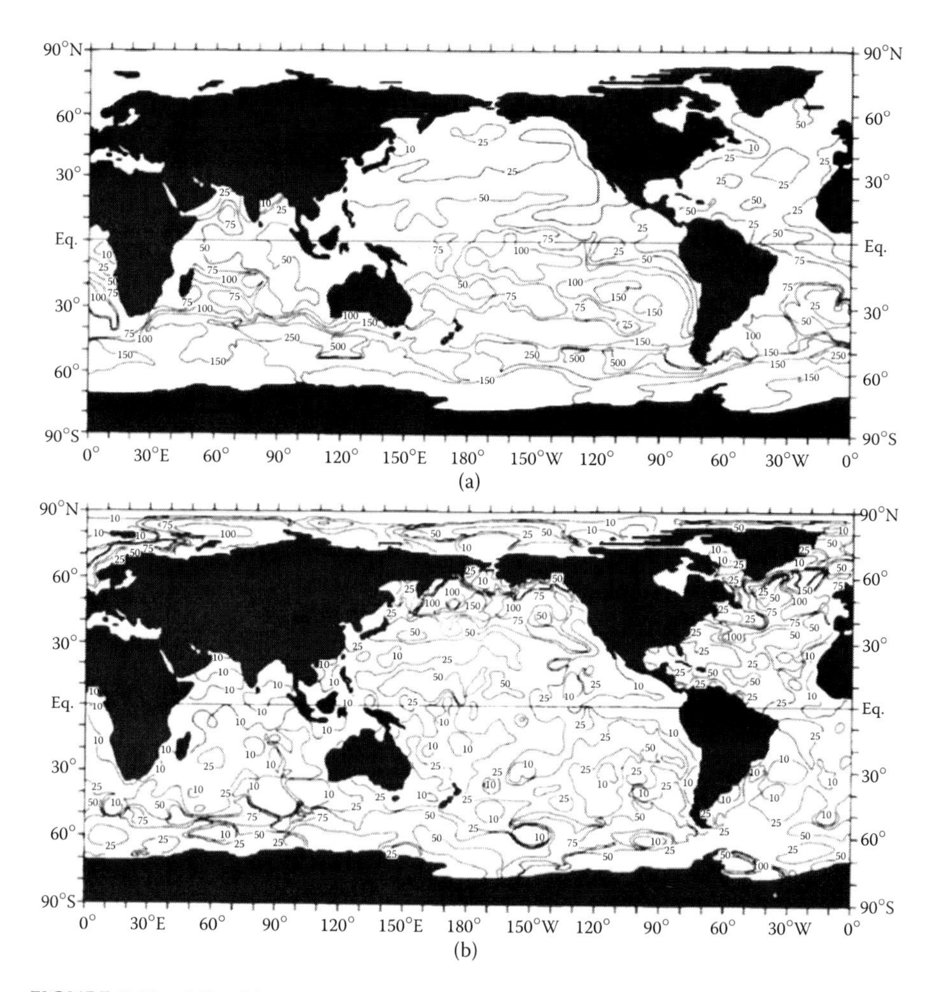

FIGURE 3.16 Mixed layer depths (m) for September based on (a) a temperature criterion of 0.5°C and (b) a density criterion of 0.125×10^{-3} g cm^{-3}. (Adapted from Levitus, S. 1982. Climatological atlas of the world ocean. NOAA Professional Paper 13.)

Traditionally, it has been assumed that the mixed layer is the vertical extent of the turbulent boundary layer. Peters and Gregg (1987) have pointed out that the turbulence initiated by the exchanges of energy, buoyancy or momentum across the air-sea interface may actually penetrate the lower boundary of the mixed layer as defined by the previous criteria. Thus, precise discussions of the mixed layer are often frustrated by imprecise terminology.

The SLD and the mixed-layer depth (MLD) often coincide since sound speed increases with depth down to the MLD whereafter a decrease in temperature typically occurs resulting in a local sound-speed maximum, referred to as the SLD (Helber et al., 2008). In reality, however, the SLD and MLD may not always be the same since the sound speed is substantially more sensitive to temperature than salinity. Since the MLD is a commonly known and studied parameter, it is often used as a proxy for the SLD in scientific and operational applications. In the boreal spring, when fresh restratification events occur, the SLD is 10 m deeper (shallower) than the MLD in 39% (7%) of the observed profiles. Three parameters (SLD, MLD_T, and MLD_{TS}) characterize the upper ocean in different ways. The SLD represents the potential of the upper ocean to trap acoustic energy in a duct near the surface. The SLD methodology requires profile pairs of temperature and salinity that are used to compute a sound speed profile. The MLD_T is the penetration depth of the most recent surface mixing that is resolved in a profile as defined by potentially small changes in near surface vertical gradients of temperature. Salinity is not used for the MLD_T. The MLD_{TS} is a density based threshold method that most closely represents the seasonal MLD and requires profile pairs of temperature and salinity.

Some regional examples will serve to clarify features of the mixed layer as evidenced by the thermal structure of the water column. For these examples, the continental shelf region off the Texas-Louisiana coast in the Gulf of Mexico (Etter and Cochrane, 1975) will be explored. Figure 3.17 presents the annual variation of water temperature down to 225 m. In this graphical representation, the seasonal variation of the depth of the mixed layer is portrayed as a uniform layer of temperature versus depth. In January, for example, the mixed layer is approximately 100 m deep while in August the MLD is seen to shallow to about 25 m. These variations are further illustrated in Figures 3.18 and 3.19, which are vertical sections across the shelf. In winter (February), for example, the isotherms tend to be vertically oriented, evidencing strong mixing and vertical (versus the more typical horizontal) stratification. In summer (July), however, the isotherms tend to be more horizontal, the result of weak mixing, with attendant horizontal stratification.

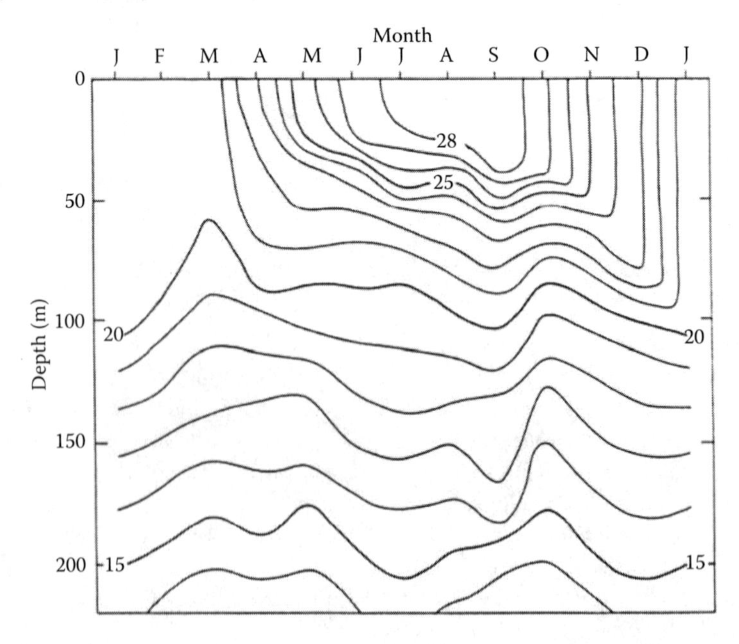

FIGURE 3.17 Temperature contours (°C) for a region of the Texas-Louisiana continental shelf in the Gulf of Mexico. (Adapted from Etter, P.C. and Cochrane, J.D., *Water temperature on the Texas–Louisiana shelf*. Texas A & M Univ., 1975. TAMU-SG-75-604.)

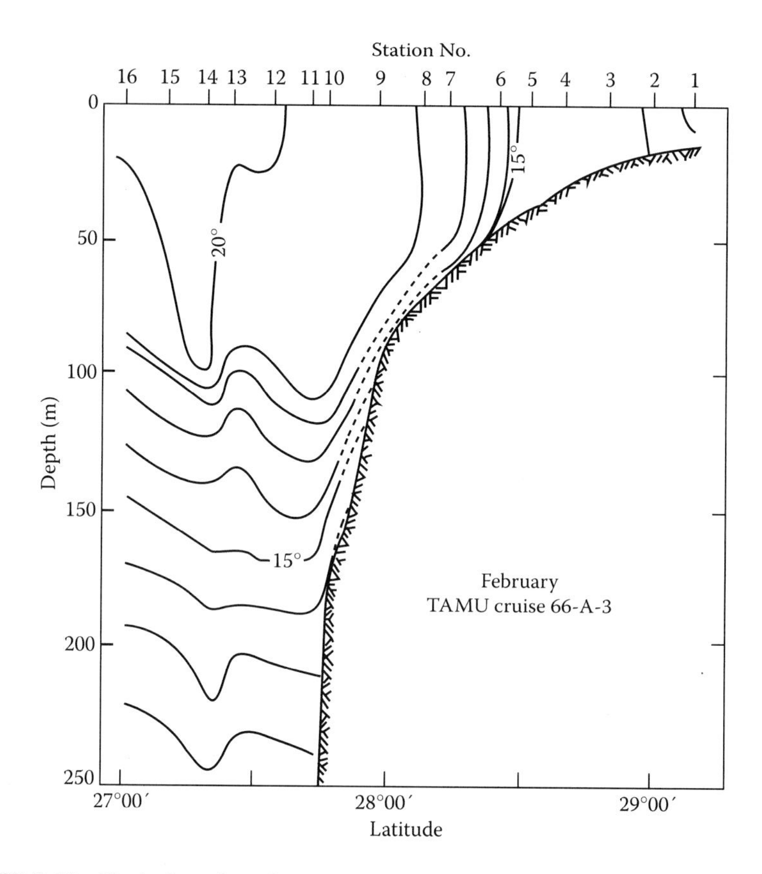

FIGURE 3.18 Vertical section of water temperatures (°C) over the Texas-Louisiana continental shelf in the Gulf of Mexico in February. (Adapted from Etter, P.C. and Cochrane, J.D., *Water temperature on the Texas–Louisiana shelf.* Texas A & M Univ., 1975. TAMU-SG-75-604.)

3.7.2 General Propagation Features

A computer-generated ray diagram for a source in a typical mixed layer is shown in Figure 3.20. Under the conditions for which the diagram was drawn, the ray leaving the source at an angle of 1.76° becomes horizontal at the base of the mixed layer. Rays leaving the source at shallower angles remain in the layer, and rays leaving the source at steeper angles are refracted downward to greater depths. A shadow zone is produced beneath the layer at ranges beyond the immediate sound field. The shadow zone is not completely devoid of acoustic energy since it is insonified by diffraction and by sound scattered from the sea surface. The rate of sound leakage out of the surface duct can be quantified in terms of an empirical leakage coefficient (α_L), which expresses the attenuation (dB km^{-1}) of sound trapped within the duct. This leakage coefficient varies with the surface roughness (wave height), duct thickness (SLD, or MLD), sound speed gradient below the layer, and acoustic frequency (Urick, 1983: 153–4).

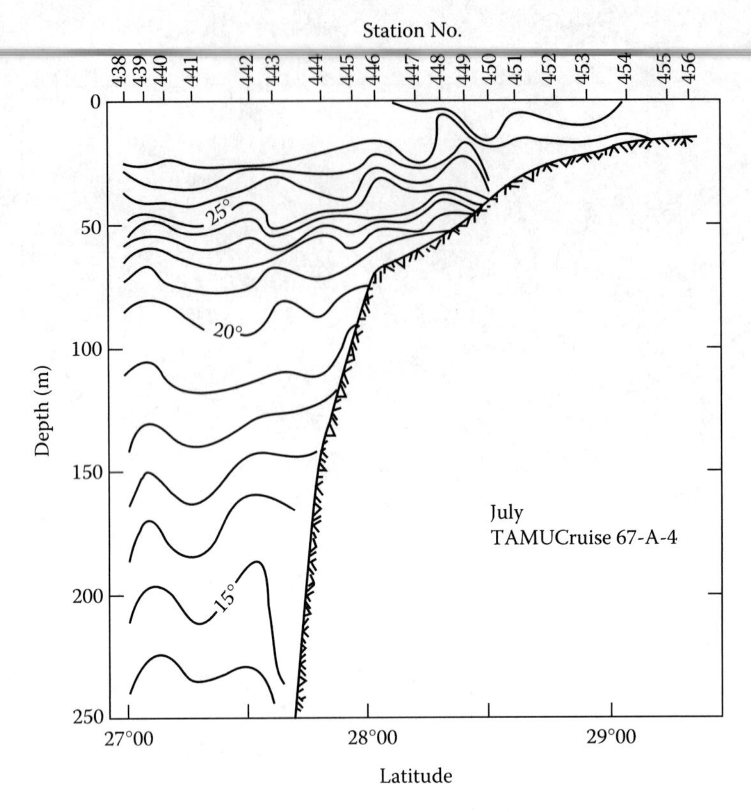

FIGURE 3.19 Vertical section of water temperatures (°C) over the Texas-Louisiana continental shelf in the Gulf of Mexico in July. (Adapted from Etter, P.C. and Cochrane, J.D., *Water temperature on the Texas–Louisiana shelf.* Texas A & M Univ., 1975. TAMU-SG-75-604.)

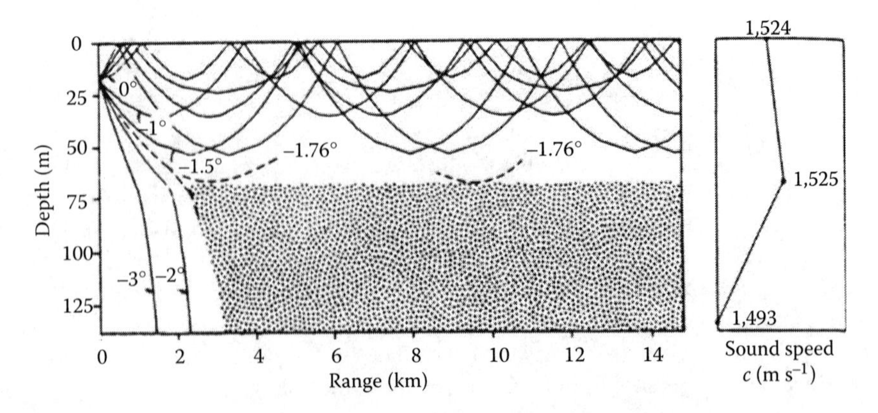

FIGURE 3.20 Ray diagram for a sound source located at a depth of 16 m, showing propagation and trapping of energy in the surface duct, with a shadow zone below 60 m, for the sound speed profile shown at the right. (Reproduced from Urick, R.J., *Principles of Underwater Sound*, 3rd ed. McGraw-Hill, New York, 1983. With permission of McGraw-Hill.)

Over short distances, or even at a fixed location over time, the thickness of the mixed layer may vary because of internal waves. These waves propagate along the density discontinuity at the base of the mixed layer (Figure 3.21). Internal waves complicate the propagation of underwater sound by causing variations in TL over short distances. Fluctuations of echoes between a source and receiver separated by the mixed (or sonic) layer depth are attributable in part to the existence of internal waves.

Urick (1982: Chapter 6) reviewed the available open-literature algorithms for estimating TL in the surface duct. In particular, an analysis of the extensive data set collected during the AMOS program (Marsh and Schulkin, 1955) has been used to characterize TL in the surface duct in the frequency range 2–8 kHz. Graphical results derived from Condron et al. (1955) are summarized in Figures 3.22 and 3.23. These figures demonstrate the importance of source-receiver geometry (relative depths) in surface duct propagation, particularly in those cases where the source and receiver are situated on opposite sides of the layer depth (referred to as cross-layer geometries).

It has been demonstrated experimentally that the shadow zone below the upper channel in a two-channel ocean waveguide can be illuminated by the scattering of signals from highly anistropic fine-structure inhomogeneities embedded in Mediterranean outflow water (Galkin et al., 2006). (See related discussions in Section 2.5.3.1.) The experiment was conducted in the Iberian Basin in the northeastern Atlantic Ocean where warm and saline Mediterranean waters intrude at depths between 600 m and 1500 m. This creates a double-channeled waveguide with the two-channel axes (sound-speed minima) situated at 450–500 m and at 2,000 m, that is, above and below the sound-speed maximum corresponding to the depth of the Mediterranean-water intrusion. (These effects are related to a submesoscale inter-thermocline lens of Mediterranean Water in the Iberian Basin of the Atlantic Ocean.) The sound source emitted a continuous pseudonoise signal in the frequency band 2.5–4.0 kHz. The ratio of the vertical scale to the horizontal scale of the inhomogeneities was approximately 0.001 or less, meaning that the inhomogeneities extended over 100 m or greater in the horizontal direction. This areal extent would provide opportunities to scatter sound into the shadow zone.

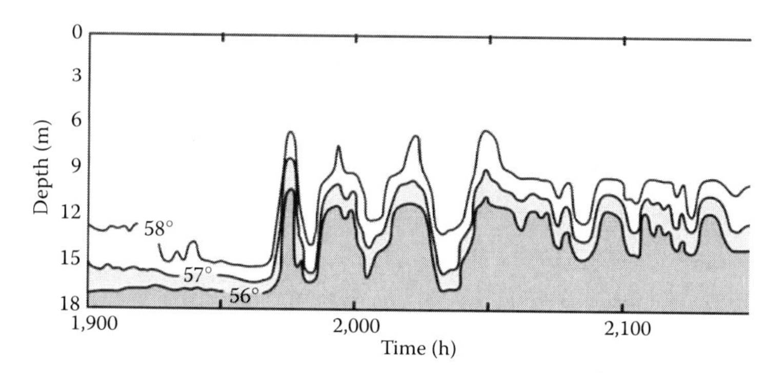

FIGURE 3.21 Fluctuations in the depth of the mixed layer caused by the passage of internal waves. Water temperature is measured in °F. The mixed layer depth is defined by the 58°F isotherm. (Reprinted from LaFond, E.C. pp. 731–751. *The Sea*, Vol. 1, *Physical Oceanography*, Interscience Publishers, New York, 1962. Copyright Wiley-VCH Verlag GmbH & Co. KGaA. With permission.)

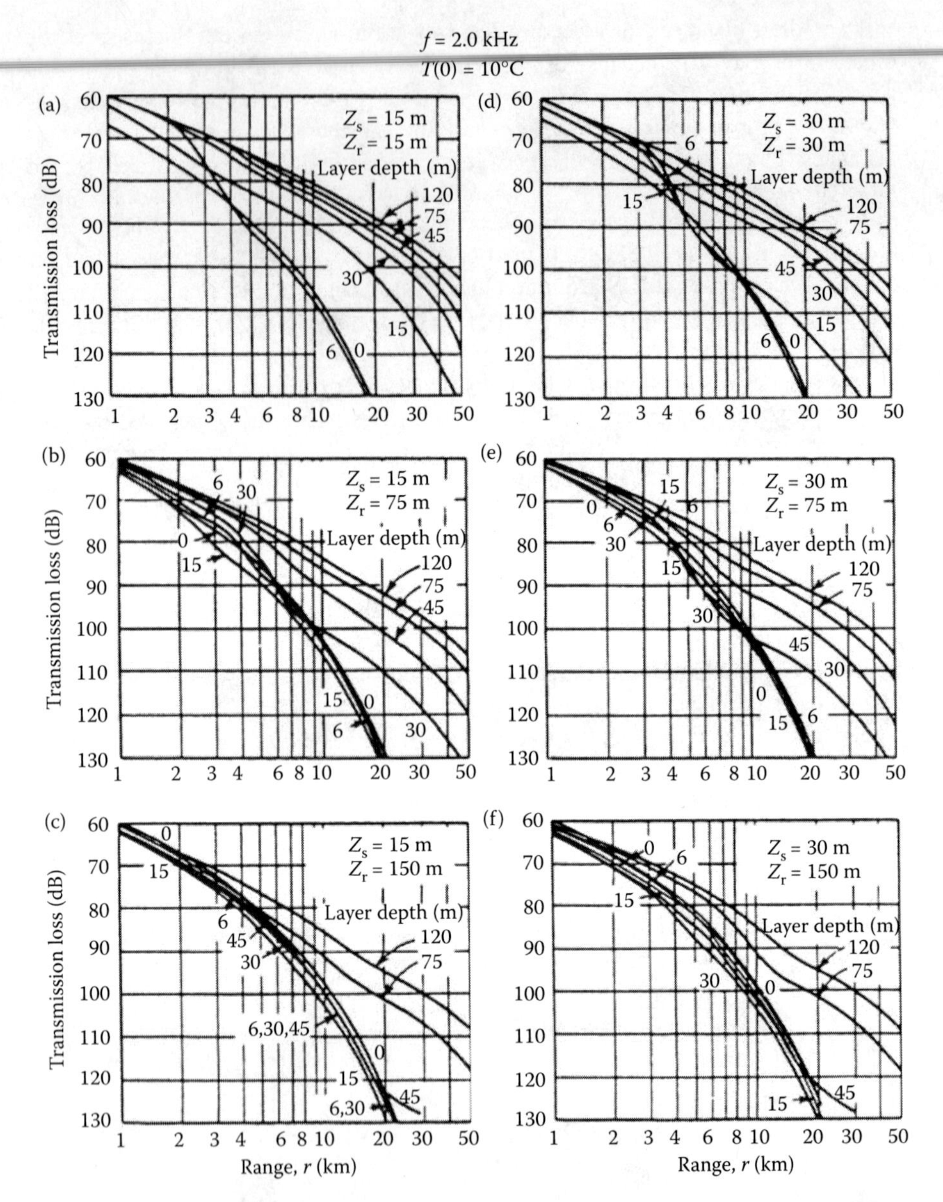

FIGURE 3.22 Surface duct transmission loss estimates generated from the AMOS empirical relationships at a frequency of 2 kHz for various combinations of MLDs, source depths (z_s) and receiver depths (z_r). (From Condron, T.P., Onyx, P.M., and Dickson, K.R. 1955. Contours of propagation loss and plots of propagation loss vs. range for standard conditions at 2, 5, and 8 kc. Navy Underwater Sound Lab., Tech. Memo. 1110-14-55, by Apel, J.R., *Principles of Ocean Physics, International Geophysics Series*, Vol. 38. Academic Press, San Diego, 1987. With permission.)

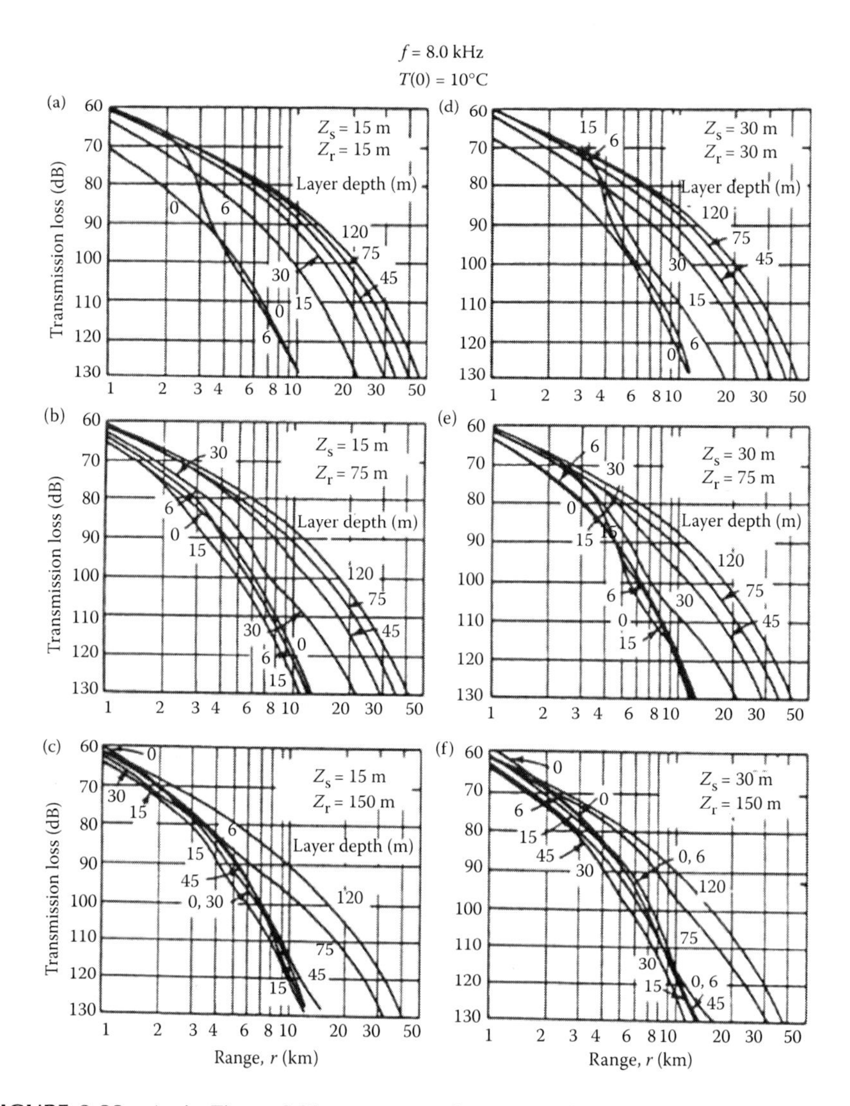

FIGURE 3.23 As in Figure 3.22, except at a frequency of 8 kHz. (From Condron, T.P., Onyx, P.M., and Dickson, K.R. 1955. Contours of propagation loss and plots of propagation loss vs. range for standard conditions at 2, 5, and 8 kc. Navy Underwater Sound Lab., Tech. Memo. 1110-14-55, by Apel, J.R., *Principles of Ocean Physics*, International Geophysics Series, Vol. 38. Academic Press, San Diego, 1987. With permission.)

3.7.3 Low-Frequency Cutoff

At very low frequencies, sound ceases to be trapped in the surface duct. The maximum wavelength for duct transmission may be derived from the theory of radio propagation in ground-based radio ducts to be (Kerr, 1951: 20)

$$\lambda_{\max} = \frac{8\sqrt{2}}{3}\int_{0}^{H}\sqrt{n(z)-n(H)}\,dz \tag{3.9}$$

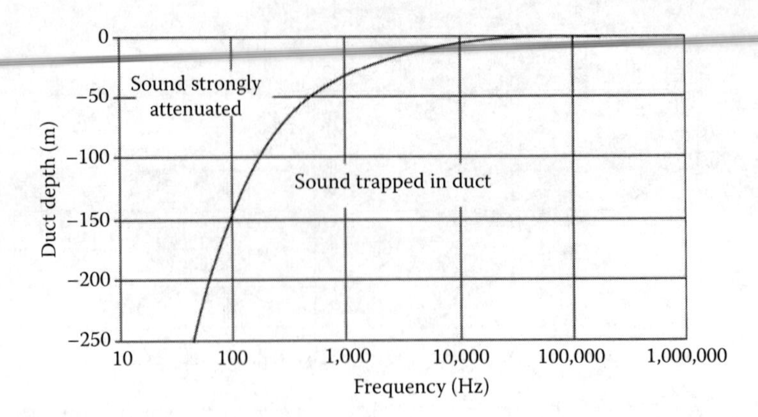

FIGURE 3.24 Cutoff frequency calculated according to Equation 3.10 assuming a sound speed of 1,500 m s^{-1}.

where $n(z)$ is the index of refraction at any depth z in the duct and $n(H)$ is the index of refraction at the base of the duct. Using values of sound speed and sound-speed gradient appropriate for sound propagation in the mixed layer, Equation 3.9 reduces to

$$\lambda_{\max} = 8.51 \times 10^{-3} H^{3/2} \tag{3.10}$$

for the maximum wavelength ($\lambda_{\max}$) in meters trapped in a mixed-layer duct of depth H in meters. For example, a mixed layer 30-m thick would trap a maximum wavelength of 1.4 m, corresponding to a frequency of approximately 1070 Hz (assuming that the sound speed is 1500 m s^{-1}). This relationship is illustrated in Figure 3.24 under the assumption that the speed of sound is fixed at 1500 m s^{-1}. Although this does not represent a sharp cutoff, wavelengths much longer (or frequencies much lower) than this are strongly attenuated. Conversely, wavelengths much shorter (or frequencies much higher) suffer losses due to absorption and leakage. Thus, for a mixed layer of a given thickness, it follows that there is an optimum frequency for propagation at which the loss of sound is a minimum (Figure 3.25).

3.8 DEEP SOUND CHANNEL

The deep sound channel, sometimes referred to as the sound fixing and ranging (SOFAR) channel, is a consequence of the sound speed profile characteristic of the deep ocean (see Section 2.3). This profile has a sound-speed minimum at a depth that varies from about 1000 m at mid-latitudes to near the surface in polar regions. This sound speed minimum causes the ocean to act like a lens: above and below the minimum, the sound speed gradient continually refracts the sound rays back toward the depth of minimum sound speed. This depth is termed the axis of the sound channel (refer back to Figure 2.7). A portion of the acoustic energy originating in the deep sound channel thus remains within the channel and encounters no losses by reflection from the sea surface or the sea floor. Sound in this channel will be diminished by the effects of absorption. The properties of the deep sound channel were first

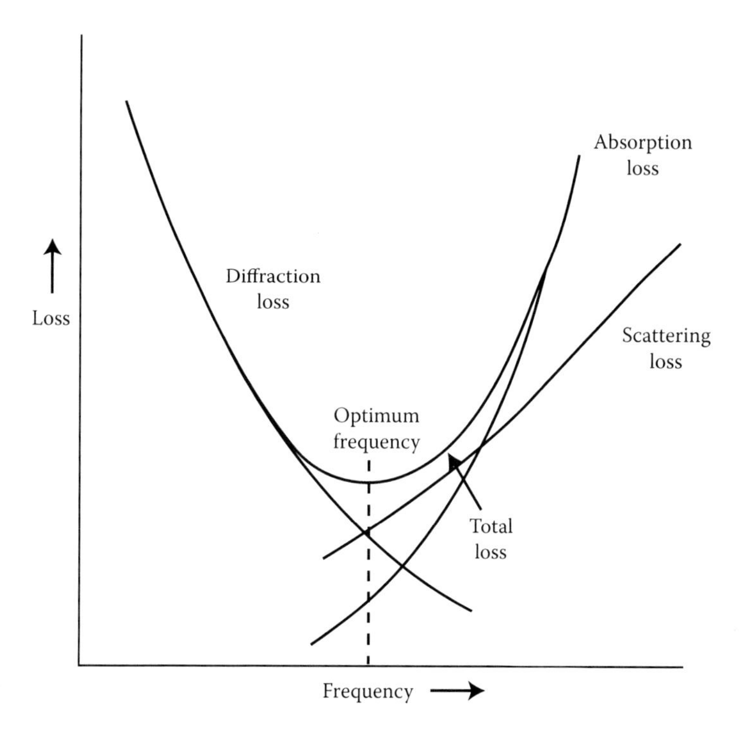

FIGURE 3.25 Loss to a fixed range in a surface duct. The optimum frequency is that at which the total loss is a minimum. (Adapted from Urick, R.J., *Sound Propagation in the Sea*. US Government Printing Office, Washington, 1979.)

investigated by Ewing and Worzel (1948). The exceptional ducting characteristics of this channel have been used to advantage by oceanographers in the design and conduct of acoustic tomography experiments (see Section 6.10).

In terms of the sound speed profile, the upper and lower limits of the channel are defined by the two (conjugate) depths of equal maximum sound speed in the profile between which a minimum exists. In Figure 3.26, these limits of the deep sound channel are the depths A and A'; the depth A' is referred to as the critical (or conjugate) depth. Different ray paths from a source in the channel exist depending on whether or not the channel extends to the sea surface or to the sea floor. In Figure 3.26(a), the sound speeds at the sea surface and sea floor are the same. All depths in the water column then lie within the channel, and sound is propagated via paths that are either refracted (path 1) or reflected (path 2). In Figure 3.26(b), the upper limit of the deep sound channel lies at the sea surface (which may happen at high latitudes). Here, in addition to paths 1 and 2, refracted-surface-reflected (RSR) paths occur (path 3) involving losses at levels intermediate to those suffered by paths 1 and 2. In Figure 3.26(c), the channel is cut off by the sea floor, and refracted-bottom-reflected (RBR) paths exist (path 4). The entirely refracted paths and the low transmission losses associated with these paths do not exist when the source or the receiver is outside the depth limits A and A' of the channel.

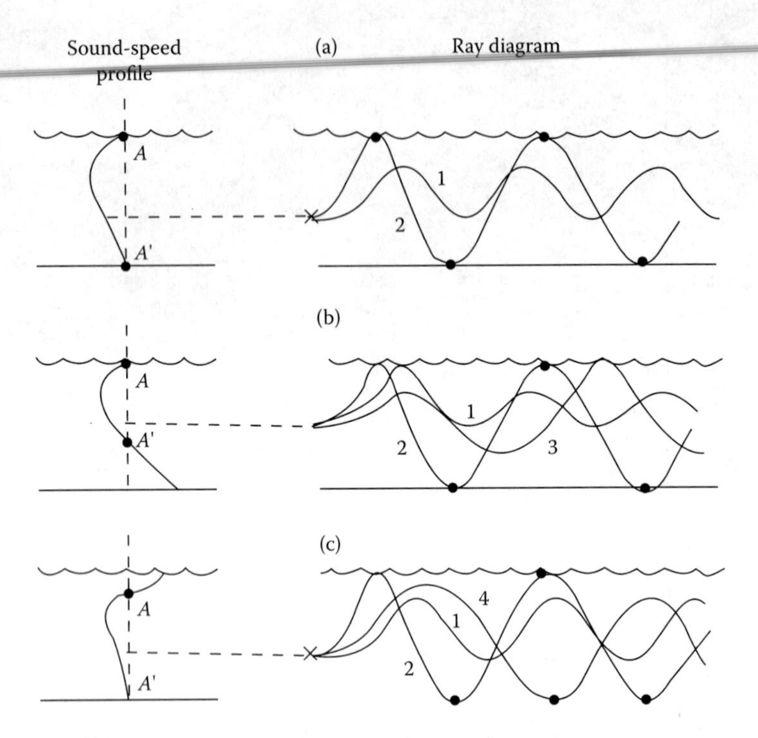

FIGURE 3.26 Ray paths for a source in the deep sound channel. In (a), the channel extends between the sea surface and the sea floor. It is cut off by the sea surface in (b) and by the sea floor in (c). The depth A' is referred to as the critical depth. (From Urick, R.J., *Principles of Underwater Sound*, 3rd ed. McGraw-Hill, New York, 1983. Reproduced with permission of McGraw-Hill Publishing Company. With permission.)

3.9 CONVERGENCE ZONES

Urick (1983: 163–8) described the formation of convergence zones in the ocean (Figure 3.26). Specifically, there must exist a refracted ray that leaves the source horizontally. If this ray is reflected at either the sea surface or the sea floor (ray types 2 or 4 in Figure 3.26), then the required caustic pattern is destroyed and no convergence is possible. Another requirement for convergence is that the water depth must be greater than the critical depth (depth A' in Figure 3.26) in order to allow the rays traveling downward to refract without striking the bottom and to later converge downrange (ray type 3). There must also be a depth excess (i.e., a vertical separation between the critical depth and the bottom) on the order of a few hundred meters. For water depths less than critical (Figures 3.26(a) and 3.26(c)), the rays that would converge if the water were deeper are cut off by the bottom and become bottom-reflected without convergence.

Because the deep waters of the ocean are of a fairly uniform low temperature (near 1°C), the speed of sound at great depths is largely a function of pressure only. Near the surface, however, the speed of sound is determined largely by the water temperature. Thus, water temperature near the sea surface and the water depth in any particular area will largely determine whether sufficient depth excess exists and therefore whether or not a convergence zone will occur. Charts of surface temperature and water depth can then be used as basic prediction tools for ascertaining the existence of convergence zones. A convergence-zone-range slide rule (TACAID 6–10) was developed in 1973 by the Naval Underwater Systems Center based on an analysis of oceanographic data performed by E.M. Podeszwa. This slide rule could be used in the North Atlantic and North Pacific Oceans, and the Mediterranean, Norwegian, and Caribbean Seas to determine CZ ranges.

In the North Atlantic Ocean, convergence zones are seen to appear at intervals of approximately 35 nmi (65 km), with zone widths of about 2 nmi (4 km). TL is significantly lower than spherical spreading within the zones, but significantly higher than spherical spreading between zones. Successive convergence zones get wider with increasing range until, at a range beyond a few hundred nautical miles (nmi), they coalesce. Beyond this range, TL increases smoothly with range and is characterized by cylindrical spreading plus attenuation.

3.10 RELIABLE ACOUSTIC PATH

When a source is located at the critical depth (depth A' in Figure 3.26), and provided sufficient depth excess exists, propagation to moderate ranges can take place via the so-called RAP, as illustrated in Figure 3.5(f). Such paths are termed "reliable" because they are sensitive neither to near-surface effects nor to bottom interaction.

3.11 SHALLOW-WATER DUCTS

There are two definitions of shallow water: hypsometric and acoustic. The hypsometric definition is based on the fact that most continents have continental shelves bordered by the 200-m bathymetric contour, beyond which the bottom generally falls off rapidly into deep water. Therefore, shallow water is often taken to mean continental-shelf waters shallower than 200 m. Using this definition, shallow water represents about 7.5% of the total ocean area.

Acoustically, shallow water conditions exist whenever the propagation is characterized by numerous encounters with both the sea surface and the sea floor. By this definition, some hypsometrically shallow-water areas are acoustically deep. Alternatively, the deep ocean may be considered shallow when low-frequency, long-range propagation conditions are achieved through repeated interactions with the sea surface and the sea floor.

Shallow-water regions are distinguished from deep-water regions by the relatively greater role played in shallow water by the reflecting and scattering boundaries. Also, differences from one shallow water region to another are primarily driven by differences in the structure and composition of the sea floor. Thus, aside from water depth, the sea floor is perhaps the most important part of the marine environment that distinguishes shallow-water propagation from deep-water propagation (Jensen, 2009).

The most common shallow water bottom sediments are sand, silt and mud (see Chapter 2), with compressional sound speeds greater than that of the overlying water.

Sediments are also characterized by shear waves, which are not present in the water column. Acoustic energy that strikes the sea floor at sufficiently small grazing angles is nearly totally reflected back into the water column. This results in a slightly lossy duct with TL approximately characterized by cylindrical spreading within the frequency range 100–1500 Hz. At low frequencies, the acoustic field can extend into the bottom with sound being returned to the water by subbottom reflection or refraction (Eller, 1984b).

The tendency toward cylindrical spreading is illustrated in Figure 3.27, which shows one-third-octave-band TL data at a center frequency of 200 Hz as a function of range. An omnidirectional hydrophone was located at a depth of 91 m in water approximately 210 m deep. The sources were set at a depth of 91 m on a track along which the water depth increased gradually from about 220 m to about 300 m. The sound speed profile was nearly constant along the track, and the bottom sediments were reported to be silty-sand near the beginning range and sand-silt-clay at greater ranges. Figure 3.27 also presents reference curves depicting spherical and cylindrical spreading (beyond 1 km). The acoustic energy is effectively trapped in the shallow-water duct at ranges less than about 40 km, beyond which the TL drops below the reference curve for cylindrical spreading.

3.12 ARCTIC HALF-CHANNEL

The acoustic waveguide in the Arctic is determined by the geometry of the ocean (type of ice cover and water depth) and by a positive-gradient sound speed profile

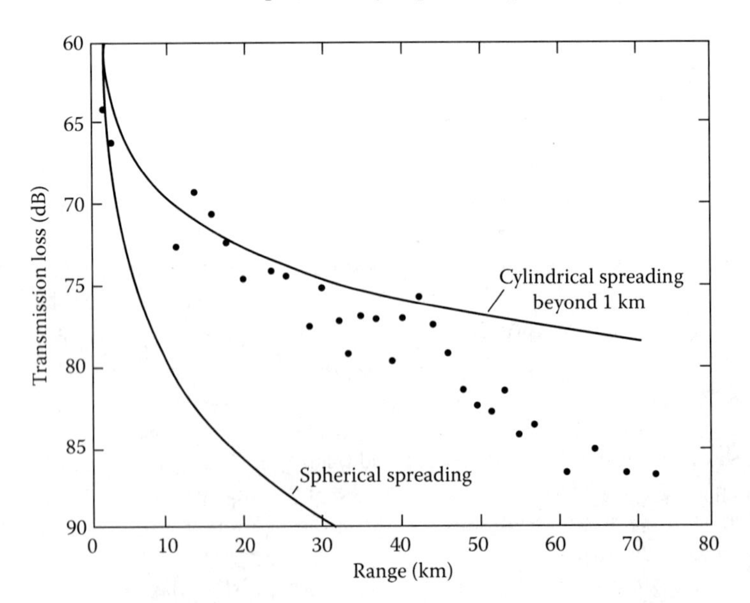

FIGURE 3.27 Example of transmission loss data in shallow water illustrating the cylindrical spreading associated with energy trapping by the waveguide. Data are for the one-third-octave band centered at 200 Hz. (Adapted from Eller, A.I. 1984b. Acoustics of shallow water: A status report. Nav. Res. Lab., Memo. Rept 5405.)

(sound speed increases with increasing depth). This waveguide forms a half-channel, that is, the lower half of the deep sound channel, with the axis of the sound channel located at the sea surface.

The continuously upward-refracting propagation conditions that generally prevail in Arctic waters (as demonstrated by the ray trace in Figure 3.28) cause repeated interactions with the ice canopy and tend to create a low-pass filter, favoring the propagation of low-frequency (<300 Hz) signals. The situation is further complicated when the geometry of the sea floor and the acoustic wavelength combine to yield shallow-water conditions with the attendant increase in bottom interaction opportunities.

Measurements of TL in the Arctic are limited since access to this region has historically been restricted to the spring season when conditions are favorable for manned camps. Consequently, there is a relatively poor understanding of the seasonal variability based on historical data. This situation has gradually improved with the introduction of autonomous sensors that can be deployed through the ice from aircraft throughout the year.

One of the principal characteristics of acoustic propagation measurements under an ice cover is the rapid increase in TL with range at frequencies above about 30 Hz (the low-pass filter effect). The loss mechanism has been attributed primarily to scattering at the ice-water interface. Other possible mechanisms include dissipative processes in the ice canopy, conversion of waterborne energy into energy traveling in and confined to the ice canopy, and increased absorption in the water column.

Buck (1968) summarized available Arctic TL data (Figure 3.29). These curves give the average measured TL in the Arctic Ocean at a number of frequencies. The standard deviation of the various data points from the smooth curve is stated for each frequency. The dashed line shows spherical spreading ($\text{TL} = 20\log_{10} r$, where r is the range measured in meters). It is evident that sound propagation in the

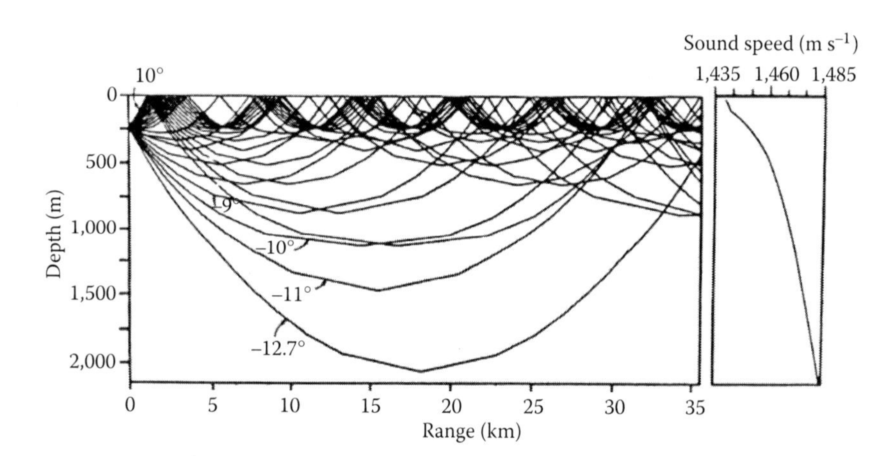

FIGURE 3.28 Typical sound speed profile and corresponding ray diagram for sound propagation in the Arctic region. (Adapted from Urick, R.J., *Sound Propagation in the Sea*. US Government Printing Office, Washington, 1979.)

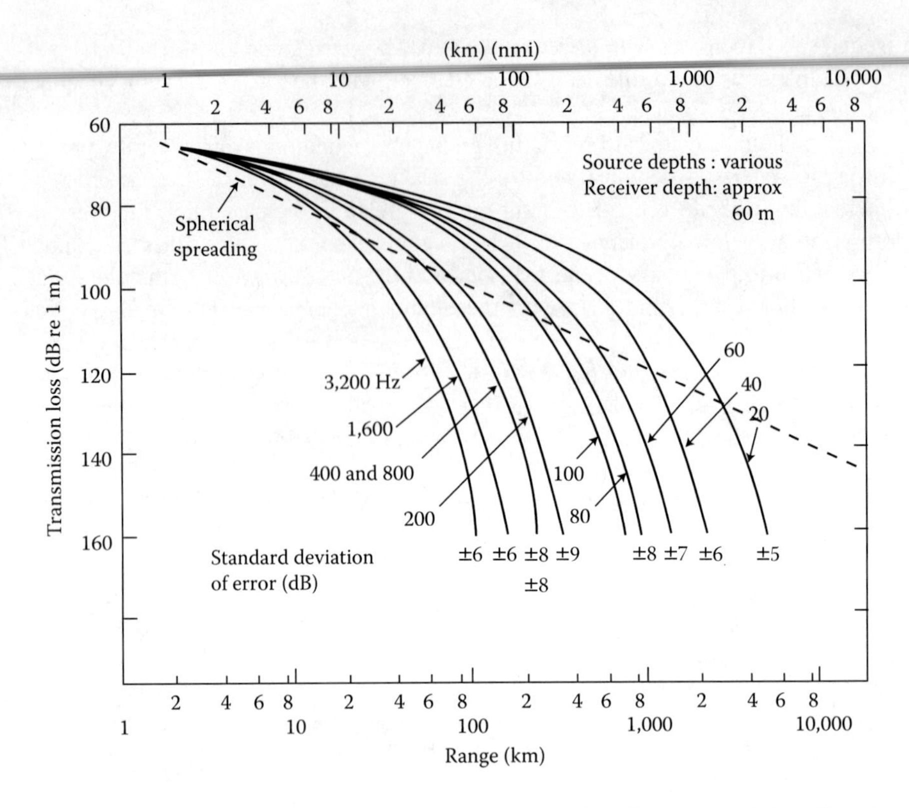

FIGURE 3.29 Average curves of Arctic TL versus range based on measured data. Note that range is plotted on a log scale. (From Buck, B.M., *Arctic Drifting Stations*, Arctic Institute of North America, Calgary, Canada, pp. 427–438, 1968. With permission.)

Arctic degrades rapidly with increasing frequency, particularly above about 30 Hz. At short-to-moderate ranges, ducting improves propagation relative to spherical spreading. At long ranges, repeated encounters with the under-ice surface degrade propagation.

3.13 COHERENCE

Coherence is defined as a measure of the phase and amplitude relationships between sets of acoustic waves. In the ocean, which is characterized by both temporal and spatial variations, the effects of the medium on small-amplitude wave propagation can be described in terms of coherence time, coherence bandwidth, spatial coherence and angular coherence. As described by Ziomek (1985: Chapter 7), for example, this information can be obtained from the generalized coherence (or autocorrelation) function.

Temporal coherence (fluctuations) refers to changes in a received signal (relative to a steady signal) over a period of time. Spatial coherence refers to the changes in the signals received at different locations in the ocean at a given time. Urick

(1982: Chapters 12 and 13) discussed additional aspects of temporal and spatial coherence.

Signal fluctuations can be caused by a number of physical processes including

1. Source or receiver motion
2. Oceanic fine-scale features
3. Sea surface motion (waves and tides)
4. Scatterers
5. Internal waves and internal tides
6. Ocean currents and eddies

4 Propagation II: Mathematical Models

4.1 BACKGROUND

Chapter 1 described a framework within which all underwater acoustic models could be categorized. It was shown that propagation models formed the foundation for the category of models classified as basic acoustic models. In turn, basic acoustic models supported the more specialized category of sonar performance models. Propagation models are the most common (and thus the most numerous) type of underwater acoustic models in use. Their application is fundamental to the solution of all types of sonar performance problems.

Chapter 3 described the observations and physical (physics-based) models that are available to support the mathematical modeling of sound in the sea. Conceptually, it was convenient to separate propagation phenomena into the categories of boundary interactions, volumetric effects and propagation paths. A similar approach will be adopted here in the description of the mathematical models.

The mathematical models will first be distinguished on the basis of their theoretical treatment of volumetric propagation. Then, as appropriate, further distinctions will be made according to specification of boundary conditions and the treatment of secondary volumetric effects such as attenuation due to absorption, turbidity, and bubbles. These secondary effects are generally accommodated by using the physical models described in Chapter 3. Special propagation paths such as surface ducts, shallow water and Arctic half-channels will be discussed in Chapter 5.

The various physical and mathematical models all have inherent limitations in their applicability. These limitations are usually manifested as restrictions in the frequency range or in specification of the problem geometry. Such limitations are collectively referred to as "domains of applicability," and vary from model to model. Most problems encountered in model usage involve some violation of these domains. In other words, the models are misapplied in practice. Therefore, considerable emphasis is placed on these restrictions and on the assumptions that ultimately give rise to them. Finally, model selection criteria are provided to guide potential users to those models most appropriate to their needs. Comprehensive summaries identify the available models and associated documentation. Brief descriptions have been provided for each model.

The emphasis in this chapter, as throughout the book, is placed on sonar (versus seismic) applications. Reviews of mathematical models of seismo-acoustic propagation in the ocean have been provided by Tango (1988) and by Schmidt (1991). Tango (1988) placed particular emphasis on the very-low-frequency (VLF) band.

4.2 THEORETICAL BASIS FOR PROPAGATION MODELING

The theoretical basis underlying all mathematical models of acoustic propagation is the wave equation. The earliest attempts at modeling sound propagation in the sea were motivated by practical problems in predicting sonar performance in support of antisubmarine warfare (ASW) operations during World War II. These early models used ray-tracing techniques derived from the wave equation to map those rays defining the major propagation paths supported by the prevailing marine environment. These paths could then be used to predict the corresponding sonar detection zones. This approach was a forerunner to the family of techniques now referred to as ray-theoretical solutions.

An alternative approach, referred to as wave-theoretical solutions, was first reported by Pekeris (1948), who used the normal-mode solution of the wave equation to explain the propagation of explosively generated sound in shallow water.

As modeling technology matured over the intervening decades, the attendant sophistication has complicated the simple categorization of ray versus wave models. The terminology is still useful in distinguishing those models based principally on ray-tracing techniques from those using some form of numerical integration of the wave equation. Occasionally, a mixture of these two approaches is used to capitalize on the strengths and merits of each and to minimize weaknesses. Such combined techniques are referred to as hybrid approaches. Related developments in propagation modeling have been reviewed by Harrison (1989), McCammon (1991), Buckingham (1992), Porter (1993), and Dozier and Cavanagh (1993). Finite element (FE) methods have also been used in underwater acoustics to treat problems requiring high accuracy (see Kalinowski [1979] for a good introduction to applications in underwater acoustics). Developments in finite-element (FE) modeling will be discussed in appropriate sections throughout this book.

In September 2013, the European Commission commissioned a project entitled Impacts of Noise and use of Propagation Models to Predict the Recipient Side of Noise under a Framework Service Contract. For Task 4 (Compile existing information on underwater sound propagation models), Wang et al. (2014) used relevant literature and research results to compile an inventory of existing models, including advantages and disadvantages. Also included were known gaps in the reliability and the information required for applying these models, together with explicit statements of model assumptions and limitations.

4.2.1 Wave Equation

The wave equation is itself derived from the more fundamental equations of state, continuity and motion. Rigorous derivations have been carried out in numerous basic texts in physics. Kinsler et al. (1982: Chapter 5) presented a particularly lucid derivation. DeSanto (1979) derived a more general form of the wave equation that included gravitational and rotational effects. Accordingly, the derivation will not be repeated here. Rather, the mathematical developments described in this book will build directly on the wave equation.

Formulations of acoustic propagation models generally begin with the three-dimensional (3D), time-dependent wave equation. Depending on the governing assumptions and intended applications, the exact form of the wave equation can vary considerably (DeSanto, 1979; Goodman and Farwell, 1979). For most applications, a simplified linear, hyperbolic, second-order, time-dependent partial differential equation is used:

$$\nabla^2 \Phi = \frac{1}{c^2} \frac{\partial^2 \Phi}{\partial t^2} \tag{4.1}$$

where ∇^2 is the Laplacian operator $[= (\partial^2/\partial x^2) + (\partial^2/\partial y^2) + (\partial^2/\partial z^2)]$, Φ is the potential function, c is the speed of sound, and t is the time.

Subsequent simplifications incorporate a harmonic (single-frequency, continuous wave) solution in order to obtain the time-independent Helmholtz equation. Specifically, a harmonic solution is assumed for the potential function Φ:

$$\Phi = \phi \, e^{-i\omega t} \tag{4.2}$$

where ϕ is the time-independent potential function, ω is the source frequency $(2\pi f)$ and f is the acoustic frequency. Then the wave Equation 4.1 reduces to the Helmholtz equation

$$\nabla^2 \phi + k^2 \phi = 0 \tag{4.3a}$$

where $k = (\omega/c) = (2\pi/\lambda)$ is the wavenumber and λ is the wavelength. In cylindrical coordinates, Equation 4.3a becomes

$$\frac{\partial^2 \phi}{\partial r^2} + \frac{1}{r} \frac{\partial \phi}{\partial r} + \frac{\partial^2 \phi}{\partial z^2} + k^2(z)\phi = 0 \tag{4.3b}$$

Equation 4.3a is referred to as the time-independent (or frequency-domain) wave equation. Equation 4.3b, in cylindrical coordinates, is commonly referred to as the elliptic-reduced wave equation.

Various theoretical approaches are applicable to the Helmholtz equation. The approach used depends on the specific geometrical assumptions made for the environment and the type of solution chosen for ϕ, as will be discussed in the following sections. To describe the different approaches effectively, it is useful to first develop a classification scheme, with associated taxonomy, based on five canonical solutions to the wave equation: ray theory, normal mode, multipath expansion, fast field, and parabolic equation (PE) techniques.

Throughout the theoretical development of these five techniques, the potential function ϕ normally represents the acoustic field pressure. When this is the case, the transmission loss (TL) can easily be calculated as

$$\text{TL} = 10 \log_{10}[\phi^2]^{-1} = -20 \log_{10} |\phi|$$

This relationship necessarily follows from Equations 3.1 and 3.2. If phases are considered, the resulting TL is referred to as coherent. Otherwise, phase differences are ignored and the TL is termed incoherent.

4.2.2 Classification of Modeling Techniques

Although acoustic propagation models can be classified according to the theoretical approach employed, the cross-connections that exist among the various approaches complicate a strict classification, or taxonomic, scheme. Consequently, as the schemes become more detailed, more cross-connections will appear. A generalized classification scheme has been constructed using five categories corresponding to the five canonical solutions of the wave equation (also see Jensen and Krol, 1975; DiNapoli and Deavenport, 1979; Weston and Rowlands, 1979).

Within these five categories, a further subdivision can be made according to range-independent and range-dependent models. Range independence means that the model assumes a horizontally stratified ocean in which properties vary only as a function of depth. Range dependence indicates that some properties of the ocean medium are allowed to vary as a function of range (r) and azimuth (θ) from the receiver, in addition to a depth (z) dependence. Such range-varying properties commonly include sound speed and bathymetry, although other parameters such as sea state, absorption and bottom composition may also vary. Range dependence can further be regarded as two-dimensional (2D) for range and depth variations, or 3D for range, depth, and azimuthal variations.

In order to illustrate the relationships among the five approaches used to solve the wave equation, the rather elegant scheme developed by Jensen and Krol (1975) will be adopted with slight modifications (Figure 4.1). According to this classification

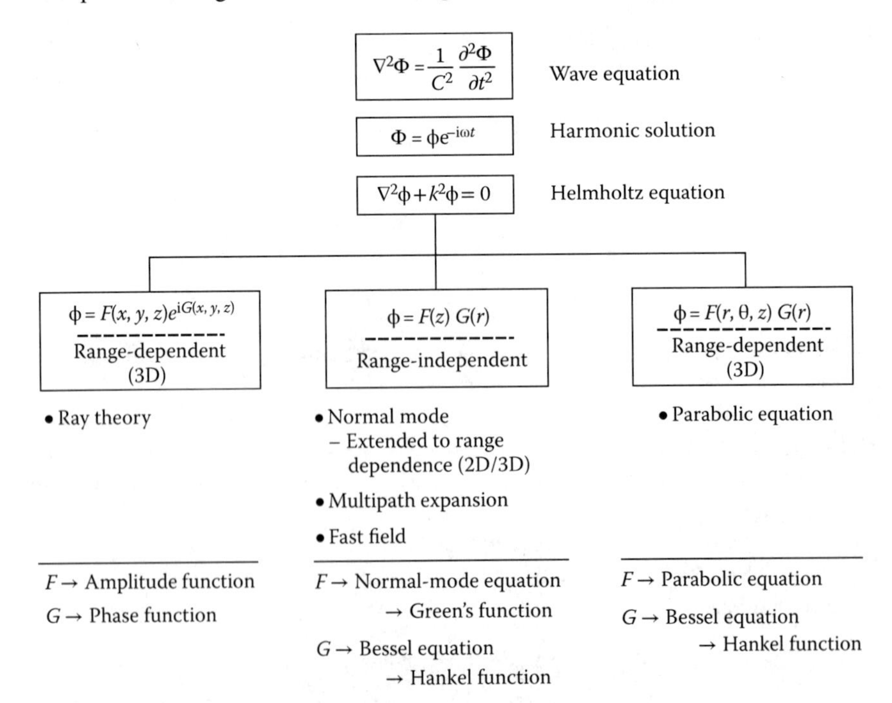

FIGURE 4.1 Summary of relationships among theoretical approaches for propagation modeling. (Adapted from Jensen, F.B. and Krol, H. *SACLANT ASW Res.* Ctr, Memo. SM-72, 1975.)

scheme, there are three avenues that connect the five basic approaches applicable to underwater acoustic propagation modeling. These five categories of propagation models will be described in detail in the following sections, and Figure 4.1 will serve as a useful road map. For convenience, the general functions and equations represented in Figure 4.1 have been identified with the letters F and G. In the discussions that follow, different symbols will be substituted to facilitate identification with relevant physical properties or with other well-known mathematical functions.

4.3 RAY-THEORY MODELS

4.3.1 Basic Theory

Ray-theoretical models calculate TL on the basis of ray tracing (National Defense Research Committee, 1946). Ray theory starts with the Helmholtz equation. The solution for ϕ is assumed to be the product of a pressure amplitude function $A = A(x, y, z)$ and a phase function $P = P(x, y, z)$: $\phi = Ae^{iP}$. The phase function (P) is commonly referred to as the eikonal, a Greek word meaning "image." Substituting this solution into the Helmholtz Equation 4.3a and separating real and imaginary terms yields

$$\frac{1}{A}\nabla^2 A - [\nabla P]^2 + k^2 = 0 \tag{4.4}$$

and

$$2[\nabla A \cdot \nabla P] + A\nabla^2 P = 0 \tag{4.5}$$

Equation 4.4 contains the real terms and defines the geometry of the rays. Equation 4.5, also known as the transport equation, contains the imaginary terms and determines the wave amplitudes. The separation of functions is performed under the assumption that the amplitude varies more slowly with position than does the phase (geometrical acoustics approximation). The geometrical acoustics approximation is a condition in which the fractional change in the sound speed gradient over a wavelength is small compared to the gradient c/λ, where c is the speed of sound and λ is the acoustic wavelength. Specifically

$$\frac{1}{A}\nabla^2 A << k^2 \tag{4.6}$$

In other words, the sound speed must not change much over one wavelength. Under this approximation, Equation 4.4 reduces to

$$[\nabla P]^2 = k^2 \tag{4.7}$$

Equation 4.7 is referred to as the eikonal equation. Surfaces of constant phase (P = constant) are the wavefronts, and the normals to these wavefronts are the rays. Eikonal refers to the acoustic path length as a function of the path endpoints. Such rays are referred to as eigenrays when the endpoints are the source and receiver positions. Differential ray equations can then be derived from the eikonal equation. Typically, four sets of eigenrays are considered (Figure 4.2): direct path (DP), refracted-surface-reflected (RSR), refracted-bottom-reflected (RBR), and refracted-surface-reflected-bottom-reflected (RSRBR). The physical models described in Chapter 3 are generally incorporated into ray models to account for boundary interaction and volumetric effects.

Weston (1998) provided concise and insightful descriptions of the basic equations used in ray (or geometrical) acoustics, discussed applications, presented expansions and extensions to ray theory, and commented on the validity of the methods. Bergman (2005) used an elegant development of the connection between the acoustic rays in a moving fluid and the null geodesics of a pseudo-Riemannian manifold to derive several well-known results used in underwater-acoustic ray theory from the underlying principle of isometry (i.e., distance-preserving mapping).

The geometrical acoustics approximation effectively limits the ray-theoretical approach to the high-frequency domain. An approximate guideline for defining high frequency is provided by the relation

$$f > 10\frac{c}{H} \tag{4.8}$$

where f is the frequency, H is the duct depth, and c is the speed of sound.

The computation of the pressure amplitude can be accomplished using the transport Equation 4.5 and by invoking the principle of conservation of energy flux for a ray bundle (Figure 4.3). Assuming constant density (Tolstoy and Clay, 1966: 57)

$$A_2 = \left[\frac{c_2\, \mathrm{d}\sigma_1}{c_1\, \mathrm{d}\sigma_2}\right]^{1/2} A_1 \tag{4.9}$$

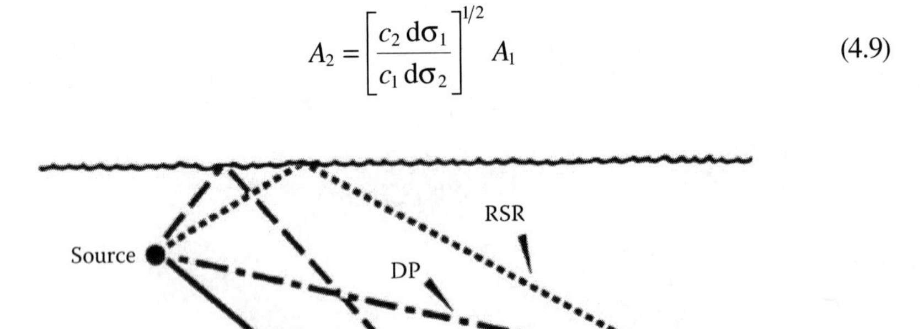

FIGURE 4.2 Four basic types of eigenrays: DP, direct path; RSR, refracted-surface-reflected; RBR, refracted-bottom-reflected; RSRBR, refracted-surface-reflected-bottom-reflected.

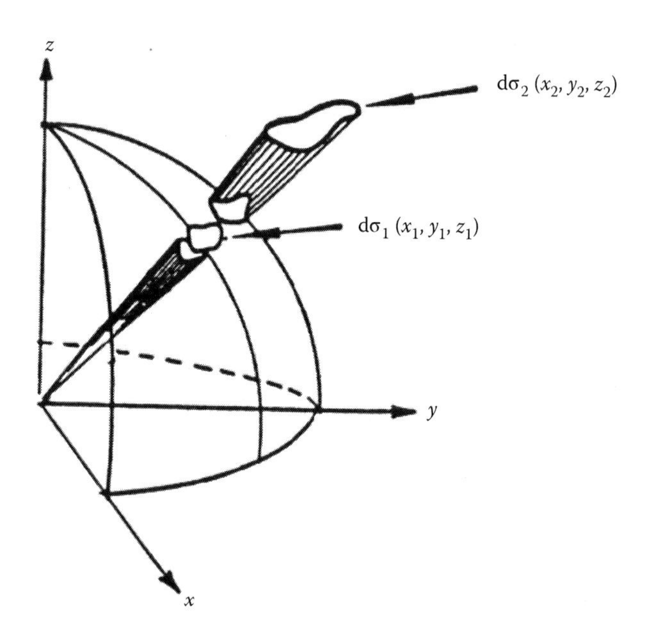

FIGURE 4.3 Geometry of a ray bundle.

where

$$A_2 = A(x_2, y_2, z_2) \quad A_1 = A(x_1, y_1, z_1)$$

are the signal amplitudes,

$$c_2 = c(x_2, y_2, z_2) \quad c_1 = c(x_1, y_1, z_1)$$

are the sound speeds, and

$$d\sigma_1 = \text{ray bundle cross-section at } x_1, y_1, z_1$$

$$d\sigma_2 = \text{ray bundle cross-section at } x_2, y_2, z_2$$

As $d\sigma_2$ approaches zero, A_2 approaches infinity. Thus, ray theory does not hold in the vicinity of focal surfaces (caustics) and focal points.

4.3.2 Caustics

Focal surfaces, or caustics, are formed when the refractive properties of the ocean environment focus a number of adjacent rays into close proximity. There are two types of caustics: smooth and cusped. A cusp is actually the intersection of two smooth caustics. Examples of smooth and cusped caustics are presented in Figure 4.4. In optics, a caustic is the envelope of light rays reflected or refracted by a curved surface

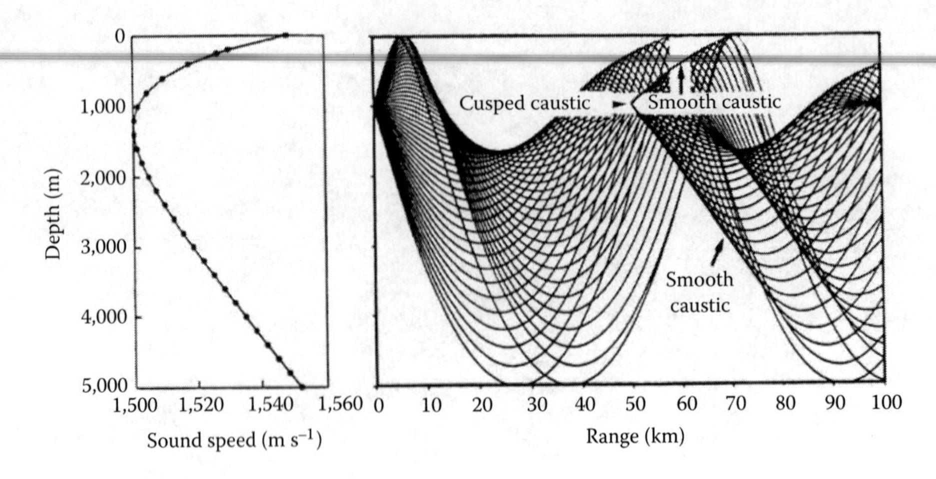

FIGURE 4.4 Sound speed profile and associated ray trace showing the formation of smooth and cusped caustics. (Adapted from Jensen, F.B., *IEEE J. Oceanic. Engr.*, **13**, 186–197, 1988. With permission.)

or object, or the projection of that envelope of rays onto another surface. Caustic can also refer to the curve to which light rays are tangent, defining a boundary of an envelope of rays as a curve of concentrated light. These shapes often have cusp singularities. In the case of sunlight, such concentration of light can burn, thus the name *caustic*. Stewart and James (1992) proposed using acoustic ray-tracing to demonstrate examples of wave propagation (versus the usual examples of light propagation) to illustrate the general wave phenomenon of refraction in a classroom environment.

In the vicinity of caustics, a higher-order approximation can be used to yield predictions on the caustic itself, and also in the nearby shadow zone. One theory, developed by Sachs and Silbiger (1971), is essentially an approximate asymptotic method that predicts a spatially oscillating field amplitude on the illuminated side of the caustic. In the shadow zone, the field is damped with increasing distance from the caustic boundary. Boyles (1984: Chapter 5) provided a lucid description of caustic formation together with appropriate corrections to ray theory. With appropriate frequency-dependent (diffraction) corrections and proper evaluation of caustics, ray theory can be extended to frequencies lower than those normally associated with the geometrical acoustics approximation. Under these conditions, the approach is commonly termed "ray theory with corrections."

For ducted propagation in a waveguide where the source and receiver are placed close to the depth of the waveguide axis, cusped caustics occur repeatedly along the axis. Interference in the vicinity of these cusped caustics results in the formation of a coherent structure (i.e., the axial wave) that propagates along the waveguide axis like a wave (Grigorieva and Fridman, 2008).

4.3.3 Gaussian Beam Tracing

Another method useful in dealing with caustics is Gaussian beam tracing (e.g., Porter and Bucker, 1987; Bucker, 1994), which has been adapted from seismic applications.

This method associates with each ray a beam with a Gaussian intensity profile normal to the ray. A pair of differential equations that govern the beamwidth and curvature are integrated along with the standard ray equations to compute the beam field in the vicinity of the central ray of the beam. This method avoids certain ray-tracing artifacts such as perfect shadows and infinite energy levels at caustics. Furthermore, this technique is attractive for high-frequency, range-dependent applications in which wave-theoretical approaches might not be practical alternatives.

4.3.4 Range Dependence

Although ray-tracing techniques are theoretically applicable to fully range-dependent (3D) problems (refer to Figure 4.1), they are rarely implemented as such. The mathematical complexity discourages 3D versions in favor of 1D or 2D versions. The 2D versions can be implemented by one of three methods: (1) by mapping rays over discrete range intervals in which the environment remains constant (Weinberg and Dunderdale, 1972; Weinberg and Zabalgogeazcoa, 1977); (2) by dividing the range-depth plane into triangular regions (Bucker, 1971; Roberts, 1974; Watson and McGirr, 1975); or (3) by allowing the environment to vary smoothly as a function of range, as through the use of cubic splines (Foreman, 1983). These methods are explained in the following paragraphs.

A significant problem that confronts range-dependent ray-tracing programs is the proper representation of the transition of sound speed profiles between adjacent measurement points in the range dimension. Two aspects of this problem are important from a practical standpoint. First, the interpolated intermediate sound speed profiles should be physically plausible. Second, it is desirable that the resulting ray trajectories be analytically computable (versus numerically integratable) in order to maximize computational efficiency.

In method (1) (Figure 4.5a), rays are traced in the first range interval. Selected rays are then mapped into the second interval, and the process continues throughout the remaining intervals. Drawbacks to this approach stem from discontinuities in ray tracing at the boundaries between adjacent intervals, and also from the potential omission of important rays in the selection process at each boundary. In particular, as the water (or duct) depth decreases, rays can properly be eliminated. As the depth increases, however, there is no valid and consistent process for adding new rays into consideration. The concept of "imaginary eigenrays" has been introduced as a way of supplementing the geometrically obvious paths with those arising from the leakage of trapped rays and from shadow-zones near caustics where geometric paths do not exist (e.g., Martin, 1995). Energy from these imaginary eigenrays has been observed in experimental data but their strength, if not their occurrence, may be more sensitive to small changes in the environment than that due to the real eigenrays. Also see the discussion of complex eigenrays by Jensen et al. (1994: 154–5).

In method (2) (Figure 4.5b), the input sound speed profiles are represented as piecewise-linear functions of depth. Then, the range-depth plane between specified profiles can be divided into triangular sections within which the sound speed varies linearly in both range and depth. In each sector, ray trajectories then correspond to arcs of circles for which analytical ray-tracing programs are available (e.g., Urick,

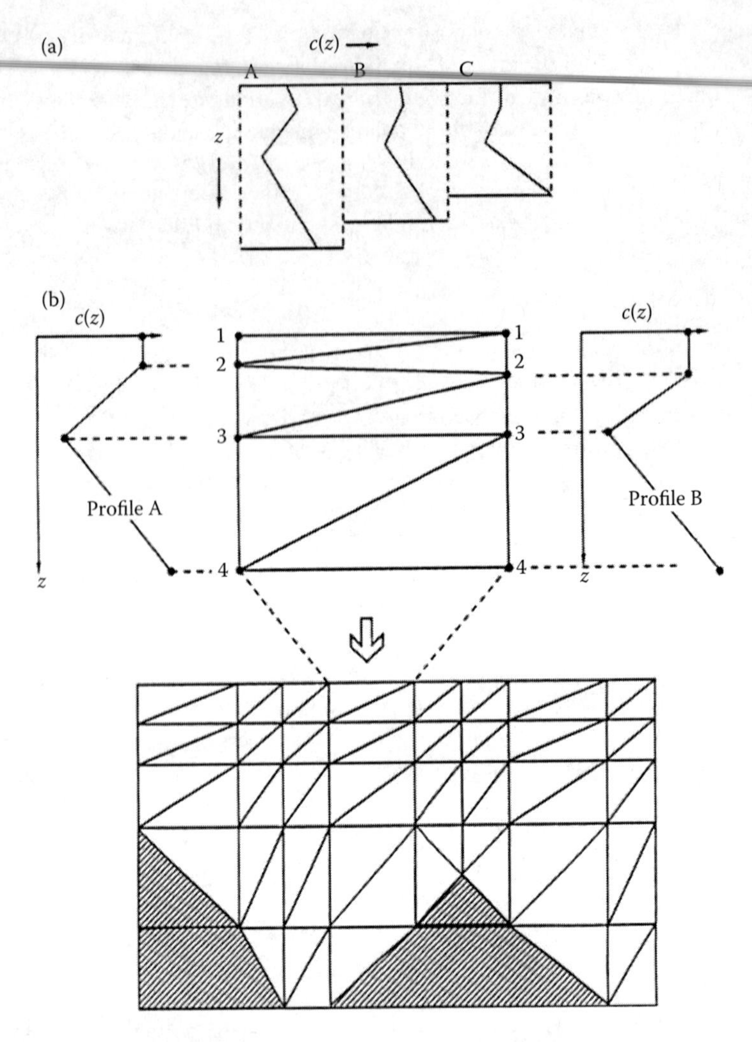

FIGURE 4.5 Range-partitioning techniques: (a) discrete range intervals; and (b) typical network of triangular regions formed by connecting corresponding features between two adjacent sound speed profiles.

1983: 124–8). Each triangular sector is selected to provide a smooth and physically realizable transition between adjacent profiles. The vertexes are commonly selected at those depths representing sound-speed minima. Otherwise, the triangular sides are selected to subtend the gradient that departs least from the gradient of the preceding triangle. These procedures generally require the expertise of an oceanographer to ensure the generation of realistic intermediate sound speed profiles. One tries to maintain continuity in features such as the sonic layer depth and the sound channel axis.

In method (3), the sound speed profiles and bathymetry are fitted with cubic splines (Solomon et al., 1968) or quadratic functions. The domains thus formed are rectangular, much like that in method (1) above. However, linear interpolations in range are performed between adjacent profiles to obtain continuous range derivatives. The differential ray geometry and amplitude equations (derived from Equations 4.4 and 4.5, respectively) are then solved numerically (Foreman, 1983).

The proper use and application of complex range-dependent models requires a great deal of planning, especially with regard go the selection and spacing of the range-dependent environmental data inputs. Improper spacing of inputs can overlook important ocean frontal systems or bathymetric features (Henrick, 1983).

Limited 3D ray-tracing techniques have been developed to account more properly for the effects of horizontal gradients of oceanic properties, which are manifested as cross-range variations in the TL patterns. Examples include the use of a hybrid approach involving horizontal rays and vertical modes (Weinberg and Burridge, 1974; Burridge and Weinberg, 1977), and the use of 3D Hamiltonian ray tracing (Jones, 1982; Jones et al., 1986). A compact, 3D ray-tracing algorithm suitable for implementation on small computers was developed by Einstein (1975). Bowlin et al. (1992) developed a versatile, range-dependent ray-tracing program (RAY). Dushaw and Colosi (1998) described a FORTRAN ray model (EIGENRAY) that closely followed the Bowlin et al. (1992) ray code (RAY), which was written in C. EIGENRAY generates fast and accurate deep-water wavefront and eigenray travel-time predictions at basin-scale ranges for application to long-range ocean acoustic tomography. Both RAY and EIGENRAY are available from the Ocean Acoustics Library (OALIB) website (see Appendix C).

Hovem and Knobles (2002) described a range-dependent propagation model based on a combination of range-dependent ray tracing and plane-wave bottom responses. The ray-tracing module of the model determined all the eigenrays between any source-receiver pairs. The received wave field was then synthesized by adding the contributions of all the eigenrays, taking into account the reflections from the bottom and the surface. The model could treat arbitrarily varying bottom topography and a layered elastic bottom as long as the layers were parallel. The bottom was modeled with a sedimentary layer over an elastic half space, but more complicated structures could be implemented easily in future versions. This new model has been tested against other models on several benchmark problems and also applied in the analysis and modeling of up-slope and down-slope propagation data recorded on a 52-element center-tapered array that was deployed at two locations about 70 miles east of Jacksonville, Florida.

4.3.5 Arrival Structure

A useful property of ray models is their ability to calculate arrival structure. Similar information can be obtained from wave-theoretical models, but only with much additional computation. Arrival-structure contours indicate the horizontal range from source to receiver that will be traversed by a ray leaving the source (or arriving at the receiver) at the specified angle for each propagation path of interest. These contours are not frequency dependent, but they do depend on the source and receiver depths.

These contours are also referred to as θ–r diagrams. An alternative way of viewing these contours is that r represents the range at which a ray leaving the source (arriving at the receiver) at angle θ crosses a horizontal line representing the depth of the receiver (source). To be properly utilized, these contours must indicate a reference to either the source or the receiver angle, and the sign conventions must be explicitly stated. For example, relative to the receiver, a negative angle (θ) usually, but not always, signifies rays arriving from above the horizontal axis (i.e., rays traveling downward), and a positive angle signifies rays arriving from below (i.e., rays traveling upward).

Arrival structure information is not just an academic curiosity. Fundamentally, it facilitates discrimination of multipath arrivals. This information is essential for the proper computation of the vertical directionality of ambient noise (see Chapter 8) and for the proper evaluation of the performance of vertically oriented hydrophone arrays. Arrival structure information is also important in the calculation of volumetric and boundary reverberation (see Chapter 10).

In practice, arrival structure diagrams are produced in conjunction with ray traces, TL curves and other diagnostic information, as will be demonstrated during discussions of the RAYMODE propagation model later in this chapter. Here, Figure 4.6 is introduced in isolation to illustrate the utility of arrival structure diagrams in identifying particular ray families, which are evidenced by distinctive sets of patterns. For example, the inverted L-shaped trace at very short range in Figure 4.6 is indicative of direct path arrivals, while the more graceful arching contours on either side of the horizontal axis are associated with bottom-bounce paths. The sets of contours (for both positive and negative arrival angles), which

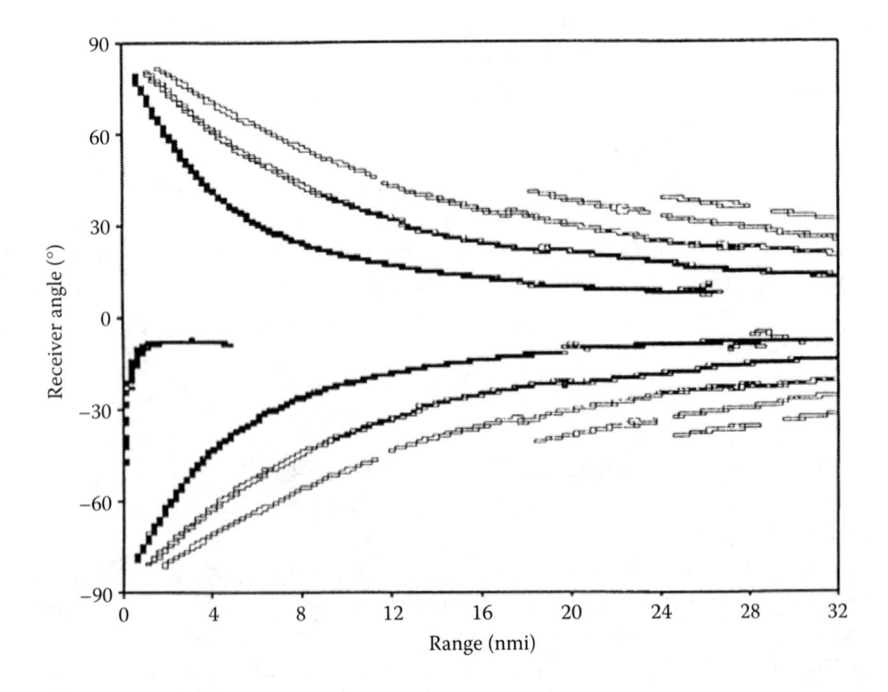

FIGURE 4.6 Example of a vertical arrival structure diagram.

represent bottom-bounce paths, are referred to as order contours. Rays with the same number of bottom reflections belong to the same order. Thus, the first set of contours represents one bottom bounce and is of order 1. The contour associated with the positive angle within this first order has traveled from the source to the receiver over a RBR path and arrives at the receiver from below the horizontal axis. The corresponding contour with the negative angle has traveled from source to receiver over a refracted-bottom-reflected-surface-reflected path and arrives at the receiver from above the horizontal axis.

These order contours can be used to identify all ray paths connecting the source and receiver that encounter the bottom at any range *r*. This information is very useful in the calculation of bottom reverberation. For example, only those rays encountering the bottom will contribute to bottom reverberation. The concept of order contours also applies to convergence zone (CZ) paths.

The θ–r contours can also be used to identify caustic formations. An instructive example was provided by Franchi et al. (1984). Figure 4.7 illustrates order contours obtained for a typical deep-ocean sound speed profile. The two rays that just graze the bottom at ranges X_B and $X_{B'}$ determine points B and B′ on the contour. The two rays that just graze the surface at ranges X_A and $X_{A'}$ occur at shallower angles and determine points A and A′ on the contour. Thus, rays with angles in the interval ($\theta_{B'}$, $\theta_{A'}$) and (θ_B, θ_A) hit the surface but not the bottom. Caustics are associated with points C and C' on these contours.

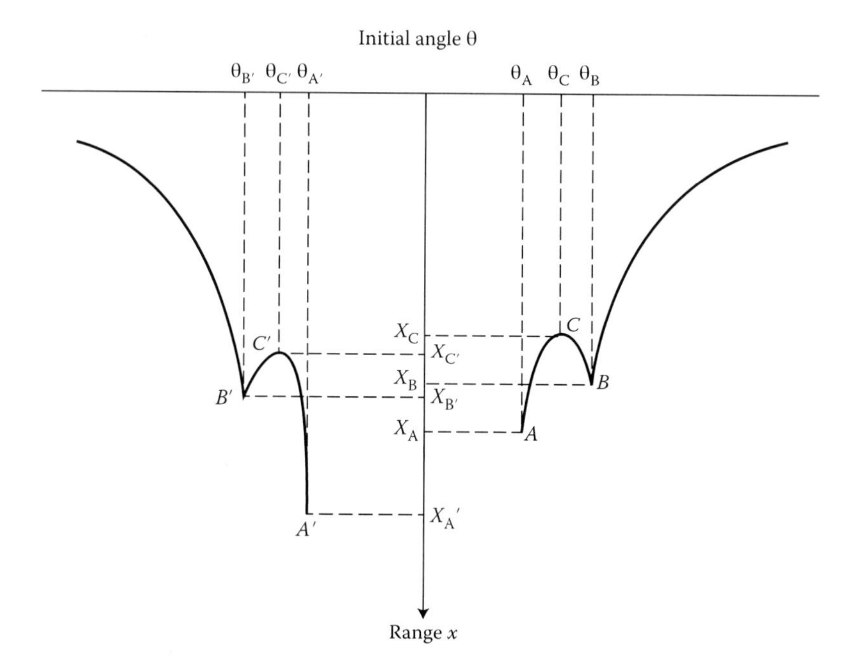

FIGURE 4.7 Smoothed order contours showing caustic behavior. (Adapted from Franchi E.R et al., *Nav. Res. Lab.*, Rept 8721, 1984.)

4.3.6 Beam Displacement

Propagation models based on ray-tracing techniques generally treat bottom reflection as specular and reduce the intensity through application of a bottom reflection loss. However, acoustic energy can be transmitted into the bottom where it is subsequently refracted, attenuated and even transmitted back into the water column at some distance down range (Figure 4.8). This spatial offset is referred to as "beam displacement." Time displacements would also be associated with this process while the ray is absent from the water column.

In their initial study of the effects of beam displacement in ray calculations, Tindle and Bold (1981) considered a simple two-fluid Pekeris model as a good first approximation to many shallow-water environments. The Pekeris model (Pekeris, 1948) consists of a fluid (water) layer of depth H, density ρ_1, and sound speed c_1 overlying a semi-infinite fluid (sediment) layer of density ρ_2 and sound speed c_2, where $c_2 > c_1$. Attenuation was neglected and only rays totally reflected at the interface were considered to propagate to a range (r) greater than the water depth (H). The vertical wavenumbers (γ_1, γ_2) in the two layers were defined as

$$\gamma_1 = \left(\frac{\omega}{c_1}\right)\sin\theta, \quad \gamma_2 = \left(\frac{\omega}{c_1}\right)\sqrt{\cos^2\theta - \frac{c_1^2}{c_2^2}}$$

where θ is the grazing angle and ω is the angular frequency of the wavefield. The lateral displacement (D) of a beam of finite width undergoing reflection at the water-sediment interface is

$$\Delta = \frac{2k_h\rho_1\rho_2\left(\gamma_1^2 + \gamma_2^2\right)}{\gamma_1\gamma_2\left(\rho_1^2\gamma_2^2 + \rho_2^2\gamma_1^2\right)}$$

where k_h is the horizontal wavenumber in the water layer

$$k_h = \left(\frac{\omega}{c_1}\right)\cos\theta$$

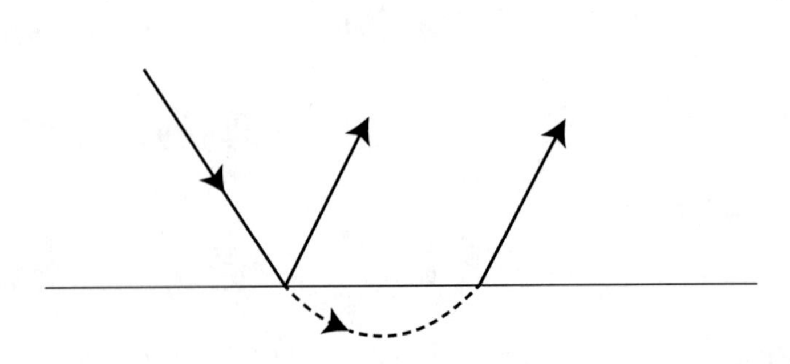

FIGURE 4.8 Geometry for beam displacement.

A typical ray path for this simple model is presented in Figure 4.9. The source and receiver depths are z_0 and z, respectively. A ray leaving the source at an angle θ relative to the horizontal travels in a straight (unrefracted) path in the water layer. There is no beam displacement at the sea surface and the ray is reflected in the conventional manner. At the bottom, the beam is displaced horizontally by an amount Δ before traveling upward again at angle θ.

Recent efforts to improve ray theory treatments of bottom attenuation and beam displacement have been conducted by Siegmann et al. (1987) and Westwood and Tindle (1987), among others. Modified ray theory with beam and time displacements has advantages over wave theory in that the interaction of the acoustic energy with the bottom can be intuitively visualized. Moreover, Jensen and Schmidt (1987) computed complete wave theory solutions for a narrow Gaussian beam incident on a water-sediment interface near the critical grazing angle. They observed that the fundamental reflectivity characteristics of narrow beams could be explained entirely within the framework of linear acoustics.

4.3.7 Waveguide Invariant

Advances in the computational efficiency of propagation codes have facilitated the practical analysis of broadband sources using the waveguide invariant approach. The "waveguide invariant" summarizes in a single scalar parameter the dispersive characteristics of the acoustic field in a waveguide. Using a ray (versus normal mode) formulation, the invariant is computed in part by varying the ray launch angle about a mean value and then computing the corresponding changes in the ray-cycle distance and the ray-cycle time. The term "invariant" derives from the fact that the (computed) dispersive character of the propagation in the environment is nearly independent of the particular ray pair selected. This interesting approach was described by Brekhovskikh and Lysanov (1982) (and earlier by S.D. Chuprov). Song et al. (1998) and D'Spain et al. (1999) described practical applications of the waveguide invariant approach in realistic environments, including matched field processing.

Harrison (2003, 2005) derived closed-form expressions for two-way propagation (and reverberation) in variable-depth ducts for isovelocity water using ray invariants and acoustic flux. These expressions provided useful benchmarks for testing other models (Harrison, 2003). These expressions were extended to the case of uniform-slope bathymetry combined with a range-independent linear sound speed and bistatic geometry. It was noted that the effects of caustics were very small in shallow water and at frequencies below about 1 kHz (Harrison, 2005).

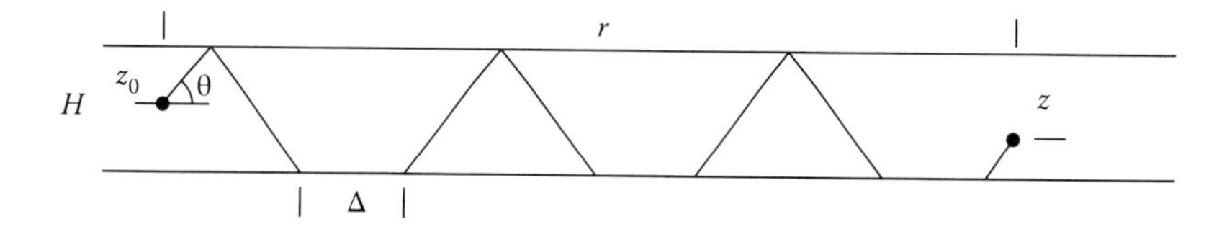

FIGURE 4.9 A typical ray path with beam displacement included. (From Tindle C.T. and Bold G.E.J. *J., Acoust. Soc. Am.*, **70**, 813–819, 1981. With permission.)

In weakly range-dependent ocean environments, it has been observed that the acoustic wavefield can be described in terms of ray-based or mode-based approaches (Brown et al., 2005). Using ray-based descriptions, distributions of the ray amplitude and phase (travel time) are largely controlled by the so-called stability parameter (α). Alternatively, using mode-based descriptions in either purely range-independent environments or in range-independent environments on which weak (range-dependent) perturbations have been superimposed, many wavefield properties are controlled by the so-called waveguide invariant (β). When β is evaluated using asymptotic mode-theoretical results (in a stratified environment), it was found that $\beta = \alpha$, which is consistent with the well-known ray-mode duality. In weakly range-dependent environments, however, α was found to control ray stability and several measures of travel-time dispersion while β controlled the spread of modal-group delays due to mode coupling. Cockrell and Schmidt (2010) also described a relationship between the waveguide invariant and wavenumber integration.

In essence, the waveguide invariant (β) summarizes the pattern of constructive and destructive interference between acoustic modes propagating in the ocean waveguide. It manifests itself as interference fringes (or striations) in a plot of frequency versus source-receiver separation. The waveguide invariant summarizes in a single scalar parameter the dispersive propagation characteristics in a waveguide.

In oceanic waveguides, acoustic propagation is described by the modal propagation theory; thus, the acoustic field can be viewed as a sum of modes (Le Gall and Bonnel, 2013). When plotted in the range (r)-versus-frequency (f) domain, the acoustic intensity $I(r,f)$ created by a broadband source exhibits a characteristic pattern stemming from interferences between the propagating modes. This interference pattern is made up of striations with a slope $\delta f/\delta r$ that is often characterized by a scalar parameter β, called waveguide invariant: $\delta f/\delta r = \beta \cdot (f/r)$. The waveguide invariant summarizes the acoustic dispersion in the waveguide, and this property is used in various applications. In shallow water, the waveguide invariant is only mildly dependent on the environmental parameters (e.g., sound speed profile, bottom properties) making it an appealing feature when the marine environment is poorly known. The waveguide invariant principle is applied: to perform source localization, to enhance target detection, to suppress the effect of dispersion, and in time reversal mirror experiments. It is usually assumed that $\beta = 1$, which is a typical value in shallow water. However, this is only an approximation and the gap between this approximation and the reality may considerably hinder applications. Moreover, the dependence of the waveguide invariant with respect to the marine environment can be used to obtain information on the environment. Hence, accurate knowledge of the waveguide invariant value is often of critical interest; thus, techniques for estimating the waveguide invariant in operational context are required. Estimating the waveguide invariant is not a simple issue since the waveguide invariant depends on frequency and on the pairs of modes considered. Thus, the waveguide invariant may be better modeled by a distribution than by a scalar.

Le Gall (2015) noted that an acoustic wave propagating in the underwater environment carries information about the source (range, depth) and about the physical properties of the oceanic environment (water depth, sound speed in the water column, and sound speed in the sea floor). This information can be obtained by solving an inverse problem, which is conventionally addressed through the minimization of an error function between the measured pressure field and the pressure field obtained from a propagation model (such as matched-field processing methods). The results of the inversion are inherently uncertain: the measurements are affected by noise, and the propagation model is imperfectly known because of the complexity of the marine environment. Le Gall (2015) dealt with performance analyses of these estimation methods: the estimation of these parameters constitutes a nonlinear problem that suffers from significant ambiguities beyond a certain noise level, and errors in the assumed propagation medium are considerably difficult. Statistical tools for predicting the performance of these inverse problems used the Ziv-Zakai bound (addresses lower bounds on signal parameter estimation) and the interval-error method. These tools were used to analyze the performance of a particular source localization problem. Furthermore, the source localization's robustness against modeling errors of the oceanic environment was considered; specifically, a Bayesian source localization method was developed to handle uncertainty in the oceanic environment. The Green's function was considered as a random vector that accounted for environmental uncertainty through its probability density. A physical *a priori* modal propagation model enhanced performance in those cases when the environment was poorly known. Moreover, the Bayesian formalism provided a quantitative measure of the confidence that can be given to each estimate of the source position, which is very valuable information in an operational context.

4.3.8 Energy-Flux Models

Some applications, such as work relating to acoustic impacts on marine mammals, do not require extremely high-fidelity model outputs. TLs averaged over depth, for example, are often adequate.

An approach referred to as energy-flux models (Weston, 1980a,b) is useful for the rapid calculation of TL where the propagation conditions are dominated by numerous boundary-reflected multipaths and when only the coarse characteristics of the acoustic field are needed.

Lurton (1992) described the range-averaged intensity model (RAIM), originally developed by Weston (1971), which is based on a probabilistic description of the intensity field in terms of cyclic ray beams. The contributions of these cyclic ray beams are functions of their probability of presence at the receiver depth (as obtained from their cycle length) and of the classical geometrical spreading factor.

In specific configurations, especially at long ranges in shallow-water environments, the transmitted field can be viewed as being composed of many paths propagating by successive reflections from the surface and bottom boundaries. Here, the acoustic energy will remain trapped between these two boundaries. Furthermore, if the acoustic frequency is high enough that the field oscillations can be considered to be random, then an average intensity can be calculated using simple algebraic formulas (related discussions will be presented in Section 5.2.1).

This concept can be extended to ocean environments where the sound speed is not constant, or where there are slight losses at the boundaries. In such cases, the transmitted field cannot be taken as a volumetric average. Rather, it has to be decomposed into its angular components and the cyclic characteristics of the various beams must be detailed (Lurton, 2002).

This approach is very useful for predicting incoherent TLs in cases where there are a large number of modes (≥6). These flux models are fast because there is no requirement to find modes or eigenrays. They can also lead to very simple and intuitive expressions. This method has recently been extended for application to reverberation (Holland, 2010).

Heinis et al. (2013) noted that the energy-flux formulation of waveguide propagation is closely related to the incoherent mode sum, and its simplicity has led to development of efficient computational algorithms for reverberation and target echo strength. However, the formulation lacks the effects of convergence or modal interference. By starting with the coherent mode sum, rejecting the most rapid interference, and retaining beats on a scale of a ray-cycle distance, it was shown that convergence could be included in a hybrid formulation requiring only minimal extra computation. Three solutions were offered by evaluating the modal intensity cross terms using Taylor expansions: (1) in the most efficient approach, the double summation of the cross terms was reduced to a single numerical sum by solving the other summation analytically; (2) a local range average; and (3) a local depth average. Favorable comparisons were made between these three solutions and the wave model Orca with, and without, spatial averaging in an upward refracting duct. It was also shown that the running range average was very close to the mode solution (excluding its fringes), given a relation between averaging window size and effective number of modes which, in turn, is related to the waveguide invariant.

Nijhof et al. (2014) proposed a new hybrid approach for modeling sound levels in the water column due to pile-driving operations. Their proposed method used a local FE model that accurately captured the source characteristics of the pile; this FE model was then coupled to an adiabatic normal-mode model for efficient evaluation of the sound propagation over large distances in range-dependent environments. Ultimately, the FE model will be coupled to the more efficient flux-based propagation model used in AQUARIUS (Heinis et al., 2013; Weston, 1971; Weston, 1976) or, alternatively, the recently developed flux based model SOPRANO (Sertlek and Ainslie, 2014).

4.3.9 Advanced Algorithms

The recursive ray acoustics (RRA) algorithm is a simple, fast, and accurate algorithm that can be used to find eigenrays and compute the position, angles of propagation, travel time, phase, and path length along a ray path as well as to draw ray-trace plots for speeds of sound that are functions of all three spatial variables (Ziomek, 1989, 1994; Ziomek and Polnicky, 1993). In addition, the RRA algorithm can compute the sound-pressure level (SPL) along individual ray paths for arbitrary, one-dimensional (1D), depth-dependent speeds of sound. SPL calculations are made along individual ray paths for 1D, depth-dependent speeds of sound using an enhanced version of the RRA algorithm. The SPL calculations are valid (i.e., finite) at turning points and focal points and do not require the use of Airy functions. The SPL calculations include the effects of frequency-dependent volume attenuation and frequency-dependent attenuation due to surface and bottom reflections. The ocean surface and bottom are treated as boundaries between viscous fluid media. Although the ocean surface is modeled as a planar boundary, the bathymetry is an arbitrary function of horizontal range. Sound-speed-versus-depth and bathymetric data are represented by orthogonal-function expansions.

Smith (1974) derived an integral formula to estimate the averaged steady-state TL for a source-receiver pair in a channel whose environmental parameters varied slowly with range. The derivation was based on the theory of ray acoustics with lossy specular reflection from the boundaries. The averaging process eliminated all but gradual changes with range of the TL, and dependence on the depth of each point was retained. By way of example, numerical results for an isogradient channel with constant bottom slope were illustrated.

Doran and Fredricks (2007) used the level-set method, which is a fixed-grid method for generating solutions to the high-frequency approximation to the wave equation. In this method, the user controls the underlying grid and thus the accuracy of the solution. Osher and Sethian (1988) developed generic computational techniques that are referred to as PSC (propagation of surfaces under curvature) algorithms.

Zielinski and Geng (1996) described a simple method for acoustic ray tracing based on the concept of a traveling wavefront (TWF), which led to a simple algorithm for ray tracing. The approach is suitable for arbitrary sound-speed profiles (which can be approximated by differentiable functions) and arbitrary sea-bottom topography. It does not require linearization and it is particularly convenient for computer simulation. It can be easily generalized to include range-dependent cases. A specific example of correcting the sea-floor depth as measured by a multibeam bathymetry system was used to illustrate its application to precise marine measurements.

Brooks (2008) approximated the Green's function from the cross-correlation of sound recorded at two locations in a shallow-water oceanic waveguide; this approach was referred to as ocean-acoustic interferometry. Active-source and ship-dominated ambient noise were both considered for application to ocean-acoustic interferometry. A stationary-phase argument was used to relate cross-correlations from active sources to the Green's function between hydrophones. A vertical line source, a horizontal line source, and a horizontal hyperbolic source were considered. The theory and simulations were found to be in agreement with those presented by previous investigators. Empirical Green's function approximations were determined from ship-dominated ocean noise cross-correlation. Direct and secondary path travel times between hydrophones were determined; these results compared favorably with simulated data. Averaging the cross-correlations between equi-spaced horizontal line array hydrophone pairs was shown to increase the SNR. Analysis of temporal variations in the cross-correlations confirmed that, at any given time, the signal was generally dominated by one or two sources. Cross-correlations obtained from data collected during a tropical storm were shown to be clearer than those collected at other times. This was thought to be due to a reduction in nearby shipping and an increase in the overall sound levels caused by increased wave action generated by the storm. Two practical applications were derived from this work: (1) diagnosis of a multichannel hydrophone array, and (2) array hydrophone self-localization.

4.4 NORMAL-MODE MODELS

4.4.1 Basic Theory

Normal mode solutions are derived from an integral representation of the wave equation. In order to obtain practical solutions, however, cylindrical symmetry is assumed in a stratified medium (i.e., the environment changes as a function of depth only). Then, the solution for the potential function ϕ in Equation 4.3b can be written in cylindrical coordinates as the product of a depth function $F(z)$ and a range function $S(r)$:

$$\phi = F(z) \cdot S(r) \tag{4.10}$$

Next, a separation of variables is performed using ξ^2 as the separation constant. The two resulting equations are

$$\frac{d^2 F}{dz^2} + \left(k^2 - \xi^2\right) F = 0 \tag{4.11}$$

$$\frac{d^2 S}{dr^2} + \frac{1}{r}\frac{dS}{dr} + \xi^2 S = 0 \tag{4.12}$$

Equation 4.11 is the depth equation, better known as the normal mode equation, which describes the standing wave portion of the solution. Equation 4.12 is the range equation, which describes the traveling wave portion of the solution. Thus, each normal mode can be viewed as a traveling wave in the horizontal (r) direction and as a standing wave in the depth (z) direction.

The normal mode equation (Equation 4.11) poses an eigenvalue problem. Its solution is known as the Green's function. The range equation (Equation 4.12) is the zero-order Bessel equation. Its solution can be written in terms of a zero-order Hankel function ($H_0^{(1)}$). The full solution for ϕ can then be expressed by an infinite integral, assuming a monochromatic (single-frequency) point source:

$$\phi = \int_{-\infty}^{\infty} G\left(z, z_0; \xi\right) \cdot H_0^{(1)}(\xi r) \cdot \xi \, d\xi \tag{4.13}$$

where G is the Green's function, $H_0^{(1)}$ is a zero-order Hankel function of the first kind, and z_0 is the source depth. Note that ϕ is a function of the source depth (z_0) and the receiver depth (z). The properties of these various functions are described in standard mathematical handbooks such as Abramowitz and Stegun (1964).

4.4.2 Normal-Mode Solution

To obtain what is known as the normal mode solution to the wave equation, the Green's function is expanded in terms of normalized mode functions (u_n). The eigenvalues, which are the resulting values of the separation constants, are represented by ξ_n. These eigenvalues (or characteristic values) represent the discrete set of values for which solutions of the mode functions u_n exist. The infinite integral in Equation 4.13 is then evaluated by contour integration:

$$\phi = \oint \sum \frac{u_n(z) \cdot u_n(z_0)}{\xi^2 - \xi_n^2} H_0^1(\xi r) \cdot \xi \, d\xi + \text{branch-cut integral} \tag{4.14}$$

The contour integral represents the trapped (or discrete) modes that propagate through the water column. The branch-cut integral is associated with the continuous mode spectrum, which represents those modes propagating through the ocean floor and which are strongly attenuated. The branch-cut integral describes the near-field conditions and corresponds in ray theory to those rays striking the bottom at angles greater than the critical angle. Thus, the contribution from the branch-cut integral is often neglected, particularly when horizontal distances greater than several water depths separate the source and receiver. For acoustic propagation problems using an impedance boundary condition at the bottom, the solution actually comprises three spectral intervals: the continuous, discrete, and evanescent. The evanescent spectrum is associated with interface waves that decay exponentially away from the boundary (Jensen et al., 1994). Interface waves that propagate along a fluid-solid boundary (i.e., water-sediment interface) are called Scholte waves. In seismology, interface waves propagating along the boundaries between solid layers are called Stoneley waves.

By neglecting the branch-cut integral, evaluating the contour integral, and replacing the Hankel function expression by its asymptotic expansion for large arguments

$$H_0^{(1)}(\xi r) \approx \sqrt{\frac{2}{\pi \xi r}} \, e^{i(\xi r - \pi/4)} \quad \text{for} \quad \xi r >> 1$$

(where $\xi r >> 1$ is the far-field approximation), a simple solution for the potential function ϕ can be obtained:

$$\phi = g(r,\rho) \sum \frac{u_n(z) \cdot u_n(z_0)}{\sqrt{\xi_n}} \exp\left[i(\xi_n r - \pi/4)\right] \exp(-\delta_n r) \tag{4.15}$$

where $g(r,\rho)$ is a general function of range (r) and water density (ρ). Each of the n-terms in Equation 4.15 corresponds to the contribution of a single normal mode of propagation. Each of these modal contributions is propagated independently of the others. Under idealized conditions, there is usually an upper limit on the number of modes to be calculated, and this number increases with increasing frequency.

The attenuation coefficient δ_n can be written in the form (Miller and Wolf, 1980)

$$\delta_n = \varepsilon \gamma_n + \varepsilon_c \gamma_n^{(c)} + \varepsilon_s \gamma_n^{(s)} + S_n^{(0)} + S_n^{(1)} + \alpha_n \tag{4.16}$$

where

ε = plane-wave attenuation coefficient in the sediment layer

ε_c, ε_s = compressional and shear plane-wave attenuation coefficients, respectively, of the basement

$\gamma_n, \gamma_n^{(c)}, \gamma_n^{(s)}$ = measures of nth mode interactions with sediment, and with basement compressional and shear wave mechanisms

$S_n^{(0)}, S_n^{(1)}$ = attenuation of the modal field due to the interaction of the nth mode with statistically rough boundaries at the pressure release boundary (sea surface) and the water-sediment boundary, respectively

α_n = attenuation due to absorption by sea water

One advantage of normal mode solutions over ray-theoretical methods is that TL can easily be calculated for any given combination of frequency and source depth (z_0) at all receiver depths (z) and ranges (r) (refer to Equation 4.14). Ray models, on the other hand, must be executed sequentially for each change in source or receiver depth (compare Figures 3.2 and 3.3).

A disadvantage associated with normal mode solutions is the degree of information required concerning the structure of the sea floor, as indicated in Equation 4.16. In order to execute effectively, this type of model generally requires knowledge of the density as well as the shear and compressional sound speeds within the various sediment layers. Figure 4.10 shows an ideal three-layer physical

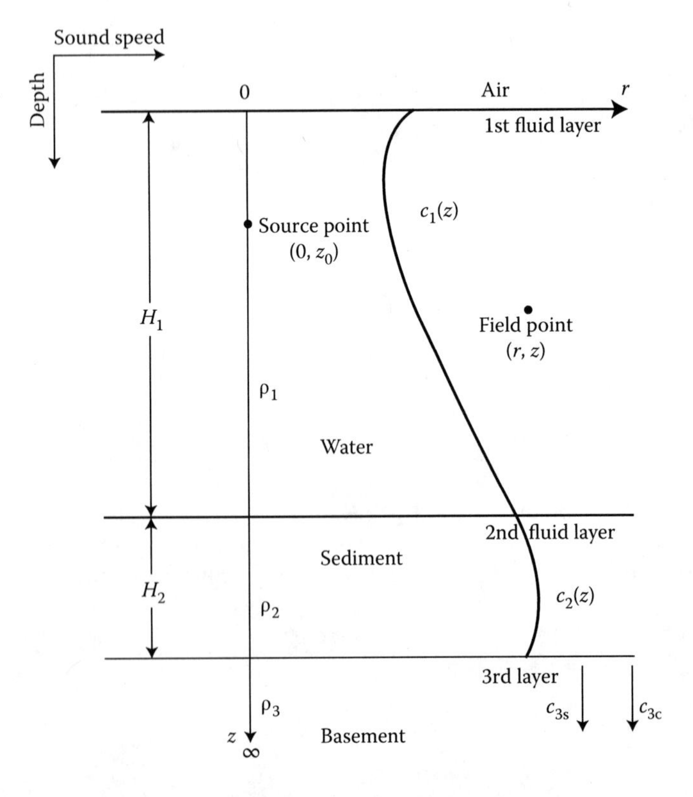

FIGURE 4.10 Simple physical model depicting an infinite half-space consisting of two fluid layers with depths H_1, H_2; densities ρ_1, ρ_2, and sound speeds c_1, c_2. A semi-infinite third layer (representing the bottom) has density ρ_3, and compressional (c_{3c}) and shear (c_{3s}) sound speeds. (Adapted from Miller, J.F. and Ingenito, F., *Nav. Res. Lab.*, 1975. Memo. Rept 3071.)

model usable in normal mode solutions. The fluid layer boundaries are defined by duct depths or by other notable features in the sound speed profile. Accordingly, more layers can be added to the physical model as the complexity of the profile increases. The computational intensity increases in proportion to this complexity. In Figure 4.10, the sediment layer is treated as a second fluid layer (below the water column) and is characterized by a compressional sound speed only, thus ignoring shear effects. The effects of sediment attenuation can be included, as noted above in Equation 4.16.

Some normal mode models (Stickler, 1975; Bartberger, 1978b) partition the sound speed profile into N layers such that the square of the index of refraction in each layer can be approximated by a straight line and the density can be assumed constant. Under these conditions, it is then possible to represent the depth-dependent portion of the pressure field in terms of Airy functions and thus improve computational efficiency. The model by Stickler (1975) is noteworthy in that it includes the continuous modes as well as the trapped modes. Fast finite difference methods have been used to accurately determine the real (versus imaginary) eigenvalues (Porter and Reiss, 1984). Errors in these eigenvalues would otherwise appear as phase shifts in the range dependence of the acoustic field.

4.4.3 Dispersion Effects

Unlike ray-theoretical solutions, wave-theoretical solutions inherently treat dispersion effects. Dispersion is the condition in which the phase velocity is a function of the acoustic frequency. If present, dispersion effects are most noticeable at low frequencies. In oceanic waveguides, dispersion depends on the characteristics and geometry of the waveguide and is referred to as geometrical dispersion. This is distinguished from intrinsic dispersion as might result from sound propagation through bubbly water layers near the sea surface (Clay and Medwin, 1977: 311–12).

4.4.4 Experimental Measurements

Ferris (1972) and Ingenito et al. (1978) summarized results from field measurements at a site near Panama City, Florida. The experiments were conducted over tracks having a range-independent water depth and sound speed profile. The sea floor was composed of hard-packed sand. The modal field distribution was calculated on the basis of Equation 4.14. If all acoustic and environmental parameters except for the receiver depth are held constant, then the dependence of the pressure amplitude (ϕ_n) of the nth-order mode on receiver depth (z) is directly proportional to the corresponding mode function (u_n):

$$\phi_n(z) \propto u_n(z)$$

If range and receiver depth are held constant, then the dependence of the pressure amplitude (ϕ_n) of the nth-order mode on source depth (z_0) is

$$\phi_n(z_0) \propto u_n(z_0)$$

Comparisons of measured and calculated pressure amplitude distributions at the receiver (using an arbitrary amplitude scale) are presented in Figure 4.11 for the first and second modes. Two frequencies (400 Hz and 750 Hz) were considered, as were three different sound speed profiles (positive gradient, negative gradient, and isospeed). The influence of the sound speed gradient on the low-order modes is evident. Specifically, the first mode is nearly symmetrically distributed about mid-depth in the isospeed case (Figure 4.11a). The sound speed gradients alter this symmetry and concentrate the energy in the low-speed portion of the water column (Figures 4.11b and c).

Boyles (1984: Chapter 6) provided a comprehensive discussion demonstrating that certain groupings of normal modes could be associated with particular ray families propagating in specific oceanic waveguides. This was accomplished by explicitly calculating the TL associated with selected subsets of modes using Equation 4.14. These calculations were then compared to the full TL curve. One example involved a typical North Atlantic Ocean sound speed profile during the winter season. This profile was characterized by a surface duct, a sound channel, and sufficient depth excess to support CZ propagation (refer back to Figure 2.7). The source and receiver were both placed within the surface duct, and a frequency of 30 Hz was selected. Only 27 modes were trapped in the water column at this frequency. Of these, only one mode (mode 11) contributed to the TL in the surface duct. Modes that uniquely contribute to surface duct propagation are referred to as "virtual modes" (Labianca, 1973). Thus, mode 11 is also referred to as "virtual mode 1." Modes 13–27 accounted for the TL associated with the convergence-zone paths. The remaining modes contributed very little to the overall TL. The association of certain normal modes with particular ray families will be explored further in discussions of the RAYMODE model later in this chapter.

4.4.5 Range Dependence

As noted previously in Figure 4.1, normal mode models assume range independence. Extensions to range dependence can be accommodated either by "mode coupling" or by "adiabatic approximation." Mode coupling considers the energy scattered from a given mode into other modes. The adiabatic approximation assumes that all energy in a given mode transfers to the corresponding mode in the new environment, provided that environmental variations in range are gradual.

Three-dimensional (3D) propagation modeling using normal mode theory has been attempted using two different approaches. The first approach employs the $N \times 2D$ technique in which the 3D problem is solved using N horizontal radials (or sectors) in conjunction with range-dependent (2D) adiabatic mode (or coupled mode) theory along each radial. The resulting quasi-3D propagation fields can then be contoured on polar plots, for example. The second approach directly includes the effects of horizontal refraction through use of the lateral wave equation. One implementation by Kuperman et al. (1988) and Porter (1991) employed Gaussian beam tracing (Porter and Bucker, 1987) to solve the lateral wave equation. In essence, the horizontal field of each modal wavenumber is translated into a horizontal sound speed field that defines the Gaussian beam environment for each mode. In related work, a

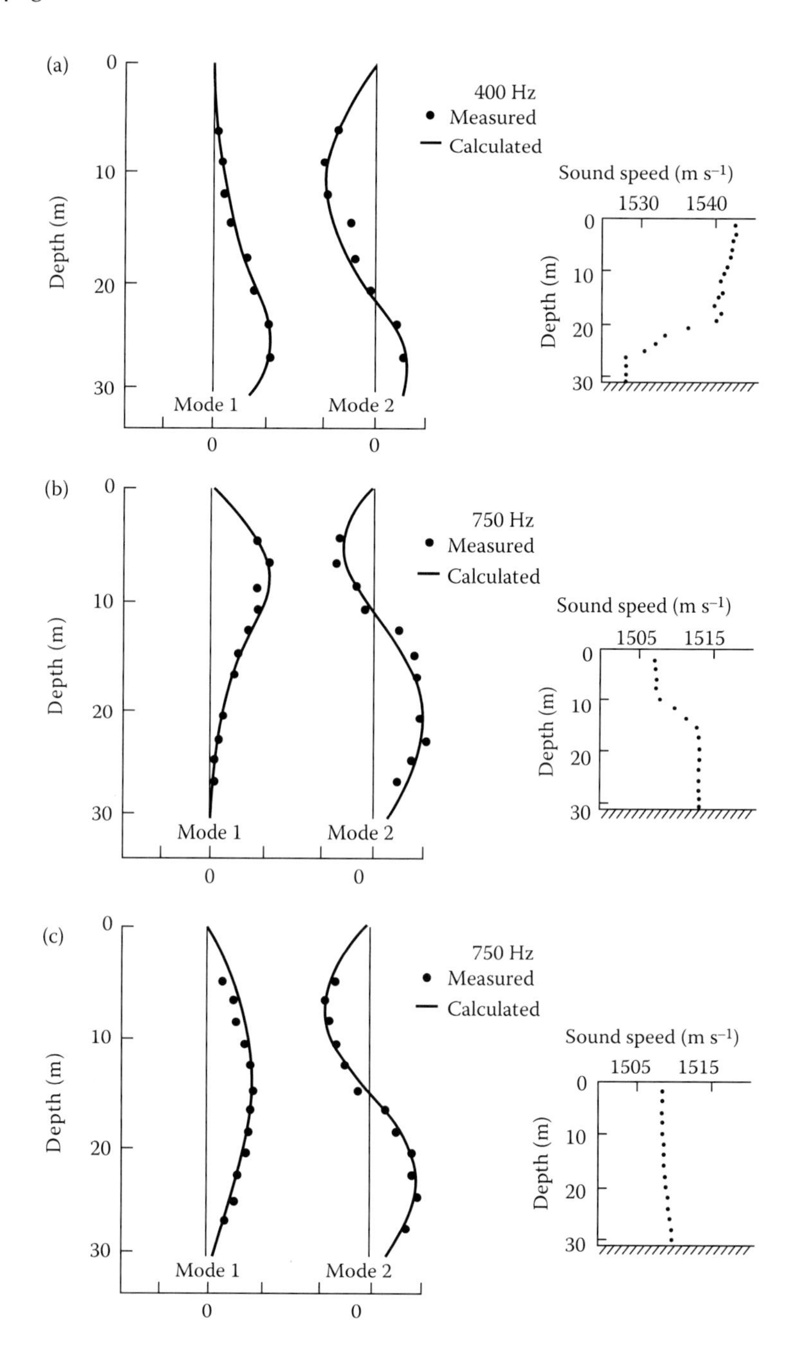

FIGURE 4.11 Comparison of measured and calculated amplitude functions (using arbitrary scales) for the first and second modes: (a) at 400 Hz with a downward-refracting (negative-gradient) profile; (b) at 750 Hz with and upward-refracting (positive-gradient) profile; and (c) at 750 Hz with a nearly constant sound speed profile. (From Ferris, R.H. *J. Acoust. Soc. Amer.*, **52**, 981–988, 1972. With permission.)

new development called WRAP (wide-area rapid acoustic prediction) precomputed local acoustic eigenvalues and normal modes for complex 3D ocean environments comprising a number of distinct local environments (Perkins et al., 1990). For mild horizontal variability, the full 3D acoustic field was constructed by adiabatic mode computations. Another 3D model called CMM3D included horizontal refraction and radial mode coupling (Chiu and Ehret, 1990, 1994). Ainslie et al. (1998a) demonstrated the importance of leaky modes in range-dependent environments with variable water depth. In this particular investigation, the bottom-interacting field was computed by mode summation. Gabrielson (1982) investigated the application of normal mode models to leaky ducts.

Higham and Tindle (2003) extended coupled perturbed-mode theory to include a range-dependent penetrable bottom. Coupled perturbed mode theory combines conventional coupled modes and perturbation theory for problems of sound-speed range dependence over slopes.

Stotts (2002) applied the differential-equation approach to solve second-order coupled-mode equations in inhomogeneous ocean environments. The model incorporated sound-speed profile points to construct depth-dependent, piecewise-linear ocean and bottom environments along a range grid. The modal solutions were evaluated in terms of Airy functions. Stotts et al. (2011) utilized a two-way integral equation coupled-mode (IECM) approach to establish a benchmark solution for the reverberation time series in an environment with a range-dependent, fine-scale, rough-bottom boundary that included mode coupling and generated a scattered field. This exact solution to the wave equation computed both the scattered-field and direct-blast (fathometer-return) components. The transition from direct-blast to scattered-field dominance could be identified in the total field time series. Energy conservation was maintained.

Stotts and Koch (2015) showed that energy conservation and the derivation of the two-way coupled mode range equations could be extended in three dimensions to complex mode functions and eigenvalues. Furthermore, the energy in the coupled mode formulation was conserved for finite-thickness fluid ocean waveguides with a penetrable bottom boundary beneath any range dependence. The derivations relied on completeness and a modified orthonormality statement. The mode-coupling coefficients were specified solely and explicitly by the waveguide range dependence. The statement of energy conservation was applied to a numerical coupled-mode calculation. It was demonstrated that, for more general fluid environment specifications than previously considered, energy conservation consistent with the Helmholtz equation was preserved for a properly formulated two-way coupled mode description. An explicit demonstration of conservation was necessary because, in the past, some coupled-mode formalisms failed this test. Previous coupled-mode developments treated various aspects of horizontal variability (e.g., bathymetry, sound speed, and density), three dimensions, penetrable-bottom boundaries, and absorption; however, none dealt with all these elements together. Furthermore, previous discussions concerning energy conservation for coupled modes assumed ideal boundary conditions involving real mode functions. The formalism presented by Stotts and Koch (2015) incorporated all elements of these various aspects with both discrete mode and continuum effects. The coupled-mode range equations for the modal amplitudes were derived for modes evaluated in the complex horizontal wavenumber plane and for a finite-depth 3D waveguide. The formulation with an impedance boundary at finite depth and continuum contributions required a modified form of orthonormality for the local depth mode functions. The resulting equations were demonstrated to be equivalent to the coupled mode range equations derived by applying mode orthogonality in the form of a product of two mode functions integrated to infinite depth. A statement of energy conservation derived from the coupled normal mode formulation was shown to be consistent with that derived directly from the Helmholtz equation. Numerical results for both flat and rough bottom environments, with both trapped and untrapped modes, were presented to illustrate the energy conservation for these solutions. It is particularly noteworthy that Stotts and Koch (2015) provided a careful historical account and analysis of the conceptual development of coupled modes, punctuated with essential references to the supporting scientific literature.

4.4.6 High-Frequency Adaptations

Normal mode approaches tend to be limited to acoustic frequencies below about 500 Hz due to computational considerations (and not due to any limitations in the underlying physics). Specifically, the number of modes required to generate a reliable prediction of TL increases in proportion to the acoustic frequency. However, by invoking some simplifying assumptions regarding the complexity of the ocean environment, upper frequency limits in the multi-kilohertz range can be achieved (Ferla et al., 1982).

4.4.7 Wedge Modes

Primack and Gilbert (1991) investigated so-called "wedge modes," which are the intrinsic normal modes in a wedge (i.e., sloping-bottom) coordinate system. Wedge modes are identical to the usual normal modes of a range-invariant waveguide except that the mode functions are referenced to the arc of a circle rather than a vertical line. This difference derives from use of a polar coordinate system, with its origin at the apex of the wedge, rather than the usual range-depth coordinate system in a wedge domain (Fawcett et al., 1995). For shallow-water acoustic propagation, the acoustic wavelength is commensurate with the water depth, but short compared to the horizontal extent of the problem. Under these conditions, a sloping bottom causes the development of normal modes having wavefronts that are curved in the vertical direction. Using simple slopes, for example, wedge modes were found to propagate with cylindrical wavefronts (Mignerey, 1995).

Fawcett et al. (1995) developed an efficient coupled-mode method based on the concept of wedge modes. Leaky modes were also included because of their importance in range-dependent waveguide geometries. In related work, Tindle and Zhang (1997) developed an adiabatic normal mode solution for a well-known benchmark wedge problem (discussed in Chapter 12) that included both fluid and solid attenuating bottom boundaries. The continuous-mode contribution was treated as a sum of leaky modes, and each trapped mode gradually transitioned into a leaky mode as the water depth decreased.

In a waveguide characterized by a wedge geometry with perfectly reflecting surface (pressure release) and bottom (either pressure release or rigid) boundaries, the acoustic field propagates as a set of normal modes with circular wave fronts centered at the apex of the wedge. In laboratory experiments (Tindle et al., 1987), the curvature of the modal wave fronts was manifested as a depth-dependent time delay in the arrival of the full signal at the range of the receivers. Other research (Yudichak et al., 2006) simulated wedge-mode propagation with a penetrable bottom using broadband adiabatic mode, coupled mode, and PE model computations. The model predictions were compared with earlier tank experiments (Tindle et al., 1987) and good agreement was found for bottom slopes up to 6°. It was concluded that the experimentally observed propagation by modes with curved wave fronts was correctly described using coupled normal-mode formalisms. Moreover, for slopes ≤3°, the predictions of an adiabatic approximation closely approximated those of the more complete coupled-mode approach. This last observation reaffirmed the applicability of adiabatic approximations of normal-mode theory to ocean environments with gradual range dependence.

Tadeu et al. (2013) implemented 2.5D and 3D Green's functions to simulate wave propagation in the vicinity of 2D wedges. All Green's functions were defined by the image-source technique, which does not account directly for the acoustic penetration of the wedge surfaces. The performance of these Green's functions was compared with solutions based on a normal-mode model, which are found not to converge easily for receivers whose distance from the apex was similar to the distance from the source to the apex. The applicability of the image source Green's functions was then demonstrated by means of computational examples for 3D wave propagation. For this purpose, a boundary element formulation in the frequency domain was developed to simulate the wave field produced by a 3D point pressure source inside a 2D fluid channel. The propagating domain could couple different dipping wedges and flat horizontal layers. The full discretization of the boundary surfaces of the channel was avoided by using the 2.5D Green's functions. The boundary-element method was used to couple the different subdomains, discretizing only the vertical interfaces between them.

4.5 MULTIPATH EXPANSION MODELS

Multipath expansion techniques expand the acoustic field integral representation of the wave equation (Equation 4.13) in terms of an infinite set of integrals, each of which is associated with a particular ray-path family. This method is sometimes referred to as the "WKB method" since a generalized Wentzel–Kramers–Brillouin (WKB) approximation is used to solve the depth-dependent equation derived from the normal mode solution (Equation 4.11). Each normal mode can then be associated with corresponding rays.

The WKB approximation (sometimes also referred to as the WKBJ or Liouville–Green approximation) facilitates an asymptotic solution of the normal mode equation by assuming that the speed of sound varies gradually as a function of depth. Advanced versions of the WKB method provide connection formulas to carry the approximation through "turning points" (i.e., depths where an equivalent ray becomes horizontal). Unlike ray-theoretical solutions, however, the WKB method normally accounts for first-order diffraction effects and caustics.

The specific implementation of this approach is accomplished by directly evaluating the infinite integral of Equation 4.13 over a limited interval of the real ξ-axis. Thus, only certain modes are considered. By using a restricted number of modes, an angle-limited source can be simulated. The resulting acoustic pressure field ϕ in Equation 4.13 is then expressed as a sum of finite integrals, where each integral is associated with a particular ray family. As implemented, this approach is particularly applicable to the modeling of acoustic propagation in deep water at intermediate and high frequencies. Multipath expansion models thus have certain characteristics in common with ray models. Moreover, the pressure field is properly evaluated in caustics and shadow zones. Weinberg (1975) provided a brief summary of the historical development of this technique. Clark (2005) extended the multipath expansion method to range-dependent environments; this extension accounted for horizontal variations in bottom depth, bottom type, and sound speed using the stationary phase approximation. This approach is explored in more detail in Section 4.8 where the RAYMODE model (which is based in part on the multipath expansion approach) is described.

A new wavefront modeling method for calculating waveforms in underwater sound propagation is based on a Hankel-transform generalized WKB solution of the wave equation (Tindle, 2002). The resulting integral led to a form of ray theory that is valid at relatively low frequencies and allows evaluation of the acoustic field on both the illuminated and shadow sides of caustics and at cusps. The integral was evaluated by stationary-phase methods for the appropriate number of stationary points. Rays of nearby launch angles having a travel time difference less than a quarter period must be considered jointly. All other ray arrivals can be described by simple ray theory. The phase, amplitude and travel time of broadband acoustic pulses were obtained directly from a simple graph of ray-travel time as a function of depth at a given range. Although the method can handle range dependence, initial illustrations were limited to long-distance propagation in deep water where the ray paths do not pass close to the surface or bottom. The method is fast (because it operates in the time domain and thus avoids the need for Fourier synthesis) and provides close agreement with normal-mode calculations. The field on the shadow side of a caustic is given properly in terms of rays with complex launch angles, but good approximations can be obtained without the need to find complex rays.

4.6 FAST-FIELD MODELS

In underwater acoustics, fast-field theory is also referred to as "wavenumber integration." In seismology, this approach is commonly referred to as the "reflectivity method" or "discrete-wavenumber method." In fast-field theory, the wave-equation parameters are first separated according to the normal-mode approach. Then, the Hankel function expression in Equation 4.13 is replaced by the first term in the asymptotic expansion (DiNapoli and Deavenport, 1979):

$$H_0^{(1)}(\xi r) \approx \sqrt{\frac{2}{\pi \xi r}}\, e^{i(\xi r)} \quad \text{for} \quad \xi r >> 1$$

Equation 4.13 can now be written as

$$\phi = \int_{-\infty}^{\infty} \sqrt{\frac{2\xi}{\pi r}}\, G(z, z_0; \xi) e^{i(\xi r)}\, d\xi \tag{4.17}$$

The infinite integral is then evaluated by means of the fast Fourier transform (FFT), which provides values of the potential function ϕ at n discrete points for a given source-receiver geometry. Evaluation of the Green's function can be simplified by approximating the sound speed profile by exponential functions. Such an approximation facilitates the matrizant solution, but complicates specification of the sound speed profiles.

Historically, models based on fast-field theory did not allow for environmental range dependence. However, two early developments introduced the possibility of range-dependent calculations of TL. First, Gilbert and Evans (1986) derived a generalized Green's function method for solving the one-way wave equation exactly in an ocean environment that varied discretely with range. They obtained an explicit marching solution in which the source distribution at any given range step was represented by the acoustic field at the end of the previous step. Gilbert and Evans (1986) further noted that their method, which they called the range-dependent fast-field program (RDFFP) model, was computationally intensive. Second, Seong (1990) used a hybrid combination of wavenumber integration and Galerkin boundary element methods (BEM), referred to as the SAFRAN model, to extend the fast field theory technique to range-dependent ocean environments (see also Schmidt, 1991). The experimental nature of these early methods imposed uncertain restrictions on their application to system performance modeling in range-dependent ocean environments. As described below, however, further research proved these methods useful in modeling range-dependent wave propagation.

One approach to range-dependent modeling partitioned the ocean environment into a series of range-independent sectors called "super elements" (Schmidt et al., 1995). Goh and Schmidt (1996) extended the spectral super-element approach for acoustic modeling in fluid waveguides to include fluid-elastic stratifications. Their method used a hybridization of FEs, boundary integrals, and wavenumber integration to solve the Helmholtz equation in a range-dependent ocean environment. It provided accurate, two-way solutions to the wave equation using either a global multiple scattering solution or a single-scatter marching solution.

Grilli et al. (1998) combined BEM and eigenfunction expansions to solve acoustic wave propagation problems in range-dependent, shallow-water regions. Their hybrid BEM technique, or HBEM, was validated by comparing outputs to analytical solutions generated for problems with simple boundary geometries including rectangular, step and sloped domains. HBEM was then used to investigate the transmission of acoustic energy over bottom bumps while emphasizing evanescent modes and associated "tunneling" effects. Related developments in boundary-element modeling in shallow water were reported by Santiago and Wrobel (2000).

The FFP approach has been modified to accommodate acoustic pulse propagation in the ocean by directly marching the formulation in the time domain (Porter, 1990).

Applications included specification of arbitrary source time series instead of the more conventional time-harmonic sources used in frequency-domain solutions of the wave equation. Other recent developments in wavenumber integration approaches are summarized in Section 4.9.

4.7 PARABOLIC EQUATION MODELS

Use of the parabolic approximation in wave propagation problems can be traced back to the mid-1940s when it was first applied to long-range tropospheric radio wave propagation (Keller and Papadakis, 1977: 282–4). Subsequently, the parabolic approximation method was successfully applied to microwave waveguides, laser beam propagation, plasma physics, and seismic wave propagation. Hardin and Tappert (1973) reported the first application to problems in underwater acoustic propagation (also see Spofford, 1973b: 14–16). Lee and Pierce (1995) and Lee et al. (2000) carefully traced the historical development of the PE method in underwater acoustics.

4.7.1 Basic Theory

The PE (or parabolic approximation) approach replaces the elliptic reduced equation (Equation 4.3b) with a parabolic equation (PE). The PE is derived by assuming that energy propagates at speeds close to a reference speed—either the shear speed or the compressional speed, as appropriate (Collins, 1991).

The PE method factors an operator to obtain an outgoing wave equation that can be solved efficiently as an initial-value problem in range. This factorization is exact when the environment is range independent. Range-dependent media can be approximated as a sequence of range-independent regions from which backscattered energy is neglected. Transmitted fields can then be generated using energy-conservation and single-scattering corrections. The following derivation is adapted from that presented by Jensen and Krol (1975).

The basic equation for acoustic propagation, Equation 4.3a, can be rewritten as

$$\nabla^2\phi + k_0^2 n^2 \phi = 0 \tag{4.18}$$

where

k_0 = reference wavenumber (ω / c_0)
$\omega = 2\pi f$ = source frequency
c_0 = reference sound speed
$c(r, \theta, z)$ = sound speed in range (r), azimuthal angle (θ) and depth (z)
n = refraction index (c_0/c)
ϕ = velocity potential
∇^2 = Laplacian operator

Equation 4.18 can be rewritten in cylindrical coordinates as

$$\frac{\partial^2\phi}{\partial r^2} + \frac{1}{r}\frac{\partial\phi}{\partial r} + \frac{\partial^2\phi}{\partial z^2} + k_0^2 n^2 \phi = 0 \tag{4.19}$$

where azimuthal coupling has been neglected, but the index of refraction retains a dependence on azimuth. Further, assume a solution of the form

$$\phi = \Psi(r,z) \cdot S(r) \tag{4.20}$$

and obtain

$$\Psi\left[\frac{\partial^2 S}{\partial r^2} + \frac{1}{r}\frac{\partial S}{\partial r}\right] + S\left[\frac{\partial^2 \Psi}{\partial r^2} + \frac{\partial^2 \Psi}{\partial z^2} + \left(\frac{1}{r} + \frac{2}{S}\frac{\partial S}{\partial r}\right)\frac{\partial \Psi}{\partial r} + k_0^2 n^2 \Psi\right] = 0 \tag{4.21}$$

Using k_0^2 as a separation constant, separate Equation 4.21 into two differential equations as follows:

$$\left[\frac{\partial^2 S}{\partial r^2} + \frac{1}{r}\frac{\partial S}{\partial r}\right] = -S k_0^2 \tag{4.22}$$

and

$$\left[\frac{\partial^2 \Psi}{\partial r^2} + \frac{\partial^2 \Psi}{\partial z^2} + \left(\frac{1}{r} + \frac{2}{S}\frac{\partial S}{\partial r}\right)\frac{\partial \Psi}{\partial r} + k_0^2 n^2 \Psi\right] = \Psi k_0^2 \tag{4.23}$$

Rearrange terms and obtain

$$\frac{\partial^2 S}{\partial r^2} + \frac{1}{r}\frac{\partial S}{\partial r} + k_0^2 S = 0 \tag{4.24}$$

which is the zero-order Bessel equation, and

$$\frac{\partial^2 \Psi}{\partial r^2} + \frac{\partial^2 \Psi}{\partial z^2} + \left(\frac{1}{r} + \frac{2}{S}\frac{\partial S}{\partial r}\right)\frac{\partial \Psi}{\partial r} + k_0^2 n^2 \Psi - k_0^2 \Psi = 0 \tag{4.25}$$

The solution of the Bessel equation (Equation 4.24) for outgoing waves is given by the zero-order Hankel function of the first kind:

$$S = H_0^{(1)}(k_0 r) \tag{4.26}$$

For $k_0 r >> 1$ (far-field approximation)

$$S \approx \sqrt{\frac{2}{\pi k_0 r}} \exp\left[i\left(k_0 r - \frac{\pi}{4}\right)\right] \tag{4.27}$$

which is the asymptotic expansion for large arguments. The equation for $\Psi(r, z)$ (Equation 4.25) can then be simplified to

$$\frac{\partial^2 \Psi}{\partial r^2} + \frac{\partial^2 \Psi}{\partial z^2} + 2i\, k_0 \frac{\partial \Psi}{\partial r} + k_0^2\left(n^2 - 1\right)\Psi = 0 \tag{4.28}$$

Further assume that

$$\frac{\partial^2 \Psi}{\partial r^2} << 2k_0 \frac{\partial \Psi}{\partial r} \tag{4.29}$$

which is the paraxial approximation. Then, Equation 4.28 reduces to

$$\frac{\partial^2 \Psi}{\partial z^2} + 2i\, k_0 \frac{\partial \Psi}{\partial r} + k_0^2 \left(n^2 - 1\right) \Psi = 0 \tag{4.30}$$

which is the parabolic wave equation. In this equation, n depends on depth (z), range (r) and azimuth (θ). This equation can be numerically solved by "marching solutions" when the initial field is known (e.g., Tappert, 1977). The computational advantage of the parabolic approximation lies in the fact that a parabolic differential equation can be marched in the range dimension whereas the elliptic reduced wave equation must be numerically solved in the entire range-depth region simultaneously. Typically, a Gaussian field or a normal-mode solution is used to generate the initial solution. Additional methods for generating accurate starting fields have been described by Greene (1984) and by Collins (1992). Collins (1999) improved the "self-starter" (a PE technique for generating initial conditions) by removing a stability problem associated with evanescent modes. Song and Peng (2006) developed a propagator for use in the PE that has no far-field limitation and can be used as a PE self-starter; its application was demonstrated in a Pekeris channel.

By using operator formalism, the PE can be derived in another fashion (Lee and McDaniel, 1987: 314–5). By assuming commutation of operators,

$$\frac{\partial}{\partial r} \frac{\partial}{\partial z} \Psi = \frac{\partial}{\partial z} \frac{\partial}{\partial r} \Psi \tag{4.31}$$

the equation

$$\frac{\partial^2 \Psi}{\partial r^2} + \frac{\partial^2 \Psi}{\partial z^2} + 2i\, k_0 \frac{\partial \Psi}{\partial r} + k_0^2 \left(n^2 - 1\right) \Psi = 0 \tag{4.32}$$

can be factored as

$$\left[\frac{\partial}{\partial r} + i\, k_0 - i \sqrt{k_0^2 + k_0^2\left(n^2-1\right) + \frac{\partial^2}{\partial z^2}}\right]\left[\frac{\partial}{\partial r} + i\, k_0 + i \sqrt{k_0^2 + k_0^2\left(n^2-1\right) + \frac{\partial^2}{\partial z^2}}\right]\Psi = 0 \tag{4.33}$$

By considering only the outgoing wave, Equation 4.33 reduces to

$$\left(\frac{\partial}{\partial r} + i\, k_0 - i \sqrt{k_0^2 + k_0^2\left(n^2-1\right) + \frac{\partial^2}{\partial z^2}}\right)\Psi = 0 \tag{4.34}$$

which can be rewritten as

$$\frac{\partial}{\partial r} \Psi = i\left(\sqrt{Q} - k_0\right) \Psi \tag{4.35}$$

where

$$Q = k_0^2 + k_0^2\left(n^2 - 1\right) + \frac{\partial^2}{\partial z^2} \tag{4.36}$$

The square-root operator (Q) can be approximated using a rational functional representation

$$\sqrt{Q} = k_0\left(\frac{A + Bq}{C + Dq}\right) \tag{4.37}$$

where

$$q = \left(\frac{Q}{k_0^2}\right) - 1 \tag{4.38}$$

Using Equation 4.37, Equation 4.35 can be rewritten as

$$\frac{\partial}{\partial r}\Psi = i\,k_0\left(\frac{A + B\,q}{C + D\,q} - 1\right)\Psi \tag{4.39}$$

For [$A\ B\ C\ D$] = [1 ½ 1 0], Equation 4.39 reverts back to that of Tappert (Equation 4.30). For [$A\ B\ C\ D$] = [1 ¾ 1 ¼], the form attributed to Claerbout (1976) is obtained.

4.7.2 Numerical Techniques

Existing PE models employ one of four basic numerical techniques: (1) split-step Fourier algorithm; (2) implicit finite-difference (IFD); (3) ordinary differential equation (ODE); or (4) FE. The original split-step algorithm developed by Tappert (1977) solves the parabolic wave equation by imposing an artificial zero bottom-boundary condition and pressure-release surface condition.

At the time Tappert introduced the PE method to the underwater acoustics community, there was a critical need for a capability to predict long-range, low-frequency sound propagation, as would occur in the vicinity of the sound channel axis. Since this type of propagation is characterized by low-angle, nonboundary interacting energy, the PE method was ideally suited to this purpose. Thus, the first introduction of the PE method to naval sonar applications was a celebrated event. Subsequent work focused on making the PE method more robust so that it could be applied to a wider range of problems in underwater acoustics.

While the split-step algorithm is an efficient method for solving a pure initial-value problem (Perkins et al., 1982), several difficulties arise when there is significant interaction with the sea floor (Bucker, 1983). These difficulties are due to density and compressional sound speed discontinuities at the water-sediment interface, strong gradients of the compressional sound speed in the sediment layers, and rigidity in the sediment layers that produces shear waves. Thus, when bottom interaction is strong, a more general-purpose solution is desirable. For this reason, IFD schemes (Lee et al., 1981; Robertson et al., 1989) and ODE methods (Lee and Papadakis, 1980) have also been developed to solve the PE. Lee and McDaniel (1987) provided an in-depth development of the IFD technique. Their development was made more useful by the inclusion of benchmark test examples together with a listing of the computer program (comprising approximately 1700 lines of FORTRAN code). A microcomputer implementation of the IFD model was reported by Robertson et al. (1991), who included a listing of their computer program. Collins (1988a) described a FE solution for the PE. Kampanis et al. (2007) described the benchmarking of two FE models based on the parabolic approximation: one for underwater object identification (FENL) and one for atmospheric sound propagation over an irregular terrain (CNP1-NL). The FENL model is also described in the Ocean Acoustics Library (refer to Appendix C).

Recent modeling developments utilizing parabolic approximations have been directed at refining and expanding the capabilities of existing techniques. Useful summaries of this progress are available in the literature (Lee, 1983; Scully-Power and Lee, 1984; Lee and Pierce, 1995; Lee, et al., 2000). Actual comparisons of computer model results for a set of four ocean acoustic test cases was documented by Davis et al. (1982). A more recent comparison of PE models involving seven test cases was described by Chin-Bing et al. (1993b). One of the most interesting test cases described by Chin-Bing et al. (1993b) involved a leaky surface duct. So-called "ducted precursors" are generated by modal energy in the surface duct that leaks out and travels as convergence-zone or bottom-bounce paths, or both, before coupling back into the surface duct down range. The importance of this problem is that a small phase error in the refracted (leaky) path can produce large changes in the predicted sound level in the duct beyond the first CZ (Porter and Jensen, 1993). This pathology was particularly evident in some wide-angle, split-step PE models. Yevick and Thomson (1994) explored this problem further and formulated a new propagation operator that retained the computational efficiency of the split-step algorithm but which was more accurate.

The UMPE, SNAP, and PROSIM models were used to study the focusing of acoustic energy over RBR paths in the Straits of Florida (DeFerrari et al., 2003). The sound-speed profile $[c(z)]$ beneath the surface mixed layer could be expressed as $c(z) = c_0 \cosh[g(1-z/D)]$, where D is the channel depth and c_0 and g are scaling coefficients. This particular profile produced focusing at all ranges. In earlier studies, Monjo and DeFerrari (1994) had noted the generation of precursors at discrete ranges where RBR and SRBR phase fronts exchanged energy with the RSR (refracted-surface-reflected) phase front in the duct. The effects of mode coupling, which were manifested as smeared arrivals, could arise from interactions with an irregular bottom or with internal waves. Judicial selection of the experiment site and source-receiver geometry minimized the effects of mode coupling. See the related analyses of precursors using the SWAMP model described by Sidorovskaia (2004) and by Tamendarov and Sidorovskaia (2004).

Mikhin (2004) developed a model called OWWE, which is based on the innovative one-way wave equation developed by Godin (1999). This equation was originally generalized by Godin to include the source terms and also to account for motion of the medium. Solutions of the differential OWWE are strictly energy conserving and reciprocal. The derivation presented for the multiterm Padé PE model is applicable to a broad class of finite-difference PE models. Brekhovskikh and Godin (1999: 448) noted that the reciprocity of PE solutions is important in solving inverse problems by the method of backpropagation. Much like energy conservation, reciprocity is known to hold within narrow-angle approximations, but commonly used wide-angle PE methods are not reciprocal in general range-dependent media.

Palmer (2002) applied the horizontal ray-acoustic approximation to obtain a PE that is energy conserving and has a spreading factor that describes the field intensification for antipodal propagation for application to global propagation of acoustic waves in the ocean or atmosphere.

Traditionally, PE models have been applied to anelastic ocean-bottom regions by treating the shear waves as an additional loss mechanism. This approximation breaks down when anelastic propagation becomes significant, as in shear conversion due to backscattered acoustic fields. In general, PE models propagate the acoustic field only in the forward direction, thus excluding backscatter. A two-way PE model (Collins and Evans, 1992) was developed to improve the computation of backscattered energy for use in reverberation simulations (e.g., Schneider, 1993). A modified version of this two-way PE model (called spectral PE) treats backscattering problems in 3D geometries (Orris and Collins, 1994). Lingevitch and Collins (1998) argued that a poro-acoustic medium is, in fact, the limiting case of a poro-elastic medium in which the shear wave speed vanishes. Collins and Siegmann (1999) extended energy-conservation corrections from the acoustic case to the elastic case.

An improved approach for handling boundaries, interfaces and continuous depth dependence with the elastic PE was derived and benchmarked by Jerzak et al. (2005). By way of example, this approach was used to model the propagation of Rayleigh and Stoneley waves. Specifically, solutions from the PE were compared with a reference solution based on a separation of variables. Depending on the choice of dependent variables, the operator in the elastic wave equation may not factor, or the treatment of interfaces may be difficult. These problems can be resolved by using a formulation in terms of the vertical displacement and the range derivative of the horizontal displacement. These two quantities, which are continuous across horizontal interfaces, permit the use of Galerkin's method to discretize in depth. This implementation extended the capability of the elastic PE to handle arbitrary depth dependence and could thus lead to improvements in range-dependent problems.

Metzler et al. (2014) presented a 2D fluid PE that treated range dependence through a scaled-mapping technique. Specifically, this approach mapped the original domain, where the bathymetry varied with range, to one where both the bathymetry and surface were flat. Propagation in the mapped domain was obtained by appropriately expanding (or contracting) the waveguide as the solution was marched through range. This procedure was then compared to reference solutions for shallow-water and deep-water environments where it was shown to be more robust than an earlier mapping approach by Collins and Dacol (2000), which distorted the waveguide only translationally.

4.7.3 Wide-Angle and 3D Adaptations

Tappert's (1977) original split-step PE method handled small-angle (≤15°) propagation paths, a restriction imposed by the basic paraxial assumption. Unless the split-step algorithm is modified for wide angles (e.g., Thomson and Chapman, 1983), large phase errors may be introduced into the solution. In order to extend the maximum angle of half-beamwidth propagation in the PE model, alternate forms of the square-root operator have been explored. For split-step PE formulations, the Thomson and Chapman (1983) wide-angle approximation has been used. The Claerbout (1976: 206) wide-angle approximation is used in some finite-difference PE formulations with values of $[A\ B\ C\ D] = [1\ \frac{3}{4}\ 1\ \frac{1}{4}]$ in Equation 4.39. Other approximations of the square-root operator have been explored including higher-order Padé forms, which have been incorporated in finite-difference and FE PE formulations to achieve near 90° half-beamwidth propagation (Chin-Bing et al., 1993b). Thomson and Wood (1987) investigated a post-processing method for correcting phase errors in the parabolic approximation approach. This method was later extended by Thomson and Mayfield (1994) to include an exact, nonlocal boundary condition at the sea floor. Since strong boundary (surface and bottom) interactions are associated with wide-angle propagation, the accurate treatment of irregular interfaces and rough boundaries assumed greater importance in PE models (Bucker, 1983; Lee and McDaniel, 1983; Dozier, 1984; Collins and Chin-Bing, 1990; Brooke and Thomson, 2000). Adaptations to under-ice environments have also been of interest to sonar modelers.

Extensions to three dimensions are generally implemented in an approximate manner (e.g., $N \times 2D$), although the primary parabolic wave equation is fully 3D (Perkins et al., 1983; Siegmann et al., 1985; Lee and Siegmann, 1986). Such 3D extensions are often coupled with wide-angle modifications to achieve maximum utility from the models (Botseas et al., 1983; Thomson and Chapman, 1983).

Godin (2002) developed a class of wide-angle parabolic wave equations for sound in a 3D inhomogeneous moving fluid. These equations provided higher accuracy than the familiar wide-angle PEs for moving media and were consistent with the law of energy conservation and with the flow-inversion theorem in the parabolic approximation. In related work, Godin (2011) developed an exact wave equation for sound in inhomogeneous, moving and nonstationary fluids.

The PE method has been further modified to accommodate solid–solid interfaces using the single-scattering solution (Küsel et al., 2007). This solution is based on an iteration formula that has improved convergence, and a transverse operator of the parabolic wave equation that is implemented efficiently in terms of banded matrices. Problems involving large contrasts across sloping stratigraphy can be handled by subdividing a vertical interface into a series of two or more scattering problems. This approach is applicable to a large class of seismic problems, and accurate solutions can often be obtained by using just one iteration.

Lu and Zhu (2007) used the perfectly matched layer (PML) technique to truncate unbounded domains in numerical simulations of wave propagation problems. In the frequency domain, the PML corresponds to a complex coordinate stretching. Specifically, the PML was modified and applied to a wide-angle PE model to solve a range-dependent benchmark problem.

Lin et al. (2013) used the split-step Fourier method in 3D PE models to compute underwater sound propagation in the forward direction. Their method was implemented in both Cartesian (x, y, z) and cylindrical (r, θ, z) coordinate systems. The Cartesian model had uniform resolution throughout the domain, but had errors that increased with azimuthal angle from the x-axis. The cylindrical model exhibited consistent validity in each azimuthal direction, but a fixed cylindrical grid of radials could not produce uniform resolution. Two different methods were presented to achieve more uniform resolution in the cylindrical PE model: (1) increase the grid points in azimuth (as a function of r) according to nonaliased sampling theory; and (2) make use of a fixed arc-length grid. A point-source starter was derived for the 3D Cartesian PE model. Results from idealized seamount and slope calculations were shown to verify the performance of these methods.

Petrov and Sturm (2016) presented a new asymptotic analytical solution for the problem of wave propagation in a 3D penetrable wedge with a small apex angle. The solution was based on the adiabatic mode parabolic equation (MPE theory), and the solution of MPEs was analytically derived using operator disentanglement identities. The resulting formula for the acoustical field in the wedge-shaped waveguide was compared with the numerical solution of a fully 3D PE based model that included a leading-order cross term correction, and very good agreement was observed in the across-slope direction. These cross terms were introduced in 3D PE models in order to reduce the phase errors inherent to any 3D PE computation. The analytical solution was valid only under the usual requirement of adiabatic theory, and there will always exist some combinations of waveguide parameters for which it is inapplicable (e.g., when the source is located close to the cut-off depth of a mode).

4.7.4 Range-Refraction Corrections

Range-refraction corrections are introduced to accommodate propagation through strong oceanic fronts. The standard parabolic approximation is known to have intrinsic phase errors that will degrade the accuracy of any PE solution for long-range propagation in the ocean (Tappert and Lee, 1984; Jensen and Martinelli, 1985). The accuracy can be improved using updated mean phase speeds as a function of range. Tolstoy et al. (1985) suggested a simple transformation of the sound speed profile to compensate for the inability of the PE approach to correctly locate the range of the signal turning points (e.g., Brock et al., 1977). Schurman et al. (1991) developed an energy-conserving PE model that incorporated range refraction. A model called LOGPE (Berman et al., 1989) used a logarithmic expression for the index of refraction in the standard PE in order to closely approximate solutions of the more exact Helmholtz equation in weakly range-dependent ocean environments.

4.7.5 High-Frequency Adaptations

At frequencies higher than about 500 Hz, PE models become impractical due to excessive execution times. Computational intensity is proportional to the number of range-interval steps. As the frequency increases, the step size decreases and more steps are thus required to achieve the desired prediction range. High-frequency approximations can be obtained by introducing a hybrid approach that combines aspects of parabolic approximations and ray theory (Tappert et al., 1984). A special-purpose microcomputer model (PESOGEN) was developed to compute the acoustic field in a fully range-dependent environment over a wide range of frequencies for both deep-water and shallow-water regions with greatly reduced execution times (Nghiem-Phu et al., 1984).

The so-called calculation frequency method (CFM) can be used to extend PE solutions to high frequencies with execution speeds typically associated with those at low frequencies (Moore-Head et al., 1989). The CFM produces range-averaged TL information by substituting a (low) calculation frequency for a (high) prediction frequency. This substitution is accomplished by sacrificing volume attenuation and boundary-loss phase information under the assumption that diffraction and interference effects are not important at the desired (high) prediction frequency.

4.7.6 Time-Domain Applications

Only frequency-domain applications of the PE have been explored thus far. However, the parabolic approximation has been adapted to consider time-domain applications. Specifically, the acoustic pressure is advanced in time by the so-called progressive wave equation (PWE). However, the PWE is, for practical purposes, limited to narrow-angle propagation paths. Collins (1988b) derived an inverse Fourier transform of the wide-angle PE referred to as the time-domain PE (TDPE). The TDPE advances the acoustic pressure field in range. Collins (1988b) used the TDPE to investigate the effects of sediment dispersion on pulse (or broadband) propagation in the ocean. Orchard et al. (1992) described the development of a TDPE model appropriate for use in 3D ocean acoustic propagation problems.

The PWE has also been adapted to model the propagation of a nonlinear acoustic pulse that is subject to refraction and diffraction. McDonald and Kuperman (1987) examined the problem of acoustic propagation of finite-amplitude pulses and weak shocks in the ocean where refraction can lead to caustic formation. Their result was a first-order nonlinear progressive-wave equation (NPE), which is the nonlinear time-domain counterpart of the linear frequency-domain PE. When the nonlinear term is omitted, the NPE reduces to the linear frequency-domain PE. The simplicity of their formulation suggested additional applications to broadband linear acoustic problems in the ocean. The NPE was also used to study the nonlinear signature of long-range acoustic propagation for high-amplitude sources in an ocean waveguide (Castor et al., 2004).

4.8 RAYMODE MODEL—A SPECIFIC EXAMPLE

The RAYMODE model, originally developed by Leibiger (1968), numerically solves the wave equation for underwater acoustic propagation using the multipath expansion approach (Section 4.5). RAYMODE is actually a hybrid approach involving elements of both ray theory and wave theory. RAYMODE is a stand-alone propagation model intended for application to passive sonar performance prediction problems in range-independent ocean environments.

Since its original development, the RAYMODE passive acoustic propagation model was widely used in the naval sonar modeling community. The US Navy later brought RAYMODE under configuration management where it was maintained as an interim standard for many years. While use of this model has subsequently been eclipsed by the development of more capable models, RAYMODE continues to provide an instructive example of acoustic model construction. Additional sources of information for RAYMODE are provided in Section 4.9.

The model accommodates beam patterns at both the source and receiver. Boundary interaction losses are also included. Sea-surface losses are computed by a user-supplied table or by internal algorithms. Surface-scattering losses assume two mechanisms: (1) a high-frequency, large-roughness loss (Beckmann and Spizzichino, 1963: 89; Clay and Medwin, 1964); and (2) a low-frequency loss (Marsh and Schulkin, 1962a). Seafloor losses are computed either through identification of bottom province types in order to access stored values or by a user-supplied table. The volume attenuation expression of Thorp (1967) is used. TL is calculated both coherently and incoherently.

The basic assumptions incorporated in RAYMODE are as follows:

1. Range-independent sound speed and bathymetry.
2. Sound speed profile fit with segments such that the index-of-refraction squared is a linear function of depth.
3. Multiple reflections due to sound speed discontinuities are ignored except in surface duct situations.
4. Plane wave reflection coefficients are assumed.
5. Only a finite number of ray cycles is considered in the multipath evaluation of pressure integrals.
6. A harmonic (single-frequency) source is assumed.
7. A water density of unity is assumed.

The specific implementation of the multipath expansion technique is explained as follows. The infinite integral in Equation 4.13 is expressed as a sum of finite integrals, each of which is associated with a particular ray path. In Equation 4.13, ϕ is taken to represent the acoustic field pressure and the integration is performed piecewise over particular ray-path regions defined by the limiting rays (R.L. Deavenport, 1978, unpublished manuscript):

$$\int_{-\infty}^{\infty} \cong \int_{\xi_3}^{\xi_0} = \underset{(\mathrm{SD})}{\int_{\xi_1}^{\xi_0}} + \underset{(\mathrm{CZ})}{\int_{\xi_2}^{\xi_1}} + \underset{(\mathrm{BB})}{\int_{\xi_3}^{\xi_2}} \tag{4.40}$$

where $\xi_i = \omega/c_i$, c_i = sound speed at vertexes, SD = surface duct region, CZ = convergence zone region, and BB = bottom bounce region.

Figure 4.12 illustrates a typical geometry. Here, z_s is the source depth, c_0 is the sound speed at the sea surface, c_1 is the sound speed at the sonic layer depth, and c_2 is the sound speed at the sea floor (at the base of the water column). The particular sound speed profile and source depth geometry illustrated here would support all three ray-path families described (i.e., surface duct, CZ, and bottom bounce). Note that

$$\xi_3 = \left[\frac{\omega}{c_s}\right]\cos\theta_s$$

where θ_s is the maximum ray angle leaving the source and c_s is the sound speed at the source depth.

The integral for each region is expanded into four parts:

$$\phi = \sum_{i=1}^{4} \phi_i$$

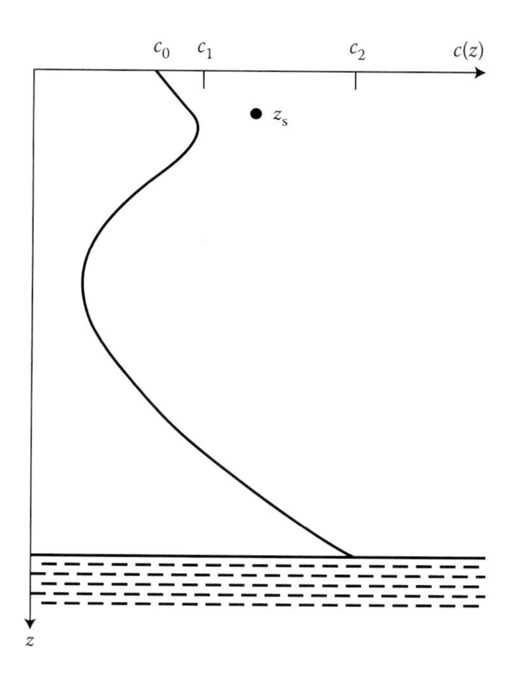

FIGURE 4.12 Typical environment for the RAYMODE propagation model showing a sound speed profile with a source located at depth z_s.

Each ϕ_i corresponds to one of the four ray types associated with upgoing and downgoing rays at both the source and the receiver. The path geometries and associated sign conventions are prescribed in the RAYMODE L-path chart illustrated in Figure 4.13. These paths may be compared with those described previously in Figure 4.2. The cycle counter (q) represents the number of bottom reflections.

Each segment is then evaluated by either wave theory or ray theory. A special option is available to the user for surface ducts in order to incorporate the effects of leakage (diffraction effects). This option is exercised only for that wavenumber segment for which it is appropriate. If the special surface duct option is not exercised, then the segment is evaluated using the normal mode approach. However, if the number of normal modes for a given segment exceeds ten, then a multipath expansion of the four parts of the integral is employed. If the segment contains bottom bounce paths, then ray-theoretical methods are used. At frequencies below 45 Hz, a normal mode solution is used to evaluate propagation through the sediments. There is also a separate high-frequency routine that is used for frequencies greater than 1 kHz. The key parameters controlling the decision for the method of solution are the acoustic frequency and the physical environment. Consequently, slight variations in the problem statement may possibly result is unexpectedly abrupt changes in TL brought about by changes in the method of solution.

The resulting acoustic pressure field is modified by a beam pattern attenuation factor characterizing the off-axis beam position of an equivalent ray. The real and imaginary pressure components (P_i) are then used to form incoherent and coherent

Resolution* of sign of source and receiver angles in RAY-MODE:

General range $\bar{R} = qR_C \pm R_S \pm R_d$ or $qR_C + P_2R_S + P_2R_d$ where q = Cycle 0, 1, 2,...λ

R_C = Cycle range
R_S = Source term
R_d = Receiver term

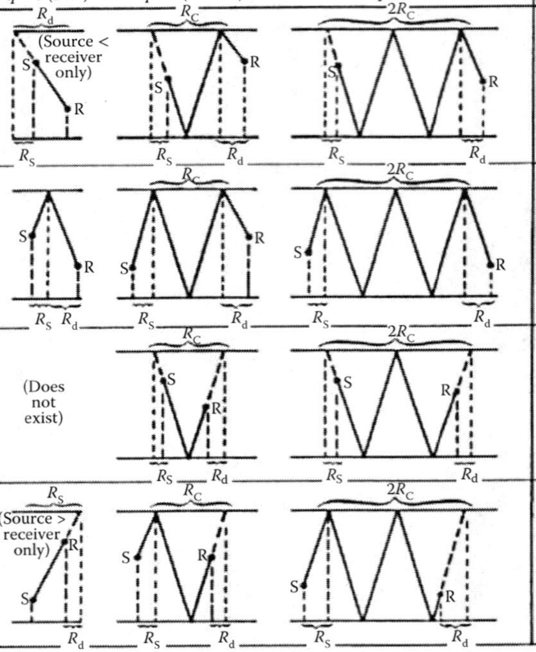

Path L	P_2	P_1	Equation $\bar{R}$	$q = 0$ (D.P.)	$q = 1$ (EX. BB)	$q = 2$	Source angle	Receiver angle
1	–	+	$qR_C - R_S + R_d$	R_d, (Source < receiver only), S, R, R_S	R_C, S, R, R_S, R_d	$2R_C$, S, R, R_S, R_d	+ Down	– Up
2	+	+	$qR_C + R_S + R_d$	S, R, R_S R_d	R_C, S, R, R_S, R_d	$2R_C$, S, R, R_S, R_d	– Up	– Up
3	–	–	$qR_C - R_S - R_d$	(Does not exist)	R_C, S, R, R_S, R_d	$2R_C$, S, R, R_S, R_d	+ Down	+ Down
4	+	–	$qR_C - R_S - R_d$	R_S, (Source > receiver only), R, S, R_d	R_C, S, R, R_S, R_d	$2R_C$, S, R, R_S, R_d	– Up	+ Down

* Sign of angle is opposite sign of range adjustment term for both the source and the receiver.

FIGURE 4.13 RAYMODE L-path chart illustrating the sign convention for source and receiver angles. (Adapted from Yarger, D.F. *Nav. Underwater Syst.* Ctr, Tech. Memo. 821061, 1982.)

intensities for each range point and each ξ-partition. TL relative to unit intensity at unit distance from the source is calculated from the incoherent and coherent intensity sums as follows:

Incoherent TL

$$-10\log_{10}\left\{\begin{array}{l}\sum_{i=1}^{4}\Big[(\mathrm{Re}\,P_i)_{\mathrm{SD}}^2+(\mathrm{Im}\,P_i)_{\mathrm{SD}}^2+(\mathrm{Re}\,P_i)_{\mathrm{BB}}^2\\ +(\mathrm{Im}\,P_i)_{\mathrm{BB}}^2+(\mathrm{Re}\,P_i)_{\mathrm{CZ}}^2+(\mathrm{Im}\,P_i)_{\mathrm{CZ}}^2\Big]\end{array}\right\}+\alpha r \tag{4.41}$$

Coherent TL

$$-10\log_{10}\left\{\begin{array}{l}\left(\sum_{i=1}^{4}\Big[(\mathrm{Re}\,P_i)_{\mathrm{SD}}+(\mathrm{Re}\,P_i)_{\mathrm{BB}}+(\mathrm{Re}\,P_i)_{\mathrm{CZ}}\Big]\right)^2\\ +\left(\sum_{i=1}^{4}\Big[(\mathrm{Im}\,P_i)_{\mathrm{SD}}+(\mathrm{Im}\,P_i)_{\mathrm{BB}}+(\mathrm{Im}\,P_i)_{\mathrm{CZ}}\Big]\right)^2\end{array}\right\}+\alpha r \tag{4.42}$$

where α is the attenuation coefficient according to Thorp (1967).

In the RAYMODE model, user-supplied input information includes the following:

1. Sound speed profile
2. Source and receiver depths
3. Frequency
4. Horizontal range determinants
5. Bottom loss (default values used if inputs not supplied)
6. Surface loss (default values used if inputs not supplied)
7. Source and receiver beam patterns (default values used if inputs not supplied)
8. Program controls (e.g., number of ray cycles to be processed, ray angle limits, mode cutoff values, calculation method options)
9. Output options

Graphical outputs from RAYMODE include the following:

1. Sound speed profile plot, and plots of bottom loss, surface loss and beam deviation loss
2. Source and receiver angle versus horizontal range (i.e., arrival structure) for selected ray paths (Figure 4.14)
3. Travel time versus horizontal range for the ray paths in item (2) above (Figure 4.15)
4. TL versus horizontal range for coherent or incoherent combination of multipaths with beam pattern attenuation (Figure 4.16)
5. Ray trace diagram (Figure 4.17)

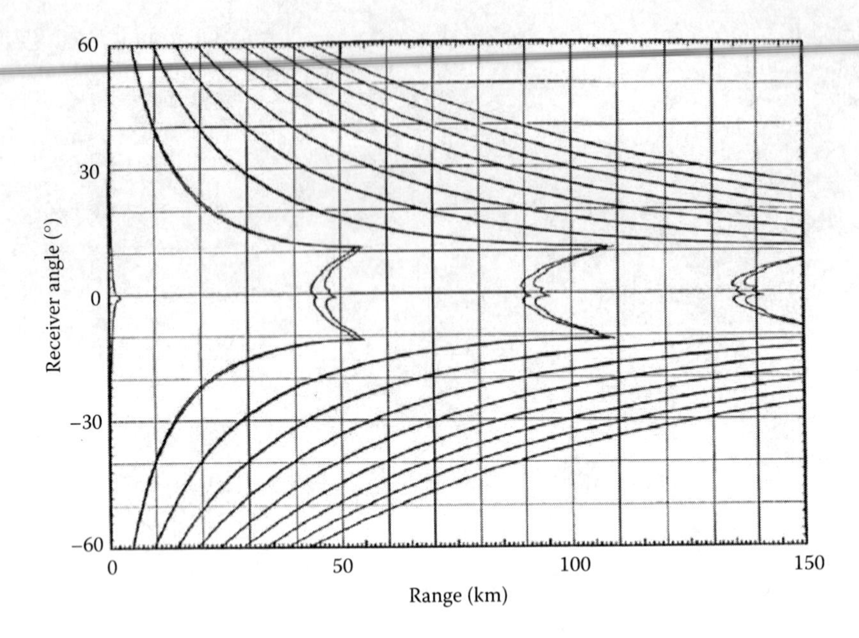

FIGURE 4.14 RAYMODE arrival structure diagram. This graph displays receiver angle versus horizontal range for selected rays representing bottom bounce (BB) and convergence zone (CZ) paths. The CZ paths are associated with the near-axis features at 45, 90, and 135 km. (Adapted from Yarger, D.F. *Nav. Underwater Syst.* Ctr, Tech. Memo. 821061, 1982.)

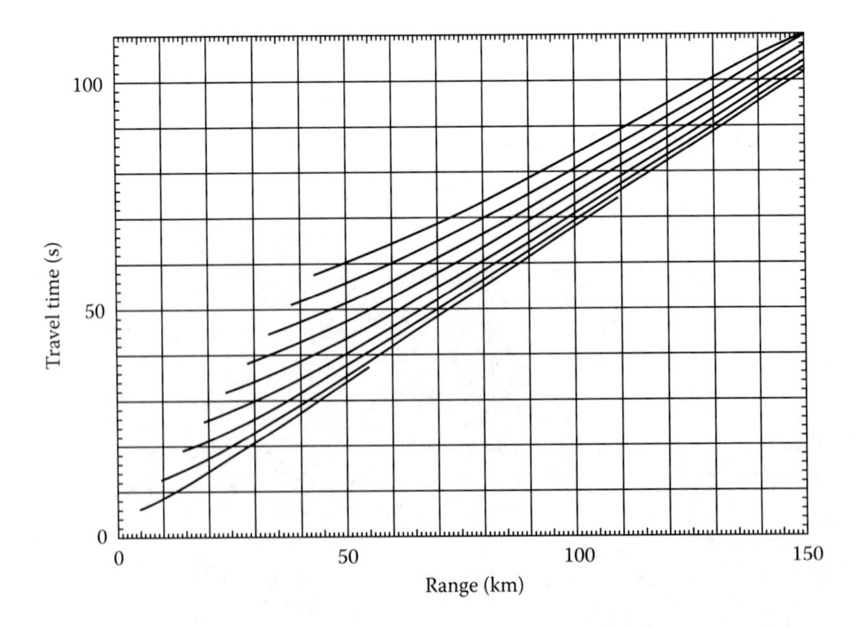

FIGURE 4.15 RAYMODE travel time versus horizontal range for the ray paths selected in Figure 4.14. (Adapted from Yarger, D.F. *Nav. Underwater Syst.* Ctr, Tech. Memo. 821061, 1982.)

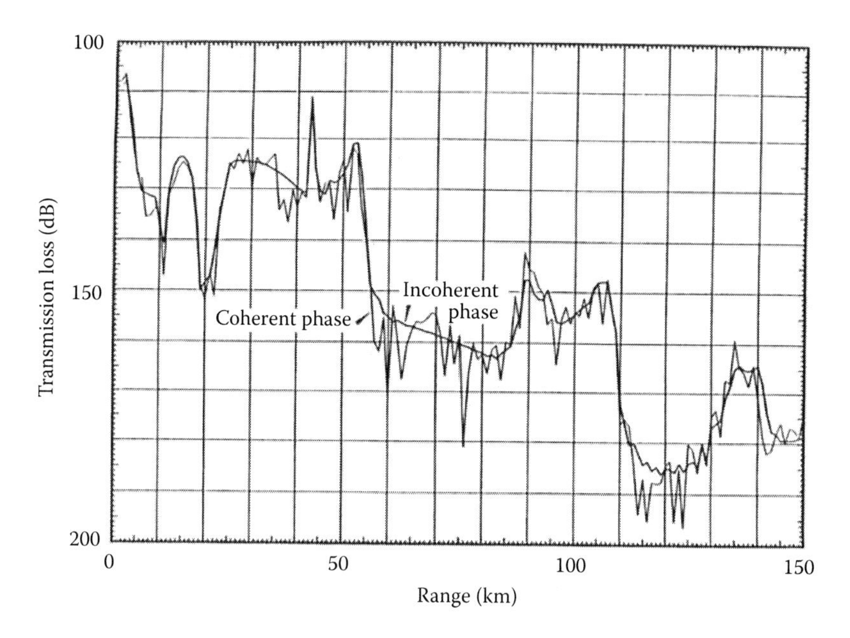

FIGURE 4.16 RAYMODE transmission loss versus range for both coherent and incoherent phase summation. (Adapted from Yarger, D.F. *Nav. Underwater Syst.* Ctr, Tech. Memo. 821061, 1982.)

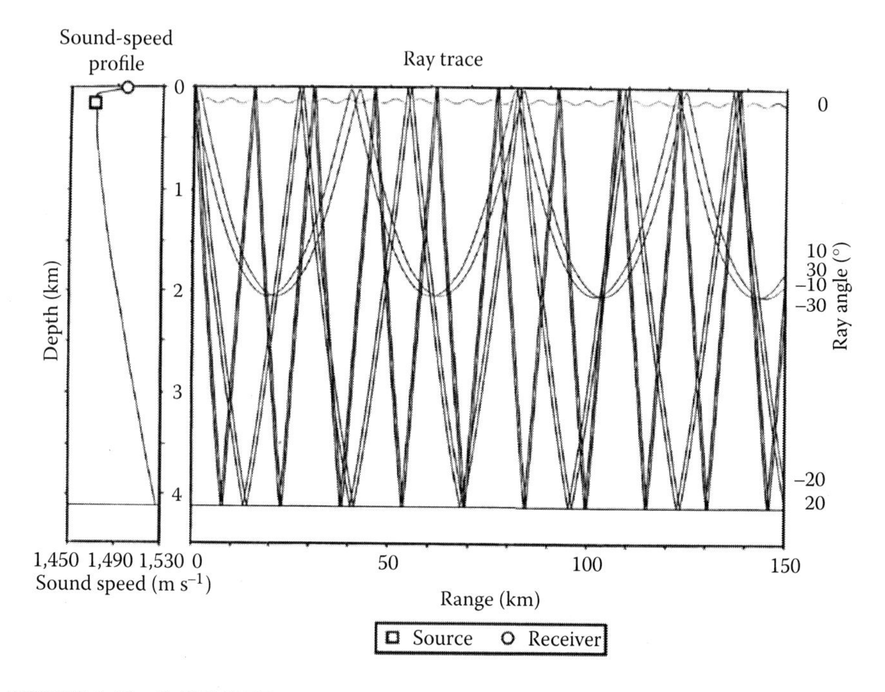

FIGURE 4.17 RAYMODE ray trace diagram. (Adapted from Yarger, D.F. *Nav. Underwater Syst.* Ctr, Tech. Memo. 821061, 1982.)

4.9 NUMERICAL MODEL SUMMARIES

This section summarizes available underwater acoustic propagation models. These models are applicable to work in sonar technology (e.g., sonar design and operation) and in acoustical oceanography (e.g., oceanographic research and data analysis). As is true of any summary of this nature, the information represents a snapshot in time of the present state of the art. While comprehensive in coverage, this summary does not claim to be exhaustive.

In order to optimize the utility of this information, the models are arranged in categories reflecting the basic modeling technique employed (i.e., the five canonical approaches) as well as the ability of the model to handle environmental range dependence. Such factors define what is termed "domains of applicability." Hybrid models occasionally compromise strict categorization, and some arbitrariness has been allowed in the classification process. The environmental range dependence considers variations in sound speed or bathymetry. Other parameters may be considered to be range-dependent by some of the models, although they are not explicitly treated in this summary.

The specific utility of these categories is further explained below. In sonar design and operation problems, for example, the analyst is normally faced with a decision matrix involving water depth (deep versus shallow), frequency (high versus low) and range-dependence (range-independent versus range-dependent ocean environments). Jensen (1982, 1984) developed a very useful classification scheme for optimizing this decision logic against the available modeling approaches and their domains of applicability. Jensen's scheme is slightly modified here to accommodate the five modeling approaches utilized in this book (Figure 4.18). The following assumptions and conditions were imposed in the construction of Figure 4.18:

1. Shallow water includes those water depths for which the sound can be expected to interact significantly with the sea floor. Typically, a maximum depth of 200 m is used to delimit shallow water regions. (A more accurate definition of shallow water would be expressed in terms of water depth and acoustic wavelength.)
2. The threshold frequency of 500 Hz is somewhat arbitrary, but it does reflect the fact that above 500 Hz, many wave-theoretical models become computationally intensive. Also, below 500 Hz, the physics of some ray-theoretical models may become questionable due to restrictive assumptions.
3. A solid circle indicates that the modeling approach is both applicable (physically) and practical (computationally). Distinctions based on speed of execution may change as progress is made in computational capabilities.
4. A partial circle indicates that the modeling approach has some limitations in accuracy or in speed of execution.
5. An open circle indicates that the modeling approach is neither applicable nor practical.

Table 4.1 identifies available stand-alone passive propagation models, which are categorized according to the five modeling approaches discussed earlier.

Model type	Applications							
	Shallow water				Deep water			
	Low frequency		High frequency		Low frequency		High frequency	
	RI	RD	RI	RD	RI	RD	RI	RD
Ray theory	○	○	◑	●	◑	◑	●	●
Normal mode	●	◑	●	◑	●	◑	◑	○
Multipath expansion	○	○	◑	◑	◑	◑	●	◑
Fast field	●	◑	●	◑	●	◑	◑	◑
Parabolic equation	◑	●	○	○	◑	●	◑	◑

Low frequency (<500 Hz) RI: Range-independent environment
High frequency (>500 Hz) RD: Range-dependent environment

● Modeling approach is both applicable (physically) and practical (computationally)
◑ Limitations in accuracy or in speed of execution
○ Neither applicable nor practical

FIGURE 4.18 Domains of applicability of underwater acoustic propagation models. (Adapted from Jensen, F.B., *MTS/IEEE Oceans 82 Conf.*, 147–154, 1982. by IEEE. With permission.)

(Stand-alone FE models are not included in this summary.) These models are further segregated according to their ability to handle range-dependent environments. Note that range-dependent models can also be used for range-independent environments by inserting a single environmental description to represent the entire horizontal range. Figure 4.18 has been further modified to indicate some range-dependent capability for multipath expansion and fast field (or wavenumber integration) models. Numbers within brackets following each model refer to a brief summary and appropriate documentation. Model documentation can range from informal programming commentaries to journal articles to detailed technical reports containing a listing of the actual computer code. Abbreviations and acronyms are defined in Appendix A.

Taken together, Figure 4.18 and Table 4.1 provide a useful mechanism for selecting a subset of candidate models once some preliminary information is available concerning the intended applications. There are many cases in which two different modeling approaches are suitable (as indicated by a solid circle in Figure 4.18). In such cases, the user is encouraged to select candidates from both categories in order to assess any intermodel differences. Even when only one modeling approach is suitable, more than one candidate model should be tried. In the event of divergences in model predictions (a very likely event), the envelope defining spreads in model outputs can be used to assess the degree of uncertainty in the predictions. Moreover, this envelope may provide diagnostic information that is useful in identifying model pathologies.

TABLE 4.1
Summary of Underwater Acoustic Propagation Models

Technique	Range Independent		Range Dependent	
Ray Theory	CAPARAY [1]		ACCURAY [9]	MPC [22]
	FACT [2]		BELLHOP-2D/3D [10]	MPP [23]
	FLIRT [3]			
	GAMARAY [4]		Coherent DELTA [11]	Pedersen [24]
	ICERAY [5]		FACTEX [12]	PlaneRay [25]
	PLRAY [6]		FeyRay [13]	PWRC [26]
	RANGER [7]		GRAB [14]	Ray5 [27]
	SMOD [8]		GRASS [15]	RAYSON [28]
			HARORAY [16]	RAYWAVE [29]
			HARPO [17]	RP-70 [30]
			HARVEST [18]	SHALFACT [31]
			LYCH [19]	TRIMAIN [32]
			MEDUSA [20]	TV-APM [33]
			MIMIC [21]	WaveQ3D [34]
				XRAY [35]
Normal Mode	AP-2/5 [36]	POPP [47]	ADIAB [51]	MOCTESUMA [63]
	BDRM [37]	PROTEUS [48]	ASERT [52]	NAUTILUS [64]
	COMODE [38]	SHEAR2 [49]	ASTRAL [53]	PROLOS [65]
	DODGE [39]	Stickler [50]	CENTRO [54]	PROSIM [66]
	FNMSS [40]		CMM3D [55]	SHAZAM [67]
	MODELAB [41]		COUPLE [56]	SNAP/C-SNAP [68]
	NEMESIS [42]		CPMS [57]	SWAMP [69]
	NLNM [43]		FELMODE [58]	WEDGE [70]
	NORMOD3 [44]		IECM [59]	WKBZ [71]
	NORM2L [45]		Kanabis [60]	WRAP [72]
	ORCA [46]		KRAKEN [61]	3D Ocean [73]
			MOATL [62]	
Multipath Expansion	FAME [74]		Integrated Mode [78]	
	MULE [75]			
	NEPBR [76]			
	RAYMODE [77]			
Fast Field or Wavenumber Integration	FFP [79]	RPRESS [84]	CORE [89]	RDOASP [93]
	Kutschale FFP [80]	SAFARI [85]	OASES-3D [90]	RDOAST [94]
	MSPFFP [81]	SCOOTER [86]	RDFFP [91]	SAFRAN [95]
	OASES [82]	SPARC [87]	RD-OASES [92]	
	Pulse FFP [83]	VSTACK [88]		
Parabolic Equation	Use Single Environmental Specification		AMPE/CMPE [96]	OWWE [120]
			Cartesian 3DPE [97]	PAREQ [121]

(Continued)

TABLE 4.1 (Continued)
Summary of Underwater Acoustic Propagation Models

Technique	Range Independent	Range Dependent	
		CCUB/SPLN/CNP1 [98]	PDPE [122]
		CDPE [99]	PE [123]
		Corrected PE [100]	PECan [124]
		CRAM [101]	PE-FFRAME [125]
		DREP [102]	PEREGRINE [126]
		FDHB3D [103]	PESOGEN [127]
		FEPE [104]	PE-SSF (UMPE/ MMPE) [128]
		FEPE-CM [105]	RAM/RAMS/ RAMGEO [129]
		FEPES [106]	RAMSURF [130]
		FOR3D [107]	RMPE [131]
		FWRAM [108]	SNUPE [132]
		HAPE [109]	Spectral PE [133]
		HYPER [110]	SPUR [134]
		IFD Wide Angle [111]	TDPE [135]
		IMP3D [112]	Two-Way PE [136]
		LOGPE [113]	ULETA [137]
		MaCh1 [114]	UNIMOD [138]
		MCPE [115]	3DPE (NRL-1) [139]
		MONM3D [116]	3DPE (NRL-2) [140]
		MOREPE [117]	3D TDPA [141]
		NSPE [118]	3DWAPE [142]
		OS2IFD [119]	

Note:

RAY THEORY

Range Independent

[1] CAPARAY simulates the effects of layering in the ocean bottom on the propagation of a broadband signal. The ocean bottom is characterized by a geoacoustic profile that includes compressional and shear velocities, density and compressional wave attenuation. Velocity profiles that allow the derivation of analytic expressions for a ray's range, travel time, and attenuation are used in the ocean and in the bottom layers. A transfer function representing the total effect of the environment is constructed from the characteristics of each eigenray. The receiver time series is generated using a linear systems approach (Westwood and Vidmar, 1987).

[2] FACT combines ray theory with higher-order asymptotic corrections near caustics to generate transmission loss in range-independent ocean environments (Spofford, 1974; Baker and Spofford, 1974; Jacobs, 1982; McGirr et al., 1984; Holt, 1985; King and White, 1986a, 1986b).

[3] FLIRT provides estimates of bottom-bounce and convergence-zone transmission loss using linear ray-tracing methods (McGirr and Hall, 1974).

(Continued)

TABLE 4.1 (Continued)
Summary of Underwater Acoustic Propagation Models

[4] GAMARAY is a multi-frequency (broadband) extension to the ray-theoretical approach (Westwood, 1992).

[5] ICERAY is an extension of CAPARAY that includes layers above the ocean surface to model propagation in the ice. Three types of ray paths are included in the model: (1) compressional rays; (2) shear vertical rays in the upper layer that are converted at the water-ice interface from (or to) compressional rays in the water; and (3) a ray form of the evanescent shear or compressional field in the upper layers (Stotts and Bedford, 1991; Stotts et al., 1994).

[6] PLRAY is a ray propagation loss program applicable to a horizontally stratified ocean with a flat bottom. It contains a surface duct model based on a simplified approximation to mode theory, which replaces ray computations when the sources and receivers are located in a surface duct (Bartberger, 1978a).

[7] RANGER is a package of acoustic ray propagation programs used as research tools for bottom-interaction studies. These programs assume an ocean environment that is horizontally stratified with respect to sound speed and water depth (Foreman, 1977).

[8] SMOD, sound modeling and ranging software, uses a suite of empirical sonar equations to determine sound-propagation paths, transmission loss, and probability of detection of signals emanating from a predefined underwater sound source. SMOD was developed by The Institute for Maritime Technology, a Division of Armscor Defence Institutes (Pty) Ltd., for use by the South African Navy. Wainman (2012) utilized SMOD to test the feasibility of deriving useful sonar input parameters from the thermal characteristics of the water column in the vicinity of the Benguela Current in the southeastern Atlantic Ocean, as deduced from a single sea-surface temperature (SST) value, provided that sufficient historic temperature-depth profiles were available for that region. Applications of SMOD are discussed in more detail in Chapter 1, Section 1.2 (measurements and prediction).

Range Independent

[9] ACCURAY uses complex ray methods to find the field due to a point source in the presence of a single plane, penetrable interface (Westwood, 1989b). This method has been applied to the Pekeris (flat, isovelocity) waveguide and to a sloping, isovelocity (penetrable wedge) waveguide.

[10] BELLHOP-2D/3D computes acoustic fields in range-dependent environments (Porter, 1991, 2011) via Gaussian beam tracing (Porter and Bucker, 1987). (Also see the updated summary of SACLANTCEN models by Jensen et al. [2001]). The 2D beam-tracing algorithms in the BELLHOP acoustic model have been extended to 3D in BELLHOP3D. This new model includes virtually all of the capabilities of the 2D model including: (1) spatially varying bottom-types; (2) eigenray calculations; (3) transmission-loss calculations; and (4) time-series calculations. The 3D model now includes both the geometric and Gaussian-beam tracing options. An optional change allows a switch between 2D and 3D calculations to assess the effects of horizontal refraction and determine when horizontal refraction is important (Porter, 2015).

[11] Coherent DELTA is an extension of a ray-theoretic algorithm developed by A.L. Piskarev, which computes acoustic intensity without calculating eigenrays or focusing factors in caustics. The original algorithm has been modified to sum ray contributions coherently in a range-dependent environment (Dozier and Lallement, 1995).

[12] FACTEX extends the FACT model to range-dependent environments by adiabatic mapping of ray families over a sequence of sound-speed profiles with associated bathymetry (Garon, 1975).

(Continued)

TABLE 4.1 (Continued)
Summary of Underwater Acoustic Propagation Models

[13] FeyRay was developed to accommodate the speed, fidelity, and implementation requirements of sonar trainers and simulators. It is a broadband, range-dependent, point-to-point propagation model optimized for computational efficiency. FeyRay utilizes the Gaussian-beam approximation, which reduces the acoustic wave equation (a partial differential equation) to a more tractable system of ordinary differential equations (Howard et al., 2000; Newman et al., 2002; Collins et al., 2002; Foreman and Speicher, 2003; Collins and Scannell, 2005).

[14] GRAB computes high-frequency (10–100 kHz) transmission loss in range-dependent, shallow-water environments. The model is based on Gaussian ray bundles, which are similar in form (but somewhat simpler) than Gaussian beams (Weinberg and Keenan, 1996; Keenan, 2000; Keenan and Weinberg, 2001). The US Navy standard GRAB model (under OAML configuration management) is a subset of CASS (Keenan et al., 1998).

[15] GRASS utilizes a ray-tracing technique involving iteration along the ray path to compute transmission loss in range-dependent ocean environments (Cornyn, 1973a,b).

[16] HARORAY is a 2D broadband propagation model based on ray theory. This code was written to provide synthetic waveforms to simulate signals used in the Haro Strait coastal-ocean processes experiment (June 1996). The model uses a single point source and a vertical array of receivers. The speed of sound in the water column and bottom layers is homogeneous in depth but variable in range (Pignot and Chapman, 2001).

[17] HARPO numerically integrates Hamilton's equations in three dimensions. Transmission loss is calculated by computing 3-D ray divergence and complex reflection coefficients at the upper and lower boundaries, and by numerically integrating absorption (Jones, 1982; Jones et al., 1982, 1986; Weickmann et al., 1989; Newhall et al., 1990; Georges et al. 1990; Harlan et al., 1991a,b; Jones and Georges, 1991).

[18] HARVEST is a general hybrid technique that solves the two-dimensional acoustic-viscoelastic equations for bottom-interacting acoustics in water depths exceeding 1 km in the frequency range 100–500 Hz. The model comprises three methods: a Gaussian-beam method is used to propagate the source wave field vertically through the water column; a viscoelastic, finite-difference grid is used to compute the complex acoustic-anelastic interaction of the incident wave field with the rough sea floor; and the backscattered wave field is extrapolated to a distant receiver array using the Kirchhoff integral (Robertsson et al., 1996).

[19] LYCH calculates transmission loss on the basis of ray tracing in an environment where both sound speed and bathymetry vary as functions of range (Plotkin, 1996).

[20] MEDUSA is a ray theoretical model intended for range-dependent, azimuthally symmetric ocean environments. The model uses modified cubic splines to fit the bathymetry and sound speed profiles with smooth curves, thus eliminating false caustics and shadow zones (Foreman, 1982, 1983).

[21] MIMIC is a wave-like ray summation model that treats propagation at low frequencies (<150 Hz) and short ranges (<CZ ranges) in range-dependent ocean environments. The environment is modeled as a water column overlying a sedimentary seabed with an acoustically hard bottom (Ocean Acoustic Developments Ltd, 1999b).

[22] MPC is a real-time model for engineering applications, particularly training simulators. Surface duct and CZ caustic calculations are performed by the FACT model while the methods of the RAYWAVE model are used to transition across ocean frontal zones (Miller, 1982, 1983).

[23] MPP is a ray-theoretical model that accommodates range-dependent environments using constant-gradient sound-speed layers with a triangular profile interpolation system (Spofford, 1973a; Jacobs, 1974).

(Continued)

TABLE 4.1 (Continued)
Summary of Underwater Acoustic Propagation Models

[24] Pedersen is a ray-theoretical model that uses multiple sound-speed profiles above the sound-channel axis, flat bathymetry and bottom loss inputs to compute transmission loss as a function of range. The individual sound speed profiles are fit by segments with continuous slopes (Pedersen et al., 1962; Gordon, 1964).

[25] PlaneRay provides a unique sorting and interpolation routine for efficient determination of a large number of eigenrays in range-dependent environments. No rays are traced into the bottom since bottom interaction is modeled by plane-wave reflection coefficients. The bottom structure is modeled as a fluid sediment layer over a solid half-space. This approach balances two conflicting requirements: ray tracing is valid for high frequencies while plane-wave reflection coefficients are valid for low frequencies where the sediment layers are thin compared with the acoustic wavelength (Hovem, 2008; Hovem et al., 2008; Hovem and Dong, 2006; Hovem and Knobles, 2002, 2003). The performance the PlaneRay ray-tracing model was assessed in a range-dependent, shallow-water environment against a parabolic equation (PE)-based model being particularly suited to this type of environments (Korakas and Hovem, 2013). Comparisons were first carried out in a simple range-independent (Pekeris-like) environment against both a wavenumber integration (WI) model and the PE-based model to validate the use of each of the models. The comparisons were carried out at four frequencies ranging from 15 Hz to 100 Hz; transmission-loss curves were compared at ranges up to 20 km. Overall, a very good agreement was observed between the ray-tracing model and the two wave-theoretic models (PE and WI) despite the fact that ray-tracing models are conventionally considered to be valid at high frequencies and/or deep water environments. In particular, the transmission-loss-versus- range curves predicted by the ray-tracing model were observed to closely track the detailed variations as a function of range at frequencies as low as 50 Hz. At lower frequencies, discrepancies were observed, although the overall transmission-loss levels were fully consistent and in good agreement with the predictions by the wave theoretic models.

[26] PWRC is a ray-based model that performs geoacoustic inversions in range-dependent ocean waveguides. The pressure field is modeled approximately by separating the ocean propagation ray paths from the layered bottom interaction. The bottom interaction is included by using a full-wave description, making PWRC a hybrid model, in contrast to a full-ray theory approach that traces rays into the bottom layers. The field contribution from the bottom interactions partially includes beam-displacement effects associated with internally reflected or refracted returns from the sediment since the complex bottom reflection coefficients are obtained from a full-wave solution. This method is comparable in accuracy to normal mode and analytic solutions (in range-dependent environments) for frequencies > 100 Hz (Stotts et al., 2004).

[27] Ray5, developed by Trond Jenserud at the Forsvarets forskningsinstitutt (Norwegian Defence Research Establishment), uses direct integration in a sound-speed field specified either analytically or by interpolation from measured data. The Ray5 program is well suited for ray-tracing calculations in acoustic fields described analytically. It needs the sound-speed values, their spatial derivatives and second derivatives at all points within the field. Hence, if these values can be given analytically as functions of the oceanographic and bathymetric parameters, no interpolation is necessary and the program reduces its computing time. Further developments in Ray5 have made it possible to calculate eigenrays. This allows phase information to be retained for a given frequency so that coherent pressure values can be summed for the rays arriving at the receiver. The actual pressure values for the individual rays are calculated by assuming the pressure distribution in the direction normal to the ray to have a Gaussian behavior (i.e. Gaussian beams). It is also possible to calculate the incoherent sound levels (Kristiansen, 2010). A separate report includes the MATLAB code for Ray5 (Olsen, 2008).

(Continued)

TABLE 4.1 (Continued)
Summary of Underwater Acoustic Propagation Models

[28] RAYSON was developed by Semantic T.S. (France) to solve the Helmholtz equation using a ray-theoretic approximation for high frequencies. In a stratified (range-independent) environment, analytic solutions are obtained for ray paths that are portions of circles. In range-dependent environments, the ray equations are numerically integrated using a fourth-order Runge-Kutta method to propagate the rays along the range axis. The bottom composition can vary with range and the state of the sea surface can vary in time as well as in range. The software is coded in C++ and is available commercially (Semantic TS, 2002; Viala et al., 2004, 2005, 2006).

[29] RAYWAVE generates long-range transmission loss values in ocean environments where sound speed, bottom depth and boundary losses may vary significantly along the transmission path. RAYWAVE computes sound speed as a function of depth and range by using triangular regions between adjacent profiles. The prototype RAVE model of Bucker (1971) was the starting point for the development of RAYWAVE (Watson and McGirr, 1975).

[30] RP-70 is a long-range ray-theoretical model that allows for changing sound speed profiles and bottom depth over the propagation track. The transmission loss routines utilize an adjacent-ray approach and provide loss values based on incoherent signal addition. An interpolation routine computes intermediate sound speed points, and a linearly segmented bottom-depth profile is used. Absorption can also be made to vary along the propagation path (Colilla, 1970; Harding, 1970).

[31] SHALFACT is a special version of the FACT model designed for use in shallow-water environments by allowing for a sloping (flat) bottom (Garon, 1976). Contributions from surface-reflected and bottom-reflected paths are made using a rapidly computed analytic expression that represents the average acoustic field.

[32] TRIMAIN is a ray-theoretical model that divides the range-depth plane into triangular regions (Roberts, 1974). Interpolation of sound speed (c) within each triangle is performed by making $1/c^2$ linear in depth and range. The rays are assumed to be parabolic in each triangle.

[33] TV-APM, the time variable acoustic propagation model, simulates underwater acoustic propagation in time-variable environments where such variability induces a strong Doppler channel spread. This is an important consideration in testing and estimating the performance of equalization algorithms. Doppler spread is usually included a posteriori over a stationary acoustic propagation model (APM). The ray-trace model TRACE was selected for the channel-frequency response modeling and to account for range-dependent water column and bottom properties (Silva et al., 2010).

[34] WaveQ3D (wavefront queue 3D) is a C++ implementation of a 3D Gaussian ray bundling model based on the same latitude, longitude, altitude coordinates used in the underlying environmental databases. It incorporates an implementation of 3D refraction, 3D interface reflection, 3D eigenray detection, and a 3D variant of the Gaussian ray bundles (GRAB) model. Instead of transforming the 3D environment into a collection of N × 2D radials, WaveQ3D maintains the 3D nature of the environment by using spherical polar coordinates (r,θ,φ) to solve the acoustic eikonal equation (Reilly et al., 2016). The impulse response of the environment is modeled as a series of acoustic wavefronts that propagate away from the source as a function of time. To improve transmission loss accuracy in the neighborhood of shadow zones and caustics, this derivation includes the development of a 3D variant of the GRAB model.

[35] XRAY combines ray tracing in a range-dependent water column with local full-field modeling of interactions with a seabed composed of multiple range-dependent layers of fluid or solid materials (Svensson et al., 2004).

(*Continued*)

TABLE 4.1 (Continued)
Summary of Underwater Acoustic Propagation Models

NORMAL MODE

Range Independent

[36] AP2 is a normal mode model based on the Pekeris branch cut in the complex plane, an infinite sum of normal modes and a branch line integral. AP5 is a normal mode model similar to AP2 but with a geophysical bottom approximated by up to five sediment layers containing sound speed and attenuation gradients overlying a homogeneous semi-infinite basement layer characterized by a shear wave speed (Bartberger, 1978b).

[37] BDRM extends the WKBZ model to include beam-displacement ray-mode theory. These extensions permit calculation of pulse propagation in shallow water as well as extensions to elastic bottom boundaries (Zhang and Li, 1999; Liu et al., 2001).

[38] COMODE is a normal mode model that treats bottom attenuation exactly using complex eigenvalues (Gilbert et al., 1983).

[39] DODGE uses the effective-depth method to determine the normal modes in a shallow-water channel. Using equations developed from simple arguments, the grazing angle and phase speed of each propagating mode are easily determined (Ward, 1989). The effective-depth method (Weston, 1960) replaces the actual seabed (at depth H) with a pressure-release surface at an effective depth (He) below the surface.

[40] FNMSS is a normal mode model that computes boundary losses due to surface scattering (Gordon and Bucker, 1984). Empirical scattering tables for Arctic pack ice are optional.

[41] MODELAB is an efficient and numerically robust algorithm for calculating acoustic normal modes in a fluid-layered ocean. Each layer has a sound speed profile for which the mode functions can be expressed analytically in terms of Airy functions. Attenuation is included as a perturbation. The form of the propagator matrices avoids the numerical instabilities associated with evanescent fields (Levinson et al., 1995).

[42] NEMESIS is a low-frequency model designed to compute eigenvalues and normal modes in a horizontally stratified deep ocean with single-channel profiles and multiple fluid sediment layers overlying a substrate. Sound speeds in the water and sediment vary with depth, although the density is constant within each layer. The last layer is a homogeneous, semi-infinite fluid or solid substrate in which the compressional and shear speeds (and density) remain constant with depth (Gonzalez and Hawker, 1980; Gonzalez and Payne, 1980).

[43] NLNM is a normal mode model for ideal layered media. User assisted mode selection permits flexibility, but requires user familiarity with mode structures (Gordon, 1979). This model is an extension of the two-layer model of Pedersen and Gordon (1965).

[44] NORMOD3 is a normal mode model that uses a finite-difference algorithm (Blatstein, 1974). This model is based on the earlier model by Newman and Ingenito (1972).

[45] NORM2L calculates the discrete normal modes and acoustic propagation loss for the Pekeris model of the ocean, a simple two-layer model in which the water and seabed have constant acoustic properties (Ellis, 1980). The user interactively enters the frequency, water depth, sound speed, and density for the two layers, and the absorption coefficient for the bottom layer.

[46] ORCA uses a normal-mode method to model propagation in acousto-elastic ocean waveguides. The model assumes horizontally stratified layers in which the sound speed (c) is either constant or varies linearly in $1/c^2$. Multiple-duct environments are handled, and short-range propagation is accurately modeled by including leaky modes in the mode summation. Seismic interface

(Continued)

TABLE 4.1 (Continued)
Summary of Underwater Acoustic Propagation Models

modes such as the Scholte and Stonely modes are also computed. The model uses analytic solutions to the wave equation in each layer: Airy functions for gradient layers and exponentials for iso-speed layers (Westwood et al., 1996). (Also see the PROSIM model, which is based in part on ORCA. Tolstoy [2001] compared the performance of the ORCA, KRAKEN and PE models against a reference solution generated by the SAFARI model. All three models were highly accurate [mostly within 1 dB of the reference solution] for range-independent environments.)

[47] POPP is a range-independent version of the PROLOS normal-mode propagation model (Ellis, 1985).

[48] PROTEUS is a multi-frequency (broadband) extension to the normal-mode approach (Gragg, 1985).

[49] SHEAR2 extends the standard Pekeris waveguide model (homogeneous layer of fluid overlying an infinite homogeneous fluid half-space of greater sound speed) to handle the case of a fluid overlying an elastic basement in which the shear speed is less than the compressional speed of sound in the fluid. This gives rise to leaky modes in which both the mode eigenfunctions and eigenvalues are complex (Ellis and Chapman, 1985).

[50] Stickler is a normal mode model that computes the continuous as well as the discrete modal contributions (Stickler, 1975; Stickler and Ammicht, 1980; Ammicht and Stickler, 1984). The determination of the continuous spectrum to the transmission loss is performed using a uniform asymptotic method that avoids costly numerical evaluation. However, the improper eigenvalues must be found.

NORMAL MODE

Range Dependent

[51] ADIAB is an adiabatic normal mode program that conserves energy using a consistent boundary-condition approximation (R. A. Koch, unpublished manuscript). Milder's (1969) criterion is used to test the amount of coupling per mode-cycle distance.

[52] ASERT uses the ASTRAL model, together with specialized databases, to generate multi-radial portraitures of ocean acoustic transmission loss (Lukas et al., 1980a).

[53] ASTRAL assumes adiabatic invariance in propagating mode-like envelopes through a range-dependent environment (Spofford, 1979; Blumen and Spofford, 1979; White et al., 1988; Dozier and White, 1988; White, 1992; White and Corley, 1992a,b).

[54] CENTRO is an adiabatic normal mode model that includes absorptive effects and elastic waves in the ocean floor (Arvelo and Überall, 1990). The source is located in the water column, while the receiver may be located either in the water or in the elastic medium.

[55] CMM3D is a three-dimensional, coupled-mode model (Chiu and Ehret, 1990, 1994) that is based on the earlier work of Pierce (1965). This model has been interfaced with data generated by an ocean circulation model to examine three-dimensional environmental effects on sound transmission through intense ocean frontal systems.

[56] COUPLE utilizes a stepwise-coupled-mode method that overcomes the failure of previous coupled mode techniques to properly conserve energy over sloping bottoms. This method of stepwise-coupled modes avoids problems associated with sloping bottoms by using only horizontal and vertical interfaces. The full solution includes both forward and backscattered energy (Evans and Gilbert, 1985; Evans, 1983, 1986).

(Continued)

TABLE 4.1 (Continued)
Summary of Underwater Acoustic Propagation Models

[57] CPMS combines a conventional coupled-mode solution with perturbation theory to provide fast and accurate solutions of transmission loss in range-dependent ocean environments. The efficiency of the coupled-mode solution is improved by removing the need to solve the depth-separated wave equation for the normal modes at each range step. Specifically, the normal modes are found by applying perturbation theory to the modes of the previous step (Tindle et al., 2000).

[58] FELMODE is a standard normal mode model that considers an ocean environment consisting of three layers: water column, sediment layer and homogeneous half-space. Density and attenuation in all layers are independent of depth, although attenuation is a function of frequency. The sound speed in the water column and sediment varies with depth while it remains constant in the half-space. Finite-difference discretization is used to solve the modal equation and its boundary conditions. Only discrete modes are calculated, omitting the continuous spectrum. Volume-attenuation losses in the water, sediment and half-space are incorporated by first-order perturbation theory. Shear in the bottom layers is ignored. Range-dependent ocean environments are handled by using the adiabatic approximation. The model was developed in MATLAB with a user-friendly GUI (Simons and Laterveer, 1995; Simons and Snellen, 1998).

[59] IECM is a two-way coupled-mode formalism that provides an exact solution to the wave equation (Stotts et al., 2011). This model was used to establish a benchmark solution that is an exact numerical solution for the reverberation time series for an environment with a range-dependent, fine-scale rough bottom boundary that induces mode coupling and generates a scattered field. The solution includes scattering effects to all orders in that it sums the infinite series of forward and backward contributions at each range point and maintains energy conservation.

[60] Kanabis is a normal mode model appropriate for low-frequency, long-range propagation in shallow-water regions (Kanabis, 1975, 1976).

[61] KRAKEN is a normal mode model that handles range dependence through either adiabatic coupling or full forward mode coupling between environmental provinces (Porter and Reiss, 1984, 1985; Porter, 1991). KRAKEN is recommended for more experienced modelers or for those requiring a 3D capability. (The original KRAKEN algorithm was incorporated into the SNAP model, whereupon SNAP was renamed SUPERSNAP. Since 1984, SUPERSNAP became a standard model at SACLANTCEN and is now simply referred to as SNAP [Porter, 1991]. KRAKENC is a special version of KRAKEN that finds eigenvalues in the complex plane and includes material attenuation in elastic media such as ice. Also see the updated summary of SACLANTCEN models by Jensen et al. [2001].)

[62] MOATL computes transmission loss on the basis of the discrete normal modes that propagate in the water column. The sound speed profile in the water is arbitrary, and the ocean bottom consists of a sediment layer with arbitrary sound speed profile and an underlying uniform fluid or solid subbottom (Miller and Ingenito, 1975; Ingenito et al., 1978; Miller and Wolf, 1980).

[63] MOCTESUMA is a coupled normal mode model developed by Thomson Sintra—Activités Sous Marines, France (Dr. Alain Plaisant). There are two versions: one for 2D environments with fluid/elastic sediments and one for 3D environments with fluid sediments (Noutary and Plaisant, 1996; Papadakis et al., 1998).

[64] NAUTILUS is a broadband, range-dependent normal-mode model based on the adiabatic formulation. This model is designed for use in shallow water and is currently under review by OAML for designation as a US Navy configuration-managed model. The mode-generating engine is derived from the MODELAB model (Koch and LeMond 2001a,b,c; Koch and Knobles, 1995).

(Continued)

TABLE 4.1 (Continued)
Summary of Underwater Acoustic Propagation Models

[65] PROLOS is a normal-mode propagation loss program that uses the adiabatic approximation to model an environment in which the sound speed profile, water depth and bottom properties vary slowly with range. PROLOS can handle up to ten layers in the bottom and used up to 500 layers to interpolate the sound speed profile in the water column. The normal mode equations are solved using a two-ended shooting technique borrowed from quantum mechanics and generalized to include the effects of density. Shooting from both ends results in a more stable algorithm and gives more accurate eigenfunctions than those obtained from conventional one-ended shooting methods (Ellis, 1985).

[66] PROSIM is a broadband adiabatic normal-mode propagation model, the kernel of which is based on the range-independent normal-mode propagation model called ORCA. Using PROSIM, the calculation of broadband transfer functions at frequencies up to 10 kHz in shallow water is attainable in a few minutes on a modern workstation (Bini-Verona et al., 2000). (The PROSIM project was partly funded by the European Commission through the Marine Science and Technology programme [MAST3]. Recent applications of the PROSIM model were reported by Siderius et al. [2001] and by Simons et al. [2001]. Also see the updated summary of SACLANTCEN models by Jensen et al. [2001].)

[67] SHAZAM is a normal mode model with both adiabatic and coupled-mode solutions for application to shallow-water environments (Navy Modeling and Simulation Management Office, 1999).

[68] SNAP is a normal mode model based on a computer program originally developed at the US Naval Research Laboratory (Miller and Ingenito, 1975; Ingenito et al., 1978). Solutions are found only for the discrete part of the spectrum corresponding to the propagating modes. Losses are introduced in a perturbational manner. Slight range dependence is handled in the adiabatic approximation (Jensen and Ferla, 1979). C-SNAP is a coupled-mode version of SNAP (Ferla et al., 1993). The numerical solution technique for one-way mode coupling was obtained from KRAKEN. When the original KRAKEN algorithm was incorporated into the SNAP model, SNAP was renamed SUPERSNAP. Since 1984, SUPERSNAP became a standard model at SACLANTCEN and is now simply referred to as SNAP (Porter, 1991). (Also see the updated summary of SACLANTCEN models by Jensen et al. [2001].) Diachok and Wales (2005) studied bio-acoustic absorptivity using the C-SNAP coupled normal-mode model modified to simulate the effects of bio-acoustic absorption layers due to fish with swim bladders and the effects of the sea floor on transmission loss. This special-purpose version of C-SNAP was referred to as BIO-C-SNAP. The absorbing layer was characterized by a horizontally uniform bio-absorption coefficient (α_B), a mid-layer depth and a layer thickness.

[69] SWAMP is a range-dependent normal-mode model that contains closed-analytical forms of the vertical mode functions, which facilitate computation of one-way mode-coupling coefficients between adjacent range-independent regions by neglecting weak backscattering components (Sidorovskaia, 2003, 2004). The model was used to understand the physics of pulse propagation in double-ducted shallow-water environments where precursors have been observed. The pulse temporal response is modeled using Fourier synthesis in the frequency domain. The model accounts for scattering events along the acoustic signal propagation path and has been extended to model acoustic pulse scattering by spherical elastic-shell targets in inhomogeneous waveguides within the T-matrix approach (Tamendarov and Sidorovskaia, 2004).

(*Continued*)

TABLE 4.1 (Continued)
Summary of Underwater Acoustic Propagation Models

[70] WEDGE calculates downslope propagation using a two-dimensional representation of the continental slope and realistic environmental parameters (Primack and Gilbert, 1991). Numerical implementation is divided into two parts: (1) pre-calculation of local modes, and (2) mode manipulation. This wedge-mode model is limited to geometries where the propagation can accurately be predicted using only trapped modes. Wedge modes are the intrinsic normal modes in the wedge coordinate system.

[71] WKBZ is an adiabatic normal-mode model based on a uniform WKB approximation to the modes (Zhang et al., 1995).

[72] WRAP uses adiabatic and coupled-mode theory for rapid computation of acoustic fields in three-dimensional ocean environments (Perkins et al., 1990; Kuperman et al., 1991).

[73] 3D Ocean expresses the solution of the reduced wave equation for an almost stratified medium in the form of an asymptotic power series. The vertical structure of the solution is expressed as a linear combination of the normal-mode eigenfunctions whose coefficients satisfy two-dimensional eikonal and transport equations (Weinberg and Burridge, 1974).

MULTIPATH EXPANSION

Range Independent

[74] FAME is based on the multipath expansion for acoustic propagation in a horizontally stratified ocean. The multipaths are expressed in terms of Fresnel integrals and effective range derivatives (Weinberg, 1981).

[75] MULE is a multi-frequency (broadband) extension to the multipath expansion method (Weinberg, 1985a).

[76] NEPBR modifies the RAYMODE model to account for frequency smearing in the narrowband spectrum (McCammon and Crowder, 1981).

[77] RAYMODE is a hybrid technique combining elements of ray theory and normal mode theory to compute transmission loss in range-independent ocean environments (Leibiger, 1968; Yarger, 1976, 1982; Medeiros, 1982a, 1985a; Almeida and Medeiros, 1985; Naval Oceanographic Office, 1991a,c). (The RAYMODE model is discussed in detail in section 4.8.)

MULTIPATH EXPANSION

Range Dependent

[78] Integrated Mode extends the multipath expansion method to range-dependent environments (Clark, 2005). This approach accounts for horizontal variations in bottom depth, bottom type and sound speed using the stationary phase approximation.

FAST FIELD or WAVENUMBER INTEGRATION

Range Independent

[79] FFP applies fast Fourier transform methods to acoustic field theory to obtain transmission loss as a function of range (DiNapoli, 1971; DiNapoli and Deavenport, 1980).

[80] Kutschale FFP is a rapid and accurate method for calculating transmission loss in the ice-covered Arctic Ocean. Input parameters include source and detector depth, frequency, ice roughness bottom topography, and the sound velocity structure as a function of depth in the ice, water and bottom. Computation is performed by direct integration of the wave equation derived from a harmonic point source located in a multilayered, interbedded liquid-solid half-space. The integration method introduced by H.W. Marsh employs the FFT for rapid evaluation of the integral solution (Kutschale, 1973).

(Continued)

TABLE 4.1 (Continued)
Summary of Underwater Acoustic Propagation Models

[81] MSPFFP, or multiple scattering pulse FFP (Kutschale, 1984), models bottom-interacting pulses corresponding to the coherent summation of many modes over a limited time interval. Specifically, MSPFFP decomposes the Kutschale (or pulse) FFP model (Kutschale, 1973; Kutschale and DiNapoli, 1977) into ray-path contributions. Each decomposed term can then be interpreted as the desired path contribution for a corresponding bottom-interacting pulse. The FFP algorithm directly integrates the full-wave solution. Temporal waveforms are computed by Fourier synthesis.

[82] OASES is a general-purpose computer code for modeling seismo-acoustic propagation in horizontally stratified waveguides using wavenumber integration in combination with the direct global matrix solution technique (Schmidt, 1999). The OASES model is essentially an upgraded version of SAFARI. Compared to SAFARI (version 3.0), OASES provides improved numerical efficiency, and the global matrix mapping has been redefined to ensure unconditional numerical stability. OASES is downward compatible with SAFARI and the preparation of input files follows the format used by SAFARI. OASES (Schmidt, 2004) supports all environmental models available in the earlier SAFARI model including any number and combination of isovelocity fluids, fluids with sound-speed gradients and isotropic elastic media. Any number of transversely isotropic layers may be specified. Other media with general dispersion characteristics can also be included. OASES (Version 3.1) is an umbrella for related modules. OASES uses graphics post-processors compatible with other acoustic propagation models employing the MINDIS graphics library. (Also see the updated summary of SACLANTCEN models by Jensen et al. [2001].)

[83] Pulse FFP is a time-marched fast-field program for modeling acoustic pulse propagation in the ocean (Porter, 1990).

[84] RPRESS uses a high-order, adaptive integration method for efficient computation of the Hankel-transform integral for the wave field in a laterally homogeneous fluid-solid medium (Ivansson and Karasalo, 1992). This model has been used to investigate frequency-dependent propagation losses in shallow water caused by shear losses in the sediment (Ivansson, 1994).

[85] SAFARI is based on the wavenumber integration technique (or fast field program) to solve the wave equation exactly for strictly range-independent environments (Schmidt and Glattetre, 1985; Schmidt, 1988). The model includes propagation of waves in fluid and solid media by discretizing the environment into layers with different acoustic properties. The Green's function is evaluated analytically within each layer and the boundary conditions are fulfilled at each layer interface by using a numerically stable direct-global-matrix approach. SAFARI consists of three independent models: FIPR to compute the plane-wave complex reflection coefficient due to an arbitrarily layered medium; FIP for general single frequency wave propagation; and FIPP for pulse propagation. An updated version of SAFARI called OASES (discussed above) is available on the web. (Also see the updated summary of SACLANTCEN models by Jensen et al. [2001].)

[86] SCOOTER is a finite-element FFP code for computing acoustic fields in range-independent environments. It is recommended for use when the horizontal range is less than ten water depths (Porter, 1990).

[87] SPARC is a time-marched FFP model that treats problems dealing with broadband or transient sources (i.e., pulses) (Porter, 1990, 1991).

[88] VSTACK (near-field wavenumber integration model) computes synthetic pressure waveforms versus depth and range for arbitrarily layered acoustic environments using the wavenumber integration approach to solve the exact (range-independent) acoustic wave equation. This model is

(Continued)

TABLE 4.1 (Continued)
Summary of Underwater Acoustic Propagation Models

valid over the full angular range of the wave equation and can fully account for the elasto-acoustic properties of the sub-bottom. VSTACK computes sound propagation in arbitrarily stratified water and seabed layers by decomposing the outgoing field into a continuum of outward-propagating plane cylindrical waves. Seabed reflectivity in VSTACK is dependent on the seabed layer properties: compressional and shear wave speeds, attenuation coefficients, and layer densities. Fundamental to the modeling is the physical constraint that the instantaneous displacements of the seabed and the water are equivalent at the interface between the two media. VSTACK computes pressure waveforms via Fourier synthesis of the acoustic transfer function in closely spaced (<1 Hz) frequency bands. In addition, VSTACK includes the ability to model distributed monopole sound sources in the water column and in the sub-bottom. VSTACK assumes range-invariant bathymetry with a horizontally stratified medium that is azimuthally symmetric about the source. Although the range-independent restriction assumes that the layering of the environment is invariant with range, it does not assume that the vibration field is invariant with range: the vibration field exhibits very complex variations with distance from the source due to interference between the multiple acoustic paths corresponding to reflection and transmission of vibration from the different layers in the model (which is accounted for in VSTACK calculations). For pile-driving scenarios, VSTACK fully accounts for vibration of the seabed interface excited by the pile and its coupling to the water column. VSTACK is thus best suited to modelling the sound field in close proximity to the pile. A vertically distributed array of sources is used to calculate both pressure and particle velocity in the nearfield region of a pile (MacGillivray et al., 2011).

FAST FIELD or WAVENUMBER INTEGRATION

Range Independent

[89] CORE is a coupled version of OASES for range-dependent environments. It belongs to the new spectral super-element class of propagation models for range-dependent waveguides. This approach is a hybridization of the finite-element and boundary-element methods. The ocean environment is divided into a series of range-independent sectors separated by vertical interfaces (Goh et al., 1997).

[90] OASES-3D target modeling framework is used to investigate scattering mechanisms of flush buried spherical shells under evanescent insonification (Schmidt et al., 1998, 2004; Schmidt, 2001, 2004).

[91] RDFFP utilizes a generalized Green's function method for solving the one-way wave equation exactly in an ocean environment that varies discretely with range. An explicit marching solution is obtained in which the source distribution at any given range step is represented by the acoustic field at the end of the previous step (Gilbert and Evans, 1986).

[92] RD-OASES is a range-dependent version of OASES (Schmidt, 1999). RD-OASES extends to fluid-elastic stratifications the development of a spectral super-element approach for acoustic modeling in fluid waveguides using a hybridization of finite elements, boundary integrals and wavenumber integration to solve the Helmholtz equation in a range-dependent ocean environment. The ocean environment is divided into a series of range-independent sectors separated by vertical interfaces. This model provides accurate, full two-way solutions to the wave equation using either a global multiple scattering solution or a single-scatter, marching solution.

[93] RDOASP is a pulse version of RD-OASES (Goh and Schmidt, 1996). RDOASP is the range-dependent version of OASP, which calculates the depth-dependent Green's function for a selected number of frequencies and determines the transfer function at any receiver position by evaluating the wavenumber integral. The frequency integral is evaluated in the post-processor. Stresses and particle velocities can be determined. The field may be produced by either point or line sources. Pulsed beam propagation can be analyzed by arranging the sources in a vertically phased array.

(Continued)

TABLE 4.1 (Continued)
Summary of Underwater Acoustic Propagation Models

[94] RDOAST refers to the specific combination of RD-OASES and VISA (Goh and Schmidt, 1996). The virtual source algorithm (VISA) uses the marching, local single-scatter approximation to the transmission and reflection problem at the sector boundaries; thus, a virtual array of sources and receivers is introduced on each sector boundary. RDOAST (OASES range-dependent transmission loss module) is a range-dependent version of OAST (OASES transmission loss module).

[95] SAFRAN uses a hybrid combination of wavenumber integration and Galerkin boundary element methods to extend the fast field theory technique to range-dependent ocean environments (Seong, 1990).

PARABOLIC EQUATION

Range Independent

Use single environmental specification with range-dependent model.

Range Dependent

[96] AMPE was developed to solve global-scale ocean acoustic problems that are too large to solve with other existing three-dimensional codes. The PE method was used to solve two-dimensional wave equations for the adiabatic mode coefficients over latitude and longitude (Collins, 1993d). CMPE is a generalization of AMPE that includes mode-coupling terms. It is practical to apply this approach to large-scale problems involving coupling of energy between both modes and azimuths (Abawi et al., 1997).

[97] Cartesian 3DPE parabolic equation program implements a split-step Fourier algorithm with a wide-angle PE approximation, and is thus a 3D variant of the PE model of Thomson and Chapman (1983). The advantage of employing Cartesian coordinates in the numerical scheme is that the model resolution is uniform over the computational domain (Duda, 2006b; Chiu et al., 2011).

[98] CCUB/SPLN/CNP1 represent a family of higher-order, finite-element (FE) PE methods developed by the Foundation for Research and Technology—Hellas, Institute of Applied and Computational Mathematics, Greece (Papadakis et al., 1998).

[99] CDPE (complex density PE) model was developed to predict acoustic transmission losses in the shallow-water environment off Sakhalin Island (Russia) in the Sea of Okhotsk. This specialized PE code was based on the RAM model (Collins, 1993a); RAM was not used directly since it did not account for shear-wave losses at the seabed. Collins also developed a modified PE code (RAMS) that treated shear waves in a robust sense; however, that model was excessively slow because it used hepta-diagonal matrices instead of the standard tri-diagonal operator matrices in RAM. Instead, a shear-wave approach was implemented that maintained the standard tri-diagonal matrix inversion scheme and used a complex density method (Zhang and Tindle, 1995). The CDPE approach was more than five times faster than the reference hepta-diagonal matrix approach and has been shown to produce results that are nearly identical to the reference approach for uniform low-shear-speed, shallow-water environments with silt and sand bottoms. This method applied a complex multiplicative factor to the seabed density; the factor was dependent on the shear-wave speed and the shear-wave attenuation coefficient parameters. CDPE has been tested extensively by direct comparisons of its transmission loss predictions with those produced by RAM and RAMS. When the shear speed was set to zero in CDPE, its transmission loss predictions were identical to those from RAM for all frequencies. For frequencies below 400 Hz, the match between CDPE and

(*Continued*)

TABLE 4.1 (Continued)
Summary of Underwater Acoustic Propagation Models

	RAMS was near perfect, even for shear speeds as high as 600 m s^{-1}. Less than 0.5-dB differences between RAMS and CDPE model results were observed at any range along a 10-km track in a shallow upslope environment starting at 30-m depth and ending at 10-m depth. At frequencies above 400 Hz, the match in TL amplitude remained good, but slight phase errors caused small range mismatches in the locations of nulls. These phase errors did not introduce significant problems for amplitude estimates because the model results were summed over many frequency bands, thereby averaging out the influence of individual nulls (Hannay and Racca, 2005).
[100]	Corrected PE solves the parabolic equation using the split-step algorithm. (Perkins and Baer, 1978; Perkins et al., 1982). By assuming cylindrical propagation symmetry, it is possible to simulate a tilted acoustic array placed in the calculated complex-valued acoustic field.
[101]	CRAM is based entirely on NSPE2.0 (a descendant of RAM developed by Collins, 1993a) coded from scratch in ANSI-standard C. The code handles all environmental (via net CDF) and source/receiver geometry input and permits use of OpenMP to optimize memory and multi-processor threading. Multiple sources-to-receivers can be run in the execution of a single command. Setup of the N × 2D propagation problem is handled automatically for desired receiver output grids in geographic coordinates. The N × 2D PE model has been extended to basin-scale problems. The assumptions inherent in the N × 2D approximation, versus full three-dimensional (3D) propagation modeling, are that horizontal refraction and out-of-plane bathymetric scattering can be neglected in the environment of interest, so that adjacent radials can be computed independently without coupling. The set of independent radials, and the range-marching within each radial, are selected such that the complex pressure for each source-receiver pair is phase-exact in the along-range direction, and approximated in the cross-range direction. This preservation of spatial coherence allows for beamforming and other post-processing operations that require high fidelity of the complex pressure output. As much of the 2D PE grid setup as possible was reused over multiple frequencies, thus allowing for a more rapid computation of broadband and time-domain pressure responses. To leverage the multiprocessor capability of modern computers, the program was parallelized over the N-independent radials as well as more limited parallelization over frequency and Padé coefficient index, without causing changes to the output. Environmental inputs are interpolated from a variety of four-dimensional (3D space plus time) ocean models and bathymetry databases as they are needed in the calculations. The model can use standard geoacoustic profiles that are range- as well as depth-dependent, but its ability to take a scalar mean grain size (φ), available from sediment cores or even from the sediment type read off a navigation chart, and convert this information into geoacoustic profiles using Hamilton's (1980) relations greatly facilitates the problem setup. Additionally, the model can output a variety of file formats including Keyhole Markup Language (an XML notation for expressing geographic annotation and visualization) format that can be imported directly into popular viewers. The RAM core of the CRAM model is based on an estimate of a solution to the acoustic wave equation, and therefore is not exact. The model does not incorporate the shear properties of the bottom, which could influence the accuracy of the model, especially with higher density bottom types. The model does account for acoustic backscatter. The CRAM acoustic propagation code was written by Richard Campbell and Kevin Heaney of OASIS, Inc., using Mike Collins' RAM program as the starting point (Helble et al., 2013).
[102]	DREP is a wide-angle parabolic equation based on an operator-splitting that permits the use of a marching type Fourier transform solution method (Thomson and Chapman, 1983; Thomson, 1990).

(Continued)

TABLE 4.1 (Continued)
Summary of Underwater Acoustic Propagation Models

[103] FDHB3D is a hybrid 3D, two-way propagation model for solving 3D backscattering problems. It is based on the implicit finite difference (IFD) parabolic equation (PE) approach (Zhu and Bjørnø, 1999).

[104] FEPE computes the complex acoustic pressure in range-dependent ocean environments. The model solves higher-order parabolic equations using a numerical solution based on Galerkin's method, alternating directions and Crank-Nicolson integration (Collins, 1988a). FEMODE is a module in FEPE that generates eigenvalues and modes. The model allows arbitrary sound-speed, density and attenuation profiles. FEPE provides several options for the initial starting field including the Gaussian PE starter (Tappert, 1977) and Greene's (1984) PE starter.

[105] FEPE-CM combines the FEPE code with the PERUSE surface scattering formulation to model the forward scattering from both periodic and single realizations of randomly rough sea surfaces. A conformal mapping technique converts the rough-surface scattering problem into a succession of locally flat-surface problems (Norton et al., 1995). Norton and Novarini (1996) used FEFE-CM to investigate the effect of sea-surface roughness on shallow-water waveguide propagation.

[106] FEPES extends the FEPE model to handle range-dependent elastic media (Collins, 1993b,c). This energy-conserving, elastic PE model is applicable to a large class of problems involving waveguides consisting of both fluid and solid layers.

[107] FOR3D implements the Lee-Saad-Schultz (LSS) method for solving the three-dimensional, wide-angle wave equation (Botseas et al., 1987; Lee et al., 1988, 1992). The model is designed to predict transmission loss in ocean environments that vary as functions of range, depth and azimuth.

[108] FWRAM (far-field waveform synthesis model) computes synthetic pressure waveforms versus range and depth for range-varying marine acoustic environments using the parabolic equation approach to solving the acoustic wave equation. This software uses the same underlying algorithmic engine as MONM (noise model) for computing acoustic propagation along 2-D range-depth transects, and takes the same environmental inputs (bathymetry, water sound speed profiles, and seabed geoacoustics). FWRAM computes pressure waveforms via Fourier synthesis of the modeled acoustic-transfer function in closely spaced frequency bands. Like MONM, FWRAM accounts for range-varying properties of the acoustic environment and is therefore capable of computing rms sound-pressure levels (SPL) at long ranges (outside the nearfield zone). Since FWRAM is a time-domain model, it is well suited to computing time-averaged rms SPL values for impulsive sources. For pile-driving scenarios, rms SPLs from marine pile driving can be computed by modeling far-field pressure waveforms along single-range depth transects (MacGillivray et al., 2011).

[109] HAPE implements the rational-linear approximation in a finite-difference formulation to provide accurate treatment of high-angle propagation to angles of about 40° with respect to the horizontal (Greene, 1984). A rational approximation to the square-root operator was first suggested by Claerbout (1976, p. 206). This approximation reduced the phase error problem inherent in the original, narrow-angle PE models that had been limited to about 15°. (Note: more recent, higher-order PE implementations have virtually eliminated the phase error problem and are accurate for propagation angles of nearly 90°.)

[110] HYPER is a hybrid parabolic equation-ray model intended for high-frequency applications (Tappert et al., 1984).

(Continued)

TABLE 4.1 (Continued)
Summary of Underwater Acoustic Propagation Models

[111] IFD Wide Angle implements an implicit finite-difference scheme for solution of the parabolic equation (PE) (Lee and Botseas, 1982; Botseas et al., 1983). This model can handle arbitrary surface boundary conditions and an irregular bottom with arbitrary bottom boundary conditions.

[112] IMP3D extends the FOR3D model by including a simplified elastic impedance bottom boundary condition (Papadakis et al., 1998).

[113] LOGPE uses a logarithmic expression for the index of refraction in the standard parabolic equation in order to closely approximate solutions of the more exact Helmholtz equation in weakly range-dependent ocean environments (Berman et al., 1989).

[114] MaCh1 is a broadband, range-dependent acoustic propagation model based on the one-way form of the Maslov integral representation of the wave field. The one-way propagation assumption may be combined with a parabolic approximation (Brown, 1994).

[115] Monte Carlo parabolic equation (MCPE) is a method of modeling acoustic fluctuations that are induced by variability in the ocean (White et al., 2013). The parabolic equation contains the relevant physics for propagation; the MCPE method provides an evaluation of the simulated propagation environment, which is composed of a background sound-speed plus perturbations. The MCPE method is commonly used in the ocean acoustics community to model acoustic signal fluctuations in 2D. The idea is to generate random instances of an inhomogeneous sound-speed field that consists of some range-independent or slowly varying background plus perturbations to this background. The NSPE was chosen because it provided test cases and was actively maintained by the US Navy. The code includes an implementation of a split-step Fourier algorithm (Tappert, 1977), and an implementation of a split-step Padé algorithm derived from the range-dependent acoustic model RAM (Collins, 1993a); MCPE utilized the split-step Padé version. The NSPE is capable of propagation through a range-dependent sound-speed field such as that of an internal-wave perturbed ocean model environment. The angle between the horizontal and the direction of acoustic paths leaving the source was well within the limits for this PE. This method was further employed by Andrew et al. (2015, 2016).

[116] MONM3D incorporates techniques that reduce the required number of model grid points. The concept of tessellation (i.e. covering the plane with a pattern in such a way as to leave no region uncovered) is used to optimize the radial grid density as a function of range, reducing the required number of grid points in the horizontal planes of the grid. The model marches the solution out in range along several radial propagation paths emanating from a source position. Tessellation, as implemented in MONM3D, allows the number of radial paths in the model grid to depend on range from the source. In addition, the model incorporates a higher-order azimuthal operator, which allows a greater radial separation and reduces the required number of radial propagation paths (Austin and Chapman, 2009, 2011).

[117] MOREPE is an energy-conserving parabolic equation that can handle relatively large variations in the index of refraction (Schurman et al., 1991).

[118] NSPE, the Navy Standard PE model, consists of two methods of solving the acoustic parabolic wave equation: split-step Fourier parabolic equation model (SSFPE); and split-step Padé (finite-element) parabolic equation (SSPPE) known as RAM.

[119] OS2IFD is a full-featured version of IFD parabolic equation that runs on personal computers (PCs) under the OS/2 operating system (Robertson et al, 1991).

[120] OWWE (Mikhin, 2004) is based on the innovative one-way wave equation developed by Godin (1999). This equation was generalized by Godin to include the source terms and also to

(Continued)

TABLE 4.1 (Continued)
Summary of Underwater Acoustic Propagation Models

account for motion of the medium. The solutions of the differential OWWE are strictly energy conserving and reciprocal. The derivation presented for the multiterm Padé PE model is applicable to a broad class of finite-difference PE models.

[121] PAREQ is a parabolic equation model based on the split-step algorithm (Jensen and Krol, 1975). This model is a modified version of the one developed by AESD (Brock, 1978). (Also see the updated summary of SACLANTCEN models by Jensen et al. [2001].)

[122] PDPE is a pseudodifferential parabolic equation model. The documentation contains a numerical algorithm for its implementation (Avilov, 1995).

[123] PE is based on the original algorithm developed by Hardin and Tappert (1973) and Tappert (1977). The original US Navy version was implemented by Brock (1978). Upgrades have been reported by Holmes (1988) and by Holmes and Gainey (1992a,b,c). A version of this model can accommodate ice-covered regions. (See NSPE for additional updates to this model.)

[124] PECan is an N × 2D/3D parabolic equation model for underwater sound propagation that was developed for matched-field processing applications (Brooke et al., 2001).

[125] PE-FFRAME applies the finite-element method to produce a full-wave, range-dependent, scalar, ocean acoustic propagation model (Chin-Bing and Murphy, 1988; Murphy and Chin-Bing, 1988). The term 'full wave' includes the continuous and discrete spectra, forward propagation and backscatter. The model can be given a long-range capability through super elements or marching frames.

[126] PEREGRINE (Heaney and Campbell, 2013) is a recoding in C of the split-step Padé PE model RAM (Collins, 1993a). This version permits the inclusion of arbitrary range-dependent sediments and surface layers. It also includes an ocean-wave induced sea surface following the methodology of Rosenberg (1999). PEREGRINE can place a realization of an ice layer (with the spatial resolution of the PE step-size) on top of a range-dependent ocean, bathymetry, and arbitrary sediments. The ice thickness is a realization of a random draw from an input probability density function of ice thickness. The high-angle PE model accurately captures forward scattering effects due to ice roughness on very short spatial scales. The ice layer is characterized by its thickness, compressional speed, density, and attenuation. The PEREGRINE PE model is a fluid-fluid model and is not expected it to capture the shear loss or the re-conversion of shear energy back into compressional energy in the water column. To approximate the loss due to shear conversion, an effective ice-parameter model is used that attenuates energy via volume attenuation. In later work, Heaney and Campbell (2016) further noted that Peregrine is a tightly integrated interface to the Seahawk split-step Padé PE marcher, based on RAM (Collins, 1993a), which is able to interpolate directly from geographically defined temperature-salinity-bathymetry inputs, includes an optional 3D azimuthal coupling operator, integrated time-domain synthesis, range and depth antialiasing, non-spheroidal Earth corrections, volume attenuation, and one- or two-parameter sediment specification (grain size and thickness), among other improvements. Both Seahawk and RAM are 2D (r and z) propagation codes. Peregrine automatically uses all available central processing units for parallel loops and task pipelining. The integrated nature of Peregrine thus allows larger problems to be tackled including 3D global scale propagation, full azimuthal dependency of tactical-scale problems, and higher frequency full-field acoustics, than might otherwise be practical on typical computing configurations.

[127] PESOGEN is a special-purpose microcomputer model used to compute the acoustic field in a fully range-dependent environment over a wide range of frequencies for both deep-water and shallow-water regions with greatly reduced execution times (Nghiem-Phu et al., 1984).

(Continued)

TABLE 4.1 (Continued)
Summary of Underwater Acoustic Propagation Models

[128] PE-SSF (UMPE/MMPE)—MMPE and its predecessors, UMPE (Smith and Tappert, 1993) and MIPE, are now collectively referred to as PE-SSF (Parabolic Equation—Split-Step Fourier algorithm). PE-SSF thus represents a wide class of PE models (Tappert, 1998).

[129] RAM/RAMS/RAMGEO—RAM incorporates an improved elastic parabolic equation (Collins, 1998) together with a stable self starter (Cederberg and Collins, 1997). A companion version called RAMS is available for acousto-elastic problems. Both RAM and RAMS use the split-step Padé solution, which is approximately two orders of magnitude faster than the Crank-Nicolson solution of the wide-angle PE (Collins, 1993a). RAMGEO is a version of RAM modified to handle sediment layers that are range dependent and parallel to the bathymetry (Fialkowski et al., 2003).

[130] RAMSURF is RAM recoded in C and vectorized for speed. RAMSURF allows for a rough sea surface. This model is available from the OALIB (Ocean Acoustics Library), as described in Appendix C. RAMSURF is similar to RAMGEO, but also inputs a file specifying the height of the top boundary of the water column as a function of range (e.g. Eggen et al., 2002; Duncan and Maggi, 2006).

[131] RMPE is a ray-mode parabolic-equation solution that is expressed in terms of normal modes in the vertical direction and mode coefficients in the horizontal direction. The model is based on the beam-displacement ray-mode (BDRM) theory and the parabolic equation (PE) method. The BDRM theory is used to analyze the local normal modes. The PE method is used to solve the wave equations for mode coefficients (Zhang et al., 2009).

[132] NUPE achieves linearization of the depth-direction operator by expansion into a multiplicative Padé approximation. Galerkin's method is used for computational efficiency to approximate the depth-direction equation. Crank-Nicolson's method is used to approximate the range-direction equation. A numerical self-starter initiates the near-field solution (Seong and Choi, 2001).

[133] Spectral PE extends the two-way parabolic equation to three-dimensional problems involving a point source in a waveguide in which the acoustic properties vary with depth and range (Orris and Collins, 1994). This model presents an efficient approach for solving backscattering problems that are separable in one of the horizontal Cartesian coordinates, including backscattering from extended features such as ridges.

[134] The synthetic pulse reception (SPUR) model is an improved version of the range-dependent acoustic model (RAM), which is based on the split-step Padé solution (Kelly, 2002).

[135] TDPE is the inverse Fourier transform of the wide-angle parabolic equation (Collins, 1988b).

[136] Two-Way PE generalizes the parabolic equation to handle backscattered acoustic energy in the ocean (Collins and Evans, 1992). The two-way PE is based on the single-scattering approximation and the approach of the two-way coupled modes in which range-dependent environments are approximated by a sequence of range-independent regions. The outgoing and incoming fields are propagated by two-way range marching.

[137] ULETA is based on Tappert's (1977) parabolic equation model. ULETA generates transmission losses for single frequency (CW) or pulse sources in range-dependent ocean environments (Palmer et al., 1988b).

[138] UNIMOD is a hybrid combination of PE (for water-borne energy) and FACT or MPP for bottom-interacting energy (R. B. Lauer, unpublished).

[139] 3DPE (NRL-1) solves the three-dimensional parabolic equation using the split-step algorithm (Perkins and Baer, 1982; Perkins et al., 1983). The input sound-speed profiles and bathymetry

(Continued)

TABLE 4.1 (Continued)
Summary of Underwater Acoustic Propagation Models

need not be specified on a regular grid. However, interpolation schemes based on user-supplied triangularizations of the region are employed. Three graphic outputs can be produced: (1) TL versus range (with depth and azimuthal angle held fixed); (2) TL versus depth (with range and azimuthal angle held fixed); and (3) TL versus azimuthal angle (with range and depth held fixed). An additional program can simulate the performance of a horizontal acoustic array at given ranges.

[140] 3DPE (NRL-2) handles wide propagation angles in depth, narrow propagation angles in azimuth, and rough boundaries (Collins and Chin-Bing, 1990).

[141] 3D TDPA uses one-way narrow-angle and wide-angle, three-dimensional, time domain paraxial approximations to simulate pulse propagation in three-dimensional ocean geometries characterized by volume attenuation and variable density (Orchard et al., 1992).

[142] 3DWAPE incorporates higher-order finite-difference schemes to handle the azimuthal derivative term in a three-dimensional (3D) parabolic equation model (Sturm and Fawcett, 2003). Broadband pulse propagation problems were solved in a 3D waveguide using a Fourier synthesis of frequency-domain solutions (3DWAPE) in a penetrable wedge-shaped waveguide (Sturm, 2002). The 3DWAPE model includes a wide-angle paraxial approximation for the azimuthal component. This version of 3DWAPE was used to investigate broadband sound pulse propagation in two shallow-water waveguides: the 3D ASA benchmark wedge and the 3D Gaussian canyon (Sturm, 2005).

5 Propagation II: Mathematical Models

5.1 BACKGROUND

Mathematical models of underwater acoustic propagation include both numerical models and empirical models. Chapter 4 addressed the theoretical development of numerical models and summarized their availability. This chapter addresses the development of empirical models applicable to special propagation paths such as surface ducts, shallow water, and Arctic half-channels. Where appropriate, comparisons are made with predictions generated by numerical models. Data support requirements for mathematical models of propagation are discussed and a select number of field experiments are described in order to highlight promising areas for future research and development.

5.2 SURFACE DUCT MODELS

Properties of the surface duct were discussed previously in Chapter 3. Both ray-theoretical and wave-theoretical solutions can be applied to propagation in the surface duct.

5.2.1 Ray-Theory Models

An expression for transmission loss (TL) in a surface duct may be obtained through simple ray-theoretical considerations (Urick, 1983). In Figure 5.1, let a nondirectional source of sound be located at P in a surface duct (or mixed layer). Also, let c_0 denote the reference sound speed in the duct. Of all the rays leaving the source, only those within a certain limiting angle 2θ remain in the duct. At a distance of 1 m, the power contained in this ray bundle is distributed over a portion of the spherical surface A_1. At a long distance r, this same amount of power (in the absence of leakage and absorption) is distributed over a cylindrical surface A_2. Because the power crossing areas A_1 and A_2 is conserved, the TL to range r, averaged over the duct thickness H, is

$$\mathrm{TL} = 10 \log_{10} \frac{A_2}{A_1}$$

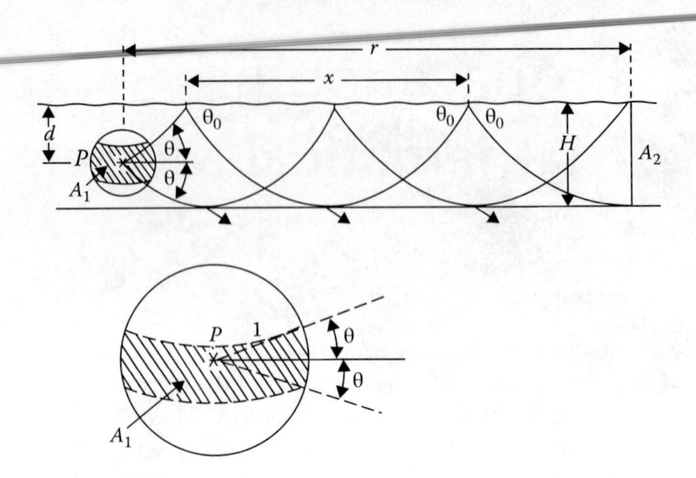

FIGURE 5.1 Propagation geometry in a surface duct. (From Urick, R.J., *Principles of Underwater Sound*, McGraw-Hill, New York, 1983. Reproduced with permission of McGraw-Hill.)

By geometry

$$A_2 = 2\pi rH \quad \text{and} \quad A_1 = 2\pi \int_{-\theta}^{\theta} \cos\theta \, d\theta = 4\pi \sin\theta$$

Therefore, for the surface duct

$$\mathrm{TL} = 10 \log_{10} \frac{rH}{2\sin\theta} = 10 \log_{10} r r_0 = 10 \log_{10} r_0^2 \frac{r}{r_0}$$

where $r_0 = H/(2 \sin \theta)$ and θ is the inclination angle at the source of the maximum trapped ray. The quantity $r_0^2\ r/r_0$ indicates that the propagation to range r may be viewed as the result of spherical spreading out to a transition range r_0, followed by cylindrical spreading from r_0 to r. The range r_0 also corresponds to the condition in which the vertical extent of the beam subtended by the angle 2θ equals the duct thickness H. Ranges greater than r_0 are generally considered to represent far-field conditions. When the attenuation due to absorption and leakage is added, the duct TL expression becomes

$$\mathrm{TL} = 10 \log_{10} r_0 + 10 \log_{10} r + (\alpha + \alpha_L) r \times 10^{-3} \tag{5.1}$$

where α is the absorption coefficient (dB km^{-1}), α_L is the leakage coefficient (dB km^{-1}) that expresses the rate at which acoustic energy leaks out of the duct, and r is the range (m). This expression is applicable to all ducts.

For a duct with a constant sound speed gradient (g), in which the rays are arcs of circles with radius of curvature R = c_0/g, the following relationships result from geometry when $R >> H$ and when $\sin \theta_0 << 1$ at source depth d (refer to Figure 5.1):

$$R = \frac{c_0}{g} = \text{radius of curvature of rays}$$

$$x = \sqrt{8RH} = \text{skip distance of limiting ray}$$

$$\theta_0 = \sqrt{\frac{2H}{R}} = \text{maximum angle of limiting ray (rad)}$$

$$\theta = \sqrt{\frac{2(H-d)}{R}} = \text{angle of limiting ray at source depth (rad)}$$

$$r_0 = \sqrt{\frac{RH}{8}}\sqrt{\frac{H}{H-d}} = \frac{x}{8}\sqrt{\frac{H}{H-d}} = \text{transition range}$$

Kinsler et al. (1982: 402–6) also presented an insightful development of these relationships.

A related approach, referred to as an energy-flux modeling (Weston, 1980a,b; Lurton, 2002: 31–2), is useful for rapid calculations of TL where the propagation conditions are dominated by numerous boundary-reflected multipaths and when only the coarse characteristics of the acoustic field are needed. Refer to Section 4.3.8 for more details on this approach.

Marsh and Schulkin (1955) developed empirical surface-duct equations based on an extensive set of data collected during Project AMOS (acoustic, meteorological, and oceanographic survey) in 1953–54. These equations have been incorporated into many ray-theoretical models to handle the special case of surface duct propagation. Graphical summaries of the AMOS data were presented in Chapter 3. Other empirically derived formulas have been reviewed by Urick (1982, Chapter 6).

5.2.2 Wave-Theory Models

Pedersen and Gordon (1965) adapted Marsh's (1950) normal mode approach to short ranges for a bilinear gradient (positive gradient overlying a negative gradient) model of a surface duct (Figure 5.2). This now classic bilinear surface duct model has been incorporated into some ray-theoretical models to augment their capabilities.

Assuming that leakage due to surface roughness can be neglected, the TL is given as a function of range by

$$\text{TL} = -10\log_{10}\left|\sum_{1}^{N} H_0^{(2)}(\lambda_n r)u_n(t)u_n(t_0)\right|^2 - 20\log_{10}\pi + \alpha r \tag{5.2}$$

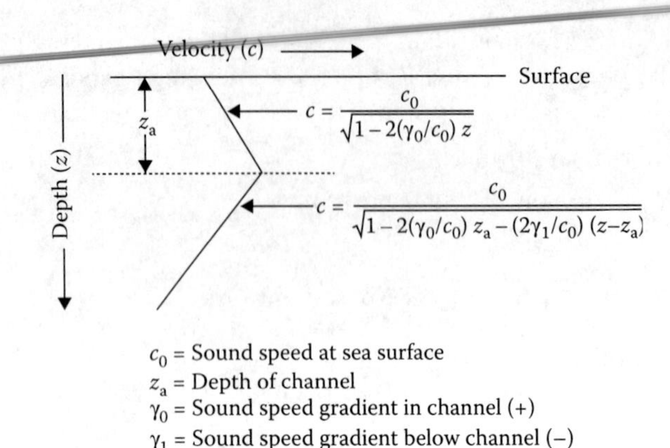

FIGURE 5.2 Geometry of bilinear-gradient model. (From Pedersen, M.A. and Gordon, D.F., *J. Acoust. Soc. Am.*, 37, 105–118, 1981. With permission.)

where

N = number of modes included in the computation
$H_0^{(2)}$ = zero-order Hankel function of the second kind
λ_n = complex wavenumber
n = mode counter index ($n = 1,2,\ldots,N$)
r = range
$u_n(\,)$ = depth function
t = ratio of receiver depth to depth of channel
t_0 = ratio of source depth to depth of channel
α = absorption coefficient

Qualitatively, the relationship in Equation 5.2 predicts cylindrical spreading as a function of range combined with the effects of modal interactions. These modes may be thought of as damped sinusoidal waves. There are two basic factors that determine the degree to which a particular mode contributes to the result. First, the product $|u_n(t)u_n(t_0)|$ depends on the source and receiver depths, but not on range. Specifically, an exponential damping factor implicit in the formulation causes the relative contributions of a mode to decrease with increasing range and with increasing mode number.

By invoking simplifying approximations, Pedersen and Gordon (1965) were able to generate analytical solutions using a variation of Equation 5.2 that agreed favorably with both ray-theoretical solutions and with experimental data. There are three important aspects of this approach that differ from other normal mode solutions: (1) the solution is valid for short ranges because the branch-cut integral is zero for this model; (2) the modes are damped since the wavenumbers are complex; and (3) there are no cutoffs in the frequency domain. Thus, higher-order modes are highly damped and only the lower-order modes need be considered for most practical problems.

5.2.3 Oceanographic Mixed-Layer Models

New developments in numerical modeling in oceanography now permit the depth of the mixed layer to be forecast in both time and location. These predicted values can then be incorporated into surface duct models to predict the corresponding acoustic TL. Such interfacing of models is referred to as coupled ocean acoustic modeling (Mooers et al., 1982). Models of the mixed layer have largely been restricted to the treatment of one-dimensional (1D) approximations, which have proved useful when horizontal advection can be neglected. In many ocean areas, however, the 1D approximations appear to be inappropriate for estimating mixed layer depths (Garwood, 1979).

Mixed layer models are of two basic types: differential and bulk. Differential models use the equations for conservation of momentum, heat, salt, and turbulent kinetic energy (TKE) in their primitive form, and are not integrated over the mixed layer. The region where the local TKE is large enough to provide a certain minimum level of vertical mixing defines the mixed layer for these models. Bulk, or integrated, models assume that the mixed layer is a well-defined layer that is uniform in temperature and salinity. The governing equations for these models are obtained by integrating the primitive equations over the depth of the mixed layer.

Mixed layer models respond to three basic types of forcing conditions: wind deepening, heating, and cooling. Wind deepening is defined to occur when the mixed layer deepens due to the erosion of the stably stratified region at its base by wind-generated turbulence. The depth of mixing is governed by a balance between the stabilizing effect of surface heating, or a positive surface buoyancy flux, and the effect of mixing due to wind-generated turbulence. This balance governs the mixed layer depth during periods when the mixed layer is shallowing. Under conditions of cooling, a net surface heat loss, or negative surface buoyancy flux, causes the mixed layer to deepen due to convection. Convection usually occurs at night and is the dominant mechanism for deepening the mixed layer in fall and winter, especially under conditions of reduced solar heating and increased evaporative cooling due to increased wind speeds.

A number of mixed layer models have been developed and some have been coupled with underwater acoustic TL models to provide input (and feedback) for the parameters important for prediction of sound propagation in the surface duct. One system that is operational for fleet applications is the thermodynamical ocean prediction system, or TOPS (Clancy and Martin, 1979). TOPS is categorized as a differential mixed layer model. Forecasts appear to agree with measurements in those ocean areas where the layer depth is dependent primarily on local conditions and not on advection from neighboring regions. A comparison of observations and predictions generated by TOPS (Figure 5.3) shows reasonable agreement. Figure 5.3a illustrates the observed wind speed. Figure 5.3b compares the observed and predicted mixed-layer temperatures while Figure 5.3c compares the observed and predicted mixed-layer depths over the same period.

Clancy and Pollak (1983) advocated coupling the TOPS synoptic mixed-layer model to an objective ocean thermal analysis system in order to produce a continuously updating, real-time, analysis-forecast-analysis system. The forecast component

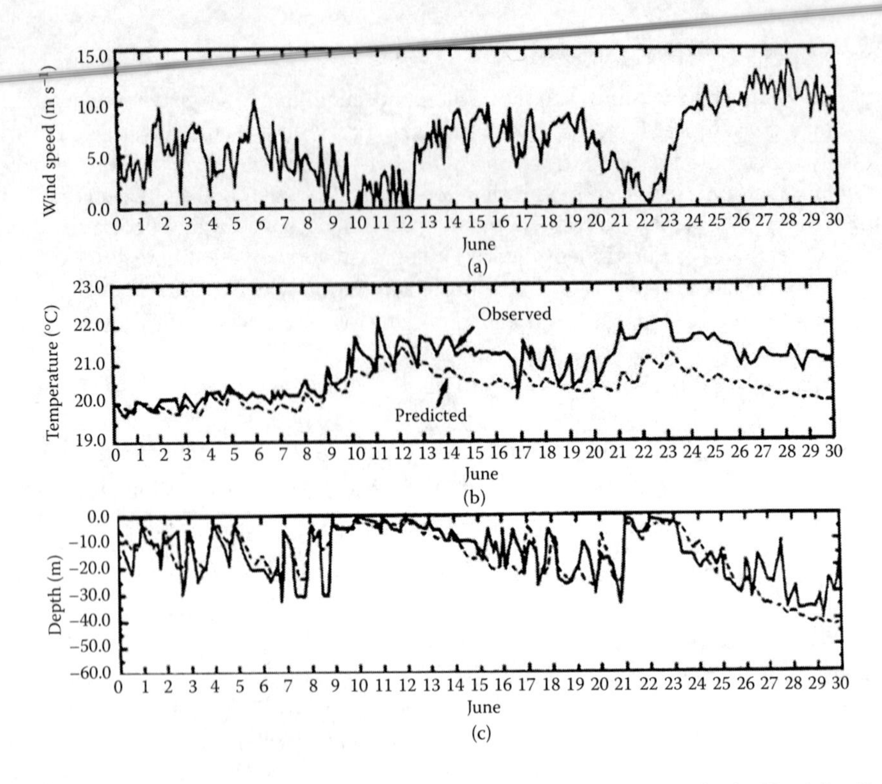

FIGURE 5.3 Observations and predictions at Ocean Station November (in the North Pacific Ocean) during June 1961. (a) Observed wind speed. (b) Observed (solid line) and predicted (dashed line) sea-surface temperatures. (c) Observed (solid line) and predicted (dashed line) mixed layer depths. (Adapted from Clancy, R.M. et al., 1981. Test and evaluation of an operationally capable synoptic upper-ocean forecast system. Nav. Ocean Res. Devel. Activity, Tech. Note 92.)

(TOPS) employs the Mellor and Yamada (1974) level-2 turbulence parameterization scheme. It includes advection by instantaneous wind drift and climatologically averaged geostrophic currents, and is forced by surface fluxes supplied by atmospheric models.

The diurnal ocean surface layer (DOSL) model, which predicts diurnal patterns of the sea-surface temperature (SST) field, has been developed to support high-frequency sonar operations against shallow targets (Clancy et al., 1991b; Hawkins et al., 1993). Diurnal variability in the mixed layer dominates short-term changes in the acoustic behavior of the surface duct. The well-known "afternoon effect" discovered during World War II denotes the loss of a surface duct due to the creation of a shallow, transient thermocline by local heating. Diurnal SST changes also affect the use and interpretation of satellite data. Specifically, satellites measure the skin temperature of the ocean, and these surface measurements may or may not be characteristic of the bulk mixed-layer temperature (the desired parameter for naval operations) depending on the amplitude and phase of the SST cycle at the time of measurement. Since the diurnal SST response is a strongly nonlinear function of the wind speed, smooth variations in the synoptic wind field can produce sharp horizontal SST gradients that might be misinterpreted as the thermal signature of ocean frontal features.

5.3 SHALLOW-WATER DUCT MODELS

5.3.1 Shallow-Water Propagation Characteristics

Coastal environments are generally characterized by high spatial and temporal variabilities. When coupled with attendant acoustic spectral dependencies of the surface and bottom boundaries, these natural variabilities make coastal regions very complex acoustic environments. Specifically, changes in the temperature and salinity of coastal waters affect the refraction of sound in the water column. These refractive properties have a profound impact on the transmission of acoustic energy in a shallow-water waveguide with an irregular bottom and a statistically varying sea surface. Thus, accurate modeling and prediction of the acoustic environment is essential to understanding sonar performance in coastal oceans.

Physical processes controlling the hydrography of shelf waters often exhibit strong seasonal variations. Annual cycles of alongshore winds induce alternating periods of upwelling and downwelling. The presence of coastal jets and the frictional decay of deep-water eddies due to topographic interactions further complicate the dynamics of coastal regions. Episodic passages of meteorological fronts from continental interiors affect the thermal structure of the adjacent shelf waters through intense air-sea interactions. River outflows create strong salinity gradients along the adjacent coast. Variable bottom topographies and sediment compositions with their attendant spectral dependencies complicate acoustic bottom boundary conditions. At higher latitudes, ice formation complicates acoustic surface boundary conditions near the coast. Waves generated by local winds under fetch-limited conditions, together with swells originating from distant sources, conspire to complicate acoustic surface boundary conditions and also create noisy surf conditions. Marine life, which is often abundant in nutrient-rich coastal regions, can generate or scatter sound. Anthropogenic sources of noise are common in coastal seas including fixed sources such as drilling rigs and mobile sources such as merchant shipping and fishing vessels. Surface weather, including wind and rain, further contribute to the underwater noise field. Even noise from low-flying coastal aircraft can couple into the water column and add to the background noise field.

Acoustic propagation in shallow water is dominated by repeated interactions with the sea floor. Generally, shallow water is restricted to consideration of the continental shelves with depths less than 200 m. Detection ranges in shallow water are severely limited both by the high attenuation that results from interaction with the bottom and by the limited water depth, which will not support the long-range propagation paths available in deep water. In a recent book, Katsnelson and Petnikov (2002) discussed results from acoustical measurements made over the continental shelves of the Barents Sea and the Black Sea. Also see the more recent book by Katsnelson et al. (2012).

Determination of source location (bearing, range, and depth) can be affected by the horizontal refraction caused by repeated boundary reflections over a sloping bottom. Doolittle et al. (1988) experimentally confirmed the horizontal refraction of CW acoustic radiation from a point source in a wedge-shaped ocean environment. A striking graphical presentation of a 3D ray trace in a complicated wedge-shaped ocean environment is illustrated in Figure 5.4 (Bucker, 1994). This ray trace vividly displays the effects of horizontal refraction caused by a sloping bottom boundary. The source (denoted by an asterisk) appears in the background. Such horizontal refractive effects complicate the determination of bearing angles between sources and receivers. Consequently, sonar detections made against targets in shallow water may need to be corrected for horizontal refraction.

It is convenient to categorize sound speed profiles into generic groupings to facilitate subsequent discussions of shallow-water propagation. Assuming linearly segmented profiles, three groupings of profiles can be distinguished by the degree of segmentation: Category I—linear, Category II—bilinear, and Category III—multiply segmented. Sub-groupings (labeled A, B …) can be formed to further distinguish these profiles according to the sound-speed gradient.

Linear profiles consist of single segments that can be further distinguished according to their gradient as: I-A—positive gradient $\left(\frac{\partial c}{\partial z} > 0\right)$, I-B—negative gradient $\left(\frac{\partial c}{\partial z} < 0\right)$ or I-C—isovelocity $\left(\frac{\partial c}{\partial z} = 0\right)$, where c is the speed of sound and z is the depth (measured positive downward).

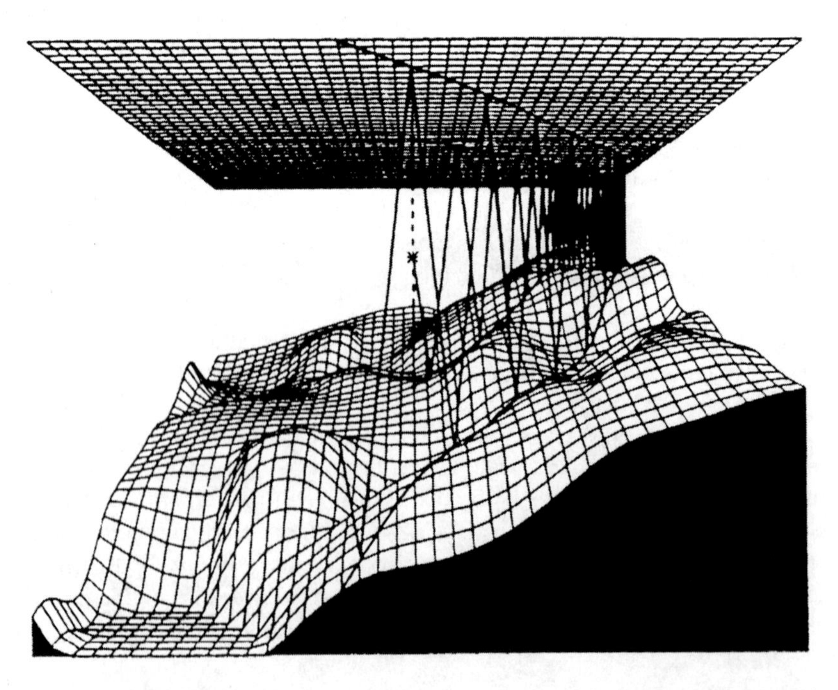

FIGURE 5.4 Typical beam trace in a shallow-water region. (From Bucker, H.P., *J. Acoust. Soc. Am.*, **95**, 2437–2440, 1971. With permission.)

Bilinear profiles consist of two segments and can be formed in two ways in the ocean. If a positive gradient overlies a negative gradient (II-A), then a surface duct is formed in the upper layer above the sound-speed maximum. If a negative gradient overlies a positive gradient (II-B), then a sound channel is formed at the juncture of the two segments (i.e., at the sound-speed minimum).

Multiply segmented profiles consist of three or more segments and can assume a variety of forms. However, the most common manifestation of this type occurs when a surface duct overlies a channel (III-A). Other manifestations typically involve multiple channels (III-B).

This classification system provides a convenient method for describing the general distribution of sound-speed profiles in shallow-water environments (with depths ≤200 m). For example, Reise and Etter (1997) used this classification system in a sonar trade study that examined representative shallow-water profiles from the Pacific and Atlantic Oceans, and the Mediterranean and Arabian Seas (Table 5.1). Based on this small but representative sampling, the most common occurrence (42% of all profiles examined) was the bilinear profile with a surface duct (II-A). This form was almost twice as likely to occur in summer as in winter (64% versus 36%, respectively). The next most common occurrence (23% of all profiles examined) was the linear positive-gradient profile (I-A). This form occurred exclusively in winter (100%). No isovelocity cases (I-C) were encountered in the study. Multiply segmented forms (III-A and III-B) represented 15% of the profiles examined and were three times more frequent in summer than in winter (75% versus 25%). The linear negative-gradient profile (I-B) represented 12% of the profiles examined and occurred exclusively in summer (100%). Finally, the bilinear sound channel (II-B) represented 8% of the profiles examined and occurred as frequently in summer as in winter (50% each). It should be noted that these results might not be representative of every shallow-water region in every season.

5.3.2 Optimum Frequency of Propagation

Understanding and predicting acoustic sensor performance in shallow water is complicated by the relatively high temporal and spatial variability of the ocean environment. Improper optimization of the frequency or the depth of operation at a particular location and time of year can degrade sonar performance.

In shallow-water environments, the optimum frequency of propagation is often the result of competing propagation and attenuation mechanisms at either end of the frequency spectrum. Jensen and Kuperman (1979, 1983) investigated this problem and concluded that the optimum frequency is strongly dependent on water depth, is somewhat dependent on the particular sound speed profile, and is only weakly dependent on the bottom type. Jensen and Kuperman (1983) also noted that shear waves in the bottom were important in determining the optimum frequency of propagation and the actual TL levels at lower frequencies. A major loss mechanism for low-frequency acoustic propagation in shallow water is the attenuation in ocean sediments. Research results (e.g., Focke, 1984) indicate that variations in attenuation as a function of sediment depth have a significant impact on propagation.

TABLE 5.1
Categorization of Shallow-Water Sound-Speed Profiles.

Group	Linear			Bilinear		Multiply Segmented	
Designation/ Subgroup	I-A Positive Gradient (%)	I-B Negative Gradient (%)	I-C Isovelocity	II-A Surface Duct (%)	II-B Sound Channel (%)	III-A Duct Over Channel (%)	III-B Multiple Channels
Relative occurrence	23	12	—	42	8	15	
Summer	—	100	—	64	50	75	
Winter	100	—	—	36	50	25	

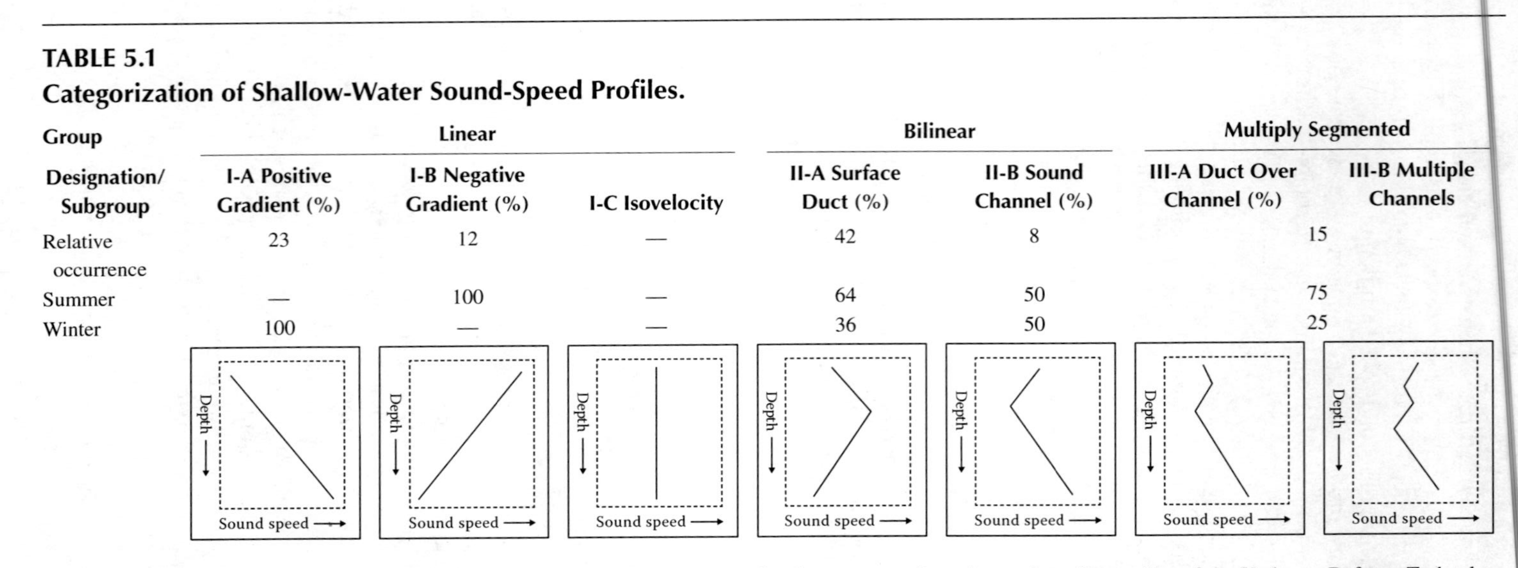

Source: Adapted from Reise, B. and Etter, P.C. 1997. Performance assessment of active sonar configuration options. Proceeding of the Undersea Defence Technology Conference (UDT Europe), pp. 408–13.

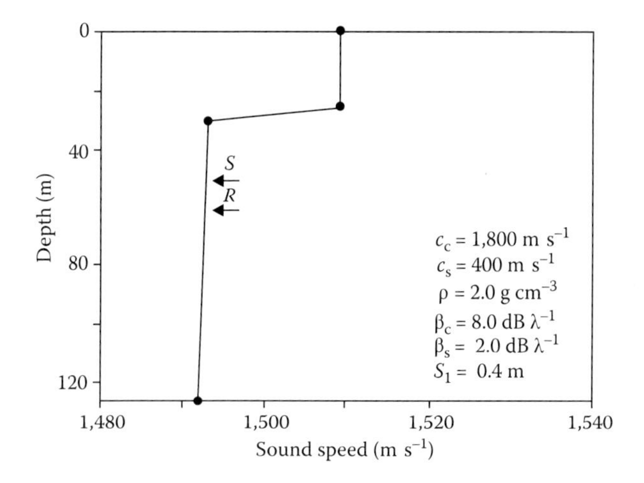

FIGURE 5.5 Sound speed profile and bottom properties for a shallow-water area located in the eastern North Atlantic Ocean: c = sound speed, ρ = density, β = attenuation, S_1 = rms bottom roughness. Subscripts "c" and "s" refer to compressional and shear waves, respectively. Source (*S*) and receiver (*R*) depths are indicated on the sound speed profile. (Adapted from Jensen, F.B. and Kuperman, W.A., Environmental acoustical modelling at SACLANTCEN. SACLANT ASW Res. Ctr, Rept SR-34.)

An important analytical tool used by Jensen and Kuperman (1979, 1983) was a frequency-range representation of TL, as illustrated in Figures 5.5 and 5.6. This type of representation has been used to characterize acoustic propagation in different waveguides (e.g., Milne, 1967). For the particular environment studied (Figure 5.5), a normal mode model was used to generate repeated transmission-loss runs versus frequency. In Figure 5.6b, the optimum frequency predicted by the model is about 200 Hz, as evidenced by the elongated axis of low-loss values. These predicted results compared favorably with experimental results (Figure 5.6a).

Shallow-water modeling techniques include both numerical models and empirical models. Eller (1984b, 1986) summarized important developments in shallow-water acoustic modeling. Katsnelson and Petnikov (2002) and Katsnelson et al. (2012) reviewed experimental results in shallow-water acoustics together with approximate approaches for modeling such phenomena.

5.3.3 Numerical Models

Both ray-theoretical and wave-theoretical approaches have been used to numerically model sound propagation in shallow water. Since shallow water environments are best approximated by range-dependent geometries (e.g., sloping bottom and high spatial variability of water-column and sediment properties), attention will be focused on the range-dependent modeling approaches. Moreover, the high-angle boundary interactions encountered in shallow water have traditionally limited consideration to two basic approaches: ray theory and normal mode solutions (with either adiabatic approximations or mode coupling). More recently, appropriately modified parabolic

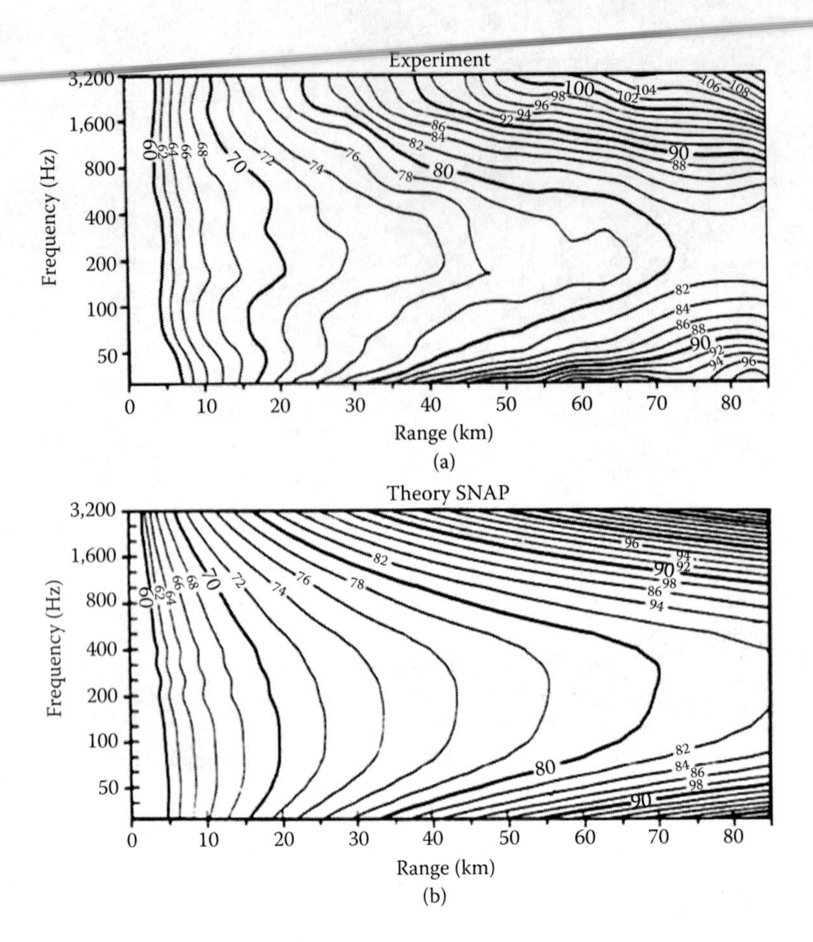

FIGURE 5.6 Transmission loss contours (in 2-dB intervals) for the environment presented in Figure 5.5: (a) experimental measurements and (b) predictions generated by the SNAP propagation model. (Adapted from Jensen, F.B. and Kuperman, W.A. 1979. Environmental acoustical modelling at SACLANTCEN. SACLANT ASW Res. Ctr, Rept SR-34.)

equation (PE) models have also been utilized successfully in shallow-water environments (Jensen, 1984; Jensen and Schmidt, 1984). Approximately 18% of the numerical modeling inventory is specifically tailored for shallow-water applications (Etter, 2001c).

Much emphasis has been placed on modeling sound propagation over a sloping bottom. This geometry is commonly referred to as a "wedge problem" and involves both upslope and downslope propagation. The direction of propagation (i.e., upslope or downslope) considerably alters the observed propagation characteristics. Consequently, this problem is of great practical interest to sonar operations in shallow water. This geometry has also been used as a benchmark problem in model evaluation (see Chapter 12).

The basic mechanisms involved in acoustic propagation in a horizontally stratified (i.e., range-invariable) waveguide are spreading loss, attenuation due to bottom-interaction effects, and intermode phasing effects. In a range-variable waveguide, an additional mechanism must be considered. This mechanism is related to changes in the acoustic energy density that occur with bathymetric changes and is often referred to as a renormalization loss, or megaphone effect. The term *renormalization* is used because the so-called megaphone effect is manifested as a change in the normalization of the normal mode depth function due to changes in the waveguide depth (Koch et al., 1983). The megaphone effect produces a gain in upslope propagation and a loss in downslope propagation.

The processes involved in upslope propagation can be better understood by using the ray-mode analogy. This analogy is a heuristic concept which states that any given mode trapped in the water column can be associated with upgoing and downing rays corresponding to specific grazing angles at the bottom (e.g., Urick, 1983: 174–6; Boyles, 1984: 197–204). As sound propagates upslope, the horizontal wavenumber associated with each mode decreases. In the ray analogy, the grazing angle at the bottom increases. For each mode, then, a point on the slope will be reached at which the grazing angles of the analogous rays will approach the critical angle at the bottom. At this point, the energy essentially leaves the water column and enters the bottom. In the ray analogy, the bottom reflection losses associated with those rays become very large. In the wave analogy, the modes transition from the trapped (waterborne) to the continuous (bottom propagating) spectrum. This point is called the "cutoff depth" for the equivalent modes. Upslope propagation is then said to exhibit a transition from a trapped to a radiative state (e.g., Arnold and Felsen, 1983; Jensen, 1984). In the case of downslope propagation, the transition is from a radiative to a trapped state.

Conventional modal formulations fail to adequately explain circumstances in which a mode suddenly disappears from the water column with its energy being radiated into and dissipated within the sea floor (Pierce, 1982). Thus, the particular case of mode coupling where discrete modes (trapped in the water column) couple into continuous modes (which propagate in the bottom) has been further explored as a matter of practical interest (e.g., Miller et al., 1986). Evans and Gilbert (1985) developed a stepwise-coupled-mode method that overcame the failure of previous coupled-mode techniques to properly conserve energy over sloping bottoms. Their method of stepwise-coupled modes avoided problems associated with sloping bottoms by using only horizontal and vertical interfaces. The full solution thus included both forward and backscattered energy (Jensen and Ferla, 1988). Also see the discussion of wedge modes in Section 4.4.7.

Collins (1990a) suggested using a rotated PE in those wedge geometries involving complicated bottom-boundary conditions. Specifically, by rotating the coordinate system, the PE could be marched parallel to the sea floor. The sea surface was then a sloping boundary with simplistic boundary conditions (pressure-release surface) that could be approximated by a sequence of range-independent regions in which the surface was specified as a series of stair steps.

The rotated PE of Collins (1990a) has been generalized to problems involving ocean-sediment interfaces of variable slope. This has been accomplished by segmenting the interface into a series of range intervals where each interval can be characterized by a different (but constant) slope (Outing et al., 2006). The original rotated

PE algorithm was used to march the field through each interval. An interpolation-extrapolation approach was used to generate a starting field at the beginning of each interval beyond the one containing the source. For the elastic case, a series of operators was applied to rotate the dependent-variable vector along with the coordinate system. The variable rotated PE should provide accurate solutions for a large class of range-dependent seismo-acoustics problems. For the fluid case, the accuracy of the approach was confirmed through comparisons with reference solutions. For the elastic case, variable rotated PE solutions were verified by comparison with energy-conserving and mapping solutions.

Eigenray formulations can be useful in determining significant propagation paths and propagation mechanisms in a wedge geometry. One particular phenomenon of interest in the upslope problem is that of backscattered eigenrays. These rays have paths that travel up the slope, past the receiver, and then back down the slope before arriving at the receiver (Westwood, 1990). The method of images has also been used to construct ray-path solutions in shallow-water environments with a sloping bottom (Macpherson and Daintith, 1967).

The Shallow Water Acoustic Modeling (SWAM) Workshop, held in Monterey, California, in September 1999, provided a forum for the comparison of single-frequency (CW) and broadband (pulse) propagation models in synthetic (i.e., virtual) environments. Test cases included up-sloping, down-sloping, flat, and 3D bathymetries. Additional cases considered the effects of internal waves and a shelf break. The goal was to determine which shallow-water environmental factors challenged existing propagation models and what details were important for constructing accurate, yet efficient, solutions. The results of this workshop, designated SWAM '99, were published in a series of papers in the *Journal of Computational Acoustics* (see Tolstoy et al., 2001).

Harrison and Nielsen (2007) observed that in shallow-water propagation, steeper ray angles are weakened the most by boundary losses. When regarded as a continuous function of angle, the sound intensity can be converted into a function of travel time to reveal the multipath pulse shape received from a remote source (via a one-way path) or from a target (via a two-way path). The simple closed-form isovelocity pulse shape could be extended to the more complicated case of upward or downward refraction. The envelope of the earliest arrivals was roughly trapezoidal with a delayed peak corresponding to the slowest, near-horizontal refracted paths. The tail of the pulse fell off exponentially (or linearly in decibels) with a decay constant that depended only on the bottom-reflection properties and water depth, irrespective of travel time. This is a useful property for both geoacoustic inversion and sonar design. The nontrivial analytical problem of inverting explicit functions of angle into explicit functions of time was solved by numerical interpolation. Thus, exact solutions could be calculated numerically. Explicit closed-form approximations were given for one-way paths. Two-way paths were calculated by numerical convolution. The wave model C-SNAP was used in several broadband cases of interest to demonstrate that these solutions correspond roughly to a depth average of multipath arrivals.

Jones et al. (2006) examined SUS (expendable underwater explosive sound signal device) waveform data collected over numerous shallow-water tracks. Comparisons with weak-shock theory were best when transmission was via surface reflections in an isothermal ocean with a nearly smooth sea surface. Kuperman and Lynch (2004) reviewed the problems associated with generating and detecting acoustic signals in shallow-water environments. These challenges were viewed as opportunities to explore adaptive processing or inversion methods (discussed in Chapter 6).

Tindle and Deane (2005) extended the wavefront modeling method of Tindle (2002) to handle range dependence in both bathymetry and sound speed. This method was used to model sound propagation in the surf zone with variable water depth (i.e., wedge geometry) and surface waves. Such propagation conditions are relevant to the study of acoustic communications systems operating in shallow water near the surf zone. Supporting experimental data showed that the underside of a wave crest acted as a curved mirror and served to focus energy and increase signal amplitude. Since the acoustic path length is longer under a wave crest, the increased intensity was also associated with a time delay. See the related work by Tindle et al. (2009).

Abawi and Porter (2007) noted that the virtual-source technique, which is based on the boundary-integral method, provided the means to impose boundary conditions on arbitrarily shaped boundaries by replacing them with a collection of sources whose amplitudes were determined from the boundary conditions. The virtual-source technique was used to model the propagation of waves in a range-dependent ocean overlying an elastic bottom with an arbitrarily shaped sea floor interface. The method was applied to propagation in an elastic Pekeris waveguide, an acoustic wedge, and an elastic wedge. For propagation in an elastic Pekeris waveguide, the results agreed very well with those obtained from the wavenumber-integral technique. For propagation in an acoustic wedge, the results agreed very well with the solution of the parabolic equation (PE) technique. For propagation in an elastic wedge, the results agreed qualitatively with those obtained from an elastic PE solution.

Jensen (1984) examined both upslope and downslope propagation using an appropriately modified parabolic equation (PE) model. These results will be discussed below. Other researchers have investigated the wedge problem utilizing normal mode solutions (Evans, 1983; Tindle and Zhang, 1997) and ray theory (Arnold and Felsen, 1983; Westwood, 1989c). Costa and Medeiros (2010) provided a summary of numerical modeling and simulation of acoustic propagation in shallow water. Jensen (2001) presented a useful summary of the notable features of sound propagation in shallow water.

Reeder (2016) described a field experiment that was carried out in the Columbia River Estuary over the period 27–29 May 2013 to observe the acoustic propagation characteristics of the estuarine salt wedge. During one flood and one ebb cycle, linear frequency-modulated acoustic signals in the 500–2000 Hz band were transmitted by an acoustic source deployed over the side of the ship and observed at a hydrophone 1.36 km distant. The 2D BELLHOP coherent Gaussian-beam acoustic propagation model was used to compute band-averaged TLs between the source and receiver for the entire acoustic transmission period during flood on May 27. In the cross-range direction, the bathymetry was modeled as planar. In the along-path direction, the environment was modeled as range-dependent in terms of water-column sound speed and bathymetry, and range-independent in terms of the geoacoustic parameters of the seabed, and a planar sea surface. These model results provided insights into the physics of the acoustic propagation through the salt wedge. Observations, together with associated acoustic modeling results, demonstrated that the salt wedge front was the dominant physical mechanism controlling the acoustic propagation in this environment. Specifically, the received signal energy was relatively stable before and after the passage of the salt wedge front when the acoustic path consisted of a single medium (either entirely fresh water or entirely salt water), and suffered a 10–15 dB loss and increased variability in a dual-media environment during salt wedge front passage due to vertical and horizontal refraction of energy up and out-of-plane from the receiver. From a phenomenological standpoint, acoustic-propagation characteristics of the estuarine environment, in terms of average energy level and variance, depended on the position of the source and receiver relative to the salt wedge, acoustic frequency, and environmental parameters, including the three-dimensional salt wedge-induced sound speed gradient, riverbed sediment acoustic properties, and morphology (e.g., bathymetry, bedforms), dynamic surface conditions with downwelling, subducted bubbles, mixing and turbulence at the front, and potential presence of migrating fish and internal waves propagating along the density contrast interface. Therefore, future work will be required to quantify the contribution of each of these physical mechanisms to estuarine acoustic propagation characteristics.

5.3.3.1 Upslope Propagation

For the wedge problem involving upslope propagation, Jensen (1984) considered the particular environment illustrated in Figure 5.7. (Note that the material presented by Jensen [1984] is also available in a report by Jensen and Schmidt [1984], which contains the papers presented by these two researchers at the same conference.) The

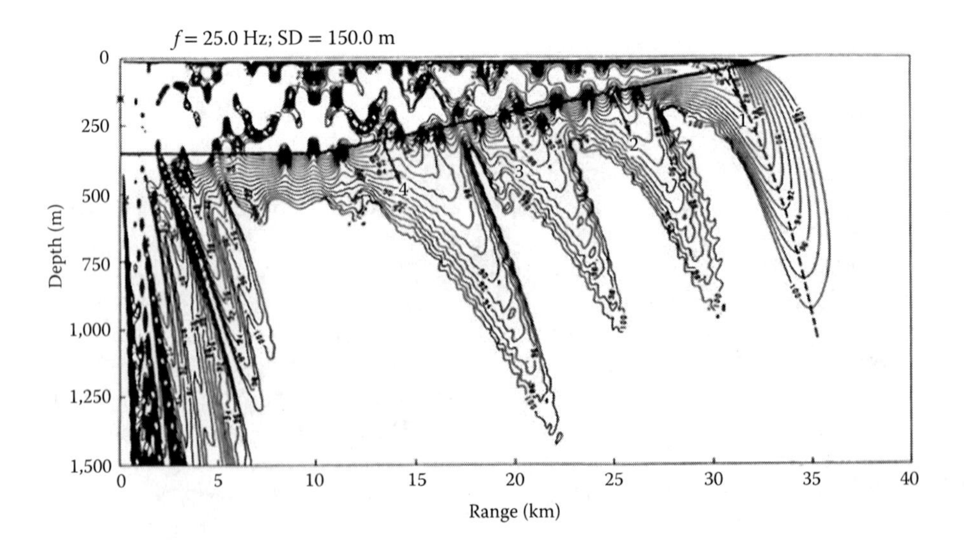

FIGURE 5.7 Upslope propagation showing discrete mode cutoffs. Results were generated using a parabolic equation propagation model. (Adapted from Jensen, F.B., *Hybrid Formulation of Wave Propagation and Scattering*. Martinus Nijhoff, Dordrecht, 1984.)

water-bottom interface is indicated on the contour plot by the heavy line starting at 350-m depth and then inclining toward the sea surface beyond a range of 10 km. The bottom slope is 0.85°. The frequency is 25 Hz and the source is located at a depth of 150 m. The sound speed is constant at 1500 m s^{-1}. The bottom has a sound speed of 1600 m s^{-1}, a density of 1.5 g cm^{-3}, and an attenuation coefficient of 0.2 dB per wavelength. Shear waves were not considered.

In Figure 5.7, Jensen displayed TL contours between 70 dB and 100 dB in 2-dB intervals. Thus, high-intensity regions (loss < 70 dB) are shown as white areas within the wedge, while low-intensity regions (loss > 100 dB) are shown as white areas in the bottom. The PE solution was started by a Gaussian initial field, and there are four propagating modes. The high intensity within the bottom at ranges less than 10 km corresponds to the radiation of continuous modes into the bottom. As sound propagates upslope, four well-defined beams (numbered 1 to 4) are seen, one corresponding to each of the four modes. This phenomenon of energy leaking out of the water column as discrete beams has been confirmed experimentally by Coppens and Sanders (1980). These points correspond to the cutoff depths associated with each mode.

5.3.3.2 Downslope Propagation

For the wedge problem involving downslope propagation, Jensen (1984) considered the environment illustrated in Figure 5.8. The initial water depth is 50 m and the bottom slope is 5°. The sound speed is constant at 1500 m s^{-1}. The sound speed in the bottom is 1600 m s^{-1} and the attenuation coefficient is 0.5 dB per wavelength. The density ratio between the bottom and the water is 1.5. The contour plot is for a source frequency of 25 Hz. The initial field for the PE calculation was supplied by a normal mode model, and only the first mode was propagated downslope. Shear waves were not considered.

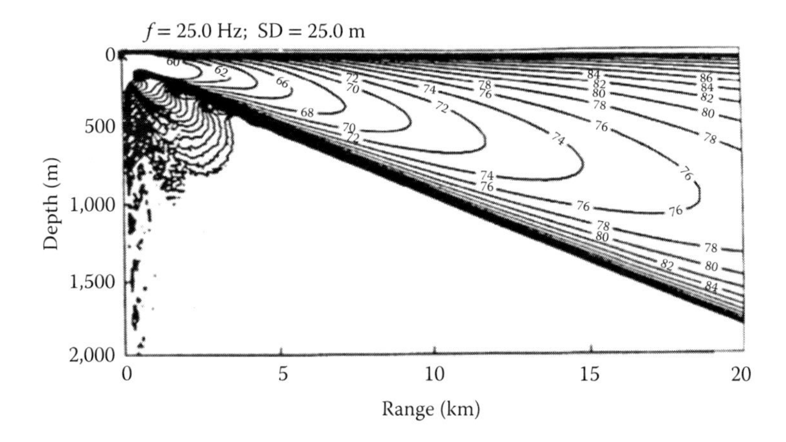

FIGURE 5.8 Downslope propagation over a constant 5° slope. Results were generated using a parabolic equation propagation model. (Adapted from Jensen, F.B. and Schmidt, H., 1984. Review of numerical models in underwater acoustics, including recently developed fast-field program. SACLANT ASW Res. Ctr, Rept SR-83.)

Figure 5.8 shows that some energy propagates straight into the bottom at short ranges. That is, it couples into the continuous spectrum. Beyond the near field, however, propagation within the wedge is adiabatic, with the one mode apparently adapting well to the changing water depth. At a range of 20 km, the energy is entirely contained in the local first mode, even though as many as 21 modes could exist in a water depth of 1800 m.

5.3.4 Empirical Models

Two noteworthy empirical algorithms have been developed for use in predicting TL in shallow water, both of which provide depth-averaged estimates for range-independent ocean environments. One model (Rogers, 1981) was derived from theoretical (physics-based) considerations. The second model (Marsh and Schulkin, 1962b; Schulkin and Mercer, 1985), also known as Colossus, was derived from field measurements obtained from a limited number of geographic areas.

5.3.4.1 Rogers Model

Rogers (1981) found that virtually all shallow-water TL curves could be described by an equation of the form

$$\mathrm{TL} = 15\log_{10} R + AR + B + CR^2 \tag{5.3}$$

where R is the range, and A, B, and C are coefficients.

For the case of a negative sound speed gradient (i.e., sound speed decreases with increasing depth), Rogers obtained the following equation:

$$\mathrm{TL} = 15\log_{10} R + 5\log_{10}\left(H\beta\right) + \frac{\beta R\theta_L^2}{4H} - 7.18 + \alpha_w R \tag{5.4}$$

where

R = range (m)
H = water depth (m)
β = bottom loss (dB rad^{-1})
θ_L = limiting angle (rad)
α_w = absorption coefficient of sea water

The term $15\log_{10} R$ represents the spreading loss for the mode-stripping regions. Thereafter, the spreading loss corresponds to cylindrical spreading ($10\log_{10} R$). The limiting angle (θ_L) is the larger of θ_g or θ_c, where θ_g is the maximum grazing angle for a skip distance (i.e., the maximum refracted-bottom-reflected, or RBR, ray) and θ_c is the effective plane-wave angle corresponding to the lowest propagating mode:

$$\theta_g = \sqrt{\frac{2Hg}{c_w}}\ (\mathrm{rad}) \tag{5.5}$$

$$\theta_c = \frac{c_w}{2fH}\,(\text{rad}) \tag{5.6}$$

where g is the magnitude of the negative sound speed gradient (s^{-1}), c_w is the maximum (sea surface) sound speed ($m\ s^{-1}$), and f is the frequency (Hz).

The bottom loss (β) was derived from the theoretical expression for the Rayleigh reflection coefficient for a two-fluid lossy interface, for small grazing angles. For most cases of interest (small values of θ_L), the bottom loss can be approximated as

$$\beta \approx \frac{0.477 M_0 N_0 K_s}{\left[1-N_0^2\right]^{3/2}}\,(\text{dB rad}^{-1}) \tag{5.7}$$

where

$N_0 = c_w/c_s$
$M_0 = \rho_s/\rho_w$
ρ_w = density of sea water
ρ_s = sediment density
K_s = sediment attenuation coefficient (dB m^{-1} kHz^{-1})

For example, at a fixed frequency of 200 Hz, Rogers (1981) considered eight different sound speed profiles for which sound speed decreased monotonically with depth (Figure 5.9). Using Equation 5.3, the coefficients A, B, and C were determined for a number of test cases between 5 km and 100 km. The maximum deviation between Equation 5.3 and the actual TL curves generated by a normal mode model was also reported (Table 5.2). The sediment properties were based on

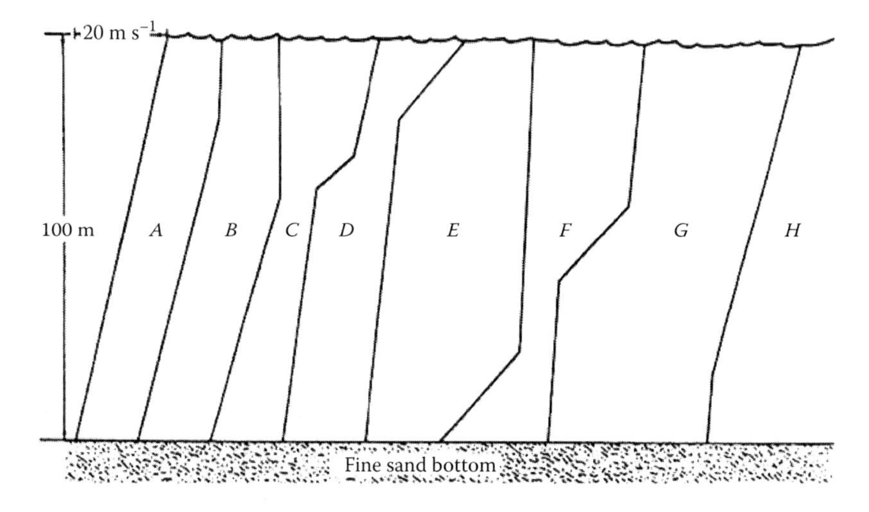

FIGURE 5.9 Eight different sound speed profiles in which the sound speed decreases with an average gradient is 0.2 s^{-1}. (Adapted from Rogers, P.H., 1981. Onboard prediction of propagation loss in shallow water. Nav. Res. Lab., Rept 8500.)

TABLE 5.2
Coefficients for the Empirical Shallow-Water TL Formula: TL = 15 $\log_{10}(R)$ + $AR + B + CR^2$ (dB). Range (R) is Measured in km, and the Coefficients A, B, and C are Valid for a Frequency of 200 Hz for the Sound Speed Profiles (A–H) Illustrated in Figure 5.9 over a Fine-Sand Bottom

		Three-Parameter Fit			Maximum
Sediment	Profile Type	A	B	$C \times 10^4$	Deviation
Fine sand	A	0.18	49.18	+1.098	0.017
Fine sand	B	0.22	49.04	+1.772	0.043
Fine sand	C	0.27	49.02	+0.461	0.042
Fine sand	D	0.149	49.04	+0.020	0.047
Fine sand	E	0.0729	49.61	−0.801	0.042
Fine sand	F	0.146	49.76	−0.587	0.051
Fine sand	G	0.252	48.80	−4.360	0.046
Fine sand	H	0.173	49.14	+0.090	0.021

Source: Adapted from Rogers, P.H., Onboard prediction of propagation loss in shallow water. Nav. Res. Lab., Rept 8500, 1981.

Hamilton (1980). The intent was to demonstrate the importance of the sound speed profile (versus the average gradient in the duct) in determining depth-averaged TL. The eight transmission-loss curves generated using the coefficients presented in Table 5.2 are plotted in Figure 5.10 to demonstrate the spread in values. At a range of 100 km, the minimum and maximum transmission-loss values differ by more than 20 dB even though the overall gradients of all eight sound speed profiles were the same. These results demonstrate the importance of the shape of the sound speed profile, and not just the overall gradient, in determining TL in a shallow-water waveguide.

5.3.4.2 Marsh-Schulkin Model

The model by Marsh-Schulkin (1962b), also referred to as Colossus, is an empirical model for predicting TL in shallow water (e.g., Podeszwa, 1969). Colossus employs several concepts: (1) refractive cycle, or skip distance; (2) deflection of energy into the bottom at high angles by scattering from the sea surface; and (3) a simplified Rayleigh two-fluid model of the bottom for sand or mud sediments. With a few free parameters (including water depth), about 100,000 measurements spanning the frequency range 100 Hz to 10 kHz were fitted within stated error bounds.

Along with the concepts of skip distance and bottom loss, the Colossus model used the AMOS results (Marsh and Schulkin, 1955) for a deep-water isothermal surface duct, but with an average thermocline appropriate for shallow water. It also used measurements of actual TLs in shallow water off the east coast of the United States as a function of frequency, categorized by bottom type (sand or mud) and by season. Two other mechanisms characteristic of shallow-water processes were also

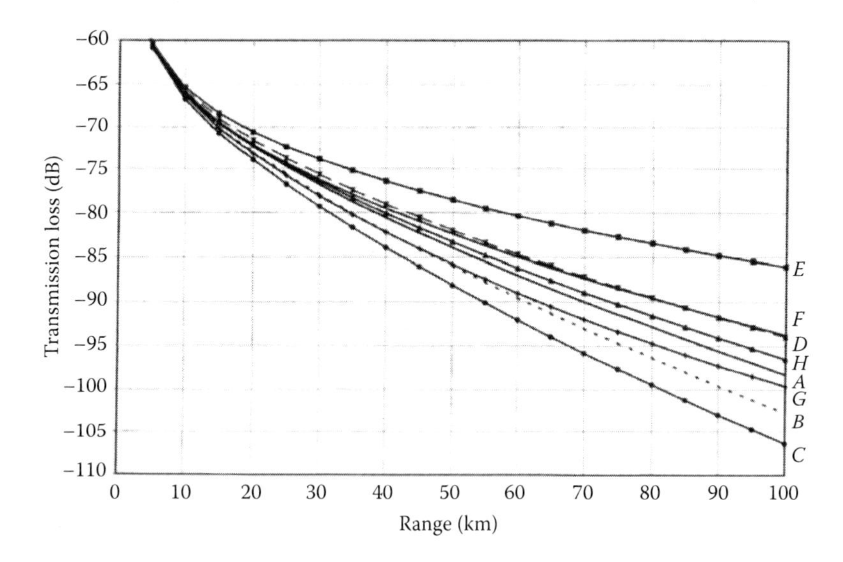

FIGURE 5.10 Composite illustration of the eight transmission-loss curves resulting from the coefficients presented in Table 5.2. (Adapted from Rogers, P.H. 1981. Onboard prediction of propagation loss in shallow water. Nav. Res. Lab., Rept 8500.)

included in the Colossus model. The first mechanism was a "near-field anomaly" correction in the direct radiation zone that included the gain due to multiple bottom and surface bounces. The second was an energy-conservation rule that was used to establish the effective shallow-water attenuation coefficient (α_t), which included the additional loss due to the coupling of energy from the wind-roughened sea surface to the bottom.

The bilinear gradient used in the Colossus model is composed of two constant, linear segments drawn toward the surface and toward the bottom from the depth of maximum sound speed (or temperature). This profile corresponds to the bilinear gradient with a surface duct (Category II-A) that was illustrated previously in Table 5.1. By placing the layer depth (maximum sound speed) at the sea surface, a linear negative-gradient profile (Category I-B) is obtained. Alternatively, placing the layer depth at the sea floor produces a linear positive-gradient sound-speed profile (Category I-A). As noted in Table 5.1, profiles from Categories I-A, I-B, and II-A collectively comprised about 77% of all sound speed profiles encountered in shallow-water regions.

In Colossus, the sound ray cycles have one upward radius of curvature (positive sound speed gradient) for surface bounces and one downward radius of curvature (negative gradient) for bottom bounces. Based on the depth of the surface layer and the water depth, a single effective skip distance (H) is formulated. Multiples of this effective skip distance are used to define a zone of direct ray paths ($20 \log_{10} R$, where R is the range), a zone of mode stripping ($15 \log_{10} R$), and a zone of single-mode control ($10 \log_{10} R$). The mode-stripping process was found to be complete at a range equal to $8H$, where H is the skip distance.

In the Marsh-Schulkin (or Colossus) model, TL is a function of sea state (or wave height), bottom type, water depth, frequency, and the depth of the positive-gradient layer. The skip distance (H) is used as a reference to define regions where wavefront spreading follows square, three-halves, and first-power laws as a function of range (R). Accordingly, three equations were developed to provide for the gradual transition from spherical spreading in the near field to cylindrical spreading in the far field:

Short range ($R < H$)

$$\mathrm{TL} = 20\log_{10} R + \alpha R + 60 - K_{\mathrm{L}} \tag{5.8}$$

Intermediate range ($H < R < 8H$)

$$\mathrm{TL} = 15\log_{10} R + \alpha R + \alpha_{\mathrm{t}}\left[\frac{R}{H} - 1\right] + 5\log_{10} H + 60 - K_{\mathrm{L}} \tag{5.9}$$

Long range ($R > 8H$)

$$\mathrm{TL} = 10\log_{10} R + \alpha R + \alpha_{\mathrm{t}}\left[\frac{R}{H} - 1\right] + 10\log_{10} H + 64.5 - K_{\mathrm{L}} \tag{5.10}$$

where

$H = [(L + D)/3]^{1/2}$ = skip distance (km)
L = mixed layer depth (m)
D = water depth (m)
R = range (km)
α = absorption coefficient (dB km^{-1})
α_{t} = effective shallow-water attenuation coefficient (dB per bounce)
K_{L} = near-field anomaly (dB)

The absorption coefficient (α) can be estimated from Figure 3.14. The effective shallow-water attenuation coefficient (α_{t}) and the near-field anomaly (K_{L}) are functions of frequency, sea state, and bottom composition. Values of α_{t} range from about 1 dB per bounce to about 8 dB per bounce. Typical values of K_{L} range from about 1 dB to about 7 dB. Complete tables of α_{t} and K_{L} are contained in the paper by Marsh and Schulkin (1962b).

According to Schulkin and Mercer (1985), the model's chief criticisms have been that it could not be adjusted for arbitrary negative sound speed gradients (it uses the same constant gradient in all cases), and that it uses empirical bottom loss values. The issue concerning the arbitrary negative sound speed gradients provided motivation for the efforts documented by Rogers (1981). Consequently, the Colossus model has been extended to accommodate arbitrary gradients in both negative and bilinear sound-speed profiles. These extensions use new general expressions for the skip distance, the near-field anomaly, and the reflection coefficients. The extended model (Schulkin and Mercer, 1985) and the model by Rogers (1981) were found to give about the same predictions when the same inputs were used.

5.3.5 Field Experiments

Four field experiments will be described below to illustrate recent attempts to measure and understand shallow-water acoustics: (1) SWAT experiments in the South China Sea; (2) SWARM experiment in the North Atlantic Ocean; (3) the littoral acoustic demonstration center (LADC) in the Gulf of Mexico; and (4) shallow water '06 in the North Atlantic Ocean. Since these are fairly recent experiments, only preliminary scientific results have been published to date.

5.3.5.1 Swat Experiments in the South China Sea

Low-frequency sound propagation features and bottom-sediment properties were studied during the SWAT (shallow water acoustic technology) experiments, which were conducted in the South China Sea over the period 1999–2001 (Ohta et al., 2005). Eigenvalues and eigenfunctions were computed using the KRAKEN model at frequencies of 25 Hz and 40 Hz over tracks of about 18 nmi in water depths of about 100 m. The range variation of geoacoustic parameters will be examined in future studies.

5.3.5.2 Swarm Experiment in the Atlantic Ocean

A theoretical model was used to explain the decorrelation of modal amplitudes on time scales of the order of 100 s observed during the SWARM (shallow water acoustics in a random medium) experiment, which was conducted off the US coast of New Jersey in 1995 (Rouseff et al., 2002; Apel et al., 1997, 2007). Packets of internal waves caused mode coupling, and their motion resulted in changing acoustic interference patterns.

Data from the SWARM experiment were used to further investigate the effects of nonlinear-internal waves on the propagation of acoustic signals (Frank et al., 2005). Results generated by the three-dimensional adiabatic mode parabolic equation (AMPE) model strongly suggested the presence of horizontal refraction when a packet of nonlinear internal waves crossed the propagation track at a high incidence angle (i.e., when the wave fronts were nearly parallel to the acoustic propagation direction), which was close to the critical angle for total internal reflection.

5.3.5.3 Littoral Acoustic Demonstration Center

Rayborn (2006) described the LADC, which was founded in 2001 as a consortium of scientists from universities and the US Navy. Initially, it consisted of the University of Southern Mississippi, the University of New Orleans, and the US Naval Research Laboratory at the Stennis Space Center located in Mississippi near the Gulf of Mexico. The University of Louisiana at Lafayette later became a part of the LADC group. Formative funding was provided by the US Office of Naval Research (ONR). The LADC was originally formed to conduct ambient-noise measurements, marine-mammal acoustic measurements, and analyses in shallow water. Research was later broadened to include airgun-calibration measurements.

Sidorovskaia et al. (2013) reviewed the LADC library of broadband passive acoustic data, collected by autonomous bottom-moored buoys, which sampled the Gulf of Mexico regional ambient noise, seismic-airgun array emissions, and marine-mammal activities six times during the past decade. The LADC acoustic data represent an opportunity to study short-term and long-term effects of environmental changes on the marine-mammal population. Environmental factors included baseline anthropogenic noise levels, passages of tropical storms, regional seismic-exploration surveys, and the 2010 Deepwater Horizon oil-spill incident. Specifically summarized were recent findings on the relationships between regional population dynamics of sperm and beaked whales, in addition to abrupt environmental changes emphasizing the 2010 oil spill. Statistically significant results of this study suggested a need to establish consistent acoustic monitoring protocols in the oceanic areas of current or potential industrial activities.

5.3.5.4 Shallow Water '06

The US ONR sponsored a multidisciplinary experiment off the US coast of New Jersey in the summer of 2006. This large-scale experiment (called *Shallow Water '06*) had three components: LEAR (littoral environmental acoustics research), NLIWI (nonlinear internal waves initiative), and AWACS (acoustic wide area coverage for surveillance) using AUVs and gliders (Tang et al., 2007).

Wan et al. (2016) utilized acoustic normal-mode dispersion curves, mode shapes, and modal-based longitudinal horizontal coherence to define a three-objective optimization problem for geoacoustic parameter estimation in the Shallow Water 2006 Experiment. This inversion scheme was applied to long-range combustive sound source data obtained from L-shaped arrays deployed on the New Jersey continental shelf during the summer of 2006. A two-layer (sand ridge overlaying a half-space basement) range-independent sediment model was utilized based on the subbottom layering structure derived from the compressed high-intensity radiated pulse reflection survey at the experimental site. When using geoacoustic inversion methods, one objective function may not result in a unique solution of the inversion problem because of the ambiguities among the unknown parameters. The ambiguities of the sound speed, density, and depth of the sand ridge layer were partially removed by minimizing these objective functions. The inverted seabed sound speed over a frequency range of 15–170 Hz was comparable to the ones from direct measurements and other inversion methods in the same general area. The inverted seabed attenuation (α_b) showed a nonlinear frequency dependence expressed as $\alpha_b = 0.26 f^{1.55}$ (dB/m) from 50 Hz to 500 Hz, or $\alpha_b = 0.32 f^{1.65}$ (dB/m) from 50 Hz to 250 Hz, where f is in kHz.

5.4 ARCTIC MODELS

5.4.1 Arctic Environmental Models

Basic descriptors of the Arctic marine environment that require specialized algorithms for use in propagation models are limited to absorption and surface (under-ice) scattering (Etter, 1987c; Ramsdale and Posey, 1987). The generation of other parameters, such as sound speed and bottom scattering, appear to be adequately supported by existing algorithms that are valid over a wide range of oceanic conditions.

Absorption is regionally dependent due mainly to the pH dependence of the boric acid relaxation. In the Arctic, the pH range is roughly 8.0–8.3 (versus 7.7–8.3 for nominal sea water), but the greatest variability occurs much closer to the sea surface than in other ocean areas. The attenuation of low-frequency sound in the sea is pH-dependent; specifically, the higher the pH (lower acidity), the greater the attenuation (Browning et al., 1988). Absorption formulas appropriate for use in the Arctic regions were presented by Mellen et al. (1987c).

Existing under-ice scattering loss models appropriate for inclusion in mathematical models of acoustic propagation were reviewed and evaluated by Eller (1985). Chapter 3 described physical models of under-ice roughness. Many under-ice scattering loss models can be incorporated directly into existing propagation models that were constructed using a modular architecture.

During the winter season, the tidal flats in the Bay of Fundy are littered with large muddy icebergs that are dense enough to sink. Submerged debris transported by tidal currents may collide with tidal turbines, and the added cost associated with subsurface collisions threatens the viability of tidal-energy projects (also see Section 6.16.2.7). Detection of sediment-laden ice could serve to protect tidal-power infrastructure. Dourado (2015) attempted to identify echoes from the interior of the ice using a broadband echosounder system. Acoustic backscattering measurements from calibration targets encased in bubble-free ice were used to locate strong targets in ice. However, scattering from a planar surface of bubble-free ice overwhelmed echoes from encased targets; hence, it was not possible to conclusively detect solid and hollow spheres in ice. The total echoes from ice with inclusions could not be differentiated from ice without inclusions. These results imply that echoes from sediment-laden ice blocks cannot be interpreted by modeling separate scattering mechanisms within the ice; therefore, future modeling should focus on echoes from sediment-laden ice surfaces.

5.4.2 Arctic Propagation Models

Developments in acoustic propagation modeling for the Arctic Ocean have been very limited. Much of the past effort on characterizing propagation in the Arctic has been devoted to gathering acoustical data and developing empirical models based on that data. While these models tend to be site and season specific with little generality, they do provide basic information on the frequency and range dependence of acoustic propagation in ice-covered regions.

In general, there are four factors peculiar to the Arctic environment that complicate the modeling of acoustic propagation: (1) the ice keels present a rapidly varying surface; (2) the reflection, transmission, and scattering properties at the water-ice interface are not well known; (3) the measurement of under-ice contours is difficult; and (4) the diffraction of sound around ice obstacles may be important. In the Arctic half-channel, ray theory may provide a useful predictive method, whereas its utility in temperate-water surface ducts may actually be quite limited. The utility of ray theory in the Arctic derives from the fact that the Arctic half-channel is between one and two orders of magnitude greater in gradient (and much greater in depth) than the temperate-water surface ducts. The strong positive thermocline and halocline produce an exceptionally strong positive sound speed gradient in the subsurface layer. This markedly shortens the ray-loop length, causing many surface reflections for rays with small grazing angles (Mobile Sonar Technology, 1983, unpublished manuscript).

The open ocean region in the Arctic environment also poses potential problems as far as existing propagation models are concerned. Surface reflection losses, which may not be significant in deep-ocean propagation, become more important in the upward-refracting environment of the Arctic since multiple surface reflections now play a dominant role. Moreover, multiple ducts are prevalent in this region and many propagation models do not adequately treat them.

Two basic modeling approaches are currently being pursued: the application of ice-scattering coefficients to existing numerical models of acoustic propagation and the development of empirical models. These two approaches are discussed in more detail below.

5.4.3 Numerical Models

Numerical models of underwater acoustic propagation specifically designed for ice-covered regions are limited. Since the Arctic half-channel acts as a low-pass filter (discriminating against higher-frequency components), and since bottom interaction is not as important as in other ocean regions (because of the upward-refracting sound speed profile), most modeling applications in the Arctic have employed either normal mode (Gordon and Bucker, 1984) or fast-field (Kutschale, 1973, 1984) approaches. These models are considered most appropriate for prediction of low-frequency (<350 Hz) propagation at long ranges (>25 nmi). Other modeling techniques (e.g., ray theory and PE) have also proved suitable for calculation of TLs once the required ice-scattering algorithms were incorporated (e.g., Chin-Bing and Murphy, 1987). For example, Stotts et al. (1994) developed a ray-theoretical propagation model called ICERAY that is valid for under-ice environments, and at least one implementation of the PE technique has been modified to include the effects of ice scattering (refer back to Table 4.1).

Shot signals in the Arctic channel propagate via low-order normal modes, a result of the constructive interference of RSR rays travelling in the upper few hundred meters of the water column. Due to scattering at the rough boundaries of the ice, only low frequencies (typically less than 40 Hz) can propagate to long distances in the Arctic channel. If the channel is sufficiently deep along the entire propagation path, as over an abyssal plain, then the received wave trains display an impulsive character corresponding to arrivals of deep-penetrating RSR rays while the latter part of the wave trains retain the nearly sinusoidal character typical of low-order modes traveling in the upper layers of the water column (Kutschale, 1984). Using the normal-mode theory of Pekeris (1948), the measured frequency dispersion of low-order modes has often been used to infer acoustic properties of sediments in shallow-water waveguides. Kutschale and Lee (1983) and Kutschale (1984) utilized a similar approach to infer acoustic properties of bottom sediments in the Arctic based on the dispersion of high-order normal modes. A normal mode model was used to interpret the frequency dispersion of bottom-interacting wave trains. The MSPFFP model was used to derive a geoacoustic model that predicted synthetic waveforms matching the dispersion profiles of the bottom-interacting signals. More information on MSPFFP can be found in Table 4.1.

Krupin (2005) used iterative algorithms based on the WKB method to achieve fast and accurate calculations of the local group velocities and attenuation coefficients for deep-water regions of the Arctic Ocean. Specifically, modal characteristics were determined using the adiabatic approximation by integrating the local group velocity and attenuation coefficient over the horizontal distance between the ends of the propagation path.

Gavrilov and Mikhalevsky (2006) analyzed data collected during an experiment called ACOUS (Arctic climate observations using underwater sound) to determine the correlation between acoustic propagation loss and the seasonal variability of sea-ice thickness. They deployed autonomous sources that transmitted tomographic signals at 20.5 Hz once every 4 days over a 15-month period. These signals were received on a vertical array deployed at a distance of 1250 km. They studied the influence of ice parameters, variations of the sound-speed profile and mode-coupling effects on the propagation losses of individual modes. This experiment demonstrated the possibility of using low-frequency, cross-Arctic acoustic transmissions to observe long-term, basin-scale changes in the Arctic ice cover. These results were compared with theoretical predictions generated by the KRAKENC model (KRAKENC is a special version of KRAKEN that finds eigenvalues in the complex plane and includes material attenuation in elastic media such as ice). According to the theory of ice scattering, coherent losses of low-frequency signals propagating under the ice cover depend primarily on the height and correlation length of the ice roughness. This relationship allows one to construct an acoustic inversion method for remote sensing of ice thickness. However, the accuracy of such inverse sensing of ice thickness needs to be examined in more detail using ice-draft profiling along the acoustic path at the time of observation.

Collins (2015) extended the PE method to handle problems involving ice cover and other thin elastic layers. PE solutions are based on rational approximations that are designed using accuracy constraints (to ensure that the propagating modes are handled properly) and stability constrains (to ensure that the nonpropagating modes are annihilated). The nonpropagating modes are especially problematic for cases involving thin elastic layers. It has been demonstrated by previous investigators that stable results may be obtained for such problems by using rotated rational approximations and generalizations of these approximations. The current approach was applied to problems involving ice cover with variable thickness and sediment layers that taper to zero thickness; for some cases, it was necessary to use a rotation angle greater than 90°. The rotated rational approximations achieved stability at the cost of moving all of the accuracy constraints far from the eigenvalues that were associated with the propagating modes.

Collis et al. (2016) introduced a range-dependent elastic PE solution for acoustic propagation in an ice-covered underwater environment. The solution was benchmarked against a derived elastic normal-mode solution for range-independent underwater acoustic propagation (OASES). Results from both solutions accurately predicted plate-flexural modes that propagated in the ice layer as well as Scholte interface waves that propagated at the boundary between the water and the seafloor. The parabolic-equation solution was used to model a scenario with range-dependent ice thickness and a water sound-speed profile similar to those observed during the

2009 Ice Exercise (ICEX) in the Beaufort Sea. Effects due to elasticity in overlying ice can be significant enough that low-shear approximations (such as effective complex density treatments) might not be appropriate. Several issues remain to be explored: (1) while some roughness in the ice layer can be included when using the elastic PE (due to its ability to handle range dependence and refined gridding), the inclusion of a rough water-ice boundary must be examined; (2) if the environment contains large ice keels, the forward scattering approximation intrinsic to the elastic PE might be violated (the use of a single-scattering approximation could improve results); the incorporation of the upper fluid-solid boundary condition into a two-way PE could also be investigated; finally, three-dimensional environments remain an important area of study for elastic PE solutions, and the inherently three-dimensional nature of the sub-ice surface represents an important problem where such solutions are relevant.

5.4.4 Empirical Models

Empirical models are inherently limited by the databases from which they were derived. Attempts to fit results from a large data set with simplistic curves generally imply large errors in the model results. Comparisons of model results with data are clearly quite limited by a lack of comprehensive data sets. A recent intercomparison of available empirical techniques applicable to the Arctic revealed large unresolved discrepancies in the manner in which under-ice scattering losses were computed (Deavenport and DiNapoli, 1982). Two of the better-known empirical models are described below.

5.4.4.1 Marsh-Mellen Model

The Marsh-Mellen Arctic TL model (Marsh and Mellen, 1963; Mellen and Marsh, 1965) is based on observations made during the summers of 1958 and 1959 between Arctic drift stations separated by 800–1200 km. The measured arrivals were found to consist of a dispersive, quasi-sinusoidal wave train in the 10–100 Hz frequency range. A half-channel model in which the higher frequencies were attenuated by under-ice scattering explained these features. Using these and other experimental data, long-range, low-frequency (<400 Hz) TL data in the Arctic were fitted with an equation of the form

$$\mathrm{TL} = 10 \log_{10} r_0 + 10 \log_{10} R + \alpha_s N_s \tag{5.11}$$

where r_0 is the skip distance for the limiting ray, N_s is the number of surface reflections, R is the range in meters ($R = r_0 N_s$), and α_s is the loss per bounce. The wind-generated ocean wave spectrum (Marsh, 1963) was taken to approximate the ice roughness.

5.4.4.2 Buck Model

The Buck Arctic TL model (Buck, 1981) consists of a short-range (10–100 nmi) and a long-range (100–1000 nmi) model for low-frequency (<100 Hz) TL in that part of the Arctic Ocean deeper than 1000 m. The crossover range at 100 nmi is where higher-order, deeper cycling modes begin to dominate the acoustic propagation. These preliminary empirical models represent linear regression fits to winter data collected in the Beaufort Sea in 1970, in the Fram Strait in 1977, and in an intermediate area in 1979 for a source depth of 244 m and a receiver depth of 30 m:

Short range (10–100 nmi)

$$\mathrm{TL} = 62.4 + 10\log_{10} R + 0.032\,f + 0.065\,R + 0.0011\,f\,R \tag{5.12}$$

Long range (100–1000 nmi)

$$\mathrm{TL} = 68.5 + 10\log_{10} R + 0.07\,f - 0.0015\,sR + 0.000487\,f\,s\,R \tag{5.13}$$

where f is the frequency (Hz), R is the range in nautical miles (nmi) and s is the standard deviation ice depth (m), also referred to as the under-ice roughness parameter. A chart of the standard deviation ice depth (s) suitable for use in Equation 5.13 is presented in Figure 5.11 (from Buck, 1985).

There is some doubt regarding the applicability of Equations 5.12 and 5.13 to summer conditions and to other source-receiver depth combinations. Therefore, these equations should be used with caution. To obtain coarse estimates, refer to Figure 3.29 where average curves of Arctic TL versus frequency have been derived from measured data (Buck, 1968).

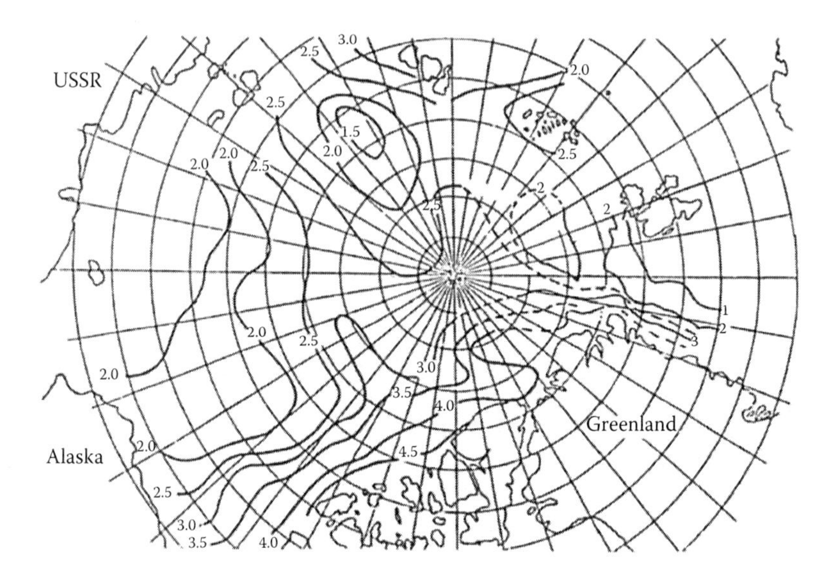

FIGURE 5.11 Estimates of under-ice roughness in standard deviation (m) about the mean ice depth. These estimates were derived, in part, from nuclear submarine sonar data analyzed by L.A. LeSchack. (Adapted from Buck, B.M. 1985. Long term statistical measurements of environmental acoustics parameters in the Arctic [AEAS Report No. 2—Low-frequency transmission loss measurements in the central Arctic Ocean]. Polar Res. Lab., TR-55.)

5.4.5 Field Experiments

Mikhalevsky et al. (2015) observed that new technologies are needed to allow synoptic *in situ* observations year-round to monitor and forecast changes in the Arctic atmosphere-ice-ocean system at daily, seasonal, annual, and decadal scales. Multipurpose acoustic networks were discussed including subsea cable components. These networks provide communication, power, underwater and under-ice navigation, passive monitoring of ambient sound (ice, seismic, biologic, and anthropogenic), and acoustic remote sensing (tomography and thermometry), supporting and complementing data collection from platforms, moorings, and vehicles. The development and implementation of regional to basin-wide acoustic networks were advocated as an integral component of a multidisciplinary *in situ* Arctic Ocean observatory. Applications of acoustic networks in the Arctic include: acoustic thermometry and tomography; underwater acoustic navigation and communication; ambient noise and bioacoustics in the Arctic; monitoring seismicity in the Arctic; and Fram Strait multipurpose acoustic network (a key location to study the impact of the Arctic Ocean on global climate). Accordingly, Mikhalevsky et al. (2015) recommended that a seafloor-cabled network be constructed in the Arctic to provide near-real-time access to data, instruments, and platforms in the Arctic environment. Although cables are expensive to install, the life-cycle costs over a climate cycle (several decades) would be substantially lower than those of alternative solutions such as autonomous seafloor moorings, ice camps, and shipborne expeditions. Cables could also provide high bandwidth connectivity to Arctic communities to enhance connectivity. An informative review of recent and relevant field experiments in the Arctic provided a useful context for these recommendations.

5.5 DATA SUPPORT REQUIREMENTS

The development of numerical models requires data with which to support model initialization and model evaluation. Initialization of propagation models requires various descriptors of the ocean environment including the water column, the sea surface, and the sea floor. Wave-theoretical models tend to be more demanding of bottom sediment information than do the ray-theoretical models. Evaluation of propagation models requires TL measurements that are keyed to descriptions of the prevailing ocean environment. Such coordinated measurements are necessary to ensure that the model is initialized to the same environment for which the TL data are valid (e.g., Hanna, 1976).

Data support requirements are further complicated by the fact that long-range TL can rarely be considered to occur in a truly range-independent environment. With few exceptions, changes in sound speed or water depth (among other parameters) can be expected not only as a function of range but also as a function of azimuth (or bearing). Consequently, data management in support of model development and operation can be formidable, and is often a limiting factor in the proper employment and evaluation of numerical models.

When working with a variety of propagation models (either different models within one category or models from different categories, as illustrated previously in Figure 4.1), it becomes obvious very quickly (sometimes painfully so) that each model not only requires somewhat different parameters but sometimes requires varying format specifications for any given parameter. In response to this situation, two related developments have occurred. First, gridded databases (i.e., those organized by latitude and longitude) have been established that contain all the required ocean environmental parameters in a standardized format with automated update and retrieval mechanisms. Second, model-operating systems (see Chapter 11) have been created to automatically interface the standard databases with the various models resident in the operating system, and to accommodate the format specifications peculiar to each of the models.

A graphic example of a retrieval from a gridded ocean database is presented in Figure 5.12. This presentation of rang-dependent sound speed and bathymetry data is suitable for use in many existing propagation models. Chapter 11 presents a summary of available oceanographic and acoustic databases.

One of the newest automated ocean data products is the generalized digital environmental model (GDEM), developed by Dr. Tom Davis at the US Naval Oceanographic Office (Wells and Wargelin, 1985). GDEM uses a digitized database of major oceanic features together with their climatic location to display temperature, salinity, and sound speed profiles on a 30-minute latitude and longitude grid. A simple parabolic fit is used to describe the deep portion of the sound speed profile. The more variable near-surface fields of temperature, salinity, and sound speed

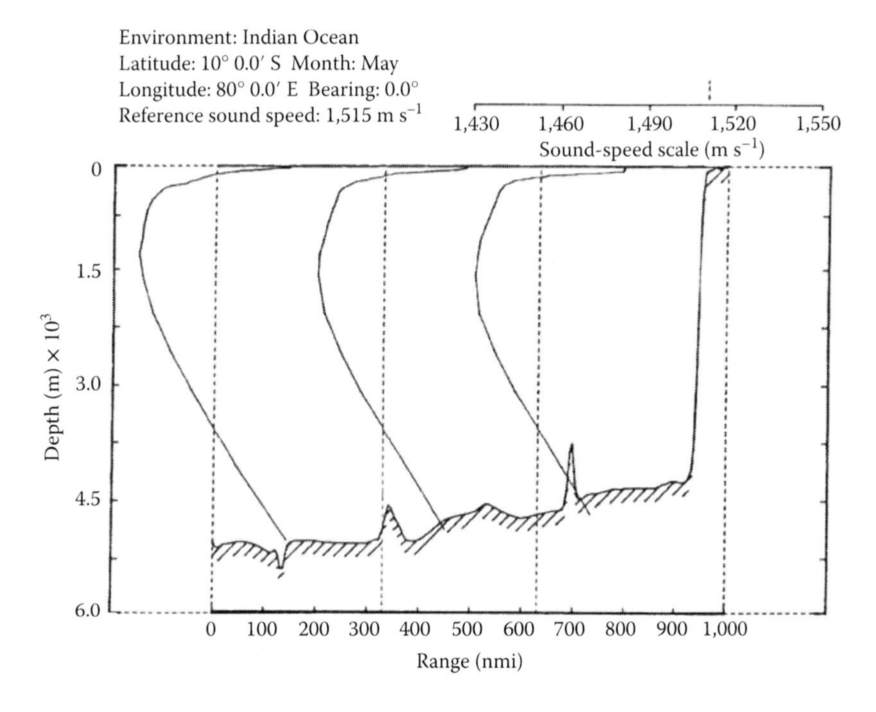

FIGURE 5.12 Example of a range-dependent ocean environment extracted from a gridded database appropriate for use in many propagation models.

are constructed using orthogonal polynomials between the surface and a depth of 800 m. GDEM has been rigorously compared with reliable climatologies (Teague et al., 1990). Refer to Chapter 11 (specifically Table 11.5) for further details regarding GDEM. The specification of sound speed profiles for initialization of propagation models is very important, and the following sections will elaborate on this topic.

The GDEM (Generalized Digital Environment Model) has served as the US Navy's global gridded ocean temperature and salinity climatology since its development in 1975. A new version of this climatology, named GDEM-V 3.0, was introduced in 2009 (Carnes, 2009). The largest change in methodology was to discontinue fitting temperature profiles to a prescribed nonlinear function and replace it by a method that horizontally interpolates temperature or salinity values separately at each depth level. The horizontal interpolation has been redesigned to avoid averaging dissimilar profiles separated by land boundaries. The vertical gradients of averaged profiles have been improved, particularly in shallow water, by application of a gradient correction algorithm based on statistics derived from the profile data set.

5.5.1 Sound-Speed Profile Synthesis

In preparation for a typical modeling run, a sound speed profile is reconstructed (or synthesized) from a tabulation of sound-speed values at discrete depths. The best functional representation of the synthesized sound speed profile varies with the type of modeling technique employed. There are basically two different approaches in use, each with its own advantages and disadvantages, as summarized in Figure 5.13.

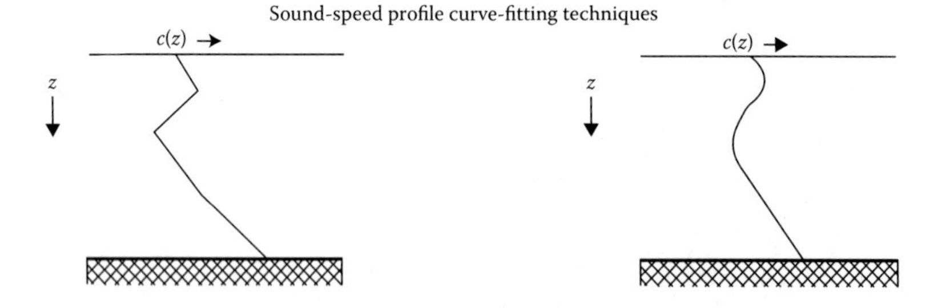

Segmented constant gradient

- Discontinuous first derivative leads to false caustics in ray tracing techniques.
- Line segments in which $1/c^2$ varies linearly with depth permits mode functions to be expressed in terms of Airy functions in normal mode techniques.

Curvilinear or continuous gradient

- Quadratic equations—fit data points to within specified tolerances.
- Cubic splines—do not result in closed-form solutions for ray path equation in ray tracing-techniques; continuous first and second derivatives.
- Conic sections and hyperbolic cosines—result in closed-form solutions for ray path equation using elliptical and hyperbolic ray paths in ray-tracing techniques.
- Exponential forms—facilitate matrizant solution in fast-field theory techniques.

FIGURE 5.13 Summary of sound speed profile curve-fitting techniques.

5.5.1.1 Segmented Constant Gradient

The sound speed profile can be constructed by connecting the discrete points with straight-line segments. Because of discontinuities in the first derivative of the resulting function, TLs calculated on the basis of ray-tracing techniques are undefined at certain ranges (see Pedersen, 1961). These regions are referred to as false caustics since they are false regions of infinite intensity.

In some propagation models, the sound speed profile is fitted with segments in which the inverse of the sound speed squared (c^{-2}) varies linearly with depth. This often permits a more efficient mathematical solution in which the mode functions are expressed in terms of Airy functions.

5.5.1.2 Curvilinear or Continuous Gradient

Curvilinear profile approximations that preserve the continuity of slope as well as sound speed have been developed by Pedersen and Gordon (1967). The sound speed profile can also be fitted with quadratic equations within specified tolerances (Weinberg, 1969, 1971).

Cubic splines can be used to approximate the sound speed profile for application to ray-theoretical techniques (Solomon et al., 1968; Moler and Solomon, 1970). However, this method does not result in closed-form solutions for the rays. Approximating profiles with conic sections and hyperbolic cosines results in closed-form solutions for the ray-path equations when using elliptical and hyperbolic ray paths (Flanagan et al., 1974).

Exponential forms are sometimes used in those modeling techniques employing fast-field theory to facilitate a matrizant solution. To simplify the mathematical treatment of long-range sound propagation in the ocean, some investigators (Munk, 1974; Flatté et al., 1979) have introduced what is termed a "canonical" model of the sound speed profile. This model has an exponential form that is valid in the vicinity of the deep sound channel axis:

$$c(z) = c_1\left[1+\varepsilon\left(\eta+e^{-\eta}-1\right)\right]$$
$$\eta = \frac{2(z-z_1)}{B} \qquad (5.14)$$
$$\varepsilon = \frac{B\gamma_A}{2}$$

where

$c(z)$ = sound speed as a function of depth
c_1 = sound speed at channel axis (z_1)
η = dimensionless distance beneath channel axis
B = scale depth
ε = perturbation coefficient
γ_A = fractional sound speed gradient for adiabatic ocean

Munk (1974) used the following typical values: c_1 = 1492 m s^{-1}, B = 1.3 km, z_1 = 1.3 km, $\gamma_A = 1.14 \times 10^{-2}$ km^{-1}, and $\varepsilon = 7.4 \times 10^{-3}$.

5.5.2 Earth Curvature Corrections

In problems involving acoustic propagation over great distances (greater than 30 nmi or 56 km), the fact that the reference plane is situated on a spherical surface can no longer be ignored. The curvature of Earth's surface is roughly equivalent to a linear sound speed profile. A linear surface duct is closely analogous to a whispering gallery.

First-order spherical Earth curvature corrections are usually applied to the sound speed profile before any curve-fitting techniques are employed. These corrections are typically of the form (Watson, 1958; Hoffman, 1976)

$$c_i = c_i'\left(1 + \frac{z_i'}{R_0}\right) \tag{5.15}$$

$$z_i = z_i'\left(1 + \frac{z_i'}{2R_0}\right) \tag{5.16}$$

where

R_0 = mean radius of Earth (6370.949 km)
c_i' = input sound speed at depth z_i'
c_i = corrected sound speed at corrected depth z_i

The effect of curvature is more than just a matter of chords versus arcs of great circles. Rather, the important feature is that the contours of constant sound speed (as represented on a range-depth plane) are actually concentric spheres instead of flat, parallel planes. The differences in ray angles as calculated by Snell's law produce effects of potential importance to sonar performance predictions. A detailed comparison between state-of-the-art ray theory and experimental TL data (Pedersen et al., 1984) revealed that Earth curvature corrections to the sound speed profile produced model predictions that were in better agreement with the data than would have otherwise been obtained. The effect of curvature was to reduce the range to the leading edge of the first convergence zone (CZ) by about 1%. Errors might be expected to increase with increasing range. Yan (1999) noted that the incorporation of spherical-Earth-curvature corrections into 2D ray equations considerably extended the applicability of these equations in certain classes of 3D problems.

Based on historical field data, Vadov (2005) noted that computed ranges to the first CZ typically overestimated experimentally observed ranges by 1–2 km. Part of this difference could be attributed to diurnal variability in the propagation conditions. Some of this diurnal variability could also be attributed to the lack of simultaneity between the measured sound speed profiles used to calculate propagation conditions and the experimentally measured CZ ranges. More substantial differences could be attributed to differences among the various algorithms used to convert observed temperature and salinity profiles into sound speed profiles used for computation of the propagation-path geometries. It was noted that there is no universal formula that can be used reliably to calculate distances to the first CZ (although eight formulas were examined in the course of the study). However, judicious selection of appropriate sound-speed conversion formulas (presumably reflecting the appropriate domains of applicability) could reduce range differences to about 0.5 km in a variety of ocean areas.

5.5.3 Merging Techniques

Propagation models generally require sound speed data from the surface of the ocean to the sea floor. When the local water depth exceeds the limits of available sensing devices such as the XBT, the required surface-to-bottom sound speed profiles are often derived from a combination of *in situ* data for the near-surface (or shallow) layer together with climatology for the deep layer (Fisher and Pickett, 1973).

A number of algorithms have been developed for automatically merging data from the shallow and deep layers. Statistical relationships based on oceanographic analyses are typically employed to associate the shallow-layer data with the most appropriate deep-layer climatology. This practice is usually satisfactory when the historical databases are characterized by high spatial and temporal resolutions. However, merging algorithms can occasionally introduce errors into the derived surface-to-bottom sound speed profile as a result of curve-fitting artifacts or incomplete data near the juncture of the two layers. Additional errors may result from the measurement and digitization processes. Careful inspection of graphical representations of the merged profiles can reduce these types of errors.

5.6 CELLULAR AUTOMATA

Kwon (2016: Chapter 4) discussed the application of cellular automata (plural of automaton) to underwater acoustics. Specifically, this included the formulation of TL and boundary conditions. These formulations were then applied to wave propagation across a flat-bottomed ocean floor, over a curved hill, and over a sloping bottom.

Cellular automaton (CA) is an alternative computational technique that can be used to model the behavioral response of dynamic systems. Since this method allows great flexibility in the application of various types of boundary conditions, it is particularly attractive for use in developing alternative propagation models for ocean acoustics. The modeling scheme created a profile of propagation losses versus range in an acoustic medium (Bailey, 2013). The discrete nature of the CA technique defined a domain within which every point was governed by a set of rules that defined finite values at each location. As time progressed in discrete time iterations, the finite value at each point changed synchronously according to a rule that was being implemented at that time step. The nature of the rules being applied to each location was dependent on the finite values contained by its associated neighbors. Overall, the beauty behind the CA technique is that it can produce a rich spectrum of complex patterns from the application of easy-to-apply rules. Consequently, the nature of these patterns captured the essence behind the behavioral response of complex systems. This particular modeling scheme created a generic profile of propagation losses in terms of decibels (dB) versus range in an acoustic medium. In addition, visualization of wave propagation in a cellular domain with complex boundaries could be animated to demonstrate their effects. The nature of all restraints on each boundary is not easily modeled by ordinary numerical techniques; however, they were easily modeled with the CA method. Furthermore, a multi-grid meshing technique was introduced to increase the computational efficiency in modeling the CA domain. It was shown that modeling complex boundary conditions could be performed easily with CA methods. As with other modeling methods, however, the computational time increased when the solution refinement increased. An alternative multi-grid modeling scheme was shown to increase the performance time of the CA method significantly. This improvement was dependent on the total number of global grid points inside the multi-grid domain. The end result showed that a multi-grid (with fewer nodal points) produced accurate results that could replicate a uniform grid that utilized a larger number of nodal points (Bailey, 2013).

6 Special Applications and Inverse Techniques

6.1 BACKGROUND

This chapter reviews specialized applications for underwater acoustic (UWA) models. Progress in these particular areas is advancing rapidly and the following sections are intended to identify areas of particular interest at the present time as well as areas that may be appropriate for future research. Topics will touch on: statistical representations of acoustic fields and consequent prediction uncertainties; advanced applications for inverse techniques and parameter estimation; acoustic impacts of oceanographic phenomena and rapid environmental assessments (REAs); developments in underwater acoustic networks (UANs), communications, and vehicles; and marine mammal protection. In many cases, results are not conclusive and more research is clearly required. Nevertheless, these seemingly dissimilar developments do reveal discernable trends in modeling applications. At the same time, deficiencies in current modeling capabilities are revealed, pointing to future opportunities (or emerging frontiers) for model enhancements.

Over the past decade, the technical and popular literature has often described changes in the ocean soundscape. A soundscape is a combination of sounds that characterize, or arise from, an ocean environment. The study of a soundscape is sometimes referred to as acoustic ecology. Changes in the soundscape have been driven by anthropogenic activity (e.g., naval-sonar systems, seismic-exploration activity, maritime shipping, and wind farm development) and by natural factors (e.g., climate change and ocean acidification [OA]). The disruption of the natural acoustic environment results in noise pollution. In response to these developments, new regulatory initiatives have placed additional restrictions on uses of sound in the ocean: mitigation of marine mammal endangerment is now an integral consideration in acoustic-system design and operation. Modeling tools traditionally used in UWAs have undergone a necessary transformation to respond to the rapidly changing requirements imposed by this new soundscape. Advanced modeling techniques now include forward and inverse applications, integrated-modeling approaches, non-intrusive measurements, and novel processing methods.

Farina (2014) provided a broad overview of soundscape ecology. Topics addressed included: underwater sound sources; marine sound signatures; acoustic masking in freshwater ecosystems; acoustic masking in marine ecosystems; ship and boat traffic impacts on marine life; effect of traffic noise on wildlife; noise as signal in marine environments; noise from multisource environments; oil spill, noise, and effects on animals; wind turbines and noise; and military noise.

A recent book by Caiti et al. (2006) summarized the results of a workshop held in Italy in 2004 to review progress in the application of inverse-acoustic methods that have the potential to provide fast and accurate characterizations of shallow-water environments. Contributions ranged from ocean-acoustic tomography to estimation of the seabed and subbottom properties, to marine biology. The editors noted that ambient noise in the ocean is increasingly being investigated as a replacement for traditional acoustic sources to make acoustic inversion less invasive and more environmentally friendly as a methodology with which to probe the ocean.

6.2 STOCHASTIC MODELING

Methods by which to investigate the stochastic (versus deterministic, or completely predictable) nature of acoustic propagation in the ocean have been refined as part of ongoing, long-term projects. Such efforts are concerned with probabilistic predictions of the environmental limits to sonar aperture designs (Perkins et al., 1984). A workshop conducted at the US Naval Research Laboratory addressed the model and database requirements for current and new highly complex sonar systems (Spofford and Haynes, 1983). Among the topics addressed were estimates of both the first and second moments (variances) or spreads in space, time, and frequency required in a variety of ocean environments to support the operation of current sonar systems and the design of advanced sonar systems. (Also see the discussions in Section 6.13 regarding prediction uncertainties in complex environments.)

A review of stochastic signal modeling efforts (Wood and Papadakis, 1985) concluded that spread function models are efficient for computing lower-order statistics, although with limited capability. General-purpose Monte Carlo models are desirable for their higher-order capabilities, despite higher computing costs.

In the so-called "model-based" approach to acoustic signal processing in the ocean, mathematical models of physical phenomena and measurement processes are incorporated into the processor. The inclusion of a propagation model in the signal-processing scheme introduces environmental information in a self-consistent manner. Furthermore, stochastic properties of the oceanic medium can be included in the model. Solutions using state-space techniques employ two sets of equations: the state equation and the measurement equation. The state equation describes the evolution in space of the modal and range functions while the measurement equation relates these states to the hydrophone array measurements. One specific implementation (Candy and Sullivan, 1992) cast the normal-mode propagation model into state-space form, extended the formulation to a Gauss-Markov model (least squares estimator), and applied the results to an ocean-acoustic signal-processing problem in the context of a horizontally stratified ocean with a known source position. A related implementation investigated the inverse reconstruction of a sound speed profile from hydrophone measurements (Candy and Sullivan, 1993).

The use of finite element (FE) techniques for computing UWA propagation has been studied as part of a computationally intensive probabilistic acoustics program, the major objective of which was to develop models for propagation of the moments of the acoustic field in regions where the ocean boundaries are random surfaces (Goldstein, 1984). The possibility of using exact techniques such as the finite-element method (FEM) for solving the wave equation have been made more attainable by the augmentation of computers with array processors. Recent developments in finite element modeling include ISVRFEM (Pack, 1986), the Finite-Element Parabolic Equation (FEPE) (Collins, 1988a), the Finite Element Ocean Acoustic Model (FOAM) (Murphy and Chin-Bing, 1989), and the Seismo-Acoustic Finite Element (SAFE) model (Murphy and Chin-Bing, 1991). Murphy and Chin-Bing (2002) described the development and validation of FEM models for application to acoustic and elastic wave propagation in complex ocean environments.

6.3 BROADBAND MODELING

Broadband (also referred to as pulse or wideband) propagation modeling is concerned with simulating the effects associated with the transmission of a signal characterized by a frequency spectrum (versus a single-frequency continuous wave). When considering such signals, the simplest approach has been to first calculate the geometric mean frequency (f_M), defined as $f_M = (f_1 f_2)^{1/2}$, where f_1 and f_2 are the lower and upper limits, respectively, of the frequency band. The propagation of an equivalent signal at frequency f_M is then simulated using one of the available propagation models. When the bandwidth is small, this approach probably generates a reasonable approximation. Otherwise, this approach may lead to substantial errors, particularly when the spectrum is not flat over the bandwidth, and other methods must be tried including Fourier synthesis and time-domain methods.

In Fourier synthesis, multiple executions of an existing propagation model are performed over the frequency range (f_1, f_2) at a number of discrete frequencies at intervals Δf, where Δf might be 1 Hz, for example. The resulting transmission losses for each frequency in the bandwidth are then combined through an appropriate weighting and averaging process (i.e., an interpolation postprocessor) to arrive at the transmission loss corresponding to the bandwidth. Examples of such multifrequency extensions include the ray-theoretical model GAMARAY (Westwood, 1992), the normal mode model PROTEUS (Gragg, 1985), and the multilayer expansion model MULE (Weinberg, 1985a).

Alternatively, the method developed by McDonald and Kuperman (1987) for modeling the propagation of a broadband linear pulse in a waveguide is one example of a broader class of techniques referred to as time-domain methods (Kuperman, 1985). In principle, the frequency domain wave equation (valid for a single-frequency continuous wave signal) can treat broadband signals by Fourier synthesis of the individual CW solutions over the frequency spectrum. In the presence of nonlinearities, however, interactions among frequency components invalidate the frequency-domain approach. In the time domain, the wave equation can be formulated using methods that remove such pathological limitations from the numerical solutions.

In related developments Porter (1990) developed a time-marched fast-field program (Pulse FFP) for modeling acoustic pulse propagation in the ocean. Collins (1988b) used the time-domain parabolic equation (TDPE) model to investigate the effects of sediment dispersion on pulse propagation. Orchard et al. (1992) developed the 3D time-domain parabolic approximation (TDPA) model for simulating pulse propagation in 3D ocean geometries.

Jensen (1988) summarized wave-theoretical techniques suitable for the practical modeling of low-frequency acoustic pulse propagation in the ocean. Jensen emphasized the computational efficiency of pulse propagation predictions using Fourier synthesis of existing CW propagation models based on normal mode and parabolic equation (PE) approaches. Jensen (1993) further explored these issues by placing particular emphasis on propagation in leaky surface ducts.

Futa and Kikuchi (2001) investigated the use of the finite-difference time-domain (FDTD) method for pulse propagation in shallow water. The FDTD method facilitates a direct analysis of the effect of sediment impedance on the time-depth pattern of the received acoustic pulse. The FDTD method was particularly efficient in analyzing the acoustic field in the vicinity of the sound source. By comparison, the normal-mode approach would have required a complicated analysis of both the discrete and continuous modes in the near field. Existing FDTD methods sometimes fail to match the required accuracy. Strategies to improve conventional methods were explored by Tsuru and Iwatsu (2010) including the application of compact finite-differences on a staggered grid with adjusted coefficients, and the usage of optimized multistep time integration. It was shown that highly accurate simulations were attainable.

Elston and Bell (2004) observed that pseudospectral time-domain (PSTD) methods are closely related to FDTD techniques, which discretize the wave equation by replacing the exact differential operators with local-difference approximations over a discrete space-time grid. A recurrence relation emerged that related the pressure at any node within the grid to the pressure at temporally and spatially adjacent nodes. Iteration of the recurrence relationship generated a numerical solution to the full spatial and temporal evolution of the acoustic field. The FDTD method was computationally intensive since sub-wavelength and sub-period discretizations were necessary to stabilize the solution. Alternatively, the PSTD method represented the infinite-order limit of the FDTD technique and remained stable (and accurate) up to the Nyquist limit. Consequently, for the same computational load, the PSTD method could simulate larger environments at higher acoustic frequencies. Moreover, the PSTD method avoided the numerical dispersion associated with FDTD techniques. In implementation, the PSTD calculated spatial derivates exactly using the global FFT while the FDTD approximated spatial derivatives using local differences. Elston and Bell (2004) successfully used the PSTD method to simulate a sidescan sonar system operating in deep water over a sand-rippled bottom modeled as a directional fractal surface. In particular, the statistical distributions that matched the properly normalized synthetic image histograms were the same distributions that described real sonar data. Also see the discussions in Riordan et al. (2005).

Piao et al. (2008) used the digital waveguide (DWG) mesh method to calculate the acoustic vector fields in a Pekeris waveguide. The UMPE model was modified to compute acoustic particle velocities. Comparisons with the KRAKEN and FFP models demonstrated that the DWG method was accurate in the near-field region of the waveguide. Numerical results showed that the DWGmesh method could be used to simulate the low-frequency, two-dimensional acoustic vector fields in shallow water; moreover, this method is easily applied to the calculation of three-dimensional acoustic fields.

The performance of four broadband models (GRAB, PROSIM, C-SNAP, and RAM) was compared in shallow-water test environments out to ranges of 10 km with signal bandwidths of 10 Hz–1 kHz (Jensen et al., 2003). It was concluded that coupled modes with wavenumber interpolation in both frequency and range were the most promising wave-modeling approach for broadband signal simulations in range-dependent, shallow-water environments. For frequencies higher than 1 kHz, ray-based methods were the most practical tools for simulating broadband signals in such environments.

Wang and Liu (2002) applied the normal-mode method to numerically simulate broadband acoustic propagation in the ocean. Their approach entailed computing the solution at the high-frequency end and then interpolating to the low-frequency end.

Solutions of wideband UWA problems with the classical normal-mode method in the frequency domain are required to solve the problem repetitively for every frequency component of the source-signal bandwidth. Sertlek and Aksoy (2013) presented an analytical time-domain normal mode method appropriate for arbitrary time-dependent acoustic sources for a single layered isovelocity waveguide. An incomplete separation of variables technique was used to solve the inhomogeneous wave equation directly in the time domain. This approach made it feasible to calculate the time-domain acoustic pressure in a single run.

6.4 MATCHED FIELD PROCESSING

The use of matched-field processing (MFP) techniques in UWAs has been explored by a number of investigators, and comprehensive overviews are available (Tolstoy, 1992, 1993; Baggeroer et al., 1993). This technique correlates the acoustic pressure field from a submerged source, as detected at each receiver in a hydrophone array, with the field modeled at the array by assuming a particular source position and ocean environment. Consequently, a high degree of correlation between the experimental and modeled pressure fields indicates a high probability that the source is located at the estimated position in the range-depth plane. The basic components of MFP are illustrated in Figure 6.1 (Tolstoy, 1993). Accordingly, this technique shows promise as a high-resolution localization tool. This technique can also be used to reconstruct prevailing oceanic conditions when the source and receiver positions are known *a priori*.

A variety of mathematical estimator functions have been utilized to perform the comparisons between the experimental and modeled pressure fields. These include linear, minimum variance, multiple constraint, matched mode, eigenvector, maximum entropy, approximate orthogonal, variable coefficient likelihood, and optimum uncertain field processors (Fizell, 1987; Tolstoy, 1993). The outputs of these processors are often presented graphically on so-called "performance surfaces" to facilitate identification of probable source positions. The ambiguity surface illustrated in Figure 6.1 is one example. The matched field technique has been found attractive for application to shallow-water geometries. For example, Bucker (1976) used a conventional cross correlation of the experimental and modeled pressure fields while Del Balzo et al. (1988) used a maximum-likelihood estimator. The maximum-likelihood estimator has also been applied to deep-water Arctic environments by Fizell and Wales (1985). Dosso et al. (1993) used matched field inversion to estimate properties of the ocean bottom. In related work, Snellen and Simons (2008) applied the downhill simplex algorithm to reduce the uncertainty in matched field inversion results.

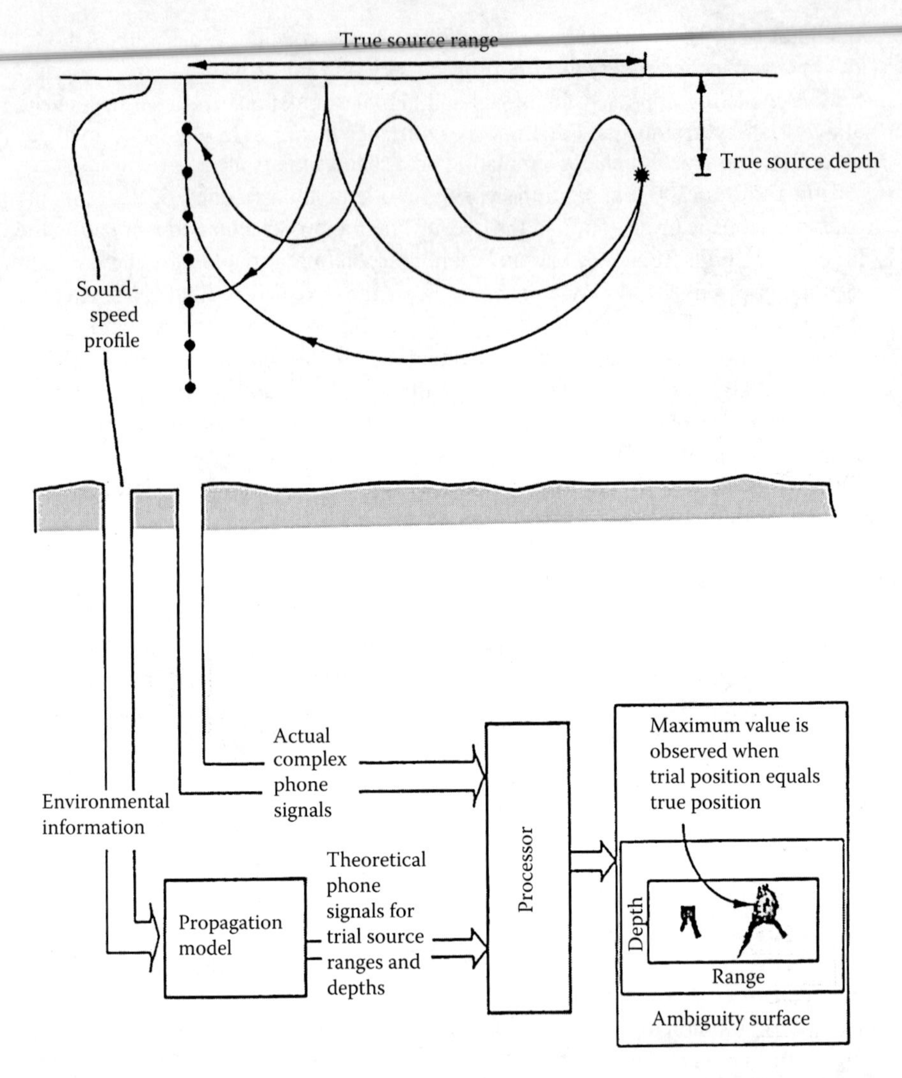

FIGURE 6.1 Illustration of the basic components of MFP. (Reproduced From Tolstoy, A., *Matched Field Processing for Underwater Acoustics*, World Scientific Publishing, Singapore, 1993. With permission.)

Tolstoy (2000) provided an overview of recent applications of MFP to inverse problems in UWAs including source localization, tomographic estimation of deep ocean sound-speed profiles, estimation of shallow-water bottom properties, evaluation of numerical model performance relative to benchmark models, and estimation of hydrophone array deformation. MFP was also utilized to perform geoacoustic inversion; specifically, the SNAP model was used to generate the synthetic acoustic field corresponding to an assumed true parameter value (Yang et al., 2004).

Westwood (1992) explored the use of broadband signals in MFP. In addition, D'Spain et al. (1999) developed an adiabatic normal-mode model to analyze broadband, MFP data collected in shallow water. The model incorporated the concept of "effective depth," which was first introduced by Weston (1960) for a Pekeris waveguide and later extended by Chapman et al. (1989) to include shear waves. In essence, the phase change associated with the reflection of a plane wave from a fluid-elastic interface at the sea floor is equal to that from a pressure-release boundary that is offset a distance below the true bottom (the effective depth). This offset, which is virtually independent of the grazing angle, can be calculated from available waveguide parameters. Thus, the normal-mode wave numbers can be provided by a closed-form expression rather than by more cumbersome numerical complex-root-finding techniques.

Aksoy et al. (2007) used the normal-mode method to compute transmission loss in a Pekeris waveguide. The analysis incorporated the effective-depth method (described above). Results from a simulation (KITMİR) compared favorably with outputs from the KRAKEN model.

Bogart and Yang (1992) used matched mode localization (MML) as an alternative to MFP and found that the matched mode ambiguity surface showed equal or improved (i.e., lower) sidelobes compared to that of MFP; moreover, it was easier to compute. Collison and Dosso (2000) documented a useful comparison of modal decomposition algorithms for matched mode processing.

Dosso (2002) examined how the accuracy of range-dependent propagation models affects the results of matched-field inversion for seabed geoacoustic parameters. Due to errors in both measurements and theory, the appropriate tradeoff between modeling accuracy and computational speed was not obvious, especially in terms of the degradation in information content for the geoacoustic parameters that resulted from inaccurate propagation modeling. The information content was quantified using the marginal posterior probability distributions of the geoacoustic parameters, as computed from a fast Gibbs sampling approach to Bayesian inversion (Gibbs sampling is discussed in more detail in Section 6.12.1). A synthetic example used the parabolic equationto model acoustic fields for a shallow-water, upslope environment in which different levels of modeling accuracy and execution speed were controlled by the range and depth step sizes of the computational grid.

Applications of MFP can be limited by sensitivities to variations in environmental parameters. Virovlyansky (2017) explored an alternative approach called matched shadow processing (MSP) based on comparisons of intensity distributions in the depth-arrival-angle plane using the formalism of coherent states borrowed from quantum mechanics. The distributions of these intensity fields in the phase space included an insonified zone in which the field intensity was relatively large, and a shadow zone where the intensity was close to zero. The parameter by which the fields were compared was the ratio of the integral intensities in the insonified and shadow zones. This approach lowered sensitivity to inaccuracies in the environmental models used for the calculations. It required that the shadow zone occupy a significant part of the phase-plane area corresponding to admissible depths and arrival angles. When working with CW signals, this condition can be satisfied only at short ranges. For pulse signals, the applicable ranges may be wider since the phase space acquires a time dimension.

6.5 TRANSMUTATION APPROACHES

A transmutation approach has been applied to UWA propagation by Duston et al. (1986). As described by Gilbert and Wood (1986), who also provided historical references to the literature, transmutation theory allows one to find an integral operator that transforms the solutions of a simpler partial differential equation into solutions of another, more complicated, partial differential equation. A third partial differential equation exists that the kernel of the integral operator must then satisfy. The advantage gained is that more freedom exists in assigning useful initial boundary conditions to the kernel. Thus, instead of separating the Helmholtz equation into two ordinary differential equations, it is separated into two related patrial differential equations, one of which is solved analytically and the other by a hybrid of symbolic and numerical computations. Koyunbakan (2009) proved the existence of a transmutation operator between two Schrödinger equations with a perturbed, exactly solvable potential.

Makrakis (2014) dealt with the problem of paraxial sound propagation in an ocean with a fluid bottom using a transmutation of nonlocal boundary conditions in ocean acoustics. This approach departed from the primitive nonlocal problem for the Helmholtz equation in the water column (which was derived by eliminating the bottom dynamics through an appropriate Neumann-to-Dirichlet [NtD] map at the water–bottom interface) and instead used the established transmutation method for taking the Helmholtz equation to the Schrödinger equation and constructing a nonlocal boundary condition for the paraxial equation. These nonlocal boundary conditions for paraxial equations facilitated efficient numerical computations for practical applications in ocean acoustics.

6.6 NONLINEAR ACOUSTICS AND CHAOS

Chaos in UWA modeling has been explored by Tappert et al. (1988) and by Palmer et al. (1988a). They observed that ray path solutions exhibit "classical chaos," that is, unpredictable and stochastic behavior. The phenomenon of chaos in UWA is presumed to be caused by the exponential proliferation of catastrophes (in the form of caustics) due to the loss of control implied by the nonseparability of variables in the eikonal equation. Consequently, there exists a prediction horizon that cannot be exceeded even when the ocean environment is known exactly.

The application of chaotic concepts to UWA propagation modeling has been further investigated by Smith et al. (1992a,b), Tappert et al. (1991), and Brown et al. (1991a,b). These studies have emphasized the practical importance of chaotic ray trajectories as limiting factors in generating deterministic predictions of acoustic propagation, particularly in the presence of mesoscale ocean structure. Collins and Kuperman (1994a) suggested that the computational difficulties associated with chaos may be overcome by solving eigenray problems with boundary-value techniques as opposed to initial-value techniques.

The deliberate use of chaotic waveforms as sonar signals has also been investigated. One theoretical study (Alapati et al., 1993) used chaotic metrics (e.g., Rasband, 1990) and conventional ambiguity functions to evaluate the performance of several nonlinear waveforms after convolution with realistic ocean impulse-response functions. These impulse-response (or Green's) functions were generated using the Generic Sonar Model (Section 11.4.4). Ziomek (1985: 4–5) discussed the mathematical formalism of impulse-response functions in a signal-processing context. Because of their amenability to signal enhancement (or noise-reduction) techniques, chaotic waveforms have been suggested for use in bistatic active sonar systems operating in shallow-water regions (Section 11.2.2).

Time-domain analysis of ocean ambient-background pressure fluctuations collected at the Atlantic Undersea Test and Evaluation Center (AUTEC) during a mine-deployment exercise (MINEX) revealed a positive Lyapunov exponent, which identified the system as chaotic. The prediction horizon was confined to a few samples. Determination of the degrees of freedom was important for the construction of physical models and nonlinear noise-reduction filters, which were based on characteristics of the observed degrees of freedom (in this case, nine) from the background acoustic source. The magnitude of the largest Lyapunov exponent provided a measure of confidence for signal-state prediction (Frison et al., 1996).

In nonseparable, range-dependent environments, ray paths can be chaotic, thus placing a fundamental limit on tracing rays by the classical *shooting* approach in which the launch angles of rays from a source point are varied until the rays intersect the receiver endpoint within specified tolerances. To circumvent this problem, Mazur and Gilbert (1997a,b) used Rayleigh-Ritz and simulated-annealing methods rather than minimizing the travel-time integral indirectly.

The effects of ocean internal waves on long-range acoustic pulse propagation were analyzed from the geometrical-optics viewpoint by Simmen et al. (1997), who also investigated the chaotic behavior of rays and the microfolding of timefronts. The extent of the region of the timefront in which strongly chaotic rays appear, and the strength of the rays' sensitivity to initial conditions, were found to depend on the average sound-speed profile, the source-to-receiver range, and the internal-wave spectral model.

Virovlyansky (2005a) developed an approximate analytical approach to describe the chaotic behavior of ray trajectories in a deep-water acoustic waveguide with inhomogeneities induced by a random field of internal waves at horizontal ranges on the order of 10^3 km.

Tappert and Tang (1996) found that groups of chaotic eigenrays tended to form *clusters* having stable envelopes. Sundaram and Zaslavsky (1999) studied the dispersion of wave packets using a parabolic approximation to the wave equation. They noted that, in a manner similar to that observed in quantum chaos, enhanced dispersion due to chaotic ray dynamics was counterbalanced by wave coherence effects.

Bjørnø (2002) reviewed developments in nonlinear UWA over the past 40 years emphasizing development and exploitation of parametric acoustic arrays. Widespread use of parametric arrays (characterized by good beam qualities) has not been realized owing chiefly to poor conversion efficiencies (<2%).

Liu et al. (2013) utilized the quadrature method to solve the wave equation with van der Pol boundary conditions. These particular boundary conditions created chaotic acoustical vibrations typical of those generated by noise signals radiated from underwater vehicles. (The van der Pol oscillator is nonconservative with nonlinear damping. Oscillations converge at the limit cycle, a state at which energy generation and dissipation balance. The van der Pol oscillator has become a common model for oscillatory processes in physics, biology, sociology, and economics. Balthazar van der Pol, who was a Dutch physicist, was a pioneer in the field of radio and telecommunications.) Also see related research by Liu et al. (2016).

6.7 THREE-DIMENSIONAL MODELING

Modeling UWA propagation in three dimensions (3D), sometimes referred to as volume acoustic field modeling (Chin-Bing et al., 1986), has assumed greater importance as sonar systems have become more complex (Jones, 1983). Ray theory and parabolic equations can theoretically treat 3D propagation, although such implementations are rarely accomplished in practice due to computational complexity (e.g., Johnson, 1984). Therefore, such modeling is generally, but not always, accomplished by extending the capabilities of existing range-dependent (2D) techniques such as PE, normal mode, and ray theory models to form composite 3D pictures. Such approaches are commonly termed N × 2D since the models are sequentially executed for *N* adjacent range-dependent (2D) radials (or sectors). Tolstoy (1996) stressed the point that N × 2D (sometimes referred to as 2½ D) approximations to full 3D modeling will fail whenever the out-of-plane energy is significant, as in the case of bottom topography (wedges, ridges, and seamounts), eddies, and fronts.

Three-dimensional propagation modeling has been used to simulate acoustic interactions with seamounts (Medwin et al., 1984) and with mesoscale oceanographic features (Kuperman et al., 1987; Tsuchiya et al., 1999). Coupled with appropriate color graphic displays, 3D modeling promises to be a powerful analytical and predictive tool. In the case involving seamounts, Medwin et al. (1984) were able to show that: (1) depending on the roughness of the sea surface, diffraction over the crest of the seamount can be the strongest contributor to the sound field in the shadow region; (2) the diffracted signal always arrives before the multiply reflected sound; (3) the diffraction loss is proportional to the square root of the acoustic frequency; and (4) a 2D model produces excessive diffraction and excessive multiple reflection signal compared to a more realistic 3D model. The 3D model used by Medwin et al. (1984) is a hybrid solution formed by combining the range-independent FACT model (to calculate acoustic propagation to and from the seamount) with 3D physical models of the frequency-dependent wave interaction at the seamount. Comparisons with experimental data obtained from the Dickens Seamount in the northeastern Pacific Ocean (Chapman and Ebbeson, 1983; Ebbeson and Turner, 1983) verified the accuracy of this model in the frequency range 50–500 Hz. A ray trace illustrating interaction with a seamount is presented in Figure 6.2.

Most seamounts have not yet been surveyed, and their numbers and locations are not well known. Yesson et al. (2011) used global bathymetric data at 30 arc-sec resolution to identify seamounts and knolls (defined in Appendix B). In all, 33,452 seamounts and 138,412 knolls were identified. Seamounts cover approximately 4.7% of the ocean floor while knolls cover 16.3%.

Evans (2006) used a coupled-mode model to study the scattering of ambient noise by a seamount. Specifically, a stepwise coupled-mode procedure was developed

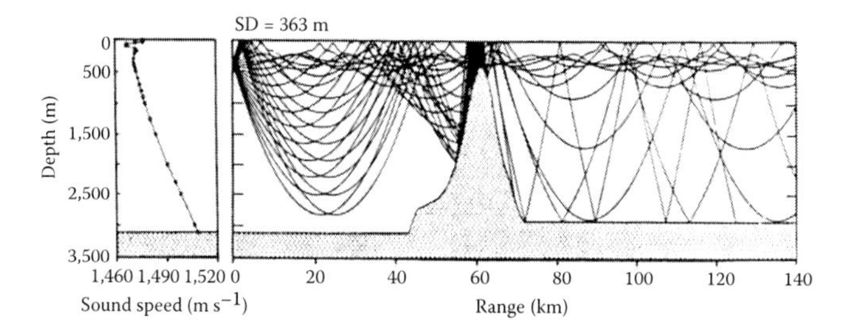

FIGURE 6.2 Ray-trace representation of sound propagation across a seamount based on data from Chapman, N.R. and Ebbeson, G.R. (*J. Acoust. Soc. Am.*, 73, 1979–1984, 1983). (From Jensen, F.B., *IEEE J. Oceanic Eng.*, **13**, 186–197, 1988. Copyright [1988] **IEEE**. With permission.)

for computing the total and scattered fields using a hybrid analytical-numerical solution. The stepwise coupled-mode calculations of ambient-noise scattering by a cylindrically symmetric representation of the Dickins seamount (in the northeast Pacific Ocean) indicated that interactions with the seamount removed energy from the higher angles. This energy was lost to attenuation through bottom interaction, but was not necessarily transferred into lower-angle propagation paths. Numerical results were obtained for a cylindrically symmetric incident noise field. Theoretical results suggested that the effect of the seamount on the vertical distribution of scattered ambient noise would remain the same with a gradual asymmetry in the incident noise field. The energy lost through bottom interaction did not always occur in all three dimensions (i.e., it might be entirely 2D, or coplanar). Thus, the same effect could be simulated with a two-dimensional range-dependent propagation model using a ridge in place of the seamount. However, the transition in the other direction (i.e., from two to three dimensions) was not as obvious.

The 3D Hamiltonian ray-tracing model, HARPO, has been used in tomographic studies of the ocean (Jones et al., 1991; Newhall et al., 1990). Both HARPO (Lynch et al., 1994) and the 3D coupled-mode model CMM3D (Chiu and Ehret, 1994) have been interfaced with sound speed fields generated by the Harvard open ocean model (HOOM). Lee and Schultz (1995) described a stand-alone 3D ocean acoustic propagation model. In related work, Perkins et al. (1993) modeled the ambient noise field in 3D ocean environments.

Work by Ballard (2012) was motivated by acoustic measurements of three-dimensional effects previously recorded at a site off the southeast coast of Florida in 2007. During that experiment, a low-frequency source was towed parallel to the shelf. The acoustic data showed the direct (i.e., nonrefracted) path arrival at the bearing of the tow ship and a second path arrival as much as 30° inshore of the direct arrival. The local shoaling of the bathymetry had already been identified as responsible for the refracted path. In this updated analysis, an acoustic propagation model was applied to predict measurements of out-of-plane propagation recorded on the southeast Florida shelf. The modeling approach was based on a 3D adiabatic mode technique for which the horizontal-refraction equation was solved using a PE in Cartesian coordinates (Collins, 1993d). The ORCA model was used to calculate the modal eigenvalues and eigenfunctions since it solved for nontrapped modes in the continuous portion of the spectrum (Westwood et al., 1996). This approach provided a consistent treatment for modes that propagated at the source and receiver locations, but which were past cutoff elsewhere in the horizontal plane. The local shoaling of the seafloor was shown to be responsible for the horizontal refraction of sound. The higher received level of the refracted path observed in the measured data was explained by range-dependent sediment properties. The arrival along the direct path, which propagated over the elastic bottom, was highly attenuated due to shear conversion; the amplitude of the arrival along the refracted path, which propagated over the sandy slope, was preserved. The modal decomposition of the acoustic field illustrated several aspects of the propagation, including mode stripping along the direct path, as well as individual modal contributions to the refracted path arrival. Additionally, through examination of the individual modal amplitudes, it was demonstrated that the uneven face of the slope was responsible for horizontal focusing of the refracted sound.

Ballard et al. (2015) formulated an approximate normal-mode/parabolic-equation hybrid model in a cylindrical-coordinate system using a separation of variables. The depth-separated Helmholtz equation was solved with a gradient half-space approach (Westwood and Koch, 1999), and leaky modes were included in the solution to account for energy dissipated into the subbottom. Modal amplitudes were calculated from the horizontally separated part of the equation using a hybrid technique in which an accounting of mode-coupling in the radial direction was accomplished with a stepwise coupled-mode technique (Evans, 1983). The PE solution provided a description of horizontal refraction in the azimuthal direction (Collins, 1994). This hybrid model was considered approximate since both mode coupling (in the azimuthal direction) and backscattering were neglected. Next, this hybrid model was used to calculate propagation over an infinitely long cosine-shaped hill where the waveguide was an isovelocity water column over an acoustic half-space having the properties of sand. The solution was calculated for a source located at a depth of 20 m and a frequency of 100 Hz. At this frequency, there were four trapped modes at the location of the source; adding six leaky modes was sufficient to obtain a convergent solution. The infinite cosine hill induced strong horizontal refraction and mode-coupling effects. These results were evaluated by comparison to solutions generated by a FEM technique (Isakson et al., 2014); for this environment, FEM solutions validated the assumption that mode coupling in the azimuthal direction was weak and therefore could be neglected. Solutions from the hybrid model were also compared against those calculated by N × 2D models (Perkins and Baer, 1982) and adiabatic-mode models to further isolate any effects of mode-coupling and horizontal refraction. Three-dimensional features of the acoustic field were identified, including a single mode interference pattern on the side of the hill toward the source, and a shadow zone of the side of the hill away from the source. Using a horizontal ray trace program (Weinberg and Burridge, 1974), it was shown that the horizontal refraction caused a shift in the modal interference pattern on the side of the hill away from the source. In addition, azimuthal angles in the horizontal plane for which the mode-coupling effects were weak were attributed to *transparent resonances* that depended on the length of the 2D projection of the hill.

6.8 OCEAN FRONTS, EDDIES, AND INTERNAL WAVES

Ocean fronts, eddies (or rings), and internal waves were discussed in Section 2.5.2 where they were classified as mesoscale oceanic features. This section will examine their acoustic impacts.

6.8.1 Fronts and Eddies

The distribution of ocean fronts and eddies was discussed previously in Chapter 2. The impacts of fronts and eddies on acoustic propagation have been intensely studied and modeled principally because of their importance to naval operations. Heathershaw et al. (1991) demonstrated the feasibility of coupling the range-dependent propagation model GRASS with a 3D eddy-resolving ocean model to study acoustic propagation through frontal systems. The effects of eddies on acoustic propagation were investigated by Vastano and Owens (1973) using ray-theoretical models. More sophisticated wave-theoretical models have since been employed in more comprehensive studies. For example, Baer (1981) utilized a primary three-dimensional version of the PE to study propagation through an eddy, and Hall and Irving (1989) used an adiabatic normal mode technique to investigate propagation through Baer's example of an eddy. All investigators indicated that eddies significantly modify major characteristics of the acoustic field relative to those obtained in the absence of eddies. Such effects include the shifting of convergence zones, altered multipath arrival sequences, and horizontal refraction, among others (Munk, 1980). In a comprehensive study of warm-core eddies, Browning et al. (1994) found that the greatest acoustic impact was obtained for a shallow source and receiver configuration. Tsuchiya et al. (1999) used a 3D wide-angle PE model to analyze acoustic propagation through an ocean populated by warm-water and cold-water masses approximating the characteristics of eddies in the Pacific Ocean. The 3D model computed horizontal and vertical refraction for comparison with 2D results. Relative to the 3D model, the 2D results underestimated the effects of eddies on acoustic propagation. Shang et al. (2001) proposed a new, less restrictive, criterion for adiabaticity in the presence of ocean fronts or internal solitary waves in shallow water.

As an illustration, the results of Vastano and Owens (1973) are used to quantify the effects of a cold-core eddy (surrounded by warmer Sargasso Sea water) on acoustic propagation. Assuming a frequency of 100 Hz, a source depth (S) of 200 m, a receiver depth (R) of 300 m, and a figure of merit (FOM) of 90 dB (see Chapter 11), the resulting transmission loss curves in Figure 6.3a would result in the detection zones in Figure 6.3b for a sonar located both inside and outside the eddy. The shaded areas denote detection opportunities under the stated conditions. The shifts of the convergence zones are readily apparent as one effect of the eddy. The particular situation illustrated in Figure 6.3 could arise, for example, if a sonar were located on the outer edge of an eddy, with the eddy on the left and the Sargasso Sea on the right.

To further explore the effects of eddies (or rings, as they are sometimes called) on sonar detection, the case of a surface ship located in the Sargasso Sea trying to

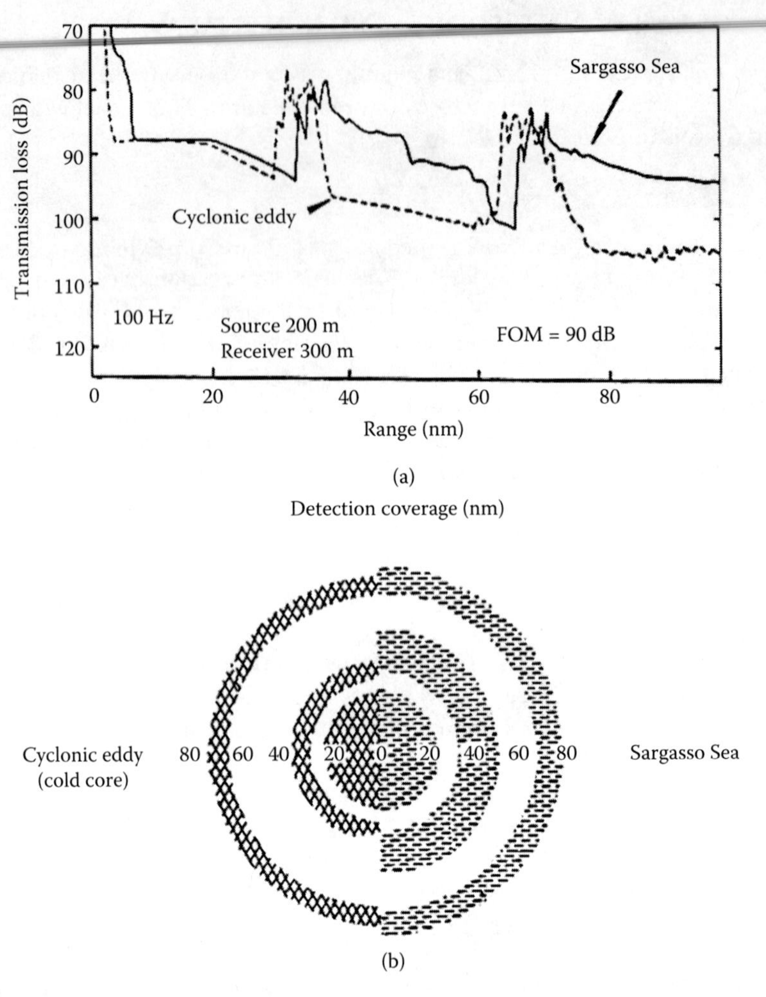

FIGURE 6.3 Effect of a cold-core eddy on acoustic propagation: (a) transmission loss versus range; and (b) corresponding detection zones assuming a figure of merit (FOM) of 90 dB. (TL curves are from Vastano, A.C. and Owens, G.E., *J. Phys. Oceanogr.*, **3**, 470–478, 1973. Copyright by the American Meteorological Society. With permission.)

detect a submarine located inside an eddy will be examined. Figure 6.4 shows contours of constant sound speed through a typical Gulf Stream eddy situated within the Sargasso Sea (Gemmill and Khedouri, 1974). Since sound speed is proportional to water temperature in the upper layers of the ocean, this figure also demonstrates this to be a cold-core eddy since the low sound speed values in the center reflect lower water temperatures relative to the surrounding waters. Also shown (by a dashed line) is the SOFAR channel axis, which indicates the depth of minimum sound speed. Next, consider Figure 6.5, which schematically portrays the corresponding isotherms. A surface ship located within the warmer Sargasso Sea is frustrated in its

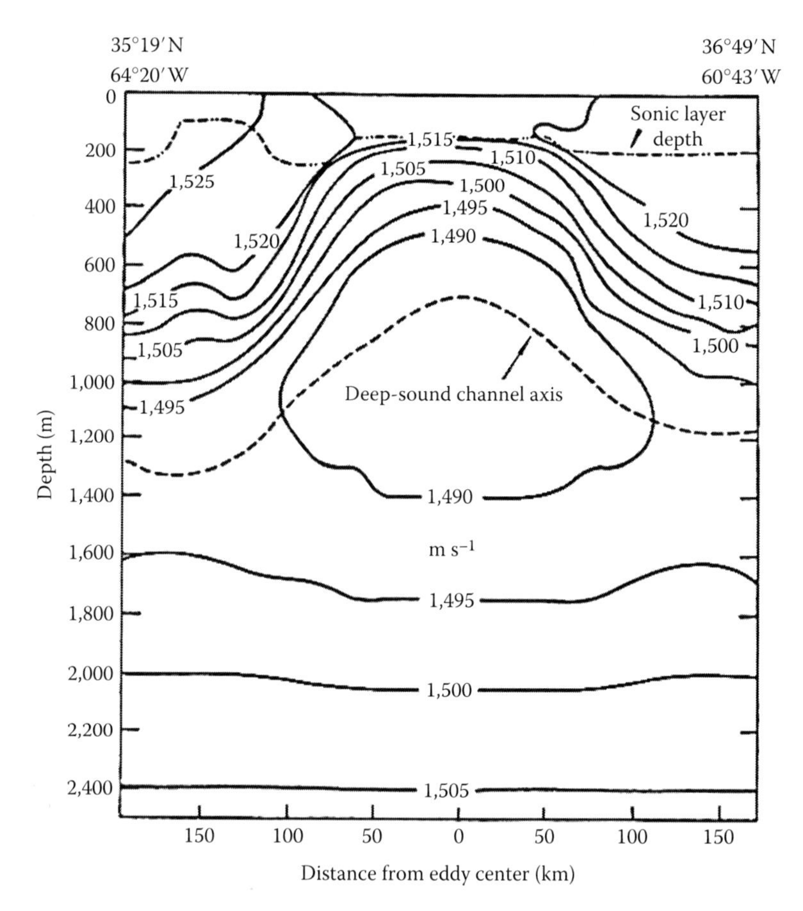

FIGURE 6.4 Sound speed contours through a cold-core eddy. (Adapted from Gemmill, W. and Khedouri, E., A note on sound ray tracing through a Gulf Stream eddy in the Sargasso Sea. Nav. Oceanogr. Off., Tech. Note 6150-21-74, 1974.)

attempts to passively detect the sound energy emitted by the submarine located in the colder interior of the eddy since the sound waves are refracted downward (away from the higher temperatures and higher sound speeds) and away from the ship's passive listening device. The submarine thus can successfully avoid detection by positioning itself inside known eddy locations.

6.8.2 Internal Waves

Zhou et al. (1991) investigated the interaction of underwater sound with internal gravity waves (specifically solitons) in an attempt to explain the anomalous behavior of low-frequency (~300–1100 Hz) acoustic propagation conditions observed in some shallow-water areas. As a result of this investigation, it was noted that acoustic measurements could be employed in an inverse fashion to remotely sense internal wave activity in the coastal zone. Rodríguez et al. (2000) used a range-dependent normal

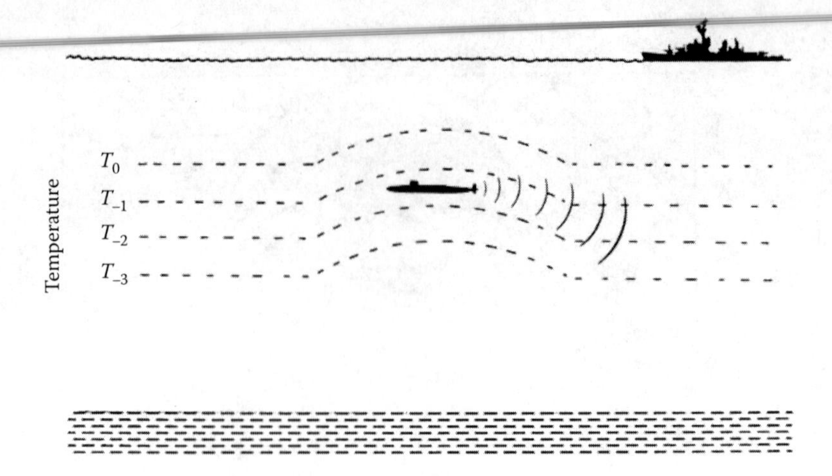

FIGURE 6.5 Schematic illustration of a submarine in a cold-core eddy avoiding detection by a surface ship equipped with passive sonar. Sound is refracted downward away from the ship by the effects of the colder water in the interior of the eddy.

mode model (C-SNAP) to simulate the propagation of acoustic signals through soliton-like fields of temperature and sound speed. The simulation reproduced experimental observations of signal enhancement attributed to focusing effects correlated with peaks in current, temperature, and surface tides. Chin-Bing et al. (1993a) numerically simulated the effects of a solitary internal wave on the low-frequency acoustic field in a shallow-water waveguide using two-way, range-dependent, FEM models (FOAM, FFRAME, and SAFE). These simulations focused on the refractive and backscatter effects of a single soliton wave packet in an effort to understand the cumulative effects of multiple resonances on acoustic propagation. The FEPE model was employed to support the claim that anomalous losses in shallow-water acoustic signals were due to acoustic-mode conversions produced by solitary internal waves (solitons) (Chin-Bing et al., 2003; Warn-Varnas et al., 2003). Additional work was reported by Warn-Varnas et al. (2009) and by Chin-Bing et al. (2009).

Fredricks et al. (2005) observed multipath scintillations from long-range acoustic transmissions on the New England continental shelf and slope during the PRIMER experiment in 1996. Specifically, intensity fluctuations for broadband, 400-Hz multipath arrivals were observed. Acoustic signals were generated by two bottom-mounted sources located on the continental slope in roughly 290-m water depth and were received on a 52-m-long vertical-line-array (VLA) located in 93-m water depth. Propagation ranges were 42.2 km and 59.6 km. Acoustic observables of point intensity, peak intensity, and integrated energy over the VLA were treated in terms of the scintillation index, log-intensity variance, and intensity probability density functions (PDFs). Variability of the observables was decomposed into high-frequency and low-frequency components with timescales less than and greater than 2 hours to facilitate correlation to ocean processes at different timescales. Numerical simulations using the RAM propagation model, in conjunction with a quasi-random undular tidal bore model, were able to reproduce many of the observed intensity fluctuations to within a factor of two. Internal tides were modeled using first-order Korteweg-de Vries (KdV) dynamics, and the internal-wave induced sound-speed fluctuations were modeled as a vertical advection of the local sound-speed structure. The KdV equation is a nonlinear, partial differential equation of third order that was first formulated as part of an analysis of shallow-water waves in canals (first published in 1895). Subsequently, this equation has been used to study a wide range of physical phenomena including those exhibiting shock waves, traveling waves, and solitons. Many different closed-form, series approximations, and numerical solutions are known for particular sets of boundary and initial conditions.

Oba and Finette (2002) simulated acoustic propagation in a shallow-water, anisotropic ocean environment at frequencies of 200 Hz and 400 Hz for horizontal ranges extending to 10 km. Their simulation employed a 3D PE code based on differential operators representing wide-angle coverage in elevation and narrow-angle coverage in azimuth. The water column was characterized by random volumetric fluctuations in the sound-speed field induced by internal gravity waves superimposed on a thermocline. There was a localized contribution from a solitary wave packet (undular bore, or solibore). The resulting simulation showed azimuthal filtering of the propagated field, with the strongest variations appearing when the propagation was parallel to the solitary wave depressions of the thermocline. The solitary wave packet was interpreted as a nonstationary oceanographic waveguide within the water column that preferentially funneled acoustic energy between the thermocline depressions.

Tang and Tappert (1997) used a broadband model (UMPE) to explain the lack of multipath replicas of a transmitted pulse in broadband acoustic experiments in the Straits of Florida. The observed single broad cluster was attributed to the effects of internal waves, which produced moving acoustic "footprints" on a rough seafloor. Tielbürger et al. (1997) investigated the acoustic field properties in a shallow-water waveguide where the sound speed had a deterministic, time-independent component and two stochastic components induced by internal-wave activity. Simons et al. (2001) used a broadband normal mode model (PROSIM) to simulate variability in a shallow-water channel resulting from fluctuations in oceanographic parameters. The acoustic band of interest was 1 kHz–8 kHz.

Recent research (Godin et al., 2006) concluded that the commonly assumed "frozen-medium" approximation (i.e., time-independent ocean) appears to be sufficiently accurate for modeling the forward scattering of sound by internal gravity waves. It was also noted that acoustic frequency shifts induced by temporal variations of the sound speed field are small. The study used a 2D version (RAY2C) of a 3D model developed by A. G. Voronovich and V. V. Goncharov (unpublished Russian manuscript, 1985). Within the ray approximation, the techniques reduce the simulation of 3D and 4D acoustic effects to the calculation of certain integrals along 2D acoustic eigenrays corresponding to a range-dependent ocean. Here, 3D acoustic effects represented the difference between the values an acoustic observable takes in a horizontally inhomogeneous ocean (i.e., weak sound-speed perturbations are superimposed on a range-dependent background) and the value of the observable in a range-dependent ocean. Furthermore, 4D acoustic effects represented the difference between the values an acoustic observable takes in a time-dependent ocean and the value of the observable in an ocean frozen in time. In the case of random inhomogeneities, statistical moments of acoustic observables are calculated as integrals of appropriate statistical moments of environmental variables along 2D eigenrays in a deterministic, range-dependent ocean. It was concluded that the propagation of statistical moments of acoustic variables provided a very efficient alternative to Monte Carlo simulations when mapping environmental fluctuations into fluctuations of the acoustic field.

Duda (2006a) developed expressions governing coherence scales of sound passing through a moving packet of nonlinear internal waves over the continental shelf. These expressions described the temporal coherence scale at a point and the horizontal coherence scale in a plane transverse to the acoustic path. Coherence scales derived from the numerical simulation of coupled-mode acoustic propagation through moving wave packets substantiated these expressions.

Long-range, low-frequency (<5 kHz) sound propagation was experimentally observed in the central region of the Baltic Sea during the summer season (Vadov, 2001). The significant excess of attenuation over predicted absorption was assumed due to the presence of internal waves.

The effects of upper-ocean stirring (*millifronts*) were added to the effects of internal waves to account for the major part of acoustic variability in the ocean (Dzieciuch et al., 2004). The critical dependence on mixed-layer processes suggested a scheme for acoustically monitoring the upper ocean from sensors deployed at the surface-conjugate depth (3–5 km).

It was demonstrated by van Uffelen et al. (2009) that oceanic internal waves could facilitate the penetration of acoustic energy into the shadow zone below cusps of timefronts, as observed by bottom-mounted receivers at multi-megameter ranges. This penetration was much larger than predicted by diffraction theory. Acoustic data collected at the bottom-mounted horizontal-receiving arrays did not provide information on the vertical structure of the shadow-zone arrivals. However, acoustic data from two vertical-line array receivers deployed in close proximity in the North Pacific Ocean, together virtually spanning the water column, showed the vertical structure of the shadow-zone arrivals for transmissions from broadband 250-Hz sources moored at the sound-channel axis (750 m) and slightly above the surface conjugate depth (3000 m) at ranges of 500 and 1000 km. Comparisons of the field data with PE simulations based on climatological sound-speed fields that did not include significant internal-wave variability showed that observed early branches of the measured timefronts consistently penetrated as much as 500–800 m deeper into the water column than predicted. Subsequent PE simulations incorporating sound-speed fluctuations consistent with the Garrett–Munk internal-wave spectrum at full strength predicted the observed energy level to within 3–4 dB rms over the depth range of the shadow-zone arrivals.

Uscinski and Nicholson (2008) compared field data with predictions from parabolic-moment equations for propagation and scattering in randomly irregular media using the standard Garrett-Munk model for internal waves. The experimental results and theoretical predictions were in good agreement. However, this comparison raised new questions about the correlation of intensity fluctuations as the acoustic transmission frequency was varied. This aspect requires further investigation.

6.9 COUPLED OCEAN-ACOUSTIC MODELING

Oceanographers have recently developed synoptic forecasting techniques appropriate for predicting the locations of frontal features such as currents and eddies (Mooers et al., 1982; Robinson et al., 1984; Robinson, 1987; Peloquin, 1992). These forecasts can be used by forces afloat to facilitate the efficient allocation of naval resources during anti-submarine warfare (ASW) operations.

These same dynamical ocean models can be used in conjunction with UWA propagation models to generate timely forecasts of sonar performance in the vicinity of highly variable frontal features. When used in this fashion, the ocean models generate input variables necessary for initialization of the acoustic models. This synergistic arrangement is referred to as coupled ocean-acoustic modeling. Such coupling has been successfully demonstrated, for example, by Botseas et al. (1989), who interfaced an implicit-finite-difference PE model with ocean forecasts generated by the HOOM.

Coupled ocean-acoustic forecast systems comprise three basic components: an oceanic forecast scheme, a coupling scheme, and an ocean acoustic propagation scheme (Robinson et al., 1994). These systems can also be used to generate nowcasts and hindcasts. Nowcasts are estimates of the present state of a system. They are based on a combination of observations and dynamical modeling. Hindcasts are *a posteriori* forecasts. They are useful in evaluating modeling capabilities based on historical benchmark data (e.g., Martin, 1993).

Requirements for oceanographic data to support coupled ocean-acoustic forecast systems often exceed observational capabilities. Therefore, data assimilation, which introduces data generated by feature models, is used to achieve accurate synoptic realizations. Feature models are statistical representations of common synoptic structures in the ocean such as fronts and eddies (Robinson et al., 1994).

The optimum thermal interpolation system (OTIS) (Clancy et al., 1991a) forms the basis for an upgraded tactical ocean thermal structure (TOTS) system (Hawkins, 1992) designed for use aboard ships equipped with the tactical environmental support system (TESS). OTIS assimilates real-time observations from multiple sources into a complete 3D representation of the oceanic thermal field. A corresponding 3D representation of the salinity field is derived using empirical techniques. To compensate for sparse subsurface measurements, OTIS supplements actual observations with synthetic data derived from an empirical orthogonal function (EOF) representation of water masses, from ocean feature models and from subjective interpretation of satellite imagery. The resulting product provides an accurate 3D representation of the temperature and salinity fields together with a realistic depiction of fronts and eddies. Such information is useful in organizing naval assets as well as in planning oceanographic field work. OTIS has also been used to generate high-quality data for the evaluation of oceanographic forecasting models (Lai et al., 1994).

The finite volume community ocean model (FVCOM) is a prognostic, unstructured-grid, finite-volume, free-surface, 3D primitive equations community ocean model (Chen et al., 2011). The current version of FVCOM is a fully coupled ice-ocean-wave-sediment-ecosystem model system with options for various turbulence mixing parameterizations, generalized terrain-following coordinates, data assimilation schemes, and wet/dry treatments with inclusion of dike and groyne structures under hydrostatic or nonhydrostatic approximation. FVCOM solves the governing equations on Cartesian or spherical coordinates in integral form by computing fluxes between nonoverlapping horizontal triangular control volumes. Either mode-split or semi-implicit schemes can be selected. This finite-volume approach combines the best of FEMs for geometric flexibility and finite-difference methods (FDM) for simple discrete structures and computational efficiency. This numerical approach also provides a much better representation of mass, momentum, salt, and heat conservation in coastal and estuarine regions with complex geometry. The conservative nature of FVCOM, in addition to its flexible grid topology and code simplicity, makes FVCOM ideally suited for interdisciplinary application in the coastal ocean.

Long et al. (2015) developed an underwater sound model to simulate sound propagation from marine-hydrokinetic energy (MHK) devices or from offshore wind-energy platforms. Specifically, they coupled a time-domain hydrodynamic model with a frequency-domain acoustic model. The acoustic model was then coupled with the 3-D ocean hydrodynamic circulation model FVCOM, which resolves variations in water density and sea-surface elevation (Chen et al., 2011).

The coupling of oceanographic mixed layer models with surface duct propagation models was discussed in Section 5.2.3. The subject of marine modeling, which properly embraces the numerical modeling of ocean physics, geology, chemistry, and biology, has been summarized by Goldberg et al. (1977) and by Kraus (1977).

6.10 ACOUSTIC TOMOGRAPHY

Ocean acoustic tomography (Munk and Wunsch, 1979) is an inverse technique that measures perturbations in travel times between fixed acoustic sources and receivers (see Figure 6.6). This technique is analogous to the medical procedure called tomography (from the Greek word for "slice" or "cut"), and also has elements in common

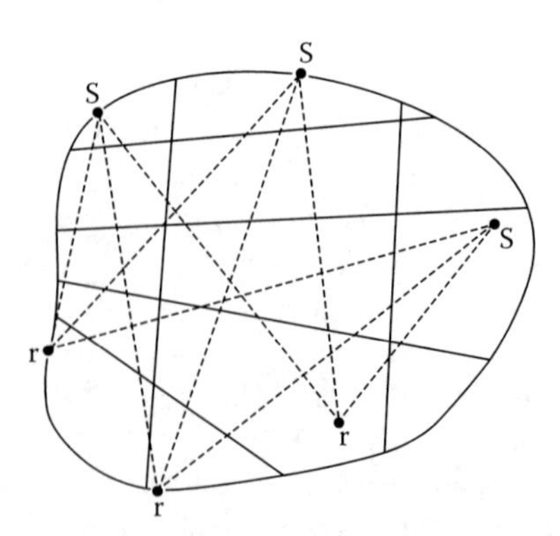

FIGURE 6.6 Schematic illustration of an ocean tomographic geometry. Perturbations in acoustic travel time from any source s to any receiver r (dashed paths) are used to estimate sound speed perturbations in an arbitrary grid area formed by the solid lines. (From Munk, W. and Wunsch, C., *Deep-Sea Res.*, **26**, 123–161, 1979. Copyright by Pergamon Press PLC. With permission.)

with conventional seismology (Menke, 1989). A recent book by Munk et al. (1995) provided a comprehensive review of the oceanography and mathematics necessary to understand and develop ocean-acoustic tomographic systems.

This technique has been proposed for large-scale monitoring of ocean basins. Specifically, the perturbations in travel times between sources and receivers can be used to deduce sound speed (and thus water density) fluctuations in the interior of the oceans. Moreover, if reciprocal transmissions are used between sources and receivers, the differences in travel times can be used to compute the mean ocean currents along the acoustic path (Spiesberger, 1989). Sea surface spectra can also be estimated using acoustic tomography (Miller et al., 1989).

The temporal resolution required to distinguish multipath arrivals has been estimated at 50 ms. Existing low-frequency (<200 Hz) broadband sources appear to satisfy these requirements. The real limiting factor will probably be the lack of spatial resolution owing to practical limitations on the number of sources and receivers. Consequently, there will usually be insufficient information from which to determine a unique solution to the inverse problem.

A field test of ocean-acoustic tomography was conducted in the North Atlantic Ocean in 1981 (Cornuelle et al., 1985). The test was conducted over a two-month period in a 300 km square centered at 26°N, 70°W using nine acoustic deep-sea moorings with seafloor transponders. The acoustic sources operated at a center frequency of 224 Hz, with a 5.4 Hz rms bandwidth. The three-dimensional sound speed field obtained by inversion techniques compared favorably with direct observations of sound speed profiles. Tomographic maps constructed at three-day intervals over the test period revealed a pattern of eddy structure in general agreement with that deduced from direct observations. Mapping errors were attributed to noise variance in travel-time measurements.

Acoustic tomography experiments have been conducted to detect climatic trends of temperature in the oceans with basin-scale resolution (Munk and Forbes, 1989; Spindel and Worcester, 1990a,b, 1991; Spiesberger and Metzger, 1991a; Spiesberger et al., 1992; Forbes, 1994). Since acoustic travel times are inversely related to water temperature, tomography provides a method for ascertaining the bulk temperature of the oceans. This bulk temperature is potentially a more reliable metric for global warming than are conventional atmospheric measurements which, for example, can be influenced by the heat island effect that is often associated with urban temperature records. Moreover, basin-scale averages of water temperature suppress the intense mesoscale eddies that would otherwise dominate the climatic variations.

The Heard Island Feasibility Test (HIFT) established the limits of phase-coded, electrically driven sound sources (instead of explosive sources) in tomographic experiments (Munk, 1994). Figure 6.7 illustrates the ability of an acoustic source situated at Heard Island to sample vast volumes of the oceans over ray paths unimpeded by bathymetric obstructions. These ray paths represent refracted geodesics along the surface containing the sound channel axis (refer back to Figure 2.8). A geodesic path is the shortest distance between two points on the surface of an elliptical (versus spherical) Earth. Geodesics would be the proper ray paths in a uniform sound-speed field; however, the refractive effects of lateral sound-speed gradients alter these paths.

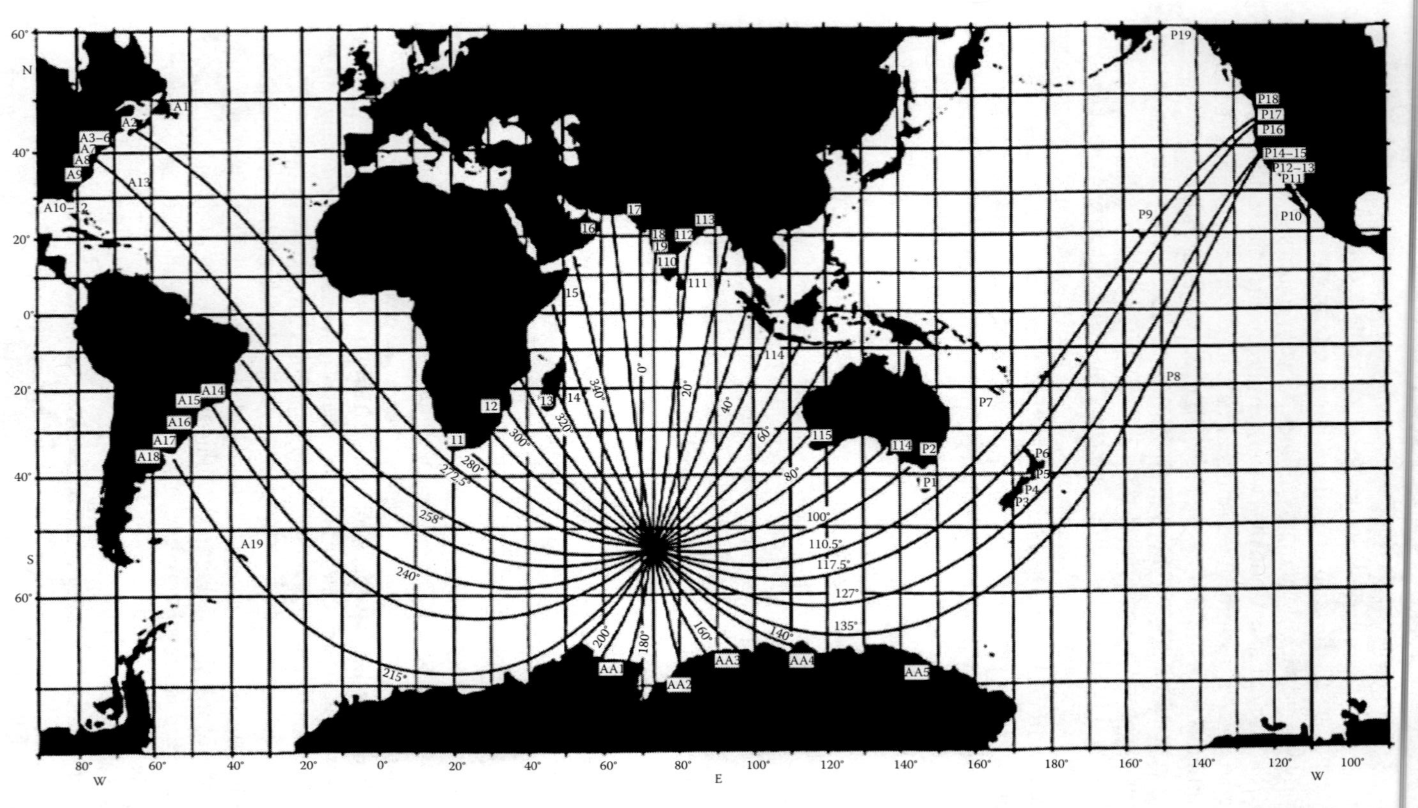

FIGURE 6.7 Illustration of bathymetrically unimpeded ray paths emanating from Heard Island with the indicated launch angles. All rays are refracted geodesics along the surface containing the sound channel axis (see Figure 2.8). White boxes indicate locations of oceanographic research stations capable of receiving acoustic signals. (From Munk, W.H. and Forbes, A.M.G., *J. Phys. Oceanogr.*, **19**, 1765–1778, 1989. Copyright by the American Meteorological Society. With permission.)

HIFT is a prerequisite to a program for acoustic thermometry of ocean climate (ATOC). The goals of ATOC are (1) to verify the acoustically measured changes in travel time by comparison with those inferred from traditional oceanographic measurements along the transmission path in the North Pacific Ocean; (2) to study the coherence of climatic scale variability in the North Pacific gyre; (3) to test long-range acoustic transmission characteristics using 40-element vertical receiver arrays; and (4) to test specially designed ATOC sound sources. The potential effects of acoustic transmissions on marine mammals have also been investigated in related efforts (National Research Council, 1994; Green, 1994). A series of 17 technical papers addressing various aspects of HIFT was recently published. These papers were introduced by Munk and Baggeroer (1994).

Long-range acoustic propagation issues relevant to tomographic investigations include bathymetric interactions. These aspects have been investigated by Munk (1991), Munk and Zachariasen (1991), and Heaney et al. (1991). Forward scatter by islands and seamounts was found to produce significant scattered arrivals (Munk and Zachariasen, 1991). Such arrivals can complicate interpretation of the recorded acoustic travel times if they are not properly differentiated from the direct arrivals.

Worcester and Spindel (2005) summarized the results of the North Pacific Acoustic Laboratory (NPAL), which is a series of individual long-range acoustic propagation experiments that have been conducted in the North Pacific Ocean during the last 15 years (including ATOC) using various combinations of low-frequency and wide-bandwidth transmitters together with horizontal and vertical line-array receivers. These measurements were undertaken to advance understanding of the physics of low-frequency, broadband propagation and the effects of environmental variability on signal stability and coherence. Moreover, NPAL acoustic thermometry is directed at using travel-time data obtained from a few acoustic sources and receivers located throughout the North Pacific to study the temperature variations of the North Pacific Ocean basin at the largest scale. Such measurements would facilitate a better understanding of climate variations by relating changes in acoustic travel times to changes in bulk water temperatures over a period of time.

Gaillard (1985) demonstrated that mobile sources or receivers can also be used in tomographic experiments, but that this mobility increases the sensitivity to noise. More efficient use of *a priori* knowledge of the ocean reduces this sensitivity by decreasing the number of degrees of freedom. Moving ship tomography (MST) is a method for obtaining high-resolution, quasi-synoptic, 3Dmaps of the oceanic temperature field over large areas (The Acoustic Mid-Ocean Dynamics Experiment Group [AMODE], 1994). All previous work in ocean acoustic tomography has been performed with fixed or moored instruments. MST is part of the largerAMODE, which has the goal of determining the measurement limits of MST. The larger number of acoustic ray paths generated by a moving receiver will improve measurements of advecting fronts and interacting eddies. This increased volumetric resolution will enable more rigorous testing of the predictive capabilities of dynamical ocean circulation models.

Methods for computing and plotting tomographic inversions in ocean environments have been described by Nesbitt and Jones (1994a,b,c). In related work, Weickmann and Jones (1994) described computer programs used to perform ocean-acoustic tomography inversions based on a nonperturbative-inversion method. Harrison et al. (1998) described a localization technique that was an efficient approximation to the maximum *a posteriori* probability (MAP) estimator intended for use in matched-field source-localization methods. Yaremchuk and Yaremchuk (2001) developed a nonlinear method for inverting ocean-acoustic tomography data. This method accounts for the quadratic term of the travel-time expansion in the powers of the reciprocal sound-speed perturbations. The higher-order correction to the travel-time model can be essential in regions of large spatial variability. This method was tested against data acquired in the Kuroshio extension region (near 32°N, 148°E). While this method does improve the quality of data processing, it increases the associated computational requirements. Dzieciuch et al. (2001) introduced a technique referred to as a "turning point filter" that permits a uniform treatment of long-range acoustic transmissions from the early ray-like arrivals to the late mode-like arrivals. This method may also be applicable to source localization problems.

Long-range sound transmissions in the ocean are controlled largely by the SOFAR waveguide. The observed agreement between measured and predicted arrival patterns has been generally excellent except for an observed extension downward by many hundreds of meters from the lower caustics into the deep sound shadow (Rudnick and Munk, 2006). This deepening of early arrivals is hypothesized to be associated with the sharp transition zone marking the lower boundary of the surface mixed layer. The base of the mixed layer is distorted by internal waves. Multiple collisions of the ray-like acoustic transmissions with the wavy mixed-layer base often lead to a mean deepening of the lower caustics. Over a 1,000-km transmission, there were typically 20 such collisions. Monte Carlo simulations yielded statistics of ray inclination, range, travel time, and lower-turning point. The resulting timefront included a deepening by several hundred meters. These results suggested the possibility that abyssal acoustic rays could be used to monitor upper-ocean processes.

Tolstoy (2008) discussed the tomographic 3D geoacoustic-inversion approach including second-order regularization. Since inverse problems are usually ill-posed, it is necessary to use some method to reduce their deficiencies. One method involved regularization by derivative matrices. Second-order regularization occurs when the matrix is formed by second-order differences. (Order zero means that the regularization matrix is the identity matrix.)

Mansour (2012) examined processes by which to extract the maximum information from both active and passive acoustic-tomographic signals. Classic algorithms were applied to deal with residual problems. New high-order statistics (HOS) estimators were proposed to analyze the spatial diversity of the original sources, including independence-discrimination criteria. Finally, blind separation of observed acoustic signals entailing independent component analysis (ICA) algorithms were used to separate the observed mixed signals. See related discussions in Section 11.5.5.

The analyses of signals for acoustic tomography often use the unrefracted geodesic path, an approximation that is justified from theoretical considerations relying on estimates of horizontal gradients of sound speed, or on simple theoretical models. For the purposes of acoustic tomography, then, the small effects arising from horizontal refraction can be ignored. To quantify the effects of horizontal refraction caused by a realistic ocean environment, Dushaw (2014) computed horizontal refractions of long-range signals using global ocean-state estimates for 2004 from the *Estimating the Circulation and Climate of the Ocean* (ECCO2) project. Basin-scale paths in the eastern North Pacific Ocean and regional-scale paths in the Philippine Sea were used as examples. At basin scales on the order of 5 Mm, refracted geodesic and geodesic paths differed by only about 5 km. Gyre-scale features had the greatest refractive influence, but the precise refractive effects depended on the path geometry with respect to oceanographic features. Refraction decreased travel times by 5–10 ms and changed azimuthal angles by about 0.2°. At regional scales on the order of 500 km, paths deviated from the geodesic by only 250 m, and travel times deviated by less than 0.5 ms. Refraction details depended only slightly on mode number and frequency.

Marin (2015) examined the feasibility of using ocean-acoustic tomography to monitor large areas of the continental shelf waters off the coast of Brazil to identify and characterize mesoscale oceanographic features. This work discussed the estimation of sound-speed perturbations in shallow-water environments using measurements of modal travel time based on the processing of data collected in INTIMATE96 (internal tide measurements with acoustic tomography experiments), which was carried out in June 1996 off the coast of Portugal. The purpose of ocean-acoustic tomography is to estimate marine environmental characteristics from measurements of acoustic propagation. For shallow-water regions, alternatives to ray tracing techniques have been found in matched-field techniques and modal tomography. Marin (2015) addressed the problem of estimating environmental parameters in shallow water based on identification of the modal structure of broadband acoustic signals measured at a hydrophone. The dispersive characteristics of shallow-water regions pointed to normal modes for effective modeling of acoustic propagation in such environments.

As summarized by Spindel (1985), acoustic tomography is particularly attractive for application to the ocean for a number of reasons: (1) images of the ocean interior can be obtained from periphery measurements; (2) large ocean areas can be measured with relatively few instruments; (3) a single source-receiver pair samples the ocean vertically due to multipath propagation effects; and (4) the integrating properties of acoustic propagation produce average quantities not obtainable with point measurements.

6.11 PHASE CONJUGATION AND TIME-REVERSAL MIRRORS

A time-reversal mirror (TRM), also referred to as phase conjugation, focuses sound from a source-receiver array (SRA) back to the source probe (SP) that had originally insonified the SRA. The SRA receives the SP pulse field, then time reverses it and uses the time-reversed data to excite a source array that is collocated with the receiving array on the SRA. If the ocean environment does not change significantly during the two-way travel time, the phase-conjugate field will refocus at the SP regardless of the complexity of the medium (unless excessive loss in the system degrades the process). Since the focus is both spatial and temporal, it undoes the multipath from the first part of the transmission. Since this process offers an approach to compensate for multipath interference and other distortion through a complex medium, it may be applicable to various adaptive sonar and communication concepts. Moreover, by putting more energy on the target and keeping it away from the ocean boundaries, these techniques promise significant increases in echo-to-reverberation level (also see Section 9.4).

Kuperman et al. (1998) experimentally demonstrated that a time-reversal mirror (or phase-conjugate array) could spatially and temporally refocus an incident acoustic field back to its origin. This work was extended by Song et al. (1998) to refocus an incident acoustic field at ranges other than that of the probe source. The basic idea of the approach was that the sound field maxima could be shifted to different ranges by appropriately increasing or decreasing the source frequency for a specific propagation environment.

Time-reversal acoustics can be applied in shallow water to focus energy back to a source location. This refocusing produces spatial intensification of the field (through removal of multipath spreading) as well as temporal convergence of the signal. These properties suggested potential applications in UWA communication systems (Abrantes et al., 1999).

An experimental method has been demonstrated to obtain coherent acoustic wavefronts by measuring the space-time correlation function of ocean noise between two hydrophones (Roux et al., 2004). This concept was subsequently used in a demonstration of noise-based TRM (Roux and Kuperman, 2005).

Broadband correlation processing for extracting time-domain Green's functions and coherent acoustic wavefronts from random ocean noise has been demonstrated recently using experiments and numerical simulations that are consistent with theoretical predictions (Roux et al., 2011). Ocean-acoustic noise processing presents additional challenges over its seismological counterpart. Mainly, the ocean environment is temporally nonstationary and it is spatially heterogeneous. Furthermore, in the frequency regime 20–500 Hz, space-time episodic shipping is the dominant noise source. Data from different publications and research groups were reviewed in an effort to demonstrate the viability and potential applications of passive-coherent array processing in the ocean.

Experimental results for UWA communications using time reversal have been discussed in the recent literature (Edelmann, 2005; Edelmann et al., 2005). Time-reversal acoustics was used to environmentally adapt the acoustic propagation effects of a complex medium and focus energy at a particular target range and depth (Heinemann et al., 2003). These experimental results suggested a high potential in data rates for the time-reversal approach in shallow-water acoustic communications.

Walker et al. (2005) noted that the traditional employment of TRM requires that a probe source first broadcast from the desired focal point. The sound can then be refocused back at the original probe source position with the assistance of model-based calculations. Methods already exist by which to refocus sound at different ranges using frequency shifting. A technique has now been developed to refocus sound at depths different from the original probe source in a shallow-water, range-independent waveguide. The ability to shift focus could find applications to MIMO communications (see Section 11.5.5). However, there is a requirement to collect data from various ranges at a single depth, as from a moving broadband radiator, over a distance sufficient to construct the relevant frequency-wavenumber (f–k) structure of the waveguide. With this information, it would then be possible to focus at an arbitrary depth at any of the ranges at which the probe source data were taken. This theory was experimentally demonstrated at ultrasonic frequencies (~600 kHz) in a laboratory setting that scaled to typical shallow-water scenarios.

A technique that can be related to time-reversal for underwater-acoustic applications is Tikhonov regularized inverse filtering. This is a fast method for obtaining a stable inverse filter design by calculating the filter in the frequency domain using the fast Fourier transform; it was originally developed for use in audio-reproduction systems. Earlier research demonstrated that the Tikhonov regularized filter design outperformed time-reversal when using Dirac impulse transmission within a simulated underwater environment. Dumuid (2012) extended this previous work by exploring implementation of Tikhonov regularized inverse filtering with communication signals in shallow-water environments. Given appropriate parameters, the Tikhonov regularized inverse filter was shown to have good or comparable performance to the best filter tested for communication systems, especially with respect to multichannel communication capability.

6.12 DEDUCTIVE GEOACOUSTIC INVERSION

Since direct measurements of seabed parameters are practical only over short distances, considerable attention has been devoted to the inversion of acoustic propagation measurements as a cost-effective alternative to measurements over large ocean areas (e.g., Diachok et al., 1995; Siderius et al., 2000). Using inverse techniques, the desired geoacoustic seabed parameters are extracted from the forward measurements of acoustic propagation in the oceanic waveguide.

Deductive geoacoustic inversion derives the desired sediment and substrate parameters from direct measurements of acoustic propagation. The desired parameters are deduced from the propagation measurements using, for example, simulated annealing or genetic algorithms (Gerstoft, 1994; Snellen et al., 2001). (Also see related discussions in Sections 6.12.1, 6.12.2, and 6.18.) The resulting geoacoustic parameters represent averages over long ranges. Direct measurements over comparably long ranges are frustrated by sparse sampling (Ainslie et al., 2001; Hamson and Ainslie, 1998). Deductive geoacoustic inversion assumes that acoustic data are sensitive to different geoacoustic parameters at different frequencies. Sediment parameters of interest include sediment thickness, sound speed (and gradient), density, attenuation, and shear speed. Substrate parameters of interest include density, attenuation, compressional sound speed, and shear speed (Ainslie et al., 2000).

Siderius et al. (2001) used experimental data collected at a shallow-water site to demonstrate the impact of ocean sound-speed fluctuations on the quality of seabed properties deduced from geoacoustic inversion. The experimental transmission loss (TL), when averaged over frequency (0.2–3.8 kHz) and time (3 days), was modeled with sufficient accuracy. In fact, the mean modeled TL was within one standard deviation of the data for both short-range (2 km) and long-range (10 km) propagation. However, the modeled standard deviation of the TL was in poor agreement with that of the data due to insufficient knowledge of the changing ocean environment. Fluctuations in the sound speed destroyed coherent processing beyond a few kilometers. Without reliable predictions of acoustic propagation, the inverted geoacoustic parameters showed erroneous time variability (erroneous since the sea-bed properties being deduced are expected to remain fixed over time). At ranges of 2 km, the temporal stability of the geoacoustic inversion indicated that the method was indeed sound. This demonstrated that, without sufficient temporal and spatial knowledge of the volumetric properties of the ocean, geoacoustic inversion at long ranges is subject to degradation.

Ainslie et al., (2001) compared results deduced from three different methods, all of which used a Bartlett processor: (1) the evolutionary search algorithm (ESA) used a PE forward model; the genetic algorithm (GA) used a normal mode forward model; and the deductive rapid environmental assessment model (DREAM) used a normal mode forward model. The results showed good agreement among the three methods, although differences in the computational procedures complicated definitive numerical assessments.

Recently, geoacoustic inversion of ambient noise has also been explored (also see Section 7.8). In particular, Harrison and Simons (2001) noted that inversion of the noise field could complement active-acoustic propagation inversion techniques in which sound speed fluctuations might impede model matching.

Huang et al. (2006) suggested a method for the statistical estimation of transmission loss based on the posterior PDF of environmental parameters obtained from the geoacoustic-inversion process. Jiang et al. (2006) described the results of geoacoustic inversion using broadband signals from an experiment conducted in the Florida Straits. They recorded M-sequence coded pulse trains at carrier frequencies centered between 100 Hz and 3.2 kHz on a vertical line array at a distance of 10 km. (M-sequences are a class of pseudorandom binary sequences used as test signals in analytical schemes. The M-sequence method has several advantages for investigating linear and nonlinear systems: ease of implementation, rapid calculation of system kernels, and a solid theoretical framework.) They carried out geoacoustic inversion to determine the feasibility of inverting the environmental parameters from this long-range experiment. The received signal at frequencies below 400 Hz consisted of a dominant water-column signal together with a secondary arrival that was delayed by 0.4 s. This secondary signal was spatially filtered by beamforming the array data. These beam data were then inverted by matched-beam processing in the time domain and combined with an adaptive simplex simulated annealing algorithm. The inverted shear-wave speed appeared to be a sensitive parameter and was consistent with the compressional wave speed.

Koch and Knobles (2005) estimated geoacoustic parameters for a typical high-noise, shallow-water region in the Gulf of Mexico using the radiated noise emitted from a cooperative surface ship. They demonstrated an efficient method for geoacoustic inversion using broadband beam cross-spectral data. Furthermore, they discussed the robust performance of cost functions involving coherent or incoherent sums over frequency and one or multiple time segments. They were able to obtain inversions for the first sediment-layer sound speed and thickness at nominal ship ranges of 20–50 water depths. The sub-aperture beams included 10 frequencies equally spaced in the 120–200 Hz band. As part of the validation process, they compared measured transmission losses with those modeled based on the inverted geoacoustic parameters.

A range-independent, normal-mode model (ORCA) was used since the selected track was close to being horizontally stratified. Simulated annealing was selected to sample the geoacoustic parameter space for the best data-model fit.

Lapinski and Chapman (2005) performed geoacoustic inversion using a matched-field inversion algorithm, which is an established technique for estimating geoacoustic parameters of the seabed. To demonstrate how parameter estimation can be affected by unknown (or ignored) random range dependence in the real environment while the inversion model assumes range independence, simulations with controlled statistics were carried out using a simple shallow-water model consisting of an isospeed water column over a homogeneous elastic half-space. The inversion parameters included water depth, compressional speed in the seabed, seabed density, and compressional wave attenuation. On average, the environment was range independent; that is, some parameters were constant while other parameters were random with range-independent means and variances. A PE model was used to calculate propagation fields for the range-dependent environment as well as fields for the range-independent inversion model. It was concluded that ignoring the variability of even a single geoacoustic parameter could lead to significant bias and variance in the estimation of all inverted parameters.

Morley et al. (2008) showed that neglecting environmental range dependence in matched-field inversion resulted in biased theory errors that led to biased geoacoustic parameter estimates using standard inversion methods. Both hard-bottom and soft-bottom environments were considered at a number of scales of lateral variability for water depth and seabed sound speed. The biases appeared to result from additional losses in range-dependent propagation, which were compensated for in range-independent inversion by adjusting geoacoustic parameters to decrease the seabed reflection coefficient. The effects of range dependence differed for different environments: the soft-bottom cases were sensitive to range-dependent sound speed; the hard-bottom cases were sensitive to range-dependent water depth.

An approach was developed to guide accurate estimation of a set of unknown geoacoustic parameters from remote acoustic measurements (Zanolin et al., 2004). The approach computed sample sizes (or signal-to-noise ratios, SNRs) necessary for estimates to have variances that asymptotically attained the Cramer-Rao lower bound (CRLB) and have CRLBs that fell within a specified design error threshold.

Choi and Dahl (2005, 2006) analyzed the relationship between a head wave and the arrival of a ground wave using PE- based simulations. They discussed parametric dependencies for distinguishing between these arrivals and the implications for geoacoustic inversion. Head waves and ground waves are strongly linked with properties of the seabed; thus, they provide a useful metric for geoacoustic inversion. For a Pekeris waveguide, the ground wave can be interpreted as a sequence of first-order (or classical) head waves, which are referred to as a head-wave sequence. The term "first-order" originates from a ray-series classification. However, the presence of a sound-speed gradient in the sediment gives rise to very different effects. Specifically, the first-order head wave is replaced by either a lower-amplitude interference head wave, whose properties are akin to a first-order head wave, or a higher-amplitude interference head wave (or related diving wave) whose properties are zeroth-order. There is also a shift in the dominant frequency of the first-arriving head wave in the transition regime between these arrivals. Synthetic data and results were obtained using the RAM PE model.

Chotiros and Isakson (2004) noted that measurements conducted over a broad range of frequencies have shown that the sound-speed dispersion is significantly greater than that predicted by the Biot-Stoll model with constant coefficients, and the observed sound attenuation does not seem to follow a consistent power law. The sound-speed dispersion may be explained by Biot plus squirt flow (BISQ) while the attenuation may be explained by BISQ plus grain contact squirt flow and viscous drag (BICSQS). It is helpful to note that in partially saturated rocks, compressional wave attenuation is much larger than in fully saturated rocks. In partially saturated rocks, cracks are partially saturated with liquid, and the compression of cracks causes liquid to flow into regions occupied by gas, resulting in viscous energy losses. The grain-shearing (G-S) model relates dispersion relationships for four wave properties (compressional-wave phase speed, compressional-wave attenuation, shear-wave phase speed, and shear-wave attenuation) to three macroscopic physical variables (porosity, grain size, and depth) in unconsolidated granular media, that is, without an elastic frame (Buckingham, 2005).

Papadakis and Flouri (2007, 2008) modeled acoustic propagation in the ocean via the parabolic approximation where the bottom boundary condition was of the form of a NtD map. An optimal control method was executed using the adjoint operator for recovering the density, sound speed, and attenuation of the bottom. The method was applied to two test cases resulting in satisfactory convergence of the inversion scheme. This inversion method can be extended to other forms of Dirichlet-to-Neumann (DtN) or NtD maps. Current limitations of the method include a horizontal bottom interface, homogeneous bottom region, and knowledge of the complex field at the source. The method can handle a range-dependent water column.

The following two subsections will discuss specialized topics relevant to data processing in geoacoustic inversion: navigating parameter landscapes and tabu search. Also see related discussions in Section 6.18 pertaining to seismo-acoustic inversion.

6.12.1 Navigating Parameter Landscapes

Fialkowski et al. (2003) developed an efficient method for performing geoacoustic inversions in range-dependent ocean waveguides. This method combined a simulated annealing search with an optimal coordinate rotation that increased the efficiency of navigating parameter landscapes for which parameter coupling was important. The coordinate rotation provided information concerning which parameters were resolvable for a particular inversion frequency and array geometry. Results from several single-frequency inversions could be combined to obtain an estimate for the sediment parameters.

It has been stressed that accurate modeling of the response of the acoustic environment is a necessary aspect of geoacoustic inversion (Battle et al., 2004). The complex effective depth (CED) concepts of Zhang and Tindle (1993) were employed, motivated by the modeling requirements of Gibbs sampling, which may necessitate on the order of 10^5 model runs for a single inversion. Gibbs sampling is a general method for probabilistic inference. It is well suited to problems involving incomplete information, although implementation comes at some computational cost. Nevertheless, understanding Gibbs sampling provides valuable insights into problems of statistical inference and can be used to train Bayesian networks with missing data. (For more information on Gibbs sampling, refer to a recent book by Train (2009). The power of Gibbs sampling derives from the fact that the joint distribution of the parameters will converge to the joint probability of the parameters, given the observed data.)

van Leijen (2010) and van Leijen et al. (2011) explored the use of ant colony optimization (ACO), which has elements in common with genetic algorithms. Both are population-based algorithms that search a discrete space and provide uncertainty analyses. The main difference is the mechanism that handles and recombines components of better candidate solutions (i.e., ant-pheromones trails versus genetic operators). ACO is further distinguished by having a form of memory (the ant pheromone trails), while genetic algorithms are without memory. Specifically, when the pheromones evaporate, identifiers of paths with above-average quality fade out. Thus, high rates of evaporation mean that only recent information can be retrieved (typical for short-term memory); alternatively, low rates of evaporation allow recollection of much older information (typical for long-term memory). When applied to inversion, the world of the ants acts as an analogy for the geoacoustic environment.

6.12.2 Tabu Search

Michalopoulou and Ghosh-Dastidar (2004) described tabu search, which has traditionally been applied to combinational optimization problems (also see the book by Glover and Laguna [1997]). The tabu search began by marching to a local minimum. This approach avoided entrapment in cycles by forbidding (tabu), or penalizing, moves that took the solution in the next iteration to points in the solution space previously visited.

6.13 PREDICTION UNCERTAINTIES IN COMPLEX ENVIRONMENTS

Finette (2005) defined "uncertainty" as a quantitative measure of our lack of knowledge of the sound-speed field and boundary conditions constituting the waveguide information necessary for simulation of the acoustic field. This uncertainty is distinct from any errors related to numerical solution of the wave equation. Existing mathematical methods typically solve a deterministic wave equation separately over many realizations, and the resulting set of pressure fields is then used to estimate statistical moments of the field. Proper sampling may involve the computation of thousands of realizations to ensure convergence of the statistics. A stochastic differential equation incorporating wave dynamics and environmental uncertainty was derived and incorporated, for demonstration, into Tappert's narrow-angle PE (although the proposed approach is not limited to the narrow-angle formulation). Sensitivity to errors in computing coefficients in the stochastic differential equation might affect robustness, and the inclusion of multiple sources of uncertainty might affect numerical convergence of the solution. This approach effectively computed all realizations simultaneously by solving a set of coupled differential equations. The benefit is that significant computational speed can be achieved relative to Monte-Carlo based methods in which realizations are computed individually. Moreover, this approach provides a more compact description of uncertainty in complex systems comprised of subsystems for which uncertainty can be transferred among the subsystems. In ocean-acoustic modeling, a two-component system might consist of the ocean environment and the pressure field propagating within that environment. Processes characterized by a short correlation length may need many terms in the spectral expansion, thus increasing the complexity and computational load. Uncertainty in the boundary conditions can also be included in the formalism. Following this line of investigation, Finette (2006) maintained that to quantify environmental uncertainty, a measure of incomplete environmental knowledge should be included as part of any simulation-based predictions linked to acoustic-wave propagation. A method was proposed for incorporating environmental uncertainty directly into the computation of acoustic-wave propagation in ocean waveguides. Polynomial chaos expansions were selected to represent uncertainty in both the environment and the acoustic field. The sound-speed distribution and the acoustic field were generalized to stochastic processes, wherein uncertainty in the field was interpreted in terms of its statistical moments. Starting from the narrow-angle parabolic approximation, a set of coupled-differential equations was derived in which the coupling term linked incomplete environmental information to the corresponding uncertainty in the acoustic field. Propagation of both the field and its uncertainty in an isospeed waveguide was considered in which the sound speed was described by a random variable. The first two moments of the field were computed explicitly and compared to those obtained from independent Monte-Carlo simulations of the deterministic PE used to describe the acoustic wave properties. This comparison showed that the truncated polynomial chaos expansions reproduced the Monte Carlo estimates of the moments.

Pace and Jensen (2002) documented the proceedings of a conference organized by the SACLANT Undersea Research Centre (now NURC) that was held in Lerici, Italy. Papers addressed the limiting effects of complex and variable littoral environments on sonar-system performance. Topics included ocean and seabed variability, measurements and models of acoustic fluctuations, sonar signal processing, and performance predictions.

Culver and Camin (2008) linked an understanding and reduction of environmental uncertainty with an estimated ocean detector (EOD). The EOD comprised a new signal-processing structure that utilized a probabilistic description of received signal parameters, thus providing improved sonar performance. (Also see related discussions in Section 6.2 regarding stochastic modeling.)

6.14 RAPID ENVIRONMENTAL ASSESSMENTS

REA provide deployed naval forces with environmental information in littoral waters within meaningful tactical timeframes (Whitehouse et al., 2004). Since its establishment as a formal requirement by SACLANT in 1995, REA has evolved from a concept to a functioning network supporting diverse military operations including ASW, mine warfare, and amphibious operations. REA product specifications, which were formalized in 2001, are distinguished by the temporal and spatial scales, environmental boundaries of littoral regions, and by the speed and format in which deployed forces require these products. A differentiation between static products (weeks to years) and dynamic products (hours to days) recognizes the different timescales associated with the prevailing ocean features found in littoral regions. REA-enabling technologies are grouped into three categories: (1) worldwide web, geographic information systems (GIS), and data communications; (2) environmental modeling and adaptive sampling techniques; and (3) sensors and platforms. Unlike open-ocean waters, littoral waters are sovereign and thus present difficulties for traditional surface-platform *in situ* (point-source) observations. Alternatively, satellites, aircraft, and shore-based stations can provide synoptic portraitures of the maritime battlespace. The REA process entails (1) data management, fusion, and display; (2) systems engineering (refer to Section 13.6.1); (3) a sensitivity to client needs; and (4) covertness.

In support of naval operations in littoral regions, acoustical oceanographers have employed ocean-acoustic models as adjunct tools that can be used to conduct REA in remote locations.

6.15 UNDERWATER ACOUSTIC NETWORKS AND VEHICLES

An UAN is composed of multiple nodes that communicate via an acoustic channel in support of underwater sensors and vehicles. This section will address three principal topics: channel models; localization methods (including a discussion of range-based and range-free schemes); and vehicles (including unmanned underwater vehicles [UUVs], autonomous underwater vehicles [AUVs], and gliders).

6.15.1 Channel Models

A model of the ocean medium between acoustic sources and receivers is called a *channel model*, and it may be digital or analog. In an oceanic channel, characteristics of the acoustic signals change as they travel from transmitters to receivers. These characteristics depend on the acoustic frequency, the distances between sources and receivers, the paths taken by the signals, and the prevailing ocean environment in the vicinity of the paths. Properties of received signals can be derived from those of the transmitted signals using channel models.

The following subsections provide further details regarding the role of channel models in UANs. Specific topics include: channel structure; network structure; channel emulators and network simulators; network performance and optimization (including denial of service); underwater communications; medium access control (MAC); and data delivery schemes.

6.15.1.1 Channel Structure

Van Walree (2013) surveyed a collection of systematic measurements to characterize shallow-water acoustic propagation channels for application to underwater communications. The objective of channel sounding is to measure the channel impulse response as a function of time and time delay. The *in situ* measurements were performed in different frequency bands, geographical areas, seasons, deployment geometries, and used different probe-signal parameters. The survey comprised measurements from northern Europe covering the continental shelf, Norwegian fjords, a sheltered bay, a channel, and the Baltic Sea. These measurements were conducted in frequency bands between 2 kHz and 32 kHz. The study compared 15 channels that differed in many ways, frustrating attempts to define a typical acoustic communication channel. Miscellaneous forward-propagation effects were also presented that were relevant to channel models suitable for the design of modulation schemes, network protocols, and simulation environments. For 13 of the 15 cases examined, the waveguides were confined by the seafloor and sea surface, or surface ducts; part or all of the received signal energy had Doppler spread due to surface interactions. Case 14 involved a communication channel that was a submerged sound channel formed by a local minimum in the sound-speed profile. Case 15 dealt with a wideband channel.

6.15.1.2 Network Structure

Underwater networks consist of variable numbers of sensors and vehicles deployed in concert to perform collaborative monitoring tasks over a given area (Yang, 2010). A sensor network comprises nodes that communicate via acoustic waves over multiple wireless hops to perform environmental monitoring, naval surveillance, and ocean exploration. Nodes in underwater sensor networks are constrained by harsh physical environments, energy limitations, long and variable propagation delays, and limited bandwidth. Data delivery schemes originally designed for terrestrial sensor networks are unsuitable for use in the underwater environment. Relatively few new schemes have been proposed for underwater use, and no single scheme has yet emerged as the *de facto* standard.

Otnes et al. (2012) presented a comprehensive overview of UAN that served both as an introduction to the subject and as a summary of existing protocols. It provided a useful background and further described the state-of-the-art for all networking facets that are relevant to underwater applications.

Partan (2006) developed a taxonomy that characterized the spatial extent of a network by comparing it to the acoustic range of the nodes. If all nodes are in direct contact it is referred to as a single-hop network with either centralized or distributed control. In networks covering larger areas, communications require multiple hops to reach destinations. When the geographic coverage is greater than the unpartitioned link-layer coverage of all nodes, routing requires techniques from disruption-tolerant networking (DTN).

Darehshoorzadeh and Boukerche (2015) described opportunistic routing as a promising approach for increasing the performance of wireless networks. In this approach, a group of candidate nodes was selected to assist as the next-hop forwarder. Each candidate that received the packet could continue forwarding the packet. By using a dynamic relay node to forward the packet, the transmission reliability and network throughput were increased. In this application, underwater sensor networks collected data from the environment and transferred them to sonobuoys on the surface for further transfer to a processing center. The acoustic channels common to underwater sensor networks have low bandwidth, high error probability, and longer propagation delay compared to radio channels. These properties make underwater channels potential candidates for employing opportunistic-routing concepts to deliver packets to the destination.

6.15.1.3 Channel Emulators and Network Simulators

Channel simulators can be grouped according to simulation method: (1) direct replay—reproduce measured channel conditions; (2) stochastic replay—generate channel conditions with statistical properties similar to field measurements; (3) model-based simulation—generate channel conditions by physical modeling based on marine-environmental information (van Walree et al., 2008). The meaning of the term *channel conditions* can vary from simulator to simulator; while it often refers to the time-varying impulse response (TVIR), it may also denote noise models or more elaborate channel-characterization methods. The channel simulator introduced by van Walree et al. (2008) for UWA channels later received the name MIME.

Otnes et al. (2013a,b) discussed validation methods for UWA communication channel simulators and validated the direct and stochastic replay of UWA communication channels as implemented in a channel simulator called MIME. MIME mimics the acoustic channel by passing the transmit signal through a channel modeled as a TVIR, where delay and time both are sampled at the sampling rate of the signal. Three simulation modes are included: mode 1 is direct replay of a measured TVIR; mode 2 is stochastic replay of a measured TVIR; and mode 3 is model-based simulation. Mode 3 does not require TVIR measurements but instead requires environmental data such as sound-speed profiles and wave spectra. Direct replay filters an input signal directly with a measured time-varying impulse response; stochastic replay filters an input signal with a synthetic impulse response consistent with the scattering function of the measured channel. The validation process used data from two sea experiments in concert with a diverse selection of communication schemes. Good agreement was found between bit error rates and packet error rates (PERs) of *in situ* transmissions and simulated transmissions. Long-term error statistics of *in situ* signaling were also reproduced in simulation when a single channel measurement was used to configure the simulator. In all except one comparison, the PER in simulation was within 20% of the PER measured on location. The implication is that this type of channel simulator can be employed to test new modulation schemes in a realistic fashion without going to sea, except for the initial data collection.

The underwater acoustic channel replay benchmark (WATERMARK) is a realistic simulation tool that is available to the underwater communications community. It is built around the validated Norwegian Defence Research Establishment (FFI) channel simulator MIME (Otnes et al., 2013a,b). The initial release comes with two test channels measured in Norwegian waters (in the 10–18 kHz band) with bottom-mounted sources and receivers (van Walree et al., 2016).

Dol et al. (2013) created a practical simulation framework, called impulse response simulator (IRSIM) that enabled the generation of impulse response time evolutions of realistic appearance for both stationary and mobile communication nodes, to be used by channel simulators such as MIME. This work was performed within the framework of the European Defence Agency project Robust Acoustic Communications in Underwater Networks (EDA–RACUN). The RACUN project is part of the European Unmanned Maritime Systems for MCM and other naval applications (EDA-UMS) program, and is funded by the Ministries of Defense of the five participating nations: Germany, Italy, The Netherlands, Norway, and Sweden.

Network simulators can be used to assess the performance of underwater-acoustic protocols and applications in complex ocean scenarios (Guerra et al., 2009). Specifically, the network simulator (NS2) has been adapted to provide a detailed reproduction of the propagation of sound in water by means of ray tracing instead of empirical relationships. This configuration provided a shared environment for the simulation of underwater networks. Available simulation frameworks were used to construct a customizable tool that included acoustic propagation, physical-layer modeling, and cross-layer specification of networking protocols. (The ISO open systems interconnection [OSI] model contains seven layers, the first of which is the physical layer [at the bottom] consisting of the basic networking hardware transmission technologies of a network. The remaining layers in the OSI model are data link, network, transport, session, presentation, and application. Any given layer serves the layer above it and is in turn served by the layer below it. Cross-layer functions [such as management] are not restricted to a given layer but can affect multiple layers.) This tool was used in a case study that compared three MAC protocols for underwater networks over different types of physical layers. The results compared the transmission-coordination approach opted by each protocol and demonstrated when it was better to rely on random access as opposed to loose or tight coordination. Baldo et al. (2010) described the world ocean simulation system (WOSS) and the multi-interface cross-layer extension (MIRACLE) of NS2 (see Appendix C for more information on WOSS). Çinar and Örencik (2009) designed and implemented an UWA channel model that accounted for multipath, fading, and shadow-zone effects by using a ray-tracing propagation method in the NS2 network simulator. Jiang et al. (2012) constructed UWA communication software using advanced interface technologies.

Sehgal et al. (2010) described the AquaTools simulation toolkit that supports simulation of UANs using either static or mobile nodes. Three different channel models provided flexibility in accommodating transmission frequency, distance between nodes, depth, ambient temperature, salinity, and acidity. AquaTools is based on the NS2 simulator and thus provides a flexible scripting interface to set up the simulations. The simulated results include detailed packet traces. A high degree of similarity in the channel characteristics predicted by AquaTools compared to numerical and published models validates the reliability of the simulator. The ability to test different protocols and systems with AquaTools makes it a valuable tool in place of at-sea testing.

Hauge and Hetland (2015), in cooperation with Kongsberg Maritime, developed the hydroacoustic channel emulator (HACE) to test hydroacoustic solutions as a total system for the purpose of reducing the number of required sea trials. The emulator replaced the transducers and sea water with a computer simulating the acoustics, an audio interface, and voltage attenuation. The principal simplifications limited the number of self-contained mathematical models: ray-trace calculations were performed external to HACE using BELLHOP. Models embedded in HACE included: reflection loss; geometric spreading; propagation delay; ambient noise; acoustic absorption; point-to-point and network communication in 3D; varying sea bed in 3D; varying sea surface; surface scatter; Doppler spread (time variation); and varying sound speed with ray tracing.

Zuba et al. (2013) introduced a new communication framework to represent the physical layer of the Aqua-Sim simulator. They integrated and tested a new attenuation model (based on Rogers [1981] model), a noise model, a modulation scheme based of orthogonal frequency-division multiplexing (OFDM) acoustic modems, and two practical real system characteristics. The proposed communication framework showed an improvement over existing simulation results. They also compared the new simulation results with the results from field experiments in the Chesapeake Bay.

Accurate acoustic channel models are critical for the study of UANs. Existing models include physics-based models and empirical approximation models: the former enjoy good accuracy, but incur heavy computational load, rendering them impractical in large networks; the latter are computationally inexpensive but inaccurate since they do not account for the complex effects of boundary reflection losses, multipath phenomenon, and ray refraction in the stratified ocean medium. Wang et al. (2013) proposed a stratified acoustic model (SAM) based on frequency-independent geometrical ray tracing, accounting for each ray's phase shift during propagation. It is a feasible channel model for large scale UAN simulation, allowing prediction of transmission loss with much lower computational complexity than the traditional physics-based models. The accuracy of the model was validated via comparisons with experimental measurements in two different oceans (Australia and Taiwan Strait). Satisfactory agreements with the measurements and with other computationally intensive classical physics-based models were demonstrated.

Qarabaqi and Stojanovic (2013) developed a statistical model for (UWA) channels that took into account physical aspects of acoustic propagation as well as the effects of inevitable random channel variations. Channel variations were classified into small-scale and large-scale based on the notion of the underlying random displacement being on the order of a few, or many wavelengths, respectively. While small-scale modeling treated random-channel variations over short displacements and short intervals of time (subsecond) during which the system geometry and environmental conditions did not change, large-scale modeling treated variations caused by location uncertainty (displacement from the nominal geometry) as well as varying environmental conditions. The proposed small-scale model described intrapath dispersion caused by scattering as complex Gaussian multiplicative coefficients with particular correlation properties in time and frequency. Specifically, it was shown that an autoregressive Gaussian displacement of scattering points led to a frequency-dependent exponential time-correlation function of the small-scale fading coefficients, while frequency correlation was dictated by the variance of the intrapath delays. In addition, motion-induced random Doppler shifting, resulting from surface waves or transmitter/receiver drifting, was shown to lead to Bessel-type autocorrelation functions. Based on this model, a computationally efficient channel simulator was proposed in which each path's small-scale coefficient was represented by an autoregressive Gaussian process itself; further provision was made to account for frequency correlation across the signal bandwidth. Large-scale modeling focused on the channel gain, which offered a measure of the received signal strength averaged locally over small-scale phenomena. A log-normal model was proposed for the large-scale gain, the mean of which followed a log-distance dependence. Experimental data from four deployment sites with varying degrees of mobility were used for a statistical analysis. Probability distributions and correlation functions of the salient small- and large-scale parameters demonstrated a good match with the theoretical models. The present model was based on the Gaussian assumption for the underlying processes, which is a starting point from a statistical point of view. Future challenges may focus on examining frequency-correlation properties of the small-scale fading in wideband experimental systems, as well as on extending the proposed models to address spatial correlation properties of the acoustic channel on both small and large scales.

Domingo (2008) surveyed ray-theory-based multipath Rayleigh underwater channel models for subsea wireless communications and summarized the research challenges for an efficient communication in this environment. These channel models were valid for shallow or deep water. They were based on acoustic propagation physics that captured different propagation paths of sound underwater and considered all the effects of shadow zones, multipath fading, operating frequency, depth, and water temperature. The propagation characteristics were assessed through mathematical analysis. Transmission losses between transceivers were investigated through simulations. Further simulations were carried out to study the bit-error rate (BER) effects and the maximum internode distances for different networks and depths, considering a 16-quadrature amplitude modulation scheme with orthogonal frequency-division multiplexing as the multicarrier transmission technique. The effect of weather and the variability of ocean environmental factors (such as water temperature) on the communication performance were also investigated. The mathematical analysis and simulations were used to guide the deployment and operation of underwater wireless communication networks.

Chitre (2007) developed a mathematical model for signal propagation through a warm, shallow-water acoustic channel. This model, which was used to study the performance of communications systems, was shown to capture the essential physics of such channels including multipath arrival structure and statistical effects such as fading and arrival-time jitter. The time correlation of the fading is defined by the Doppler spread, which has to be measured to calibrate the model; this might impact simulations where the data-packet length is much larger than the coherence time of the arrival-time jitter.

Cheng et al. (2013) discussed the time variability of an UWA channel induced by the Doppler effect. They proposed a time-varying channel simulator based on the BELLHOP ray-tracing model. The new channel model incorporated many factors of realistic underwater environments so that it fully reflected the attenuation, multipath-propagation, and time-variation of the channel. The problem was further analyzed by examining the issue of signal-frequency shift and time expansion (or compression) due to the Doppler effect.

Zărnescu (2014) developed environmental and acoustic models for an underwater communication channel located in the northwestern part of the Black Sea.

Mean seasonal sound-speed profiles, bathymetric data, geophysical properties of the seafloor sediments, and wind speed at the sea surface were used to generate simulations that were both accurate and comparable to field measurements in the marine environment. Two algorithms from the acoustic toolbox user interface and post processor (AcTUP; see Appendix C) were used to obtain the seasonal attenuation in the channel in the frequency band 0.1–99.9 kHz for three medium-range transmission distances: Bounce (bottom-reflection coefficient calculation for layered media), and BELLHOP (ray and Gaussian-beam tracing model with range-dependent bathymetry). The average transmission losses and ambient-noise levels were computed as functions of frequency. The optimal transmission frequencies for each season and three transmission distances were then derived. A comparison of the simulated results with the observed mean oceanic communication channel (MOCC) revealed that the optimal transmission frequency for the winter and spring seasons were approximately equal to those of the MOCC. In the summer season at a range of 0.5 km, and in the autumn season at a range of 1 km, it was observed that the optimal transmission frequency exceeded that of the MOCC. This indicated that the transmission communication speed could increase and, therefore, a larger bandwidth would be available. In the autumn season at ranges of 0.5 km and 2 km, the optimal transmission frequency was smaller than that of the MOCC, indicating that the transmission speed decreased with an associated decrease in bandwidth.

Khan et al. (2013) evaluated the performance of an OFDM-based scheme for underwater-acoustic communications using simulated channel models. System performance was assessed under the effects of seasonal variations in the northwestern region of the Arabian Sea off the coast of Ormara, Pakistan. Channel models were simulated using the BELLHOP ray-tracing program. The results of these simulations demonstrated that OFDM-based communication is a robust technique capable of mitigating the seasonal effects of underwater multipaths. Furthermore, these results could be used to determine the maximum ranges at which the transmitter and the receiver could communicate at any given source level. This study formed the basis for an expanded feasibility study of an UWA sensor network and telemetry solution to be deployed in that area of the Arabian Sea. Dynamic channel-condition modeling under the effects of wind and platform motion will be used in future studies to mitigate Doppler-induced signals.

The second order and fourth order moment method (M2M4) is an adaptive algorithm for estimating SNR through second-order and fourth-order moment operations based on characteristics of the modulation system. For an FM modulation system, Gao et al. (2013) demonstrated that M2M4 is the optimal algorithm for UWA channels modeled by a Rayleigh channel, a Rician channel, and a channel modeled by the BELLHOP propagation model.

6.15.1.4 Network Performance and Optimization

Control and optimization of acoustic sensors can improve network performance in variable undersea environments (Cai et al., 2007). However, the design of optimal control systems requires tractable models of acoustic-wave propagation phenomena. High-fidelity acoustic models that capture the influence of environmental conditions on wave propagation involve partial-differential equations that are computationally intensive. Moreover, for given boundary and initial conditions, partial differential equations do not provide closed-form solutions for propagation loss. A simple Bayesian-network model of acoustic propagation was developed for sonar control. The performance of this model compared favorably with that of a radial-basis function neural network. The sensor dependency on spatial and temporal coordinates could be estimated and utilized to compute optimal sonar control strategies.

In many emerging applications, UANs are integrated into larger networks comprising terrestrial, radio, and satellite networks (Dong et al., 2014). A cross-disciplinary project called coastal and Arctic maritime operations and surveillance (CAMOS) sensor networks, developed at Norwegian University of Science and Technology (NTNU), has been developing a robust communication framework that integrates underwater, terrestrial radio, and satellite communications in a resilient infrastructure with specific attention given to denial-of-service (DoS) attacks. In this way, a multitude of applications can be supported with specific attention given to sensor networking in the Arctic region. DoS attacks are designed to prevent UAN nodes from utilizing all or part of their network connectivity. DoS attacks may extend to multiple layers of the protocol stack: (1) physical layer—DoS attacks can be launched against the physical layer by using communication-jamming devices to compromise service availability; (2) data-link (or MAC) layer—adversaries may only need to induce a collision in one octet of a transmission to disrupt an entire packet; and (3) network (or routing) layer—DoS attacks against the network layer can be accomplished by (a) congesting a network through flooding; (b) maliciously dropping packets to deny services; (c) stopping packets in the network by refusing to forward messages; (d) malicious nodes broadcasting themselves as the optimal node to select for data forwarding and then dropping packets to deny service; (e) compromising protocols; (f) using an authenticated device to disrupt the system; (g) using a malicious device to fabricate multiple identities, behaving as if it were a larger number of nodes. Countermeasures comprise two basic approaches: (1) each node monitors the packet-forwarding activities of its neighbors and rates the transmission reliability of all alternative routes to a particular destination node; and (2) the neighbors of any single node collaborate in rating the node according to how well the node executes the functions requested of it.

Ensuring the required reliability and energy efficiency is an essential issue in UANs. Many schemes have been developed to improve reliability and energy efficiency in such networks. However, most of the existing schemes are based on the assumption that the noise is uniformly distributed in the underwater surrounding and the noise attenuation is not considered. Xu et al. (2013) took the noise attenuation into account and proposed a new asymmetric multipath division communications (AMDC) mechanism to improve reliability and energy efficiency in UANs. To this end, an asymmetric multiple layer division (AMLD) scheme was developed to divide the underwater communication space for the purpose of constructing tree-based multipaths. The problem of energy efficiency of the AMDC was then formulated as a distributed optimization problem and was solved to achieve a set of feasible solutions. Finally, simulation experiments were conducted to evaluate the performance of the proposed AMDC. The results revealed that the AMDC outperformed the existing multipath transmission scheme in UANs in terms of energy efficiency and the total PER.

6.15.1.5 Underwater Communications

Stojanovic and Preisig (2009) noted that an acoustic communication system is inherently wideband in the sense that the bandwidth is not negligible with respect to its center frequency. The channel can have a sparse impulse response, where each physical path acts as a time-varying low-pass filter, and motion introduces additional Doppler spreading and shifting. Surface waves, internal turbulence, fluctuations in the sound speed, and other small-scale phenomena contribute to random signal variations. There are presently no standardized models for the acoustic channel fading, and experimental measurements are often made to assess the statistical properties of the channel in particular deployment sites.

Baggeroer (2012) updated earlier reviews of acoustic communications (Acomms) dating back to 1984. Specifically addressed were the topics of (1) channel models; (2) equalizers; (3) time reversal and passive phase conjugation; (4) OFDM and MIMO; (5) diversity; (6) acquisition and synchronization; (7) coding; (8) networks; and (9) stealth. Emphasis was placed on the physical layer of Acomms systems since this layer is fundamental to the successful functioning of networks.

UWA communications are influenced by spreading loss, noise, multipath discrimination, Doppler spread, and high and variable propagation delays. Moreover, UWA channels normally have low data rates and time-varying fading. These factors determine the temporal and spatial variability of the acoustic channel and make the available bandwidth of the ocean channel both limited and dependent on range and frequency. Challenges due to fading, multipath, and refractive properties of the sound channel necessitate the development of precise underwater-channel models. Some existing channel models are simplified and do not consider multipath or fading. Multipath interference due to boundary reflection in shallow-water acoustic communications poses major obstacles to reliable, high-speed underwater communication systems (Su et al., 2010). Socheleau et al. (2010) developed a time-varying UWA channel model based on the principle of maximum entropy. Their model relied only on the available knowledge of the environment in addition to a few channel parameters (e.g., channel average power and Doppler spread). They demonstrated through fading statistics and bit error rate measurements that accurate channel impulse responses could be obtained for communications applications.

The Centre for Maritime Research and Experimentation (CMRE) has developed, tested, and promoted JANUS, which is a robust and simple modulation-and-coding scheme intended as the first standard to support interoperability in digital underwater communications. The initial frequency band for JANUS was derived from the attractiveness of the 9–14 kHz band for a range of typical operational communication scenarios. The JANUS waveform is fully scalable, with all implementation parameters depending solely on center frequency and bandwidth. Two key mechanisms have been built into JANUS to handle the medium access: a channel-reservation feature;

and a MAC. JANUS-based services of relevance for operational communities include underwater AIS and underwater METOC services. The idea of using JANUS to transmit the AIS picture to submarines navigating at depth had been proposed as a mechanism to reduce the probability of accidents between surface ships and submerged assets (Alves et al., 2016).

Ocean-bottom sensor nodes are used for oceanographic data collection, pollution monitoring, offshore exploration, tactical surveillance applications, and REA (Akyildiz et al., 2004, 2005). Factors that determine the temporal and spatial variability of the acoustic channel also limit the available bandwidth of the ocean channel and make it dependent on range and frequency. Specifically, long-range systems (~10 km) have bandwidths of a few kilohertz while short-range systems (~0.1 km) have bandwidths on the order of a hundred kilohertz. A moored-buoy ocean observatory system comprising oceanographic sensors was linked by acoustic communications to retrieve data from sensors in the water column at ranges of approximately 3 km (Frye et al., 2006). The observatory was deployed off Vancouver Island in the northeastern Pacific Ocean in May 2004 (for a period of 13 months) to study the correlation of seismicity and fluid flow in a seep area along the Nootka fault.

While many applications may require long-term monitoring of the deployment area, battery-powered network nodes limit the lifetime of these networks. Shallow-water acoustic channel characteristics (such as low available bandwidth, highly varying multipath, and large propagation delays) restrict the efficiency of underwater-acoustic networks. A useful survey of existing network technologies and their applicability to underwater-acoustic channels was provided by Sozer et al. (2000).

Ebihara et al. (2009) applied a chip-interleaved multiple-access (CIMA) method to underwater communications. Although CIMA is commonly used for wireless radio communication, it was shown to be useful in reconstructing acoustically transmitted data in multipath ocean environments. Ebihara et al. (2010) simulated the performance of an UWA communication system using orthogonal signal division multiplexing (OSDM) in the presence of intersymbol interference and Doppler shift. The results of this simulation were compared with an existing communication system that used coherent modulation with an adaptive filter. It was shown that OSDM is robust to heavy intersymbol interference and it was expected to be suitable for a rapidly time-varying channel. However, it was also determined that OSDM is more sensitive to the Doppler effect compared with the existing communication system; hence, it would be necessary to apply frequency-offset compensation.

UWA communications are complicated by doubly-spread channels (i.e., large time spreads resulting from multipath and Doppler). This ill-conditioned problem exists for conventional OSDM, which employs a single transducer in the receiver. The introduction of a multichannel receiver was found to be effective against the ill-condition problem. Ebihara and Mizutani (2014) proposed a UWA communication system using OSDM that measured the multipath profile without an adaptation or interpolation process to achieve stable communication in doubly-spread channels. Specifically, they experimentally compared the performance of OSDM and existing communication schemes (single-carrier with decision feedback equalizer [DFE] and OFDM) in a test tank with respect to communication quality, data rate, frame length, and calculation complexity. It was shown that OSDM with a multichannel receiver was attractive in terms of communication quality. Moreover, OSDM achieved a far better BER performance compared to the other schemes in both static and dynamic channels with various input signal-to-noise ratios. However, the complexity was less than that achieved with single-carrier DFE. They suggested that OSDM may be a reliable communication alternative for UWA under conditions of multipath and Doppler spread (such as shallow water) with practical complexity.

Cooperative transmission is a new wireless communication technique in which diversity gain can be achieved by utilizing relay nodes as virtual antennae. These transmission techniques were investigated for UWA communications. First, the performance of several cooperative transmission schemes was studied in an underwater scenario. Second, by taking advantage of the relatively low propagation speed of sound in water, a new wave cooperative transmission scheme was designed. In this scheme, the relay nodes amplified the signal received from the source node and then forwarded the signal immediately to the intended destination. The goal was to alter the multipath effect at the receiver. Third, the upper bound of performance was derived for the proposed wave cooperative transmission scheme. The simulation results showed that the proposed wave cooperative transmission had significant advantages over both the traditional direct transmission and the existing cooperative transmission schemes originally designed for radio wireless networks (Han et al., 2008).

Linton (2016) examined the application of iterative- and adaptive-processing techniques to multiple-access interleave-division multiple access (IDMA) systems. Optimization of the iterative detection process was achieved through power allocation, forward-error-correction code allocation, and perfect space-time coding. In addition, IDMA systems (with iterative receivers) were applied to UWA communications; variance transfer charts were used to analyze the iterative-receiver performance. Since the UWA channel is a challenging environment characterized by long delay-spreads and limited bandwidth, an OFDM-IDMA system was posed as a solution. An iterative receiver for IDMA using a nonlinear Kalman filter was used to perform joint decoding and channel equalization for doubly-spread UWA channels. The nonlinear Kalman filter utilized low-rank basis expansion models to track the temporal variation of the channel. OFDM was combined with an IDMA overlay to develop a multiple-access communications system that provided robust performance in the presence of large time-delay spread and other impairments presented by the shallow-water acoustic channel. A low-complexity iterative decoding algorithm based on the turbo-decoding concept was developed for the OFDM-IDMA system receiver, and experimental results demonstrated good performance. The UWA channel was extended to the doubly-spread case. The relative motion between the transmitter, receiver, and scattering objects imparted each path with a unique Doppler shift, so that multipath propagation also induced a frequency-domain spreading effect on the information signal. Such channels are both delay- and Doppler-spread and are thus referred to as *doubly-spread*.

6.15.1.6 Medium Access Control

Tan and Seah (2007) showed that an efficient and effective medium-access-control (MAC) scheme is required to coordinate access to shared communications channels. Typical terrestrial MAC protocols are unable to handle the long propagation delays encountered in UWA environments. Existing underwater MAC schemes are generally centralized in nature and therefore are not scalable to large sensor networks; moreover, they may have high control overheads. (See also Makhija et al. [2006] and Mahdy [2008a,b].) Typically, MAC protocols can be classified as deterministic or nondeterministic. The latter is also known as random-access protocols that are contention-based in nature, meaning that the nodes compete to transmit data at various times, and access to the channel is not guaranteed. As such, contention-based protocols are not able to provide the quality-of-service guarantees required by real-time data transmissions. Tan and Seah (2007) proposed a distributed MAC protocol for long-latency access networks (or PLAN) that utilizes code-division-multiple-access (CDMA) as the underlying multiple-access technique to minimize the multipath and Doppler effects that are inherent in underwater physical channels. The proposed MAC protocol involved a three-way handshake (RTS-CTS-DATA) that collated the request-to-send (RTS) from multiple neighboring nodes before sending a single clear-to-send (CTS). Simulation results showed that the proposed scheme outperformed the so-called Aloha protocol (with retransmissions) and multiple-access-with-collision-avoidance (MACA) in terms of higher throughput while incurring lower overheads.

Linton et al. (2008, 2009) presented a novel multiple-access communications scheme that was shown to be robust in shallow-water acoustic networks where large time-delay spreads are commonly encountered. The scheme was based on OFDM with an Interleave-Division Multiple Access (IDMA) overlay. The IDMA is a variant of CDMA. This approach was found to outperform other common multiple-access schemes in shallow-water environments.

Based on typical sound-speed profiles for the Taiwan Strait (representative of shallow-water conditions), Tao and Xu (2007) adopted the FFP SPARC to simulate the delay time and amplitude fluctuations of multipaths under different channel conditions. These simulation results revealed relationships between delay, amplitude, and typical channel parameters such as sound-speed gradient, distance, and channel depth, which provided valuable guidance for the practical design of wireless acoustic communication systems in a shallow-water environment. In related work, the Gaussian beam model was used to compute eigenrays and transmission losses for flat and sloping bottoms using representative sound-speed profiles for the Taiwan Strait including positive, negative, and isovelocity gradients (Xu et al., 2006). The resulting simulations provided guidance for the practical design of underwater wireless communication systems including development of more robust schemes and selection of the most appropriate schemes for any given environmental conditions. Further efforts will be needed to improve the precision and robustness of the Gaussian beam model for application to shallow-water wireless communication systems.

Chitre et al. (2008) provided an overview of the key developments in point-to-point communication techniques and underwater networking protocols. MACA-based contention protocols, time-division multiple access (TDMA), or CDMA-based contention-free protocols can be used in an underwater local-area network depending on the exact requirements and constraints. Multi-modem adaptive MAC protocols for AUV networks were proposed to provide a unified interface to higher layers. *Ad hoc* on-demand distance vector (AODV) and dynamic source routing (DSR)-based lightweight routing protocols were also proposed for underwater use. Chitre et al. (2008) concluded that efficient routing for *ad hoc* mobile underwater networks still remains an open research challenge; furthermore, standardization for underwater networking is required to provide interoperability and ease of operation to accelerate field research.

Climent et al. (2014) presented a comprehensive technology assessment of the current state-of-the-art in UWA sensor networks organized according to physical layer, MAC (or data link) layer, and routing (or network) layer.

6.15.1.7 Data Delivery Schemes

Lee and Seah (2007) compared two data delivery schemes: vector based forwarding (VBF) and the multipath virtual sink (MVS) architecture. The MVS architecture presents a simple but effective approach. To further optimize performance, it may be beneficial for sources to choose only a few paths among those available, rotating among different paths so as to avoid any collisions among spatially diverse but interleaving paths. The main drawback to MVS lies in its fixed-path forwarding method. Fortunately, MVS is versatile and other forwarding mechanisms can be used with MVS. The VBF architecture offers a location-based scheme that is built on a limiting assumption: the nodes are equipped with signal distance and angle-of-arrival detection capabilities. In particular, accurate angle-of-arrival information is difficult to achieve in the presence of multipath fading, even with directional hydrophones. To enhance VBF, optimal performance values might be dynamically changed to suit network conditions. The adaptation algorithm should also mitigate the effect of overlapping routing.

6.15.2 Localization Methods

Localization algorithms are relevant to underwater sensor networks, but there are challenges in meeting requirements imposed by emerging applications for such networks, particularly in offshore engineering (Chandrasekhar et al., 2006; Tan et al., 2011). Localization algorithms can be broadly categorized into *range-based* and *range-free* schemes, as discussed below.

6.15.2.1 Range-Based Schemes

Range-based schemes use precise distance or angle measurements to estimate the location of nodes in a network. These schemes use time-of-arrival (ToA), time-difference-of-arrival (TDoA), angle-of-arrival (AoA), or received-signal-strength-indicator (RSSI) methods. Traditional range-based schemes have fixed anchor nodes whose locations are known; however, some newer methods do not require anchor nodes or beacon signals. ToA or TDoA schemes require tight time synchronization between the transmitter and receiver clocks. RSSI schemes need to account for multipath effects.

Myagotin and Burdinsky (2010) examined an acoustic navigation network for AUV positioning based on one-way signal transmissions. They used phase-manipulated M-sequences (refer back to Section 6.12) in acoustic transmissions to improve the accuracy of time-of-flight (ToF) measurements. Simulations of different network cell geometries revealed that a random distribution of buoys was preferable to triangular and rectangular grids due to a larger visibility coefficient (i.e., the number of buoys visible from the current AUV location).

In acoustics-based positioning systems where it is important to know accurately the time-of-flights of the acoustic signals, pseudorandom codes may provide some immunity to the physical variability of the environment (Aparicio et al., 2013). A statistical study of the influence in range detection of common effects in UWA, such as noise and wind speed, was presented. The performance of a relative positioning system was studied for two different scenarios: a shallow-water environment, and a surface-duct environment. A multidimensional-scaling technique was employed to obtain buoy positions. This algorithm required as inputs the distances between the buoys. In these simulations, the real distances between the buoys were contaminated with a Gaussian distribution of errors, whose parameters varied depending on the environment.

In acoustic positioning systems, accurate detection of arrival times is critical for the accurate estimation of distances between nodes. ToA accuracy can be improved by employing acoustic signals coded using pseudorandom noise; however, these signals are still affected by underwater-channel phenomena. Aparicio et al. (2015) analyzed the detection of spread-spectrum modulated signals in channels that were affected by multipath and reverberation. A spread-spectrum signal consisting of a modulated Kasami code (binary sequences) was transmitted through two different pools to a receiver after following several line-of-sight and nonline-of-sight paths. Next, a correlation process was performed to obtain information regarding the arrival times that comprised the multipath structure. These ToFs were next compared against those generated by an underwater acoustic propagation model (APM) to validate the performance of the APM and its ability to predict signal-detection outcomes in underwater environments characterized by strong multipath and reverberation components.

Scour holes tend to make bridge foundations weaker and thus prone to collapses. Dahal et al. (2013) simulated the utility of the received signal strength (RSS) method for measuring bridge scour depth versus ToF or ToA methods for range measurements in underwater environments. Simulations showed that the estimated scour depth tended to approach the real depth as the number of acoustic sensors employed was increased. Also, since erosion of sediments leads to the formation of a dense water layer over the water bottom due to suspended sediments, an effort was made to compare the effect of the layered nature of the water bottom on scour depth to that of nonlayered approach; it was shown that the layered approach gave more accurate values for the scour depth. For example, the simulations indicated that a sediment layer depth of 1.5 m resulted in a total scour depth of 2.8 m; this value increased with the thickness of the sediment layer. In contrast, when a single-layered bottom was considered, the resultant scour depth was about 2 m. Since a single-layered bottom assumption is fairly unrealistic, this latter result could be misleading.

6.15.2.2 Range-Free Schemes

Range-free schemes are simpler than range-based schemes, but they only provide a coarse estimate of a node's location. Range-free schemes can be classified according to *hopcount-based* and *area-based* schemes:

Hopcount-based schemes place anchor nodes at the corners or along the boundaries of a square grid. *Hopcount* is the number of network devices between the starting node and the destination node through which data must pass (this number is used in determining position uniqueness, as in *n*-hop multilateration). Examples include the centroid scheme, DV-hop (where DV stands for distance vector) and density-aware hop-count localization (DHL). DV-hop is one of the most basic range-free schemes, and it first employs a classical DV exchange so that all nodes in the network get distances, in number of hops, to the anchor nodes. DV-hop performs well only in networks that have uniform and dense node distributions. *Iterative mulitlateration* is the process in which unknown nodes become anchor nodes after having estimated their respective locations.

Area-based schemes approximate the area in which a node is located in very large and densely populated sensor networks. Examples include the area-localization scheme (ALS) and the approximate point-in-triangle (APIT) scheme. In the APIT scheme, a node chooses three anchors from all audible anchors (i.e., anchors from which beacon signals were received) and tests whether it is inside the triangular region formed by these three anchors using RSSI information from the beacon signals.

Seah and Tan (2006) proposed a virtual sink architecture for UWA sensor networks that achieved robustness and energy efficiency under harsh marine conditions. They utilized multipath data delivery to overcome the long-range propagation delays and adverse link conditions. Moreover, a virtual sink design avoids contention near the sink. Sun and Seah (2007) developed a data-delivery scheme based on a multi-sink underwater wireless sensor network (UWSN) architecture to achieve fast and reliable data delivery under harsh oceanic conditions. The scheme dynamically redirects data packets when temporal link failures are encountered; this is done without requiring network-state information to be updated.

6.15.3 Vehicles

Underwater networking is an enabling technology for the operation of AUVs. In particular, *ad hoc* networks entail wireless communications for mobile hosts called nodes. In these networks, there is no fixed infrastructure. Mobile nodes that are within range communicate directly via wireless links, while those that are far apart rely on other nodes to relay messages as routers. Node mobility in an *ad hoc* network causes frequent changes of the network topology. Since *ad hoc* networks can be deployed rapidly with relatively low cost, they are attractive for military, emergency, commercial, and scientific applications (Akyildiz et al., 2004, 2005). Wang et al. (2010) developed a channel simulator for testing the performance of UUV communications.

AUVs, or UUVs, constitute part of a larger group of undersea systems known as UUVs, a classification that includes nonautonomous remotely operated vehicles (ROVs) that are controlled and powered from the surface by an operator (or pilot) via an umbilical connection.

Graver (2005) noted that underwater gliders actually constitute a new class of autonomous underwater vehicles that glide by controlling their buoyancy and attitude using internal actuators. Gliders have useful applications in oceanographic sensing and data collection because of their low cost, autonomy, and capability for long-range, extended-duration deployments. They serve as adjuncts to ship-based hydrographic casts, towed sensors, UUV/AUV and satellite-based sensors, but they also present challenges in communications common to all untethered subsurface sensors.

Guizzo (2008) identified three variants of gliders (Seaglider, Spray, and Slocum) that are contending for funding under the US Navy Littoral Battlespace Sensing, Fusion, and Integration (LBSF&I) program. Klamper (2007) described the planned fielding of a mobile, autonomous underwater surveillance network comprising gliders, drifters, and bottomed sensors in shallow-water environments.

Goldhahn et al. (2014) proposed a method for data fusion and Bayesian target tracking for a network of multiple AUVs in an ASW scenario; the port-starboard ambiguity problem associated with horizontally towed arrays was specifically addressed. This method accounted for the hypothesized probability of detection as a function of target position in bistatic geometries using the associated environmental parameters of the waveguide. The acoustic model for bistatic SNR was based on the work of Harrison (2005) wherein a closed-form solution existed when the bathymetry was range independent and the sound speed was isovelocity. The target strength was assumed to be constant with respect to target-aspect angle. The proposed method was shown to be optimal given one target in the area of surveillance. If the sound speed were depth-dependent and/or range-dependent, then numerical models such as ARTEMIS or BELLHOP could be used to achieve better fidelity.

Developments in the use of underwater swarm sensor networks (USSNs) require new approaches for short-range mobile acoustic underwater communications essential for (1) coordinated maneuvering of closely spaced operating vehicles; and (2) data transmission within the swarm. Implementation of vehicle swarms can improve the current ability of single vehicles to survey and explore the undersea environment. Burrowes et al. (2014) speculated that the application of an USSN, which has many vehicles in a dense topology, would impact the reverberation channel since the vehicles themselves become sound-scattering objects. Their simulations demonstrated that the closer the vehicles were operating to each other, the higher were the swarm-scattering values. This new type of swarm-scattering reverberation exhibits a strong relationship between range (propagation time) and packet length (transmission time). They further investigated the impact that reverberation levels, in particular swarm and sea-surface scattering, would have on the signal to noise-plus-interference ratio (SNIR), as well as the influence that transmitter power would have on this ratio for autonomous vehicles operating at close ranges in a swarm-like fashion. In their simulations, Burrowes et al. (2014) utilized the following parameters: swarm size = 5 vehicles; average range between vehicles = 20–50 m;

center frequency = 40 kHz; and bandwidth = 10 kHz. A new USSN MAC layer protocol was proposed: adaptive space time-time division multiple access. This protocol effectively used a single channel broadcast acoustic environment while incorporating a method for handling the unique characteristics of long propagation delay and low bandwidths underwater by utilizing the space-time diversity available in the channel. The broadcast approach was used since it provided rapid dissemination of information within the swarm. It was shown that knowing the relationship between transmitter power, reverberation levels, and SNIR can be beneficial in a vehicle for predicting optimal transmitter power levels in a way similar to the feedback utilized in terrestrial networks. Optimizing the energy levels used to send packets would be beneficial for both reducing energy consumption as well as increasing packet-delivery success (i.e., delivered without error). The OpNet Modeler was used to simulate operations of a MAC protocol specifically designed for an USSN and to analyze the impact of these reverberation levels on SNIR.

Braca et al. (2014) considered a multistatic network of AUVs where a collaborative multi-sensor data fusion scheme moved beyond the limitations of individual sensors. They further explored the cognitive paradigm wherein individual AUV units optimized their path planning vis-à-vis the intended purpose of the network using accumulated data. A multistatic configuration of the platforms and a corresponding acoustic model were considered in order to derive a proper Bayesian model. Using the information contained in the Bayesian full posterior, cognitive detection and tracking algorithms were designed. The information contained in the Bayesian posterior was used to demonstrate the benefit of the cognitive paradigm in a practical scenario.

Blouin et al. (2015) identified five challenges to full autonomy of underwater systems: (1) networking—the development and testing of an underwater networking protocol incorporating and exploiting the environmental knowledge to make routing robust and faster; (2) collaboration—the identification of adequate metrics to determine the amount and nature of data-sharing occurring between underwater network nodes for increasing the success likelihood of their collective mission; (3) data aggregation—the identification of conditions leading to stable and convergent data aggregation schemes, along with the identification of potential degrees-of-freedom accelerating or improving the accuracy of such convergence; (4) communication—the development and testing of a sufficiently accurate acoustic propagation model running on-board modems, and advanced acoustic modem hardware to improve the reliability of low-power and longer transmission range; and (5) adaptation—the identification of key adaptation schemes, their interplay, and robustness in terms of the overall performance of a network of underwater systems.

6.16 MARINE MAMMAL PROTECTION

This section will address five areas pertinent to marine mammal protection. The first relates to regulatory initiatives and measurement programs. The second deals with the rising levels of underwater noise due to increases in shipping, OA, and wind farm development. The third concerns seismic operations and their impacts on whales. The fourth topic addresses modeling efforts. Finally, ASW training ranges and mitigation techniques are reviewed.

Over the past several decades, the soundscape of the marine environment has responded to changes in both natural and anthropogenic influences. The disruption of the natural acoustic environment results in noise pollution. The field of UWA enables us to observe quantitatively and predict the behavior of this soundscape and the response of the natural acoustic environment to noise pollution.

The soundscape baseline is defined by ambient noise, which is the prevailing, background of sound at a particular location in the ocean at a given time of the year. It does not include transient sounds such as the noise of nearby ships and marine organisms, or of passing rain showers. In practice, ambient noise excludes all forms of self-noise, such as the noise of current flow around the sonar. For sonar processing, however, it is the background of noise (including interfering sounds), typical of the time, location, and depth against which an acoustic signal must be detected. New regulatory initiatives have placed additional restrictions on uses of sound in the ocean: mitigation of marine mammal endangerment is now an integral consideration in acoustic-system design and operation.

Dahl et al. (2007) compared ambient sound levels in air and underwater using units of intensity spectral density (Wm^{-2} Hz^{-1}) over the frequency range 10 Hz–100 kHz. The intensity spectral density ranged from 10^{-16} to 10^{-4}. It was found that the intensity spectral density of a quiet residential environment (with distant traffic influence) exceeded that of nominal high-level underwater ambient noise conditions. In air acoustics, a reference sound pressure of 20 μPa is used, while the current choice in UWA is 1 μPa. Therefore, pressure measurements of equal pressures in air and water will differ by 26 dB (i.e., $20 \log_{10}$ [20 μPa/1 μPa]), being 26 dB higher in water than in air due solely to differences in the reference sound pressures (e.g., see Richardson et al., 1995: 18). Moreover, the characteristic impedance of water is about 3,600 times that of air, so the conversion factor for the intensity of sounds of equal pressure in air versus water is 36 dB (i.e., $10 \log_{10}$ [3600]). Therefore, intensity measurements of equal pressures in air and water differ by 62 dB (= 36 dB + 26 dB).

A set of procedures has been developed to allow preliminary estimates to be made of underwater noise and its effects on marine species (Hazelwood and Connelly, 2005). Heathershaw et al. (2001) discussed the growing problem of "noise pollution" in the marine environment and described techniques for assessing the potentially adverse effects of underwater sound on marine life.

6.16.1 Regulatory Initiatives and Measurement Programs

McCarthy (2004) examined the issue of anthropogenic sound in a global context and considered the need for new regulatory initiatives to deal with the conflicting uses of ocean space related to noise. She identified the existing legal, economic, and political barriers to the creation and implementation of a new international regime designed to manage anthropogenic noise in the ocean.

The Committee on Potential Impacts of Ambient Noise in the Ocean on Marine Mammals was charged by the Ocean Studies Board of the US National Research Council to assess the state of our knowledge of underwater noise and recommend research areas to assist in determining whether noise in the ocean adversely affects marine mammals (National Research Council, 2003a). One of the findings of this committee was that models describing ocean noise are better developed than are models describing marinemammal distribution, hearing, and behavior. The biggest challenge lies in integrating the two types of models. The National Research Council (2005) also examined what constitutes "biologically significant" in the context of level B harassment as used in the latest amendments to the US Marine Mammal Protection Act (MMPA). The MMPA separates harassment into two levels. Level A harassment is defined as "any act of pursuit, torment, or annoyance which has the potential to injure a marine mammal or marine mammal stock in the wild." Level B harassment is defined as "any act of pursuit, torment, or annoyance which has the potential to disturb a marine mammal or marine mammal stock in the wild by causing disruption of behavioral patterns, including, but not limited to, migration, breathing, nursing, breeding, feeding, or sheltering." The MMPA, enacted in 1972, was the first legislation that called for an ecosystem approach to natural-resource management and conservation; it specifically prohibited the take (i.e., hunting, killing, capture, and/or harassment) of marine mammals.

Reflecting the importance of standards, a process has been advocated for communicating the relevance of data-quality standards to marine mammal acoustic-effects models (Tozzi et al., 2006). Southall et al. (2007) advanced initial scientific recommendations for marine mammal noise exposure criteria.

The US Department of Energy (2009) focused on potential impacts of marine and hydrokinetic technologies to aquatic environments, fish and fish habitats, ecological relationships, and other marine and freshwater aquatic resources. Most considerations of the environmental effects have been in the form of predictive studies and environmental assessments (EA) that have not yet been verified. While these assessments cannot predict what, if any, impact a given technology may have at a given site, they have been instructive in identifying several common elements among the technologies that may pose a risk of adverse environmental effects including noise during construction and operation. For some environmental issues, it will be difficult to extrapolate predicted effects from small to large numbers of units because of complicated, nonlinear interactions between the placement of the devices and the distribution and movements of aquatic organisms. Assessment of these cumulative effects will require careful environmental monitoring as the projects are deployed. There is no conclusive evidence that marine and hydrokinetic technologies will actually cause significant environmental impacts.

Robinson et al. (2014) provided guidance on best practices for *in situ* measurements of underwater sound, for processing the data, and for reporting the measurements using appropriate metrics. It was noted that measured noise levels are sometimes difficult to compare because different measurement methodologies or acoustic metrics were used, and results can take on different meanings for each different application, leading to a risk of misunderstandings between scientists from different disciplines. Acoustic measurements are required for applications as diverse as acoustical oceanography, sonar, geophysical exploration, underwater communications, and offshore engineering. More recently, there has been an increased need to make *in situ* measurements of underwater noise for the assessment of risk to marine life. Although not intended as a standard, these guidelines addressed the need for a common approach, and the desire to promote best practices.

Popper and Hawkins (2012) edited a collection of 153 papers presented at the August 2010 Second International Meeting on the Effects of Noise on Aquatic Life that took place in Cork, Ireland. (The first meeting was held in Nyborg, Denmark in 2007.) The material was organized into 10 parts: introduction; sound detection by aquatic animals; sound production by aquatic animals; physiological effects of sound; anthropogenic sounds and behavior; population effects; anthropogenic sound sources and their measurement; science, regulation, and sound exposure criteria; monitoring, management, and mitigation; workshops; and concluding remarks. As a follow on, Popper and Hawkins (2016) edited a collection of 162 papers presented at the Third International Conference on the Effects of Noise on Aquatic Life in Budapest, Hungary (August 2013). The themes of this third meeting were chosen to cover the principal subjects of current interest: hearing abilities of aquatic animals; communication by means of underwater sound; the description of aquatic soundscapes; different sound sources and their characteristics; the effects of sound on behavior; and assessing, mitigating, and monitoring the effects of aquatic noise. A fourth meeting was conducted in Dublin, Ireland in 2016; presenters were encouraged to document their work in the *Proceedings of Meetings in Acoustics (POMA)*, an archival and open-access journal of the Acoustical Society of America.

Dekeling et al. (2014a,b,c) created a three-volume report providing information needed to begin the ocean-noise monitoring required to implement the marine strategy framework directive (MSFD). (The Marine Directive was adopted in June 2008 and aims to protect the resource base on which marine-related economic and social activities depend. The directive established European marine regions and subregions on the basis of geographical and environmental criteria. The directive lists four European marine regions: Baltic Sea, Northeast Atlantic Ocean, Mediterranean Sea, and Black Sea.) This study focused on ambiguities, uncertainties, and other shortcomings that may hinder monitoring initiatives, provided solutions, and described a methodology for monitoring both impulsive and ambient noise in such a way that the information needed for management and policy formulation could be collected in a cost-effective way. Part I (Dekeling et al., 2014a) contained the executive summary for policy and decision makers responsible for the adoption and implementation of MSFD at the national level; it provided the key conclusions and recommendations that would enable assessment of the current level of underwater noise. Part II (Dekeling et al., 2014b) provided specifications for the monitoring of underwater noise, with dedicated sections on impulsive noise and ambient noise; it is designed for those responsible for implementation of noise monitoring, modeling, and noise registration. Part III (Dekeling et al., 2014c) provided additional information, examples, and references that supported the monitoring guidance specifications. Discussions included methods for producing seasonal and statistical noise maps for each geographical location. Such efforts included BIAS (Baltic Sea information on the acoustic soundscape); STRIVE (science, technology, research, and innovation for the environment); LIDO (Listening to the Deep-Ocean Environment); and QUONOPS© (ocean noise anthropogenic forecasting platform).

6.16.2 Rising Levels of Underwater Noise

Over the past four decades, field observations have shown that the levels of underwater noise are rising. Identifiable causes and potential mitigation issues are explored below: increased shipping levels; OA; MHK devices; wind turbines; pile-driving noise; wave-energy devices; tidal turbines; noise-reduction methods; and passive acoustic monitoring (PAM).

6.16.2.1 Increased Shipping Levels

Shipping lanes, a term used to indicate the general flow of merchant traffic between two ports (refer to Section 8.7), are routes that historically have been optimized for shortest distances and travel times, and which are modified to avoid extreme weather events (Carey and Evans, 2011). Noise from distant shipping generally occupies the frequency band 20–500 Hz (see Section 7.2.2).

A comparison of time-series measurements of ocean ambient noise in the northeast Pacific Ocean over two periods (1963–1965 and 1994–2001) revealed that noise levels from the latter period exceeded those of the earlier period by about 10 dB in the frequency ranges 20–80 Hz and 200–300 Hz, and by about 3 dB at 100 Hz (Andrew et al., 2002). The observed increase was attributed to increases in shipping. Ambient noise measurements collected at the same site but separated by an interval of nearly four decades (1964–66 and 2003–04) revealed an average noise increase of 2.5–3 dB per decade in the frequency band 30–50 Hz (McDonald et al., 2006, 2008). These results are similar to those obtained by Andrew et al. (2002).

Jasny (1999) drew attention to the emerging risks to marine life associated with the rise in underwater noise; this report was later updated by Jasny et al. (2005). Kumagai (2006) provided a very readable overview of the impact of ocean noise on marine mammals together with current efforts to mitigate adverse effects.

Merchant (2013) noted that long-term trends in the levels of underwater noise have been studied at some open-ocean sites while in shallower coastal regions, the high spatial and temporal variability of the noise field present substantial methodological challenges. In response, Merchant (2013) introduced new measurement techniques that combined multiple data sources for ship-noise assessments in coastal waters. These data sources included AIS ship-tracking data, shore-based time-lapse footage, meteorological data, and tidal data. Specifically, two studies were described. First, AIS data and acoustic recordings from Falmouth Bay in the western English Channel were combined using an adaptive threshold, which separated ship passages from background noise in the acoustic data. These passages were then cross-referenced with AIS vessel track data, and the noise exposure associated with shipping activity was subsequently determined. Second, at a site in the Moray Firth, Scotland, this method was expanded to include shore-based time-lapse footage, which enabled visual corroboration of vessel identifications and the production of videos integrating the various data sources. Finally, a new technique termed spectral probability density (SPD) was introduced for the statistical analysis of long-term passive acoustic datasets. It was demonstrated that the SPD technique could reveal characteristics such as multimodality, outlier influence, and persistent self-noise, which may not be apparent using conventional techniques. Merchant (2013) indicated that these new methods could promote efforts to standardize underwater noise measurements and further assist investigations of the effects of shipping noise on marine life.

Erbe et al. (2013) examined UWA recordings of six floating production storage and offloading (FPSO) vessels moored off Western Australia. Monopole-source spectra were computed for use in environmental impact assessments (EIAs) of underwater noise. Given that operations on the FPSOs varied over the period of recording, and were sometimes unknown, a statistical approach was used to estimate the noise levels. No significant or consistent aspect dependence was found for the six FPSOs. Noise levels did not scale with FPSO size or power. The 5th, 50th (median), and 95th percentile source levels (broadband, 20–2500 Hz) were 188, 181, and 173 dB re 1 μPa @ 1 m, respectively.

The Bureau Veritas (2014) rule note dealt with underwater radiated noise (URN) emitted by any self-propelled ship. It aimed to control and limit the environmental impact on marine fauna in both shallow and deep waters. The measurement procedures focused on measurement uncertainty with the goals to improve measurement repeatability and to control measurement uncertainty. Three classes of vessels were addressed: controlled, advanced, and specified. The maximum URN levels corresponding to the notation *URN—controlled vessel* were the lowest (most stringent), with levels being progressively relaxed for advanced and specified vessels. Regarding bioacoustic impacts, the introduction of underwater noise into the marine environment was linked to the impact on the marine fauna. In reference to the European Directive 2008/56/EC and Commission Decision 2010/477/EU, the Descriptor 11.2 "Continuous low frequency sound" was to be assessed. (Directive 2008/56/EC of 17 June 2008 established a framework for community action in the field of marine environmental policy [MSFD]; Commission Decision 2010/477/EU of 1 September 2010 established criteria and methodological standards on good environmental status of marine waters.) For that purpose, a dedicated assessment of the pressure levels of the ⅓-octave bands centered at 63 Hz and 125 Hz were to be performed. The measurement procedure and post-processing described in this Rule Note implicitly enabled this assessment. For specific cases of vessels operating in a restricted and defined area, an additional report on the species of underwater noise concern could be requested.

Tennessen and Parks (2016) used acoustic-propagation modeling (specifically, MMPE) to predict how underwater anthropogenic noise might impair communication ranges between mother-and-calf pairs of endangered North Atlantic right whales, and further to illustrate how vocal compensation strategies commonly employed by marine mammals could improve the range over which communication signals could be detected. Attention was focused on the Bay of Fundy because of its dual status as an area with substantial shipping activity and as a critical habitat for right whales, especially mother-and-calf pairs that spent increasing time apart and communicated over distance. Point-source noise from a transiting container ship substantially limited upcall detection ranges (upcalls are tonal sounds with an upsweep in frequency over the duration of the call, and are produced by both mothers and calves during separation events). Increasing upcall amplitude and frequency greatly increased upcall detection ranges during episodes of point-source noise. Model results suggested that the documented 30 Hz increase in average upcall minimum frequency that occurred during the last half of the twentieth century (coinciding with increases in low-frequency ambient noise) reduced upcall transmission loss and increased detection range. These results suggested that the documented increase in upcall frequency may be an adaptive response by right whales to the globally increasing levels of ocean noise.

Shapiro et al. (2014) noted that underwater noise is now classed as pollution in accordance with the European Union's MSFD (Directive 2008/56/EC dated 17 June 2008). Noise from shipping is a major contributor to the ambient noise levels in the ocean, particularly at low (<300 Hz) frequencies. Shapiro et al. (2014) studied patterns and seasonal variations of underwater noise in the Celtic Sea by using a coupled ocean model (Proudman Oceanographic Laboratory Coastal Ocean Modeling System—POLCOMS) and an acoustic model (HARCAM) in the year 2010. POLCOMS is a three-dimensional baroclinic Arakawa B-grid model (Arakawa and Lamb, 1977) designed for the study of shelf sea processes and ocean-shelf interaction. Two sources of sound were considered: (1) noise representing a typical large cargo ship, and (2) noise from pile-driving activity. In the summer, when the source of sound is on the onshore side of the boundary front, the sound energy is mostly concentrated in the near-bottom layer. In winter, the sound from the same source is distributed more evenly in the vertical. The difference between the sound level in summer and winter at 10 m depth is as high as 20 dB at a distance of 40 km. When the source of sound is on the seaward side of the boundary front, the sound level is nearly uniform in the vertical. The transmission loss is also greater (~ 16 dB) in the summer than in the winter for a shallow source, while it is up to ~ 20 dB for a deep source at 30 km.

Chen et al. (2017) noted that shipping noise is a threat to marine wildlife. In particular, grey seals are benthic foragers and are exposed to the potential impacts of shipping noise throughout the water column. To assess the noise exposure of grey seals along their tracks, Chen et al. (2017) used ship-track data from the Celtic Sea, seal-track data, and a coupled ocean-acoustic modeling system to assess the noise exposure of grey seals along their tracks. The coupled ocean-acoustic modeling system combined a 3D oceanographic model (POLCOMS) with an acoustic propagation model (HARCAM). They found that the animals experienced step changes in sound levels up to ~20 dB at a frequency of 125 Hz, and ~10 dB on average over the frequency band 10–1000 Hz when they dove through the thermocline, particularly during summer. Their results showed large seasonal differences in the noise level experienced by the seals.

6.16.2.2 Ocean Acidification

Climate change also affects the ocean soundscape. The emission of carbon into the atmosphere through the effects of fossil-fuel combustion and industrial processes increases atmospheric concentrations of carbon dioxide (CO_2). Ocean acidification, which occurs when CO_2 in the atmosphere reacts with sea water to create carbonic acid (H_2CO_3), is increasing.

The attenuation of low-frequency sound in the sea is pH-dependent; specifically, the higher the pH, the greater the attenuation. Thus, as the ocean becomes more acidic (lower pH) due to increasing CO_2 emissions, the attenuation will diminish and low-frequency sounds will propagate farther, effectively making the ocean noisier.

According to Hester et al. (2008), OA resulting from the introduction of CO_2 from fossil fuel consumption will result in significant decreases in ocean sound absorption (α) for frequencies lower than about 10 kHz. This effect is due to known pH-dependent chemical relaxations. Under reasonable projections of future fossil fuel CO_2 emissions and other sources, a pH change of 0.3 units or more can be anticipated by mid-century, resulting in a decrease in α (dB km^{-1}) by almost 40%. Thus, ambient noise levels in the ocean within the frequency ranges critical for environmental and military interests are set to increase significantly.

Recent investigations modeled what effect the increasing acidity of the ocean would have on ambient-noise levels in shallow water in the presence of internal waves (Rouseff and Tang, 2010). This model assumed an isotropic distribution of noise sources. Exploring a scenario typical of the East China Sea, the noise at 3 kHz was predicted to increase by 30%, or about one decibel, as the pH decreased from 8.0 to 7.4. These results are representative of other contemporaneous investigations into this matter.

The oceanic carbonate system can be understood and probed through four key parameters: total alkalinity (TA), dissolved inorganic carbon (DIC), pH, and partial pressure of CO_2 (pCO_2). In principle, knowledge of any two of these four is sufficient to solve the carbonate system equations; however, overdetermination (measuring at least three parameters) is advantageous. Two parameters have been identified as useful for monitoring temporal and spatial variations in OA: pH (measure of acidity), and calcium carbonate ($CaCO_3$) mineral saturation state (Land et al., 2015). As *in situ* data continue to accumulate, attempts have been made to use available hydrographic data and remotely sensed data to provide proxy indicators of the condition of the carbonate system. The increased availability of *in situ* data has created a substantial dataset with which to develop and test the capabilities of satellite-derived products; moreover, the recent availability of satellite-based salinity measurements may provide new insights for studying and assessing OA from space. The ratio between ions (the constituents of salinity) tends to remain constant throughout the global oceans, resulting in a strong relationship between TA and salinity. Unfortunately, a universal relationship between TA and salinity does not apply in certain regions, as in areas influenced by freshwater outflows from rivers, or in areas where calcification or $CaCO_3$ dissolution occurs. It is critical to gain additional regional knowledge since different rivers will have different ionic concentrations (and therefore different TA concentrations), depending on the surrounding geology and hydrology. While it has proved difficult to use remote sensing directly to monitor and detect changes in seawater pH and their impact on marine organisms, satellites can measure sea-surface temperature (SST), sea-surface salinity (SSS), and surface chlorophyll-a, from which carbonate-system parameters can be estimated through the use of empirical relationships derived from *in situ* data. Although surface measurements may not be representative of important biological processes (e.g., fish or shellfish), observations at the surface are particularly important for OA because changes in carbonate chemistry due to atmospheric CO_2 occur first at the surface [see related discussions on coral-reef fishes in Section 7.2.3). Thus, satellites have great potential as a tool for assessing changes in the carbonate chemistry. Only recently has a satellite-based capability for measuring SSS existed. Specifically, an increase in salinity decreases the emissivity of seawater, thus resulting in changes in the microwave radiation emitted at the water surface. Regular mapping of the SSS field with unprecedented temporal and spatial resolution at global scale is now possible from satellites. The impact of using satellite SSS for carbonate-system algorithms can now be tested, where previously there was a reliance on climatology, *in situ* data, or models. This provides the means for studying the impact that freshwater influences (sea ice melt, riverine inputs, and rain) can have on the marine-carbonate system. The use of satellite SSS data may also allow evaluation of the impact of the inter- and intra-annual variations in SSS on the carbonate system.

6.16.2.3 Marine-Hydrokinetic Energy Devices

Copping and O'Toole (2010) noted that the effects of underwater noise from MHK devices on receptors such as marine mammals and fish include: physical auditory damage; behavioral changes; avoidance of area; chronic stress; altered acoustic sensitivity; and mortality. Monitoring of acoustic baseline data at the sites and levels of noise emitted by MHK devices could be undertaken with passive listening devices. Models have been developed for use in understanding the potential acoustic impacts of a number of devices or arrays. Existing (and planned) test centers and platforms could be used to develop technologies that can help industry conduct EA and quantify environmental impacts. Test sites could also provide facilities to test new devices and models, as well as measure acoustic sources.

Despite significant barriers, marine renewable energy (MRE) has made recent advances toward commercialization. These barriers include the high cost of MRE, which leaves it uncompetitive relative to more established renewable-energy technologies. A substantial proportion of this cost differential comes from operations and maintenance (O&M) activities. Such O&M activities can be reduced through the use of condition-based maintenance scheduling. In offshore environments, the submerged location of most MRE devices enables the use of underwater acoustic emission (AE), which is a relatively new condition-monitoring technique that combines acoustics (used for environmental monitoring of MRE influence on noise levels) together with AE condition monitoring (as used in air). (AE is defined by ASTM International E1316—16a as "the class of phenomena whereby transient stress/displacement waves are generated by the rapid release of energy from localized sources within a material, or the transient waves so generated.") Walsh et al. (2016) assessed the practicality of the AE approach in complex ocean environments through detailed sound propagation modeling studies using the BELLHOP propagation model available in the AcTUP toolbox (see Appendix C). Their results demonstrated that acoustic propagation was very sensitive to the variations in the shallow-water environments considered in their study. Concerning acoustic sensor placement, multipath interference in shallow-water environments means that the locations of the measuring sensors need to be carefully considered, even though such placements might not anticipate all environmental variations over the time period necessary (typically several months) for accurate long-term monitoring. These environmental variations may also restrict the range of acoustic frequencies that can be modeled accurately, thus limiting monitoring of the received levels. The results presented by Walsh et al. (2016) represented the first steps toward optimizing AE sensor positions and AE measuring strategies for arrays of MRE devices.

Etter (2017) described the utilization of UWA models for the evaluation of marine-system noise impacts associated with the installation and operation of MHK devices, particularly in coastal oceans.

In practice, noise modeling efforts in support of EIA are often carried out using simplistic UWA models, with limited environmental data, and with little or no field measurements to ground-truth the model predictions. In some cases, practitioners have developed proprietary models, the inner workings of which are not disclosed to regulators. This confronts regulatory decision makers with considerable uncertainty regarding the prediction of possible impacts; moreover, this uncertainty is often times not apparent. In an effort to better inform regulators, stakeholders, and developers of the factors that may lead to uncertainty in noise assessments, Farcas et al. (2016) provided concrete examples of how different modeling procedures can affect predictions. Raising awareness of these issues can help promote best practice in noise-impact assessments and enable more informed EIA processes for noise-generating developments. Also see Robinson et al. (2014) regarding a good practice guide for underwater noise measurement. To further explore this aspect, Farcas et al. (2016) used measurements of impact pile-driving noise that were made simultaneously at two locations in the Cromarty Firth, Scotland. Different acoustic models were then used to calculate the source level of pile-driving noise. This exercise served to illustrate that, although there is considerable uncertainty in the relationship between noise levels and impacts on aquatic species, the science underlying noise modeling appears to be well understood. Farcas et al. (2016) further observed that UWA models that are currently applied in EIA consider only the sound-pressure component of sound, which is the means by which mammals hear; however, the primary mechanism by which fish and invertebrate species detect sound is through the particle-motion component of sound.

6.16.2.4 Wind-Turbine Noise

Wind power, as an alternative to fossil fuels, is plentiful, renewable, widely distributed, clean, and produces no greenhouse gas emissions during operation. A wind farm, which is a group of wind turbines in the same location used for production of electric power, may be located offshore. The installation of ocean wind farms requires medium water depths (<30 m) and construction logistics such as access to specialized vessels to install the turbines. Economic wind generators require wind speeds of 16 km h^{-1} or greater.

The US Department of Labor (2016) projected that the employment of wind turbine service technicians (*windtechs*) would grow 108% over the 10-year period 2014–2024. This is a much faster rate than the average for all occupations; however, because this is a small occupation, the fast growth will result in only about 4,800 new jobs over the 10-year period.

Nedwell and Howell (2004) noted that a concerted effort is being made by industry to minimize any undesirable effects relating to wind farm development and operation. One potential effect of offshore wind farm development is the creation of underwater noise. Knowing the length of time the marine environment is exposed to an underwater noise source is useful when assessing environmental effects. To help in this assessment, the life cycle of an offshore wind farm was split into four phases:

- *Preconstruction.* Background noise may be used as a benchmark against which to assess the environmental impact of new sources. Activities that occur before construction begins on a wind farm include geophysical and geotechnical surveys, meteorological mast installation, and an increase in vessel traffic.
- *Construction.* One of most significant activities during wind farm construction is foundation installation. Dredging and rock laying may be undertaken during wind farm construction. Applications include scour protection, cable protection, and modifying nonideal bathymetry. Other construction activities include cable laying, turbine and turbine-tower installation, and ancillary structure installation (such as offshore transformers).
- *Operation.* By far the longest phase of a wind farm's life cycle is the operational phase. Two measurements of offshore wind turbine noise show low-frequency sound levels with a maximum of 153 dB re 1 μPa @ 1 m at 16 Hz. These measurements are of individual turbines of a relatively low power (less than 1 MW). Despite the low-level and low-frequency nature of the sound, behavioral reactions have been observed in response to the reproduction of wind-turbine noise.
- *Decommissioning.* The final stage of a wind farm's life cycle is its decommissioning, the majority of which will be a reflection of the installation process. However, the wind turbine foundation decommissioning process is unclear since options for pile-foundation removal include jet and explosive cutting below the seabed.

Abkar and Porté-Agel (2015) developed an analytical model to parameterize the effect of wind farms in large-scale atmospheric models. Wind turbines in a wind farm were parameterized as elevated sinks of momentum, and as sources of turbulence. The analytical approach estimated the turbine-induced forces as well as the turbulent kinetic energy (TKE) generated by the turbines inside the atmospheric boundary layer (ABL). The model accounted for the effects of wind-farm density, wind-farm layout, and wind direction. Model performance was tested with large-eddy simulations of ABL flows over very large wind farms with different turbine configurations. The results showed that the model was able to accurately predict the turbine-induced forces as well as the TKE generated by the turbines inside the ABL.

6.16.2.5 Pile-Driving Noise

Bailey et al. (2010) noted that a soft start (gradual increase in intensity) to pile-driving demonstrated that it successfully resulted in the gradual increase of sound pressure, which could potentially have alerted animals before levels became harmful and enabled them to swim away. Additional mitigation measures such as the use of bubble curtains can reduce the radiated sound levels of piling in shallow waters, particularly at 400–6400 Hz. Bubble-curtain systems could not be used due to the complexities of the installation operation in water depths of 42 m. Improvements in enclosed bubble curtains means they may have application for pile-driving in deep water in the future and should be investigated in areas of high cetacean activity.

Lippert and von Estorff (2014) investigated parameter uncertainty modeling in the context of pile-driving noise associated with offshore wind farm construction sites. The goal of global uncertainty analysis is to quantify potential errors in predicted quantities (x). This was accomplished by the variation of all major influencing factors within a certain range by means of Monte Carlo simulations. The resulting probability distributions of these simulations were then compared with measured data at a number of wind farm sites. An interval was then established defined by standard deviations ($x \cdot \sigma$) around the mean value μ of the Monte Carlo simulations, wherein all measured quantities could be found. When performed at a sufficiently large number of sites, a reliable prediction for future sites could be made on the assumption that, varying the parameters in the same manner as before, the quantities would again lie within a determined interval ($\mu \pm x \cdot \sigma$). Alternative uncertainty methods were considered (e.g., *fuzzy arithmetic*), but were abandoned in favor of Monte Carlo simulations using the Latin hypercube sampling approach (a statistical method that generates a quasi-random sample of parameter values from a multidimensional distribution). The basic idea was to partition each uncertainty distribution into N intervals of equal probability and to randomly generate one sampling point in each of those intervals. Subsequently, N computations were performed using random sampling-point combinations under the restriction that each point was used only once. This led to a significantly enhanced convergence rate.

The task of numerical pile-driving noise prediction involves splitting the problem into a near-field model (to accurately model the pile vibrations and its close environment) and a far-field model using a standard UWA propagation method. Lippert et al. (2014) proposed using a hybrid approach combining FE and wavenumber-integration methods together with a point-source array for the coupling between the two sub-models.

At the COMPILE workshop (a generic benchmark case for predictions of marine pile-driving noise) in June 2014, seven different modeling approaches were presented to predict the sound for the prescribed benchmark case: Hamburg University of Technology (TUHH, Hamburg, Germany); Organization for Applied Scientific Research (TNO, The Hague, The Netherlands); CMST, Curtin University (Perth, W.A., Australia); Bundeswehr Technical Centre for Ships and Naval Weapons, Maritime Technology and Research (WTD 71, Kiel, Germany); JASCO Applied Sciences (Victoria, BC, Canada); University of Southampton together with the National Physical Laboratory (UoS/NPL, UK) (only the close-range region was modeled); and Seoul National University (SNU, Seoul, South Korea). Most numerical models used to predict pile-driving noise are split into close-range and far-range models to reduce the computational effort. In the close-range region, the distinct arrivals of the different Mach wave fronts were correctly predicted by all models. In the far-range region, three different models of sound propagation in the UWA channel were used: wave-number integration method, PE approximation method, and normal-mode model. The numerical predictions of the sound exposure and peak pressure levels by the six different far-range models were in general agreement, although divergences of the results with growing distance from the sound source were observed. These divergences were likely due to numerical problems arising from the modeled sound attenuation resulting from bottom interaction, from differences in the coupling procedures, or from differences in the modeling assumptions and their implementation (Lippert et al., 2016).

Fricke and Rolfes (2015) described an approach for predicting underwater noise caused by impact pile driving, which was validated based on *in situ* measurements. This model was divided into three submodels: (1) a submodel based on the FE method was used to describe the vibration of the pile and the resulting acoustic radiation into the surrounding water and soil column; the model was solved for frequencies ranging from 12 to 1120 Hz at increments of 2 Hz (to cover a maximum impulse length of 0.5 s); calculation of the sound field using the FE model was effectively limited to a radius <100 m due to the selected frequency band and available computer hardware; (2) another submodel estimated the mechanical excitation of the pile by the piling hammer using an analytical approach that took into account the large vertical dimension of the ram (the important parts directly involved in the piling process are the ram and the anvil); (3) the last submodel, based on the split-step Padé solution of the PE, was used to compute long-range propagation out to 20 km; the seafloor soil was handled as a fluid sediment. In order to specify realistic environmental properties for model validation, a geoacoustic model was derived from spatially averaged geological information obtained from the investigation area (soil properties comprised compressional wave speed and attenuation, shear wave speed and attenuation, and density, each of which was defined as a function of depth and range). While the developed model tended to underestimate the sound exposure level (SEL) at low frequencies and large radii, the validation results provided reasonable confidence that the overall approach was appropriate with respect to the considered frequency range. In future work, it was noted that the propagation submodel should be able to handle the soil as an elastic medium to allow for the radiation of shear waves into the soil.

Dawoud et al. (2016) employed the empirical model developed by Rogers (1981) to assess underwater noise fields resulting from driving 96-m long hollow-steel piles (using a Kobe-80 diesel hammer) off the Jeddah Coast of Saudi Arabia in a water depth of 16 m. The acoustic model inputs included Red Sea bathymetry, temperature, and salinity. It was found that behavioral disturbances to marine mammals could occur within a distance of 1000 m from the pile location; threshold shifts could occur at ranges between 30 and 50 m.

6.16.2.6 Wave-Energy Device Noise

Austin et al. (2009) provided wave-energy developers in Oregon with fundamental information on the principles, methods, and equipment involved in conducting environmental noise assessments related to the permitting of such projects. In the absence of any documented ambient-noise measurements for the near-shore environment off the Oregon coast, characterizations of the environmental components that contribute to the overall ambient noise field were provided instead. Transmission losses were generated at three sites along the Oregon coast considered to be representative of the expected range of propagation conditions. These results provided indications of the rates at which sound levels could be expected to decay as a function of distance from potential wave-energy converter development sites in the Oregon coastal environment. The marine operations noise model (MONM) was used to compute transmission losses for arbitrary three-dimensional, range-varying acoustic environments using a parabolic-equation solution to the acoustic wave equation. Specifically, MONM computed approximate 3D acoustic fields by modeling transmission losses along evenly spaced radial traverses covering a 360° swath from the source. The modeling took into account a number of environmental parameters including bathymetry, sound-speed profile in the water column, and geoacoustic properties of the seafloor. A 50-m range step was used in the spatial sampling of the acoustic environments along each model traverse. Frequency dependence of the sound propagation characteristics was treated by computing acoustic transmission losses at the center frequencies of ⅓-octave bands between 10 Hz and 2 kHz and summing the transmission-loss estimates assuming a flat-spectrum source.

Ikpekha et al. (2014) developed a computer model that simulated low-frequency (<1000 Hz) acoustic signals produced by a wave-energy device in coastal environments. They analyzed these signals with the aid of audiograms of marine mammals, in this case the harbor seal. This enabled them to estimate the levels of acoustic noise experienced by marine mammals due to the presence of ocean-deployed devices. Propagation of the UWA signals was modeled using the FE method with appropriate boundary conditions at the sea surface and the sea floor. Using an audiogram of the harbor seal, it was deduced that animals at least 51 m distant from the sound source would not be affected.

6.16.2.7 Tidal-Turbine Noise

Li and Çalişal (2010) presented a preliminary study of four principal characteristics of tidal-current turbines: power output, torque, induced velocity, and AE. Numerical models were developed to predict these characteristics for tidal-current turbines. It was proposed that these same models could also be used to develop standards. The resulting hydrodynamic noise intensity (AE) was evaluated at three locations downstream from the subject turbine. The frequencies corresponding to the first peak (main noise frequency) at the three locations were all around 4 Hz. Successively smaller amplitude peaks were also observed at 18 Hz and at 31 Hz.

Lloyd et al. (2011) modeled underwater noise sources associated with horizontal-axis tidal turbines and their potential impact on shallow-water marine environments. The requirement for device-noise prediction as part of EIAswas considered in light of the limited amount of measurement data available. Noise sources included self noise, interaction noise, and hydro-elastic noise; in future studies, machinery (generator) noise and cavitation noise also need to be considered. The dominant flow-generated noise sources were modeled using empirical techniques. The predicted sound pressure level (SPL) due to inflow turbulence for a typical horizontal-axis tidal turbine was estimated to generate ⅓-octave-bandwidth pressure levels of 119 dB re 1 μPa at 20 meters from the turbine at individual frequencies. This preliminary estimate revealed that this noise source alone would not be expected to cause either permanent or temporary threshold shift (TTS) in typical marine animals of the North Sea: cod, harbor seal, and harbor porpoise.

Spiga (2015a) reviewed ocean acoustics modeling for EIAs. This review was carried out as a MRE knowledge-exchange fellowship through the UK natural environment research council (NERC) concerned with developing and testing models of fish behavior around tidal turbines. The overarching aim of this project was to provide an evidence-based tool to forecast the effects of anthropogenic noise on marine fish for EIAs. Specifically, Spiga (2015a) reviewed 18 projects (spanning the period 2000–2015) that used sound propagation modeling to assess the impacts of anthropogenic noise on marine biota. According to Spiga (2015a), it is important to note that most fish and invertebrates are sensitive to particle motion, which is likely to be important for behavioral responses near the acoustic source or at reflecting boundaries. With few exceptions, impact assessments have disregarded the particle-motion component of the sound, instead measuring only sound pressure. Since no audiograms based on this component are available, our knowledge is limited regarding the effects of noise on particle-motion-sensitive animals; moreover, the development of conclusive criteria for these species is also limited. Spiga (2015b) further developed and tested models of fish behavior around tidal turbines. The HAMMER model was implemented with the inclusion of measurements of sound levels from MRE and fish-behavioral data in response to noise exposure.

6.16.2.8 Noise-Reduction Methods

Todd et al. (2015) described noise-reduction methods and acoustic-mitigation devices. Noise-reduction methods included acoustic-isolating materials and bubble curtains that reduced initial sound output or reduced sound intensity along a propagation path. Acoustic-mitigation devices included acoustic-harassment devices (or pingers) that encouraged animals to move away from high-risk operational areas.

Würsig et al. (2000) described the use of a perforated rubber hose to produce a bubble curtain (screen) around pile-driving activity in water depths ranging from 6 to 8 m near western Hong Kong. The percussive hammer-blow sounds of the pile driver were measured on two days at distances of 250, 500, and 1000 m. Broadband-pulse levels were reduced by 3–5 dB by the bubble curtain. Sound intensities were measured from 100 Hz to 25.6 kHz, and the greatest sound reduction by the bubble curtain was evident from 400 to 6400 Hz. Sound conduction probably occurred through the substrate under the bubble curtain for at least the lowest frequencies of 100–200 Hz. Background noises from adjacent shipping channels and industrial centers probably also complicated the measurements. Although bubble screening appeared to show promise for reducing anthropogenic sounds underwater, it is only one of several potential mitigation tools.

Tsouvalas and Metrikine (2016) proposed a model for the investigation of sound reduction during marine piling operations when an air-bubble curtain was placed around the pile. The model consisted of the pile, the surrounding water and soil media, and the air-bubble curtain, which was positioned at a select distance from the pile surface. The solution approach was semi-analytical and was based on a combination of the dynamic sub-structuring technique and the modal decomposition method. Two main results were obtained: (1) a new model was proposed that could be used for predictions of the noise levels in a computationally efficient manner; and (2) an analysis was presented of the principal mechanisms responsible for the noise reduction due to application of the air-bubble curtain in marine piling operations. Understanding these mechanisms may be crucial for exploitation of the maximum efficiency of the noise-reduction system. Specifically, the principal mechanism for noise reduction depended strongly on the frequency content of the radiated sound and the characteristics of the bubbly medium. For large-diameter piles that radiate most of the acoustic energy at relatively low frequencies, the noise reduction was mainly attributed to the mismatch in the acoustic impedances between the seawater and the bubbly layer. For smaller piles, and when the radiated acoustic energy was concentrated at frequencies close to (or higher than) the resonance frequency of the air bubbles, the sound absorption within the bubbly layer became critical.

In December 2005, construction work was started to replace a harbor wall in Kerteminde Harbor, Denmark. Using an air-bubble curtain, the mean levels of sound attenuation over a sequence of 95 consecutive pile strikes were 14 dB (standard deviation 3.4 dB) for peak-to-peak values, and 13 dB (standard deviation 2.5 dB) for SEL values (Lucke et al., 2011).

Ramp-up (or soft-start) procedures employ a gradual increase in source level in order to mitigate the effects of sonar transmissions on marine mammals. Von Benda-Beckmann et al. (2014) investigated the effectiveness of ramp-up procedures in reducing the area within which changes in hearing thresholds can occur. They modeled the level of sound that killer whales (*Orcinus orca*) were exposed to from a generic sonar operation preceded by different ramp-up schemes. In their model, ramp-up procedures were shown to reduce the risk of killer whales receiving sounds of sufficient intensity to affect their hearing. The effectiveness of the ramp-up procedure depended strongly on the assumed response threshold and differed with ramp-up duration, although extending the duration of the ramp up beyond 5 minutes did not add much to its predicted mitigating effect. The main factors that limited effectiveness of ramp up in a typical ASW scenario were high source level, rapid-moving sonar source, and long silences between consecutive sonar transmissions. It was suggested that this exposure-modeling approach could be used to evaluate, and optimize, mitigation procedures.

The Annex IV Project is an international effort initiated by the International Energy Agency's Ocean Energy Systems Implementing Agreement (OES-IA) to examine the environmental effects of marine energy devices and environmental research studies from around the world and to disseminate information to marine energy researchers, regulators, developers, and stakeholders (Copping et al., 2013). Led by the US Department of Energy, the Annex IV project concluded its first phase in 2013 with a Final Annex IV Report that used the best available science and information to examine three case studies of specific interactions of marine energy devices with the marine environment: 1) the physical interactions between animals and tidal turbines; 2) the acoustic impact of marine energy devices on marine animals; and 3) the effects of energy removal on physical systems. Tethys is a knowledge-management system that actively gathers, organizes, and provides access to information on the environmental effects of marine energy, offshore wind energy, and land-based wind energy development. In essence, Tethys is a clearinghouse for information and metadata associated with the Annex IV project.

Monitoring of acoustic-baseline data at the sites and levels of noise emitted by MHK devices could be undertaken with passive listening techniques such as PAM. Models have been developed for use in understanding the potential acoustic impacts of a variety of MHK devices or arrays. Existing (and planned) test centers and platforms could be used to develop technologies that can help industry conduct EA and quantify environmental impacts. Test sites could also provide facilities to test new devices and models, as well as measure acoustic sources.

6.16.2.9 Passive Acoustic Monitoring

Passive acoustic technologies have advanced significantly in terms of hardware and software. Archival and real-time passive acoustic arrays now constitute economical approaches for mesoscale monitoring of marine areas; these arrays can be used to monitor vocal marine life in areas otherwise difficult to survey by traditional visual methods. Fixed autonomous passive acoustic arrays sample continuously for prolonged periods of time in all weather conditions, thus allowing for assessment of seasonal changes in both the distribution and acoustic behavior of individuals without the disturbance of survey vessels or aircraft. Although these techniques have primarily been used with cetaceans, the potential exists for studying many other marine animals such as pinnipeds (often generalized as seals), sirenians (commonly referred to as sea cows), and fishes.

PAM using towed, static, or attached tags is most useful when employed in the context of the acoustic behavioral ecology of the animals, and applied in regionally and seasonally appropriate contexts. The utility of PAM is improved when adequate information is available regarding individual, group, population, and species sound-level usage (van Parijs et al., 2009). Utilization of PAM is guided by best practices for *in situ* measurements of underwater sound and is further informed by recent marine mammal (TTS) experiments.

In related work, Au and Lammers (2016) assembled 15 contributed chapters describing aspects of PAM for detecting aquatic life.

6.16.3 Seismic Operations and Protection of Whales

Marine seismic surveys are used to assess the location of hydrocarbon resources, including gas and oil. There are two acquisition methods: 2D and 3D. The 2D method tows a single seismic cable (or streamer) behind the seismic vessel, together with a single source. The reflections from the subsurface of the sea floor are assumed to lie directly below the path (sail line) of the vessel. In a 3D survey, groups of sail lines (or swathes) are used to acquire orthogonal or oblique lines relative to the acquisition direction. By utilizing more than one source together with many parallel streamers towed by the seismic vessel, the acquisition of many closely spaced subsurface 2D lines can be achieved by a single sail line. Computationally intensive processing is necessary to produce a 3D image of the subsurface of the sea floor. The source arrays are powered by high-pressure air that is compressed onboard the seismic vessel. These compressors are capable of recharging the airguns rapidly and continuously, enabling the airgun source arrays to be fired at approximately 10-second intervals for periods of up to 12 hours. Typical towing depths range from 4 to 5 meters for shallow, high-resolution surveys or 8–10 meters for deeper penetration, lower-frequency targets in open waters. Typical source outputs are approximately 220 dB re 1 μPa Hz^{-1} @ 1 m. Other types of seismic sources include water guns and marine vibrators.

In 2002, the International Association of Geophysical Contractors (IAGC) hosted an informal meeting to discuss future acoustic research relevant to seismic operations and the effects of seismic exploration on sperm whales in the Gulf of Mexico (Jochens et al., 2006). The IAGC offered its support for sperm-whale research through the contribution of a seismic-source vessel for controlled-exposure experiments. In response, a proposed sperm whale seismic study (SWSS) was approved by the Minerals Management Service (MMS) in 2002. In subsequent years, IAGC was joined by a number of oil and gas companies to form the Industry Research Funders Coalition (IRFC) that has continued to provide contributions in support of SWSS studies.

Long-term (monthly to seasonal) movements and distributions of sperm whales were studied using satellite-tracked radio telemetry tags (S-tags). Short-term (hours) diving and swimming behavior and vocalizations of sperm whales were examined using recoverable digital-recording acoustic tags (D-tags) that logged whale orientation (i.e., pitch, roll, heading) and depth, as well as the sounds made by the whale and received at the whale from the environment. Diving depths and movements were examined using 3D passive acoustic tracking techniques (Jochens et al., 2006).

Remote sensing fields of sea surface height and ocean color provided information on dynamical currents (as might be generated by eddies and fronts) and on chlorophyll-rich surface waters that might create locally favorable feeding conditions for the vertically-migrating prey of sperm whales. *In situ* data such as temperature, salinity, fluorescence, chlorophyll, currents, and acoustic backscattering data enabled further characterization of the epipelagic environment (Jochens et al., 2006).

To examine potential changes in the behavior of sperm whales when subjected to seismic airgun sounds, controlled exposure experiments (CEEs) were conducted using the D-tags in conjunction with a seismic-source vessel. The location and level of airgun sounds delivered at the tagged sperm whales were controlled by the science team. These CEEs provided data on the immediate and short-term (hours) response of sperm whales to airgun sounds. Longer-term avoidance or displacement behaviors of sperm whales to seismic vessel airgun sounds were examined using location data from the S-tags and from proprietary commercial seismic shot data (Jochens et al., 2006).

The 3D tracking method requires at least two widely separated hydrophones to obtain the horizontal range and depth of acoustically active sperm whales, and would thus be suited for eventual use on a standard seismic vessel where the passive acoustic arrays (streamers) can be over a kilometer long. Instead of relying on four hydrophones deployed as a three-dimensional array (which would be difficult to deploy and process), the method used here exploited surface multipath (or "ghosts") to reduce the number of required hydrophones to three and further permitted the phones to be deployed along a single towed cable. The horizontal separation between the widely-spaced hydrophones needed to be at least 200 m in order to obtain adequate range and depth resolution at 1–km horizontal ranges. The method did not require the use of multipath from the ocean bottom, but when such bottom returns were detected they could provide an independent confirmation of these tracking procedures (Thode et al., 2002).

Another assumption used in 3D tracking is that ray-refraction effects caused by depth-dependent sound speed profiles can be neglected for 3D tracks within 1 km. This assumption may not always be valid. Using the ray-tracing program BELLHOP, the effects of the sound-speed profile on various ToA and AoA measurements could be modeled. In future experiments, these corrections can be pre-computed and then applied to the tracking routine (Jochens et al., 2006).

Lawson (2009) examined the effects of uncertainty in the modeling of anthropogenic impacts and suggested a precautionary approach to regulation. It was further noted that due to the complex patterns of sound propagation encountered in diverse shelf regions, some marine mammals may not necessarily encounter the average sound exposure conditions predicted for any given seismic survey.

Modeling sound propagation in the ocean is an essential tool to assess the potential risk of airgun shots on marine mammals (Breitzke and Bohlen, 2010). Based on a 2½D finite-difference code, a full waveform modeling approach was presented that determined both sound-exposure levels of single shots and cumulative sound-exposure levels of multiple shots fired along a seismic line. Source signatures were modeled using the NUCLEUS™ source modeling package available from Petroleum Geo-Services (PGS). Band-limited point-source approximations of compact airgun clusters deployed in polar regions were used as sound sources. Marine mammals were simulated as static receivers. Applications to deep- and shallow-water models, including constant and depth-dependent sound velocity profiles of the Southern Ocean, showed dipole-like directivities in the case of single shots, and tubular cumulative SEL fields beneath the seismic line in the case of multiple shots. Compared to a semi-infinite model, an incorporation of seafloor reflections enhanced the seismically induced noise levels close to the sea surface. Refraction due to sound velocity gradients and sound channeling in near-surface ducts were evident, but affected only low-to-moderate levels. Therefore, exposure-zone radii (r) derived for different hearing thresholds were almost independent of the sound-velocity structure. With decreasing thresholds, r increased according to a spherical 20 $\log_{10} r$ law in case of single shots and according to a cylindrical 10 $\log_{10} r$ law in case of multiple shots. A doubling of the shot interval diminished the cumulative SELs by 3 dB and halved the radii. The ocean bottom properties only slightly affected the radii in shallow waters if the normal incidence reflection coefficient exceeded 0.2.

Prideaux and Prideaux (2015) proposed template guidelines detailing the information that should be gathered to support robust and defensible decisions involving EIAs for offshore petroleum exploration seismic surveys. The Convention on Migratory Species (CMS) guidelines on EIA for marine noise-generating activities were developed to present the best available techniques and best environmental practices (Prideaux, 2016). The document was structured to stand either as a complete unit or to be used as discrete modules tailored for national and agreement approaches. The nine modules were structured to cover species areas:

- *Module A: Executive Summary*: The complete unit and the discrete modules are online at: *cms.int/guidelines/cms-family-guidelines-EIAs-marine-noise.*
- *Module B: Sound in Water is Complex*: provided an insight into the characteristics of sound propagation and dispersal. This module was designed to provide decisions-makers with the necessary foundation knowledge to interpret the other modules in these guidelines and any impact assessments that are presented to them for consideration.

- *Module C: Expert Advice on Specific Species Groups*: presented 12 separate detailed sub-modules covering each of the CMS species groups, focusing on species' vulnerabilities, habitat considerations, impact of exposure levels, and assessment criteria. The species groups covered in the following sub-modules were: inshore odontocetes; offshore odontocetes; beaked whales; mysticetes; innipeds; polar bears; sirenians; marine and sea otters; marine turtles; fin-fish; elasmobranchs; and marine invertebrates.
- *Module D: Decompression Stress*: provided important information on bubble formation in marine mammals, source of decompression stress, source frequency, level and duration, and assessment criteria.
- *Module E: Exposure Levels*: presented a summary of the current state of knowledge about general exposure levels.
- *Module F: Marine Noise-Generating Activities*: provided a brief summary of military sonar, seismic surveys, civil high powered sonar, coastal and offshore construction works, offshore platforms, playback and sound exposure experiments, shipping and vessel traffic, pingers, and other noise-generating activities. Each section presented current knowledge about sound intensity level, frequency range, and the activities general characteristics.
- *Module G: Related Intergovernmental or Regional Economic Organization Decisions*: presented the series of intergovernmental decisions that have determined the direction for regulation of anthropogenic marine noise.
- *Module H: Principles of EIAs*: established basic principles including strategic EIAs, transparency, natural justice, independent peer review, consultation, and burden of proof.
- *Module I: Frameworks EIA Guidelines for Marine Noise-Generating Activities*: presented advisory notes that should be considered in all EIA together with eight separate guideline tables for appropriate assessments of each activity area (as presented in Module F).

Multi-beam echo sounders designed for seafloor-mapping applications are a common tool for ocean exploration and monitoring. Concerns have been expressed about their impacts on marine mammals even though their inherent characteristics (12 kHz multi-sector system using short signals and narrow transmitting lobes) would tend to minimize this possibility. To address this issue, Lurton (2016) proposed an analysis of multi-beam echo sounder radiation characteristics including pulse design, source level, and radiation directivity patterns. Metrics for acoustical-impact assessment comprised maximum SPL and cumulative SEL, given by the integration of received intensity over the exposure time. In all cases, the predicted SPL and SEL values fell in a range that made them of little concern in terms of the marine mammal threshold levels in common use. A detailed radiation model included transmission through directivity sidelobes and was applied to three typical multi-beam echo sounder radiation configurations. A simplified radiation model was defined to extend the analysis to the case of cumulative insonification by a system moving along a survey line. An approximated analytical model was proposed for the accumulated intensity and showed good agreement with the complete simulation of insonification. The computation of ranges corresponding to impact thresholds accepted today showed that impacts in terms of injury were negligible for both SPL and SEL. However, behavioral response impacts cannot be excluded and should be explored further.

Zykov (2013) described the performance of an acoustic modeling study to estimate source levels, beam configurations, and sound-exposure levels generated by low-energy equipment used in geophysical surveys including single-beam echo sounders, multi-beam echo sounders, side-scan sonars, subbottom profilers, and seismic boomers. The acoustic impacts of the survey equipment were based on distances to specific thresholds for per-pulse SPL and SEL. Two ocean bottom types were considered: sandy bottom and exposed bedrock. Underwater sound propagation values were predicted using MONM, which utilized RAM for frequencies ≤1 kHz, and BELLHOP for frequencies ≥2 kHz. Volumetric assessments were obtained using the N × 2D approximation. The potential for anthropogenic noise to impact marine species was derived using marine mammal frequency weighting (M-weighting) for low-frequency (7 Hz–22 kHz) cetaceans (baleen whales), mid- (150 Hz–160 kHz) and high-frequency (200 Hz–180 kHz) cetaceans (toothed whales), and also seals, sea lions, and walrus (75 Hz–75 kHz). Extensive tabulations were included to summarize the resulting marine mammal impacts.

MacGillivray (2006) predicted underwater noise levels using an integrated modeling approach incorporating (1) an airgun array source model; (2) a broadband N × 2D propagation model based on a modified version of RAM; and (3) environmental databases comprising high-resolution bathymetry, historical CTD casts, and geoacoustic properties of the seabed. An airgun array source signature model was developed for use in predicting the acoustic source levels of seismic arrays; this model was based on the physics of the oscillation and radiation of airgun-generated bubbles and was validated against actual airgun data. Broadband sound propagation was approximated by modeling transmission losses at ⅓-octave band center frequencies. Noise levels were then computed by subtracting the transmission losses from the respective ⅓-octave band airgun array source levels. Seasonal variations in the sound-speed profiles were derived using principal-component analysis of a large collection of historical CTD data. The resulting noise level estimates were used to define impact zones around survey vessels where marine mammals could be expected to exhibit disturbance reactions to airgun noise.

6.16.4 Modeling Efforts

This section discusses four topics relating to modeling in support of marine mammal endangerment research: the acoustic integration model (AIM), the effects of sound on the marine environment (ESME), marine mammal movement models, and collision avoidance.

6.16.4.1 Acoustic Integration Model

The AIM combines an animal movement simulator with UWA models to predict (and thus minimize and mitigate) the potential effect of sound on marine mammals (Frankel et al., 2002). Simulated sound sources and animals are programmed to move in location and depth over time in a realistic fashion. AIM is a Monte Carlo statistical model based on a whale movement and tracking model and an

UWA backscattering model for a moving source. Currently, AIM incorporates the BELLHOP and PE propagation models in addition to the ETOP05 bathymetry and GDEM sound-speed profile databases. The ETOP05 is the highest resolution topographic data set with global coverage that is publicly available, with elevations posted every 5 arc minutes (approximately 10 km) for all land and sea floor surfaces. ETOP05 is distributed without restriction by the National Oceanic and Atmospheric Administration (NOAA) through its National Geophysical Data Center (NGDC)—refer to Appendix C. For details regarding GDEM, refer to Section 11.6, specifically Table 11.5.

6.16.4.2 Effects of Sound on the Marine Environment

ESME is a multidisciplinary research and development effort to explore the interactions between anthropogenic sounds, the acoustic environment, and marine mammals (Shyu and Hillson, 2006; Siderius and Porter, 2006). The "ESME workbench" models the entire sound path including the sound sources (impulsive or explosive), the medium (water column and seafloor), and the TTS models of the marine mammals. (TTS refers to a temporary increase in the threshold of hearing, that is, the minimum intensity needed to hear a sound at a specific frequency, but which returns to its pre-exposure level over time.) The goal is to predict impacts of anthropogenic sounds on marine mammals. This entails three elements: (1) accurate estimates of the sound field in the ocean; (2) accurate estimates of the cumulative sound exposure of the marine mammals; and (3) reliable predictions of the incidence of TTS for the species of interest given the estimated cumulative exposure.

For the period 1996–2015, Finneran (2015) reviewed progress in the methods employed by groups conducting marine mammal TTS experiments. Specifically, Finneran (2015) summarized the relationships between the experimental conditions, the noise exposure parameters, and the observed TTS. Major findings were synthesized across experiments to provide the current state-of-knowledge for the effects of noise on marine mammal hearing. The most critical gaps involved the manner in which exposure frequency affected the resulting patterns of TTS growth and recovery. TTS growth curves at various frequencies were needed for representative species so that effective weighting functions could be developed to predict the onset of TTS and establish upper safe limits to prevent permanent threshold shift (PTS) for various noise frequencies. The noise sources of greatest concern, such as military sonars and seismic airguns, involved acute exposures to high-intensity, intermittent sounds; however, significant questions remained regarding the rate of TTS growth and recovery after exposure to intermittent noise and the effects of single and multiple impulses. At present, data are insufficient to construct generalized models for recovery and to determine the time necessary to treat subsequent exposures as independent events. More information is needed on the relationship between auditory evoked potentials (AEPs) and behavioral measures of TTS for various stimuli. Finally, data on noise-induced threshold shifts in marine mammals are available for only a few species, and for only few individuals within these species. Questions still remain about the most appropriate methods for extrapolation to other species.

The flexibility and computational efficiency of the ESME Workbench will be enhanced in the future by merging the Navy Acoustic Effects Model (NAEMO) and ESME approaches into the "One Navy Model," which is intended to serve as the standard simulation system for use in predicting impacts of anthropogenic sound sources on marine life for environmental compliance purposes.

The ocean biogeographic information system (OBIS) is an online worldwide atlas for accessing, modeling, and mapping marine biological data in a multidimensional geographic context (Grassle, 2000).

6.16.4.3 Marine Mammal Movement Models

Houser (2006) described a method for modeling marine mammal movement and behavior (3MB) for use in EIAs. Estimating the impact of anthropogenic sound on marine animals entails consideration of animal location and behavior at the time of sound exposure. The ESME model incorporates 3MB to provide fine-scale control over simulated marine-animal (animat) movement and behavior, where the animats serve as virtual dosimeters. Control over the animats is scaleable to the information available regarding the species of concern. Movement and behavior are stochastically determined by sampling from distributions describing rates of movement in the horizontal and vertical planes, direction of travel, time at the surface between dives, time at depth, and time in and transition between behavioral states. Influence of behavior over each of the other distributions is also permitted.

An environmental impact statement (EIS), under US environmental law, is a document required by the National Environmental Policy Act (NEPA) for certain actions significantly affecting the quality of the environment. EISs can be prepared using two different analysis methods to estimate the impact of sound on marine mammals (Schecklman et al., 2011): (1) dynamic marine mammals (animat method); or (2) static distributions of marine mammals (static-distribution method). The static-distribution method was found to underestimate the number of behavioral harassments compared with the animat method. It was concluded that repeating many simulations with the animat method would provide a robust risk assessment, provide a measure of variability, and allow the probability of spurious events to be estimated.

Patterson et al. (2008) noted that animal movement is a fundamental but poorly understood population process. Although population ecology has traditionally concentrated on understanding temporal fluctuations in abundance, the recent focus has shifted to spatially explicit approaches; this shift has led to a greater appreciation of the importance of animal movement. Furthermore, some important population phenomena depend not only on spatial changes in average population density but also on individual movement behavior. Characterization of animal movement is affected by errors in the observation process; however, separating real biological signals from observation errors in the data remains challenging. Methods that admit uncertainty in movement data when estimating dynamical movement models have generally been lacking. The state-space model enables this by coupling a statistical model of the observation method with a model of the movement dynamics, which can include effects owing to behavior and to the environment. A state-space model consists of coupled stochastic models: a process model (describing the state of an animal in position and time) and an observational model (describing the observed position and inferred behavior). Both coupled models contain error parameters: inherent randomness in movement in the process model, and sampling errors in the observational model. (Also see the earlier discussion of state-space models in Section 6.2.)

6.16.4.4 Collision Avoidance

Cetaceans are prone to collisions with fast-moving vessels. In areas of high cetacean and vessel density, the sperm whale (*Physeter macrocephalus*) is of great concern. Sperm whales are highly vocal and can be localized with passive sonar; however, when at or near the surface (i.e., when they are at greatest risk of collision), they tend to stop vocalizing. Techniques employing active sonars have proved inefficient due to short detection ranges and high closing speeds. Delory et al. (2007) evaluated the efficiency of a passive-sonar solution that used vocalizing whale clicks (at depth) as acoustic sources to detect silent whales. This solution could serve as a noninvasive complement to a more complex passive localization and collision-avoidance system. A wideband N × 2D (range-and-azimuth-dependent) ray model was used to simulate a passive solution comprising an arbitrary number of active acoustic sources (vocalizing whales), an illuminated object (silent whale), and a passive-sonar receiver, all positioned in three-dimensional space with arbitrary bathymetry. Both curved-line and straight-line ray solutions were implemented, with the latter providing greater computational speeds at the expense of temporal and angular fidelity. The simulation recreated the resultant mixture of direct, reverberated, and target-backscattered signals arriving at the receiver for any array configuration, any number of sources, and one target. In the vicinity of the Canary Islands, the simulations demonstrated the applicability of the concept with a maximum detection range on the order of 1 km.

6.16.5 ASW Training Ranges and Mitigation Techniques

At issue here is operational naval training with active sonars. High-power, multistatic sonars have become more important in the face of improved diesel-electric submarine threats operating in complex coastal environments. At the same time, however, these high-power sonars have posed risks to marine mammals.

The US Navy has explored the environmental consequences of installing and operating an undersea warfare training range (USWTR) in conjunction with appropriate coordination and consultation with the National Marine Fisheries Service (NMFS) and in compliance with applicable laws and executive orders including the MMPA, the Endangered Species Act (ESA), NEPA, and the Coastal Zone Management Act (CZMA).

This balance of this section discusses environmentally adaptive sonars and frequency diversity as two viable techniques for mitigating the potentially harmful effects to marine mammals of mechanically generated noise during ASW training experiments at sea.

6.16.5.1 Environmentally Adaptive Sonars

As part of the NATO Undersea Research Centre (NURC) broadband environmentally adaptive concept, the SUPREMO sonar performance model was incorporated within a feedback mechanism for sonar-parameter optimization for any specified ocean environment (Haralabus and Baldacci, 2006). The feasibility of using active sonars to monitor the movement of marine mammals was investigated in order to recommend stand-off distances within which an acoustic source should not be deployed (Ward et al., 2004).

In complex littoral regions, which are often characterized by abundant marine life, Wang and Gong (2007) demonstrated that recursive adaptive beamforming could be used to enhance detection of objects on the ocean bottom in the presence of strong reverberation and false targets.

In "acoustic probing," transmit signals can be adapted according to *in situ* estimates of channel conditions. Waveform characteristics could be set using returns from the previous transmission as an indicator of channel response. Rapidly modifying processing parameters by analyzing previously received signals addresses the problem of fluctuating acoustic channel response owing to variations in environmental factors. Such optimal matching to the estimated channel response would improve target-echo stability (McHugh et al., 2005).

6.16.5.2 Frequency Diversity

A new framework was suggested for sonar design and operation in littoral waters where possible harm to marine life, particularly mammals, has increased (McHugh et al., 2005). Approaches such as environmentally adaptive sonars and frequency diversity were explored.

Since sound level, frequency, and duration are critical parameters in causing physical damage to marine life, transmitting over a wider frequency spectrum using spread-spectrum techniques or noise-like signals could avoid tissue damage by reducing the time spent (or energy transmitted) in particular frequency bands. Spread-spectrum techniques use a pseudo-random spreading code to produce a wideband source signature. Signals must possess high range and Doppler resolution and minimize cross-talk between different users.

Noise-like signals emulate ambient noise by covertly adapting to the fluctuating ambient noise field. Designers could model pseudo-random signals based on bio-acoustic noise; surf noise when high-frequency broadband transients could be useful (Deane and Stokes, 2010); or shipping noise when near-CW signals would be more useful.

6.17 THROUGH-THE-SENSOR PARAMETER ESTIMATION

Advances in sonar technologies have rendered modern sonar systems useful for *in situ* measurements of the ambient marine environment. For example, through-the-sensor measurements of the ocean impulse response have enabled modern sonars to perform collateral functions as tactical environmental processors.

In shallow water, interactions of the acoustic fields with the sea bed require an understanding of the sedimentary structure of the bottom to a level of detail that is usually not required in deep-water environments. In the forward-propagation case, this means that a significant amount of information is necessary to properly characterize the bottom boundary to ensure the generation of high-fidelity model outputs. This generally requires a good understanding of the physics of bottom-interacting acoustics in diverse ocean environments.

Brown and Barlett (2005) described a through-the-sensor remote sensing technique for acoustic-parameter estimation referred to as sonar active boundary loss estimation (SABLE). Bottom scattering strengths were derived from active hull-mounted naval sonar data by using eigenray paths modeled by CASS and GRAB to associate sonar-signal attributes with specific propagation paths.

The potential for an *in situ* technique to infer seafloor reflectivity at shallow grazing angles was assessed (Jones et al., 2002) as a supplement to the Royal Australian Navy sonar prediction tool for range-dependent environments, TESS 2.

6.18 SEISMO-ACOUSTIC INVERSION

At frequencies below several hundred hertz, research in UWA has overlapped with the spectral domain of marine seismologists. This area of overlapping interests has been recognized as a subdiscipline of both communities and is referred to as ocean seismo-acoustics.

A global inversion package called seismo-acoustic inversion using genetic algorithms (SAGA) utilizes a global optimization based on a direct Monte Carlo search and on genetic algorithms (Gerstoft et al., 2003; Gerstoft, 2004). The package can be used to infer geometry in underwater scenarios (source-target localization) and geoacoustic properties of the seabed. The determination of the unknown environmental properties in SAGA is performed by a systematic change of input parameters to a numerical propagation model where the results from the modeling are subsequently compared to the observed data. The measure of the mismatch between model and data (an objective function) depends on the type of data considered, but SAGA includes a suite of objective functions allowing for inversion of a wide range of acoustic data. The environmental input data for the transmission loss model that results in the best match between modeled and observed field data is the final result of the inversion. SAGA has been applied to single-frequency and multifrequency acoustic data received on vertical and horizontal hydrophone arrays, coherent and incoherent transmission loss data, reverberation data, and bottom reflection-loss data. SAGA contains several forward-propagation models including OASES (OAST and OASR modules), SNAP, PROSIM, POPP, and tropospheric parabolic equation model (TPEM). The result of an inversion can be analyzed in terms of *a posteriori* probability distributions, which give an estimate of the importance and uniqueness of each of the environmental parameters that was searched, thus permitting an assessment of the uncertainty in the solution.

6.19 SEISMIC OCEANOGRAPHY

Seismic oceanography (SO) employs low-frequency marine seismic reflection data to image thermohaline fine-structure in the oceanic water column (Holbrook, 2013). Reflection seismology can image such fine structure over large areas based on oceanic temperature contrasts of only a few hundredths of a °C. The resulting images illuminate diverse oceanic phenomena including fronts, water-mass boundaries, internal-wave displacements, internal-tides, eddies, turbulence, and lee waves (Holbrook and Fer, 2005). These low-frequency reflections can be further processed to produce quantitative estimates of sound speed (and thus ocean temperature), turbulence dissipation, and vertical-mode structure over full ocean depths, provided oceanic fine-structure reflections are present. During the first decade of SO utilization, this new technique only slowly gained acceptance as a standard tool for physical oceanographers. This slow acceptance may be attributed partly to the disciplinary boundaries between oceanography and seismology, and partly to the high expense and logistical challenges associated with seismic-data acquisition (Holbrook, 2013). (Also see related discussions in Section 2.5.3.1.)

Multichannel seismic (MCS) reflection imagery of oceanographic structures allows high-resolution teledetection of the thermohaline structure of the ocean (Holbrook et al., 2003). First implementations to study the ocean-surface boundary layer have recently been achieved (Ker et al., 2016), but remain very challenging due to the weakness and shallowness of such seismic reflectors. Ker et al. (2016) developed multifrequency seismic analyses of hydrographic datasets, collected in a seasonally stratified midlatitude shelf by a network of ARGO floats, to assess the detectability issue of shallow thermoclines. These analyses allowed characterization of both the depth and the frequency dependency of the dominant reflective feature of such complex structures. This approach provided the first statistical distribution of the range of variability of the frequency-dependent seismic reflection amplitudes of midlatitude seasonal thermoclines.

Ker et al. (2016) introduced a new parameter (fashioned after the familiar sonar equation) to quantify the overall capability of a MCS setup to detect shallow thermoclines: these parameters included the source strength, the fold (number of shot-receiver pairs that illuminate a given point of the medium), and the ambient-noise level. Ricker wavelets (a zero-phase wavelet, either the second derivative of the Gaussian function or the third derivative of the normal-PDF) were used to approximate the seismic-source signals. These quantitative guidelines aided in the design of seismic experiments targeting such oceanic reflectors. For shallow midlatitude seasonal thermoclines, it was shown that detectability was optimal for seismic peak frequencies between 200 Hz and 400 Hz. This meant that airgun and sparker sources would not be well suited, and that significant improvements in source devices would be necessary before seismic imaging of ocean-surface boundary layer structures could be reliably attempted.

Buffett et al. (2009) applied seismic-reflection profiling to the study of large-scale physical oceanographic processes in the Gulf of Cadiz and along the western Iberian coast. This location coincided with the path of the Mediterranean Undercurrent. The multichannel seismic reflection method provided clear images of thermohaline fine structure with a horizontal resolution approximately two orders of magnitude higher than CTD casting. The seismic data were compared with co-located historical oceanographic data. Three distinct seismic reflectivity zones were identified: North Atlantic Central Water, Mediterranean Water, and North Atlantic Deep Water. Seismic evidence for the path of the Mediterranean Undercurrent was found in the near-slope reflectivity patterns, with rising reflectors between about 500 m and 1500 m. However, the core of the undercurrent was largely transparent. Seismic images showed that the central and (particularly) intermediate Mediterranean Waters had fine structure that was coherent over horizontal distances of several tens of kilometers. However, the intensity of the reflectors, and their horizontal coherence, decreased downstream. This observed change in seismic reflectivity probably resulted from the diminished vertical thermohaline contrasts between adjacent water masses; thus, the double-diffusion processes were unable to sustain the thermohaline staircases. This observation strongly suggested a causal relationship between the intensity of double-diffusion processes and, therefore, fine structure and true seismic amplitude.

Dagnino et al. (2016) demonstrated the feasibility of 2D time-domain, adjoint-state acoustic full-waveform inversion to retrieve high-resolution models of ocean physical parameters, including sound speed, temperature, and salinity. The proposed method was applied to pre-stack MCS data acquired in the Gulf of Cadiz (southwest Iberia) in 2007 within the framework of the geophysical oceanography project. The inversion strategy flow included specifically designed data preconditioned for acoustic noise reduction, followed by the inversion of sound speed in the shot-gather domain. Dagnino et al. (2016) showed that the final sound-speed model had a horizontal resolution of approximately 70 m; this was two orders of magnitude better than that of the initial model constructed with coincident XBT data, and close to the theoretical resolution on the order of a wavelength. Temperature and salinity were retrieved with the same lateral resolution as sound speed by combining the inverted sound-speed model with the thermodynamic equation of seawater together with a local, depth-dependent T-S relationship derived from regional CTD measurements of the NOAA database. The comparison of the inverted temperature and salinity models with the XBT and CTD casts deployed simultaneously with the MCS acquisition showed that the thermohaline contrasts were resolved with an accuracy of 0.18° C for temperature and 0.08 psu for salinity. The combination of oceanographic and MCS data into a common, pseudo-automatic inversion scheme permitted quantitative resolution of submesoscale features that could be incorporated into larger-scale ocean models of oceans structure and circulation. These results demonstrated the potential of MCS data to cover the observational gaps that exist in oceanographic measurements at horizontal scales of 10 m to 1 km. Obtaining information on the structures and processes occurring at these scales would contribute to better understanding of the mechanisms driving the energy transfer between the internal waves and turbulent sub-ranges, and how it influences mixing.

Wang et al. (2014) observed that guided seismic waves can obscure reflections in marine seismic data; however, they also carry information about the seafloor. These guided waves can be observed in physical models. In particular, the dispersion curves from the physical models match theoretical calculations. The shape of the guided-wave dispersion curve is largely determined by the shear-wave velocity of the seafloor and is not sensitive to other physical parameters. Wang et al. (2014) were able to extract the shear-wave velocity of the seafloor from the guided waves using a least-square-based curve-fitting method. Existing filtering techniques may have difficulty separating the normal modes energy from other events; therefore, they developed a dispersion-curve filter that was successfully tested on two physical models in different water depths. Along this line of investigation, Wang et al. (2016) conducted several ultrasonic physical modeling experiments to observe marine guided waves. The guided-wave dispersion curves from these surveys fit theoretical calculations very well. Next, a new method was developed to extract the subbottom S-wave velocity and density from water column guided waves using least-squares inversion. A dispersion-curve filter was developed in the velocity-frequency domain to attenuate the guided waves. These techniques were then applied to the physical modeling data, which have different water depths and different subbottom materials. The extracted results (S-wave velocity, density, and water depth) matched the actual values well. The dispersion-domain filter clarified reflections by attenuating the guided waves, which could benefit further processing and interpretation.

7 Noise I: Observations and Physical Models

7.1 BACKGROUND

Ambient noise is the prevailing, unwanted background of sound at a particular location in the ocean at a given time of the year. It does not include transient sounds such as the noise of nearby ships and marine organisms, or of passing rain showers. It is the background of noise, typical of the time, location, and depth against which a signal must be detected. Ambient noise also excludes all forms of self-noise, such as the noise of current flow around the sonar. Thus, ambient noise is the residual sound level remaining after all identifiable, transient noise sources have been removed (Urick, 1983: Chapter 7). Levels of noise sources are commonly specified as the root-mean square (rms) sound pressure level in a 1-Hz band (referred to as "spectrum level"). Different units indicate how the noise levels were derived (see Pierce, 1989: Chapter 2, for an in-depth discussion). Units commonly encountered in the technical literature include: spectrum level, dB re 1 μPa; dB re 1 $\mu Pa^2\ Hz^{-1}$; and dB re $\mu Pa\ Hz^{-1/2}$ (where "re" is an abbreviation for "relative to"). The discussion by Carey (1995) is relevant here regarding the potential for confusion when using decibels to estimate spectral quantities.

7.2 NOISE SOURCES AND SPECTRA

Figure 7.1 is a hypothetical example of the spectrum of ambient noise in the open ocean (Urick, 1983: Chapter 7). This spectrum is composed of segments of different slope, each exhibiting a different behavior. A number of frequency bands in the spectrum can be associated with readily identifiable noise sources. Five such frequency bands are indicated in Figure 7.1. Band I, lying below 1 Hz, is associated with noise of hydrostatic origin (tides and waves) or with seismic activity (Kibblewhite and Ewans, 1985). Valid measurements in this band (and in Band II) are extremely difficult to make because of the self-noise of the hydrophone and its supporting structure caused by currents (e.g., cable strumming). Band II is characterized by a spectral slope of –8 dB to –10 dB per octave (about –30 dB per decade). The most probable source of noise in deep water appears to be oceanic turbulence. In Band III, the ambient noise spectrum flattens out and the noise appears to be dominated by distant shipping traffic. Band IV contains the Knudsen spectra (Knudsen et al., 1948) having a slope of –5 dB to –6 dB per octave (about –17 dB per decade) in which the noise originates at the sea surface near the point of measurement. Band V is dominated by thermal noise originating in the molecular motion of the sea and is uniquely characterized by a positive spectrum having a slope of +6 dB per octave. Hildebrand (2009) presented a useful tabulation and summary of typical sources of anthropogenic noise in the oceans.

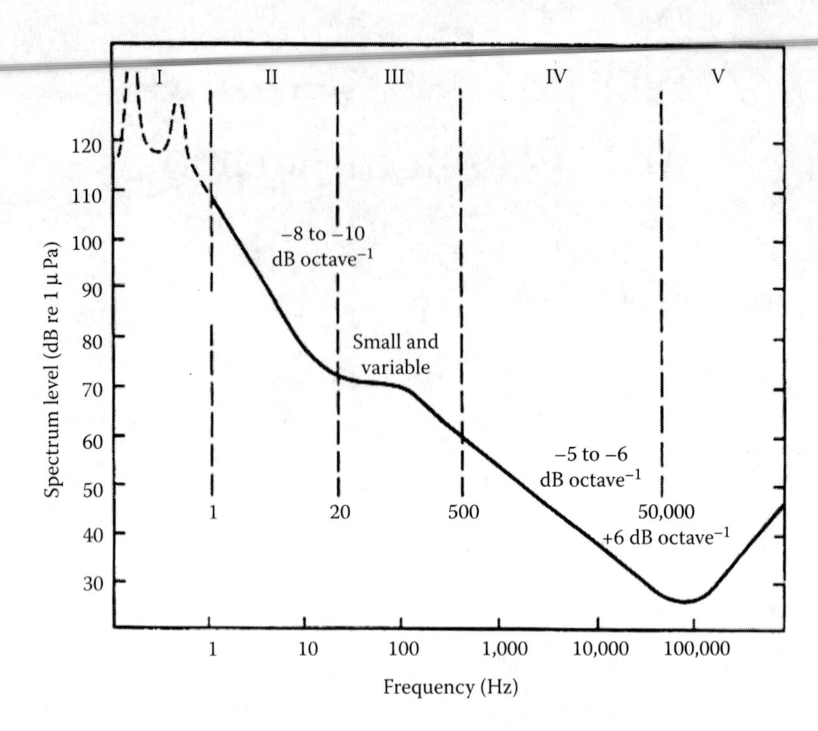

FIGURE 7.1 General spectrum of deep-sea noise showing five frequency bands of differing spectral slopes. The slopes are given in decibels (dB) per octave of frequency. (Reproduced from Urick, R.J., *Principles of Underwater Sound*, McGraw-Hill, New York, 1983. With permission of McGraw-Hill.)

For some prediction purposes, average representative ambient-noise spectra for different conditions are adequate. Such average working curves are shown in Figure 7.2 for different conditions of shipping and wind speed. These curves are adapted from Wenz (1962); consequently, they are often referred to as Wenz curves. In the infrasonic region below 20 Hz, only a single line is drawn. The ambient noise spectrum at any location and time is approximated by selecting the appropriate shipping and wind curves and connecting them at intermediate frequencies. When more than one source of noise is present (e.g., shipping and wind noise), the effective noise background is obtained by summing the intensities of the contributing sources. When using noise levels (specified in units of spectrum level, dB re 1 μPa), this summation process is easily accomplished using the "power summation" operator, denoted by the symbol ⊕, and defined as

$$\oplus = 10 \log_{10} \sum_{i=1}^{n} 10^{L_i/10} \tag{7.1}$$

where L_i is the level of the ith noise source (dB) and n is the number of contributing noise sources. This operation effectively converts the noise levels (L_i) to units of intensity, sums the intensities, and then converts the sum back to units of dB.

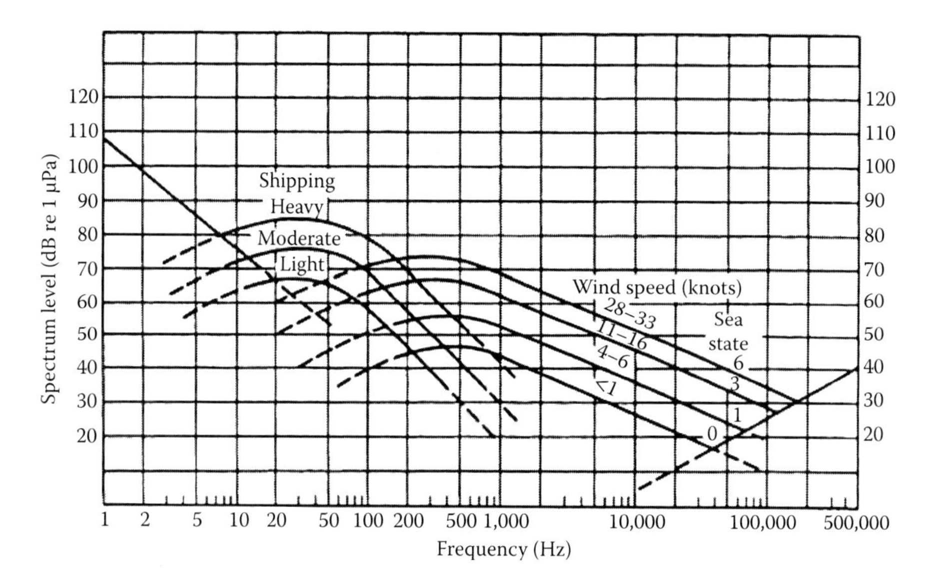

FIGURE 7.2 Average deep-sea ambient noise spectra. (Reproduced from Urick, R.J., *Principles of Underwater Sound*, McGraw-Hill, New York, 1983. With permission of McGraw-Hill.)

For example, in Figure 7.2 at a frequency of 100 Hz under conditions of moderate shipping and sea state 6, the individual noise levels are about 69 dB and 71 dB, respectively. In effect, however, the noise level obtained using Equation 7.1 is

$$69 \text{ dB} \oplus 71 \text{ dB} = 10\log_{10}\left[10^{6.9} + 10^{7.1}\right] = 73 \text{ dB}$$

or 2 dB higher than the level due to surface weather alone.

7.2.1 Seismo-Acoustic Noise

The term *seismo-acoustics* is used broadly in reference to low-frequency noise signals originating in Earth's interior and the oceans. In the frequency range below 3 Hz, Orcutt (1988) defined three specific frequency bands distinguished by the physics of the noise sources:

1. Microseism band (80 mHz–3 Hz) contains high-level microseism noise resulting from nonlinear wave-wave interactions.
2. Noise-notch band (20–80 mHz) contains noise controlled by currents and turbulence in the boundary layer near the sea floor.
3. Ultralow-frequency (ULF) band (<20 mHz) contains noise resulting from surface gravity waves.

Wilson et al. (2003) demonstrated that nonlinear surface-wave interactions are the dominant mechanism for generating deep-ocean noise in the frequency band

0.2–0.7 Hz. Ardhuin et al. (2013) noted that the generation of ultra-low frequency acoustic noise (0.1–1 Hz) by the nonlinear interaction of ocean surface gravity waves has been well established. However, the quantitative theories that attempt to predict the recorded noise levels and their variability remain controversial. A single theoretical framework was used to predict the noise level associated with propagating pseudo-Rayleigh modes and evanescent acoustic-gravity modes. The latter are dominant only within 200 m from the sea surface, in either shallow or deep water. At depths greater than 500 m, the comparison of a numerical noise model with hydrophone records from two open-ocean sites near Hawaii and the Kerguelen Islands revealed that (1) deep-ocean acoustic noise at frequencies ranging from 0.1–1 Hz is consistent with the Rayleigh wave theory in which the presence of the ocean bottom amplified the noise by 10–20 dB; (2) the local maxima in the noise spectrum supported the theoretical prediction for the vertical structure of acoustic modes in agreement with previous results; and (3) noise level and variability were well predicted for frequencies up to 0.4 Hz. Above 0.6 Hz, the model results were less accurate, probably due to the poor estimation of the directional properties of wind-waves with frequencies higher than 0.3 Hz.

Previous research observed that ocean-surface winds play a key role in the generation of underwater ambient noise. Of particular interest is the infrasonic (VLF) band from 1–20 Hz. In this spectral band, wind-generated ocean surface waves interact nonlinearly to produce acoustic waves that couple into the seafloor to generate microseisms. Nichols and Bradley (2016) examined long-term data sets in the VLF portion of the ambient-noise spectrum collected by the hydroacoustic systems of the comprehensive nuclear-test ban treaty organization (CTBTO) in the Atlantic, Pacific, and Indian Oceans. Three properties of the noise field were examined: (1) the behavior of the acoustic spectrum slope from 1–5 Hz; (2) the correlation of noise levels and wind speeds; and (3) the autocorrelation behavior of both the noise field and the wind. Results of these analyses indicated that the spectral slope was site dependent and that there was a high correlation between the wind and the noise field in the 1–5 Hz band. It was further noted that wind-dependent noise between 6 and 16 Hz in the sound channel was masked by other noise sources. This analysis confirmed that long-term VLF (1–20 Hz) underwater noise measured by a total of six hydrophones in three different ocean basins was clearly related to surface wind speeds for frequencies up to about 5 Hz. Below 5 Hz, earlier studies have demonstrated a noise saturation effect, where noise levels increased with wind speed, but only to an acoustic equilibrium spectrum. The noise equilibrium range corresponds to twice the frequency range of the ocean surface wave equilibrium range, which is defined by the condition of surface waves being unable to sustain further growth, even though the wind speed continued to increase.

7.2.2 Shipping Noise

Shipping noise can exhibit both spatial and temporal variability. The spatial variability is largely governed by the distribution of shipping routes in the oceans. Temporal variability can be introduced, for example, by the seasonal activities of fishing fleets.

The noise generated by coastal shipping and by high-latitude shipping can contribute to the noise field in the deep sound channel in tropical and subtropical ocean areas. Specifically, coastal shipping nose is introduced into the deep sound channel through the process of downslope conversion. High-latitude shipping noise is introduced through the latitudinal dependence of the depth of the sound channel axis. These mechanisms are explained below.

Wagstaff (1981) used ray-theoretical considerations to illustrate the mechanisms involved in the downslope conversion process. The following hypothetical arrangement was assumed (Figure 7.3): (1) the continental shelf extends from the coastline to the shelf break (approximately 11.4 km from shore), with an inclination angle of about 1° from the horizontal; (2) the continental slope extends from the shelf break seaward to a depth in excess of 1,000 m, with an inclination of about 5°; and (3) the deep sound channel axis is located at a depth of 1,000 m. Then, in an 11-km band extending seaward from the shelf break, downward-directed radiated noise from surface ships can enter the sound channel by direct reflection off the seafloor.

Kibblewhite et al. (1976) demonstrated the importance of the latitudinal dependence of the sound channel axis depth in introducing high-latitude shipping noise into the deep sound channel. In Figure 7.4, the sound speed structure in the North Pacific Ocean is related to the local water masses. Also shown are the critical depth and the bathymetry. Figure 7.4 vividly illustrates the shoaling of the sound channel axis as it approaches the Arctic region. Thus, low-frequency noise from shipping sources near latitude 50°N will be refracted into the (relatively) shallow sound channel and then propagate with little attenuation to lower latitudes.

Tappert et al. (2002) used the UMPE propagation model to simulate a range-dependent phenomenon that occurs over a sloping bottom when a source is located on the sea floor in shallow water with a downward-refracting (negative-gradient) sound-speed profile. Sound waves propagate downslope with small grazing angles until they reach the depth of the sound-channel axis in deep water where they then detach from the bottom and propagate within the sound channel. Tappert et al. (2002)

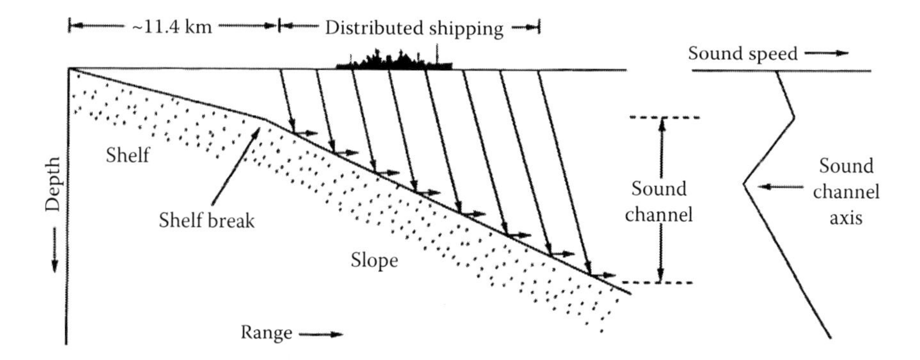

FIGURE 7.3 Illustration of the conversion of coastal shipping noise, represented by high-angle rays, to noise in the deep sound channel, represented by horizontal rays. (Adapted from Wagstaff, R.A., *J. Acoust. Soc. Am.*, **69**, 1009–1014, 1981. With permission.)

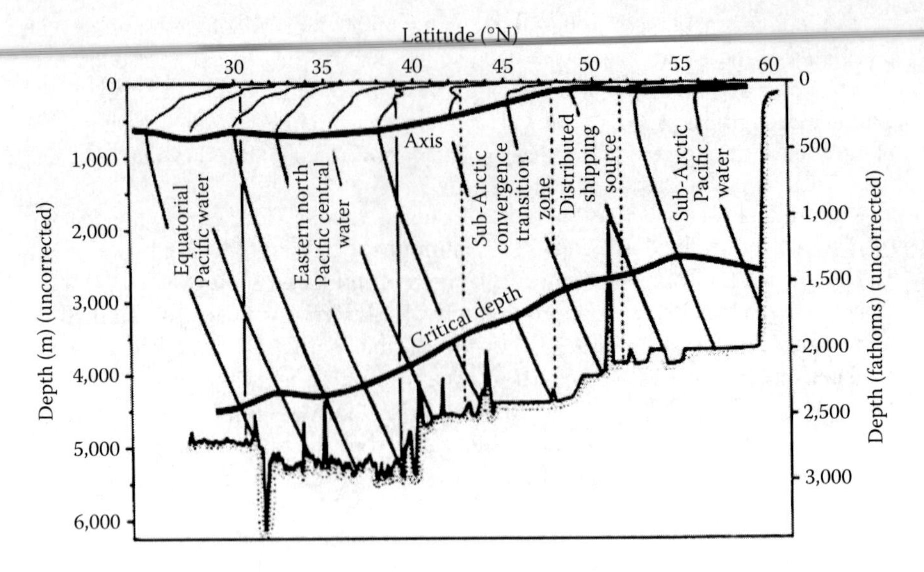

FIGURE 7.4 Bathymetric and sound-speed structure in the North Pacific Ocean. The noise from distributed shipping sources at high latitudes can enter the sound channel and propagate with little attenuation to lower latitudes. Relationships between the sound speed structure and the prevailing water masses are also illustrated. (From Kibblewhite, A.C. et al., *J. Acoust. Soc. Am.*, **60**, 1040–1047, 1976. With permission.)

refer to this as the "mudslide" effect, which they characterize as one of a few robust and predictable acoustic propagation effects that occur in range-dependent ocean environments.

Baggeroer et al. (2005) examined the statistical and directional properties of ambient noise in the 10–100 Hz frequency band from the North Pacific Acoustic Laboratory (NPAL) array. Levels reached ~150 dB re 1 μPa, suggesting that levels 60 dB greater than the mean ambient level were common in the NPAL data sets due to passing ships.

Cato and McCauley (2002) noted that lower levels of shipping-traffic noise distinguish Australian waters from those around North America and Europe, where most of the existing ambient-noise prediction methods have been developed. Thus, noise due to marine animals and breaking waves at the sea surface take on greater importance in Australian waters.

Very-low-frequency sounds 1–100 Hz propagate over long distances in the oceanic sound channel. Weather conditions, earthquakes, marine mammals, and anthropogenic activities influence sound levels in this frequency band (Wilcock et al., 2014). Weather-related sounds result from interactions between waves, bubbles entrained by breaking waves, and the deformation of sea ice. Earthquakes generate sound in geologically active regions, and earthquake T-waves propagate throughout the oceans. Blue and fin whales generate long-duration sounds near 20 Hz that can dominate regional ambient noise levels seasonally. Anthropogenic sound sources include ship propellers, energy extraction, and seismic airguns; these sources have been growing steadily. The increasing availability of long-term records of ocean sound now provide new opportunities for a deeper understanding of natural and anthropogenic sound sources and potential interactions among them.

Jones and Marten (2016) presented an overview of known sound-source levels for various dredging equipment and activities. It described a method useful for extrapolating source levels based on the pump power when a source level has not been directly measured. Furthermore, the article explained how information can be used to predict the propagation of sound from dredging activities (using numerical modeling tools) for environmental impact assessments.

7.2.3 Bioacoustic Noise

Marine bioacoustic signal sources are typically transient in nature and exhibit diverse temporal, spatial, and spectral distributions. The main contributors to bioacoustic signals include certain shellfish, fish, and marine mammals. Of the marine mammals, whales are the most notable contributors. Cummings and Holliday (1987), Au et al. (1987), and Watkins et al. (1987) described whale signal characteristics and distributions. A paper by Watkins and Schevill (1977) included a phonograph recording of actual whale codas. Medwin (2005) assembled a useful collection of papers relating to sounds in the sea.

Richardson et al. (1995: Chapter 7) provided comprehensive summaries of marine mammal sounds in the form of tabulations that included frequency ranges and associated source levels for each species. Specifically included were the sounds produced by baleen whales (bowhead, right, gray, humpback, fin, blue, Bryde's, sei, and minke), toothed whales (Physeteridae [sperm whale], Ziphiidae [bottlenose whale], Monodontidae [beluga and narwhal], Delphinidae [killer whale and dolphin], Phocoenidae [porpoise], and river dolphins), phocid (hair) seals, eared seals (sea lions and fur seals), walruses, sea otters, and sirenian (manatees and dugongs).

The so-called "boing" sound was first observed in US Navy submarine recordings made in the 1950s off San Diego, California, and off Kaneohe, Hawaii. Despite much attention, however, the source of the sound remained a mystery until recently. During a cetacean survey of the US waters surrounding the Hawaiian Islands (Rankin and Barlow, 2005), the probable source of the mysterious "boing" sound of the North Pacific Ocean was identified as a minke whale, *Balaenoptera acutorostrata*.

As discussed by Richardson et al. (1995: Chapters 8–11) and the National Research Council (2000), understanding of both the sensitivity of marine mammals' hearing and the reactions of marine mammals to various noise sources has advanced through additional fieldwork. This work provides relevant guidance to the design and operation of high-intensity sources, especially in multistatic and tomographic experiments. Furthermore, this research aids in the development of meaningful acoustical regulations to ensure the health and safety of marine mammals (see related discussions in Section 6.16).

A reef soundscape can be characterized by vocalizations from the indigenous marine life. This biological sound may be used for orientation by fish and crustaceans recruited to the reef after a larval stage spent in open water (Piercy, 2015).

Because of its role during this critical life stage, sound has been identified as a possible driver of reef population dynamics including detection, retention, and orientation to reefs. The auditory periphery of fishes is used by settlement-stage larvae of coral-reef fishes to localize reef sounds. Coral-reef sounds enable nocturnal navigation by some reef-fish larvae. A reef can be modeled as an extended sound source, which serves to increase the predicted range at which reef noise may be heard by fish larvae (known as the "reef effect") (Radford et al., 2011).

Fishes select novel habitats by responding to multiple cues. These cues, which are complicated by the structural complexity of coral-reef ecosystems and oceanographic transport corridors for pelagic larvae, include (Simpson et al., 2011):

- Polarized light sensitivity and orientation in post-larvae fish including vertical zonation in the pelagic zone and vertical migrations of larvae in relation to both light and prey distribution.
- Chemically mediated behavior of recruiting corals and fish, including olfactory cues; coral reefs are affected by ocean acidification, which appears to impair the auditory responses of some fishes; specifically, an increased growth of otoliths (earbones) observed in fishes reared in highly CO_2-enriched concentrations may compromise the processing of sensory information (see discussion in Section 6.16.2.2).

Simpson et al. (2013) described how studies of fish hearing and physical sound propagation models can be used to predict the detection distance of reefs for settling larval fish, and the potential impact of anthropogenic noise.

7.2.4 Wind and Rain Noise

Kerman (1988, 1993) and Buckingham and Potter (1996) provided updated summaries of sea-surface sound. The established relationships between surface weather phenomena and noise levels have been used to advantage by oceanographers. Shaw et al. (1978) demonstrated that surface wind speeds derived from measured noise spectra could be used to calculate the wind stress over the oceans. Synoptic wind stress information is required for the dynamic modeling of wind-driven ocean currents. Scrimger et al. (1987) and Lemon and Duddridge (1987) described the development and operation of weather observation through ambient noise (WOTAN) systems. These systems operate over the frequency range 0.5–30 kHz and have been successfully used to infer wind speeds that were highly correlated with buoy-mounted anemometer measurements. Vagle et al. (1990) made further measurements in support of WOTAN.

Zedel et al. (1999) modified an acoustic Doppler current profiler (ADCP) to record ambient sound in the frequency range 1–75 kHz. The resulting instrument package called ocean ambient sound instrument system (OASIS) inferred wind speeds and directions from these acoustic measurements that were determined to be in good agreement with direct observations made at Ocean Weather Station Mike in the Norwegian Sea.

Felizardo and Melville (1995) concluded that ambient noise correlated well with wind speed (in the Knudsen range) but correlated poorly with significant wave height. The poor correlation with wave height was attributed to the disproportionate effect of swell on the frequency of breaking waves, which are considered the primary source of wind-dependent noise in the ocean.

The noise attributable to rainfall over the oceans has also been used in an inverse fashion to provide estimates of oceanic precipitation (Nystuen, 1986). The underwater noise spectrum generated by rain has a unique spectral shape that is distinguishable from other noise sources by a broad peak at about 15 kHz. Moreover, the relationship between spectral level and rate of rainfall is quantifiable. Scrimger et al. (1987) made measurements of the underwater noise generated by rain in a lake located on Vancouver Island, British Columbia (Figure 7.5). These data illustrate the peak in noise levels at 15 kHz. Pumphrey and Crum (1990) determined that the major cause of rain-generated sound is the production of bubbles on water-drop impact at the surface. These bubbles then oscillate with small amplitude and radiate as a dipole.

The effects of bubbles and rain as noise-generating mechanisms have been investigated further by Buckingham (1991) and Laville et al. (1991), respectively. As previously noted, there are two sources of rainfall sound: raindrop impacts on the water surface and air bubble resonances. Bubble resonances were found to be responsible for the spectral peak near 13–15 kHz, while raindrop impacts were associated with a broadband spectrum having a negative slope (Laville et al., 1991). The underwater sound due to rainfall can be distinguished further according to raindrop diameters (Medwin et al., 1992): small drops (0.8–1.1 mm) radiate primarily from bubble resonances near 15 kHz; mid-size drops (1.1–2.2 mm) radiate only broadband impact

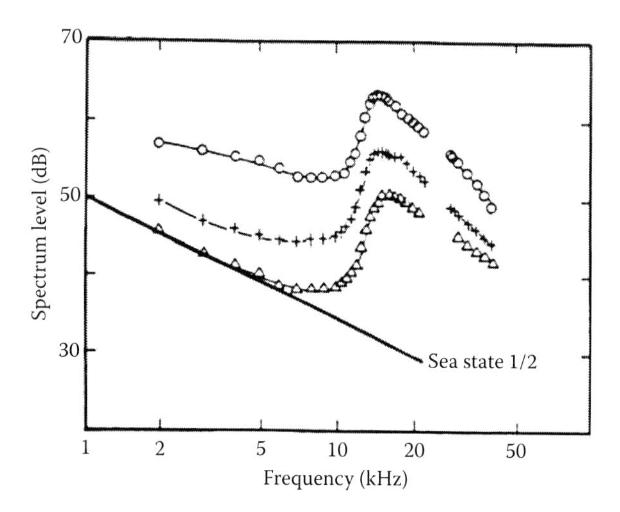

FIGURE 7.5 Rain noise spectra observed at moderate wind speeds for rain rates of 0.5 mm h^{-1} (Δ), 0.8 mm h^{-1} (+) and 1.3 mm h^{-1} (O) compared to the Knudsen curve for sea state ½. (From Scrimger, J.A. et al., *J. Acoust. Soc. Am.*, **81**, 79–86, 1987. With permission.)

sound; and large drops (>2.2 mm) radiate both impact and resonance sounds. Nystuen and Medwin (1995) proposed a new bubble-entrapment mechanism to account for a missing component in the modeling of underwater sound levels produced by raindrops. In related investigations, Rohr and Detsch (1992) attributed the effect of films on suppressing high-frequency ambient noise to the opposition of bubble-producing events by these films. Crum et al. (1992) and Pumphrey (1994) reviewed the nature of precipitation sounds underwater, while Prosperetti and Oğuz (1993) reviewed the physics of drop impacts on liquid surfaces. Collectively, these studies have improved on earlier investigations such as those reported by Heindsmann et al. (1955), Franz (1959), and Bom (1969).

Indirect measurements of rainfall are important since roughly 80% of Earth's precipitation occurs over the oceans and lakes where only about 10% of the weather stations are located (e.g., islands and buoys). This information allows oceanographers and meteorologists to improve their understanding of the oceanic heat and freshwater budgets (e.g., Etter et al., 1987, 2004) that, in turn, can provide measures of both short-term and long-term fluctuations in the global climate.

Nystuen (1994) developed an acoustic rainfall analysis (ARA) algorithm consisting of detection of rainfall in the presence of other underwater noise sources; classification of rainfall type based on drop-size distribution; and acoustic quantification of rainfall rate. The classification of rainfall type (e.g., according to stratiform or convective) will permit meteorologists to infer the vertical distribution of atmospheric latent-heat release in support of global climate studies. See Section 7.9 for a more detailed discussion of acoustic rain gauges (ARGs).

Oğuz (1994) developed a theoretical model to predict the ambient noise levels arising from bubble clouds generated by breaking waves and resultant whitecap formation. Cloud geometries were modeled by inverted hemispherical shapes within which solutions to the wave equation could be obtained analytically. Empirical relationships between wind speed and whitecap occurrence permitted calculation of noise levels as a function of frequency and wind speed.

Wilson and Makris (2006) quantified the relationship between local wind speed and underwater acoustic intensity for use in classifying hurricane strength, given eye-wall passage over the receiver.

Deane and Stokes (2010) presented a model for the underwater noise of whitecaps. The noise from a few hundred hertz up to at least 80 kHz was assumed to be due to the pulses of sound radiated by bubbles formed within a breaking wave crest. The total noise level and its dependence on frequency were a function of bubble-creation rate, bubble-damping factor, and a so-called "acoustical skin depth" associated with scattering and absorption by the bubble plume formed within the crest. (The concept of an "acoustical skin depth" defines a volume of audibility around the bubble-plume edge.) A closed-form analytical expression for the wave noise showed a −11/6 power-law dependence of noise level on frequency, which was in good agreement with the −10/6 scaling law commonly observed in the open ocean.

Cato and Tavener (1997) concluded that geostrophic wind, and thus forecasts of wind speeds, would be useful in forecasting sea noise. Supporting observations made from two regions near Australia suggested that while geostrophic-wind forecasting would predict the broad trends, it would provide poorer prediction of detail than would forecasts based on real-time local data. In general, however, the need is to forecast broad trends, such as approaching fronts that are typically accompanied by significant changes in wind speed and which would, consequently, produce substantial changes in sea noise. The drag-corrected geostrophic winds generally followed the trends of the local winds but with slight temporal displacements in peaks and troughs in wind speed. This suggested that major sea-noise effects could be forecast, but with some uncertainty about their actual timing. Forecasting the fact that major changes in wind speed are imminent may be more important than forecasting their timing accurately. Specifically, as the event approaches, the timing can be refined using other information such as reports from local weather stations.

Poikonen (2012) conducted shallow-water ambient noise measurements in a brackish-water environment where hydrophones were located at depths of 15–20 m. The measurement site was well isolated from traffic noise which made it possible to study wind-generated effects at lower frequencies. Due to the near-field conditions, the ambient-noise levels were not significantly distorted by propagation effects. The measured ambient noise spectra showed a distinctive bandpass structure characteristic of the dipolar source distribution formed by bubbles in breaking waves. The observed sharp spectral declines below 500 Hz were most likely caused by the resonances of oscillating bubble clouds created by breaking waves. The low-frequency range of the declines may be attributed to the larger bubble sizes in fresh and brackish waters compared to saline water. High-frequency ambient noise spectra exhibited a dual-slope spectral pattern above 1 kHz due to increased attenuation above 10 kHz at intermediate and high wind speeds. The study demonstrated that absorption in brackish and fresh water, unlike in ocean water, tends to decrease above a frequency of 10 kHz due to the low proportion of small bubbles in a bubbly mixture created by breaking waves. The excess high-frequency attenuation in the spectra cannot therefore be directly attributed to the effects of absorption in a bubbly mixture. Measured ambient noise spectra were modeled as a cumulative power spectrum of individual resonating bubbles distributed in a radius range of 0.01–3.3 mm using a model for the sound generated by breaking waves. The dual-slope pattern observed in the brackish water spectra was mostly explained with a bubble-size distribution that had a distinctive maximum at radii between 0.1–0.3 mm, and a relative drop in bubble density below a radius of 0.1 mm. A physical explanation for this is the fact that small bubbles have a tendency to coalesce more in fresh and brackish water than in sea water. The best fit to the average deep-water spectrum having a spectral slope of 5.7 dB/octave (19 dB/decade) was obtained with a bubble size distribution that is proportional to the bubble radius to the power of $-3/2$. A typical slope range of 5 to 6 dB/octave, reported in literature for oceanic ambient noise spectra, corresponds to the bubble size distribution power factors of -1.7 and -1.4, respectively. These brackish-water results should not be generalized due to the inherent complexity of coastal environments. Bubble densities are known to have considerable spatial variability depending on seasonal, biological, and even weather conditions.

7.3 DEPTH DEPENDENCE

Measurements of the depth dependence of low-frequency ambient noise were made by Morris (1978) in the northeastern Pacific Ocean. Hydrophones were suspended

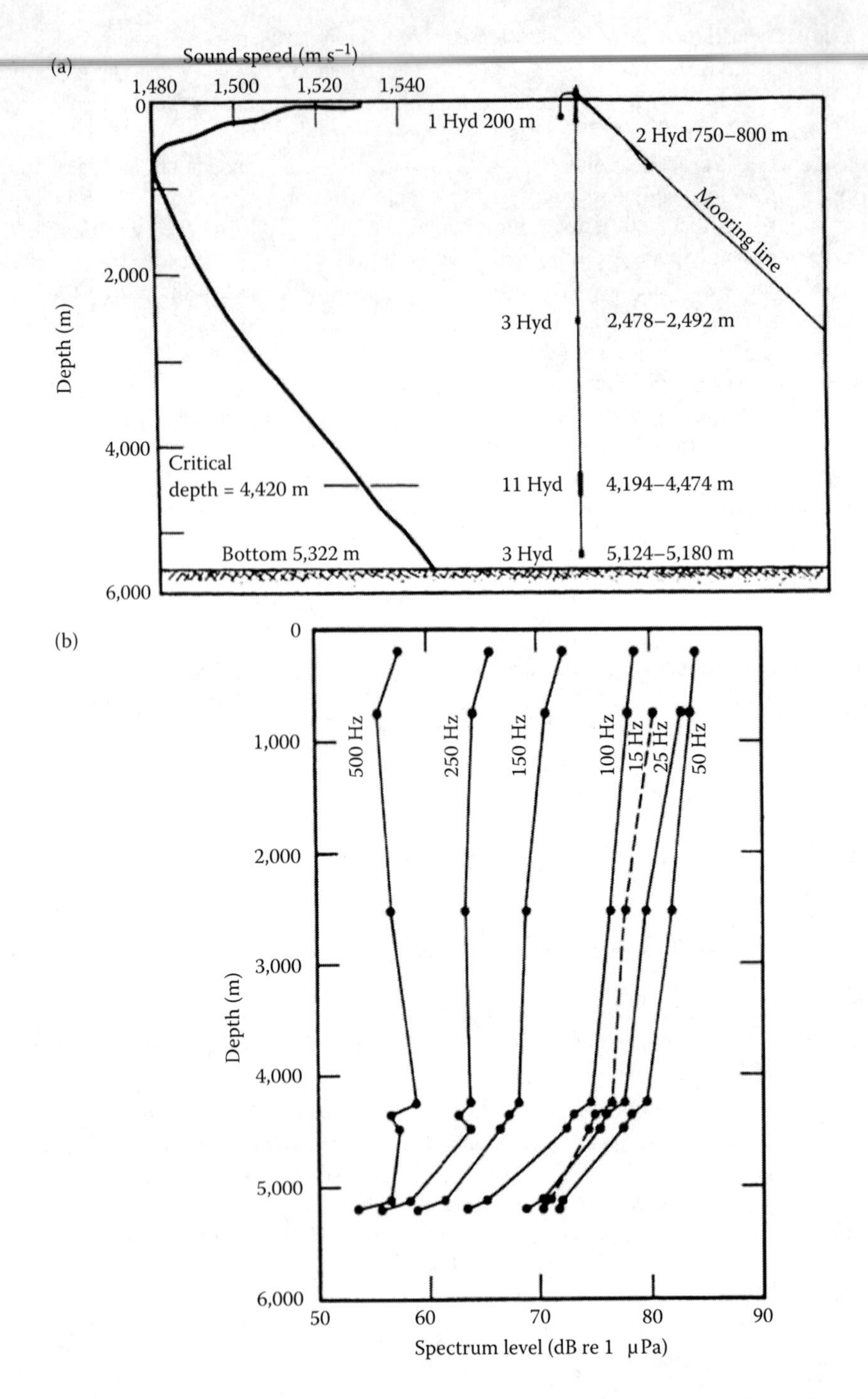

FIGURE 7.6 Northeast Pacific Ocean ambient noise measurements: (a) sound speed profile and hydrophone depths; and (b) measured noise profiles in one-third-octave bands. (From Morris, G.B., *J. Acoust. Soc. Am.*, **64**, 581–590, 1978. With permission.)

from the research platform FLIP (floating instrument platform). Figure 7.6 shows the sound speed profile, hydrophone depths, and average noise profiles in one-third-octave bands for this experiment. There is a decrease of noise with increasing depth at low frequencies, with a smaller decrease with depth at 500 Hz as wind noise overcomes the dominance of shipping noise. Below the critical depth, the fall-off with depth is steeper as the bottom is approached. This is the result of the loss of refracted sound energy through the effects of bottom interaction.

Urick (1984: Chapter 4) demonstrated that ambient noise at frequencies greater than 10 kHz is rapidly attenuated with increasing depth due to the effects of absorption.

7.4 DIRECTIONALITY

As a first-order approximation, the noise field in the ocean might be considered to be isotropic in nature, that is, uniform in all directions, both horizontal and vertical. Measurements have shown that this is not the case.

Axelrod et al. (1965) made measurements of the vertical directionality of ambient noise at frequencies of 112 Hz and 1,414 Hz. Figure 7.7 presents polar plots of the ambient noise intensity per unit solid angle $N(\theta)$ arriving at a bottomed hydrophone as a function of the vertical angle θ. At 112 Hz, more noise arrives at the hydrophone from the horizontal than from the vertical. This effect diminishes with increasing wind speed. At 1,414 Hz, the opposite is true in that more noise arrives from overhead than horizontally. This effect increases with increasing wind speed.

This directional behavior is consistent with the view that low-frequency noise originates at great distances and arrives at the measurement hydrophone principally via horizontal paths, suffering little attenuation. Alternatively, high-frequency noise originates locally at the sea surface overhead (Urick, 1984: Chapter 5).

The horizontal (or azimuthal) directionality of ambient noise can be highly variable, particularly at low frequencies. Shipping traffic is the dominant source of noise at low frequencies, and the temporal and spatial variations in shipping densities explain much of the observed azimuthal variation.

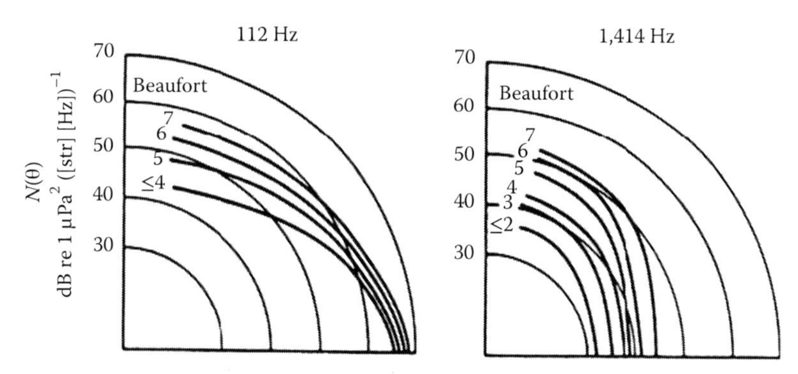

FIGURE 7.7 Distribution of ambient noise in the vertical plane at a bottomed hydrophone at two frequencies. (From Axelrod, E.H. et al., *J. Acoust. Soc. Am.*, **37**, 77–83, 1965. With permission.)

7.5 SURF NOISE

Bass and Hay (1997) noted that ambient noise in the surf zone in the frequency range 120 Hz – 5 kHz is dominated by breaking waves. Deane (1997) developed a "moving bubble sheet" source model. In this model, the production of sound within the source region is associated with the formation of bubble plumes. Above 500 Hz, the sound is consistent with radiation from individual bubble oscillations.

Fabre and Wilson (1997) provided estimates of surf-generated ambient noise, which has been characterized empirically as a source level density per meter of beach as a function of wave height and frequency. Surf noise depends on the source level characteristics of different surf types and the unique propagation conditions from the surf zone seaward. This dependence must be examined in order to develop models capable of providing accurate ambient noise levels in shallow water. An inverse method for determining ambient noise due to surf as a function of beach slope, sediment type, wave type, frequency, and wave height was developed. The surf source level density as a function of surf type can be used for any shoreline. Thus, it is possible to use this model to estimate surf conditions on a beach based on noise measurements taken offshore (Fabre et al., 1997; Wilson et al., 1997).

According to Preisig and Deane (2004), a statistical characterization of experimental data collected in the surf zone showed that acoustic interaction with surface gravity waves resulted in focused surface-reflected arrivals whose intensity often exceeded that of the direct arrival. This focusing and caustic formation adversely impacted the performance of an impulse-response estimation algorithm. Furthermore, the resulting caustics presented challenges to the reliable operation of phase-coherent underwater acoustic communications systems that must track fluctuations in the impulse response.

Means (2004) presented an analytical model for the generation of sound by individual breaking waves within the surf zone. Specifically, this model treated the entrained bubble cloud as a prolate hemispheroid versus the spherically shaped bubble clouds often used in modeling open-ocean breaking waves. Two orientations, "long" and "tall," were suggested. The "long" orientation of the bubble cloud positioned the elongated axis of the spheroid along the wave crest of the breaker. The "tall" orientation was proposed as a means of modeling the multiple subsurface bubble plumes beneath an individual breaking wave as observed in published underwater photographs.

Melville and Matusov (2002) observed that the distribution of the length of breaking wave fronts per unit area of the sea surface is proportional to the cube of the wind speed. An intense breaking surf can contribute significantly to the ambient noise in fairly deep continental shelf waters over the frequency range 20–700 Hz (Wilson et al., 1985).

Deane and Stokes (2002) measured bubble size distributions inside breaking waves in the laboratory and in the open ocean to provide a quantitative description of bubble formation mechanisms. They noted that two distinct mechanisms control the size distribution depending on bubble radius:

- Radius > 1 mm—turbulent fragmentation determines bubble size distribution resulting in a bubble density proportional to the bubble radius to the power –10/3

- Radius < 1 mm—bubbles are created by jet and drop impact on the wave face with a –3/2 power-law scaling.

Continuing along this line of investigation, Deane and Stokes (2006) measured the sound of bubbles fragmenting in fluid shear and determined the frequency, amplitude, and decay rate of the acoustic emissions from 1.8-mm-radius bubbles fragmenting between opposed fluid jets. A wide band of frequencies (1.8–30 kHz) was observed with peak pressure amplitudes in the range 0.03–2 Pa. The frequency dependence of the decay rates was consistent with the sum of thermal and acoustic radiation losses. In related work, energy dissipation rates were calculated based on observed bubble-size spectra (Garrett et al., 2000).

Rouseff and Tang (2010) modeled what effect the increasing acidity of the ocean would have on ambient-noise levels in shallow water. Because most noise sources are near the surface, high-order acoustic modes are preferentially excited. Linear internal waves, however, can scatter the noise into the low-order, low-loss modes that are most affected by the changes in ocean acidity. Their model used transport theory to couple the modes, further assuming an isotropic distribution for the noise sources. Exploring a scenario typical of the East China Sea, the noise at 3 kHz was predicted to increase by 30%, or about one decibel, as the pH decreased from 8.0 to 7.4. Consequently, a mid-frequency noise model that ignored internal-wave effects might underestimate the actual noise levels. Also see the related discussions in Section 6.16.2.2.

7.6 ARCTIC AMBIENT NOISE

The noise environment under the Arctic ice is different from that of any other ocean area. Shipping noise is extremely low due to the lack of surface traffic. The ice cover itself affects the ambient noise field significantly. It can decouple the water from the effects of the wind and produce ambient noise conditions that are much quieter than a corresponding sea state zero in the open ocean. The ice itself may produce noises as wind, waves, and thermal effects act on it (e.g., Milne, 1967). Other sources of noise in the Arctic include seismic and biological activity.

The character of the ice cover is different in areas of shore-fast pack ice, moving pack ice, and the marginal ice zone (MIZ). The under-ice noise levels, directionality, spectrum shape, and temporal character are very different in each of these regions. Noise originating within the ice stems principally from its state of stress, which gives rise to fracturing. Noise measurements under the pack ice span the frequency range from 3 Hz to more than 1 kHz (Dyer, 1984, 1988).

Pack ice is very dynamic and its characteristics are highly variable in space and time (Makris and Dyer, 1986). Nevertheless, the lack of wind-wave interaction and the absence of local shipping can lead to noise levels 10 dB lower than those encountered in the open ocean.

Noise levels in the MIZ are typically higher than those in either the pack ice regions or the open ocean regions (Diachok and Winokur, 1974; Diachok, 1980). Figure 7.8 illustrates the variations in median ambient noise levels that occur across

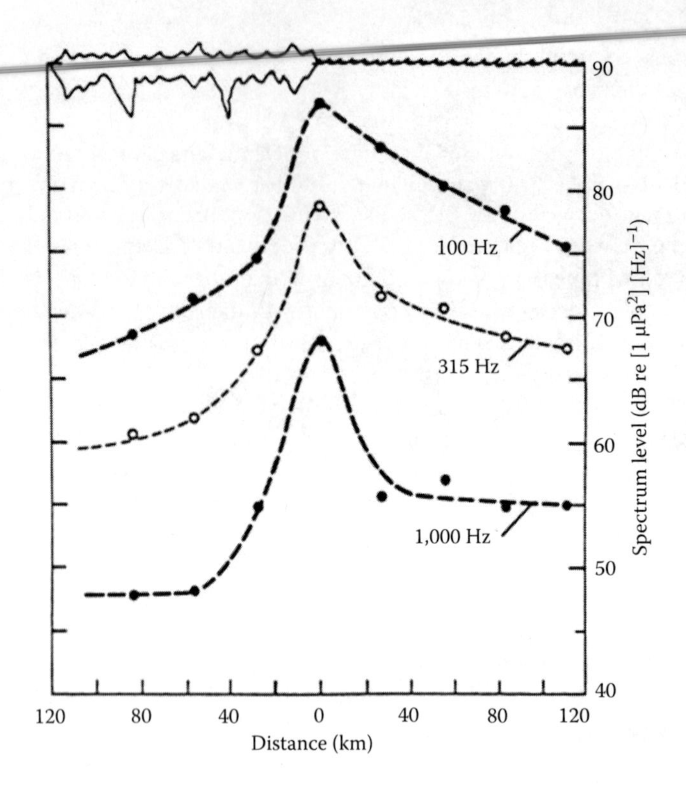

FIGURE 7.8 Variation of median ambient noise sound pressure spectrum levels with distance from a compact ice edge for frequencies of 100 Hz, 315 Hz, and 1000 Hz in sea state 2. (From Diachok, O., *Cold Regions Sci. Technol.*, **2**, 185–201, 1980. With permission.)

a compact ice-water MIZ area. The relative magnitudes of the noise levels generated at the ice-water boundary are a function of the rate of change of ice concentration with distance. Thus, the relative maximum noise level measured at a diffuse ice-water boundary would be smaller than that measured at a compact ice edge. Ambient noise levels in the MIZ also depend on such variables as sea state, water depth, and dominant ocean-wave period. The last variable is hypothesized as being related to the efficiency of coupling ocean-wave energy into the ice. Makris and Dyer (1991) demonstrated that surface gravity wave forcing was the primary correlate of ice-edge noise in the MIZ.

Pritchard (1990) developed an ambient noise model to simulate the time history of under-ice noise generated by dynamical ice movement, stress, and deformation. The model included noise contributions from local and distant sources as well as a transmission loss module appropriate for the Arctic environment. Time series data were simulated for a frequency of 31.5 Hz.

Lewis and Denner (1988) reported observations of higher-frequency (1 kHz) ambient noise data in the Arctic Ocean. These noises were attributed to the thermal fracturing of ice. Related aspects of ice-generated noise have been reviewed by Sagen et al. (1990).

7.7 ACOUSTIC DAYLIGHT

Ambient noise in the ocean can be used to form pictorial images of underwater objects. An analogy can be drawn between the natural optical (daylight) field in the atmosphere and the radiating ambient noise field in the ocean: both fields consist of random, incoherent radiation propagating in all directions. Thus, the concept of imaging with ambient noise has been referred to as "acoustic daylight" (Buckingham et al., 1992). A reasonable operating frequency range for the acoustic daylight system is 5–50 kHz; the lower limit is determined by angular resolution considerations while the upper limit is determined by the onset of thermal noise. A general introduction to imaging underwater objects with ambient noise was presented by Buckingham et al. (1996b).

Buckingham (1993) analyzed the incoherent acoustic imaging of a spherical target in order to quantify the contrast ratio (acoustic contrast) under various degrees of anisotropy. In the case of isotropic noise, the contrast (acoustic visibility) was found to have a maximum value of about 4 dB. This result was consistent with field measurements when the angle of view of the acoustic lens matched the angle subtended by the target at the phase center of the measurement array.

An acoustic daylight ocean noise imaging system (ADONIS) was constructed. The system was designed to operate over the frequency range 15–75 kHz with a corresponding minimum wavelength of 20 mm, formed 126 individual beams of nominal width of 0.76° (at the upper frequency), and operated over the spatial range 10–200 m. Simulations developed to predict the performance of ADONIS demonstrated that near perfectly reflecting objects were unlikely to be imaged in volume isotropic noise, except perhaps in the near field (Potter, 1994). However, the ocean was expected to exhibit considerable anisotropy, if only in the vertical direction, thus improving performance over the conjectured isotropic baseline. Buckingham et al. (1996a) described the results of an experiment with ADONIS, which in practice operated in the frequency range 8–80 kHz and relied on ambient noise to provide the acoustic contrast between look angles on-target and off-target. Epifanio et al. (1999) described results from the ORB (research platform) experiments, which were conducted with targets at ranges between 20 m and 40 m using ADONIS's 126 receive-only beams spanning the vertical and horizontal. Makris et al. (1994) conducted a careful analysis of this noise-imaging concept and concluded that it pressed the limits of current technology. Furthermore, they traced similar approaches back to 1985 when the possibility of detecting submarines solely by their noise absorbing and scattering properties ("acoustic contrast" versus "acoustic glow") had been investigated by Flatté and Munk.

Potter and Chitre (1999) extended the concept of "acoustic daylight" (which uses the mean intensity of backscattered ambient-noise energy to produce images of submerged objects, and is thus analogous to "vision") by exploring the information contained in higher moments. Specifically, information embodied in the second temporal and spatial moments of intensity, for which there are no visual analogs like "acoustic daylight," was referred to as ambient noise imaging (ANI), a broader imaging approach.

Bistatic sonars can detect targets outside an ellipse surrounding the source and receiver. However, when a submerged object intercepts the area between the source and receiver, it is not easily detected using traditional processing techniques because its signal is obscured by the intense direct blast of the incident field. Lei et al. (2014) noted that forward scattering of an object intercepting the source–receiver line can introduce acoustic field aberration at the receiver, although this aberration is difficult to distinguish by individual hydrophones in a littoral waveguide. This transient aberration results from interference between the incident and forward scattering fields from the object, but is difficult to predict in a multipath environment. The field aberration is generated regardless of an object's stealth ability, so it may be exploited for intruder detection. However, the aberration is weak relative to the direct blast, and is consequently difficult to detect. The potential detection of forward scattering was demonstrated in a lake experiment, and the second principal component was extracted from the beam output waveform. This study confirmed that forward scattering may be detected by principal component extraction of the beam output waveform. When the object was beyond the Fresnel zone, the field aberration could be improved to 10 dB above the background field.

7.8 GEOACOUSTIC INVERSION

In shallow water, the spatial structure of the ambient-noise field is strongly influenced by multiple interactions with the sea floor. Consequently, both the vertical directionality and coherence of the shallow-water noise field are determined primarily by the geoacoustic properties of the seabed rather than by any temporal variations in source distributions. Several investigators have deduced geoacoustic parameters through inversion of the ambient noise field in range-independent and azimuth-independent shallow-water environments in which the noise sources were uniformly distributed. For example, Buckingham and Jones (1987) determined critical angles, Carbone et al. (1998) determined compressional and shear wave speeds, and Aredov and Furduev (1994) determined reflection losses.

For nonuniformly distributed noise sources in range-dependent environments, sophisticated vertical hydrophone arrays are required to resolve the arrival structure of the noise field. Furthermore, the experimental data must be analyzed using detailed noise models. For simpler geoacoustic parameters such as reflection loss, however, Harrison and Simons (2001) argued that detailed models are not required in the inversion analysis, although a densely populated vertical array is required to resolve the arrival structure. Harrison and Simons (2002) report additional experimental results.

Between 2002 and 2004, six experiments using a drifting vertical array were conducted in the Mediterranean Sea south of Sicily to investigate bottom-reflection properties by inversion of ambient noise (called GAIN) and also to investigate sub-bottom layering (called SUPRA-GAIN) at frequencies ranging from a few hundred Hz to 5 kHz (Harrison, 2004b).

A technique based on noise correlations produced an effective passive fathometer to identify subbottom layers (Siderius et al., 2006). By cross-correlating ambient noise time series received on the upward and downward steered beams of a drifting vertical array, it was possible to obtain a subbottom layer profile. Strictly, the time differential of the cross correlation is the impulse response of the seabed. It has been shown theoretically and by simulation that completely uncorrelated surface noise results in a layer profile with predictable amplitudes proportional to those of an equivalent echo sounder at the same depth as the array (Harrison and Siderius, 2008).

Siderius et al. (2010) improved on a recently developed technique to image seabed layers using the ocean ambient noise field as the sound source (the so-called passive fathometer technique), which exploits the naturally occurring acoustic sounds generated at the sea-surface, primarily from breaking waves. The method is based on the cross-correlation of noise from the ocean surface with its echo from the seabed, which recovers the travel times to significant seabed reflectors. To limit averaging time and thus make this practical, beamforming was used with a vertical array of hydrophones to reduce interference from horizontally propagating noise. The initial development used conventional beamforming, but significant improvements have been realized using adaptive techniques. Adaptive methods allow a larger bandwidth to be included, which gives better time resolution. Further, adaptive methods suppressed the horizontal sound that interferes with the surface noise, and this allowed weaker subbottom layers to be better resolved as compared with conventional processing.

7.9 ACOUSTIC RAIN GAUGES

Surface measurements of precipitation in oceanic environments have been made using ARGs that monitor precipitation through interpretation of the underwater sound field over the frequency range 500 Hz to 50 kHz (Nystuen et al., 2000). Nystuen (2001) constructed a diagram showing the relationship between equivalent radar reflectivity and drop size. This diagram was made more useful by partitioning it acoustically to show those regions that were occupied by rainfall containing specific drop-size populations.

Mani and Pillai (2004) investigated the acoustic signals produced by raindrops impacting on the water surface. This impact generated low-frequency, damped pressure waves in the water. The low-frequency spectrum of this signal was found to be useful for measuring the kinetic energy of the raindrops, from which the drop-size distribution and the rain intensity could be estimated.

Ma and Nystuen (2005) developed a single-frequency (acoustic) rainfall-rate algorithm based on comparisons with rain-gauge data. Wind-generated and rain-generated ambient sounds from the sea surface were used to form a baseline of ocean noise (Ma et al., 2005). Five divisions of the sound spectra associated with different sound-generating mechanisms could be identified: (1) wind-only spectra levels, decreasing linearly with increasing frequency; (2) sound generated by large raindrops, invariant with wind speed; (3) sound generated by small raindrops between 8 kHz and 25 kHz, highly sensitive to wind speed; (4) sound produced by a combination of small and large raindrops; and (5) sound-masking effect during high winds or extreme rainfall rates, due to a layer of bubble clouds that form just below the sea surface. Ambient noise data collected from the intertropical convergence zone (ITCZ) were used to construct prediction algorithms that were subsequently tested on data from the western Pacific warm pool. This physically based, semi-empirical model could be used to predict ambient noise spectra over the frequency range 0.5–50 kHz at rainfall rates ranging from 2–200 mm h^{-1} and at wind speeds ranging from 2–14 m s^{-1}. Detailed equations and tables of coefficients were also provided.

8 Noise II: Mathematical Models

8.1 BACKGROUND

Chapter 7 discussed aspects of ambient noise measurements in the ocean. In particular, Figure 7.2 portrayed the average deep-sea ambient noise spectra as originating from shipping traffic at low frequencies (~5 Hz to ~200 Hz) and from surface weather at high frequencies (~200 Hz to ~50 kHz). In the range 50–500 Hz, both mechanisms (shipping and weather) contribute to the observed noise levels. The significance of these dual mechanisms has influenced the course of noise-model developments.

Mathematical models of noise in the ocean can be segregated into two categories: ambient noise models and beam-noise statistics models. Ambient noise models predict the mean levels sensed by an acoustical receiver when the noise sources include surface weather, biologics, and such commercial activities as shipping and oil drilling. Beam-noise statistics models are more specialized in that they predict the properties of low-frequency shipping noise for application to large-aperture, narrow-beam passive sonar systems. The latter models use either analytic (deductive) or simulation (inductive) techniques to generate statistical descriptions of the beam noise. In this context, beam noise is defined as the convolution of the receiver beam pattern with the sum of the intensities from the various noise sources. The analytic models calculate statistical properties directly from the components (e.g., source level, propagation loss) while the simulation models use Monte Carlo techniques.

8.2 THEORETICAL BASIS FOR NOISE MODELING

Mathematical models of noise in the ocean predict both the level and directionality (vertical and horizontal) of noise as a function of frequency, depth, geographic location, and time of year. Both categories of noise models (ambient noise and beam-noise statistics) consist of two components: a transmission loss component, and a noise level and directionality component. In principle, the transmission loss can be computed internal to the noise model or it can be input externally from other (standalone) model predictions or from field measurements.

Ambient noise models treat noise sources as variable densities distributed over large areas. This approximates the generation of wind noise and distant shipping noise. Consequently, the transmission loss calculations in ambient noise models can be range-averaged (as opposed to point-to-point). This greatly relaxes the accuracy to which transmission loss must be known. Alternatively, beam-noise statistics models treat noise sources (individual ships) as discrete sources, and thus require point-to-point representations of transmission loss.

Empirical regression formulas can sometimes satisfy low-fidelity modeling requirements. The loss of fidelity stems principally from a lack of directionality information (vertical and horizontal) and also from a lack of temporal and spatial resolution. Moreover, the use of regression formulas to estimate the ambient noise levels (but not directionality) presumes that the noise levels can be considered independently of the sonar system characteristics. Rigorous noise modeling convolves the system beam pattern (i.e., receiver response) with the calculated noise field levels and directionalities (Wagstaff, 1982).

Sadowski et al. (1984) reviewed regression formulas appropriate for the estimation of average ambient noise spectra below 100 kHz including noise sources arising from ocean turbulence, shipping traffic, surface weather (both wind and rain), and molecular agitation. Wagstaff (1973) also presented regression formulas that were incorporated into an ambient noise model that was valid over the frequency range 10–500 Hz. Ross (1976: Chapter 8) reviewed regression formulas appropriate for shipping noise levels. For a brief history of such work, see Ross (1993). Bjørnø (1998) summarized the general characteristics of ambient noise in littoral waters.

Using radiated-noise measurements collected from 272 ships over the period 1986–1992, Wales and Heitmeyer (2002) updated the classical merchant ship radiated noise regression formulas utilized by Ross (1976). These classical regression formulas postulated that the source spectrum for an individual ship was proportional to a baseline spectrum whose constant of proportionality was determined by power-law exponents for the ship speed (sixth power) and ship length (second power). The reanalysis by Wales and Heitmeyer (2002) over the frequency range 30–1200 Hz now represented the individual ship spectra by a modified rational spectrum. At high frequencies (400–1200 Hz), most of the individual spectra showed a simple power law dependence on frequency with exponents concentrated around a mean value of about 2. At low frequencies (30–400 Hz), many of the source spectra exhibited a more complex dependence on frequency with greater spectral variability across the ensemble.

Bradley and Bradley (1984) developed an elaborate empirical model called the geophysics ambient noise model. This model provided estimates of seasonal deep-water ambient noise levels and azimuthal directionalities over the frequency range 25 Hz–15 kHz. This model was based on a comprehensive empirical database of shipping and wind noise, but not noise due to biologics or industrial activity. The model further assumed a nominal receiver depth of 100 m, thus ignoring any depth dependence. Hamson (1997) reviewed techniques for modeling shipping and wind noise over the frequency range 50–3000 Hz, concentrating mainly on work performed after 1980. Noise level, horizontal and vertical directionality, and the noise responses of arrays were used to characterize the ambient noise field. Harrison (1996) used a simple ray approach to approximate the full-wave treatment of noise levels and coherence in range-independent ocean environments. Alvarez et al. (2001) used an approach based on genetic algorithms to study the physical characteristics of measured underwater ambient noise in the frequency range 10 Hz to 2000 Hz. The resulting predictability of the recorded signals was attributed, in part, to the contributions of shipping noise.

8.3 AMBIENT-NOISE MODELS

A simple model of ambient noise in the ocean would consist of an infinite layer of uniform water with a plane surface over which the sources of noise (shipping and weather) were uniformly distributed. In this model, the ambient noise level would be independent of depth (Urick, 1983: Chapter 7). Of course, a more realistic model would include volume absorption in addition to the effects of refraction and boundary reflections over long-range paths (Talham, 1964).

To compute the low-frequency component of noise due to distant shipping, three inputs are required: (1) the density of shipping as a function of azimuth and range from the receiver; (2) the source level of the radiated noise for each generic type of merchant ship; and (3) the transmission loss as a function of range between the near-surface sources (ships) and the depth of the receiver. Then, contributions from successive range rings centered about the receiver can be summed to obtain the level of shipping noise as a function of azimuth at the receiver. The theory behind this kind of modeling, in addition to some observational data on the density of shipping traffic in the North Atlantic, has been described by Dyer (1973), among others.

The high-frequency component of noise due to surface weather is usually computed on the assumption that it is locally generated and isotropic. Thus, only the weather conditions (sea state or wind speed) prevailing in the immediate vicinity of the receiver need be considered, in addition to any localized rain shower or biologic activity.

Modeling the vertical directionality of deep-water ambient noise can be approached by means of the simple model mentioned earlier. Consider a bottomless, uniform ocean without refraction or attenuation, having a surface covered with a dense, uniform distribution of noise sources. Furthermore, let each unit area of the surface radiate with an intensity $I(\theta)$ at a distance of one meter. Then, at point P (the receiver) in Figure 8.1a, the incremental intensity $\mathrm{d}I$ produced by a small circular annulus of area $\mathrm{d}A$ at horizontal range r is (Urick, 1983: Chapter 7)

$$\mathrm{d}I = \frac{I(\theta)\,\mathrm{d}A}{l^2} = \frac{I(\theta)\,2\pi r\,\mathrm{d}r}{l^2} \tag{8.1}$$

But $r = h \tan\theta$, so that $\mathrm{d}r = h \sec^2\theta\,\mathrm{d}\theta$. Also, $l = h \sec\theta$. On substituting and rearranging terms, Equation 8.1 reduces to

$$\mathrm{d}I = 2\pi I(\theta)\tan\theta\,\mathrm{d}\theta \tag{8.2}$$

If ψ is a solid angle, then $\mathrm{d}\psi = 2\pi\sin\theta\,\mathrm{d}\theta$, so that the noise intensity per unit solid angle, $N(\theta)$, becomes

$$N(\theta) = \frac{\mathrm{d}I}{\mathrm{d}\psi} = I(\theta)\sec\theta \tag{8.3}$$

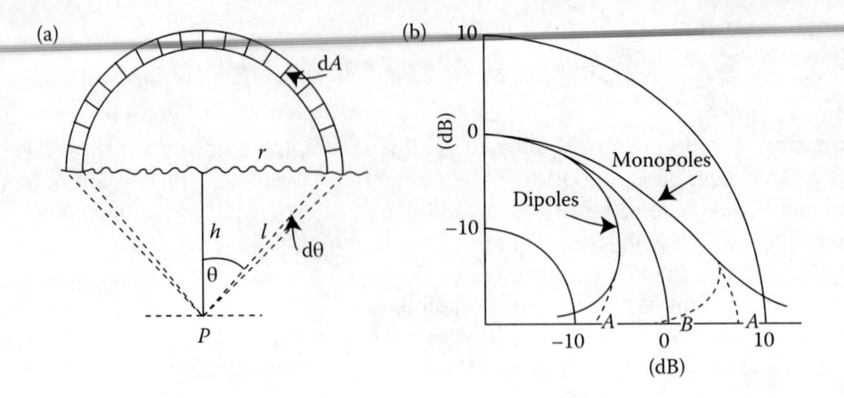

FIGURE 8.1 Simple model for the vertical directionality of ambient noise: (a) geometry with straight-line propagation paths; and (b) directional patterns for surface distribution of monopoles and dipoles. The dashed segments near the horizontal show the effect of attenuation and refraction near the seafloor (*A*) and near the deep sound channel axis (*B*). (From Urick, R.J., *Principles of Underwater Sound*, McGraw-Hill, New York, 1983. With permission.)

When $I(\theta) = I_0$, which represents a nondirectional source, Equation 8.3 yields

$$N(\theta) = I_0 \sec\theta \tag{8.4a}$$

which is the ambient noise intensity per unit solid angle for a surface distribution of monopole sources. Using a distribution of dipole sources for which the intensity radiated at an angle θ is $I(\theta) = I_0 \cos^2 \theta$, the beam pattern of the noise received at a depth below the surface is

$$N(\theta) = I_0 \cos\theta \tag{8.4b}$$

Equations 8.4a and 8.4b are plotted in polar coordinates in Figure 8.1b. At angles near the horizontal (where θ = 90°), the effects of attenuation, refraction, and boundary multipaths prevent the curves from going to either zero or infinity. Using this simple model, a receiver located within the deep sound channel would not be expected to receive noise arriving along paths near the horizontal since range-independent ray tracing would show that such ray paths do not exist. The beam pattern of the noise from monopole sources would therefore have a maximum at an oblique angle above and below the horizontal (typically ±10° to ±15° off the horizontal axis in temperate, deep-sea regions), with a minimum at θ = 90°. This simple conceptual picture is not always valid in realistic (range-dependent) environments. Specifically, more sophisticated models would consider the effects of range-dependent refraction, bottom reflection, and multipath arrivals. This aspect will be discussed further under the topic of the noise notch (Section 8.5).

For high-frequency noise sources (>1000 Hz), many investigators (e.g., Anderson, 1958; Becken, 1961; Von Winkle, 1963) have suggested a function of the form $I(\theta) = I_0 \cos^m \theta$, where I_0 is the intensity radiated by a small area of the sea surface in the downward direction ($\theta = 0°$) and where m is an integer. Values of $m = 1$, 2, or 3 have been obtained, depending upon conditions and methods of measurement. Most measurements center roughly about $m = 2$, a value consistent with the hypothesis of a dipole source formed by the actual source and its image in the sea surface. This model adequately describes surface weather noise.

For low-frequency noise sources (<500 Hz), the agreement between available observations (see Figure 7.7) and the findings of this simple model (Figure 8.1) further suggests that a distribution of monopoles ($m = 0$; thus, $I(\theta) = I_0$) adequately models distant shipping noise.

Pertinent examples from the ANDES noise model (Renner, 1995b) will illustrate how the horizontal and vertical noise directionalities are computed. The fundamental output from ANDES is the directional noise intensity per unit solid angle [$N_S(\theta,\phi)$]. The horizontal noise directionality [$N(\phi)$] is calculated from [$N_S(\theta,\phi)$] as

$$N(\phi) = \int_{-\pi/2}^{\pi/2} N_S(\theta,\phi)\cos\theta \, d\theta \tag{8.5}$$

The vertical noise directionality [$N(\theta)$] is calculated from [$N_S(\theta,\phi)$] as

$$N(\theta) = \frac{1}{2\pi}\int_{0}^{2\pi} N_S(\theta,\phi)\, d\phi \tag{8.6}$$

The omnidirectional noise level (N) is then calculated as

$$N = \int_{0}^{2\pi}\int_{-\pi/2}^{\pi/2} N_S(\theta,\phi)\cos\theta \, d\theta \, d\phi \tag{8.7a}$$

By noting the relationship with Equation 8.5 and 8.7a can be simplified as

$$N = \int_{0}^{2\pi} N(\phi)\, d\phi \tag{8.7b}$$

The horizontal angle (ϕ) is measured positive clockwise from true north while the vertical angle (θ) is measured positive upward from the horizontal plane. Note that no receiver beam patterns were convolved with the noise levels in Equations 8.5 and 8.6.

8.4 RANDI MODEL—A SPECIFIC EXAMPLE

The research ambient noise directionality (RANDI) model is an instructive example of an ambient noise model. RANDI calculates the vertical and horizontal directionalities of low-frequency (10–500 Hz) ambient noise for a selected ocean environment (Wagstaff, 1973).

8.4.1 Transmission Loss

RANDI utilizes one of three sources of transmission loss inputs:

1. Self-contained linear ray-trace routine
2. Input data consisting of transmission loss versus range in addition to arrival angle arrays
3. Deep sound channel propagation algorithm

8.4.2 Noise Sources and Spectra

RANDI considers six sources of isotropic and anisotropic surface and volumetric noise: (1) shipping, (2) wind-wave, (3) biological, (4) sea state zero, (5) rain, and (6) distant sources. A seventh source, that of a target, is also included as an option (Figure 8.2). Figure 8.3 illustrates the modeled noise field. The surface noise is generated by an infinite number of point sources distributed along a horizontal noise source plane located at a depth of about 6 m, corresponding to the nominal drafts

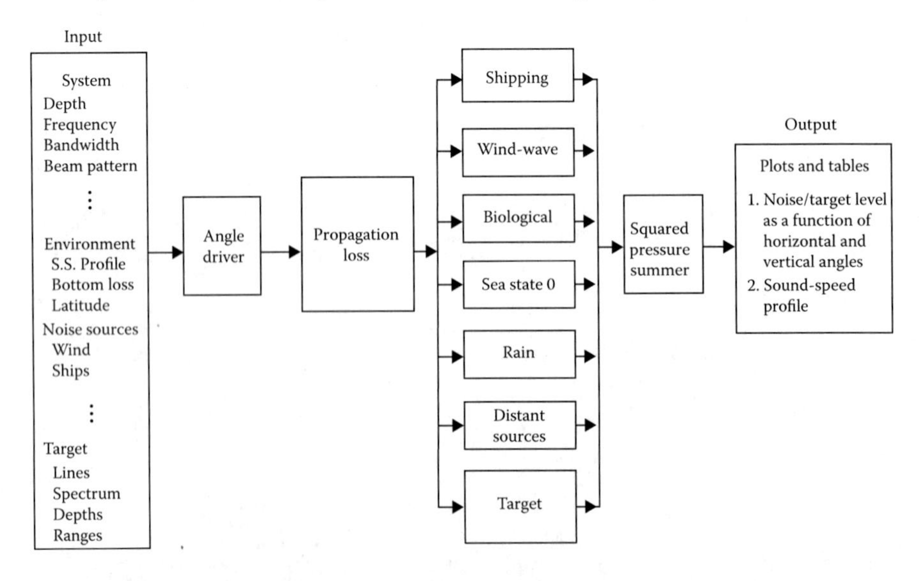

FIGURE 8.2 Block diagram of the RANDI noise model. (Adapted from Wagstaff, R.A., RANDI: Research ambient-noise directionality model. Nav. *Undersea Ctr, Tech.* Pub. 349, 1973.)

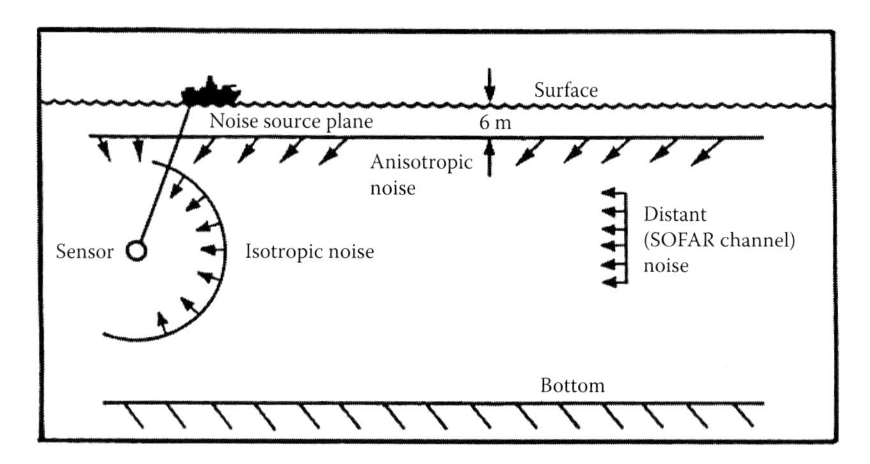

FIGURE 8.3 The noise field used in the RANDI model. (Adapted from Wagstaff, R.A., RANDI: Research ambient-noise directionality model, *Undersea Ctr, Tech*. Pub., 349, 1973.)

of surface vessels. Empirical algorithms are used to calculate squared pressure spectrum levels for each of the seven noise generation mechanisms. All of the noise (and target) squared pressure spectrum levels are then integrated over a user-specified bandwidth using an input frequency-response function. Noise level in this context refers to the mean squared spectrum level.

8.4.3 Directionality

To determine the horizontal directionality of the noise field, the ocean is divided into n wedge-shaped regions (called sectors) with the receiver at the center. The vertical thickness of the sectors is equal to the ocean depth at the receiver location. Noise calculations are performed independently in each sector defined by the user. The total squared noise pressure, as measured by an omnidirectional hydrophone, is obtained by summing the squared noise pressures for the n sectors. An example of horizontal directionality of the noise field is presented in Figure 8.4 (note that the units for noise level are referenced to degrees).

The vertical directionality of the noise field is determined for a differential vertical angle by first calculating the area defined by the intersections of the corresponding ray bundles with the anisotropic noise field plane (i.e., shipping noise). Next, these areal extents are multiplied by the local effective squared noise pressure levels. The resulting levels are reduced by the appropriate transmission loss and then summed. At this point, the contribution of the isotropic noise sources (i.e., weather noise) is considered. These new levels are then convolved with the vertical response of the receiver array. This whole process is then repeated for each new differential angle. An example of vertical directionality of the noise field is presented in Figure 8.5. Note that the units for noise level in Figure 8.5 are referenced to steradians.

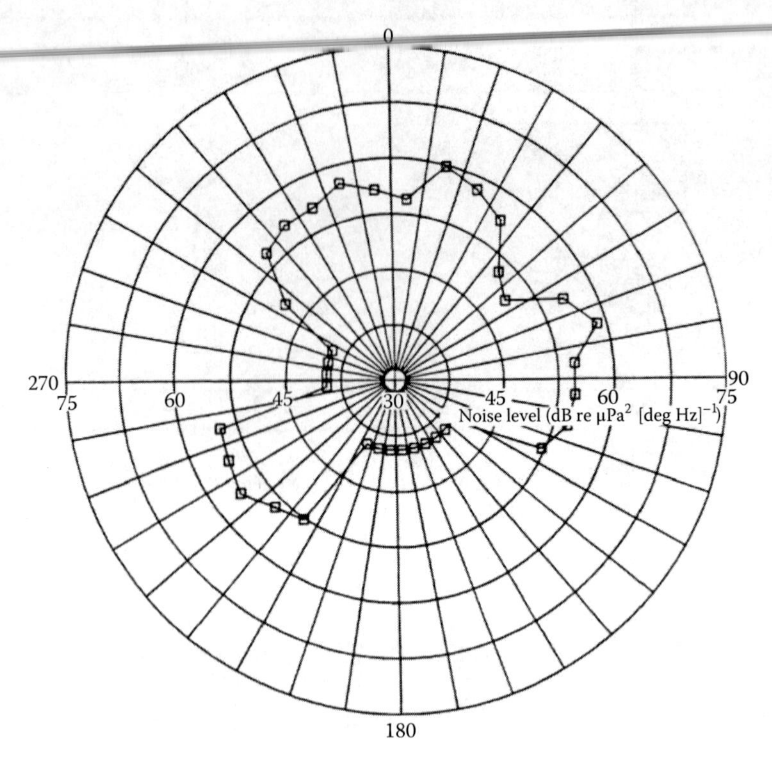

FIGURE 8.4 Per-degree horizontal directionality of ambient noise in 10° sectors generated by the RANDI noise model for a frequency of 100 Hz at a depth of 91 m in the North Pacific Ocean. (Adapted from Wagstaff, R.A., RANDI: Research ambient-noise directionality model. *Nav. Undersea Ctr, Tech. Pub.*, 1973. 349.)

8.4.4 Recent Developments

The latest version of this model (RANDI III) was developed for application to shallow water and coastal areas in the low-to-mid frequency range (~10–300 Hz). The receiver can be either a horizontal or a vertical line array (Breeding, 1993). The dominant sources of noise are assumed to be shipping, distant storms, and local winds. The ships are treated as discrete sources, and the shipping noise is propagated over great-circle routes to the receiver elements using a wide-angle, finite-element parabolic equation (FEPE) model. The arriving contributions from all ships are added coherently at the receiver, and the array response is determined by beamforming with various shading schemes. Environmental and shipping information is automatically extracted from US Navy standard databases: historical temporal shipping (HITS) for determination of shipping densities; digital bathymetric database (DBDB) for bathymetry; generalized digital environmental model (GDEM) for sound-speed profiles; and low-frequency bottom loss (LFBL) for bottom-interaction parameters (refer to Chapter 11, specifically Table 11.5).

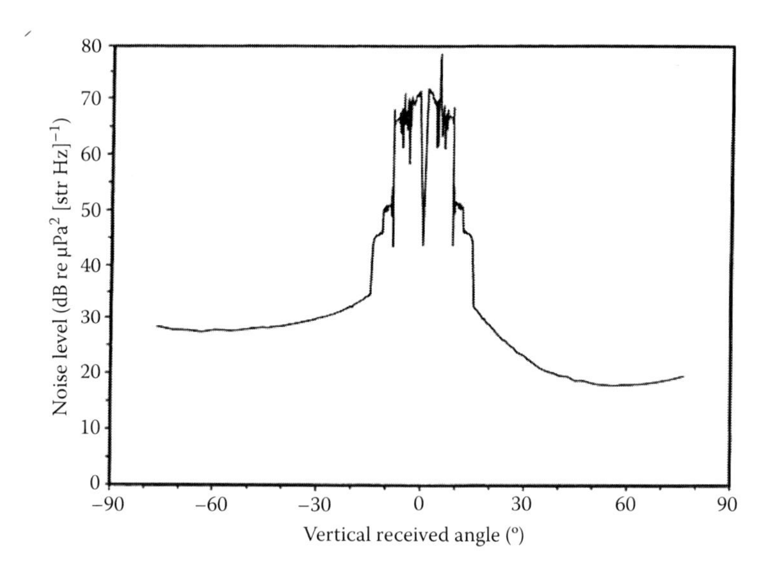

FIGURE 8.5 Ambient noise vertical directionality for the 10° horizontal sector centered at 15° in Figure 8.4. Rays with positive angles arrive at the receiver from below the horizontal plane. A ray with an angle of −90° arrives from the surface directly over the receiver. (Adapted from Wagstaff, R.A., RANDI: Research ambient-noise directionality model. *Nav Undersea Ctr, Tech.* Pub. 349, 1973.)

8.5 THE NOISE NOTCH

The so-called noise notch is often manifested in the vertical directionality of low-frequency noise fields generated by those noise models utilizing range-independent transmission loss inputs. Figure 8.6 presents a vertical noise directionality diagram generated by the fast ambient noise model (FANM). This particular example is for a low-frequency (50 Hz) case under the assumption of a range-independent ocean environment. Consistent with the conceptual picture discussed in Chapter 7 and earlier in this chapter, most of the low-frequency noise arrives near the horizontal. Furthermore, slightly higher levels are observed in the direction of the sea surface than in the direction of the seafloor. This result is also expected in that the bottom-reflected rays are attenuated more than are the corresponding surface-interacting rays. Figures 8.5 and 8.6 both show this effect. The feature of real interest, however, is the horizontal notch (or null) that is predicted by range-independent ray theory. Since the notch feature is not always observed in measurements at sea, this situation presents a paradox. The generation of notch features is not limited to range-independent ray-tracing techniques. Carey et al. (1987), for example, were able to produce similar notch features using a parabolic equation (PE) model under similar range-independent environmental assumptions.

Anderson (1979) made measurements of the vertical directionality of noise in the North Pacific Ocean in September 1973. He did not observe the horizontal null

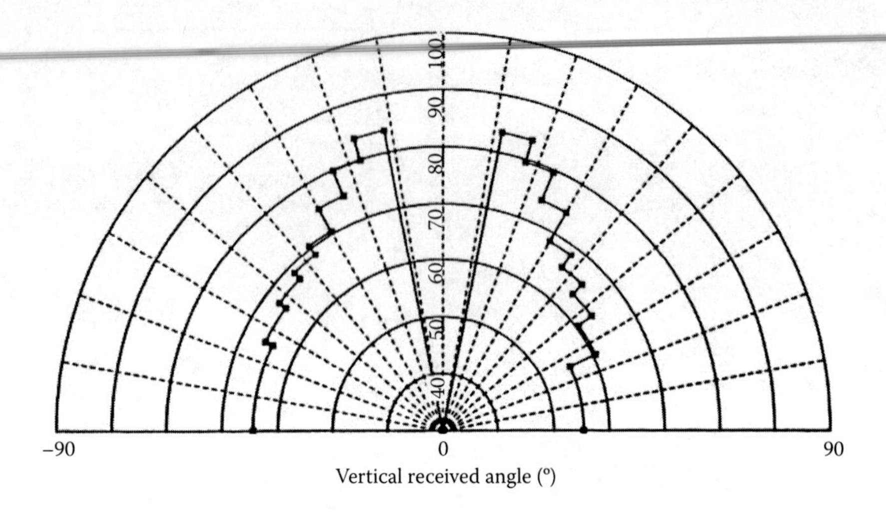

FIGURE 8.6 Vertical noise directionality (dB re μPa^2 Hz^{-1} str^{-1}) computed by the FANM noise model at a frequency of 50 Hz. Rays with positive angles arrive at the receiver from below the horizontal plane.

(notch) in the directional spectrum of the noise that is predicted by ray theory for a horizontally homogeneous (range-independent) ocean.

This aspect can be explored further by examining basic ray-theoretical considerations in typical deep-ocean, range-independent environments. Specifically, this notch (or null) is the result of ray arrivals being excluded from a region that is centered on the horizontal axis of the receiver and limited in angular width to $\pm\theta_R$, where θ_R defines the limiting ray as

$$\theta_R = \cos^{-1}\frac{c_R}{c_S} \tag{8.8}$$

where

θ_R = limiting ray angle at the receiver
c_R sound speed at the receiver position
c_S sound speed at the source position

This corresponds to those rays leaving the source horizontally ($\theta_S = 0°$). Equation 8.8 then follows directly from Snell's law. Figure 8.7 illustrates the geometry for defining the limiting ray. The sound speed profile is based on measured data presented by Anderson (1979). Assuming that the noise source plane is near the surface, and that the receiver is located at 1200 m (just below the deep sound channel axis), then Equation 8.8 yields

$$\theta_R = \cos^{-1}\left[\frac{1485\,\mathrm{m\,s^{-1}}}{1530\,\mathrm{m\,s^{-1}}}\right] = \cos^{-1}(0.971) \approx 14°$$

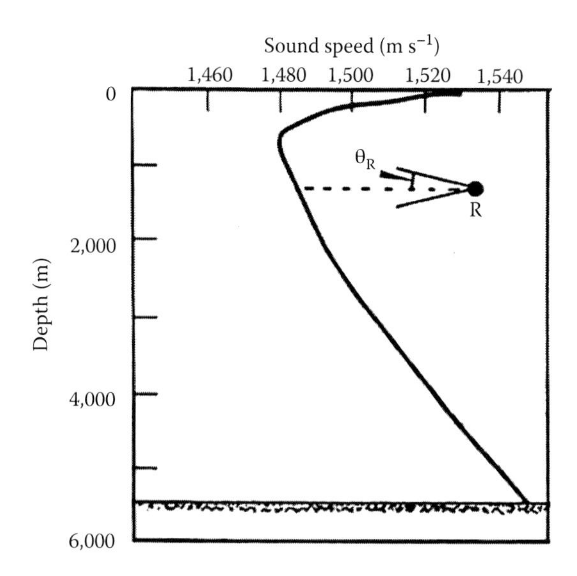

FIGURE 8.7 Geometry for calculation of limiting rays. Noise source plane is located near the sea surface. Receiver (R) is located at a depth of 1,200 m. In this example, limiting ray angles (θ_R) are ± 14° off the horizontal axis of the receiver. (Sound profile is from Anderson, V.C., *J. Acoust. Soc. Am.*, **66**, 1446–1452, 1979. With permission.)

Thus, ray-theoretical considerations indicate that noise from the surface cannot arrive within about 14° of the horizontal axis of the receiver, at least under the environmental and geometrical conditions assumed here. These conditions, however, are typical of many deep-ocean environments that support deep sound channel propagation.

The apparent paradox created by the discrepancy between range-independent ray theory and observations can be explained in terms of range-dependent propagation effects. In particular, two effects can contribute to filling the noise notch through conversion of near-surface shipping and weather noise into shallow-angle propagation paths near the sound channel axis. These mechanisms are slope conversion and horizontal sound-speed variations (e.g., Anderson, 1979). These mechanisms were addressed previously in Section 7.2.2. Slope conversion considers the reflection and scattering of sound into the deep sound channel from ships transiting the continental shelves and slopes of the ocean basins. Horizontal sound-speed variations versus latitude cause the axis of the sound channel to shoal at high latitudes, thus allowing far northern (Kibblewhite et al., 1976) or far southern (Bannister, 1986) shipping and weather noise to enter the deep sound channel and to propagate for long distances. Dashen and Munk (1984) examined this problem and suggested that diffusion may also play a role.

Carey and Wagstaff (1986) reviewed low-frequency (<500 Hz) physical noise models and measurements. They confirmed that coherent signals from surface ships were a dominant characteristic of the horizontal ambient noise field. The vertical directionality of the noise showed a broad angular distribution centered about the horizontal axis at lower frequencies (<200 Hz), and a dual-peaked (±10° to ±15° off the horizontal) distribution at frequencies between 300 Hz and 500 Hz. In the frequency range 20–200 Hz, the data exhibited a smooth variation with frequency. However, the tonal nature of surface ship radiated noise spectra would suggest a spiky spectral variation. Smooth spectral variations in the vertical directionality of the noise field are also characteristic of low-shipping areas. Carey and Wagstaff (1986) thus suggested that environmental noise sources such as wind-driven noise, in addition to shipping, might be required to explain the broad angular and frequency characteristics observed in the data. Carey et al. (1990) conducted additional simulations using a PE model. These simulations confirmed the role of downslope conversion as a low-pass filter in determining the vertical noise field at mid-basin.

Hodgkiss and Fisher (1990) made a series of direct measurements that confirmed the contribution of the downslope conversion mechanism to the near-horizontal noise distribution. They noted that the effects of absorption appeared to diminish the near-horizontal energy with increasing distance from the coast, and that these effects were more pronounced at higher frequencies.

The vertical directivity pattern of the ambient-noise field observed in shallow water typically contains a noise notch in the horizontal plane that develops since downward refraction steepens all rays emanating from near the sea surface (Rouseff and Tang, 2006). However, variability in the environment has the potential to redistribute the noise into shallower angles and thereby fill the notch. A model for the width and depth of the ambient noise notch was developed. Transport theory for acoustic propagation was combined with a shallow-water internal wave model to predict the average output of a beamformer. Ambient noise data from the East China Sea were analyzed in the frequency band 1–5 kHz. Good agreement between the model and the data for both the width and depth of the ambient noise notch was obtained at multiple frequencies, suggesting that internal wave effects are significant in filling the noise notch in this frequency band.

8.6 BEAM-NOISE STATISTICS MODELS

Models of beam-noise statistics use a statistical approach to model the low-frequency ambient noise field in the ocean. To be of practical use to large-aperture, narrow-beam passive sonar systems, these models must include sonar-specific beam pattern characteristics in addition to point-to-point representations of transmission loss.

The probability measures of beam noise depend on array configuration, orientation, location, and season. For detection predictions, the measure of interest is the total power in selected frequency bands. For prediction of false alarm rates, the desired measure is a characterization of the narrowband components of shipping noise (Moll et al., 1979). Only detection predictions will be addressed here.

Using the formalism developed by Moll et al. (1979), an expression for the total noise power in a specified band at the beamformer output can be presented. As a consequence of the principle of superposition of the instantaneous pressures of sound from multiple point sources, the averaged noise power at the beamformer output (Y) can be expressed as

$$Y = \sum_{i=1}^{m} \sum_{j=1}^{n} \sum_{k=1}^{A_{ij}} S_{ijk} Z_{ijk} B_{ijk} \tag{8.9}$$

where

m = number of routes in the basin
n = number of ship types
A_{ij} = number of ships of type j on route i (a random variable)
S_{ijk} = source intensity of the kth ship of type j on route i (a random variable that is statistically independent of the source intensity of any other ship)
Z_{ijk} = intensity transmission ratio from ship ijk to the receiving point
B_{ijk} = gain for a plane wave arriving at the array from ship ijk

The probability density function for Y can then be obtained from its characteristic function.

Heitmeyer (2006) developed a probability law for the noise field and the beam-noise time waveforms based on the assumption that breaking-wave occurrences can be described by a space-time Poisson process and that the breaking-wave waveforms are independent Gaussian processes. As an illustration, examples were presented of the first-order probability density and the correlation function for a noise field observed on a vertical array operating in shallow water. For smaller elevation angles and deep phones, where the energy is dominated by distant breaking waves, the observed noise waveforms were essentially Gaussian processes. For beams pointed toward the surface and for shallow phones, where the noise was dominated by a small number of breaking waves, the noise waveforms were not Gaussian. The elevation angles and phone depths for which the Gaussian approximation is not valid were also identified.

8.7 DATA SUPPORT REQUIREMENTS

The utilization of noise models requires a specialized database containing information on shipping routes, shipping density by merchant vessel type, and radiated noise levels by vessel type. Vessel types are usually differentiated according to freighters, tankers, and fishing craft. Further distinctions can be made according to gross tonnage. A ship-count database that can be automatically accessed by ambient noise models is the HITS database, which is incorporated in AUTOSHIPS (see Chapter 11). The resolution of HITS is 1° latitude-longitude squares by ship type. Spatial coverage is essentially worldwide (Estalote et al., 1986).

The complexity of the databasing task can be better appreciated by examining Figure 8.8, which shows generalized shipping routes from Solomon et al. (1977). The routes are generalized in the sense that the lines connecting the various ports are not necessarily indicative of the exact paths followed by the merchant traffic. (The numbering of the various routes in Figure 8.8 served as a bookkeeping mechanism for tracking port destinations.) Factors such as seasonal climates

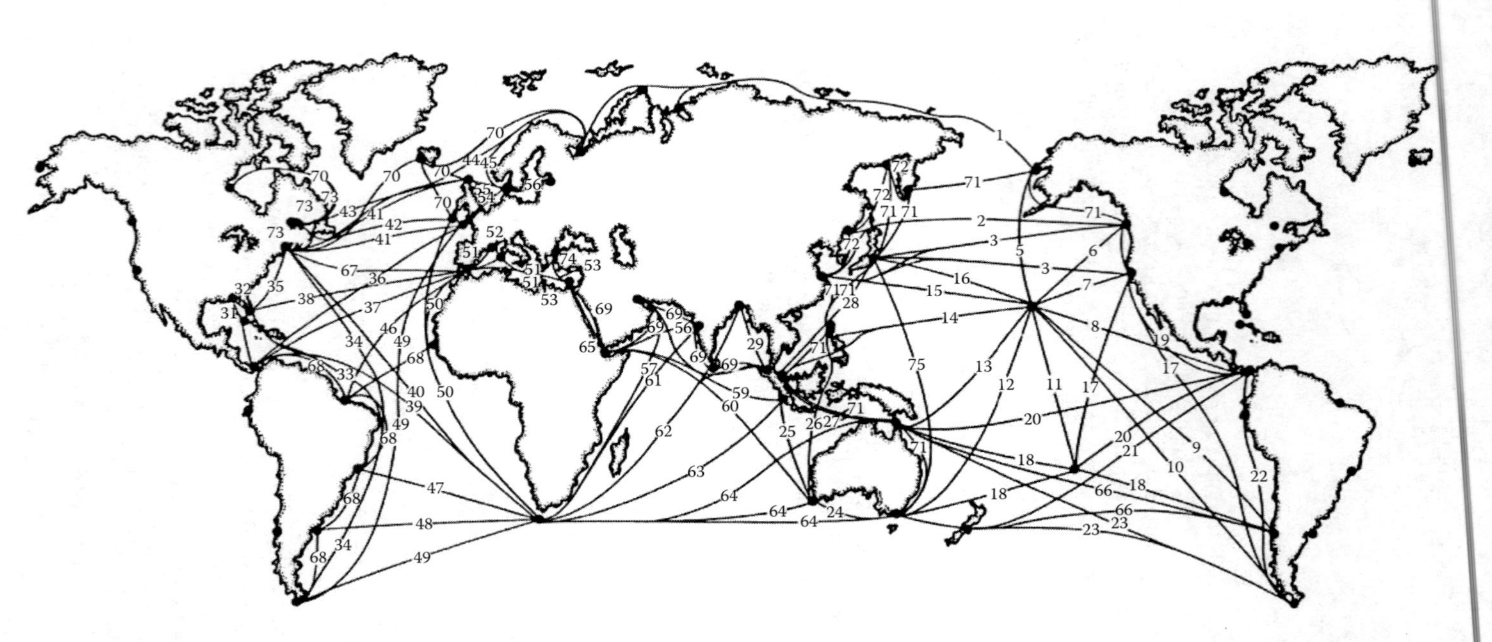

FIGURE 8.8 Generalized shipping routes. (Adapted from Solomon, L.P. et al., Ocean route envelopes. Planning Syst., Inc., Technical Report TR-036049, 1977.)

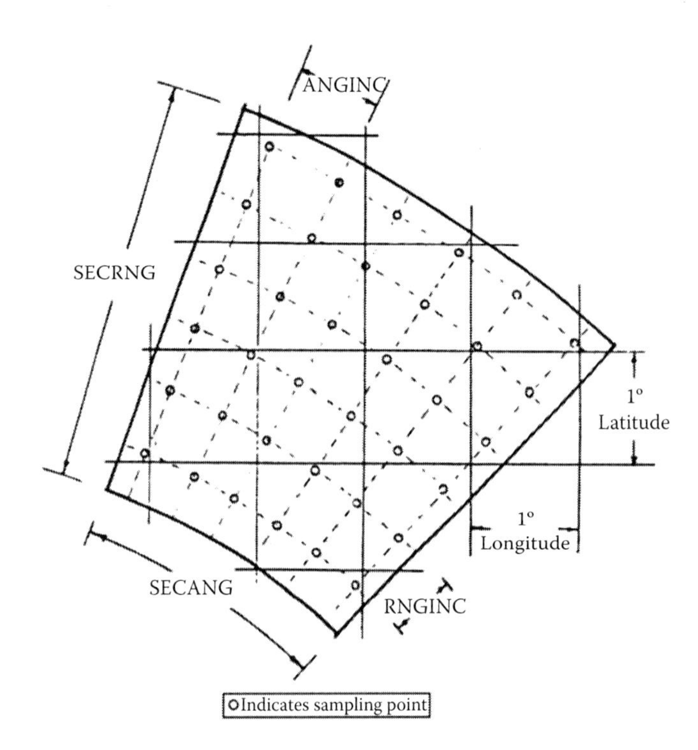

FIGURE 8.9 Mapping transformation from a rectangular to a polar coordinate system. Shipping density data are usually represented on a Cartesian coordinate system while noise models utilize a polar (sector) reference frame. This particular example is from the FANM noise model. (Adapted from Baker, C.L. 1976. FANIN Preprocessor for the FANM-I Ambient Noise Model. Ocean Data Syst., Inc.)

and severe weather patterns can alter the exact routes followed by the ships at any particular time.

Mapping transformations between rectangular (for the ship positions and routes) and polar (for noise model sector geometry applications) further complicate model implementation. Figure 8.9 illustrates one particular implementation that relates polar sector geometries to Cartesian latitude and longitude divisions. Here, a sector is defined in terms of sector angle (SECANG), sector range (SECRNG), angle increment (ANGINC), and range increment (RNGINC).

An example of the spatial variability of ambient noise levels generated by a noise model is presented in Figure 8.10 for the Mediterranean Sea. At low frequencies (<100 Hz), most noise is due to shipping traffic. The regions of high noise levels (denoted by H) coincide with established shipping routes (compare with Figure 8.8). Noise variations of as much as 20 dB can occur over relatively short distances.

8.8 NUMERICAL MODEL SUMMARIES

A summary of available noise models is presented in Table 8.1. This summary segregates the models according to the categories of ambient noise and beam-noise statistics. Models in the category of beam-noise statistics are further segregated according

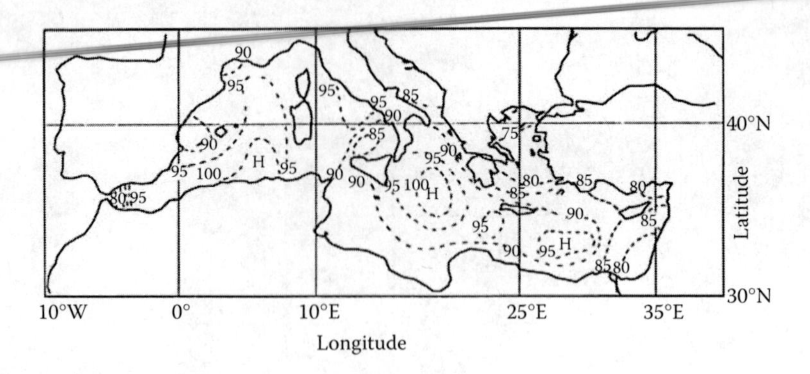

FIGURE 8.10 Ambient noise level (spectrum level, dB re 1 μPa) variations in the Mediterranean Sea as simulated by a noise model. Simulation is valid for low frequency (<100 Hz) noise for a shallow receiver depth during the winter season. The contour interval is 5 dB and H signifies areas of high noise levels.

to analytic and simulation approaches. Numbers within brackets following each model refer to a brief summary and appropriate documentation. Model documentation can range from informal programming commentaries to journal articles to detailed technical reports containing a listing of the actual computer code. Abbreviations and acronyms are defined in Appendix A. This summary does not claim to be exhaustive.

TABLE 8.1
Summary of Underwater Acoustic Noise Models

	Beam Noise Statistics	
Ambient Noise	**Analytic**	**Simulation**
AMBENT [1]	BBN Shipping Noise [18]	BEAMPL [23]
ANDES [2]	BTL [19]	DSBN [24]
ARAMIS [3]	PSSEL [20]	NABTAM [25]
CANARY [4]	Sonobuoy Noise [21]	SIAM–I/II [26]
CNOISE [5]	USI Array Noise [22]	
DANES [6]		
DANM [7]		
DINAMO [8]		
DUNES [9]		
FANM [10]		
HAMMER [11]		
INSPIRE [12]		
ISAAC [13]		
MONM [14]		
Normal Mode Ambient Noise [15]		
Quonops© [16]		
RANDI–I/II/III [17]		

(Continued)

TABLE 8.1 (Continued)
Summary of Underwater Acoustic Noise Models

Notes:

AMBIENT NOISE

[1] AMBENT calculates the ambient noise level for a cylindrically symmetric beam generated by a uniform distribution of surface radiators. Propagation effects are computed by NISSM (McConnell, 1983; Robinson and McConnell, 1983). AMBENT is intended to determine the noise level due to wind or rain, or both.

[2] ANDES (Version 4.2) addresses issues related to shallow-water ambient-noise modeling including upgrades to the shipping-density and sound-speed databases, in addition to a new capability to model fluctuations in noise directionality due to changes in wind speed and the movement of discrete sources through the transmission-loss field (Renner, 1986a,b; 1988; 1995a,b).

[3] ARAMIS consists of a number of FORTRAN, C++, and MATLAB codes that integrate US Navy standard databases with user-provided sonar system parameters to assess the performance of passive spatial processors (Arvelo, 1998).

[4] CANARY is a ray-based model of ambient noise and noise coherence that is used to estimate the performance of sonars in range-dependent and azimuth-dependent environments. CANARY treats noise sources as surface distributions rather than points (Harrison, 1997a,b; Harrison et al., 1999; Harrison et al., 2001). (Also see the DINAMO model.)

[5] CNOISE, a simple analytical ambient noise model, was developed to address a sampling problem that resulted in slow convergence of noise levels predicted by a Monte Carlo low-frequency underwater acoustics noise model (SIAM II). Inputs to CNOISE are azimuthal sector geometry about a receiver, ship counts in range-sector bins, transmission loss curves for each sector, and ship source levels. Outputs from CNOISE include horizontally directional and omnidirectional ambient noise levels. The major advantages of this model include: significantly increased execution speed relative to Monte Carlo noise models, analytical solutions that lend themselves to straightforward analysis and implementation, and the avoidance of the sampling problems that appear to plague Monte Carlo models. Comparisons of noise levels predicted by other models (BEAMPL and FANM) have shown very good agreement (Cornyn, 1980). Transmission loss versus range files must be generated externally (Estalote, 1984).

[6] DANES generates noise levels and horizontal directionality estimates for shipping traffic and wind noise (Osborne, 1979; Lukas et al., 1980b).

[7] DANM (dynamic ambient noise model) is the successor to ANDES. DANM predicts the azimuthal dependence of noise in the 25–5,000 Hz band (National Research Council, 2003a). The DANM provides a realistic simulation of the temporal noise field in which a passive receive array operates. The total noise field is obtained by separately calculating wind and shipping noise. The temporal variability of the noise field is simulated by moving merchant ships along major shipping lanes. Shipping databases provide seasonal information about shipping lanes between the world's major ports, as well as the type and number of ships that move in the lanes.

[8] DINAMO models three-dimensional (3D) noise directionality and array performance for operational applications (Harrison, 1998). DINAMO is closely related to CANARY, which was designed for research use. While CANARY first calculates the correlation matrix for the array and then sums these terms to form the array response, DINAMO performs a straightforward integral over all solid angles of the calculated noise directionality multiplied by the array's beam pattern.

(*Continued*)

TABLE 8.1 (Continued)
Summary of Underwater Acoustic Noise Models

[9] DUNES provides estimates of omnidirectional, vertical, horizontal, and 3D directional noise versus frequency. The model includes high-latitude and slope-enhanced wind noise effects. The model emphasizes calculation of noise due to the natural environment. Therefore, shipping contributions are entered explicitly and not via extensive shipping databases (Bannister et al., 1989).

[10] FANM uses a simplified (range-independent) ocean environment together with shipping and wind speed databases to predict ambient noise at a fixed receiver location (Cavanagh, 1974a,b; Lasky and Colilla, 1974; Baker, 1976; Long, 1979).

[11] HAMMER (hydroacoustic model for mitigation and ecological response) was developed by HR Wallingford Ltd. (2016), together with Loughborough University. Further development of this tool is underway in collaboration with Bristol and Exeter Universities to determine behavioral response of fish species to underwater noise. The model is based on the PE method (Collins, 1993a). HAMMER also incorporates an ecological response model to investigate the impact of underwater noise on species behavior. Individuals are represented as a particle and the model predicts movement patterns, given initial input parameters of swimming-speed limits, direction, source level intensity, and frequency. The model predicts the propagation of noise from the source point in a two-dimensional (2D) vertical slice, which determines the noise levels throughout the water column in a line out from the source, covering a full 360°. This allows sound maps of underwater radiated noise to be created so that the noise levels can be understood and interpreted for environmental impact assessments (EIA). Using agent-based modeling, HAMMER is able to predict the response of a target species to underwater noise if enough behavioral data are available for the target species. Model input data include: hearing ability, migration route, swim speeds, size of individuals, temperature and salinity tolerances of the target species, schooling behavior, and swimming depth.

[12] INSPIRE (impulse noise sound propagation and impact range estimator) model was developed by Subacoustech Ltd. specifically to model the propagation of impulsive broadband underwater noise in shallow waters (Mason, 2013). It uses a combined geometric energy-flow and hysteresis-loss model to conservatively predict propagation in relatively shallow coastal water environments. It has been tested against actual results from a large number of offshore wind-farm piling operations. The model is able to provide a wide range of physical outputs, including the peak pressure, M-Weighted sound-exposure level (SEL), and the *dBht(species)*. Transmission losses are calculated by the model on a fully range- and depth-dependent basis. The INSPIRE model imports electronic bathymetry data as a primary input to determine the transmission losses along transects extending from the pile location which has been input in addition to other simple physical data such as tide height. Thompson et al. (2013) noted that offshore wind-farm developments may impact protected marine mammal populations and thus require appropriate assessment under the EU Habitats Directive. They described a framework for assessing population-level impacts of disturbance from piling noise on a protected harbor seal population in the vicinity of a proposed wind-farm development in the North Sea off northeast Scotland. The predicted propagation of noise resulting from the piling operations (that were required to install the wind turbine foundations) was modeled using INSPIRE.

[13] ISAAC uses a Gaussian ray-tracing approach to determine the acoustic ray paths between source and target, including those reflected from the sea surface and sea-bed. The ray paths may refract with changes in bathymetry, water density, salinity, temperature, and seabed type (Barker, 2004).

(Continued)

TABLE 8.1 (Continued)
Summary of Underwater Acoustic Noise Models

ISAAC allows sensitive marine areas such as marine mammal locations, migratory routes, and fisheries to be displayed in the GIS alongside acoustic propagation results. Noise impacts on individual species can be assessed by comparing the sound pressure levels generated from anthropogenic activities with sensitivity thresholds to perform environmental risk assessments (ERA). The system has been configured specifically for use by offshore industries, environmental agencies, regulators, and others to help assess the environmental impact of underwater noise. The *dBht(species)* approach provides a measurement of sound that accounts for interspecies differences in hearing ability by passing the sound through a filter which mimics the hearing ability of the species. The *dBht(species)* metric is a pan-specific metric incorporating the concept of "loudness" by using a frequency-weighted curve based on the species' hearing threshold as the reference unit for a dB scale. A large number of both field and controlled-laboratory measurements have been made of the avoidance of a range of idealized noises, using fish with greatly different hearing as a model. All data, irrespective of source or species, indicate a dependence of avoidance reaction on the *dBht(species)* level. The data indicate three regions: no reaction below 0dBht (i.e., below the species' threshold of hearing), a cognitive avoidance region where increasing numbers of individuals will avoid the noise from 0 to 90 dBht, and instinctive reaction at and above 90 dBht where all animals will avoid the noise. This probabilistic model allows the behavioral impact of any noise source to be estimated (Nedwell et al., 2005). This approach provides an indication of the noise level that will be received for the species at various distances from the noise source. Nedwell et al. (2007) offered validation of a frequency-weighted scale, the *dBht(species)*, as a metric for the assessment of the behavioral and audiological effects on underwater animals of anthropological underwater noise. These values can then be compared to published data to indicate distances at which a species will demonstrate a strong avoidance reaction, a temporary elevation of hearing threshold, or a permanent elevation of hearing threshold (BMT Cordah EIA and Environmental Services, 2010).

[14] Marine operations noise model (MONM) incorporates a range-dependent, split-step PE acoustic model including a shear-wave computation capability. MONM has been used for precise estimation of noise produced by sub-sea construction noise, marine facilities operation, and seismic exploration, particularly in complex coastal regions. The core algorithm in MONM computes frequency-dependent acoustic transmission loss parameters along fans of radial tracks originating from each point in a specified set of source positions. The modeling is performed in individual one-third-octave spectral bands covering frequencies from 10 Hz to several kHz, which covers the overlap between the auditory frequency range of marine mammals and the spectral region in which sound propagates significantly beyond the immediate vicinity of the source. The MONM software makes use of geo-referenced databases to automatically retrieve the bathymetry and acoustic-environmental parameters along each propagation traverse, and incorporates a tessellation algorithm that increases the angular density of modeling segments at greater ranges from a source to provide more computationally efficient coverage of the area of interest. The grid of transmission-loss values produced by the model for each source location is used to attenuate the spectral acoustic output levels of the corresponding noise source to generate absolute received sound levels at each grid point. These are then summed across frequencies to provide broadband levels. A further step of Cartesian resampling and summing of the received noise levels from all the sources in a modeling scenario yields the aggregate noise level for the entire operation on a regular grid from which contours can be drawn on a GIS map. The model can either

(Continued)

TABLE 8.1 (Continued)
Summary of Underwater Acoustic Noise Models

	generate contours at evenly spaced levels or draw boundaries representing biologically significant threshold levels (Laurinolli et al., 2005). At frequencies ≤2 kHz, MONM computes acoustic propagation via a wide-angle PE solution to the acoustic wave equation based on a version of the US Naval Research Laboratory's Range-dependent Acoustic Model (RAM), which has been modified to account for an elastic seabed. MONM-RAM accounts for the additional reflection loss at the seabed due to partial conversion of incident compressional waves to shear waves at the seabed and subbottom interfaces, and it includes wave attenuations in all layers. MONM-RAM incorporates the following site-specific environmental properties: a modeled area bathymetric grid, underwater sound speed as a function of depth, and a geoacoustic profile based on the overall stratified composition of the seafloor (Matthews and Zykov, 2012; Zykov, 2013). At frequencies ≥2 kHz, MONM employs the BELLHOP Gaussian beam ray-trace propagation model and accounts for increased sound attenuation due to volume absorption at these higher frequencies. MONM-BELLHOP accounts for the source directivity specified as a function of both azimuthal angle and depression angle. MONM-BELLHOP incorporates the following site-specific environmental properties: a bathymetric grid of the modeled area and underwater sound speed as a function of depth. In contrast to MONM-RAM, the geoacoustic input for MONM-BELLHOP consists of only one interface, namely the sea bottom. This is an acceptable limitation because the influence of the subbottom layers on the propagation of acoustic waves with frequencies above 2 kHz is negligible. MONM computes acoustic fields in 3D by modeling transmission loss (via BELLHOP or RAM) within two-dimensional (2D) vertical planes aligned along radials covering a 360° swath from the source, an approach commonly referred to as N × 2D. These vertical radial planes are separated by an angular step size of $\Delta\theta$, yielding $N = 360°/\Delta\theta$ number of planes. MONM treats frequency dependence by computing acoustic transmission loss at the center frequencies of ⅓-octave bands. An adequate number of one-third octave bands, starting at 10 Hz, are modeled to include the majority of acoustic energy emitted by the source. At each center frequency, the transmission loss is modeled via BELLHOP or RAM within each vertical plane (N × 2D) as a function of depth and range from the source. One-third octave band-received SELs are computed by subtracting the band transmission loss values from the directional source level (SL) in that frequency band. Composite broadband received SELs are then computed by summing the received one-third octave band levels.
[15]	Normal Mode Ambient Noise calculates the relative noise level versus depth in addition to the coherence of the noise field at any two points. This calculation is performed using a normal-mode representation of the acoustic field (Kuperman and Ingenito, 1980).
[16]	Quonops© is a forecasting system for anthropological noise developed by Quiet-Oceans (Guelton et al., 2013). It relies on the acoustic propagation models RAM and BELLHOP. Starting with a given environmental situation that summarizes oceanographic and meteorological information, Quonops© computes the acoustic field induced by anthropological sources such as ships, windmills, or pile driving for the area of interest. The accumulated acoustic pressure in a 3D mesh is then obtained for the selected area. The computation may involve very large areas and several hundred acoustic sources. Fields are generated offline for case studies, or online for area monitoring. Offline studies require a large amount of computation involving thousands of simulations due to the use of a Monte-Carlo model. Online studies are less computationally intensive, but require tight scheduling. In both cases, high performance computing (HPC) enables the delivery of comprehensive reports or real-time monitoring. Quonops© uses three layers of parallelism via

(*Continued*)

TABLE 8.1 (Continued)
Summary of Underwater Acoustic Noise Models

distributed computing employing Caparmor III, a cluster hosted at French Research Institute for Exploitation of the Sea (Ifremer), to handle multiple simultaneous simulations. By using multi-threading, the time constraint for online simulations becomes achievable. Moreover, the manual vectorization of the computation kernels yields additional speedups that benefit both situations.

[17] RANDI–I/II/III—The original version, RANDI–I (Wagstaff, 1973) calculates and displays the vertical and horizontal directionality of ambient noise in the frequency range 10 Hz–10 kHz. RANDI–II (Hamson and Wagstaff, 1983) was constructed at the SACLANTCEN to account for the special nature of ambient noise in shallow-water environments. RANDI–III (Version 3.1) predicts ambient-noise levels and directionalities at low-to-mid frequencies in both shallow and deep water. Shipping noise can be calculated for highly variable environments using either the finite-element (FE) or split-step PE method. Local wind noise is computed using the range-independent theory of Kuperman-Ingenito, including both discrete normal modes and continuous spectra. US Navy standard and historical databases are used to describe the environment (Schreiner, 1990; Breeding, 1993; Breeding et al., 1994, 1996). Version 3.3 is a modified version of RANDI 3.1 for use in shallow water. This version provides the user with the option to supply the model with measured or estimated environmental information in areas where the US Navy standard databases may not provide coverage (Pflug, 1996). (The RANDI model is discussed in detail in Section 8.4.)

BEAM NOISE STATISTICS

Analytic

[18] BBN Shipping Noise calculates probability density functions of the beam noise power envelope using acoustic source level data for classes of surface ships, shipping routes, and traffic density along those routes (Mahler et al., 1975; Moll et al., 1977, 1979).

[19] BTL provides statistical descriptions of shipping noise for low-frequency, horizontally beamed systems (Goldman, 1974).

[20] Gervaise et al. (2015) proposed a new method for rapid computation of high-resolution shipping SEL in a single mapping step. The probabilistic shipping sound exposure level (PSSEL) modeling approach incorporates a framework for simulating the statistical distribution of SEL generated by shipping traffic, which is often required in assessing the impact of anthropogenic sounds on marine life. The SEL distribution is intrinsically computed by PSSEL without Monte Carlo simulations. Rather, the PSSEL framework is based on classical shipping-noise simulation tools using shipping-traffic descriptions derived from the automatic identification system (AIS), transmission-loss computing codes, and ship source-level models. The PSSEL outputs include a map of the proportion of time a SEL threshold criterion is exceeded. This output can be directly incorporated into marine spatial planning tools by nonacoustic specialists and then combined with other environmental layers in support to management strategies and regulations.

[21] Sonobuoy Noise was developed for sonobuoy applications (McCabe, 1976). It considers both the temporal correlation of ship-generated noise and the spatial correlation of average intensities for distributed sensors.

[22] USI Array Noise numerically estimates the ensemble and time-averaged, one-dimensional statistical probability density function of beam noise (Jennette et al., 1978).

Simulation

[23] BEAMPL computes random ambient noise time series within a user specified beam by statistically taking into account the motion of ships along user specified routes (Estalote, 1984).

(Continued)

TABLE 8.1 (Continued)
Summary of Underwater Acoustic Noise Models

[24] DSBN generates beam-noise time series from component submodels for surface ships, transmission loss, and receiver (Cavanagh, 1978a,b).

[25] NABTAM computes the response of a linear array of hydrophones to wind-sea interactions, surface ships, and designated target vessels (W. Galati, E. Moses, and R. Jennette, unpublished manuscript).

[26] SIAM II (S.C. Wales, unpublished manuscript) was designed to provide many replications of surface-ship noise for horizontal array systems, particularly narrow-beam systems, but could also be used for omnidirectional systems. Its predecessor, SIAM I (Marshall and Cornyn, 1974a,b), predicted ship-generated noise over the band 20–120 Hz by generating many replications so that ensemble statistics could be examined. A report by Science Applications, Inc. (1977) provides more details on SIAM (I/II) in addition to other legacy beam-noise statistics models. Long (1979) described plotting packages for SIAM.

9 Reverberation I: Observations and Physical Models

9.1 BACKGROUND

Reverberation is defined as that portion of the sound received at a hydrophone that is scattered by the ocean boundaries or by volumetric inhomogeneities. Accordingly, reverberation-producing scatterers in the sea can be grouped into three classes: sea surface, seafloor, and ocean volume. Surface and bottom reverberation both involve a two-dimensional (2D) distribution of scatterers and therefore can be considered jointly as boundary reverberation. Volume reverberation is produced by the marine life and inanimate matter distributed within the sea, and also by fine-scale features of the ocean itself. Useful reviews of oceanic scattering and reverberation have been provided by Andersen and Zahuranec (1977), Farquhar (1970), Ellis et al. (1993), and Pierce and Thurston (1993). A collection of papers dealing with high-frequency acoustics in shallow water (Pace et al., 1997) addressed issues relating to scattering and reverberation in shallow water. Love et al. (1996) noted that variability is the principal feature of volume reverberation in littoral waters. Hamilton et al. (2002) reviewed physical scattering mechanisms and applications in fish-stock assessments, fish migration, seabed and habitat characterization, inferences of turbidity, measurements of waves and currents, and detection of objects. A research monograph by Jackson and Richardson (2007) is part of a new book series sponsored by the US Office of Naval Research (ONR) on the latest research in underwater acoustics. This volume provides a critical evaluation of the data and models pertaining to high-frequency acoustic interaction with the seafloor.

Reverberation has several features that distinguish it from noise (Bartberger, 1965; Moritz, 1982). Principal among these is the fact that reverberation is produced by the sonar itself. Thus, the spectral characteristics of reverberation are essentially the same as the transmitted signal. The intensity of reverberation varies with the range of the scatterers and also with the intensity of the transmitted signal.

Based on its characteristic temporal and spatial correlation properties, oceanic reverberation can be segregated according to diffuse and facet components (Gerstoft and Schmidt, 1991). Diffuse reverberation results from scattering by the small-scale, stochastic structure of the oceanic waveguide (e.g., surface and bottom roughness and bottom inhomogeneity). Facet reverberation results from abrupt changes in the bathymetric and subbottom features of the ocean (e.g., seamounts and faults). Due to its stochastic nature, diffuse reverberation is characterized by a relatively low correlation. Alternatively, the deterministic nature of oceanic facets gives rise to "signal-like" reverberation from seamounts and faults.

The fundamental ratio on which reverberation depends is called the scattering (or backscattering) strength (Urick, 1983: Chapter 8). It is the ratio (in dB) of the intensity of sound scattered by a unit area (for boundary reverberation) or volume (for volume reverberation), referred to a distance of 1 m, to the incident plane-wave intensity:

$$S_{b,v} = 10 \log_{10} \frac{I_s}{I_i} \tag{9.1}$$

where

$S_{b,v}$ = scattering strength for boundary (b) or volume (v) reverberation
I_s = intensity of sound scattered by a unit area or unit volume of water
I_i = intensity of incident plane wave

The computation of volume and boundary reverberation levels based on scattering strengths is discussed in Chapter 10.

9.2 VOLUME REVERBERATION

The major source of volume reverberation in the sea has been established as biological (e.g., Johnson et al., 1956). Different marine organisms affect different bands of the active sonar spectrum. At frequencies in excess of 30 kHz, the scatterers are zooplankton. At frequencies between 2 kHz and 10 kHz, the dominant scatterers are the various types of fish that possess a swim bladder (an air-filled sac that enables fish to maintain and adjust their buoyancy). Acoustically, the bladder amounts to an internal air bubble that becomes resonant at a frequency depending on the size and depth of the fish. Love (1978) reviewed the current understanding and modeling of swim bladder resonant acoustic scattering.

Saenger (1984) developed a volume scattering strength model that is valid over the frequency range 1–15 kHz. The depth variability is a function of seasonal bio-acoustic constants (organized by ocean province), the mean water density profile and its gradient, and the acoustic frequency. The absolute level of the volume scattering strength depends on the seasonal standing crop of scatterers, which may exhibit interannual variability.

Love (1975) developed a volume reverberation model based on fish distribution data. The use of fishery data to predict volume reverberation was explored further by Love (1993). In 1988 and 1989, volume reverberation measurements were made in the Norwegian Sea and North Atlantic Ocean in the frequency range 800 Hz to 5 kHz. Below 5 kHz, volume reverberation is usually caused by scattering from the swim bladders of relatively large fish. The scattering strength (S_L) of a layer of dispersed, nonacoustically interacting fish is

$$S_L = 10 \log_{10} \sum_{i=1}^{n} \sigma_i(f) \times 10^{-4}$$

where

n = number of fish in the layer

σ = acoustic cross section of an individual fish (cm^2) at any given frequency (f)

The parameter σ represents a swim bladder-bearing fish as a spherical shell enclosing an air cavity in water. Love (1993) concluded that a possible pitfall of relying solely on fishery data to predict low-frequency volume reverberation is that fishery research concentrates on species of present or potential commercial value. Species of no commercial value could be ignored in such data, despite their potential contribution to low-frequency volume reverberation.

Raveau and Feuillade (2016) investigated resonance scattering from schools of fish (with gas-filled swim bladders) as a function of frequency and azimuth. Calculations were performed using both the effective medium method and the coupled differential equation model incorporating both multiple scattering between fish and wave interference interactions of their scattered fields. A comparison with forward scattering data showed very good agreement for both approaches, thus indicating a method for estimating fish abundance. For backscattering data, the effective medium method diverged strongly when the wavelength $\lambda < 4s$, where s is the average nearest neighbor fish separation.

9.2.1 Deep Scattering Layer

Measurements of the depth variation of volume scattering strengths show an overall decrease with increasing depth. This is consistent with the general distribution of biological organisms within the sea. However, there is often a well-marked increase at certain depths. The depth of such increased volume scattering is called the deep scattering layer (DSL).

The DSL generally exhibits a diurnal migration in depth, being at greater depth by day than by night, and with a rapid change near sunrise and sunset. The depth of the DSL can be expected to lie between 180 m and 900 m by day in midlatitudes, and to be shallower by night. In the Arctic, the DSL lies just below the ice cover.

Greene and Wiebe (1988) reported on the use of high-frequency (420 kHz), dual-beam acoustics to measure target strength distributions of zooplankton and micronekton. Their results are consistent with the present picture of the diel vertical migration of sound-scattering layers in the ocean. At this frequency, the most abundant targets (off the northeast coast of the United States) corresponded to animals the size of mature krill with target strengths between –71 dB and –62 dB. The use of acoustic techniques to estimate biomass has also been explored (e.g., Penrose et al., 1993; Wiebe et al., 1995).

From an examination of the relationship between the scattering layer and the mixed layer in shallow waters (<150 m) of the Arabian Sea in December 2003 under conditions of varying thermocline strength, it was concluded that the scattering layer was best defined when the thermocline was strongest (Kumar et al., 2005). Strong thermoclines restricted the transfer of scatterers below this zone, thus forming a strong scattering layer just above the thermocline. When the thermocline was weak, the scatterers penetrated below the thermocline and the resulting scattering layer was weaker and more diffuse.

Benoit-Bird and Au (2002) validated the relationship between acoustic backscattering strength and biomass. Moreover, they demonstrated an ability to directly convert acoustic energy from a class of animals to organic-resource units without having knowledge of the size distribution of the population being studied. Diachok et al. (2001) used measurements of bioacoustic absorptivity to estimate fish number densities.

9.2.2 Column or Integrated Scattering Strength

Some kinds of scatterers in the sea, such as the DSL or a layer of air bubbles just below the sea surface, lie in layers of finite thickness instead of being diffusely distributed throughout an irregular volume. The resultant layered reverberation is more easily considered as a form of boundary reverberation. When the layer is made infinitely great, the quantity S_b in Equation 9.1 becomes the scattering strength of the entire water column and is called the column, or integrated, scattering strength. It can be readily measured in the field by means of explosive sound signals and sonobuoys, and it is a convenient single number to characterize the total amount of volume reverberation existing at the time and location the data were obtained.

Measurements of volume reverberation have been obtained, for example, by Vent (1972) and by Gold and Renshaw (1978). Figure 9.1 compares day and night profiles of volume scattering versus depth during January 1970 in the eastern North Pacific Ocean at three frequencies: 3.5 kHz, 5 kHz, and 12 kHz (Vent, 1972). The

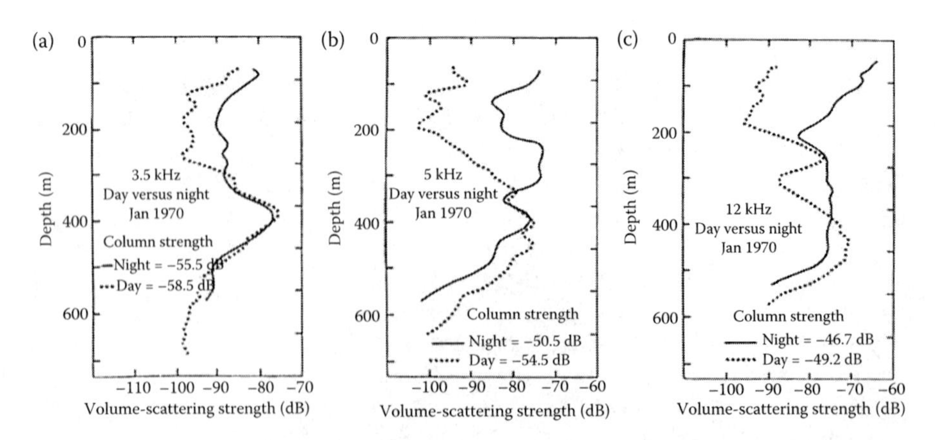

FIGURE 9.1 Comparison of day and night profiles of volume scattering strength versus depth during January 1970 in the eastern Pacific Ocean: (a) 3.5 kHz, (b) 5 kHz, and (c) 12 kHz. Column strengths are calculated by integrating the volume scattering strengths over the entire profile. (From Vent, R.J., *J. Acoust. Soc. Am.*, **52**, 373–382, 1972. With permission.)

corresponding column strengths are also indicated. Note the large differences in scattering strengths between night and day. It is important to note that different reverberation models require different characterizations of volume scattering strengths: some require input as a profile versus depth while others require input as a column scattering strength.

9.2.3 Vertical-Scattering Plumes

Pelagic organisms tend to favor horizontal orientations, thus forming what are termed scattering layers. In certain areas of the oceans, the seepage of gas or liquids from the seabed produces vertical scattering plumes that are detectable over the frequency range 10–1000 kHz (Hovland, 1988). A special case of particular interest to underwater acousticians concerns the plumes formed over some seamounts. A scattering plume over the Hancock Seamount in the Hawaiian chain was investigated using an echo sounder operating at 38 kHz. This plume was detectable over a vertical extent of approximately 300 m directly above the seamount. These seepages appear to attract various marine organisms that might further complicate the scattering dynamics in the water column overlying seamounts.

Auner (2015) inspected water-column backscatter data and identified more than 200 possible seep locations on the northwestern Gulf of Mexico continental slope. High-resolution 30 kHz multibeam echosounder data were used to detect bubble plumes floating up through the water column. Seeping hydrocarbons may originate from biogenic or thermogenic source gases. Seeps occurred throughout the range of depths surveyed (approximately 300 m to 1600 m), most commonly along the edges of kilometer-scale seafloor features including mini-basins, ridges, and domes. Seep distribution also appeared to be correlated with areas of elevated seafloor backscatter. These trends in seep distribution were consistent with a model of hydrocarbon migration facilitated by salt-related faults and the use of elevated seafloor backscatter as an indicator of possible methane seeps.

The relative importance of suspended particles and turbulence as backscattering mechanisms within a hydrothermal plume located on the Endeavour Segment of the Juan de Fuca Ridge was determined by comparing acoustic backscatter measured by the cabled observatory vent imaging sonar (COVIS) with model calculations based on *in situ* samples of particles suspended within the plume (Xu et al., 2017). The primary component of COVIS was a state-of-the-art imaging sonar that has two transmitting/receiving pairs: a 396 kHz pair used in imaging and Doppler modes for three-dimensional (3D) plume imaging and flow-rate quantification, respectively; and a 200 kHz pair used in diffuse-flow mode for 2D mapping of lower-temperature hydrothermal discharge. Analysis of plume samples yielded estimates of the mass concentration and size distribution of particles, which was used to quantify their contribution to acoustic backscatter; the number of particles with radii much smaller than 1 μm far exceeded the number of larger particles. The result showed negligible effects of plume particles on acoustic backscatter within the initial 10-m rise of the plume. The theoretically estimated backscatter from plume particles based on their estimated size distribution and mass concentration was approximately two orders of magnitudes smaller than the observed backscatter. This suggested that turbulence-induced temperature fluctuations were the dominant backscattering mechanism within lower levels of the plume. Furthermore, inversion of the observed acoustic backscatter for the standard deviation of temperature within the plume yielded a reasonable match with the *in situ* temperature measurements made by a conductivity-temperature-depth (CTD) instrument. This finding showed that turbulence-induced temperature fluctuations were the dominant backscattering mechanism and further demonstrated the potential of using acoustic backscatter as a remote-sensing tool to measure the temperature variability within a hydrothermal plume.

9.3 BOUNDARY REVERBERATION

9.3.1 Sea-Surface Reverberation

The roughness of the sea surface and the presence of trapped air bubbles make the sea surface an effective but complex scatterer of sound. Scattering can occur out-of-plane as well as within the vertical plane containing the source and receiver.

Sea-surface scattering strengths have typically been measured using nondirectional (mostly explosive) sources and receivers as well as directional sonars in which a sound beam is formed so as to intercept the sea surface at a desired angle. The scattering strength of the sea surface has been found to vary with grazing angle, acoustic frequency, and the roughness of the surface. Sea-surface roughness is usually characterized either by the near-surface wind speed or by wave heights. The measured scattering strengths show a strong variation with frequency at low frequencies and low grazing angles, and little variation with frequency at high frequencies and high grazing angles.

A series of measurements conducted by Chapman and Harris (1962) were analyzed in octave bands between 0.4 kHz and 6.4 kHz, and the results were fitted by the empirical expressions

$$S_s = 3.3\beta \log_{10}\frac{\theta}{30} - 42.2\log_{10}\beta + 2.6 \tag{9.2}$$

$$\beta = 158\left[vf^{1/3}\right]^{-0.58}$$

where

S_s = surface scattering strength (dB)
θ = grazing angle (deg)
v = wind speed (knots)
f = frequency (Hz)

Chapman and Scott (1964) later validated these results over the frequency range 0.1 kHz to 6.4 kHz for grazing angle below 80°. Figure 9.2 presents curves computed using Equation 9.2 at several frequencies with wind speed as a parameter.

McDaniel (1993) reviewed recent advances in the physical modeling of monostatic sea-surface reverberation in the frequency range 200 Hz to 60 kHz (also see the extensive review by Fortuin, 1970). Three sources of surface reverberation were

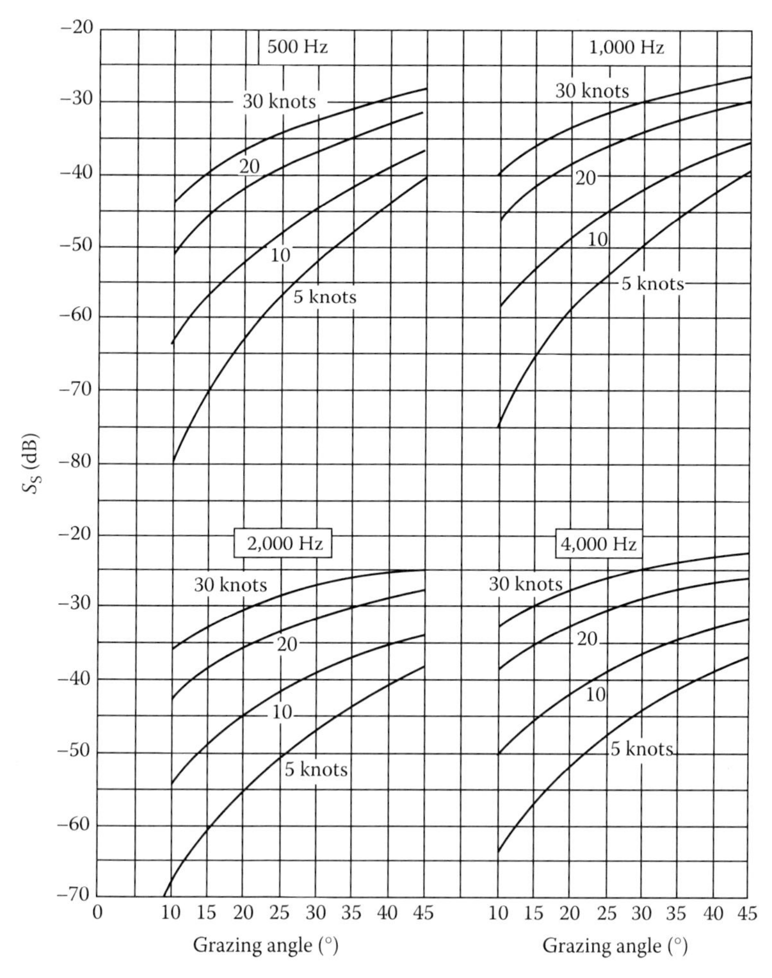

FIGURE 9.2 Low-frequency scattering strengths of the sea surface (S_S) computed from the empirical expressions given by Chapman and Harris (1962). (From Urick, R.J., *Principles of Underwater Sound*, McGraw-Hill, New York, 1983. Copyright Wiley-VCH Verlag GmbH & Co. KGaA. With permission.)

considered: rough-surface scattering, scattering from resonant bubbles, and scattering from bubble clouds (or plumes).

In the absence of subsurface bubbles, rough-surface scattering is adequately explained by the composite-roughness model (e.g., Thorsos, 1990). This model partitions treatment of surface scattering into two regimes according to the wind-wave spectrum: large-scale waves (using the Kirchhoff approximation) and small-scale waves (using a modified Rayleigh approximation). McDaniel (1993) summarized recent advances in surface reverberation modeling according to acoustic frequency.

High-Frequency Reverberation (3–25 kHz). Backscattering at grazing angles above 30° is in agreement with rough-surface scattering theories. At lower grazing angles, anomalously higher backscattering strengths are believed due to scatter from resonant microbubbles. Backscattering in coastal waters is higher (by an order of magnitude) than in the open ocean under similar wind conditions. This effect is attributed, in part, to the greater generation of microbubbles in coastal waters. McDaniel (1993) also observed that many recent backscattering-strength measurements are lower than would be predicted using the Chapman and Harris (1962) empirical model.

Low-Frequency Reverberation (<1 kHz). Anomalous scatter similar to that observed at higher frequencies is evident, apparently due to entrained air. However, the nature of the physical processes governing the scattering at these lower frequencies is not evident from an examination of the available data.

Extensive measurements of low-frequency (70–950 Hz) sea-surface backscattering strengths were made during the critical sea test (CST) experiments for grazing angles ranging from 5°–30°, and for wind speeds ranging from 1.5–13.5 m s^{-1} (Ogden and Erskine, 1994a). Analyses of these measurements revealed several regimes in the frequency-versus-wind speed (f–U) domain corresponding to at least two different scattering mechanisms. Perturbation theory (Thorsos, 1990) was found to provide adequate descriptions of the data at high frequencies for calm seas, and at lower frequencies for all wind speeds, where air-water interface scattering is the dominant mechanism. The Chapman-Harris empirical relationship adequately described surface backscattering for rougher seas at higher frequencies where scattering from bubble clouds is presumed to dominate the scattering process. In the transition region where these two effects are competing, the scattering strengths depended on the details of the surface and wind characteristics.

Ogden and Erskine (1994a) proposed a formula for computing the total scattering strength at the sea surface (S_{total}) as a combination of perturbation theory (S_{pert}) and the Chapman-Harris empirical relationship (S_{CH}):

$$S_{\text{total}} = \alpha S_{\text{CH}} + (1-\alpha) S_{\text{pert}} \tag{9.3}$$

$$S_{\text{pert}} = 10 \log_{10} \left[1.61 \times 10^{-4} \tan^4 \theta \exp\left(-\frac{1.01 \times 10^6}{f^2 U^4 \cos^2 \theta} \right) \right] \tag{9.4}$$

where

S_{CH} = Chapman-Harris empirical formula (Equation 9.2)

$$\alpha = \frac{U - U_{\text{pert}}}{U_{\text{CH}} - U_{\text{pert}}} \tag{9.5}$$

This formula is valid for grazing angles (θ) less than 40°, for wind speeds (U) less than 20 m s^{-1} (measured at a height of 19.5 m above the sea surface) over the frequency range (f) 50–1000 Hz.

For practical applications, the algorithm used a minimum wind speed of 2.5 m s^{-1} since, at lower wind speeds, swell is likely to dominate the scattering process. Also, an arbitrary cutoff of 1° has been specified for the grazing angle. In the hatched area of the f–U domain of Figure 9.3, only the Chapman-Harris formula (S_{CH}) should

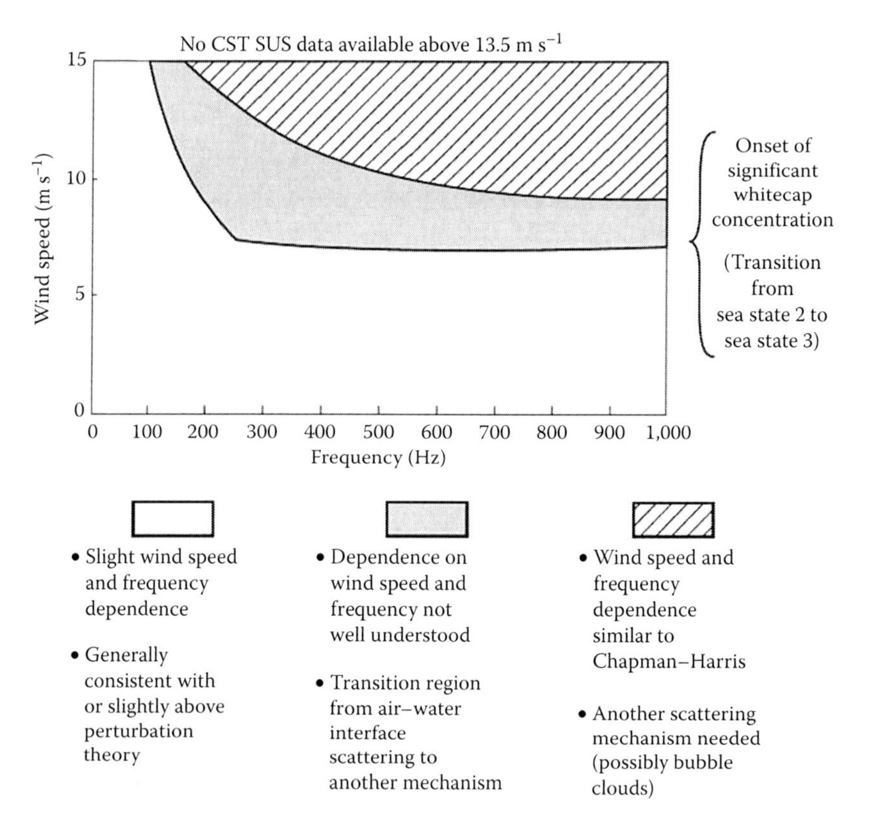

FIGURE 9.3 Frequency-wind speed (f–U) domain for sea-surface scattering strengths. (From Ogden, P.M. and Erskine, F.T., *J. Acoust. Soc. Am.*, **95**, 746–761, 1994a. With permission.)

be used (i.e., $\alpha = 1$). In the blank area of the f–U domain, only the perturbation theory formula (S_{pert}) should be used (i.e., $\alpha = 0$). In the stippled area corresponding to the transition region, the full equation (S_{total}) should be used where α is evaluated at the input frequency (f) and input wind speed (U). Figure 9.3 can be used to determine the wind speeds at the perturbation theory (lower) boundary (U_{pert}) and at the Chapman-Harris (upper) boundary (U_{CH}). For convenience, Ogden and Erskine (1994a) put these boundaries into analytical form. The lower boundary was approximated by two line segments: from 240–1000 Hz, the segment was defined by $y = 7.22$, while from 50–240 Hz the segment was given by $y = 21.5 - 0.0595x$. The upper boundary was approximated by the cubic equation

$$y = 20.14 - 0.0340x + 3.64 \times 10^{-5}x^2 - 1.330 \times 10^{-8}x^3$$

Ogden and Erskine (1994b) extended the range of environmental parameters (principally wind speed) used in modeling sea-surface backscattering strengths in the CST experiments. Related work (with minimal analysis) summarized bottom-backscattering strengths that had been measured during the CST program over the frequency range 70–1500 Hz for grazing angles ranging from 25°–50° (Ogden and Erskine, 1997). Nicholas et al. (1998) extended the analysis of surface-scattering strengths that were measured during the CST experiments over the approximate frequency range 60–1000 Hz. Unexplained variations between measured and modeled scattering strengths were attributed to an incomplete parameterization of subsurface bubble clouds.

McDaniel (2008) estimated gravity-capillary wavenumber spectra at high acoustic frequencies by applying rough-surface scattering theory to measurements of sea-surface backscatter. The ensemble-averaged scattering cross sections predicted by small-slope expansions were evaluated to examine the potential inversion of acoustic data assuming Bragg scatter. The ratio of the full fourth-order small-slope and Bragg predictions was found to exhibit a minimum value of ~2 dB at moderate angles of incidence. At such angles, the corrections to perturbation theory depend weakly on acoustic frequency and environmental conditions. This latter finding indicated that only a modest effort would be required to monitor sea-surface conditions to estimate the necessary correction. Corrections to Bragg predictions increase rapidly with increasing incidence angle. At high angles, the fourth-order contributions of the small-slope and extended small-slope expansions differ. This finding casts some doubt on the applicability of small-slope approximations to predict scattering at high-incidence angles.

Siderius and Porter (2008) described time-domain modeling approaches to sea-surface scattering problems. In particular, they developed a ray-based formulation of the Helmholtz integral equation with a time-domain Kirchhoff approximation. This approach facilitated a determination of how each ray path is modified by interaction with a rough, time-evolving sea surface. Uscinski and Stanek (2002) developed analytical expressions for the mean field of an acoustic wave scattered by a rough sea surface. The method was based on the integral equation with the single approximation that the scattering was at low angles in the forward direction. Therefore, the results were more general than the Kirchhoff approximation in this region.

Ghadimi et al. (2015a) investigated sound scattering from the sea surface in the Persian Gulf region using Chapman-Harris and Ogden-Erskine empirical relations coupled with perturbation theory. Ghadimi et al. (2016) developed the sea-surface acoustic simulator (SSAS) model based on an optimized Helmholtz–Kirchhoff–Fresnel method. This model was capable of simulating the scattering properties of the sea surface under the combined effect of sea-surface roughness and subsurface bubbles. Model outputs included scattered sound-pressure levels, sound intensity, and scattering coefficients under different environmental conditions for two different sound sources: (1) sound generated by an explosive source (model results validated using CST experiments); and (2) hydroynamic pressure generated by a moving submerged object. 3D scattered acoustic pressure levels (dB) were calculated by SSAS for both cases.

9.3.2 Under-Ice Reverberation

The detection, localization, and classification of targets with active sonars under an ice canopy is limited by reverberation from the rough under-ice surface and by the false targets presented by large ice features such as ice keels. Sea ice is the dominant cause of reverberation in Arctic regions. For undeformed first-year ice, reverberation levels at frequencies above about 3 kHz are approximately equivalent to that expected from an ice-free sea surface with a 30-knot wind.

Because of the variation in the under-ice surface, scattering strength measurements as a function of grazing angle reveal different characteristics. When the under-ice surface is relatively smooth, the scattering strength increases with grazing angle as in the open ocean. When ridge keels are present, low-grazing-angle sound waves strike them at a near-normal incidence and substantial reflection occurs. Measurements of the scattering strength of the ice-covered sea for two Arctic locations at different times of the year have been summarized by Brown (1964) and by Milne (1964). Both sets of data show an increase of scattering strength with increasing frequency and grazing angle. Using highly directional sources and receivers to exploit optimum ray paths can minimize reverberation under the ice. Such paths could provide time discrimination between target echoes and reverberation for those targets spatially separated from the sea surface (Hodgkiss and Alexandrou, 1985).

9.3.3 Sea-Floor Reverberation

The seafloor, like the sea surface, is an effective reflector and scatterer of sound. Scattering can occur out-of-plane as well as within the vertical plane containing the source and receiver. A correlation of scattering strength with the size of the particles in a sedimentary bottom has consistently been observed (e.g., McKinney and Anderson, 1964). The seafloor can then be classified according to sediment composition (sand, clay, silt) and correlated with the scattering strength. For example, mud bottoms tend to be smooth and have a low impedance contrast to water, while coarse sand bottoms tend to be rough, with a high impedance contrast. There can be large spreads in the measured data for apparently the same bottom type. This may be due in part to the refraction and reflection of sound within the subbottom sediment layers.

A Lambert's law relationship (Urick, 1983: Chapter 8) between scattering strength and grazing angle appears to provide a good approximation to the observed data for many deep-water bottoms at grazing angles below about 45°. Lambert's law refers to a type of angular variation that many rough surfaces appear to satisfy for both the scattering of sound and light. According to Lambert's law, the scattering strength varies as the square of the sine of the grazing angle. Mackenzie (1961) analyzed limited reverberation measurements at two frequencies (530 Hz and 1030 Hz) in deep water. The scattered sound responsible for reverberation was assumed to consist of nonspecular reflections obeying Lambert's law:

$$S_B = 10 \log_{10} \mu + 10 \log_{10} \sin^2 \theta \tag{9.6}$$

where

S_B = bottom scattering strength (dB)
μ = bottom scattering constant
θ = grazing angle (deg)

The term $10 \log_{10} \mu$ was found to be constant at –27 dB for both frequencies. This value has been largely substantiated by other measurements over a broad range of frequencies. The bottom scattering strength (S_B) has been graphed in Figure 9.4 according to Equation 9.6, using a value of –27 dB for $10 \log_{10} \mu$.

As part of the US Navy's acoustic reverberation special research program (ARSRP), sponsored by the ONR, an area on the northern Mid-Atlantic Ridge has been designated as a natural laboratory. This area is a natural reverberation site because its steep rock inclines, deep subsurface structures, and sediment structure provide opportunities to study many aspects of bottom reverberation. The resulting data have permitted detailed comparisons with numerical codes of bistatic reverberation (e.g., Smith et al., 1993).

Ellis and Crowe (1991) combined Lambert's law scattering with a surface-scattering function based on the Kirchhoff approximation to obtain a new functional form that allowed a reasonable extension from backscattering to a general, 3D scattering function useful in bistatic-reverberation calculations. This new functional form was tested in a bistatic version of the generic sonar model (GSM) and was shown to be an improvement over two other commonly used methods, neither of which included azimuthal dependence: the separable approximation, and the half-angle approximation.

In long-range acoustic propagation problems in the deep ocean, interactions with the seafloor generally occur at subcritical grazing angles. Under conditions of smooth basaltic surfaces and low grazing angles, most waterborne energy should be internally reflected. Furthermore, the scattering functions should be monotonic with respect to grazing angle, as would be predicted by Lambert's law. However, in ocean areas characterized by a flat, basaltic seafloor, the observed scattering functions are not always monotonic with respect to angle. Rather, peaks in the scattering functions sometimes occur at angles corresponding to compressional

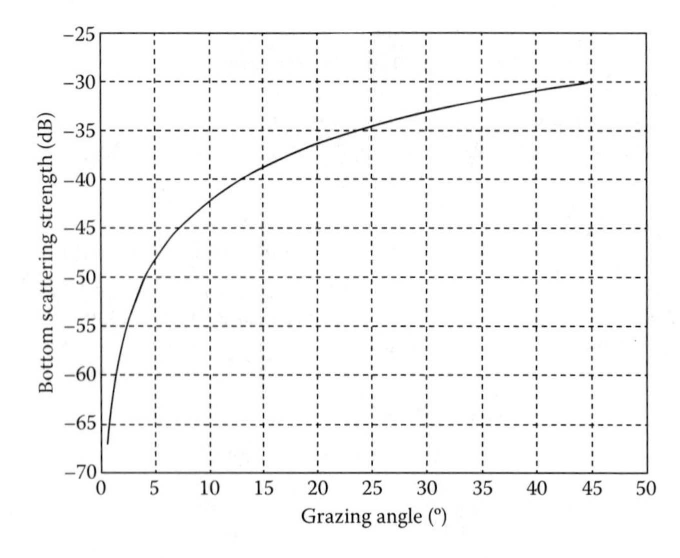

FIGURE 9.4 Bottom scattering strengths (S_B) calculated according to Equation 8.6 using a value of –27 dB for the term $10 \log_{10} \mu$.

and shear head (or interface) wave propagation. In order to investigate this seemingly anomalous behavior, Swift and Stephen (1994) generated scattering functions using a model that also included volumetric heterogeneities below a flat, basaltic seafloor. They executed the model over a wide range of length scales for the embedded scatterers when ensonified by a Gaussian pulse beam at a grazing angle of 15°. At this angle, a truly homogeneous seafloor composed of basaltic rocks would produce total internal reflection. However, they found that bottom models containing 10% velocity perturbations (representing the volumetric heterogeneities) produced significant levels of upward-scattered energy. In reality, energy leaks below the seafloor as an evanescent phase existing only when the incident energy is subcritical, and this energy only penetrates to depths of a few wavelengths. Interactions between this energy and those volumetric heterogeneities just below the surface excite interface waves (as well as compressional and shear waves in the bottom), in agreement with the observed scattering functions. Swift and Stephen (1994) also found that the scattering functions were influenced by the presence of velocity gradients (both compressional and shear) below the seafloor. Specifically, in the absence of velocity gradients, bottom-propagated energy is not refracted upward. Thus, it interacts with yet deeper velocity anomalies and produces greater scattering.

Greaves and Stephen (1997) determined that a seafloor dip on the scale of a few hundred meters influenced, but did not determine, scattering strength. This suggested that other characteristics of steeply dipping areas, such as subsurface properties or smaller-scale surface features, strongly affected the level of backscattered signals. Greaves and Stephen (2000) advanced this work in an extended analysis of ARSRP data to examine reverberation at low-grazing angles from rough and heterogeneous seafloors.

McDaniel (2003) successfully applied coupled-mode theory to a cylindrically symmetric wedge geometry in which backscatter was due primarily to progressive ray steepening and reversal. The lower boundary within the seabed (Z_B) must be chosen to obtain adequate angular resolution as well as to ensure that the assumption of weak coupling is valid. Partial modal reflection was not an important backscattering mechanism in the examples considered. Thus, with a careful choice of Z_B, estimates of reverberation could be obtained. Also see related work by McDaniel (2008).

Siemes et al. (2008) used a multibeam echosounder to provide both depth information and seafloor classification. The classification approach was model-based, employing the backscatter data. It discriminated between sediments by applying the Bayes decision rule for multiple hypotheses and implicitly accounting for ping-to-ping variability in the backscattering strength. The resulting geoacoustic estimates were validated using sediment samples collected during the experiment south of Elba in the Mediterranean Sea in 2007.

Harrison (2002a) noted that a determination of the angular dependence of bottom scattering strengths is problematic in shallow-water environments. If the bottom-scattering law is a separable function of an incoming and outgoing angle, then the resulting reverberation should contain separable incoming and outgoing propagation terms. Consequently, the returning multipaths from a scattering patch are weighted by the outgoing part of the scattering law. Comparisons of reverberation and propagation angle-dependence on a vertical receiving array have the potential to reveal the scattering law directly.

In shallow-water waveguides, the size of the ensonified reverberation patch increases with time due to multipath effects. It can be shown that this effect causes a trend toward Rayleighness in the reverberation pressure envelope. It can also be shown that when the correlation length scale of the scatterers is much greater than the size of the ensonified patch, the effects of multipath time spread on the reverberation statistics is much reduced and non-Rayleighness can persist (LePage, 2010). Specifically, it was shown that the non-Rayleighness of shallow-water reverberation with non-Gaussian chi-squared roughness is controlled by both the degree of non-Gaussianity and the ratio of the ensonified scatterer patch size to the correlation length scale. The scintillation index was used, in part, to evaluate the importance of the non-Gaussianity of the roughness height distribution on the tails of the reverberation.

The presence of plant life on the seafloor can complicate the scattering processes at the bottom boundary of the ocean. In a theoretical study, Shenderov (1998) treated acoustical scattering by algae as the diffraction of sound waves on a random system of 3D, bent, elastic bodies. This approach considered the statistical properties of algae. McCarthy and Sabol (2000) characterized submerged aquatic vegetation in terms of military and environmental monitoring applications.

Seagrass meadows are ecosystems of great ecological and economical value, and their monitoring is an important task within coastal environmental management (Paul et al., 2011). An acoustic mapping technique using a profiling sonar was applied to three different sites with meadows of *Zostera marina* (common eelgrass), *Zostera noltii* (marine eelgrass), and *Posidonia oceanica* (Neptune Grass or Mediterranean tapeweed) respectively, with the aim to test the method's applicability. From the backscatter data, the seabed could be identified as the strongest scatterer along an acoustic beam. An algorithm was developed to compute water depth, seagrass-canopy height, and seagrass coverage and to produce maps of the survey areas. Canopy height was estimated as the distance between the bed and the point where backscatter values decreased to water values. Seagrass coverage was defined as the percentage of beams in a sweep where the backscatter 5–10 cm above the bed was higher than a threshold value. (This threshold value was dynamic and depended on the average backscatter value throughout the water column.) The method is applicable in a range of turbidity conditions. Analysis of these data showed that each seagrass species had a characteristic canopy height and spatial coverage distribution. These differences could be used to perform preliminary species identification since each species had a typical canopy height and preferred depth range. Furthermore, the results showed that these differences could be used to track boundaries between species.

The acoustic properties of kelp forests are not well known, but are of interest for the development of environmental remote sensing applications. Wilson et al. (2013) examined the low-frequency (0.2–4.5 kHz) acoustic properties of three species of kelp (*Macrocystis pyrifera*, *Egregia menziessi*, and *Laminaria solidungula*) using a one-dimensional acoustic resonator. Acoustic observations and measurements of kelp morphology were then used to test the validity of the multiphase effective medium model of Wood (1930) in describing the acoustic behavior of the kelp. For *Macrocystis* and *Egregia*, the two species of kelp possessing pneumatocysts, the change in sound speed was highly dependent on the volume of free air contained in the kelp. The volume of air alone, however, was unable to predict the effective sound speed of the multiphase medium using a simple two-phase (air + water) form of Wood's model. A separate implementation of this model (frond + water) successfully yielded the acoustic compressibility of the frond structure for each species (*Macrocystis* = $1.39 \pm 0.82 \times 10^{-8}$ Pa^{-1}; *Egregia* = $2.59 \pm 5.75 \times 10^{-9}$ Pa^{-1}; *Laminaria* = $8.65 \pm 8.22 \times 10^{-9}$ Pa^{-1}). This investigation demonstrated that the acoustic characteristics of kelp are species-specific, biomass-dependent, and differ between species with and without pneumatocyst structures.

9.4 INVERSION TECHNIQUES

The utility of the reverberation field as an inverse sensing technique is analogous to that of the ambient noise field (Section 7.7). Makris (1993), for example, inverted the reverberation field to image the seafloor. This was accomplished by using simultaneous inversions of multiple reverberation measurements made at different locations. The outputs of the inversion can include optimally resolved reverberation, scattering coefficients or physical properties of the seafloor. Through simulation, Makris (1993) demonstrated the utility of this method for determining bottom scattering strengths using a monostatic-observation geometry together with operational parameters obtained from bottom-reverberation experiments sponsored by the US ONR special research program. First, a synthetic ocean basin was created using a representation of scattering coefficients. Next, the true scattering coefficients were estimated from the observed reverberation maps using a global inversion method. This method is applicable to many monostatic and bistatic experimental geometries, although it cannot be used arbitrarily since a series of separate inversions is necessary to determine the angular dependence of the bottom reverberation. Specifically, for each inversion the separation between observations must be small enough to maintain a similar orientation relative to the imaged region. It may be possible to plan an optimal set of experimental observations by considering measurement-resolution constraints, platform maneuverability, and time and observation sites by solving what W.A. Kuperman referred to as the "traveling acoustician problem" (Makris, 1993: 992).

In a special issue of *JASA Express Letters* on the topic of "Overview of Shallow Water," Lynch and Tang (2008) organized 19 papers into two categories: (1) acoustic propagation and scattering and (2) geoacoustic inversion. Papers on acoustic propagation and scattering addressed the impact of linear and nonlinear internal waves on sound propagation, the spatial and temporal coherence of the underwater sound fields, and the sensitivity of propagation to environmental parameters. Papers on geoacoustic inversion focused on development of an elasto-acoustic model of the bottom (including shear and compressional wave speeds, attenuation, and density) using available acoustic data.

Periodic reverberation peaks and valleys caused by Lloyd-mirror effects, aside from their implications on detection performance, would appear to possess characteristics that can be exploited for environmental adaptation (Cole et al., 2004).

If unaccounted for, these effects could introduce errors in the values of parameters extracted from inverse measurements and could lead to incorrect assumptions relating to reverberation statistics. Reverberation-amplitude fluctuations generated by GSM complex reverberation-pressure time series containing the interference effect compared favorably with the amplitude fluctuations observed in reverberation measurements. A version of CASS was similarly modified to contain the coherent reverberation formulation.

Reverberation data were inverted using ray-mode theory to estimate sea-bottom scattering coefficients (Liu et al., 2003). The inverted results compared well with reverberation data collected during the Asian sea international acoustic experiment (ASIAEX) in the East China Sea in 2001 at the same site but for different sea states. Also see Oikawa et al. (2007).

Time-reversal ocean acoustic experiments have been conducted at 3.5 kHz (Song et al., 2004). The focusing capability of a time reversal mirror (TRM) without *a priori* knowledge of the environment to achieve active target detection in reverberation-limited environments can be accomplished in two ways: focus acoustic energy on a target while shadowing the boundaries above and below the waveguide; or, enhance active target detection by nulling reverberation (Song et al., 2005). The concept of environmentally adaptive reverberation nulling using a TRM was demonstrated experimentally at 850 Hz and 3500 Hz using data from a monostatic shallow-water experiment conducted off the west coast of Italy in April 2003. The active transmission of a seafloor spatial null from a vertical source array was shown to attenuate prominent reverberation features by 3–5 dB, thereby reducing their levels to that of the more diffuse reverberation background. Backscattering from a rough water-bottom interface can serve as a surrogate probe source in time reversal experiments. A time-gated portion of the reverberation was refocused to the bottom interface at the corresponding range. Thus, prominent reverberation features could be attenuated using the active transmission of a seafloor spatial null.

In related work, echo-to-reverberation enhancement was successfully demonstrated using a TRM (Kim et al., 2004). A TRM was also used to measure scattering and reverberation in a temporal window corresponding to a desired focusing range from the source-receiver array (Lingevitch et al., 2002b). The potential application of phase-conjugated (or time-reversal acoustic) arrays to generate a spatio-temporal focus of acoustic energy at the receiver location was examined. This application would eliminate any distortions introduced by shallow-water channel propagation (Smith et al., 2003). Moreover, this technique is self-adaptive and automatically compensates for environmental effects and array imperfections without the need to explicitly characterize the environment.

The ability to measure seabed scattering strengths and reflection coefficients independently is an important step for the advancement of inverse methods using reverberation, such as rapid environmental assessments (REAs), since it provides the means for quantitatively measuring the robustness of those methods (Holland, 2006, 2007).

Ocean acoustic reverberation tomography was developed for 3D volumetric imaging of the ocean using wide-angle direct arrival and reverberation acoustic travel time data (Dunn, 2015). A linearized iterative approach was taken to solve the nonlinear tomography problem. It was shown that the reverberation tomography method can produce images of ocean sound speed from seafloor to sea surface over distances of tens of kilometers on a scale much smaller than traditional acoustic tomography. Because acoustic properties are measured, rather than their gradients, reverberation tomography can be complementary to the methods of seismic oceanography (see Section 6.19). The theoretical resolving length of this method is only a few tens of meters, depending on the frequency content of the acoustic waves, but practical resolution is larger and controlled by station and source spacing as well as data uncertainties. Seismic wide-angle imaging uses ship-towed sources and networks of ocean bottom seismographs rather than towed streamers. The proposed method bridges the gap between relatively low-resolution traditional tomography, and high-resolution seismic oceanography methods that measure impedance contrasts in the water column.

Godin (2012) studied the possibility of using acoustic reverberation produced by a source-of-opportunity for measuring water-column parameters by means of noise interferometry. It was first assumed that the medium was stationary. Then, changes to these results were examined in the presence of currents. Godin (2012) was not interested in the correlation function itself, but in the possibility of extracting information on the water column from it. Using general assumptions about the sound propagation and roughness of the ocean surface and seafloor, it was shown that the two-point correlation function of the acoustic reverberation produced by a source-of-opportunity carried information on the travel times of the rays as well as the phases of the modes of sound propagation in both directions between selected points. If a net of spatially separated receivers existed, then that information could, in principle, be used to obtain the sound speed and current velocity in the fluid, similar to passive acoustic tomography of the ocean based on the correlation of ambient noise. However, a number of important problems remain to be solved. Specifically, further studies are required to account for the motion of the sound source and the ocean surface, along with the horizontal inhomogeneity of the waveguide in the case of low-frequency sound. It will also be necessary to extend estimates of the signal accumulation times onto reverberation fields, thus allowing one to replace the statistical-mean values with time-averaged values. It will also be necessary to estimate the accuracy of passive measurements of mode phases and ray travel times, which can be achieved in practice in the ocean.

As described by Jagannathan et al. (2009), the ocean acoustic waveguide remote sensing (OAWRS) approach for studying marine life was first demonstrated in 2003 in the Mid-Atlantic Bight off the northeast US coast. Using a single 1-s duration transmission of a linear frequency modulated (LFM) waveform, OAWRS surveyed an area as large as the state of Connecticut. The imaging was effectively instantaneous because the entire region was surveyed in less time than it would take a marine organism to traverse a single OAWRS resolution cell. The OAWRS approach was used again with the National Marine Fisheries Service (NMFS) annual herring survey of the Gulf of Maine and Georges Bank to study herring group behavior associated with spawning in September–October 2006. In both experiments, a vertical source array transmitted sound in the frequency range 390–1400 Hz, which is near the swim bladder resonance frequency for many fish species in the survey regions.

Echoes scattered from the fish were received by a towed horizontal receiving array. Instantaneous snapshots of the ocean environment over thousands of km^2 were then formed by charting acoustic returns in horizontal range and bearing by temporal matched filtering and beamforming. A detailed technical description of the OAWRS approach was provided in a set of six appendices.

Risch et al. (2012) concluded that songs of humpback whales (*Megaptera novaeangliae*) in the Stellwagen Bank National Marine Sanctuary, in the Gulf of Maine, were reduced concurrent with transmissions of an OAWRS experiment approximately 200 km away during an 11-day period in the autumn of 2006. They compared the occurrence of songs for 11 days before, during, and after the experiment with songs over the same 33 calendar days two later years. Using a quasi-Poisson generalized linear model, they demonstrated a significant difference in the number of minutes with detected songs between periods and between years. The lack of humpback-whale songs during the OAWRS experiment was the most substantial signal in the data. They concluded that their findings provided the greatest published distance over which anthropogenic sound has been shown to affect vocalizing baleen whales, and the first time that active acoustic fisheries technology has been shown to have this effect. They submitted that the suitability of OAWRS technology for *in situ*, long-term monitoring of marine ecosystems should be reconsidered in light of its possible effects on nontarget protected species.

Gong et al. (2014) disputed the findings of Risch et al. (2012). Specifically, they showed that humpback-whale vocalization behavior was synchronous with peak annual Atlantic herring spawning processes in the Gulf of Maine. With a passive, wide-aperture, densely-sampled, coherent hydrophone array towed north of Georges Bank in an autumn 2006 OAWRS experiment, vocalizing whales could be instantaneously detected and localized over most of the Gulf of Maine ecosystem in a roughly 400-km diameter area by introducing array gain of 18 dB, orders of magnitude higher than previously available in acoustic sensing of whales. With humpback-whale vocalizations consistently recorded at roughly 2000 d^{-1}, they showed that a constant humpback-whale song occurrence rate indicated that the OAWR transmissions had no effect on humpback songs. Gong et al. (2014) suggested, in part, that the song occurrence variation reported in Risch et al. (2012) was consistent with natural causes other than sonar.

10 Reverberation II: Mathematical Models

10.1 BACKGROUND

Mathematical models of reverberation generate predictions of boundary and volumetric reverberation using the physical models of boundary and volumetric scattering strengths developed in Chapter 9. The development of reverberation models has proved to be formidable for two reasons (e.g., Moritz, 1982; Goddard, 1993). First, there are theoretical difficulties in solving complex boundary value problems for which analytical tools are poorly developed. Second, there are practical difficulties in identifying and measuring all the parameters affecting the reverberation process.

10.2 THEORETICAL BASIS FOR REVERBERATION MODELING

10.2.1 Basic Approaches

The scattering of sound by volumetric inhomogeneities in the ocean can be described either by Rayleigh's law or by geometrical acoustic scattering (Albers, 1965: 121). Rayleigh's law applies when the size (d) of the scattering particle is much smaller than the wavelength (λ) of the incident sound. Geometrical acoustic scattering is valid when d is much larger than λ. An intermediate condition applies when d is approximately equal to λ. Specifically, the following relationships hold:

1. If $d \ll \lambda$, the pressure of the scattered sound is proportional to f^2 (where f is the acoustic frequency) and to the volume of the scatterer, regardless of its shape (Rayleigh's law).
2. If $d \approx \lambda$, the pressure is a complicated function of frequency and also varies with the acoustic properties of the scatterer together with the characteristics of the ocean medium or boundaries.
3. If $d \gg \lambda$, the scattering is independent of frequency and depends only on the acoustic properties of the scatterer and its cross section (geometrical acoustic scattering).

This situation has, in part, encouraged the development of two different approaches to the modeling of reverberation in the ocean: cell-scattering models and point-scattering models.

Cell-scattering models assume that the scatterers are uniformly distributed throughout the ocean. Thus, the ocean can be divided into cells, each containing a large number of scatterers. Summing the contribution of each cell yields the total average reverberation level as a function of time after transmission. A scattering strength is used per unit area or volume, as appropriate. This approach is the most commonly used in sonar modeling.

Point-scattering models are based on a statistical approach in which the scatterers are assumed to be randomly distributed throughout the ocean. The reverberation level is then computed by summing the echoes from each individual scatterer. This approach is particularly well suited, for example, to under-ice modeling.

Reverberation is a time-domain problem. Reverberation levels can be represented as a function of range from the receiver by identifying the reverberation level corresponding to the time after transmission that the leading edge of the backscattered signal arrives at the receiver.

The transmission loss component of many reverberation models is based on ray-theoretical considerations. Models based on wave theoretical considerations are usually not well suited to calculations at the high frequencies associated with traditional sonar applications because of the required computational time (Section 4.9). However, for modern low-frequency applications, wave-theoretical techniques have been used with some success. For example, normal mode techniques have been used to model reverberation in oceanic waveguides (Bucker and Morris, 1968; Zhang and Jin, 1987; Ellis, 1993), and a two-way parabolic equation (PE) model that computes backscattered energy (Collins and Evans, 1992; Orris and Collins, 1994) has been used in reverberation simulations (e.g., Schneider, 1993). LeMond and Koch (1997) developed a normal-mode scattering formulation that was useful in computing single-frequency bottom reverberation for bistatic and monostatic scattering geometries in both shallow-water and deep-water environments. Other developments utilizing wave-theoretical techniques are addressed in Sections 10.2.2 and 10.7.

Reverberation models are typically combined with environmental, propagation, noise, and signal-processing models to form a new class of models referred to as active sonar models (see Chapter 11). In the analysis of sonar performance, reverberation and ambient noise are jointly considered as the background masking level against which a signal must be detected. In many situations, it is useful to understand which contributor (noise or reverberation) is most responsible for creating the interfering background. Such diagnostic information can be used to improve new sonar designs or to optimize the performance of existing sonars.

Figure 10.1 illustrates a graphic product generated by a long-range reverberation model based on the cell-scattering approach. The reverberation level corresponds to the time the leading edge of the signal arrives at the receiving array after scattering off the target at the specified range. The plotted reverberation level is averaged over pulse length and corrected for processing gain. This corresponds to reverberation in the receive bandwidth at the output of the signal processor before the thresholding device. Each of the bottom, surface, and volume components of reverberation is the incoherent sum of the respective intensities derived from ray tracing. The noise threshold displayed on the reverberation plot is the combined ambient noise and self-noise in the receiver bandwidth at the output of the signal processor. In this example, the noise threshold is calculated by correcting spectrum level noise for array gain, bandwidth, and processing gain. Thus, the range (time) at which the background masking level changes from reverberation-limited to noise-limited can be determined.

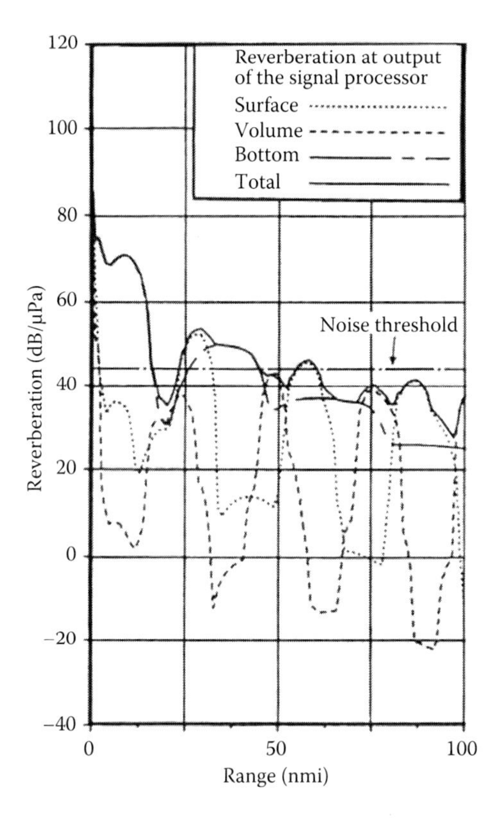

FIGURE 10.1 Averaged reverberation (surface, bottom, volume, and total) levels versus range as predicted by a reverberation model. The noise threshold is also plotted for comparison. (Adapted from Hoffman, D.W., LIRA: A model for predicting the performance of low-frequency active sonar systems for intermediate surveillance ranges. Nav. Ocean Syst. Ctr, Tech. Doc. 259, 1979.)

10.2.2 Advanced Developments

A numerical method known as transmission line matrix modeling (TLM) has been used to model the scattering of acoustic waves off underwater targets. This method models the target and propagation space using meshes, rendering the problem discrete in both space and time. The mesh method models target features in considerable detail, operates readily in both two and three dimensions, and is capable of dealing with complicated scattering processes (Orme et al., 1988). The method has already found use in volume scattering modeling, and in modeling time-varying propagation phenomena in multipath channels. It is now also being applied to transducer structure modeling (Coates et al., 1990). Scott and de Cogan (2008) applied TLM to the problem of nonuniform bounding walls by incorporating boundary-conforming Cartesian meshes into TLM schemes for acoustic propagation in the two-dimensional (2D) ideal-wedge benchmark test of Buckingham and Tolstoy (1990).

A combination of the boundary element method (BEM) and wavenumber integration (WI), referred to as the hybrid BEM-WI approach, has been used extensively in the seismic community to model elastic scattering problems. WI reduces the 2D Helmholtz equation to a one-dimensional (1D) ordinary differential equation in a range-independent environment. BEM combines an integral representation of the wave field in a volume with a point representation of stresses and displacements on the boundary between fluid and elastic media (Schmidt, 1991).

Keiffer and Novarini (1990) demonstrated that the wedge assemblage (or facet ensemble) method offers a unique description of sea-surface scattering processes in the time domain by allowing for separation of the reflected and scattered components. Jackson et al. (1986) applied the composite roughness model (Section 9.3.1) to high-frequency bottom backscattering. Lyons et al. (1994) enhanced this model by incorporating the effects of volume scattering and scattering from subbottom interfaces. A new bottom-scatter modeling approach was proposed by Holland and Neumann (1998) to account for artifacts observed in field data when the subbottom plays a role in the scattering process.

Gostev and Shvachko (2000) considered kinematic models of volume reverberation caused by scattering from different types of inhomogeneities in the caustic zones of a surface waveguide. This reverberation component is important because the scattered field contains information on the statistical parameters of the inhomogeneities, including the location of caustics in the insonified zones.

Ellis (1995) built upon the method of Bucker and Morris (1968) for computing shallow-water boundary reverberation using normal modes to calculate the acoustic energy propagating from the source to the scattering area and back to the receiver. Ray-mode analogies and empirical scattering functions were used to compute the scattered energy at the scattering area. Continuing along this line of investigation, Desharnais and Ellis (1997) developed the bistatic normal-mode reverberation model OGOPOGO, which is a further extension of the method of Bucker and Morris (1968). The propagation was described in terms of normal modes computed by the normal-mode model PROLOS. Travel times of the reverberation signals were derived from the modal-group velocities. Volume reverberation from either the water column or the subbottom is not currently included, but boundary reverberation is computed using empirical scattering functions and ray-mode analogies. The OGOPOGO model was used to interpret reverberation measurements from shallow-water sites in the frequency range 25–1000 Hz.

Preston (1999) documented the results of a workshop that identified current scientific issues relating to shallow-water reverberation, scattering mechanisms, associated reverberation experiments, and modeling efforts. The frequency range of interest was 50 Hz to 6 kHz.

It has been demonstrated that reverberation can be modeled as a four-dimensional (4D) dynamical system with a positive maximum Lyapunov exponent (Cai et al., 2002). This result may suggest new approaches to process reverberation time series based on chaos.

Abraham and Lyons (2004) developed techniques for simulating non-Rayleigh (cell-scattering) reverberation within the context of the finite-number-of-scatterers representation of distributed reverberation. This model effectively bridged the gap between the point-scatterer and Rayleigh-envelope models. Approximate methods for generating distributed random variables were developed to permit an efficient numerical implementation and to avoid the computational effort associated with simulating the response of individual scatterers. The effects of acoustic propagation through a shallow-water environment were modeled by a piece-wise stationary finite-impulse response (FIR) filter and a frequency-domain implementation and normalization. This combination allowed statistical control of the reverberation envelope and incorporation of multipath propagation at the sonar-resolution cell level, as opposed to the scatterer level for point-scatterer simulators. This approach significantly reduced the required computational effort. Shaping of the reverberation power spectrum was also accomplished by a FIR filter selected according to the transmit waveform and the Doppler effect induced by source–receiver motion and wide-band, Doppler-matched filtering. Because the spectral shaping filter alters the statistics of the reverberation, a pre-warping of the K-distribution (a standard model for radar clutter) parameters was developed to control the statistics of the filter output. The fundamental limitation and benefit of this simulation method is that the reverberation is generated after beamforming and matched filtering. The disadvantages of this method lay in the restriction to simulating reverberation arising from scatterers that are smaller than the resolution cell of the sonar, an inability to include beam-to-beam correlation, and a lack of control over individual scatterers. The advantage lies in the ability to quickly generate non-Rayleigh reverberation following a statistical distribution parameterized by the sonar system (beamwidth and bandwidth) and the environment (scatterer size and density) while correctly accounting for multipath propagation and spectral shape.

LePage (2004) noted that non-Rayleigh (or long-tailed) reverberation envelopes lead to significantly higher false-alarm rates in naval sonar systems. A high-fidelity, time-domain, shallow-water reverberation model (originally designed for the prediction of the coherent aspects of reverberation in multipath environments) was used to confirm that sufficient multipath causes reverberation from non-Gaussian distributions of scatterers to appear nearly Rayleigh, in accordance with the central limit theorem. These findings were later corroborated by Abraham (2007), who equated these results to an increase in the K-distribution shape parameter.

In other research, a two-way (PE) was developed to account for multiple scattering in a range-dependent medium that was divided into a sequence of range-independent regions (Lingevitch et al., 2002a). This method coupled the incoming and outgoing fields at a series of vertical interfaces by imposing continuity in pressure and horizontal-particle velocity. The equations were solved with the PE method by iteratively sweeping the range-dependent region. This approach would be useful for solving problems involving scattering from waveguide features and compact objects.

The traveling-wave expansion, obtained by a series expansion of a depth-dependent Green's function, has been suggested as a way to extend the useful field decomposition of ray theory to finite frequencies in an exact way (Ivansson and Bishop, 2003). Related hybrid ray-mode and ray-wavenumber-integration techniques and codes are widely used. The traveling-wave expansion works well when the pertinent wavenumber integrals can be truncated to the propagating regime. However, the traveling-wave expansion fails in isolating grazing-ray diffraction beyond boundary-produced caustics.

Here, the evanescent regime of wavenumbers is important. Consequently, reverberation levels in shallow water governed by a downward-refracting sound-velocity profile cannot be modeled correctly using the traveling-wave method. For single grazing-ray diffraction, the shadow-zone field can be represented as a residue series whose terms can be interpreted physically either as ray shedding by diffraction rays or as creeping waves. These results allow a generalization to multiple grazing-ray diffraction for generalized sound-speed profiles. Physically, ray contributions arise from wavenumbers of stationary phase while ray shedding in the shadow zones arises from leaky modes. By revisiting the traveling-wave expansion, it was shown that the pertinent wavenumber integrals do not, in general, converge at infinity because of exponential growth. To obtain convergent integrals with corresponding physically sound field components, it is necessary to recombine the integrand components. These recombinant traveling-wave expansions can be used for exact modeling of single (i.e., only one interaction with the boundary), as well as multiple grazing-ray diffraction.

Sonar clutter, particularly in shallow-water environments, introduces false targets that change the statistics of the reverberation signal. Specifically, clutter increases the probability of false alarm for a given probability of detection. This is because clutter adds to the length of the tails of the reverberation-envelope probability distribution function (PDF), moving the statistics away from the Rayleigh canonical form. Clutter can be caused by target-like features, either natural or man-made, or by non-Gaussian distributions of the scatterers. Typically, high-bandwidth or highly directive systems (or both) have more problems with clutter since, as the size of the scattering patch is reduced, the PDF of the generally non-Gaussian scatterer distributions becomes resolved by the system (LePage et al., 2006).

Lee et al. (2013) reexamined the classical ray approach for monostatic ocean boundary reverberation based on a geometrical ray-bundle concept. In this new formulation, the impulse response for the averaged scattering intensity was expressed by a simple function consisting of continuous ray-bundle quantities with respect to the time, which could be regarded as a generalized function for ray-based reverberation in a boundary cell. To numerically evaluate this impulse response, a zeroth- and a first-order polynomial interpolation method was applied to approximate the ray-bundle quantities. Then, the impulse function was reduced to the forms of the delta function and the rectangular function. They developed an explicit numerical scheme for the monostatic reverberation based on the split-step marching algorithm for the range. This numerical scheme provided high accuracy (even when using larger range steps) and gave reasonable results for numerical examples of ocean waveguides with isovelocity, summer, and winter sound-speed profiles.

Zhou and Zhang (2013) integrated the energy-flux (or angular-spectrum) method for shallow-water reverberation, based on the WKB approximation to the normal-mode solution, with physics-based rough-bottom scattering and sediment-volume scattering models. This integration resulted in a simple relationship between the classic seabed scattering cross sections and the modal scattering matrix in waveguides. The resultant shallow-water reverberation expression in the angular domain could be used for predicting reverberation in a Pekeris waveguide. Moreover, a corresponding expression in the modal domain could be used for predicting shallow-water reverberation in a non-Pekeris waveguide.

10.3 CELL-SCATTERING MODELS

Simplifying assumptions are often necessary in order to make reverberation modeling feasible. These assumptions may seem to restrict the results to purely idealized situations. However, the resulting expressions for reverberation have been found to be practical for many sonar design and prediction purposes. The assumptions necessary are the following (Urick, 1983: Chapter 8):

1. Straight-line propagation paths, with all sources of attenuation other than spherical spreading neglected. The effect of absorption on the reverberation level can be accommodated.
2. A homogeneous distribution of scatterers throughout the area or volume producing reverberation at any given time.
3. A sufficiently high density of scatterers to ensure that a large number of scatterers occurs in an elemental volume (dV) or area (dA).
4. A pulse length short enough for propagation effects over the range extension of the elemental volume or area to be neglected.
5. An absence of multiple scattering (i.e., the reverberation produced by reverberation is negligible).

In the following sections, basic theoretical relationships will be developed for volume and boundary reverberation, as adapted from Urick (1983: Chapter 8).

10.3.1 Volume-Reverberation Theory

Let a directional projector in an ideal medium (as implied by the above assumptions) insonify a volume of water containing a large number of uniformly distributed scatterers. The beam pattern of the projector can be denoted by $b(\theta, \phi)$, and the axial intensity at unit distance will be I_0. The source level (SL) is then given by SL = 10 $\log_{10} I_0$. The intensity at 1 m in the (θ, ϕ) direction is $I_0 b(\theta, \phi)$ by the definition of the beampattern function. Let there be a small volume dV of scatterers at range r. The incident intensity at dV will be $I_0 b(\theta, \phi)/r^2$. The intensity of the sound scattered by dV at a point P distant 1 m back toward the source will be $(I_0 b(\theta, \phi)/r^2)\, s_V\, dV$, where s_V is the ratio of the intensity of the scattering produced by a unit volume, at a distance of 1 m from the volume, to the intensity of the incident sound wave. The quantity 10 $\log_{10} s_V$ is the scattering strength for volume reverberation and is denoted by the symbol S_V. In the region near the source, the reverberation contributed by dV will have the intensity $(I_0/r^4)\, b(\theta, \phi)\, s_V\, dV$ and will produce a mean-squared voltage output of

$$R^2\left(\frac{I_0}{r^4}\right)b(\theta,\phi)b'(\theta,\phi)s_V\,dV$$

at the terminals of a hydrophone having a receiving beam pattern $b'(\theta, \phi)$ and voltage response R. By assumption, the elemental volumes dV can be made sufficiently small that their total contribution can be summed up by integration, and the coefficient s_V is then a constant. Furthermore, the equivalent plane-wave volume reverberation level (RL_v) can be defined as the intensity (in dB) of an axially incident plane wave producing the same hydrophone output as the observed reverberation, in which case the following relationship is obtained:

$$RL_v = 10\log_{10}\left[\frac{I_0}{r^4} s_v \int_v b(\theta,\phi)b'(\theta,\phi)\,dV\right] \tag{10.1}$$

In order to proceed further, the elemental volumes dV must be examined more closely. As shown in Figure 10.2, take dV to be an infinitesimal cylinder of finite length with ends normal to the incident direction. The area of the end-face of this cylinder dV may be written r^2 dΩ, where dΩ is the elemental solid angle subtended by dV at the source. Using a pulsed sonar, the extension in range of dV is such that the scattering produced by all portions of dV arrive back near the source at the same instant of time. Another way of viewing this is that the scattering of the front end of the pulse by the rear scatterers in dV will arrive back at the source at the same instant as the scattering of the rear of the pulse by the front scatterers in dV. The extension in range is thus $c\tau/2$, where τ is the pulse length and c is the speed of sound (Urick, 1983: 241–2). The elemental volume can then be expressed as

$$dV = r^2\frac{c\tau}{2}\,d\Omega \tag{10.2}$$

and the volume reverberation level (RL_v) is

$$\begin{aligned} RL_v &= 10\log_{10}\left[\frac{I_0}{r^4} r^2\frac{c\tau}{2} s_v \int_0^{4\pi} b(\theta,\phi)b'(\theta,\phi)\,d\Omega\right] \\ &= SL - 40\log_{10} r + S_v + 10\log_{10}\left[r^2\frac{c\tau}{2}\int_0^{4\pi} b(\theta,\phi)b'(\theta,\phi)\,d\Omega\right] \end{aligned} \tag{10.3}$$

where the range extension ($c\tau/2$) is small compared to the range r.

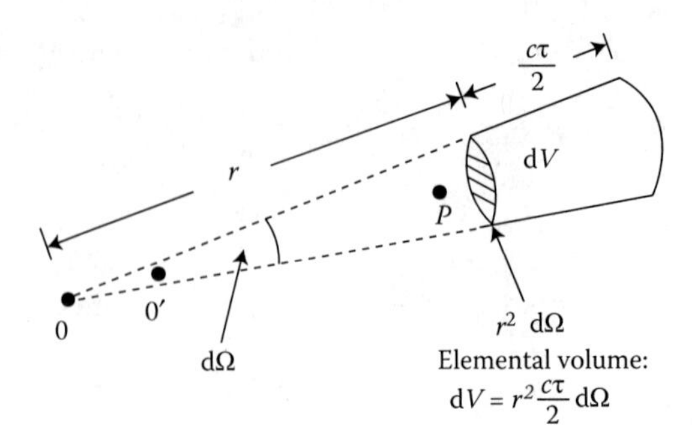

FIGURE 10.2 Geometry for volume scattering.

10.3.2 Boundary-Reverberation Theory

Boundary reverberation pertains to the reverberation produced by scatterers distributed over the sea surface or the seafloor. The derivation of an expression for the equivalent plane-wave level of boundary reverberation proceeds in a manner similar to that for volume reverberation. The result is

$$\mathrm{RL_b} = 10\log_{10}\left[\frac{I_0}{r^4} s_\mathrm{b} \int b(\theta,\phi)b'(\theta,\phi)\,\mathrm{d}A\right] \tag{10.4}$$

where dA is an elemental area of the scattering surface and 10 log s_b is the scattering strength (S_b) for boundary reverberation. If dA is taken as a portion of a circular annulus in the plane of the scatterers with the center directly above or below the transducer, then

$$\mathrm{d}A = \frac{c\tau}{2} r\,\mathrm{d}\phi \tag{10.5}$$

where dϕ is the subtended plane angle of dA at the center of the annulus. The reverberation level then becomes

$$\begin{aligned}\mathrm{RL_b} &= 10\log_{10}\left[\frac{I_0}{r^4} r \frac{c\tau}{2} s_\mathrm{b} \int_0^{2\pi} b(\theta,\phi)b'(\theta,\phi)\,\mathrm{d}\phi\right] \\ &= \mathrm{SL} - 40\log_{10} r + S_\mathrm{b} + 10\log_{10}\left[r\frac{c\tau}{2}\int_0^{2\pi} b(\theta,\phi)b'(\theta,\phi)\,\mathrm{d}\phi\right]\end{aligned} \tag{10.6}$$

10.4 REVMOD MODEL—A SPECIFIC EXAMPLE

The reverberation spectrum model (REVMOD), originally developed by C.L. Ackerman and R.L. Kesser at the Applied Research Laboratory, The Pennsylvania State University, has been expanded and further documented by Hodgkiss (1980, 1984). REVMOD is a set of computer programs that models surface, bottom, and volume reverberation. It is based on the cell-scattering approach.

REVMOD considers the effects of motion of the sonar platform, transmit signal windowing, transmit and receive beam patterns, scatterer velocity distributions, and sound absorption. Model constraints include specification of a constant sound speed profile and nonreflection of sound at the surface and bottom boundaries.

Geometrically, REVMOD divides the ensonified volume of the ocean into cells and evaluates scatterer motion relative to the sonar platform for a measure of the spectral shifting and spreading due to the environment. Acoustically, REVMOD determines the backscattering level for each cell. The contributions from each cell are then summed to compute the total reverberation power spectra received at the sonar.

REVMOD is composed of three major software modules:

1. RVMDS—computes the scattering function resulting from the combined effects of the environment, vehicle dynamics, and transmit and receive beam patterns.
2. RVMDT—convolves the scattering function with the transmit signal energy spectrum to yield the reverberation power spectrum.
3. RVMDR—convolves the reverberation power spectrum with the receiver impulse response energy spectrum.

Module RVMDS computes the scattering function for reverberation and describes the effects of the ocean medium on a transmit signal with carrier frequency ω_c. Module RVMDT completes the reverberation model by combining a detailed description of the transmit signal (pulse length, envelope shape, and SL) with the environmental characterization provided by module RVMDS. Module RVMDR creates a model of a receiver operating in a reverberant ocean environment. A matched filter envelope detector uses the reverberation spectra generated by module RVMDT. Optional software can also be included to model a matched filter envelope detection receiver appropriate for post-processing the reverberation spectra. Thus, the effects of utilizing a receiver impulse response window, which differs from that of the transmitter waveform, can be investigated.

A mathematical description of the scattering function (RVMDS) will be provided here. Calculations of the reverberation power spectrum (RVMDT) and subsequent convolution with the receiver impulse response energy spectrum (RVMDR) were discussed by Hodgkiss (1980, 1984). REVMOD generates a scattering function description of the ocean medium that does not require a detailed description of the transmit signal. Thus, the model is appropriate for investigations of the influence of the medium on various transmit-signal types.

Acoustic backscatter can be modeled as the passage of a transmit waveform through a linear, time-varying filter:

$$\tilde{s}_r = \sqrt{E_t} \int_{-\infty}^{\infty} \tilde{f}(t-\lambda)\tilde{b}\left(t-\frac{\lambda}{2},\lambda\right) d\lambda \tag{10.7}$$

where

$\tilde{s}_r(t)$ = received backscattered signal
E_t = transmit signal energy
$\tilde{f}(t-\lambda)$ = transmit waveform
t = time
λ = time delay
$\tilde{b}(t-\lambda/2, \lambda)$ = time-varying impulse function of the filter

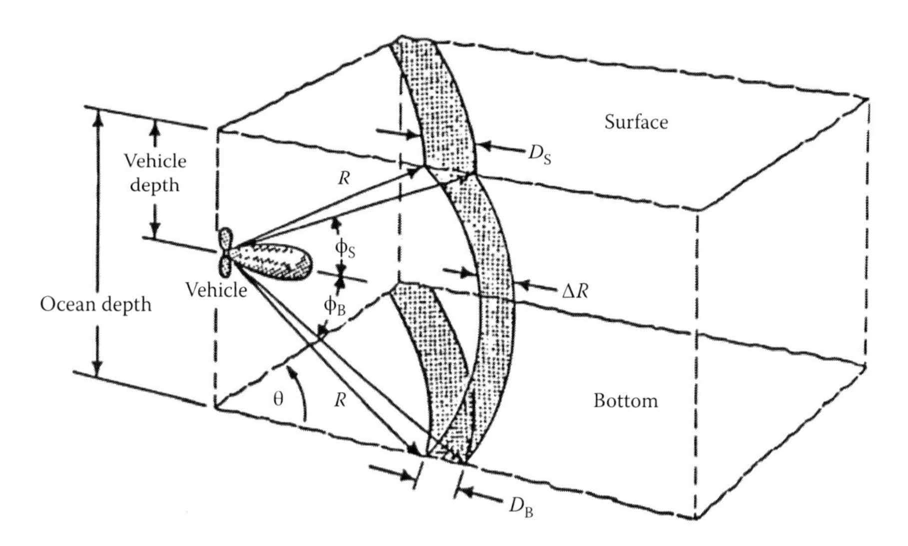

FIGURE 10.3 REVMOD reverberation model geometry. (From Hodgkiss, W.S. Jr., An oceanic reverberation model. *IEEE J. Oceanic Eng.*, **9**, 63–72, 1984. Copyright by IEEE. With permission.)

The basic geometry of REVMOD consists of a spherical shell representing that portion of the ocean medium ensonified by the signal wavefront corresponding to a range (R) after transmit (refer to Figure 10.3). The shell thickness is governed by the variables D_S, D_B, and ΔR for surface, bottom, and volume scattering, respectively. The spherical shell is further subdivided into a grid of cells, each of which contributes to surface, bottom, or volume backscatter. The location of each cell is specified in terms of an azimuth-elevation angle pair (θ_j, ϕ_i). Using this geometry, REVMOD calculates the scattering functions corresponding to the surface, bottom, and volume backscatter of a transmit waveform at specified ranges of interest. Specifically, the normalized attenuation ($\hat{A}_{i,j}$) of the transmitted signal (including propagation and beampattern effects) is calculated for each surface, bottom, and volume grid cell as follows:

Surface scattering

$$
\begin{aligned}
10\log_{10}(\hat{A}_{i,j}) &= 10\log_{10}\left[D_S(R\Delta\theta)\right] + S_S - 40\log_{10}(R) - (2\alpha R) \\
&\quad + 10\log_{10}\left[\hat{P}_T(\theta_j,\phi_S)\hat{P}_R(\theta_j,\phi_S)\right] + 10\log_{10}\left[\frac{c}{(2\Delta R)}\right]
\end{aligned}
\tag{10.8}
$$

Volume scattering

$$
\begin{aligned}
10\log_{10}(\hat{A}_{i,j}) &= 10\log_{10}\left[\Delta R(R\Delta\theta)(R\Delta\phi)\right] + S_V - 40\log_{10}(R) - (2\alpha R) \\
&\quad + 10\log_{10}\left[\hat{P}_T(\theta_j,\phi_i)\hat{P}_R(\theta_j,\phi_i)\right] + 10\log_{10}\left[\frac{c}{(2\Delta R)}\right]
\end{aligned}
\tag{10.9}
$$

Bottom scattering

$$10\log_{10}(\hat{A}_{i,j}) = 10\log_{10}\left[D_B(R\Delta\theta)\right] + S_B - 40\log_{10}(R) - (2\alpha R)$$
$$+ 10\log_{10}\left[\hat{P}_T(\theta_j,\phi_B)\hat{P}_R(\theta_j,\phi_B)\right] + 10\log_{10}\left[\frac{c}{(2\Delta R)}\right] \quad (10.10)$$

where

$\hat{A}_{i,j}$ = scattering level (normalized to 1 s in range) from the (i, j)th grid cell (s^{-1})
D_S = lateral dimension of surface-scattering patch (m)
D_B = lateral dimension of bottom-scattering patch (m)
R = range (m)
ΔR = increment in range over which scattering is averaged (m)
$\Delta\theta$ = angular grid cell width in azimuth (rad)
$\Delta\phi$ = angular grid cell height in elevation (rad)
S_S = surface-scattering coefficient (dB m^{-2})
S_V = volume-scattering coefficient (dB m^{-3})
S_B = bottom-scattering coefficient (dB m^{-2})
α = sound absorption coefficient (dB m^{-1})
$\hat{P}_T(\theta_j,\phi_i)$ = transmit beam pattern (normalized to 1 at $\theta_j = \theta_T$ and $\phi_i = \phi_T$)
$\hat{P}_R(\theta_j,\phi_i)$ = receive beam pattern (normalized to 1 at $\theta_j = \theta_R$ and $\phi_i = \phi_R$)
θ_j = azimuth to center of jth grid column (rad)
ϕ_i = elevation to center of ith grid row (rad)
(θ_T,ϕ_T) = transmit beam vector
(θ_R,ϕ_R) = receive beam vector
ϕ_S = elevation angle to surface (rad)
ϕ_B = elevation angle to bottom (rad)
c = sound speed (m s^{-1})

Hodgkiss (1984) provided sample results from REVMOD for the case of a bottom-mounted transducer with a steerable, axially symmetric beam pattern (Figure 10.4). Wind-driven surface waves and a surface current both contributed to spectral broadening and Doppler shifting of the transmit spectrum upon backscattering from the sea surface. The volume-backscattering component contributed only to spectral broadening of the transmit spectrum since the water column was assumed motionless on average. In this particular example, bottom backscattering was not a factor owing to the bottom-mounted transducer geometry.

For the purposes of discussion here, only a downwind transducer orientation is considered, although Hodgkiss (1984) also addressed upwind and crosswind orientations. This downwind case considers only the reverberation received at a range of 400 m. A quantized Hanning transmit envelope (Figure 10.5) and a Hanning receive envelope (Figure 10.6) are used in conjunction with the beam pattern shown previously in Figure 10.4. The transmit beam vector is horizontal, and the receive beam vector is inclined toward the surface 25° off the horizontal. The transmit and

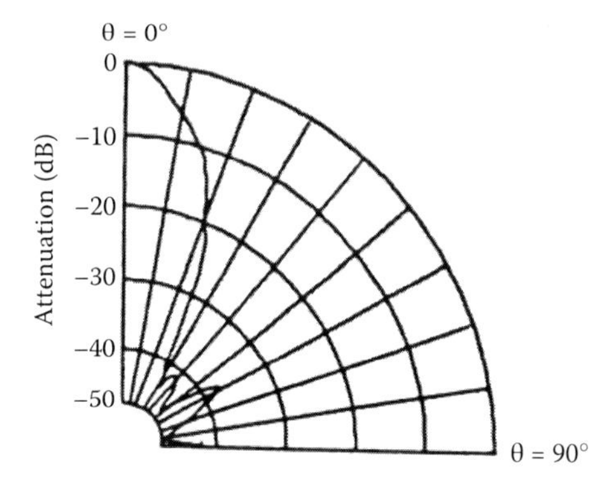

FIGURE 10.4 Transmit and receive beam pattern used in an example from the REVMOD reverberation model. (From Hodgkiss, W.S. Jr., An oceanic reverberation model. *IEEE J. Oceanic Eng.*, **9**, 63–72, 1984. Copyright by IEEE. With permission.)

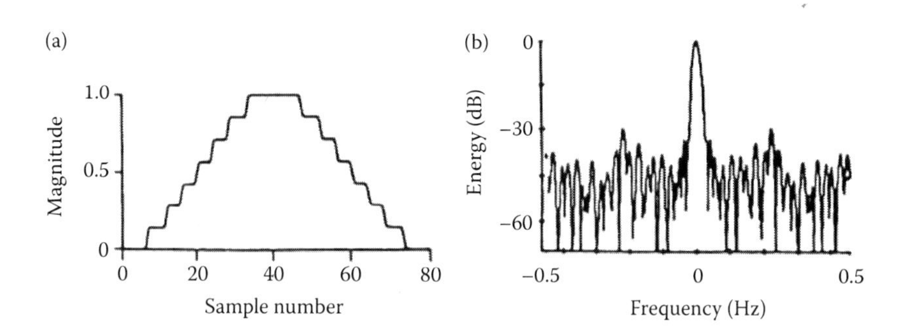

FIGURE 10.5 Sampled envelope function (a) and energy spectra (b) for quantized Hanning transmit envelope used in an example from the REVMOD reverberation model. (From Hodgkiss, W.S. Jr., An oceanic reverberation model. *IEEE J. Oceanic Eng.*, **9**, 63–72, 1984. Copyright by IEEE. With permission.)

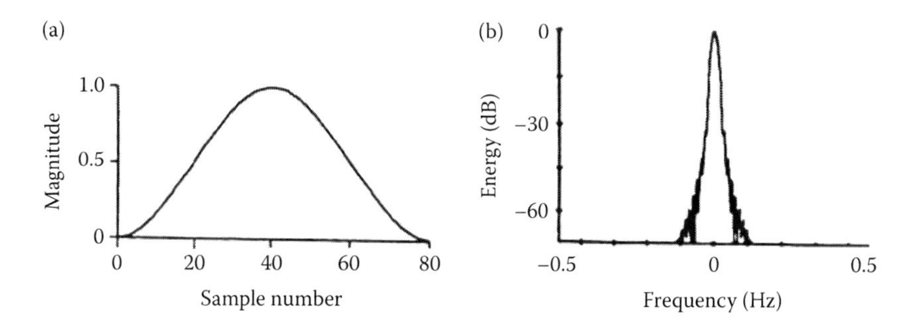

FIGURE 10.6 Sampled envelope function (a) and energy spectra (b) for Hanning receive envelope used in an example from the REVMOD reverberation model. (From Hodgkiss, W.S. Jr., An oceanic reverberation model. *IEEE J. Oceanic Eng.*, **9**, 63–72, 1984. Copyright by IEEE. With permission.)

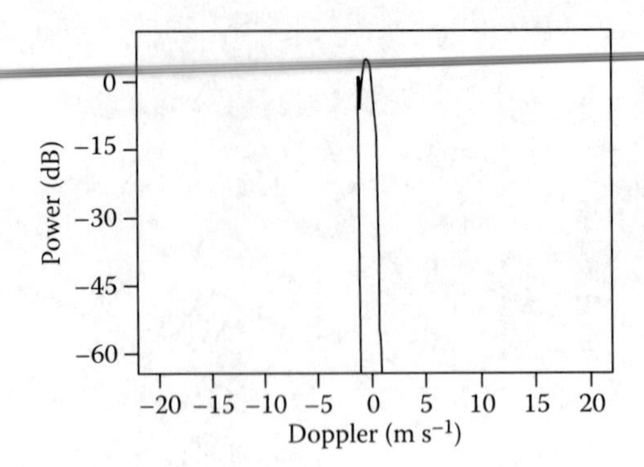

FIGURE 10.7 Example from the REVMOD reverberation model showing the surface-scattering function for a downwind case. (From Hodgkiss, W.S. Jr., An oceanic reverberation model. *IEEE J. Oceanic Eng.*, **9**, 63–72, 1984. Copyright by IEEE. With permission.)

receive envelope durations were 102 ms and 86 ms, respectively. See Ziomek (1985: Chapters 3 and 4) for a discussion of complex apertures and directivity functions.

The Doppler of the surface component of the scattering function reflects a complex interaction with this moving boundary (specifically, a negative Doppler ridge), as shown in Figure 10.7. This ridge corresponds to the Doppler-shifting effects of the surface wave and current velocity vectors.

10.5 BISTATIC REVERBERATION

10.5.1 Computational Considerations

Most traditional active sonars are configured in what is termed a monostatic geometry, meaning that the source and receiver are at the same position. In some sonar systems, however, the source and receiver are separated in range or depth, or both, in what is termed a bistatic configuration. Bistatic geometries are characterized by a triangle of source, target and receiver positions, and by their respective velocities (Cox, 1989). Such geometries are commonly employed in sonobuoy applications (Bartberger, 1985) and also in active surveillance applications (Franchi et al., 1984). Geometries involving multiple sources and receivers are termed multistatic.

A bistatic sonar must detect the target signal against the background noise level that is the intensity sum of the ambient noise; the surface, bottom, and volume reverberation; and the energy propagating directly from the source to the receiver over a set of one-way paths referred to as direct arrivals (Figure 10.8). These direct arrivals represent a significant difference between bistatic and monostatic sonars. Since the direct (one-way) paths are very intense, they must be carefully considered in predicting the performance of bistatic sonars (Bartberger, 1985). Each different propagation path may require separate predictions of transmission loss. Bistatic reverberation models are commonly confronted with substantial bookkeeping requirements associated with sorting the various reverberation and noise contributions. In the monostatic situation, reverberation is due to backscatter. In bistatics, forward and out-of-plane scattering are important (Cox, 1989).

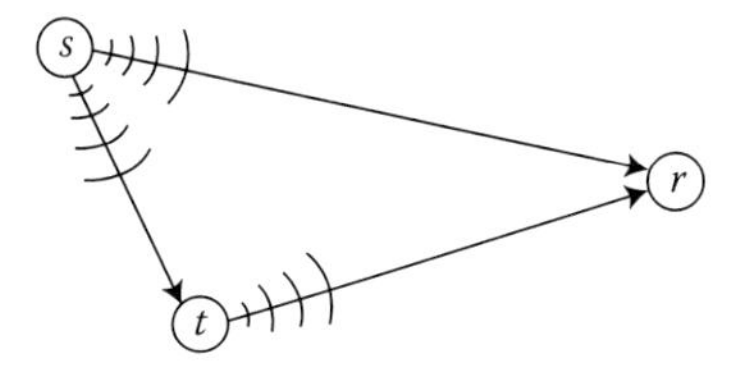

FIGURE 10.8 Geometry for bistatic reverberation showing relative positions of the source (s), target (t), and receiver (r) in either the vertical or the horizontal plane.

The bistatic problem is further complicated by the presence of certain bathymetric features, particularly seamounts, which can introduce a discrimination problem between acoustic returns from these geologic features and from potential targets of interest. Eller and Haines (1987) described seamount statistics as well as the characteristics of acoustic reverberation from seamounts. A real-time bathymetric reflection and scattering prediction system was developed to support at-sea acoustical surveys (Haines et al., 1988). The system utilized gridded databases to generate 3D environmental scenarios in conjunction with the ASTRAL propagation model and the ANDES noise model. As configured, the system operated on a HP-9020 desktop computer and provided color graphic displays of predicted ocean bottom reverberation as well as echoes from seamounts.

A computer model called ocean refraction and bathymetric scattering (ORBS) was developed to calculate the directional distribution of bottom-scattered acoustic energy received at an array (Baer et al., 1985; Wright et al., 1988). ORBS incorporated the refractive effects of a varying sound speed structure in the ocean volume as well as the effects of out-of-plane scattering caused by rough bathymetry. A seamount could be considered where it made sense to consider only a single encounter with the sea floor. The numerical techniques used included the split-step solution to the PE away from the boundary, propagation of the coherence function near the boundary, and a modified Kirchhoff formulation to incorporate scattering. Smith et al. (1996) used field measurements and model results to correlate reverberation events with bathymetric features.

Concurrent detection and classification concepts for application to mine countermeasures in shallow water were explored during the GOATS'98 experiment. One of the principal objectives of GOATS'98 was to explore the possibility of basing a detection-and-classification concept on the measurement of 3D acoustic scattering by buried objects (Schmidt, 2001). The OASES-3D target modeling framework was used to investigate bistatic scattering from buried targets in shallow water. In particular, unexpected subcritical excitation of elastic waves in buried shells and their radiation back into the water column was investigated. Schmidt et al. (2004) examined the temporal and spatial structure of 3D acoustic scattering from buried targets using autonomous underwater vehicles (AUVs) as bistatic receiver platforms. The analysis led to the development of a hybrid modeling framework (OASES-3D) for scattering from partially and completely buried elastic targets in shallow-water waveguides. Schmidt and Kuperman (1995) developed spectral representations of rough interface reverberation in stratified ocean waveguides. Li and Liu (2002) compared their model against data for bistatic reverberation in shallow water.

10.5.2 Bistatic Acoustic Model—A Specific Example

The bistatic acoustic model (BAM) predicts the performance of a sonar system consisting of a sound source and a receiver separated in both range and depth (Bartberger, 1991a; Vendetti et al., 1993a). BAM computes the echo-to-background ratios for targets located at a set of bipolar grid locations in a horizontal plane containing a specific target. The bipolar grid points are defined by the intersections of circles of constant range about the source and receiver. There are three separate versions of BAM corresponding to three different sonar system configurations: vertical line array, horizontal line array, or volumetric array. BAM consists of four computer programs:

1. BISON computes the ray arrival structure; the surface, bottom, and volume reverberation time histories; and the vertical distribution of ambient noise.
2. BISTAT computes the echo-to-background ratios on a bipolar grid within a specified circle about the receiver.
3. CTEST generates the time series of the direct arrivals; the surface, bottom, and volume reverberation; and the combined interference of all of these components.
4. PLOSS calculates the source-to-target transmission loss and the target-to-receiver transmission loss as a function of range.

A sample output product from BAM is illustrated in Figure 10.9, which shows computed echo-to-background levels. These levels are numerically coded to facilitate printer output. The digital pattern in Figure 10.9 indicates the potential for target detection under the hypothetical problem geometry and acoustic conditions:

- Frequency = 2 kHz
- Source depth = 300 m (4-element omni source array)
- Receiver depth = 500 m (6-element omni receiver array)
- Target depth = 350 m (no Doppler)
- Source-receiver range = 10 km
- Maximum target-receiver range = 50 km
- Pulse length = 5 s

The problem area is delimited by a circle centered about the receiver with a radius equal to the maximum target-receiver range. Also shown is the cumulative area coverage (km^2) contained within specified contours of echo-to-background level. In this example, the maximum area coverage possible would be $\pi \times (50 \text{ km})^2 \approx 7854 \text{ km}^2$. Thus, for echo-to-background levels $\geq$–10 dB, only 63% (4942 km^2/7854 km^2) of the specified search area would be covered effectively. Clearly, this information is useful in assessing the effectiveness of candidate sensors and search tactics. For example,

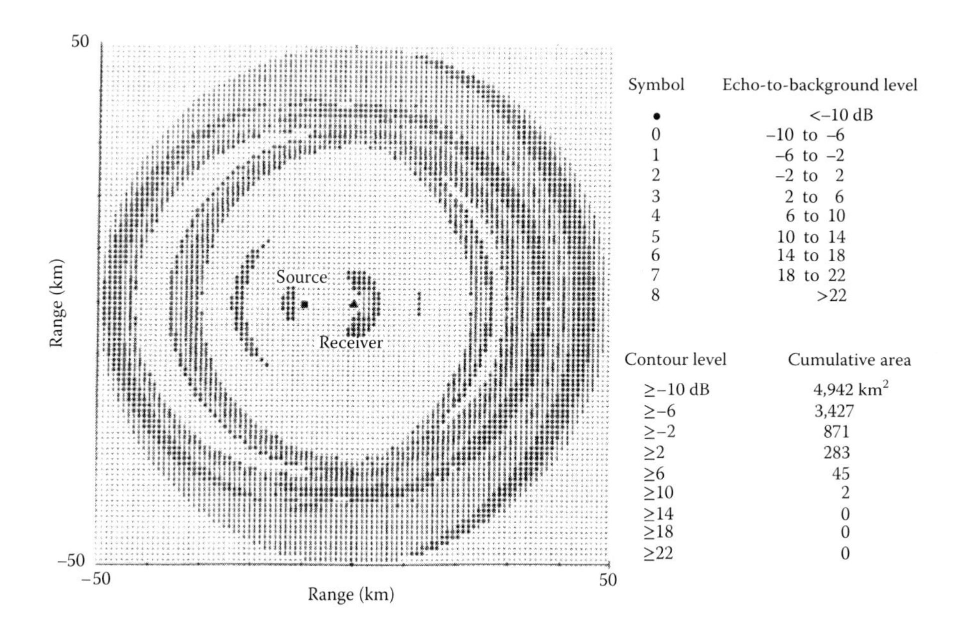

FIGURE 10.9 Sample output from the BAM showing echo-to-background levels. Also shown is the cumulative area coverage contained within specified contours of echo-to-background level.

optimum search tactics could be developed by varying selected parameters and then assessing the resulting changes in area coverage.

The elliptically shaped region of poor detection (<–10 dB) located immediately between the source and receiver is a consequence of the direct arrivals, which represent the transmission of energy directly from the source to the receiver. These direct arrivals, also referred to as the "direct blast," consist of a number of multipath arrivals extending over a time interval that is determined by geometry, environment, and system characteristics. The direct blast often masks target echoes of interest. The extent of this masking region is approximately determined by the inequality (Cox, 1989):

$$\tau_1 + \tau_2 < \tau_3 + \tau_b$$

where

τ_1 = travel time, source to target
τ_2 = travel time, target to receiver
τ_3 = travel time, source to receiver
τ_b = effective masking time of direct blast

The effects of the direct blast can be mitigated to some extent in one of three ways: (1) by controlling the vertical beam pattern of the receiver or source, or both, to minimize the contribution of steep-angle (boundary-interacting) arrivals; (2) by nulling the receiver in the direction of the source; or (3) by using Doppler to discriminate between direct-blast energy and target echoes.

10.6 POINT-SCATTERING MODELS

10.6.1 Computational Considerations

Point-scattering models are based on a statistical approach that assumes the scatterers are randomly distributed throughout the ocean. The echoes from each individual scatterer are then summed to compute the reverberation level. This approach is not as widely used as the cell-scattering technique. When dealing with scatterers whose dimensions are comparable to the acoustic wavelength, the point-scattering approach may be the preferred alternative. This situation arises in two common problems: high-frequency sonars and under-ice scattering. A point-scattering model suitable for high-frequency applications was described by Princehouse (1977). A model appropriate for under-ice scattering is briefly described below.

10.6.2 Under-Ice Reverberation Simulation Model—A Specific Example

Bishop (1987, 1989a,b) and Bishop et al. (1986, 1987) developed a bistatic, high-frequency (≥2 kHz), under-ice acoustic-scattering model to evaluate the scatter produced by a pulse, originating from an arbitrarily located source, as detected by an arbitrarily located receiver. This model, referred to as "under-ice reverberation simulation," uses measured 2D under-ice acoustic profile data and empirical results relating geometric parameters of the large-scale under-ice relief features (such as ice keels) to construct a 3D bimodal under-ice surface consisting of first-year ice keels and sloping flat-ice regions. A first-year keel is modeled as an ensemble of randomly oriented ice blocks on a planar surface inclined at some slope angle with respect to a horizontal plane at sea level. The keel is characterized by length, draft, width, ice thickness, and aspect angle. A region of flat ice is modeled as a smooth planar surface whose slope angle is less than some critical angle that serves to distinguish a flat ice feature from an ice keel. The Kirchhoff approximation is used to evaluate the target strength of a facet of an ice block. This approximation assumes that reflection coefficients appropriate for an infinite plane wave at an infinite plane interface can be used in the local scattering geometry of a rough surface (in some texts, this approximation is referred to as the tangent-plane method). The target strength of a keel is calculated in range increments as the coherent sum of the backscatter from all scattering facets contained within one-half the pulse length projected onto the keel. The model has been used to show the effects of various ice and acoustic parameters on reverberation and target-strength frequency distributions.

The under-ice reverberation simulation model was used to calculate monostatic backscattering and reverberation for one realization at a site near the North Pole, a site in the Chuckchi Sea, and a site in the Beaufort Sea (Bishop, 1989a,b). To parameterize the large-scale surface roughness at these sites, the keel frequency and mean keel relief were calculated. The Chuckchi Sea site was the roughest of the three sites with a keel frequency of 16.0 keels km^{-1} and an average keel relief of 4.1 m. The North Pole site had a keel frequency of 13.7 keels km^{-1} and an average keel relief of 4.0 m. The Beaufort Sea site was the least rough of the sites with a keel frequency of 4.5 keels km^{-1} and an average keel relief of 3.4 m.

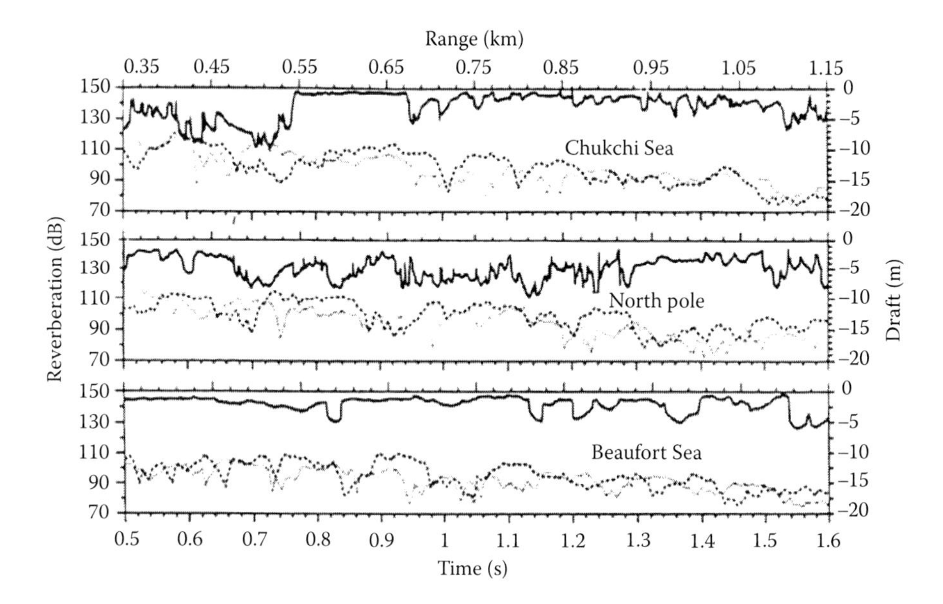

FIGURE 10.10 Under-ice profiles (solid line), measured (dashed line), and modeled (dotted line) under-ice monostatic reverberation time series for three Arctic sites. (From Bishop, G.C., *J. Acoust. Soc. Am.*, **85**, 1912–24, 1989b. With permission.)

Figure 10.10 illustrates the results obtained for these three sites. The under-ice profiles for 1.15-km transects at each site are indicated by a heavy solid line. The measured and modeled under-ice monostatic reverberation levels are indicated by the dashed and dotted lines, respectively. The timescale (in seconds) is indicated at the bottom of the figure. These results support the intuitive notion that relatively high reverberation levels are associated with areas of high ice roughness. More specifically, average keel draft determines the average scattering facet surface area (which determines the average amplitude of a specular return) and the keel frequency determines the total number of scattering facets (which determines the probability of scatter occurrence). Differences between measured and modeled results are thought to be due to significant scatterers present in the under-ice surface but not present in either the measured or modeled data.

10.7 NUMERICAL MODEL SUMMARIES

A summary of stand-alone underwater acoustic reverberation models is presented in Table 10.1. The models are segregated according to cell-scattering or point-scattering approaches. Numbers within brackets following each model refer to a brief summary and appropriate documentation. Model documentation can range from informal programming commentaries to journal articles to detailed technical reports containing a listing of the actual computer code. Abbreviations and acronyms are defined in Appendix A. This summary does not claim to be exhaustive.

TABLE 10.1
Summary of Underwater Acoustic Reverberation Models

Cell Scattering		Point Scattering	
Monostatic	**Bistatic**	**Monostatic**	**Bistatic**
C-SNAP-REV [1]	ARTEMIS [17]	REVGEN [29]	BORIS-SSA [31]
DOP [2]	BAM [18]	RITSHPA [30]	Under-Ice Reverberation Simulation [32]
EIGEN/REVERB [3]	BiKR [19]		
HYREV [4]	BiRASP [20]		
MAM [5]	BISAPP [21]		
NOGRP/ROSELLA [6]	BISSM [22]		
PAREQ-REV [7]	BISTAR [23]		
PEREV [8]	MURAL [24]		
PERM-2D [9]	OGOPOGO [25]		
REV3D [10]	RASP [26]		
REVMOD [11]	RUMBLE [27]		
REVPA [12]	S-SCARAB [28]		
REVSIM [13]			
R-SNAP [14]			
TAMAR [15]			
TENAR [16]			

CELL SCATTERING

Monostatic

[1] C-SNAP-REV computes reverberation using the C-SNAP range-dependent normal-mode model. Range dependence of the environment is treated as a one-way coupled mode solution. Surface and bottom reverberation is obtained by integrating the received intensity over the area insonified by the emitted pulse and that contributes to the reverberation at a given time. An average sound speed is assumed for all the paths (i.e. no group-velocity dependence). The scattering is described by a mode-coupling matrix that is equivalent to the plane-wave scattering function evaluated at discrete angles corresponding to the modes. The incident and scattered mode angles are modified to take into account the local slope, and are given by the local phase velocity and the sound speed at the interface. Coherent summation of mode contributions is used to correctly model the effects of deep-water convergent zones. The empirical scattering function is based on Lambert's rule. The model deals mainly with the monostatic case, though the technique is extendable to bistatic geometries. (Unpublished notes by D.D. Ellis, SACLANT Undersea Research Centre, La Spezia, Italy, various dates.)

[2] DOP divides the ocean into time-Doppler cells, sums the received energy incoherently, and produces a spectrum for the surface, volume or bottom reverberation at a given time after transmission (Marsh, 1976).

[3] EIGEN/REVERB is a series of programs used to calculate ambient noise, reverberation-versus-time signals, and transmission-loss values (Sienkiewicz et al., 1975). These programs are based on NISSM.

[4] HYREV is a high-frequency, monostatic reverberation model suitable for shallow-water environments. Arrival times and transmission losses from the source to scatterers are obtained from the appropriate eigenrays. The composite-roughness theory is used to predict the boundary scattering (Choi et al., 2002).

(Continued)

TABLE 10.1 (Continued)
Summary of Underwater Acoustic Reverberation Models

[5] MAM was designed as a monostatic companion to the bistatic BAM model. For monostatic operation, it is required that the source and receiver be located at the same horizontal location, although they may differ in depth (Bartberger, 1991b; Vendetti et al., 1993b).

[6] NOGRP/ROSELLA first runs the normal-mode program POPP (a variant of PROLOS), which calls the normal mode subprogram MODES and writes out a binary file of mode information that is then read by the monostatic reverberation code. ROSELLA, an extension of NOGRP, is used to execute the reverberation calculations with beam patterns (Ellis, 2007, 2008).

[7] PAREQ-REV is a range-dependent wave-theory model based on the parabolic approximation of the wave equation. The numerical method uses the split-step Fourier marching solution with automatic interpolation of environmental data with range. The code allows a choice of either the standard Tappert-Hardin parabolic equation or the wide-angle equation of Thomson-Chapman. Several choices of starting fields are provided, including a Gaussian source beam of varying width and tilt with respect to the horizontal. Reverberation from the ocean boundaries is computed using standard scattering laws: Lambert's rule for bottom backscatter, and either Chapman-Harris curves or Lambert's rule for sea-surface backscatter. The computational scheme uses reciprocity of propagation to compute the reverberation field for arbitrary receiver depths at the source range. (Unpublished notes by H.G. Schneider, SACLANT Undersea Research Centre, La Spezia, Italy, various dates.)

[8] PEREV (Tappert's PE reverberation model), together with the UMPE propagation model, are described by Smith et al. (1993, 1996).

[9] PERM-2D computes backscatter from range-dependent bathymetry in the oceanic waveguide (Lingevitch, 2008). This technique extends the approach of Collins and Evans (1992) to problems involving small-scale and large-scale boundary roughness. The PERM-2D model subdivides the oceanic waveguide into range-independent regions and applies the single-scattering approximation (that is, multiple forward and backward coupling in the scattered fields is neglected) to formulate a scattering problem for the reflected and transmitted pressure field at each range step. Forward-scattering loss, which can be significant at long ranges and very rough surfaces, is included in the solution. Wide-angle operators, which are accurate for the propagating and evanescent spectrum, are applied to yield stable and convergent iteration formulas for the reflected and transmitted fields (Milinazzo et al., 1997). Unlike perturbative methods that are restricted to small-roughness amplitudes, the PERM-2D model is valid for arbitrary roughness subject to the single-scattering approximation.

[10] REV3D is a deterministic version of the 3D model MOC3D. A reverberation trace can be built up by summing modified ray intensities for incoming and outgoing paths separately (Ivansson et al., 2009). REV3D was used successfully to analyze the impacts of out-of-plane scattering on the impulse responses observed during a short-range, shallow-water experiment conducted at 25 kHz. Specifically, the manifested long decaying spiky tail would be have been difficult to explain using only 2D propagation models (Ivansson and Karlsson, 2016).

[11] REVMOD calculates reverberation power spectra (C.L. Ackerman and R.L. Kesser, 1973, unpublished manuscript; Hodgkiss, 1980, 1984). (The REVMOD model is discussed in detail in Section 10.4.)

[12] REVPA is a parabolic-equation reverberation model intended for shallow-water applications (Bouchage and LePage, 2002).

[13] REVSIM generates coherent, multi-beam, non-stationary reverberation time series (Chamberlain and Galli, 1983).

(Continued)

TABLE 10.1 (Continued)
Summary of Underwater Acoustic Reverberation Models

[14] R-SNAP is a coherent monostatic reverberation model employing the range-dependent propagation model SNAP (LePage, 2003; Prior et al., 2002).

[15] TAMAR (towed array model of acoustic reverberation) is a 3D global reverberation model that predicts reverberation power versus time after beam-forming and matched-filter processing (Jenserud and Knudsen, 2004). As a ray model, TAMAR handles weakly range-dependent environments. It includes local models for scattering and reflection from the sea surface and sea floor. Local bottom scattering models have several levels of complexity ranging from simple phenomenological models to complex physics-based models.

[16] TENAR is a subroutine that uses the sonar equation to calculate underwater target echoes, reverberation and noise (Luby and Lytle, 1987).

CELL SCATTERING

Bistatic

[17] ARTEMIS (adiabatic reverberation and target echo mode incoherent sum) is a general-purpose numerical model of bistatic target echo level and surface and bottom reverberation for bistatic arrangements in an arbitrary range-dependent environment with arbitrary sound-speed variation (Harrison, 2008). The model minimizes computation time while retaining a reasonably accurate power envelope. The approach is based on the adiabatic normal-mode approximation, but with the modal series treated as a continuum and with WKB mode amplitudes (excluding the oscillatory modulation). Outputs are three dimensional and they can be presented in map form as target echo, reverberation or signal-to-reverberation-ratio. Given sparse environmental data, the trade-off between accuracy and speed is negotiated by intelligent interpolation. This is done by constructing quantities (functions of the desired variables, such as cycle distance, ray angles, and so on) that are more or less linear in space or in mode number. These are converted back to the original variables after linear interpolation. Harrison (2013) modified the computationally efficient energy-flux approach to include focusing, ray convergence, and caustic-like behavior. This derivation started with the coherent normal mode sum but retained only terms that interfered on a scale of a ray cycle distance. By starting with the adiabatic-mode sum, Harrison (2015) extended this formulation to a slowly varying range-dependent environment and applied it to the ARTEMIS target-echo and reverberation model.

[18] BAM predicts the performance of a sonar system consisting of a sound source and a receiver separated in both range and depth (Bartberger, 1985, 1991a; Vendetti et al., 1993a). The model computes the echo-to-background ratios for targets located at a set of bipolar grid locations in a horizontal plane at a specified target depth. The bipolar grid points are defined at the intersections of circles of constant range about the source and receiver. (BAM is discussed in detail in Section 10.5.2.)

[19] BiKR is a bistatic reverberation model (Fromm, 1999) based on the KRAKEN propagation model.

[20] BiRASP extended the RASP model to handle arbitrary (bistatic) source and receiver configurations in a three-dimensional, range-dependent environment (Fromm et al., 1996). RASP had been previously modified to predict range-dependent, monostatic reverberation at higher frequencies (up to 10 kHz) and in water shallower than originally intended. This modification was referred to as the Shallow Water RASP Upgrade (Fulford, 1991).

[21] BISAPP uses an integral fast eigenray model to calculate point-by-point echo levels to facilitate analysis of bistatic and multistatic scenarios (Pomerenk and Novick, 1987).

(*Continued*)

TABLE 10.1 (Continued)
Summary of Underwater Acoustic Reverberation Models

[22] BISSM computes bistatic bottom scattering strengths (Caruthers et al., 1990; Caruthers and Novarini, 1993; Caruthers and Yoerger, 1993).

[23] BISTAR is a bistatic, range-dependent reverberation model based on the method of coupled normal modes. The environment is discretized into range-independent segments and the outward propagating field is coupled at the interfaces under the single-scatter hypothesis. The propagation theory and implementation are those of C-SNAP. In order to implement the reverberation prediction, the CW field estimate of C-SNAP has been augmented with a narrowband time-series estimator. The time-domain estimates for the range-dependent environment are available separately and are also used to obtain the reverberation estimates. The scattering process itself can be modeled either as a parametric scattering strength such as Lambert's law or via perturbation theory. The model includes coherent propagation to and from the scattering patch (LePage, 1999, 2002; LePage and Harrison, 2003a,b). (Also, unpublished notes by K.D. LePage, SACLANT Undersea Research Centre, La Spezia, Italy, various dates.)

[24] MURAL, the multistatic reverberation algorithm, supports trainer development and can be used with any propagation model that produces range-sampled grids of transmission loss, travel time, launch and grazing angles. MURAL contains functions that calculate propagation, scattering and beam patterns. The operation of MURAL couples the algorithm controls to the requested resolution of the prediction with the goal of self-optimizing its performance for the requested resolution (Fromm, 2011).

[25] OGOPOGO is based on the Bucker-Morris method for computing shallow-water boundary reverberation using normal modes to calculate the acoustic energy propagating from the source to the scattering area and back to the receiver. Ray-mode analogies and empirical scattering functions are used to compute the scattered energy at the scattering area (Ellis, 1995). The normal-mode model PROLOS computes the propagation loss. Travel times of the reverberation signals are derived from the modal-group velocities. Volume reverberation from either the water column or the subbottom is not currently included, but boundary reverberation is computed using empirical scattering functions and ray-mode analogies. Both monostatic and bistatic geometries can be handled, and horizontal or vertical arrays can be specified for the source and receiver. OGOPOGO was used to interpret reverberation measurements from shallow-water sites in the frequency range 25–1000 Hz (Desharnais and Ellis, 1997). Results from the 2006 ONR reverberation modeling workshop were discussed by Ellis (2008).

[26] RASP is a sequence of computer programs using multipath propagation and scattering processes to predict the long-range, low-frequency boundary reverberation and target returns that would be received in real ocean environments (Franchi et al., 1984; Palmer and Fromm, 1992).

[27] RUMBLE calculates reverberation as a function of time for bistatic sonars (Bucker, 1986; Kewley and Bucker, 1987).

[28] S-SCARAB is a range-independent raytracing model for calculating both forward propagating and reverberant acoustic fields. Features include improved algorithms to calculate the coherent forward propagating field, the inclusion of upward refracting rays in the sediment, definition of bi-static source-receiver geometries and the possibility of specifying 3-D beam patterns of the source and receivers. The forward and back propagating acoustic fields are calculated by tracing rays both in the water column and sediment. Only contributions from the ocean bottom are considered in calculation of the reverberation. The local scattering properties of the seabed are described by known power-law expressions for both the interfaces of stratified sediment layers and volume inhomogeneities in the sediment. These scattering kernels are dependent on physical

(Continued)

TABLE 10.1 (Continued)
Summary of Underwater Acoustic Reverberation Models

descriptors of the bottom such as seabed roughness. S-SCARAB is computationally efficient compared to other reverberation tools based on normal-mode or parabolic equation approaches, particularly at higher frequencies (Marconi et al., 2004).

POINT SCATTERING

Monostatic

[29] REVGEN produces digital baseband samples of transducer or beamformer signals for use by active or passive sonars (Princehouse, 1977).

[30] RITSHPA is a reverberation module intended for use in high-resolution, wideband sonar simulators (Lam et al., 2006) in which reverberation is stochastic (non-Rayleigh) and follows the K-distribution (Abraham and Lyons, 2004). Hybrid multipaths, in which the return path is different from the transmit path, are also considered. RITSHPA assumes spherical propagation loss (i.e. isovelocity water column) and uses widely known formulas to compute reflection and scattering at boundaries.

POINT SCATTERING

Bistatic

[31] BORIS-SSA (bottom reverberation from inhomogeneities and surfaces—small-slope approximation) simulates time series resulting from acoustic scattering off various seafloor types involving various source-receiver geometries (Pouliquen et al., 1998; Canepa and Berron, 2006; Soukup et al., 2007; Staelens, 2009; Blondel and Pace, 2009). This package is an upgrade of BORIS-3D. The model parameters characterize the sonar directivity and pulse shape, the geometrical configuration of the scattering problem and the geophysical characteristics of the seafloor, the sea surface or other surfaces. These surfaces can have various statistical behaviors or can be obtained from deterministic data based on measured surface heights.

[32] Under-Ice Reverberation Simulation is a bistatic, high-frequency (≥2 kHz), under-ice acoustic-scattering model to evaluate the scatter produced by a pulse, originating from an arbitrarily located source, as detected by an arbitrarily located receiver (Bishop et al., 1986, 1987; Bishop, 1987; Bishop, 1989a, 1989b). (This model is discussed in detail in Section 10.6.2.)

11 Sonar Performance Models

11.1 BACKGROUND

The ultimate purpose of sonar performance (SP) modeling is twofold. First, advanced sonar concepts can be optimally designed to exploit the ocean environment of interest. Second, existing sonars can be optimized for operation in any given ocean environment.

In the case of naval sonars, performance prediction products can be tailored to individual sonar systems by providing the sonar operators with on-scene equipment mode selection guidance. When combined with current tactical doctrine, this information product is commonly referred to as a tactical decision aid (TDA). These decision aids are used by the force commanders to optimize the employment of naval assets in any particular tactical environment at sea. Performance prediction products help both the force commanders and sonar operators to better understand and thus exploit the ocean environment from a tactical standpoint. SP models also support various naval underwater acoustic surveillance activities.

TDAs are in the realm of engagement modeling, which will be discussed more fully in Chapter 13. In essence, engagement models use sonar detection performance data (either measured or predicted by SP models) to simulate integrated system performance. These simulations are usually conducted in the context of a naval force that is set in opposition to a hypothetical threat force. The output data typically include exchange ratios, which are useful in determining force level requirements and in developing new tactics.

SP models use active and passive sonar equations to generate performance predictions. Mathematical models of propagation, noise, and reverberation generate the input variables required for solution of these equations. This hierarchy of models was previously illustrated in Figure 1.1. SP models can logically be separated into active sonar models and passive sonar models, as would be suggested by the distinction between active and passive sonar equations.

The complexity of the SP-modeling problem is a natural consequence of the naval operations that these models support. In modern naval battle group operations, for example, transmission loss (TL) must be calculated along a multiplicity of paths connecting widely separated sources and receivers. Noise interference from nearby consort vessels must be factored into SP calculations together with distant merchant shipping and local weather noises. Furthermore, scattering and reverberation from bathymetric features and volumetric false targets must be efficiently and realistically modeled. The demands placed on computational efficiency and database management are enormous. Underwater acoustic modelers face challenges that severely tax existing mathematical methods. Moreover, with the advent of multistatic scenarios involving multiple sets of separated sources and receivers, true three-dimensional (3D) modeling is no longer a theoretical luxury but rather a practical necessity.

This situation has become even more complex with the heightened interest in shallow-water sonar operations. Acoustic interactions with highly variable seafloor topographies and compositions further compound already intensive scattering and reverberation calculations. In addition, the sound speed field in shallow-water areas is often characterized by high spatial and temporal variability. As a result, statistical approaches have been explored in order to obtain meaningful predictions of active SP in shallow-water environments. In the MOCASSIN model, for example, Schneider (1990) provided the option to specify a stochastic sound speed field in order to approximate the horizontal variability typical of coastal regions.

11.2 SONAR EQUATIONS

11.2.1 Monostatic Sonars

The sonar equations are simple algebraic expressions used to quantify various aspects of SP. These equations vary according to active or passive sonars, and within active sonars they further differentiate according to noise or reverberation limited conditions. The various system and environmental parameters that make up the sonar equations are listed in Table 11.1 together with reference locations and brief descriptions. A condensed statement of the equations is presented as follows (Urick, 1983: Chapter 2):

Active sonars (monostatic)

Noise background

$$\mathrm{SL} - 2\mathrm{TL} + \mathrm{TS} = \mathrm{NL} - \mathrm{DI} + \mathrm{RD}_{\mathrm{N}} \tag{11.1}$$

Reverberation background

$$\mathrm{SL} - 2\mathrm{TL} + \mathrm{TS} = \mathrm{RL} + \mathrm{RD}_{\mathrm{R}} \tag{11.2}$$

Passive sonars

$$\mathrm{SL} - \mathrm{TL} = \mathrm{NL} - \mathrm{DI} + \mathrm{RD} \tag{11.3}$$

In this application, recognition differential (RD) replaces detection threshold (DT) as used by Urick (1983: Chapter 12). Dawe (1997) distinguished the modern usage of RD in sonar modeling from its obsolescent usage in relation to auditory detection. The concepts of DT and RD are related to each other through the level at which a signal can be detected for a given combination of probability of detection (P_{D}) and probability of false alarm (P_{F}). The difference between DT and RD relates to the location in the information-processing chain at which the threshold signal-to-noise ratio (SNR) is effectively measured. For DT, the SNR is measured at the receiver input terminals. For RD, the SNR is measured at the display.

TABLE 11.1
Sonar Parameter Definitions and Reference Locations

Parameter	Symbol	Reference	Definition
Source level	SL	1 m from source on its acoustic axis	$10\log_{10}\dfrac{\text{intensity of source}}{\text{reference intensity}^{a}}$
Transmission loss	TL	1 m from source and at target or receiver	$10\log_{10}\dfrac{\text{signal intensity at 1 m}}{\text{signal intensity at target or receiver}}$
Target strength	TS	1 m from acoustic center of target	$10\log_{10}\dfrac{\text{echo intensity at 1 m from target}}{\text{incident intensity}}$
Noise level	NL	At hydrophone location	$10\log_{10}\dfrac{\text{noise intensity}}{\text{reference intensity}^{a}}$
Receiving directivity index	DI	At hydrophone terminals	$10\log_{10}\dfrac{\text{noise power generated by an equivalent nondirectional hydrophone}}{\text{noise power generated by actual hydrophone}}$
Reverberation level	RL	At hydrophone terminals	$10\log_{10}\dfrac{\text{reverberation power at hydrophone terminals}}{\text{power generated by signal of reference intensity}^{a}}$
Recognition differential	RD	At display	$10\log_{10}\dfrac{\text{signal power to just perform certain function}}{\text{noise power at display}}$

Source: Urick, R.J., *Principles of Underwater Sound*, McGraw-Hill, New York, 1983. Reproduced with permission of McGraw-Hill.

[a] The reference intensity is that of a plane wave of rms pressure 1 μPa.

Furthermore, RD has been subscripted in the case of active sonars to make it clear that this parameter is quantitatively different for noise (N) and reverberation (R) backgrounds. The sonar equations are valid for the condition in which the signal excess is zero. Beampattern functions are implicitly included in the TL term. The apparent simplicity of the individual components of the sonar equations can be misleading. The complexity of these parameters will be better appreciated through the discussions to be presented in Sections 11.3 and 11.4. These equations are valid for sound levels in a 1-Hz wide frequency band (referred to as spectrum level). For frequency bands greater than 1 Hz, the sound levels in Equations 11.1 through 11.3 must be corrected for bandwidth (e.g., Kinsler et al., 1982: 409–17).

It is convenient in practical work to assign separate names to different combinations of terms in the sonar equations. A summary of commonly used terms is presented in Table 11.2. The combination of terms labeled figure of merit (FOM) is the most useful because it combines environmental, sonar, and target parameters into one convenient quantitative measure of SP. Since the FOM quantitatively equals the TL under the conditions specified in the sonar equation, the FOM can give an immediate indication of the range at which active or passive sonars can detect targets provided that corresponding TL curves are available. When using the FOM in this manner, it should be stressed that the TL curves must match the ocean environment, acoustic frequency, sonar depth, target depth, and other sonar parameters used in computing the FOM value. In the case of active sonars that are reverberation (versus noise) limited, the FOM is not constant but varies with range (time) and thus fails to be a useful indicator of active SP.

TABLE 11.2
Terminology of Various Combinations of Sonar Parameters

Name	Parameters	Remarks
Echo signal level	SL – 2 TL + TS	The intensity of the echo as measured in the water at the hydrophone.
Noise-masking level	NL – DI + RD_N	Another name for these two combinations (noise-masking level and reverberation-masking level) is minimum detectable echo level.
Reverberation-masking level	RL + RD_R	
Signal excess	SL – 2 TL + TS – (NL – DI) – RD_N	Detection just occurs, under the probability conditions implied in RD_N, when the signal excess is zero. This expression pertains to a noise-limited active sonar.
Performance figure	SL – (NL – DI)	Difference between the SL and the NL measured at the hydrophone terminals.
Figure of merit (FOM)	SL – (NL – DI) – RD	Equals the maximum allowable one-way TL in passive sonars, or the maximum allowable two-way TL in active sonars when TS = 0.

Source: Urick, R.J., *Principles of Underwater Sound*, McGraw-Hill, New York, 1983. Reproduced with permission of McGraw-Hill.

The parameters listed in Table 11.1 can be related to the functions of basic acoustic models. Specifically, propagation models generate estimates of TL, noise models predict noise levels (NL), and reverberation models provide estimates of the reverberation level (RL). For passive sonars, the sonar equipment will define directivity index (DI) and RD while the target will define source level (SL); target strength (TS) is not considered in purely passive sonar models. For active sonars, the sonar equipment will define SL, DI, and RD, and the target will define TS.

Useful overviews of sonar signal processing have been assembled by Knight et al. (1981) and by Nielsen (1991). Vaccaro (1998) edited a collection of papers that reviewed past progress and future challenges in underwater acoustic signal processing. Specific applications included sonar signal processing, time-delay estimation, and underwater acoustic communications.

11.2.2 Bistatic Sonars

Bistatic geometries are characterized by a triangle of source, target, and receiver positions. Bistatic sonar equations, which differ from their monostatic counterparts, are required. Cox (1989) carefully developed a set of bistatic sonar equations using energy forms to anticipate the wide variety of waveforms likely to be used in such bistatic systems. The following nomenclature is introduced to facilitate subsequent discussions (also refer to Figure 11.1):

c = speed of sound
R_M = monostatic range from target (T) to collocated source-receiver (S–R)
TL_M = monostatic TL from target (T) to collocated source-receiver (S–R)
R_1 = range from source (S) to target (T)
R_2 = range from target (T) to receiver (R)
R_3 = range from source (S) to receiver (R)
TL_1 = TL from source (S) to target (T)
TL_2 = TL from target (T) to receiver (R)
TL_3 = TL from source (S) to receiver (R)
τ_1 = travel time from source (S) to target (T)
τ_2 = travel time from target (T) to receiver (R)
τ_3 = travel time from source (S) to receiver (R)
V_T = target velocity
β = bistatic angle ($\angle$STR)
θ = monostatic aspect angle
θ_E = equivalent (bistatic) aspect angle
θ_{TS} = relative bearing of source (S) from target (T)
θ_{TR} = relative bearing of receiver (R) from target (T)
θ_{TB} = angle from target heading to bisector of β

By reference to Figure 11.1, the equivalent bistatic aspect angle (θ_E) is defined as

$$\cos\theta_E = \frac{\cos\theta_{TS} + \cos\theta_{TR}}{2} = \cos\theta_{TB}\cos\left(\frac{\beta}{2}\right) \tag{11.4}$$

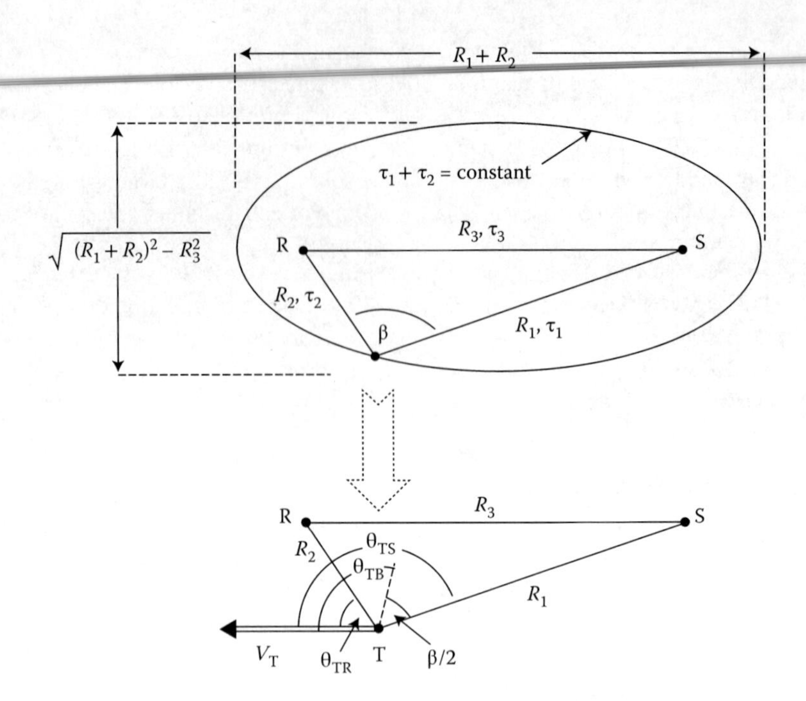

FIGURE 11.1 Bistatic geometry. (Adapted from Cox, H., *Underwater Acoustic Data Processing, NATO Advanced Science Institutes Series E: Applied Sciences*, Kluwer Academic Publishers, Dordrecht, 3–24, 1989. With permission.)

The energy source level (ESL) is related to the intensity source level (SL) as

$$\mathrm{ESL} = \mathrm{SL} + 10\log_{10} T \tag{11.5}$$

where T is the duration of the transmitted pulse. The echo energy level (EEL) received from the target at a hydrophone on the receiver array is then

$$\mathrm{EEL} = \mathrm{ESL} - \mathrm{TL}_1 - \mathrm{TL}_2 + \mathrm{TS} \tag{11.6}$$

where TS is the target strength. The noise-limited signal excess ($\mathrm{SE_N}$) can be defined as

$$\mathrm{SE_N} = \mathrm{ESL} - \mathrm{TL}_1 - \mathrm{TL}_2 + \mathrm{TS} - N_0 + \mathrm{AG_N} - \Lambda - L \tag{11.7}$$

where N_0 is the noise spectral level, $\mathrm{AG_N}$ is the array gain against noise, Λ is the threshold on the SNR required for detection, and L is a loss term to account for time spreading and system losses.

Cox (1989) noted that the problem of estimating reverberation-limited bistatic SP is more complicated than the monostatic case since it involves summing the contributions of a large number of scatterers ensonified by numerous propagation paths that differ in angle of incidence and position on the beampatterns (BPs) of the source

TABLE 11.3
Comparison of Monostatic and Bistatic Sonar Parameters

Parameter	Monostatic	Bistatic
Time of arrival	circle: $\frac{2R_M}{c}$	ellipse: $\frac{(R_1 + R_2)}{c}$
Bistatic angle	0	β
Aspect angle	θ	θ_E
Doppler	$2\frac{V_T}{c}\cos\theta$	$2\frac{V_T}{c}\cos\theta_E$
Target strength	$f(\cos\theta)$	$f(\cos\theta_E)$
Transmission loss	$2TL_M$	$TL_1 + TL_2$

Source: Cox, H., *Underwater Acoustic Data Processing, NATO Advanced Science Institutes Series E: Applied Sciences*, Kluwer Academic Publishers, Dordrecht, 3–24, 1989. With permission.

and receiver. Let R_0 represent the reverberation spectral level, then the signal excess can be written to anticipate both noise-limited and reverberation-limited cases as

$$SE = ESL - TL_1 - TL_2 - [(N_0 - AG_N) \oplus R_0] + TS - \Lambda - L \tag{11.8}$$

where $\oplus$ represents power summation (Chapter 7, see Equation 7.1).

Equations 11.7 and 11.8 do not include the effects of the direct blast. Analogies to the direct blast in monostatic sonars are the fathometer returns, which are the multiple surface-reflected and bottom-reflected arrivals following each transmission. Comparisons of selected monostatic and bistatic parameters are presented in Table 11.3.

These results can be generalized to multistatic geometries by observing that each source-target-receiver trio forms a bistatic set. It follows that additional localization opportunities can be obtained when the same target encounters multiple bistatic sets.

11.2.3 Multistatic Sonars

Multistatic scenarios involve multiple sets of separated sources and receivers. Harrison (2002b) noted that multistatic sonars introduce new problems in operational use such as mutual interference from multiple sources and variable target aspects in a background of range-dependent bistatic reverberation. To help design better trials for multistatic equipment and to investigate new tactics and tactical displays, it is essential to model these effects. Typically, however, existing propagation, reverberation, and target models are inadequate for the overall task. The multistatic sonar performance model SUPREMO includes propagation, target echoes, reverberation, and noise, all plotted as a function of delay time and bearing using an equivalent-map projection. A modular propagation section makes it possible to separate the effects of propagation modeling (e.g., theoretical basis and range-dependence) from those of scattering computations (e.g., computational efficiency, bottom slope, and scattering law). Special attention was paid to problems of interference from multiple sources firing in sequence, target-aspect dependence from multiple receivers, mixed FM and CW signals, mismatched source and receivers, multiple displays (one for each bistatic pair), and presentation of results.

Predicting the search effectiveness of a distributed, multistatic-sensor field is highly conditioned on uncertain tactical information (Incze and Dasinger, 2006). It is useful to have a method for assessing the military utility of sensor systems to inform decisions concerning optimal employment tactics and system procurement options. The combination of Monte-Carlo simulation methods and Bayesian-fusion techniques provides a robust approach for modeling the effects of uncertainty on the distribution of likely tactical outcomes.

11.3 NISSM MODEL—A SPECIFIC EXAMPLE

The Navy interim surface ship model (NISSM) II, developed by Weinberg (1973), is a computer program designed to predict the performance of active monostatic sonar systems using ray-theoretical techniques. The measure of performance is expressed in terms of (P_D) versus target range for a given false-alarm rate. Selected intermediate results can also be displayed, including ray traces, TL, and boundary and volume reverberation. This model is applicable to range-independent ocean environments.

NISSM uses a cell-scattering model for reverberation. The most time-consuming phase in the execution of NISSM is the volume reverberation computation. The volume scattering strength is expressed in terms of a column (or integrated) scattering strength, and a careful integration of the volume reverberation integral is performed, which considers multipath structure.

Because the NISSM II model encompasses all the modeling categories addressed thus far, and because it has been widely used in the sonar-modeling community, it is considered to be representative of the more general class of active sonar models. Examples of selected NISSM outputs are also presented in this section.

11.3.1 Propagation

The depth-dependent sound speed $c(z)$ and the inclination angle (θ) at a point on a ray in the x–z plane can be related by Snell's law (see Figure 11.2) as

$$\frac{c(z)}{\cos\theta} = c_v$$

where $c(z)$ is the sound speed as a function of depth (z), θ is the inclination angle, and c_v is the vertex sound speed. Snell's law implies that the range (x) and travel time (T) along the ray can be calculated as

$$x = x_a + \int_{z_a}^{z} \frac{c(z)|dz|}{\sqrt{c_v^2 - c(z)^2}} \tag{11.9}$$

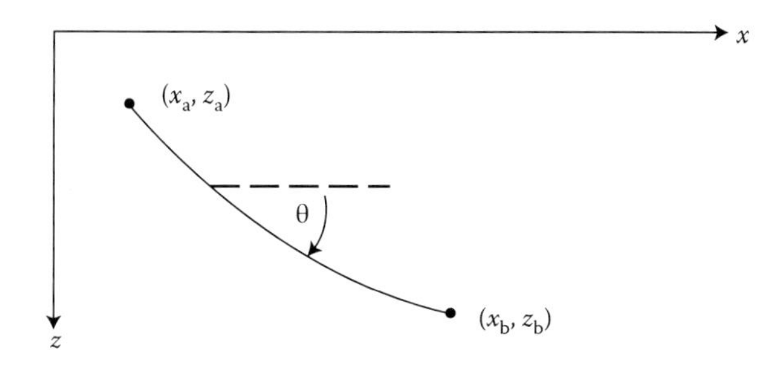

FIGURE 11.2 Ray geometry showing inclination angle along a ray, as used in the NISSM active sonar model. (Adapted from Weinberg, H., Navy interim surface ship model [NISSM] II. Nav. Underwater Syst. Ctr, Tech. Rept 4527 [also, Nav. Undersea Ctr, Tech. Pub. 372], 1973.)

$$T = T_a + \int_{z_a}^{z} \frac{c_v \left| dz \right|}{c(z)\sqrt{c_v^2 - c(z)^2}} \tag{11.10}$$

Officer (1958: Chapter 2) provided a rigorous development of these equations. The range derivative ($\partial x/\partial c_v$) is given by

$$\left.\frac{\partial x}{\partial c_v}\right|_{z=\text{const}} = \left.\frac{\partial x_a}{\partial c_v}\right|_{z=z_a} - \int_{z_a}^{z} \frac{c(z)c_v \left| dz \right|}{\left[c_v^2 - c(z)^2\right]^{3/2}} \tag{11.11}$$

(which equals zero at caustics) and is used in computing the geometrical spreading loss factor (η_{sp}):

$$\eta_{sp} = \left| p \right| = \left| c_v \tan\theta \tan\theta_0 \, x \frac{\partial x}{\partial c_v} \right|^{-1/2} \tag{11.12}$$

where θ_0 is the inclination angle of the ray at a point source of unit magnitude situated at $x_a = 0$, $z_a = z_0$.

The relative geometrical acoustic pressure field at field point (x, y) at time (t) is

$$p = \eta_{sp}\, e^{-i\omega(t-T)} \tag{11.13}$$

where $\omega = 2\pi f$ and f = frequency.

Since the range, travel time, and range-derivative integrals are symmetric with respect to the initial and final depths, it follows that the pressure satisfies the law of reciprocity. This law states that the acoustic pressure will remain unchanged if the source and receiver positions are interchanged. The sound speed profile is approximated by a continuous function of depth (Weinberg, 1971) using Leroy's (1969) sound speed formula. Earth curvature corrections are also applied.

The relative acoustic pressure along a ray is found by multiplying the relative geometrical acoustic pressure (p) by the surface, bottom, and absorption loss factors:

$$p = \eta_{sp}(\eta_{sur})^{n_{sur}}(\eta_{bot})^{n_{bot}} e^{-i\omega t + i\omega T + i\Phi + \alpha S} \tag{11.14}$$

and the TL is given by

$$\mathrm{TL} = -20\log_{10}|p| \tag{11.15}$$

where

$N_{sp} = -20 \log_{10} \eta_{sp}$ is the geometrical spreading loss (dB)
$N_{sur} = -20 \log_{10} \eta_{sur}$ is the surface loss per bounce (dB/bounce)
n_{sur} = number of surface bounces
$N_{bot} = -20 \log_{10} \eta_{bot}$ is the bottom loss per bounce (dB/bounce)
n_{bot} = number of bottom bounces
$A = -20 \log_{10} (\alpha)/\log_e (10)$ is the absorption coefficient (dB km^{-1})
α = absorption in nepers per unit distance
t = elapsed time (s)
T = travel time (s)
F = accumulated phase shift (rad)
S = arc length (km)
$\omega = 2\pi f$ = radian frequency (rad s^{-1})
f = acoustic frequency (Hz)

The surface loss may be input as a table of surface loss per bounce versus angle of incidence. The bottom loss may either be input as a table of bottom loss per bounce versus angle of incidence, or internally computed either by marine geophysical survey (MGS) data or by an empirical equation.

The absorption coefficient can be represented in one of three ways: (1) input as a table of absorption per unit distance versus frequency; (2) calculated by Thorp's (1967) equation; or (3) calculated by the equation attributed to H.R. Hall and W.H. Watson (1967, unpublished manuscript).

Eigenrays are found by tracing a preselected fan of test rays to the target depth. When two adjacent rays of the same type bracket a target range, a cubic spline interpolation is performed to determine the eigenray. The principal ray types were defined previously in Figure 4.2.

Shadow zone propagation is characterized according to Pekeris (1946) as

$$\begin{gathered} |p^2| = |p_b^2| \frac{x_b}{x} \exp[-\alpha(x - x_b)] \\ \alpha = \frac{5.93}{c_{sur}} g_{sur}^{2/3} f^{1/3} \end{gathered} \tag{11.16}$$

where

p = pressure in the shadow zone at (x, z)
p_b = pressure on the shadow boundary at (x_b, z)

c_{sur} = surface sound speed (km s^{-1})
g_{sur} = surface sound speed gradient (s^{-1})
f = frequency (Hz)

A uniform asymptotic expansion at a caustic is used to calculate the pressure in the vicinity of a caustic (Ludwig, 1966). Phase shifts are accumulated as follows: π radians decrease in Φ at the (pressure-release) surface; π/2 radians decrease in Φ at caustics; and zero phase shift at the (rigid) bottom. The computation of surface-duct propagation by ray theory does not allow for leakage from the surface layer into the underlying medium. Therefore, the AMOS equations are used (Marsh and Schulkin, 1955).

11.3.2 Reverberation

The basic theory for reverberation modeling was described in Chapter 10. Referring to Figure 11.3, the reverberation pressure at the end of the closed path is determined in NISSM as

$$|p_{rev}|^2 = \int_R \left| d_{tra}\eta_{tra}(\theta_{a,1})p_1 p_2 \eta_{rec}(\theta_{a,2})\right|^2 \mu \, dR \tag{11.17}$$

where

$d_{tra} = 20 \log_{10} d_{tra}$ is the reference level of the transmitting array
$P_1 = -20 \log_{10} |p_1|$ is the TL of the incident ray
$P_2 = -20 \log_{10} |p_2|$ is the TL of the backscattered ray
$\eta_{tra} = -20 \log_{10} \eta_{tra}$ is the transmitter response (dB)
$\eta_{rec} = -20 \log_{10} \eta_{rec}$ is the receiver response (dB)
R = region containing the scatterers corresponding to rays with travel time (t):
$T - \tau \leq t \leq T$
τ = pulse length (s)

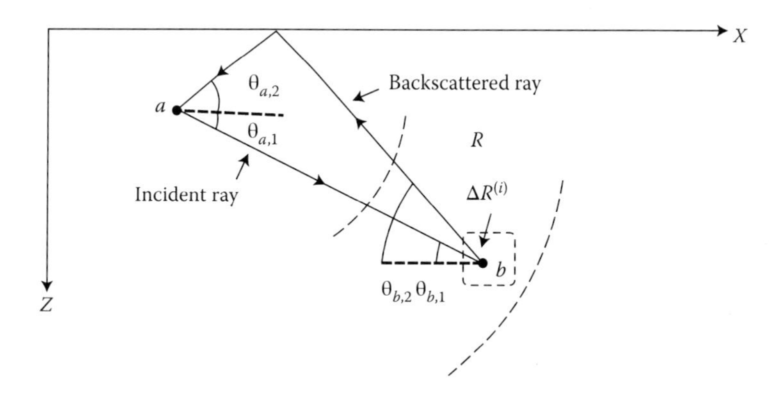

FIGURE 11.3 Closed ray path used in reverberation calculations in the NISSM active sonar model. (Adapted from Weinberg, H., Navy interim surface ship model [NISSM] II. Nav. Underwater Syst. Ctr, Tech. Rept 4527 [also, Nav. Undersea Ctr, Tech. Pub. 372], 1973.)

$T = t_0 + t_1 + t_2;\ 0 \le t_0 \le \tau$
t_0 = initial time
t_1 = travel time of incident ray
t_2 = travel time of backscattered ray
μ = backscattering strength (volume, surface, or bottom)

NISSM usually computes eight paths to each representative scatterer [$\Delta R^{(i)}$]. Thus, there may be as many as 56 (i.e., $8^2 - 8$) differently oriented closed paths per scatterer.

For volume reverberation, the expression for p_{rev} is replaced by the double summation (refer to Figure 11.4 for an illustration of the ensonified region)

$$\left|p_{\text{vol}}\right|^2 = \sum_{i,j} \left| d_{\text{tra}} \eta_{\text{tra}} \left(\theta_{a,1}^{(i,j)}\right) p_1^{(i,j)} p_2^{(i,j)} \eta_{\text{rec}} \left(\theta_{a,2}^{(i,j)}\right) \right|^2 \mu_{\text{vol}}^{(j)} x^{(i)} \Delta x^{(i,j)} \Delta z^{(j)} \Delta\phi \qquad (11.18)$$

where
$\Delta\phi$ = horizontal beamwidth (rad)
$x^{(i)}$ = range of the (i, j)th scatterer (km)
$z^{(j)}$ = depth of the (i, j)th scatterer (km)
$U_{\text{vol}}^{(j)} = 10 \log_{10} \mu_{\text{vol}}^{(j)}$ is the volume scattering strength per unit volume of ocean.

The surface reverberation is calculated using an expression similar to that for p_{rev} (Equation 11.17), except now

$U_{\text{sur}} = 10 \log_{10} \mu_{\text{sur}}$ is the surface scattering constant per unit area of sea surface (dB) (Chapman and Harris, 1962).

Bottom reverberation is calculated in a manner analogous to that for surface reverberation, but where

$U_{\text{bot}} = 10 \log_{10} \mu_{\text{bot}}$ is the bottom scattering constant per unit area of seafloor (dB) (Mackenzie, 1961).

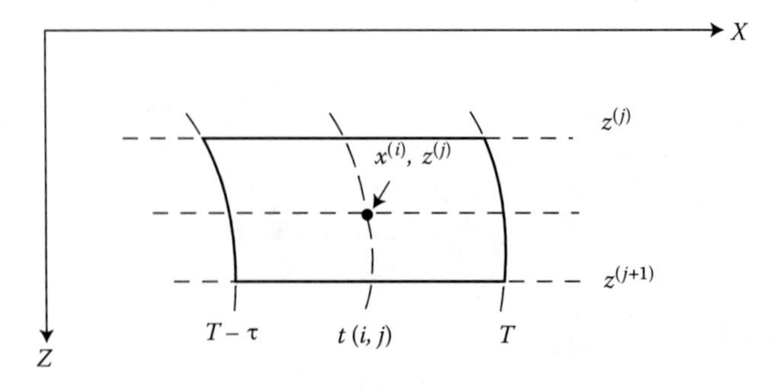

FIGURE 11.4 Ensonified region used in reverberation calculations in the NISSM active sonar model. (Adapted from Weinberg, H., Navy interim surface ship model [NISSM] II. Nav. Underwater Syst. Ctr, Tech. Rept 4527 [also, Nav. Undersea Ctr, Tech. Pub. 372], 1973.)

The total reverberation is assumed to be the random-phase addition of surface and bottom echoes from directly above and beneath the sonar (also referred to as fathometer returns), together with volume, surface, and bottom reverberation.

11.3.3 Target Echo

Target echo (or echo signal level [ESL]) at a particular time is determined by

$$\left|p_{\text{echo}}\right|^2 = \sum_i \left|d_{\text{tra}}\eta_{\text{tra}}\left(\theta_{a,1}^{(i)}\right)p_1^{(i)}p_2^{(i)}\eta_{\text{rec}}\left(\theta_{a,2}^{(i)}\right)\right|^2 \mu_{\text{tar}}^{(i)} \tag{11.19}$$

where $U_{\text{tar}} = 10 \log_{10} \mu_{\text{tar}}$ is the TS. The summation is taken over all closed paths with round-trip travel times between $T - \tau$ and T.

11.3.4 Noise

Noise is separated into two components: self-noise and ambient noise. Assuming isotropic noise fields:

$$P_{\text{noise}} = 10\log_{10}\left[\eta_{DI}\left(p_{\text{self}}^2 + p_{\text{amb}}^2\right)\right] \tag{11.20}$$

where

$DI = -10 \log_{10} \eta_{DI}$ is the directivity index (dB)
$P_{\text{self}} = 20 \log_{10} p_{\text{self}}$ is the self-noise spectrum density
$P_{\text{amb}} = 20 \log_{10} p_{\text{amb}}$ is the ambient noise spectrum density

11.3.5 Signal-to-Noise Ratio

Target echo to masking background, for a narrowband process, may be approximated by the SNR (s/n):

$$s/n = \frac{\Delta f_{\text{echo}} p_{\text{echo}}^2}{\Delta f_{\text{noise}} p_{\text{noise}}^2 + \Delta f_{\text{rev}} p_{\text{rev}}^2} \tag{11.21}$$

where

$P_{\text{echo}} = 20 \log_{10} p_{\text{echo}}$ is the target spectrum density at time T (dB)
$P_{\text{rev}} = 20 \log_{10} p_{\text{rev}}$ is the reverberation spectrum density at time T (dB)
$P_{\text{noise}} = 20 \log_{10} p_{\text{noise}}$ is the noise spectrum density at the beamformer output (dB)
Δf_{echo} = equivalent bandwidth for the received echo (Hz)
Δf_{rev} = equivalent bandwidth for the received reverberation (Hz)
Δf_{noise} = equivalent bandwidth for the received noise (Hz)

Assuming that the target echo is Gaussian distributed and centered about the receiving (Doppler shifted) frequency with a standard deviation of $\tau/2$, then

$$\Delta f_{\text{echo}} = \phi(\tau\Delta f) - \phi(-\tau\Delta f) \quad (11.22)$$

where Δf is the receiving bandwidth (Hz), τ is the pulse length (s), and where the function $\phi(x)$ is defined as

$$\phi(x) = \frac{1}{\sqrt{2\pi}} \int_{-\infty}^{x} e^{-y^2/2}\, dy \quad (11.23)$$

is the normal probability function (e.g., Burdic, 1991; Hassab, 1989; Ziomek, 1995).

Reverberation is also assumed to be Gaussian with a standard deviation of $\tau/2$, but is centered about the transmitting frequency. Therefore, reverberation energy falling outside the echo band is $\Delta f_{\text{rev}} p_{\text{rev}}^2$, where

$$\Delta f_{\text{rev}} = \phi(2\tau\Delta f_{\text{Dop}} + \tau\Delta f) - \phi(2\tau\Delta f_{\text{Dop}} - \tau\Delta f) \quad (11.24)$$

and where

$$\Delta f_{\text{Dop}} = 2f\frac{V_{\text{cl}}}{c_{\text{s}}} \quad (11.25)$$

where

Δf_{Dop} = Doppler shift (Hz)
f = frequency (Hz)
V_{cl} = closing speed (km s^{-1})
c_{s} = sound speed (km s^{-1})

Note that for passive systems the factor of 2 in Equation 11.25 is removed. If the noise-spectrum density can be considered constant over the receiving band, then

$$\Delta f_{\text{noise}} = \Delta f \quad (11.26)$$

11.3.6 Probability of Detection

P_{D} is modeled using a narrowband, square-law envelope detector (Figure 11.5). The input $x(t)$ is assumed to be either a stationary zero-mean Gaussian signal, $s(t)$, and noise, $n(t)$, or noise alone. It is assumed that the signal spectrum has the same bandwidth (B) as the narrowband filter and is centered on it, and that the noise is flat over the frequency interval. Samples of the squared-envelope, which are taken every $1/B$ seconds, are accumulated for an observation time of T seconds. The threshold (Ω) is fixed.

If P_{F} is a given probability of a false alarm, a quantity A is determined by inverting the following expression:

$$P_{\text{F}} = e^{-A} \sum_{m=0}^{M-1} \frac{1}{m!} A^m \quad (11.27)$$

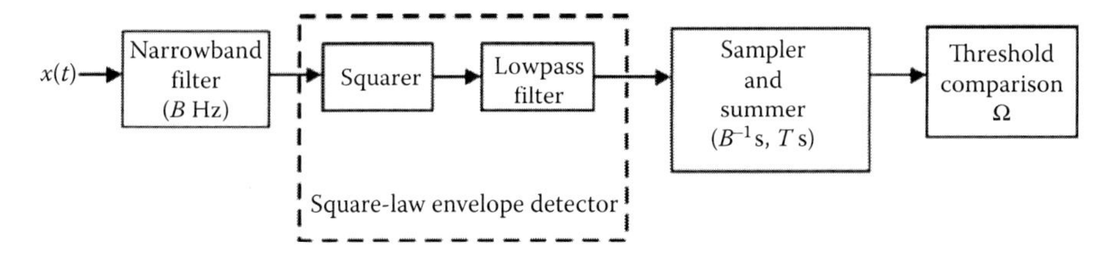

FIGURE 11.5 Narrowband detector employed in the NISSM active sonar model. (Adapted from Weinberg, H., Navy interim surface ship model [NISSM] II. Nav. Underwater Syst. Ctr, Tech. Rept 4527 [also, Nav. Undersea Ctr, Tech. Pub. 372], 1973.)

where $M = BT + 1$. Once A is known, the P_D is given by

$$P_D = e^{-A/(1+S/N)} \sum_{m=0}^{M-1} \frac{1}{m!} \left(\frac{A}{1+S/N} \right)^m \tag{11.28}$$

where (S/N) is the maximum value with respect to time of (s/n) at a particular target.

11.3.7 Model Outputs

NISSM II can generate data appropriate for graphical presentation of ray diagrams, TL versus range, reverberation (surface, bottom, volume, and total) versus time, SNR versus range, and (P_D) versus range. Sample outputs are illustrated for reverberation (Figure 11.6), SNR (Figure 11.7), and (P_D) (Figure 11.8). Representative illustrations of ray diagrams and TL have already been presented elsewhere in this book. These examples are presented for illustrative purposes only and are not intended to portray a complete model run.

The reverberation-versus-time plots illustrated in Figure 11.6 are valid for a deep-water environment with a shallow sonic layer depth, a critical depth, and sufficient depth excess to support convergence zone (CZ) propagation, as well as surface duct and bottom bounce paths. At about 11 s, a strong bottom bounce return is evident in the bottom reverberation curve. At about 63 s, a strong CZ return is evident in both the volume and surface reverberation curves (but not in the bottom reverberation curve).

The SNR plot in Figure 11.7 indicates a high (positive) ratio at ranges corresponding to direct path (0–2 km) and CZ (near 50 km) detections. These sonar detection opportunities are vividly demonstrated in Figure 11.8. Here, high probabilities of detection are associated with direct-path and CZ ranges.

It is sometimes useful to plot different combinations of active sonar parameters on the same graph. From Table 11.2, for example, ESL can be plotted together with the noise and reverberation masking levels (RMLs), as in Figure 11.9. With such information, signal excess as a function of range can be computed, as in Figure 11.10, which provides a direct indication of sonar detection performance. Specifically, all areas under the signal excess curve, but above the horizontal zero signal excess line, are associated with detection opportunities. That is, detection

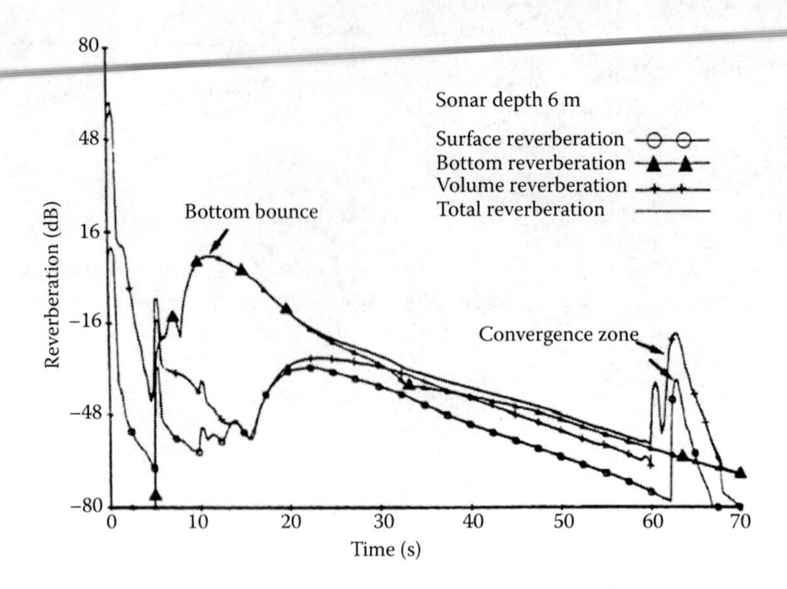

FIGURE 11.6 Sample reverberation versus time plot generated by the NISSM active sonar model. The bottom bounce contribution is readily visible in the bottom reverberation curve. The CZ contribution is evident in both the surface and volume reverberation curves. (Adapted from Weinberg, H., Navy interim surface ship model [NISSM] II. Nav. Underwater Syst. Ctr, Tech. Rept 4527 [also, Nav. Undersea Ctr, Tech. Pub. 372], 1973.) (The reverberation levels are reported here in dB re: 1 dyne cm^{-2}. To convert to levels in dB re: 1 μPa, add 100 dB to the levels shown in the figure.)

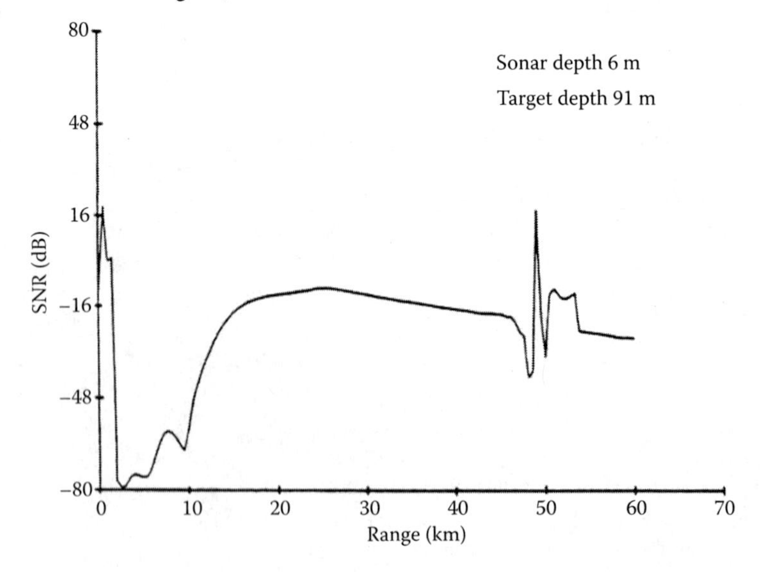

FIGURE 11.7 Sample SNR plot generated by the NISSM active sonar model based in part on the data presented in Figure 11.6. (Adapted from Weinberg, H., Navy interim surface ship model [NISSM] II. Nav. Underwater Syst. Ctr, Tech. Rept 4527 [also, Nav. Undersea Ctr, Tech. Pub. 372], 1973.)

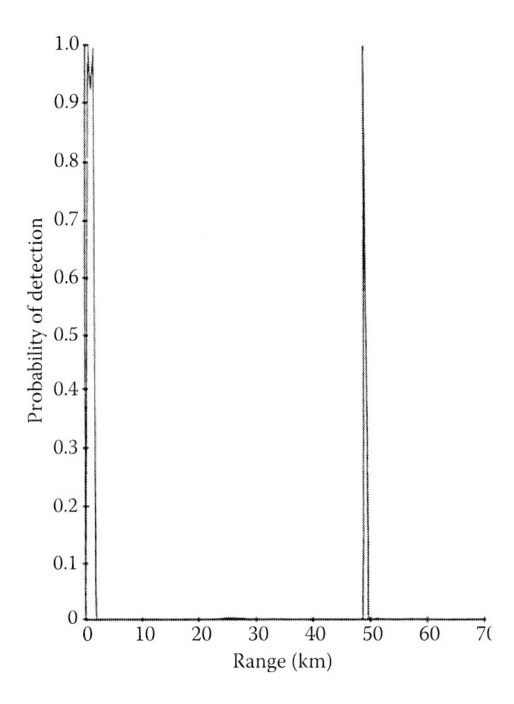

FIGURE 11.8 Sample (P_D) plot generated by the NISSM active sonar model based in part on the data presented in Figure 11.7. (Adapted from Weinberg, H., Navy interim surface ship model [NISSM] II. Nav. Underwater Syst. Ctr, Tech. Rept 4527 [also, Nav. Undersea Ctr, Tech. Pub. 372], 1973.)

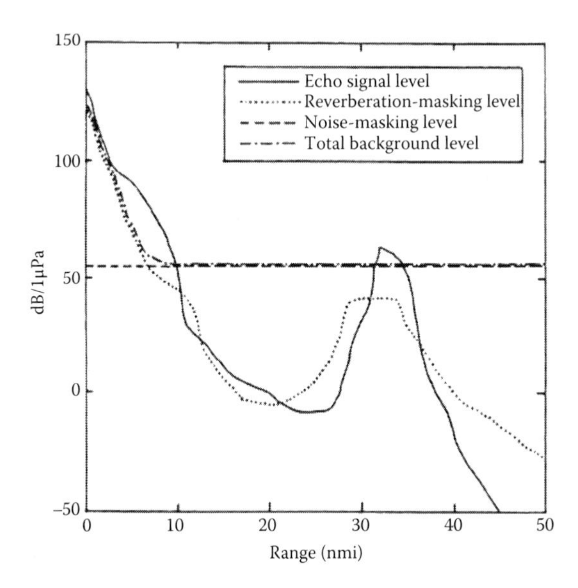

FIGURE 11.9 Sample graphical combination of selected sonar parameters. See Table 11.2 for an explanation of terms.

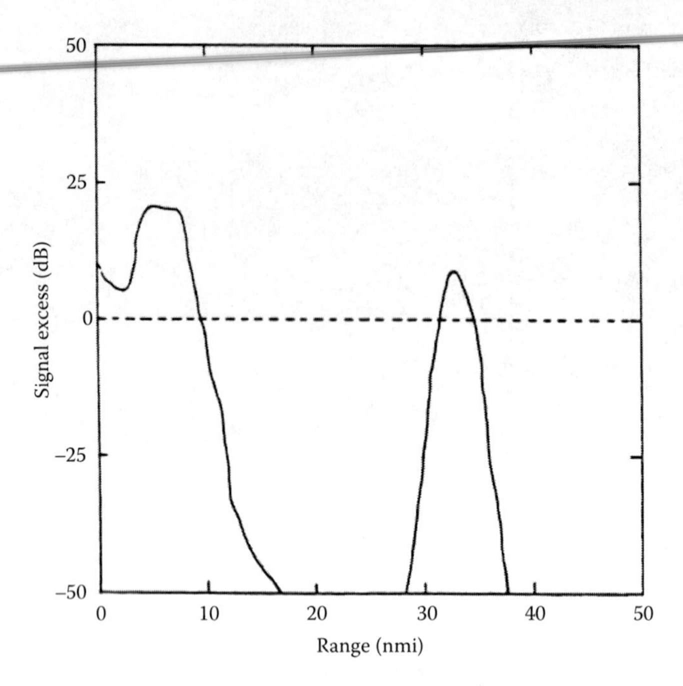

FIGURE 11.10 Signal excess as a function of range based on the data presented in Figure 11.9. Active sonar detection ranges are indicated by regions of positive signal excess.

is accomplished when the signal excess is greater than, or equal to, zero. Thus, NISSM and similar models can provide graphic assessments of active SP under the stated environmental and system-specific conditions. The intermediate quantities provide diagnostic tools for analyzing particular aspects of active sonar design and operation problems.

11.4 MODEL OPERATING SYSTEMS

The trend toward modular designs in the programming of underwater acoustic models (UAMs) is evidenced by continued interest in what are here termed model operating systems (MOS) (e.g., Eller, 1990; Holmes et al., 1990). These systems provide a framework for the direct linkage of data-management software with computer-implemented codes of acoustic models. Such systems relieve model operators of much of the tedium associated with data entry operations.

MOS systems facilitate comparative model evaluations (see Chapter 12) by standardizing the hardware and software configurations of different modeling techniques. The resulting uniformity also encourages a higher degree of configuration management and thus assists in the process of model evaluation. Sonar trainers now commonly utilize sophisticated range-dependent acoustic models to generate realistic training environments for the operators of sonar equipment (Miller, 1982, 1983). Requirements for near-real time program execution are frequently satisfied by using modular approaches similar to those employed in the construction of MOS.

In general, the development of sonar MOS is based on several considerations (Locklin and Webster, 1980): (1) it is assumed that the complex SP modeling problem can be decomposed into a set of generic functions, each of which solves a well-defined and bounded portion of the overall problem (decomposition is a powerful technique to break down large problems into manageable units); (2) it is assumed that the problem can be solved on a computer through the use of an automated data system and that the development of this system can be accomplished such that its computational structure bears a precise relationship to the problem's decomposed functions; and (3) it is assumed that the functions implemented on a computer have the means to communicate with each other in a prescribed manner.

11.4.1 System Architecture

MOS fall into two general classes of software architecture: bundled and executive. The principal distinction between these two architectures concerns the packaging of the software and databases to meet specific modeling requirements. Bundled systems are tailored to a narrow range of modeling applications and, in general, have only one method implemented to perform a generic modeling function. Executive systems are tailored to respond to a broader range of modeling applications by allowing the user to interactively control the hierarchical selection of functions. Furthermore, executive systems allow for several methods of performing a single generic function, thus providing a dynamic formulation of a modeling application.

Two different modeling environments can be addressed by the MOS: research and production. These two environments are not different with respect to what computing functions are performed, but rather how the analyst uses the computing functions. This difference in usage should be reflected in the MOS architecture in order to provide the analyst with an optimized capability to perform both types of activities, if necessary. These environments are briefly discussed below to provide potential users with a perspective and a choice of the appropriate style of computing suitable to their specific needs.

The research environment entails activities such as experimentation, iterative problem solutions, computation of intermediate model results, and model input sensitivity studies, among others. Practical applications include sonar system design and environmental-acoustics research. This implies that the MOS must be highly interactive with respect to the interface between the analyst and the MOS, and usually suggests an executive architecture.

The production environment is typified by operations that are definable, structurally organized, and ordered. To a large extent, these activities, once defined, have specific end products, solve a complete problem, and are performed repetitively using the same (or slightly modified) functions. Practical applications include routine SPpredictions and sonar operational trainers. This implies that the computing tools can be organized and structured beforehand, and then used repetitively thereafter. It further indicates that a more optimal composition of functions may be needed, with less user interactive capability, and less flexibility with respect to the computing function composition. This usually suggests a bundled architecture.

The functional organization of the MOS features a uniform way of viewing the system structure for solution of both the passive and active sonar equations. While these two problems are distinctly different, this need not imply that the active MOS and passive MOS must be constructed as two independent systems. Rather, the redundancy in system control software and in certain lower level functions common to both the active and passive sonar problems makes separate approaches less attractive. For illustrative purposes, a hypothetical functional decomposition of both the passive and active sonar equations will be described below.

11.4.2 Sonar Modeling Functions

A representative decomposition of the sonar equations into generic modeling functions is shown in the system data flows in Figure 11.11 for the passive sonar equation (Locklin and Webster, 1980) and in Figure 11.12 for the active sonar equation (Locklin and Etter, 1988). These particular decompositions are appropriate for incorporation into an executive architecture.

Although no decisionary logic options or looping controls are shown in these figures, a prerequisite processing sequence is nonetheless established. This sequence starts with an extraction process operating on databases that results in the generation of data sets specific to the analyst's application. In the case of passive sonars (Figure 11.11), subsequent calculations use the extracted data to develop the data sets required by either the beam noise (BN) or beam signal (BS) functions, which contribute to the final SP function. This particular conceptual picture identifies three methods for generating BN, and two methods for generating BS, although a greater or lesser number of methods could be accommodated. This reflects the various levels of complexity possible in noise modeling and passive sonar system parameterization, respectively.

In the case of active sonars (Figure 11.12), the three key processes involve calculation of the ESL, the RML, and the noise-masking level (NML) for a specific active sonar configuration. The signal level/spectra (SL) function for target characteristics describes the nature of the signal reflected from the target and considers the effects of TS, as specified in the active sonar equation. Furthermore, the BP function is implicit in the system characteristics (SC) function. The final SP function would then contain the algorithms necessary to generate signal excess and (P_D) outputs.

An important consideration for Figures 11.11 and 11.12 involves the usage of "EXTRACT" and "CALCULATE," and the symbols denoting application functions and data. "EXTRACT" implies such operations as database access, retrieval, reformatting, and output in a standard form, and includes the computational algorithms necessary to prepare the required data. "CALCULATE" implies that the primary emphasis is placed on some computational model or algorithm, with all input and output handled by standardized file interface utilities. The application functions shown in Figures 11.11 and 11.12 are the highest level of modularization envisioned in these particular representations. A further decomposition may be possible depending upon the particular requirements of the MOS. Each identified function would be associated with families of software application programs and associated data storage facilities.

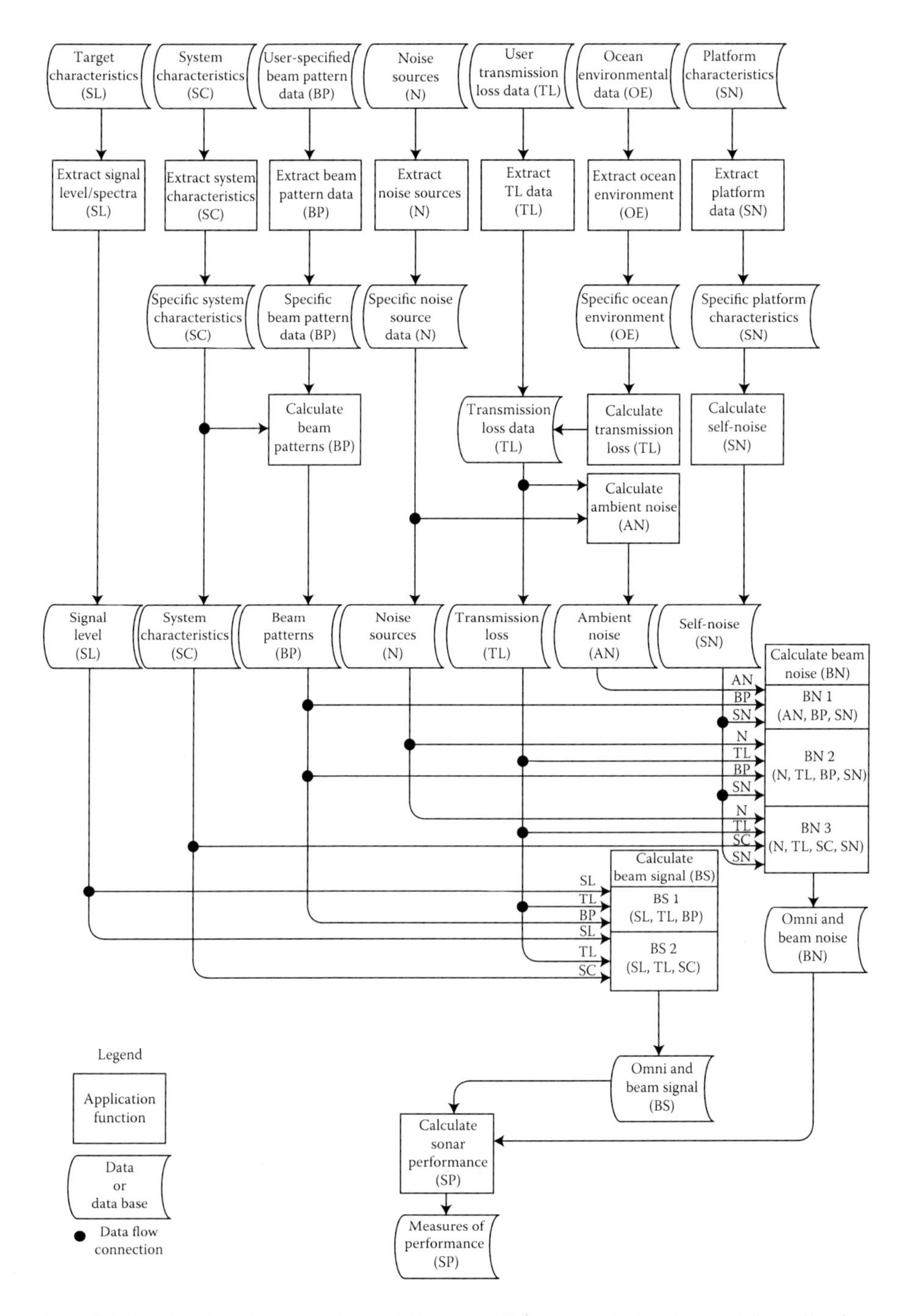

FIGURE 11.11 Sample acoustic modeling data flow concept showing MOS application functions for the passive sonar equation. (Adapted from Locklin, J. and Webster, J., NORDA Model Operating System Functional Description and FY 80 Implementation Plan. Ocean Data Syst., Inc, 1980.)

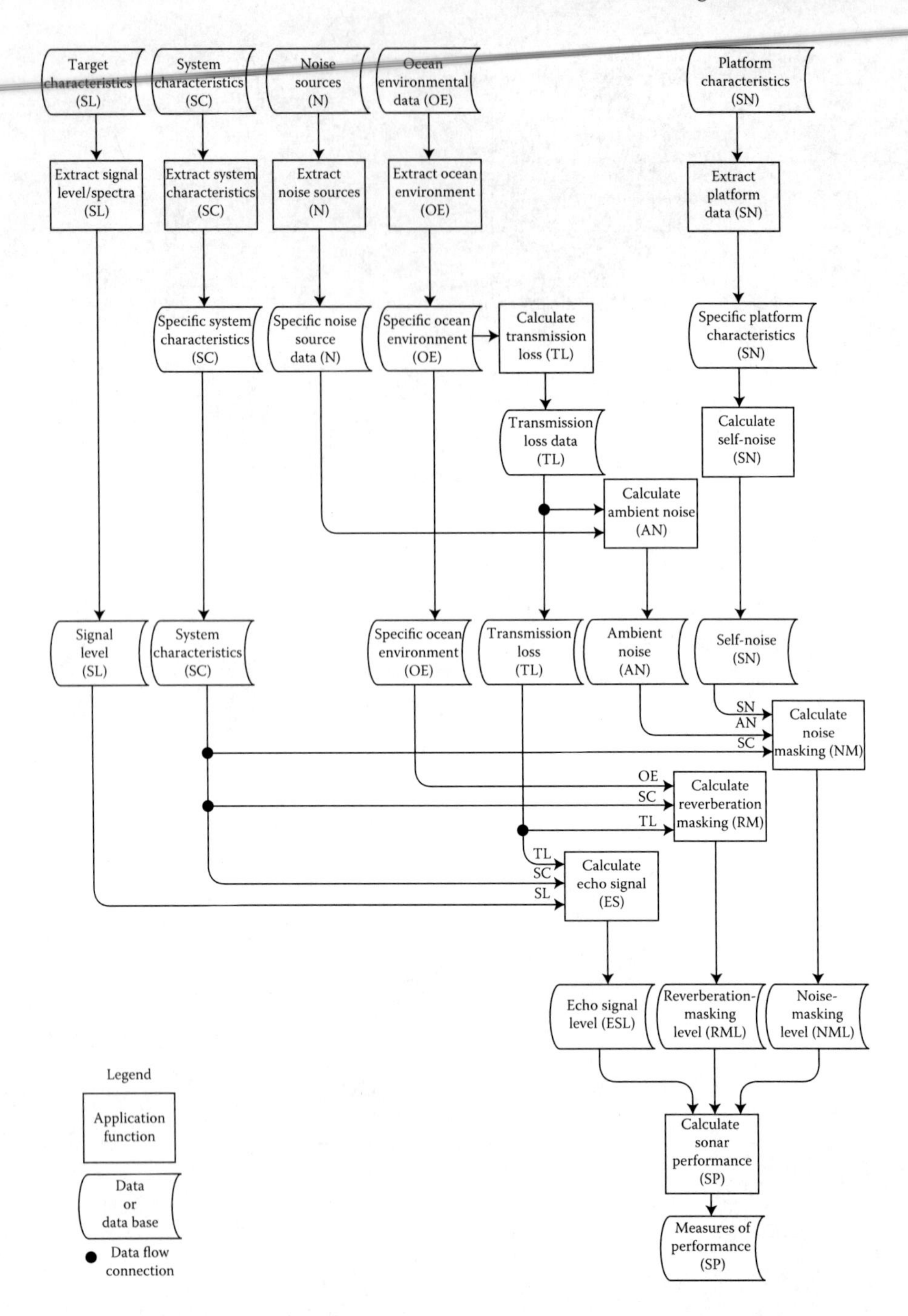

FIGURE 11.12 Sample acoustic modeling data flow concept showing MOS application functions for the active sonar equation. (Adapted from Locklin, J.H., and Etter, P.C., Sonar Model Operating System Functional Requirement. SYNTEK Engr. Comp. Syst., Inc, 1988.)

11.4.3 System Usage

MOS, by design, facilitate and thus encourage the generation of multiple solutions for any given SP problem. Divergent answers obtained from different models can alert the user to potential problems in either the problem specification or in the underlying physics of the model (i.e., a violation of the domains of applicability or an unexpected modeling pathology). Unexplained divergences can be graphically represented by envelopes illustrating the spread in predictions.

Many criteria drive the design of any particular MOS. Design decisions will likely entail trade-offs between cost and ease of use, and between accuracy and understanding. Some degree of compromise will be involved. There is always the danger of "creeping elegance" in which product embellishments not contained in the original design specifications add to the life cycle costs and delay the implementation schedule. There is also a common desire expressed by potential MOS users to obtain "all the information" generated by the models. While noble in principle, what is usually more important in practice is how the information is presented. Creative visualization techniques that logically combine essential information on a very few charts are highly desirable. The generation of high data volumes in 3D modeling, for example, necessitates the development of data fusion and animation techniques that permit the rapid assimilation and comprehension of complex acoustic interactions with dynamically changing ocean environments. Finally, documentation of MOS design and operation is essential for system maintenance and operator training.

An important aspect of the MOS design process entails the proper assessment of user requirements (also see Sections 12.7 and 13.6.1). Such requirements must be carefully specified early on in the program so that the evolving design can incorporate these requirements in addition to any derived requirements that have been flowed down to lower-level system designs. It is against these requirements that the performance of the built system must be evaluated to verify satisfactory completion of the project.

11.4.4 Generic Sonar Model—A Specific Example

The generic sonar model (GSM) was designed to provide sonar system developers and technologists with a comprehensive modeling capability for evaluating the performance of monostatic and bistatic sonar systems and for investigating the ocean environment in which these systems operate (Weinberg, 1981, 1985b). The development approach used a modular design, adhered to a strict programming standard, and implemented off-the-shelf software when practical. GSM is presently restricted to an ocean environment that is independent of range and time. In particular, the oceanic sound speed profile is layered, the surface and bottom are assumed to be horizontal reflecting boundaries, and sonar beampatterns are assumed constant over all azimuthal angles.

From a sonar system design standpoint, the significant capabilities of GSM include the modeling of passive signal excess versus range or frequency, active signal excess versus range, range and bearing errors versus range, low-frequency analysis and recording (LOFAR) diagrams, and autocorrelation and cross-correlation functions.

The next lower level of investigation involves computation of the signal, noise, reverberation, target echo, and DT. These, in turn, require inputs of ambient noise, self-noise, scattering strength profiles, scattering spectra, sonar system responses, SL, DI, filter equalizers, and eigenrays.

11.4.5 Comprehensive Acoustic System Simulation—A Specific Example

A range-dependent 3D version of GSM was developed by Dr. Henry Weinberg, formerly with the Naval Undersea Warfare Center in Newport, Rhode Island (USA). This new system is designated comprehensive acoustic system simulation (CASS) (Weinberg et al., 1997; Weinberg, 2000).

The CASS model simulates the performance of active sonar systems operating in the frequency range 600 Hz–100 kHz. Weinberg (2000) traced the evolution of CASS from its genesis in CONGRATS, through NISSM, to GSM. The GSM used integral (or multipath-expansion) approaches such as fast multipath expansion (FAME) instead of the range-derivative and divided-difference algorithms that had been used in the earlier CONGRATS and NISSM models. Like CONGRATS and NISSM before it, GSM assumed a range-independent ocean environment.

In CASS, all environmental parameters can vary as functions of range, azimuth, and frequency. Weinberg (2000) tabulated the 30 environmental-acoustics submodels, the 26 system submodels, and the 22 sonar analysis submodels comprising CASS. The CASS model computes reverberation in the time domain by accounting for the leading and trailing scattering-cell reverberation times for all possible combinations of eigenrays. Signal excess is computed at each range step by first mapping the signal into the time domain and then selecting the peak signal-to-masking level, where the masking level represents the power summation of noise and reverberation. This approach implies that the selected signal level may not always represent the time-integrated pressure level (Keenan, 2000).

The propagation component of CASS is Gaussian ray bundles (GRAB), which computes high-frequency (10–100 kHz) TL in range-dependent, shallow-water environments. Favorable modeling results have been obtained for frequencies as low as 600 Hz. The GRAB model is based on Gaussian ray bundles, which are similar in form (but somewhat simpler) than Gaussian beams (Weinberg and Keenan, 1996; Keenan, 2000; Keenan and Weinberg, 2001). Eigenrays are found by power averaging test rays having the same path history (Keenan, 2000). The GRAB model is under configuration management by OAML (Keenan et al., 1998).

Subsequent developments relating to the GRAB model are described below. Scattering theory was used to determine coherent surface reflection coefficients (CSRCs) for incorporation into ray calculations (Williams et al., 2004). Accurate determination of these CSRCs for shallow grazing angles is essential for determination of long-range, shallow-water propagation using ray models. Three classes of CSRCs were developed for testing in the GRAB model: small slope, perturbation, and Kirchhoff approximations. The Beckmann-Spizzichino model resident in GRAB was also investigated. The one-dimensional (1D) rough-surface parabolic equation (PE) code developed by Rosenberg (1999) was used as a benchmark in these evaluations. Rosenberg's PE code is an extension of RAM, but has been modified to account for the effects of a rough sea surface. Rosenberg referred to this code as Rrsfc, for RAM-rough-surface-fine-coarse. A simplified version of an isotropic two-dimensional (2D) Pierson-Moskowitz surface wave spectrum was used. The small-slope approximation provided the most accurate results, although the Kirchhoff approximation is probably sufficiently accurate for some purposes and would be simpler to implement as a module within GRAB. Perhaps due to domain mismatch, the Beckmann-Spizzichino (high-frequency, large roughness) submodel resident in GRAB did not give accurate predictions of the coherent field at long ranges. Jones et al. (2009) examined different implementations of the Beckmann-Spizzichino surface-loss model embedded in transmission-loss models and concluded that, when compared with at-sea transmission-loss data, no one surface-loss model could be considered superior. This work was further extended by Jones et al. (2010) where it was concluded that accurate surface-loss modeling will require the combination of the effects of surface roughness together with the effects of seawater bubbles.

11.5 ADVANCED SIGNAL PROCESSING ISSUES

11.5.1 Background

This section discusses a number of topics related to the mathematical modeling of signal-detection processes. Specific subjects addressed in this section include adjoint methods, stochastic resonance, pulse propagation, multiple-input/multiple-output (MIMO), clutter environments, vectors and clusters, and high-frequency acoustics.

In a tutorial-style treatment of fundamental concepts, Zoubir et al. (2012) addressed statistical robustness, which deals with deviations from distributional assumptions. Many practical engineering problems rely on the Gaussian distribution of the data, which in many situations is well justified. This enables a simple derivation of optimal estimators. However, this nominal situation is useless if the estimator was derived under distributional assumptions on the noise and signal that do not hold in practice. Measurement campaigns in both radar and sonar have confirmed the presence of impulsive (heavy-tailed) noise, which can cause optimal signal processing techniques, especially the ones derived using the nominal Gaussian probability model, to be biased or break down. Zoubir et al. (2012) defined measures of robustness, such as the influence function (IF), the breakdown point (BP), and the maximum bias curve (MBC). They further encouraged the use of robust estimators, such as the M-estimator, which can resist outliers without pre-processing the data. Collectively, robust estimators are useful in locating parameters instead of the conventionally used intuitive sample mean. Parameter estimation in linear regression was considered because of its importance in modeling many practical problems such as linear regression in impulsive interference. Moreover, the multichannel data case is becoming increasingly important in signal processing practice.

11.5.2 Adjoint Methods

Adjoint models are powerful tools for many studies that require an estimate of sensitivity of model output (e.g., a forecast) with respect to input. Actual fields of sensitivity are produced directly and efficiently, which can then be used in a variety of applications, including data assimilation, parameter estimation, stability analysis, and synoptic studies. Errico (1997) described the use of adjoint models as a tool for sensitivity analysis using simple mathematics.

When quantitative estimates of sensitivity are desired, a mathematical model of the phenomenon or relationship is required. While models have been used to assess the impacts of perturbations, and thus estimate sensitivity, a more efficient approach is to use the model's adjoint to determine optimal solutions. The adjoint operates backward in the sense that it determines a gradient (with respect to input) from another gradient (with respect to output). In a temporally continuous model, this would appear as integration backward in time. If there are no numerical instabilities associated with irreversible processes in the tangent linear model acting forward in time, there will be none in the adjoint acting backward in time (a tangent linear model provides a first-order approximation to the evolution of perturbations in a nonlinear forecast trajectory). The greatest limitation to the application of adjoints is that the results are useful only when a linearized approximation is valid. The adjoint operator (matrix transpose) back-projects information from data to the underlying model. Geophysical modeling calculations generally use linear operators that predict data from models. The usual task is to find the inverse of these calculations, that is, to find models (or make maps) from the data. The adjoint operator tolerates imperfections in the data and does not demand that the data provide full information.

The concept of adjoint modeling was introduced in shallow water acoustics for solving inverse problems (Hermand et al., 2006). Analytical adjoints were derived for normal modes and for both the standard PE and Claerbout's wide-angle approximation (WAPE). The application of a semiautomatic adjoint approach has been applied successfully for multidimensional variational data assimilation in meteorological and climate modeling. Starting from a modular graphic representation of the underlying forward model, a programming tool facilitates the generation and coding of both the tangent-linear and the adjoint models. The potential of this numerical adjoint approach for the physical characterization of a shallow water environment was illustrated with two applications for geoacoustic inversion and ocean-acoustic tomography (OAT) using Claerbout's WAPE in combination with nonlocal boundary conditions. Furthermore, the adjoint optimization was extended to multiple frequencies and it was shown how a broadband approach could enhance the performance of the inversion process. For a sparse array geometry, the generalization of the adjoint-based approach to a joint optimization across multiple frequencies was necessary to compensate for the lack of vertical sampling of the propagation modes. Results with test data synthesized from geoacoustic inversion experiments in the Mediterranean Sea showed that the the acoustic field, the sound speed profile in the water column together with the bottom properties could be retrieved efficiently with the numerical adjoint approach.

Meyer and Hermand (2006) reviewed the concept of backpropagation in the context of adjoint modeling. The different implementations of this concept were compared and discussed in the framework of experimental acoustic inversion in shallow water with application to source localization, OAT, geoacoustic inversion, and underwater communications. Well-established inversion (or focalization) methods based on matched-field processing (MFP), model-based matched filter, and time-reversal mirror are related to lesser known techniques such as acoustic retrogation and other variants of backpropagation. In contrast to the latter, adjoint-based, variational inversion approaches make use of the adjoint of a forward model to backpropagate the model-data mismatch at the receiver toward the source.

An adjoint model was derived from a forward-propagation model (e.g., normal mode or PE) to propagate data-model misfits at the observation point back through the medium to the site of those medium perturbations that were not accounted for in the forward model and which gave rise to the observed data-model misfits (Hursky et al., 2004). This property makes adjoint models attractive for use in acoustic inversion experiments.

11.5.3 Stochastic Resonance

Stochastic resonance refers to a phenomenon that is manifested in nonlinear systems whereby weak signals can be amplified in the presence of noise (Gammaitoni et al., 1998). Three components are necessary: a threshold, a periodic signal, and a source of noise. The response of the system undergoes resonance-like behavior as the NL is varied. Also see the related work reported by Zhang et al. (2008).

Stochastic resonance was applied to enhance the detection of target signals masked by shallow-water reverberation (Xu et al., 2007). Specifically, parameter-induced stochastic resonance was used to tune system parameters to recover the spatial signal that was corrupted by Gaussian noise. This method also has applicability when processing spatial signals that are corrupted by K-distributed envelope noise (a standard model for radar clutter).

11.5.4 Pulse Propagation

Pulse (or wideband) propagation modeling is concerned with simulating the effects associated with the transmission of a signal characterized by a frequency spectrum (versus a single-frequency continuous wave). Note that related discussions of broadband propagation modeling were presented previously in Section 6.3.

Makrakis and Skarsoulis (2004) investigated pulse propagation in range-independent oceans using approaches based on broadband and narrowband time-domain asymptotics. The performance of these two approximations was studied numerically:

- The broadband approximation was obtained by applying the stationary-phase method to the Fourier transform of the Green's function, which is expressed in terms of normal modes. The broadband approximation improved with increasing bandwidth.
- The narrowband approximation was obtained by incorporating the shape function of the emitted signal (assumed to be Gaussian) into the phase term and then applying the steepest-descent method. The roots of the frequency-derivative of the phase were located in the complex plane by using a second-order expansion of the eigenvalues. The narrowband approximation improved with decreasing bandwidth.

Both approximations were found to improve with increasing range. It was concluded that these approaches generated time-domain results more efficiently than the alternative Fourier synthesis approach.

Tseng and Cole (2007, 2009) developed a fast hybrid de-noising algorithm to enhance the performance of wideband acoustic signal detection in a reverberation-limited environment. By making use of this hybrid algorithm, the active sonar echolocation detector was able to estimate the motion parameters (radial-range and velocity) of a moving target in an effective and efficient manner with a very low signal-to-reverberation ratio. The hybrid algorithm was composed of two parts: adaptive-noise reduction based on an adaptive intelligent fuzzy system in the continuous wavelet transform domain, and target-motion estimation based on a recursive fast wavelets transform. In the adaptive-noise-reduction operation, which served to pre-filter the noisy signal, the signal-to-reverberation ratio of a noisy wideband signal was drastically improved by adopting the technique of an adaptive neuro-fuzzy inference system. The pre-filtered signal was transformed to the continuous-wavelet-transform domain and then processed using the recursive target-motion estimation operation, a combination of discrete wavelet de-noising and continuous-wavelet-transform techniques. Simulation results demonstrated that the proposed hybrid algorithm was not only effective in accurately predicting the motion parameters, but also was more efficient in terms of computational-time than a fuzzy detector previously developed on the basis of the adaptive-noise-reduction operation.

Borejko (2004) solved the 3D acoustic wave field produced by an impulsive point source in a wedge geometry with mixed Dirichlet-Neumann boundaries. Since the apex angle was set at 10° (an integer submultiple of $\pi/2$), the diffraction component of the field was zero. The resulting solution was exact and complete for the image component of the field for all ranges and depths since no approximation was introduced in evaluating the solution.

11.5.5 Multiple-Input/Multiple-Output

Benchekroun and Mansour (2006) applied a blind separation of sources (BSS) to the solution of MIMO channel problems in OAT, specifically a variant of OAT called passive acoustic tomography (PAT) in which a cooperative source is replaced by a noncooperative noise source such as a ship of opportunity. BSS refers to a method for retrieving unknown mixed independent sources from their observed mixture. Typically, researchers use independent-component-analysis (ICA) algorithms based on the independence assumption of the sources. For example, Xu et al. (2010) used blind-source separation to segregate sonar data into a reverberation component and a target component. Also see the work reported by Zhang et al. (2000).

11.5.6 Clutter Environments

Sonar clutter, particularly in shallow-water environments, introduces false targets that change the statistics of the reverberation signal. Specifically, clutter increases the (P_F) for a given (P_D). This is because clutter adds to the length of the tails of the reverberation-envelope probability distribution function (PDF), moving the statistics away from the Rayleigh canonical form. Clutter can be caused by target-like features, either natural or man-made, or by non-Gaussian distributions of the scatterers. Typically, high bandwidth or highly directive systems (or both) have more problems with clutter since as the size of the scattering patch is reduced, the PDF of the generally non-Gaussian scatterer distributions becomes resolved by the system (LePage et al., 2006).

A system of computer algorithms called UAIM was designed to predict multibeam SP and image the effects of variations in bathymetry, clutter objects, and bottom type, particularly in complex shallow-water environments (Wagstaff et al., 1997). Vosbein (2002) utilized UAIM to investigate the effects of out-of-plane scattering and small-scale bathymetric roughness on the backscattered acoustic intensity. It was suggested that these effects should be modeled statistically.

11.5.7 Vectors and Clusters

This section addresses replica vectors and ray clusters as techniques applicable to inverse-acoustic sensing.

11.5.7.1 Replica Vectors

Baggeroer et al. (1993: 401) discussed the concept of replica (or steering) vectors in the context of matched-field methods in ocean acoustics. Specifically, MFP refers to array-processing algorithms that exploit the full-field structure of the signals propagating in an ocean waveguide. Such algorithms may be used for source detection and localization as well as for the estimation of the waveguide parameters themselves. The replica vector is derived from the Green's function (i.e., spatial point source response) of the ocean medium. MFP uses replica vectors that are appropriate for the specific waveguide and for coherence. The multipaths and modes in the waveguide are coherent, particularly at low acoustic frequencies; this coherence can be used to advantage to enhance beamforming and to reduce low-frequency noise.

Zurk and Tracey (2005) proposed using "guide sources" to measure the transfer function between source and receiver arrays. The transfer function could then be used to reduce the troublesome mismatch commonly observed in shallow-water applications of MFP due to the absence of precise information on the underwater channel. This procedure, referred to as adaptive MFP (or AMFP), shifts the depth of the guide-source response using a vertical line array (VLA). With knowledge of the guide-source location, this vector could be shifted in depth to provide a steering vector for beamforming to alternate depths. This replica vector could be obtained without the need for environmental information or the use of a propagation model. These study results were simulated using the KRAKEN model.

An approach was described for modeling data-error correlations in matched-field geoacoustic inversion for application to vertical-array data (Huang et al., 2007). These data-error correlations originated from inhomogeneities that were not included in the environmental model. Such inhomogeneities could be due to a random environment or merely result from the chosen parametrization of the environment. In the presence of error correlation, the smooth pattern of the error vector could be mistaken as a component of the signal. This could influence the posterior probabilities if not taken into consideration. It was noted that multifrequency inversions might have a mitigating effect on the geoacoustic inversion results; however, the environmental perturbations might result in cross-frequency correlation.

11.5.7.2 Ray Clusters

In the presence of weak fluctuations in the sound-speed field, it has been observed that ray trajectories in ocean waveguides exhibit a chaotic behavior in which the travel times along eigenrays form compact clusters, the center of which is close to the arrival time of an unperturbed ray with similar geometry.

Virovlyansky (2005b) modeled the deep-water acoustic waveguide by an unperturbed sound-speed profile [$c_0(z)$] combined with weak fluctuations of sound speed [$\delta c(r,z)$] caused by a random field of internal waves conforming to the Garrett-Munk spectrum. Formally, ray equations in a random inhomogeneous medium can be considered as stochastic equations whose parameters are determined by a random function δc. Such nonlinear equations can be solved in the framework of ordinary perturbation theory, but only for short ranges. Instead, the ray structure was analyzed by using the Hamiltonian formalism in terms of action-angle canonical variables whose variation was small under conditions of weak perturbations. It was estimated that ray arrival times at a distance r occupied an interval whose average width increased in proportion with $r^{1.5}$, in agreement with previous investigators. Moreover, for $r < 10^3$ km, the ray trajectories forming the cluster deviated only slightly from the unperturbed ray trajectory. For $r > 10^3$ km, the width of the cluster grew in proportion with r^2.

Makarov et al. (2004) studied chaotic and stochastic nonlinear ray dynamics in underwater acoustic waveguides with longitudinal variations in the sound speed caused by internal waves. They adopted the model of a "frozen" medium in which the temporal variations in the environment were neglected and only the spatial variations due to the comparatively small propagation time of sound in the ocean were considered. Coherent ray clusters were observed in which large fans of rays with close initial conditions preserved close current dynamical characteristics over long distances. The cluster structure could be considered to consist of statistical and coherent parts. Rays belonging to the statistical part propagated in the same areas of phase space with the same value of the Langrangian, but did not correlate with each other and demonstrated exponential sensitivity to initial conditions. Rays belonging to the coherent part did not demonstrate sensitivity to initial conditions. This coherent clusterization might be a useful property for acoustic tomography in terms of determining spatio-temporal variations in the hydrological environment under conditions of ray chaos. Also see related work by Virovlyansky and Zaslavsky (2007) and Makarov et al. (2010).

An efficient and reliable method was proposed for performing the inversion of a neural-network underwater acoustic model to obtain seafloor parameters (Thompson et al., 2003). Two different versions of a modified particle-swarm optimization were used: two-step gradient approximation and hierarchical cluster-based approximation. Both approaches worked well.

11.5.8 High-Frequency Acoustics

A compendium of papers presented at a 2004 conference that surveyed all aspects of current research in high-frequency acoustics has been assembled in book form (Porter et al., 2004). This conference was motivated by recent interest in mine hunting, marine mammal tracking, and communications (e.g., the undersea internet). Specific areas examined included sediment acoustics, acoustic communications, boundary scattering and volume fluctuations, nonlinear bubble dynamics, marine mammals, experimental and measurement techniques, target modeling, and systems and applications.

Jackson and Richardson (2007) produced a research monograph as part of a new book series sponsored by the US Office of Naval Research (ONR) on the latest research in underwater acoustics. This particular volume provided a critical evaluation of the data and models pertaining to high-frequency acoustic interactions with the seafloor.

Chadwick and Bettess (1997) considered using finite-element methods to model high-frequency (short-wavelength) wave propagation. Usually, the domain is discretized such that there are about 10 nodal points per wavelength. Such a procedure, however, is computationally expensive and impractical for short wavelengths (i.e., high frequencies). Instead of modeling the wave potential, the wave envelope and phase were modeled. This process was applied to 2D problems. An iterative procedure was used to estimate the phase, and the resulting finite-element calculation provided a better estimate for the phase. The iterated values for the phase and wave envelope converged to the expected values for the test progressive-wave examples considered.

11.6 DATA SOURCES AND AVAILABILITY

Databases with which to support model development, evaluation, and operation are available in a variety of formats. Many have automated data-retrieval features that make them attractive for application to MOS. Available automated oceanographic and acoustic databases have been identified and summarized by Etter et al. (1984). Two categories of data banks were described: primary data banks and modified data banks. Primary data banks contain original or modestly processed data (e.g., National Oceanographic Data Center, 1992). Modified data banks are distinguished by the fact that they modify or extrapolate data derived from primary data banks in order to satisfy operational requirements for compactness or for ease of handling (e.g., Naval Oceanographic Office, 1999). Parameter summaries for primary data banks are provided in Table 11.4. Not all of these data banks are fully documented, and some may not be available for use outside the custodian's facilities. This summary does not claim to be exhaustive.

The US Navy has established sets of databases as standards for use in sonar modeling. These databases are maintained in the oceanographic and atmospheric master library (OAML), which is chartered to provide fleet users with standard models and databases (Willis, 1992; Naval Oceanographic Office, 1999, 2004). A subset of these databases is available through the tactical oceanography wide area network (TOWAN). Selected TOWAN databases are summarized in Table 11.5 together with

TABLE 11.4
Summary of Primary Data Banks

Data Bank	Custodian	Representative Database Parameters (Consult Websites for Details)			
FNMOC Data Files http://www.fnmoc.navy.mil/	FNMOC	Marine meteorology • Solar radiation • Clouds • Surface pressure	• Precipitation • Surface air temperature	• Surface marine winds • Total heat flux	• Sensible and evaporative heat fluxes
		Oceanography • Wave direction, period, and height • Combined sea height, direction, and period • Significant wave height • Swell direction, period, and height • White caps	• Insolation, reflected radiation • Sensible and evaporative heat flux • Total heat flux • Bathymetry • Shipping density	• Operational bathythermograph observations • MBT, XBT, STD, hydrocasts • Sea-surface temperature • Sea-surface temperature anomaly	• Potential mixed layer depth • Primary layer depth • Temperature at the top of the thermocline • Thermocline gradient
DoD Bathymetric Data Library https://www.nga.mil/ProductsServices/Pages/Aeronautical-Charts-and-Publications.aspx	NIMA	• Ocean floor depth			
NAVOCEANO Data Files http://www.usno.navy.mil/NAVO	NAVOCEANO	• Bathymetry • Climatology • Temperature/salinity/oxygen versus depth • Computed sound speed, sigma-t, specific volume	• Conductivity • Transmission loss • Ambient noise • Volume reverberation • Bottom loss • Surface currents • Subsurface currents	• Ice type/thickness • Core samples • Sediment samples • Geomagnetics • Seismic profiles • Gravity	• Dangerous marine animals • Boring/fouling organisms • Plankton • Bioluminescence

(Continued)

TABLE 11.4 (Continued)
Summary of Primary Data Banks

Data Bank	Custodian	Representative Database Parameters (Consult Websites for Details)			
NCDC Marine Climatic Data Files https://www.ncei.noaa.gov/	NOAA/NCDC	• Air temperature • Pressure • Waves	• Dew point temperature • Sea-surface temperature	• Low clouds • Total clouds	• Wind • Visibility
NGDC Marine Geology and Geophysics Data Files https://www.ncei.noaa.gov/	NOAA/NGDC	• Airborne magnetic survey (elements D,I,F) • Marine magnetic survey (total intensity F only)	• Megascopic core description, marine geological sample index, grain size analysis	• Digital hydrographic survey • Summary bathymetric and topographic files	• Marine bathymetry • Seismic profiles • Marine gravity
NODC Data Files https://www.ncei.noaa.gov/	NOAA/NODC	• Temperature • Salinity • Computed sound speed	• Marine meteorological parameters	• Marine chemical parameters	• Surface currents

Note: Abbreviations and acronyms are defined in Appendix A. Also see Appendix C for additional relevant web addresses.

TABLE 11.5

Databases from the US Navy Tactical Oceanography Wide Area Network (TOWAN)

Database	Contents	Temporal Resolution	Spatial Coverage/Resolution
High-Frequency Bottom Loss (HFBL)	Bottom loss vs. grazing angle (1.5 kHz $< f <$ 4 kHz).	Fixed	Global/5 minutes latitude and longitude
Low-Frequency Bottom Loss (LFBL)	Reflective and refractive characteristics of ocean bottom (bottom loss vs. grazing angle) ($f <$ 1 kHz).	Fixed	Global/5 minutes latitude and longitude
Generalized Digital Environmental Model (GDEM)	Steady-state digital model of ocean temperature and salinity. Gridded sets of coefficients are used with 1D linear or cubic spline interpolation in time to generate vertical profiles from the surface to the bottom where $z_B \geq$ 100 m.	Seasonal	Global/30 minutes latitude and longitude
Historical Wind Speed (HWS)	Ocean surface wind-speed and direction statistics based on analysis of historical data collected from 1946–1986. Data derived from NCDC Historical/Marine data. Data prior to 1980 derived from COADS.	Monthly	Global/1 degree latitude and longitude
Digital Bathymetric Data Base—Variable Resolution (DBDB-V)	Combination of different resolution gridded bathymetric databases.	Fixed	Global/variable resolution (0.5, 1, 2, or 5 minutes latitude and longitude)
Volume Scattering Strength (VSS)	Column scattering strengths at selected frequencies.	Fixed	Northern hemisphere/0.75 degree latitude and longitude
Shipping Noise—High Resolution (SNHR)	Estimated omnidirectional shipping noise and spectra for a receiver depth of 100 m at a nominal frequency of 50 Hz. Spectral range is 10–1,000 Hz for predominantly under-ice areas and 25–15,000 Hz for predominantly open-water areas.	Seasonal	Northern hemisphere/5 minutes latitude and longitude
Shipping Noise—Low Resolution (SNLR)	Estimated omni/horizontal directional shipping noise, spectra and statistics for receiver at 100 meters at a nominal frequency of 50 Hz. Ambient noise below 300 Hz is derived from industrial and natural sources (e.g., oil rigs, ice).	Seasonal	Northern hemisphere/5 degrees latitude and longitude
Historical Temporal Shipping (HITS)	Shipping densities (fishing vessels, merchants, tankers, large tankers, and super tankers).	Monthly	Global/1 degree latitude and longitude
ICECAP	Ice thickness characteristics and ice keel distribution.	Seasonal (spring/fall)	Arctic/60-nmi grid cells

Source: Tactical Oceanography Wide Area Network (TOWAN). http://www7180.nrlssc.navy.mil/homepages/TOWAN/towanpgs/TOWANDataIndex.html

Note: f = frequency; z_B = bottom depth; nmi = nautical mile.

brief descriptions of their contents, temporal resolution, and spatial coverage and resolution. Related information on websites is provided in Appendix C. It should be noted that the US Navy strictly controls dissemination of the OAML databases.

The accessibility of many of these databases has been facilitated by the US Navy's technology transfer program. One such example is naval environmental operational nowcasting system (NEONS) developed by the Naval Research Laboratory (NRL). NEONS manages three types of environmental data: observations, images, and gridded data (Schramm, 1993). The NRL designed NEONS for compatibility with computer-industry and international data-exchange standards. NEONS supports civilian distribution of FNMOC data products via the Navy/NOAA oceanographic data distribution system (NODDS).

Ocean Acoustic Developments Ltd. (1999a) created the WADER global ocean information system, which is based on DBDB5 bathymetry (an OAML database) and the Levitus temperature-salinity data (a NOAA dataset discussed previously in Chapter 2).

Clancy (1999) and Clancy and Johnson (1997) provided useful overviews of naval operational ocean data products. In particular, Clancy and Johnson (1997) identified and briefly described meteorological models (NOGAPS, NORAPS, LABL), ocean thermal models (OTIS, MODAS), ocean thermodynamics and circulation models (TOPS, SWAFS), ocean waves and surf models (WAM, SWAPS), a sea-ice model (PIPS), and a tide and surge model (TSPS). The meteorological models drive the ocean models, which, in turn, synthesize useful operational products from dissimilar and spatially sparse oceanographic data. The oceanographic models can then drive UAMs to simulate SP.

The US NRL's eddy-resolving global ocean nowcast and forecast system uses satellite observations and an ocean model to enhance real-time knowledge of the marine environment in which naval submarines and ships must operate (Wallcraft et al., 2002).

Available ocean-forecast schemes often cannot account for high-frequency ocean phenomena (sub-mesoscale) due to uncertainties in the forcing fields and initial conditions (Coelho et al., 2005). Reliance on oceanographic data collection is expensive and difficult to sustain. Alternatively, embedded-feature modeling can be used to improve local representations of sub-mesoscale dynamics and produce accurate, short-term oceanographic-field estimates to accommodate tactical modeling requirements.

Synthetic environment data representation and interchange specification (SEDRIS) is concerned with the representation of environmental data and the interchange of datasets for terrain, atmosphere, ocean, or space (see Appendix C for more details).

Interoperability in combined allied operations has been vastly improved by the allied environmental support system (AESS), which is similar to the US tactical environmental support system (TESS). TESS is an interactive data fusion system that receives, stores, processes, displays, and disseminates meteorological and oceanographic data and products. Common use of AESS will ensure that all participants have access to the same meteorological and oceanographic products during combined operations. AESS is utilized at shore-based METOC stations in support of NATO and several non-NATO allied navies.

The Navy operational global atmospheric prediction system (NOGAPS) was replaced in March 2013 by the Navy global environmental model (NAVGEM) as the US Navy's operational atmospheric-forecast system (Metzger et al., 2013). As part of this transition, both global ocean forecast system (GOFS) and Arctic cap nowcast/forecast system (ACNFS) switched from using NOGAPS to NAVGEM for atmospheric forcing. Calibrations to the wind velocities and net-heat flux were performed, wind velocities were calibrated against satellite scatterometer data, and heat-flux values were calibrated using 5-day forecast SST errors.

The bottom sediment type (BST) database describes sediment-type provinces by identifying an integer value that has a one-to-one relationship with a particular sediment type (Naval Oceanographic Office, 2003). These descriptors consider the grain size, origin, and placement of bottom deposition. The BST data are provided at high- and low-spatial resolution. The high-resolution set derives from multiple sources including the analyses of grabs and cores collected during surveys conducted by the NOO and from National Imagery and Mapping Agency charts, side-scan imagery, seismic, bathymetry, and public literature. The low-resolution set is assembled from various high-level sources, including maps, atlases, and regional studies of ocean basins. The gathered sediment information exists in a variety of diverse and sometimes incompatible formats including sample point data, areal coverage, and acoustic data. The 5-minute resolution database provides an estimate of the most likely sediment to be found on a broad, regional basis. Although the database resolution is 5 minutes, the underlying source-data resolution is uneven and far lower, typically by an order of magnitude. Therefore, the database cannot guarantee that the sediment category identified in any cell is absolutely correct, only that the category represents the prevailing sediment within the defined region.

Burnett et al. (2002, 2014) have assembled a useful tabulation of operational oceanographic Navy models that are either currently available or else are undergoing operational evaluation (Table 11.6).

The ECCO project (estimating the circulation and climate of the ocean) was established in 1998 as part of the world ocean circulation experiment (WOCE) with the goal of combining a general circulation model (GCM) with diverse observations in order to produce a quantitative depiction of the time-evolving global ocean state (Menemenlis et al., 2008). Such combinations, referred to as data assimilation, are important because remotely sensed and *in situ* observations are sparse and incomplete compared to the scales and properties of ocean circulation. These combinations also provide rigorous consistency tests for both models and data. In contrast to numerical weather prediction (which also combines models and data), ECCO estimates are physically consistent; specifically, ECCO estimates do not contain discontinuities when and where data are ingested. First-generation ECCO solutions are now available and widely used for numerous science applications; however, the coarse horizontal grid spacing and the lack of Arctic Ocean and sea-ice data (for the first-generation solutions) limit the ability to describe the real ocean. To address these shortcomings, the follow-on ECCO Phase II (ECCO2) project plans to produce a best-possible global, time-evolving synthesis of most available ocean and sea-ice data at resolutions that admit ocean eddies.

TABLE 11.6
US Navy Environmental Prediction Systems, 2002 versus 2014

Type	2002	2014
Data Assimilation	• Multivariate Optimal Interpolation (MVOI) • Modular Ocean Data Assimilation System (MODAS)	• 3D Variational Data Assimilation (3DVAR) • 4D Variational Data Assimilation (4DVAR)
Circulation	• Thermodynamic Ocean Prediction System (TOPS) • Navy Layered Ocean Model (NLOM) • Navy Coastal Ocean Model (NCOM) • Princeton Ocean Model (POM) • Advanced Circulation Model (ADCIRC)	• Hybrid Coordinate Ocean Model (HYCOM) • Navy Coastal Ocean Model (NCOM)
Waves, Surf, Tides	• Wave Action Model (WAM) • WAVEWATCH III (WW3) • Steady State Wave (STWAVE) Model • Navy Standard Surf Model (NSSM) • HYDROMAP™ • PCTides	• WAVEWATCH III (WW3) • Navy Standard Surf Model (NSSM) • PCTides • DELFT3D
Ice	• Hibler Ice Model/Cox Ocean Model	• Community Ice Code (CICE)/ Hybrid Coordinate Ocean Model (HYCOM)
Atmosphere	• Navy Operational Global Atmospheric Prediction System (NOGAPS) • Coupled Ocean Atmosphere Prediction System (COAMPS) • Geophysical Fluid Dynamics Laboratory Navy Tropical Cyclone (GFDN TC)	• Navy Global Environmental Model (NAVGEM) • Coupled Ocean Atmosphere Prediction System (COAMPS) • Coupled Ocean Atmosphere Prediction System – Tropical Cyclone (COAMPS-TC)

Source: Burnett et al., *Oceanography,* **27**, 24–31, 2014. Table 1, doi:10.5670/oceanog.2014.65.
Note: Abbreviations and acronyms are defined in Appendix A.

11.7 NUMERICAL MODEL SUMMARIES

Available SP models (subcategorized as active sonar models, model-operating systems, and TDAs) are summarized in Table 11.7. Numbers within brackets following each model refer to a brief summary and appropriate documentation. Model documentation can range from informal programming commentaries to journal articles to detailed technical reports containing a listing of the actual computer code. Abbreviations and acronyms are defined in Appendix A. This summary does not claim to be exhaustive.

The majority of the active sonar models listed in Table 11.7 are intended for use in ASW scenarios, although four of these models (CASTAR, MINERAY, SEARAY, and SWAT) were designed for use in mine-hunting scenarios.

TABLE 11.7
Summary of Sonar Performance Models Including Active Sonar Models, Model-Operating Systems, and Tactical Decision Aids

Active Sonar Models		Model Operating Systems	Tactical Decision Aids
Active RAYMODE [1]	MINERAY [14]	CAAM [27]	ASPECT [36]
ALMOST [2]	MOCASSIN [15]	CALYPSO [28]	IMAT [37]
ASPM [3]	MOC3D [16]	CASS [29]	MSTPA [38]
CASTAR [4]	MODRAY [17]	ESME/NEMO [30]	NECTA [39]
CONGRATS [5]	MSASM [18]	GSM—Bistatic [31]	ODIN [40]
ESPRESSO [6]	NISSM—II [19]	HARCAM [32]	SAKAMATA [41]
GASS [7]	SEARAY [20]	HydroCAM [33]	
HODGSON [8]	SONAR [21]	PRISM [34]	
INSIGHT [9]	SST [22]	SPPS [35]	
INSTANT [10]	SUPREMO [23]		
LIRA [11]	SWAMI/DMOS [24]		
LORA [12]	SWAT [25]		
LYBIN [13]	UAIM [26]		

SONAR PERFORMANCE

Active Sonar Models

[1] Active RAYMODE, which is based on the passive RAYMODE propagation model, also considers surface, bottom and column scattering strengths to calculate reverberation (Medeiros, 1982b, 1985b; Smith, 1982; Naval Oceanographic Office, 1991b,d).

[2] ALMOST, which was developed for the Royal Netherlands Navy, is a complete SP prediction model for active and passive systems. ALMOST contains three modules: PROPLOSS for TL; REPAS for passive sonar range predictions; and REACT for active sonar range predictions. The transmission-loss component is based on range-dependent ray tracing (Schippers, 1995). As part of the European MAST III project, an ambient noise time signal model was implemented in the ALMOST software package. The ambient noise is assumed to be generated at the surface by wind, rain or shipping. Biological sources are not included (Schippers, 1998). When sediment conditions warrant, bottom propagation (versus bottom reflection loss) can be incorporated by using the MULTIPAD model (Schippers, 1996). Further enhancements to ALMOST include bistatic calculations of reverberation and echo level. Doppler receiver-band and target-structure highlights are also available as inputs to monostatic and bistatic calculations (P. Schippers, 2002, private communication). (Dreini et al. [1995] compared the transmission-loss component of ALMOST with similar models.)

[3] ASPM is a system of prediction tools used to predict the performance of active and passive acoustic systems over a wide range of environmental conditions, including littoral regions, in support of system concept evaluation, advance deployment planning, and at-sea operational or exercise support (Berger et al., 1994).

[4] CASTAR predicts the performance of mine-hunting sonars (Naval Oceanographic Office, 1999).

[5] CONGRATS is an active sonar model constructed in several segments: I—ray plotting and eigenray generation (Weinberg, 1969); II—eigenray processing (Cohen and Einstein, 1970); and III—boundary and volume reverberation (Cohen and Weinberg, 1971).

[6] ESPRESSO is a minehunting SP assessment tool developed as a NATO standard for interfacing with NATO planning and evaluation TDAs (Davies, 2002). It uses BELLHOP as a propagation

(Continued)

TABLE 11.7 (Continued)
Summary of Sonar Performance Models Including Active Sonar Models, Model-Operating Systems, and Tactical Decision Aids

submodel, which has also been modified to calculate beam-based, high-frequency reverberation (Meyer and Davies, 2002). Espresso exists in two versions: one intended for scientific use and the other for military use. The scientific version of Espresso provides greater flexibility than the military version, including the ability to select sub-models and view the results of any sub-model. A user guide addresses the user interface for Espresso and describes the underlying software models and data output options available within Espresso (Davies and Signell, 2006a). There is a separate user guide for the military version, Espresso(m), which provides greater tailoring of the user interface, including the ability to customize parameters (Davies and Signell, 2006b).

[7] GASS is a simulator/stimulator for air ASW systems trainers. GASS contains an environmental-acoustics (EVA) server that provides underwater acoustic propagation, noise and reverberation characteristics based on US Navy standard acoustic models (ASTRAL and ANDES) and environmental databases from OAML. These characteristics are provided in the form of parameters that control the generation and propagation of time-series signals in other parts of GASS. The models used are range-dependent, and seamlessly support a wide operating envelope in frequency, range, and water depth. The EVA server was designed to be responsive to real-time systems and therefore should be suitable for a wide range of other acoustic prediction, simulation and modeling applications (US Navy Air ASW Project Office, 1996a,b).

[8] HODGSON, which was originally developed by Lt.Cdr. J.M. Hodgson of the Royal Navy, treats a fully range-dependent (sound speed and bathymetry) ocean environment. The transmission-loss component is based on range-dependent ray tracing (UK Ministry of Defence, 1995). The UK Ministry of Defence has formally validated the propagation model for both shallow and deep water. HODGSON also contains a reverberation module that computes surface and bottom reverberation. This model is available commercially from Ocean Acoustic Developments Ltd. (Dreini et al. [1995] compared the transmission-loss component of HODGSON with similar models.)

[9] INSIGHT is a PC-based model with a GUI interface. It computes signal excess for both active and passive sonars in range-independent ocean environments. INSIGHT incorporates a full ambient noise model for wind, shipping and rain noise, including propagation effects. The model has recently been upgraded with improved calculations of reverberation for both monostatic and bistatic sonar configurations. INSIGHT does not compute echo level and reverberation in the time domain. Rather, it calculates the associated energy flux density levels and displays signal excess versus target position or other user-selected parameter (Harrison et al., 1990; Packman et al., 1992; Ainslie et al., 1994, 1996; Ainslie, 2000; Mountain et al., 2001).

[10] INSTANT computes TL in range-dependent ocean environments using a hybrid of ray and mode concepts. The formulation is based on the conservation of energy flux and the exploitation of the ray invariant to model weak range dependence (Packman et al., 1995, 1996). INSTANT calculates echo and RLs for monostatic and bistatic sonar configurations, although the bistatic option is restricted to range-independent environments. Volume reverberation is not computed in INSTANT. A version of the CANARY noise model (referred to as CANARD) is used to estimate range-dependent noise (Ainslie, 2000).

[11] LIRA is an extension of LORA to predict the performance of low, mid and high frequency active sonars (Hoffman, 1979). The source and receiver may be separated in depth.

(Continued)

TABLE 11.7 (Continued)
Summary of Sonar Performance Models Including Active Sonar Models, Model-Operating Systems, and Tactical Decision Aids

[12] LORA is an extension of FAST NISSM, the utility version of the NISSM model. LORA predicts the performance of monostatic active sonar systems, either hull-mounted or towed (Hoffman, 1976).

[13] LYBIN is a range-dependent, ray-theoretical model developed by Svein Mjølsnes of the Norwegian Defence Logistic Organization (Mjølsnes, 2000; Hjelmervik and Sandsmark, 2008; Dombestein et al., 2010). Range-dependent environmental inputs include bottom type and topography, volume backscatter, sound speed, temperature, salinity, wind speed, and wave height. Choices of calculation outputs include ray trace, TL, reverberation (surface, volume, and bottom), noise, signal excess, (P_D), travel time, and impulse response. The transmission-loss module was evaluated by NURC (Ferla et al., 2001). LYBIN is available commercially from the Forsvarets forskningsinstitutt (FFI). LYBIN 6.2 consists of a COM module (LybinCom) for the Windows platform and a graphical user interface (GUI) which can be used together with LybinCom in order to build the stand-alone executable application. The GUI in LYBIN 6.0 was totally redone; it is programmed in C# and new functionality regarding range dependence was added. In LYBIN 6.2 the GUI was improved further based on user feedback and bugs was fixed both in the GUI and the calculation kernel (Dombestein and Ektvedt, 2014). LYBIN 6.2 2200 contains new target-strength tables; moreover, the editors for rang-dependent backscatter, volume backscatter, and bottom loss have been redesigned to improve user friendliness (Dombestein, 2017).

[14] MINERAY was initially developed in the 1970s to predict the performance of submarine mine-hunting sonars. There are three distinct generations of the MINERAY model. The first generation (1970s) was appropriate for modeling high-frequency sonars in deep-water environments. The second generation (mid-1980s) was extended to allow multipath sound propagation via bottom and surface bounces (Jaster and Boehme, 1984). The third generation (mid-1990s) has been extended to support modeling in littoral environments (Bailey et al., 1997).

[15] MOCASSIN predicts the performance of active sonars operating in range-dependent shallow waters that are characterized by highly variable sound-speed conditions (Schneider, 1990). Stochastic sound-speed variability in the horizontal is modeled by a diffusion approximation. Reverberation is computed and stored separately for the sea surface and ocean bottom.

[16] MOC3D (Ivansson, 2006) is a 3D model developed from the 2D model MOCASSIN (Schneider, 1990). MOC3D was used to investigate the importance of out-of-plane sound propagation in a shallow-water experiment in the Florida Straits (Sturm et al., 2008).

[17] MODRAY was developed in conjunction with DSTO (Australia) to simulate the propagation of sound through the underwater environment (Frith, 2003; Ebor Computing, 2005). MODRAY uses classical ray-tracing theory to produce sound-pressure time series at one or more receivers. The marine environment is range-independent. Seafloor composition can be specified, the sound-speed profile can be arbitrary, noise includes wind, rain, biological and shipping sources, and scattering by marine organisms is included. MODRAY can model an arbitrary number of sound sources, reflectors and receivers stationed on moving platforms. MODRAY has been used extensively to model the effectiveness of underwater communications algorithms (Alksne, 2000; McCammon, 2000).

[18] MSASM assesses the effectiveness of air-deployed, multistatic-acoustic sonobuoy fields (Navy Modeling and Simulation Management Office, 1999).

(Continued)

TABLE 11.7 (Continued)
Summary of Sonar Performance Models Including Active Sonar Models, Model-Operating Systems, and Tactical Decision Aids

[19] NISSM—II computes propagation and reverberation (Weinberg, 1973). (The NISSM model is discussed in detail in Section 11.3.)

[20] SEARAY models the acoustic environment by using ray-tracing techniques to determine sound paths in a horizontally stratified water column. The SNR along each path is determined by calculating directivity, absorption, spherical spreading loss and the effects of various noise sources, and then applying the active sonar equation (Tuovila, 1989). SEARAY is derived from the MINERAY model.

[21] SONAR is designed to compute eigenrays in ocean environments where the bottom profiles consist of a series of plane segments with slopes entirely in the vertical plane (Marsh and Poynter, 1969; Bertuccelli, 1975). Reverberation is computed using the cell scattering formulation.

[22] SST is a set of object-oriented software components and software-development tools for building sonar simulators (virtual oceans) that "sound" like a real ocean to an existing or proposed sonar system. SST-based simulators produce a digital representation of the predicted signal, which includes random fluctuations with controlled statistical properties (Goddard, 1989, 1994). This updated report emphasizes the science, mathematics, and algorithms underlying the sonar simulation toolset (SST). The SST is a computer program that produces simulated sonar signals, enabling users to build an artificial ocean that sounds like a real ocean (Goddard, 2008). Such signals are useful for designing new sonar systems, testing existing sonars, predicting performance, developing tactics, training operators and officers, planning experiments, and interpreting measurements. The signals simulated by SST include reverberation, target echoes, discrete sound sources, and background noise with specified spectra. Externally generated or measured signals can be added to the output signal or used as transmissions. Eigenrays from the GSM or the CASS can be used, making all of GSM's propagation models and CASS's Gaussian ray bundle (GRAB) propagation model available to the SST user. A command language controls a large collection of component models describing the ocean, sonars, noise sources, targets, and signals. The software runs on several UNIX computers, Windows, and Macintosh OS X. The primary model documentation is the SST Web (a large HTML website distributed with the SST software), supported by a collection of documented examples.

[23] SUPREMO is a multistatic SP model that includes propagation, target echoes, reverberation, and noise, all plotted as a function of delay time and bearing using an equivalent map projection. A modular propagation section makes it possible to separate the effects of propagation modeling (e.g., theoretical basis and range-dependence) from those of scattering computations (e.g., computational efficiency, bottom slope and scattering law). Special attention is paid to problems of interference from multiple sources firing in sequence, target aspect dependence from multiple receivers, mixed FM and CW, mismatched source and receivers, multiple displays (one for each bistatic pair), and presentation of results (Prior et al., 2002; Baldacci and Harrison, 2002; Harrison, 2002a,b; Harrison, 2003). Predictions of acoustic reverberation and target echo intensity made by the SUPREMO SP model were compared with measured data gathered in the Malta Plateau region of the Mediterranean Sea. The observed model-measurement agreement demonstrated the suitability of SUPREMO for use with an environmentally adaptive, low-frequency, active sonar system (Prior and Baldacci, 2005). Version 2.0 of SUPREMO has been documented (Prior, 2007a,b).

[24] SWAMI/DMOS models range-and-azimuth dependence (N × 2D) via adiabatic modes (Theriault and Ellis, 1997). SWAMI has been used to support the towed integrated active-passive sonar (TIAPS),

(Continued)

TABLE 11.7 (Continued)
Summary of Sonar Performance Models Including Active Sonar Models, Model-Operating Systems, and Tactical Decision Aids

which was developed as a technology demonstrator for the Canadian Forces. Software development utilized a system test bed (STB) comprising a collection of scalable, portable, and reusable components for constructing sensor-based applications. The toolset contains modules to produce predictions of TL, reverberation (MONOGO), signal excess, and (P_D). SWAMI includes the capability to model various source and receiver configurations including omnidirectional arrays, line-arrays (both horizontal and vertical) and volumetric arrays (Theriault et al., 2006). DRDC Atlantic Model Operating System (DMOS) is an evolution of the SWAMI suite of programs that enables a user to model TL, reverberation, signal excess, and (P_D) for active sonars (Calnan, 2006). Originally, the suite was based on normal-mode theory (PMODES); however, normal-mode theory is best suited to shallow water and low acoustic frequencies, and users occasionally need to model reverberation and other parameters under other conditions. DMOS was enhanced to include a Gaussian-beam acoustic propagation model, BELLHOP, as an alternate propagation engine (McCammon, 2010). A DRDC-extended version that included a range-dependant capability was chosen. DMOS may now be used to model both active and passive sonars in shallow or deep water.

[25] SWAT was developed to support mine-countermeasure (MCM) sonars (Sammelmann, 1998). SWAT actually comprises two models: one for detection sonars and one for classification sonars. The detection model is hosted on a personal computer and is designated PC SWAT. The classification model is designed to run on a workstation and is referred to as SWAT. Inputs and commands are menu driven in PC SWAT. Surface and bottom reverberation are computed by considering the multipath contributions, which are important in shallow and very-shallow littoral environments. A 3D, coherent acoustic scattering model of mines is also incorporated. Both PC SWAT and SWAT include the latest high-frequency environmental models (Applied Physics Laboratory, University of Washington, 1994).

[26] UAIM is a system of computer algorithms designed to predict multi-beam SP and image the effects of variations in bathymetry, clutter objects and bottom type, particularly in complex shallow-water environments (Wagstaff, et al., 1997). The model addresses perceived weaknesses in existing mine-countermeasure (MCM) SPmodels. The propagation code in UAIM was derived from RASP while the reverberation and signal excess codes were derived from GSM. Modifications were made to these codes to accommodate high-frequency, range-dependent applications. A utility package called SoundGuide assists the operator in creating the large input file, executes UAIM and plots the resulting data.

Model Operating Systems

[27] CAAM is a flexible R&D tool for sonar technologists (Navy Modeling and Simulation Management Office, 1999). It integrates the OAML environmental databases (Naval Oceanographic Office, 1999) together with selected propagation models including PE, ASTRAL, and RAYMODE.

[28] CALYPSO (ΚΑΛΥΨΩ) is an integrated computer environment that was developed for the analysis of underwater acoustic detection systems (Kalogerakis et al., 2004). The communication language is Greek. It can treat passive detection, broadband or LOFAR, as well as active detection, monostatic or multistatic. The system contains compact databases for environmental (coastline, bathymetry, oceanographic, geological) and operational (system parameters, target characteristics) data. Acoustic propagation calculations are performed using normal-mode, parabolic approximation

(Continued)

TABLE 11.7 (Continued)
Summary of Sonar Performance Models Including Active Sonar Models, Model-Operating Systems, and Tactical Decision Aids

and ray-theoretical codes supporting broadband calculations in range-dependent environments. The results include TL, RLs, DTs, and probabilities of detection for a variety of user-defined operational scenarios. (This work was supported by the Greek MOD.)

[29] CASS is a model architecture developed to support passive and active sonars with moderate fidelity in wide frequency bands (Weinberg, 2000). (CASS is discussed in Section 11.4.5.)

[30] ESME/NEMO (effects of sound on the marine environment) is a multidisciplinary research and development effort to explore the interactions between anthropogenic sounds, the acoustic environment and marine mammals (Shyu and Hillson, 2006; Siderius and Porter, 2006). The "ESME workbench" models the entire sound path including the sound sources (impulsive or explosive), the medium (water column and seafloor), and the temporary threshold shift (TTS) models of the marine mammals. (TTS refers to a temporary increase in the threshold of hearing, that is, the minimum intensity needed to hear a sound at a specific frequency, but which returns to its pre-exposure level over time.) The goal is to predict impacts of anthropogenic sounds on marine mammals. This entails three elements: (1) accurate estimates of the sound field in the ocean; (2) accurate estimates of the cumulative sound exposure of the marine mammals; and (3) reliable predictions of the incidence of TTS for the species of interest given the estimated cumulative exposure. The flexibility and computational efficiency of the ESME Workbench will be enhanced by merging the Navy acoustic effects model (NAEMO) and ESME approaches into the "One Navy Model," which is intended to serve as the standard simulation system for use in predicting impacts of anthropogenic sound sources on marine life for environmental compliance purposes. NAEMO, the US Navy's latest approach for analyzing the effects of sound on marine mammals, has seven basic components: scenario builder, environment builder, acoustic builder, marine mammal distribution, scenario simulator, post processor, and report generator.

[31] GSM is a modularized computer program that calculates passive and active sonar system performance in range-independent ocean environments (Weinberg, 1982, 1985b). GSM has been extended to include a bistatic active signal excess model (Powers, 1987). Version G (updated through December 1996) removed unsupported propagation models (RAYMODE, FACT, and MULE) and added bistatic scattering strength tables, among other features. (GSM is discussed in Section 11.4.4.)

[32] HARCAM (HODGSON and RAM composite acoustic model) is a dedicated acoustic propagation-loss system designed to produce accurate data over the frequency band 10 Hz–500 kHz. All data input and output is based on ASCII text files to produce graphical outputs suitable for quality control and logging purposes. The model is designed to aid users who need to produce large quantities of propagation-loss data using a computer that can run unattended. HARCAM is supplied with the same supporting data sets as WADER32 (a sonar range prediction and global ocean information system developed by Ocean Acoustic Developments Ltd); data availability is limited in littoral waters, but users are able to input their own high-resolution data. HARCAM uses the RAM and HODGSON models to produce accurate propagation-loss data at low and high frequencies, respectively; in addition, HARCAM utilizes HODGSON to generate surface-loss and absorption-loss data to correct the RAM output in the 250–1000 Hz range where these parameters are important but where RAM is unable to provide the necessary data. These two models form a symbiotic relationship to produce accurate and consistent propagation-loss data from a single user interface over a frequency band that covers virtually all operational sonars.

(Continued)

TABLE 11.7 (Continued)

Summary of Sonar Performance Models Including Active Sonar Models, Model-Operating Systems, and Tactical Decision Aids

The system is configured to allow a large number of individual propagation-loss calculations (~104) to be completed using a single set of input files; the primary output mechanism is ASCII for downstream processing. The system can provide graphical outputs including ray trace, PODgrams (probability-of-detection diagrams), and propagation-loss curves. (Ocean Acoustic Developments Ltd, 2013).

[33] HydroCAM was developed to predict the detection and localization performance of hydroacoustic monitoring networks in support of the international monitoring system (Farrell and LePage, 1996; Upton et al., 2006).

[34] PRISM is an interactive SP model used to evaluate the acoustic performance of mobile sonar systems in various ocean environments (Chaika et al., 1979). The original version of PRISM contained range-independent TL models including the FACT, RAYMODE, Normal Mode (Newman and Ingenito, 1972) and LORA models.

[35] The sonar performance prediction system, SPPS, (developed in Germany), evaluates the performance of sonar systems in various naval-warfare scenarios and assists in sonar-system design efforts. The model accommodates active and passive sonar detection problems involving both broadband and narrowband signals using a variety of sonar antennas. The integrated ocean-acoustic propagation model spans the frequency range 10 Hz–1 MHz by using a hybrid combination of PE, coupled-mode, and ray propagation sub-models. Environmental databases describe the ocean environment, targets, platforms, and sonar-system characteristics. The reverberation model predicts volume, surface, and bottom components. Noise sources include ambient, biological, rain, shipping, ice, seismic exploration, and self noise (Applied Radar & Sonar Technologies GmbH, 2002).

Tactical Decision Aids

[36] ASPECT, the active system performance estimate computer tool, is a multistatic TDA. It computes estimates of system performance for active underwater acoustic sensors. ASPECT was originally designed to satisfy the requirements of mission planning software for the improved extended echo ranging (IEERO) system. However, virtually any multistatic or monostatic active acoustic system can be modeled using this software package. ASPECT uses the FAME model for range-independent calculations of TL and active system performance model (ASPM) for range-dependent TL and reverberation. These computations are then fed into MSASM, which is capable of estimating the performance of active sonar systems for multiple sources, receivers, and targets. It simulates target motion including such features as normal and uniform probability distribution, various speed and course distributions, as well as target evasion modeling to a limited degree. ASPECT version 2.0 (and beyond) supersedes the MSASM Interactive Execution and Optimization System (MINEOS), which was the first version of mission planning software developed for the IEER program. ASPECT also exists as a stand-alone version (Applied Hydro-Acoustics Research, 2001) (AHA is now Adaptive Methods).

[37] IMAT was developed to integrate training, operational preparation, tactical execution, and post-mission analysis into a seamless support system (Ellis and Parchman, 1994; Wetzel-Smith et al., 1995; Wetzel-Smith and Czech, 1996; Foret et al., 1997; US Department of the Navy, 1999; Beatty, 1999).

[38] The multistatic tactical planning aid (MSTPA) is a tool used to model the performance of a given multistatic sensor network in terms of the (P_D) of a submarine, the ability to hold a track, and whether such a track could be correctly classified as such. The tool considers the entire chain

(Continued)

TABLE 11.7 (Continued)
Summary of Sonar Performance Models Including Active Sonar Models, Model-Operating Systems, and Tactical Decision Aids

	of events from an initial calculation of signal excess, the generation of a contact considering localization errors, followed by the subsequent tracking and classification process. The tool may be used to plan a particular multistatic scenario through operational analysis of many Monte Carlo simulations. A number of generic decision-support techniques were introduced that may be wrapped around the MSTPA tool. The acoustic performance metric that drives decisions is subject to uncertainty relating to environmental measurements and extrapolations (Strode et al., 2012).
[39]	NECTA supports oceanographic and environmental data analysis as well as sensor performance predictions. The open and modular design of the system allows the ready inclusion of additional environmental data and tactical guidance to meet changing demands (BAeSEMA Ltd, 1998).
[40]	ODIN provides an advanced capability to model the complete maritime engagement scenario at the unit and force level. The model has a consistent, integrated approach to performance assessment and algorithm development. It models the detailed interaction between individual elements (e.g., countermeasures, ships, submarines, or seabed objects) while retaining an execution speed sufficient for both real-time operations and detailed studies. ODIN provides a whole-system approach, thus enabling modeling of diverse applications using one tool. It has the flexibility to model arbitrary entities in both single-shot and Monte Carlo modes (Atlas Elektronik UK, 2015; Robinson, 2001).
[41]	SAKAMATA (sea animal kind area-dependent mitigated active transmission aid) is a naval exercise planning tool that provides the operator with techniques for careful mission planning, implementation of marine mammal monitoring, and implementation of sonar ramp-up schemes (Benders et al., 2004). The SAKAMATA software package is commercially available from TNO (www.tno.nl). When a scenario is defined, SAKAMATA will determine the most vulnerable species for this operation. The maximum allowed sound pressure level for the sonar frequency in question is derived from validated hearing-sensitive curves. Alternatively, a fixed sound pressure level can be used for each marine mammal group. Propagation loss for the given sonar in the given area is computed by the SPmodel ALMOST. This information is used to compute the received level as a function of the range and depth of the marine mammal.

MOS provide a framework for the direct linkage of data-management software with computer-implemented codes of acoustic models, thus facilitating the construction of versatile simulation capabilities. MOS are further distinguished from stand-alone active SP models by virtue of their ability to conduct sensitivity analyses by computing components of the active-sonar equation using alternative solution techniques. Since sonar MOS normally utilize existing UAMs and standard oceanographic databases, these systems are unique only in the sense of the number and types of models and databases included, and the particular architectures, graphical user interfaces (GUIs), and other features employed.

TDAs represent a form of engagement-level simulation that blends environmental information with tactical rules. These decision aids can also guide ocean scientists in planning acoustic experiments and allocating resources by exploiting knowledge of the operating environment, especially in the presence of marine mammals and other aquatic life. Such decision aids assist planners and operators in conforming to environmental regulations pertaining to underwater sound transmissions.

12 Model Evaluation

12.1 BACKGROUND

Model evaluation is defined as the systematic gathering and promulgation of information about models in order to determine model limitations and domains of applicability. Model evaluation should be viewed as a process rather than a specific result.

The intent of this chapter is not to furnish a compendium of evaluation results but rather to describe the evaluation process itself. Such a description can provide useful insights into the benefits and shortcomings of model evaluation. The fundamental steps involved in the evaluation process will be described and only a select number of examples will be used to further illustrate the process.

At first glance, the process of model evaluation might appear to be very straightforward. When examined more closely, however, a number of factors are seen to complicate and frustrate the process. To be credible, the evaluations must be based on statistically comprehensive comparisons with benchmark data. Such benchmark data can include field measurements, outputs from other models, or closed-form analytical solutions. Each of these data sources has limitations. Field measurements are quite limited with regard to temporal, spatial, and spectral coverage. Furthermore, these data sets are sometimes classified on the basis of national security concerns. Outputs from other models are useful only when those models themselves have been properly evaluated. Analytical solutions can provide very accurate, albeit idealistic, comparisons for a relatively small number of useful problems.

Model development is an evolving process. Any particular version of a model quickly becomes outdated and the attendant evaluation results become obsolete. Therefore, comparisons of model accuracy are valid only for the particular model version tested. To ignore progress in model developments gained through previous evaluation experiences is to do an injustice to those models. Accordingly, only the most recent and authoritative evaluation results should be considered when assessing models.

Rivalries among different governmental, academic, and industrial laboratories frequently add a political dimension to the model evaluation process. Many of the well-known models have gained their notoriety, in part, through the zealous advocacy of their developers and sponsors.

Clearly what is needed then is an autonomous clearinghouse to provide for both the unbiased evaluation of models and the timely promulgation of the results. An intermediate step in this direction was taken by the naval sonar modeling community through the establishment of configuration management procedures. These procedures provided a disciplined method for controlling model changes and for distributing the models (together with selected test cases) to qualified users. Configuration management comprises four major activities: (1) configuration identification and use of a product baseline; (2) configuration change control; (3) status accounting and documentation of all product changes; and (4) reviews, audits, and inspections to promote access to information and decision-making throughout the software life cycle.

The US Chief of Naval Operations established the oceanographic and atmospheric master library (OAML) in 1984. The OAML is chartered to provide fleet users with standard models and databases while ensuring consistent, commonly based environmental service products (Willis, 1992).

12.2 PAST EVALUATION EFFORTS

Comprehensive model evaluation efforts in the past have been very limited. Notable efforts have been sponsored by the US Navy, but these efforts have focused largely on propagation models. None of these efforts is presently active and their results are therefore only of limited academic interest. All of these efforts were successful in addressing the immediate concerns at the time, but were unsuccessful in the longer term due to the evolutionary nature of model developments and the lack of widely accepted evaluation criteria. It is instructive, nonetheless, to review these past efforts.

The model evaluation program (MEP) was administered by the (now defunct) acoustic environmental support detachment (AESD) under Office of Naval Research (ONR) sponsorship. Emphasis was placed on model-to-model comparisons and the evaluation results were not widely disseminated.

A methodology for comparing acoustic propagation models against both other models and measured data was developed by the panel on sonar system models (POSSM). POSSM was administered by the Naval Underwater Systems Center (NUSC) and was sponsored by the US Naval Sea Systems Command (Lauer and Sussman, 1976, 1979). Some of these results will be discussed later in this chapter. Among the many observations made by POSSM was the lack of documentation standards for acoustic models (Lauer, 1979).

In an effort to promote standardization, the US Navy established the acoustic model evaluation committee (AMEC) in 1987. The specific charter of AMEC was to develop a management structure and administrative procedures for the evaluation of acoustic models of propagation, noise, and reverberation, although initial attention was restricted to propagation models. Specific evaluation factors included model accuracy, running time, core storage, ease of effecting slight program alterations, and available ancillary information. The activities of AMEC culminated in evaluation reports for two propagation models: FACT and RAYMODE. As a result of AMEC's activities, RAYMODE was designated the (interim) Navy standard model for predicting acoustic propagation in the ocean. Subsequently, AMEC was disbanded.

More recent publications have proposed various theoretical and practical procedures for model evaluation (McGirr, 1979, 1980; Hawker, 1980; Pedersen and McGirr, 1982). Additional evaluation studies have been described for both propagation models (Hanna, 1976; Eller and Venne, 1981; Hanna and Rost, 1981) and reverberation models (Eller et al., 1982). Spofford (1973b), Davis et al. (1982) and Chin-Bing et al. (1993b) reported practical test cases suitable for all types of propagation models. In May 1994, the ONR conducted a reverberation workshop in Gulfport, Mississippi

(USA). The purpose of this workshop was to assess the fidelity of low-frequency (30–400 Hz) scattering and reverberation models. Six test cases comprising either field-measurement data sets or analytic problems with known reference solutions were used to establish the accuracy and state of current reverberation model development. Uniform plot standards facilitated intercomparison of test-case results.

Two workshops on reverberation modeling were conducted in November 2006 and March 2008 to evaluate recent progress in reverberation-related research efforts. The first workshop focused on reverberation from the environment, while the second workshop emphasized system characteristics. At the first workshop, 15 different reverberation models were represented and extensive comparisons were carried out before, during, and after the workshop (Perkins and Thorsos, 2007). The reverberation problems were posed in both two and three dimensions and included cases with different levels of boundary roughness, different sound speed profiles, and, in some cases, range dependence. Two important reverberation-modeling issues became evident: (1) the role of coherent effects in determining reverberation structure at short ranges, and (2) the role of boundary reflection loss in affecting the reverberation level at long ranges (Thorsos and Perkins, 2007). Results from these two workshops were summarized in more detail by Thorsos and Perkins (2008):

- The problems for the first workshop utilized three center frequencies (f_c): 250, 1000, and 3500 Hz. In all cases, the bandwidth was $f_c/20$. Cases with a rough bottom had two versions of bottom roughness referred to as a typical or rough. Problems 1–4 were isovelocity two-dimensional (2D) geometries with a rough bottom, a rough surface, or both. The remaining problems (5–20) were three-dimensional (3D). Problems 5–10 had either a rough surface or a rough bottom, and for each of these cases the sound speed profile was isovelocity or seasonal (winter or summer), where the seasonal profiles were approximated by a constant sound speed gradient. Problems 11–13 used a Lambert's law bottom scattering model with the same three sound speed profiles. The final problems (14–20) were all isovelocity with bottom roughness and were bistatic in range, or range dependent in depth, or both. For the 2D problems, the water depth was 50 m; for the 3D problems, the water depth was 100 m.
- The problems for the second workshop were more complex and divided into two groups. The first group introduced several important reverberation features in idealized settings including scattering from a target sphere, from a fish school, from bubble clouds near the sea surface, and from heterogeneity within a mud layer on the bottom. (A reverberation clutter problem is currently under development.) In the second group, data sets were examined for additional reverberation features of interest, and problems were formulated with fully defined environments that were expected to reproduce particular features of interest including reverberation due in part to scattering from mud volcanoes, reverberation exhibiting strong coherent and acoustic focusing effects, reverberation that included scattering from cylindrical targets in a bottom-bounce environment, reverberation in multistatic geometries, and reverberation with clutter due to scattering from seamounts. In

each case, physics-based bottom scattering models were provided in addition to generalized Lambert's-law fits made over several angular ranges. Only preliminary results were available for the second workshop. It is intended that the reverberation models and their solutions will be described in detail in forthcoming reports.

Thorsos and Perkins (2008) made two general observations regarding the evaluation of reverberation models: (1) full numerical solutions capable of providing ground-truth results were not generally practical; (2) measured reverberation data sets were not suitable for ground truth because environmental uncertainties were typically too great for that purpose. Furthermore, for evaluation purposes, the modeling should be physics-based and be capable of generating realistic field statistics, even for those cases that yielded significant clutter. Thorsos and Perkins (2008) also noted that a reliable method was needed by which to generate synthetic reverberation data with which to test inverse techniques for geoacoustic inversion.

12.3 ANALYTICAL BENCHMARK SOLUTIONS

In the absence of comprehensive experimental data, underwater acousticians have explored the use of analytical benchmark solutions to assess the quality of numerical models (Felsen, 1986; Jensen, 1986; Jensen and Ferla, 1988; Robertson, 1989). Such benchmarks emphasize idealized but exactly solvable problems. In the case of propagation models, two benchmark problems have been investigated in detail: (1) upslope propagation in a wedge-shaped channel; and (2) propagation in a plane-parallel waveguide with range-dependent sound speed profile (Felsen, 1987). These two problems are illustrated in Figures 12.1 and 12.2, respectively (Jensen and Ferla, 1988, 1990).

A special series of papers coordinated by the Acoustical Society of America described the results of a concerted effort to apply standard benchmark problems (with closed-form solutions) to the evaluation of propagation models. The propagation models selected for evaluation were based on different formulations of the wave equation (Felsen, 1990; Jensen and Ferla, 1990; Buckingham and Tolstoy, 1990; Thomson, 1990; Thomson et al., 1990; Stephen, 1990; Collins, 1990b; Westwood, 1990). The results of this evaluation provided several important insights (Jensen and Ferla, 1990):

1. One-way (versus two-way) wave equation solutions do not provide accurate results for propagation over sloping bottoms.
2. The COUPLE model provides a full-spectrum, two-way solution of the elliptic-wave equation based on stepwise-coupled normal modes. This code is ideally suited for providing benchmark results in general range-dependent ocean environments. However, the solution technique is computationally intensive and is therefore impractical at higher frequencies.
3. IFDPE provides a limited-spectrum, one-way solution of the parabolic approximation to the full wave equation. The implicit finite-difference solution technique is computationally efficient, and accurate one-way results are provided for energy propagating within ±40° of the horizontal axis.

4. PAREQ provides a narrow-angle, one-way solution of the parabolic approximation to the full wave equation. The split-step Fourier solution technique is computationally efficient, and accurate one-way results are provided for energy propagating within ±20° of the horizontal axis.

The term "pathological" is sometimes used in reference to those test cases that prove to be particularly troublesome to models. In the evaluation of propagation models, for example, ocean environments exhibiting double sound speed channels near the surface make especially good pathological test cases. Ainslie and

Parameters common to all three cases

Wedge angle: θ_0 = arctan 0.05 ≈ 2.86°
Frequency: f = 25 Hz
Isovelocity sound speed in water column: c_1 = 1,500 m s^{-1}
Source depth: 100 m
Source range from the wedge apex: 4 km
Water depth at source position: 200 m

Pressure-release surface

Case 1: *Pressure-release bottom.* This problem should be done for a line source parallel to the apex, i.e., 2D geometry.

Case 2: *Penetrable bottom with zero loss.*
Sound speed in the bottom: c_2 = 1700 m s^{-1}
Density ratio: ρ_2/ρ_1 = 1.5
Bottom attenuation: 0 dB λ^{-1}
This problem should be done for a point source in cylinderical geometry.

Case 3: *Penetrable lossy bottom.* As in case 2 except with a bottom loss of 0.5 dB λ^{-1}

Receiver depths

Case 1: 30 m
Cases 2 and 3: 30 m and 150 m

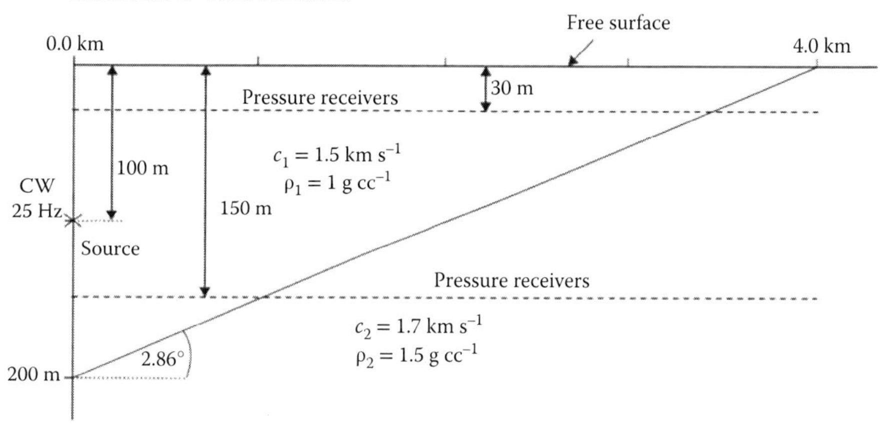

FIGURE 12.1 Analytical benchmark problem: wedge geometry for cases 1, 2, and 3. (Adapted from Jensen, F.B. and Ferla, C.M., Numerical solutions of range-dependent benchmark problems in ocean acoustics. SACLANT Undersea Res. Ctr, Rept SR-141, 1988.)

$$\frac{C(r,z)}{C_0} = \left\{ 1 + \left(\frac{\pi l_1}{L}\right)^2 e^{-2\pi r/L} + \left(\frac{2\pi l^2}{L}\right)^2 e^{-4\pi r/L} - \frac{2\pi l_1}{L}\left[1 - \left(\frac{2\pi l_2}{L}\right) e^{-2\pi r/L}\right] \cos\left(\frac{\pi z}{L}\right) e^{-\pi r/L} \left(\frac{4\pi l_2}{L}\right) \cos\left(\frac{2\pi z}{L}\right) e^{-2\pi r/L} \right\}^{-1/2}$$

where $C_0 = 1500\ \mathrm{ms}^{-1}$ is a reference sound speed, z = the depth below the pressure-release surface, L = the channel depth, and r = the range from the source; l_1 and l_2 are parameters with the values $l_1 = 0.032\mathrm{L}$ and $l_2 = 0.016L$.

Case 4: *Low-frequency, shallow water*
f = 25 Hz (frequency)
L = 500 km (channel depth)
R = 4 km (range coverage)

Case 5: *High-frequency, deep water*
f = 100 Hz (frequency)
L = 3 km (channel depth)
R = 20 km (range coverage)

Note:
The field should be computed for both the source and the receiver at depth $z = L/2$.

FIGURE 12.2 Analytical benchmark problem: plane-parallel waveguide for cases 4 and 5. (Adapted from Jensen, F.B. and Ferla, C.M. Numerical solutions of range-dependent benchmark problems in ocean acoustics. SACLANT Undersea Res. Ctr, Rept SR-141, 1988.)

Harrison (1990) proposed using simple analytic algorithms as diagnostic tools for identifying model pathologies. They developed simple analytic expressions for computing the intensity contributions from standard propagation paths such as bottom-reflected paths, bottom-refracted paths, and Lloyd's Mirror interference. Ainslie and Harrison (1990) further demonstrated the utility of these analytical tools in assessing the performance of numerical models of acoustic propagation. They accomplished this by combining the intensity contributions from the individual propagation paths appropriate for any particular ocean environment and sonar system geometry. This approach allowed a path-by-path analysis of numerical model performance and illuminated any modeling pathologies in an instructive fashion. These tools were subsequently organized into a highly interactive sonar model called INSIGHT (Packman et al., 1992). (INSIGHT is described in more detail in Chapter 11, specifically Table 11.7). The ability to alter a variety of environmental or sonar system parameters interactively and then rapidly visualize the resulting impacts on system performance makes INSIGHT especially attractive for instructional purposes.

12.4 QUANTITATIVE ACCURACY ASSESSMENTS

Quantitative accuracy assessments of propagation models can be accomplished according to two procedures: difference techniques and figure-of-merit (FOM) techniques (McGirr, 1979).

Difference techniques measure the distance between the model prediction and a standard (which can comprise field measurements or outputs from other models) in terms of dB differences at a given range, or over a set range interval (Figure 12.3). These techniques are best suited to comparative model evaluations conducted in research environments.

FOM techniques are essentially an inverted distance measure wherein a specified dB level (or FOM) is selected and the corresponding ranges (or sonar detection zones) are determined and compared with a standard. The FOM equation was previously discussed in Chapter 11. These techniques are best suited to sensitivity analyses that have as their objective a determination of the operational (versus scientific) impacts of model prediction errors. Such techniques place importance on the location of sonar detection zones relative to a standard. These techniques thus recognize the ultimate application of the model predictions to naval operations. A hypothetical example of this technique, used in combination with the difference technique, is illustrated in Figure 12.4 (Leibiger, 1977).

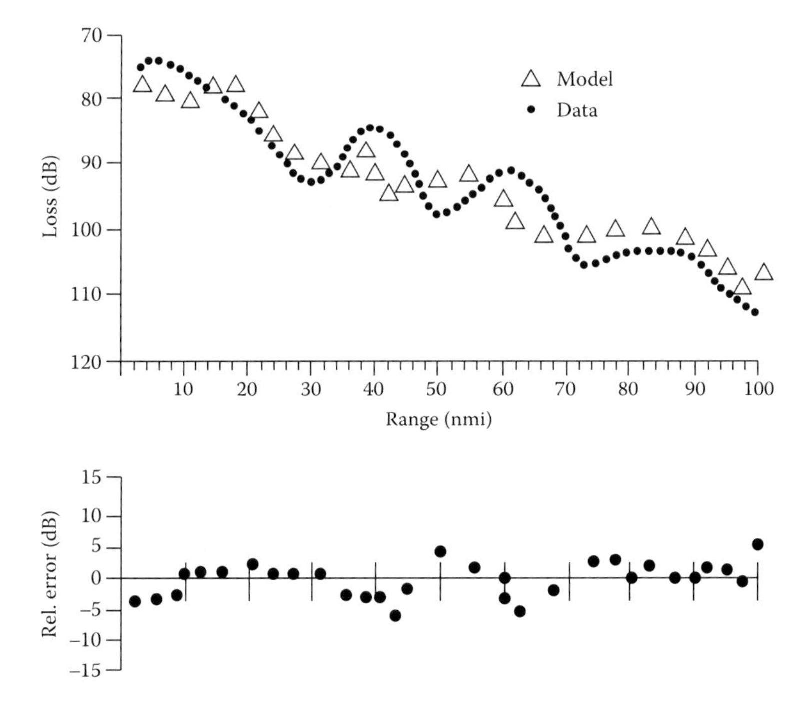

FIGURE 12.3 Sample outputs from the AESD model evaluation program (MEP) described by R.C. Cavanagh (1974, unpublished). (Adapted from McGirr, R.W. Acoustic model evaluation procedures: A review. Nav. Ocean Syst. Ctr, Tech. Doc. 287, 1979.)

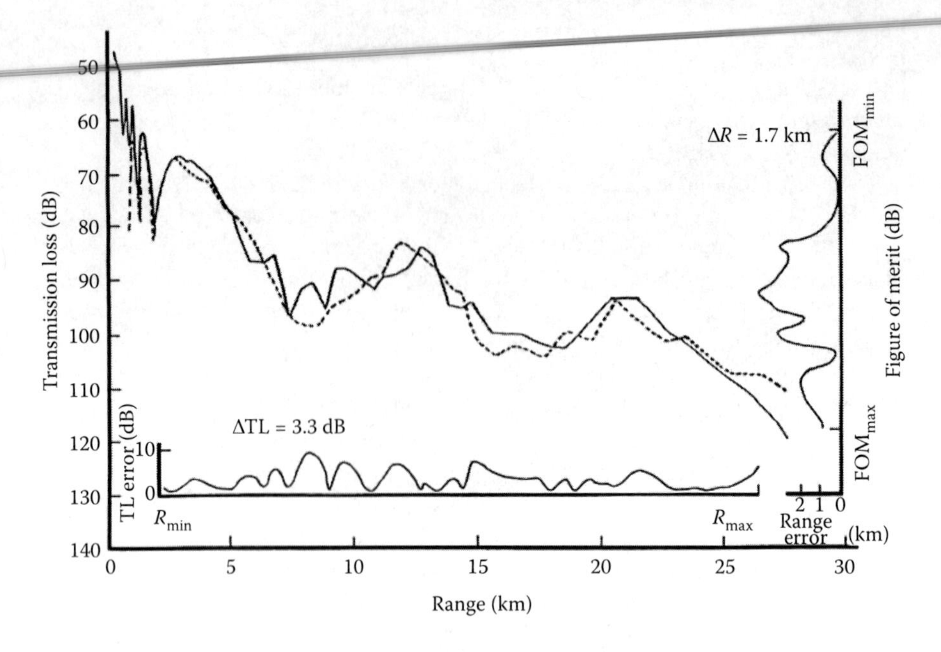

FIGURE 12.4 Transmission loss comparison concept that combines elements of both the difference technique and the FOM technique. This schematic only illustrates the concept and is not meant to be numerically exact. (Adapted from Leibiger, G.A. Criteria for propagation loss model assessment for APP application. Nav. Underwater Syst. Ctr, Tech. Memo. 771245, 1977.)

A computer model can have a number of context-specific inputs that define a particular situation within which the model is intended for use. When the values of one or more of these context-specific inputs are unknown, observations can be used to characterize them. This process is referred to as calibration. Such plug-in predictions treat the context-specific inputs as if they were known, when in reality they are only estimated and the residual uncertainty about these inputs need to be recognized in subsequent predictions of the model. Kennedy and O'Hagan (2001) developed a Bayesian approach to calibrate computer models. The computer codes considered were deterministic in that running the same code with the same inputs always produced the same outputs. Using observed data, they derived the posterior distribution of the parameter vector, which then quantified the residual uncertainty. This uncertainty was fully accounted for when using the computer models for subsequent predictions by obtaining a predictive distribution.

Another aspect of statistical model evaluation entails an assessment of just what a particular model is capable of estimating (Canepa and Irwin, 2005). All models are a compromise in which physical processes are chosen for explicit treatment and which ones are excluded. More than one model is typically selected for assessment in a statistical evaluation exercise. Often, there is already an "accepted" model, and the purpose of the evaluation is to prove whether a candidate model's performance is significantly better than the "accepted" model. Models differ in the physical processes characterized, the sophistication of input data required, and the numerical processing required. If several models can be shown to have statistically similar performance, then a potential user might select from this subset a model that best meets available resources in input data, computer expertise, and processing time. Simplicity of assumptions is a desired trait in modeling. As illustrated in Figure 12.5, as the model physics increases in complexity (to explicitly treat more physical processes), the number of input variables is increased, thereby increasing the likelihood of degrading the model's performance due to input data errors and model parameter uncertainty. Underlying the model physics and input data uncertainty, there is the natural or stochastic uncertainty that the model does not characterize ("noise"). Hanna (1989) noted that while models attempt to characterize how the ensemble mean varies, these same models may not necessarily be fashioned to explain more of the physical processes and thus characterize more of the explainable variations; this would reduce the amount of unexplained natural variations.

Alternatively, one might select from a group of models (having similar performance) a model that is known to handle a specific process. However, for testing specific processes, there may be very few databases suitable for use in a rigorous evaluation. Thus, part of model evaluation entails judicious use of sparsely available field data. Consequently, familiarity with the available field data that could possibly be used, and knowing the strengths and weaknesses of each dataset, are definite assets in model evaluation.

Taylor (2000) introduced a diagram that can provide a concise statistical summary of how well patterns (observed or modeled) match each other in terms of their correlation, their root-mean-square difference, and the ratio of their variances. These depictions are now commonly referred to as "Taylor diagrams" in the literature (Figure 12.6). While the form of the diagram is general, it may be particularly useful in evaluating complex models, such as those used to study

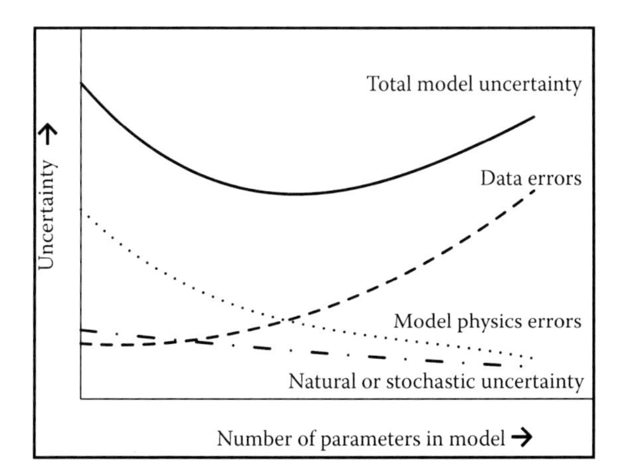

FIGURE 12.5 Illustration of the variation of model uncertainty components with number of parameters in model. (From Hanna, S.R., *Encyclopedia of Environmental Control Technology*, Gulf Publishing Company, Houston, 1989. With permission.)

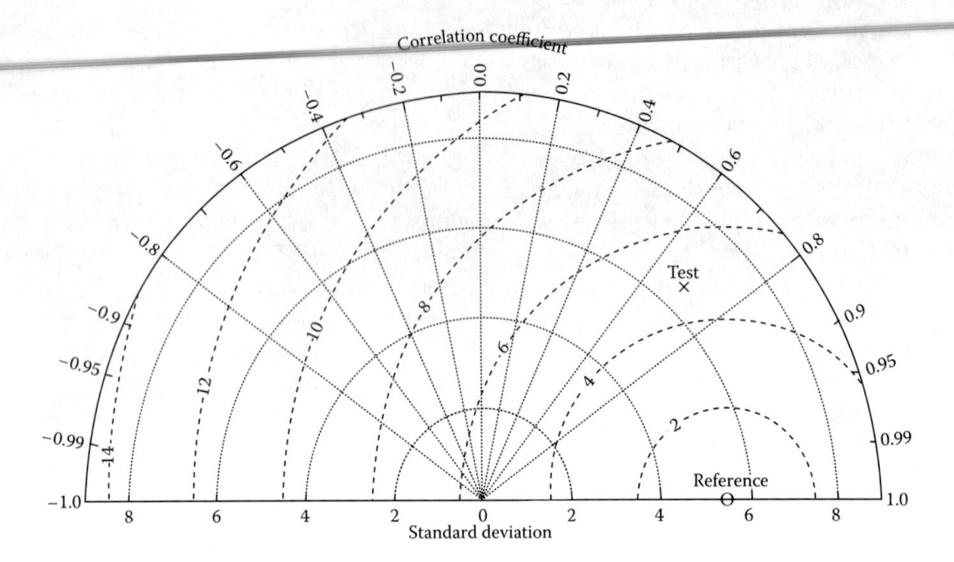

FIGURE 12.6 Sample Taylor diagram for displaying pattern statistics. The radial distance from the origin is proportional to the standard deviation of a pattern. The rms difference between the test and reference field is proportional to their distance apart (in the same units as the standard deviation). The correlation between the two fields is given by the azimuthal position of the test field. (Adapted from Taylor, K.E. Summarizing multiple aspects of model performance in a single diagram. Program for Climate Model Diagnosis and Intercomparison (PCMDI), University of California, Lawrence Livermore National Laboratory, Report No. 55 (UCRL-JC-138644, 2000).)

geophysical phenomena. Taylor included examples showing how the diagram can be used to summarize the relative merits of a collection of different models or to track changes in the performance of a model as it is modified. Methods were suggested for indicating on these diagrams the statistical significance of apparent differences and the degree to which observational uncertainty and unforced internal variability limit the expected agreement between simulated and observed behaviors. The geometric relationships among the plotted statistics also provide guidance for devising skill scores that appropriately weight among the various measures of pattern correspondence. Taylor (2001) provided specific examples to better illustrate practical applications of this diagram.

Referring to Figure 12.6, if the model under evaluation (*test*) exactly reproduced the benchmark data (*reference*), it would lie at the *reference* point (marked by "o"). The benchmark data can represent observations or output from a trusted model. The distance between the *reference* point and the *test* point (marked by "×") represents the rms error. The dashed arcs on the diagram represent lines of constant rms error. The correlation coefficient is represented on the outer arc of the diagram with increasing correlation with the angle from the y-axis. The normalized standard deviation is represented as the distance to the origin (value 0.0 on the x-axis); if the *test* point is closer to the origin than the *reference* point, then the model has lower variance than the benchmark.

12.5 POSSM EXPERIENCE—A SPECIFIC EXAMPLE

The deliberations of the POSSM are considered to be representative of past evaluation efforts, and aspects of these results will be discussed. The methodology developed and implemented by POSSM is summarized in Figure 12.7 (DiNapoli and Deavenport, 1979; Lauer, 1979). Only propagation models were evaluated and one model was selected as the standard against which to compare the other models. More than 10 different model versions were evaluated using an ocean environment described by a single sound speed profile and a flat bottom (i.e., range-independent environment).

All subject models were executed at four frequencies ranging from 35 Hz–200 Hz. The resulting transmission loss data were modestly smoothed and then compared with the standard model. These comparisons were conducted within variable range intervals specifically chosen to correspond to common sonar detection zones: direct path (DP); first and second bottom bounce (BB) regions; and first, second, and third convergence zones (CZ). The means (μ) and standard deviations (σ) of the differences in transmission loss between the standard and the subject models were then calculated within each interval. A zero mean indicated that the standard model and the subject model produced transmission loss values that were identical when averaged over the interval. The resulting means and standard deviations were appropriately weighted and averaged to obtain the cumulative accuracy measures (CAMs). Mathematically, the CAMs for the mean and standard deviation were expressed as (Lauer and Sussman, 1979)

$$\mathrm{CAM}_{\mu} = \frac{1}{\sum_{i=1}^{N_c} N_{R_i}} \sum_{i=1}^{N_c} \sum_{j=1}^{N_{R_i}} |\mu_{ij}| W_{ij} \tag{12.1}$$

$$\mathrm{CAM}_{\sigma} = \frac{1}{\sum_{i=1}^{N_c} N_{R_i}} \sum_{i=1}^{N_c} \sum_{j=1}^{N_{R_i}} \sigma_{ij} W'_{ij} \tag{12.2}$$

where

i = case index
j = range interval index
N_c = total number of cases
N_{R_i} = total number of range intervals for the ith case
W_{ij} = weights in each case and range interval bin $(\Delta R)_{ij}$ to be applied to mean values of transmission loss differences
W'_{ij} = weights applied to standard deviations of transmission loss differences

In the event that all values of W_{ij} and W'_{ij} were chosen to be unity, CAM_{μ} became the "grand mean" taken over all cases and range intervals, and CAM_{σ} became the "grand standard deviation" taken over all cases and range intervals. The weighting functions (W_{ij}, W'_{ij}) were functions of the acoustic frequency (f), source depth (Z_S), receiver depth (Z_R), and range interval (ΔR).

For the present discussions, only four specific versions of propagation models have been selected. These models represent four of the five different modeling approaches described earlier in Chapter 4. The only modeling approach not investigated by POSSM was the parabolic equation (PE) approach. At the time of POSSM's activities, PE models were not yet widely used. Moreover, fast-field program (FFP) models were considered to be the most accurate available and were commonly used as standards for comparison. The four approaches and corresponding models represented are

Approach	Model
Ray theory (with corrections)	FACT
Multipath expansion (hybrid)	RAYMODE
Normal mode	NLNM
Fast-field program	FFP

In the comparisons described by Lauer and Sussman (1979), and by DiNapoli and Deavenport (1979), the FFP model was used as the standard. Table 12.1 summarizes the computer program sizes and execution times for each model in the context of the standard problem selected by POSSM. A quantitative comparison of model accuracies was reported by Lauer and Sussman (1979) based on the methodology summarized in Figure 12.7. Only the four particular models identified above will be discussed here. Representative results from one test case are summarized in Table 12.2. It must be stressed that these results pertain only to the configuration of these models as they existed at the time of the tests, and these results are only valid for the particular test case run. Different results might be obtained using more recent configurations of the same models or with other test scenarios and geometries, or even with other measurement standards or host computers.

The CAM were also computed (Table 12.3). These results support the intuitive notion that model accuracy improves with increasing model complexity and computational intensity. Accuracy alone, however, should not be the sole evaluation criterion. A broader view of the model evaluation process is presented in the next section.

TABLE 12.1
Example Program Sizes and Execution Times for Selected Propagation Models

Model	Program Size (Decimal Words)	Execution Time (s)
FACT	21,855	2.5
RAYMODE	14,115	19.2
NLNM	53,000	30.6
FFP	51,572	373.0

Source: Adapted from Lauer, R.B. and Sussman, B., A Methodology for the Comparison of Models for Sonar System Applications—Results for Low Frequency Propagation Loss in the Mediterranean Sea, Vol. II. Nav. Sea Syst. Command, SEA 06H1/036-EVA/MOST-11, 1979.

Note: The standard problem consists of a single sound-speed profile overlying a flat bottom executed at a low frequency (≤200 Hz) on a UNIVAC 1108 computer.

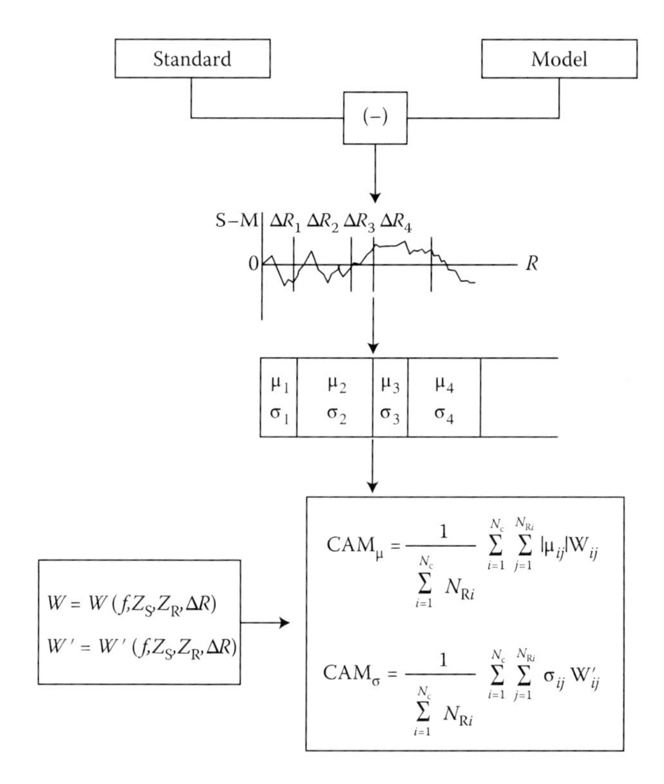

FIGURE 12.7 Summary of the POSSM model evaluation methodology. (Adapted from Lauer, R.B., Acoustic model evaluation: Issues and recommendations incorporating the experience of the panel on sonar system models (POSSM). Nav. Underwater Syst. Ctr, Tech. Rept 6025, 1979.)

TABLE 12.2
Means (μ) and Standard Deviations (σ) of the Differences between the FFP Propagation Model and Candidate Model Results for One Particular Case Over Direct Path (DP), Bottom Bounce (BB), and Convergence Zone (CZ) Regions

	Range Interval													
	DP		BB1		CZ		BB2		To 100 km		To 150 km		To 200 km	
Model	μ	σ	μ	σ	μ	σ	μ	σ	μ	σ	μ	σ	μ	σ
FACT	–	–	1.4	3.6	0.9	2.2	0.1	3.0	−2.2	2.0	−2.0	1.3	−1.9	1.5
RAYMODE	–	–	1.3	3.7	0.5	2.0	−0.4	2.7	−0.3	1.1	−0.4	1.0	−0.8	1.6
NLNM	–	–	0.6	2.0	0.4	2.2	−0.8	1.9	0.7	1.9	−0.3	2.6	−1.0	2.5
FFP	–	–	–	–	–	–	–	–	–	–	–	–	–	–

Source: Adapted from Lauer, R.B. and Sussman, B., A Methodology for the Comparison of Models for Sonar System Applications—Results for Low Frequency Propagation Loss in the Mediterranean Sea, Vol. II. Nav. Sea Syst. Command, SEA 06H1/036-EVA/MOST-11, 1979.

TABLE 12.3
Cumulative Accuracy Measures (CAMs): Averages of Means (μ) and Standard Deviations (σ) Over All Cases and Range Intervals Using FFP Propagation Model Results as the Standard

Model	$|\mu|$	σ
FACT	1.3	2.5
RAYMODE	0.9	2.4
NLNM	0.6	2.3
FFP	–	–

Source: Adapted from Lauer, R.B. and Sussman, B., A Methodology for the Comparison of Models for Sonar System Applications—Results for Low Frequency Propagation Loss in the Mediterranean Sea, Vol. II. Nav. Sea Syst. Command, SEA 06H1/036-EVA/MOST-11, 1979.

12.6 EVALUATION GUIDELINES

Model evaluation entails the systematic gathering and promulgation of information about models to determine model limitations and domains of applicability. There is no evaluation procedure appropriate for all models. The procedures must be tailored on the basis of model structure, documentation and other information available to the evaluators. Criteria for model evaluation can be segregated into five basic categories (US General Accounting Office, 1979): documentation, verification, validation, maintainability, and usability.

12.6.1 Documentation

Documentation of computer models is important for two reasons: (1) to ensure that the model is thoroughly understood and can be operated and maintained in the present as well as in the future, and (2) to facilitate independent evaluation of the model, particularly by someone other than the model developer or initial user.

Elements that should be included in model documentation are (Gass, 1979)

1. A precise statement of what the model is supposed to do
2. The mathematical and logical definitions, assumptions, and formulation of the problem being modeled
3. A complete set of current input and output, and test cases that have been run
4. A complete set of flow charts of the computer program
5. A set of operating instructions for the computer operator
6. An explanation of the various options available in using the model
7. The computer program itself (listing), with comments about various operations in the program

Further guidelines for documentation are provided later in this chapter.

12.6.2 Verification

Verification entails an examination of the model to ensure that the computer program accurately describes the model and that the program runs as expected. In order to do this, the following factors must be examined: (1) consistency of mathematical and logical relationships; (2) accuracy of intermediate numerical results; (3) inclusion of important variables and relationships; and (4) proper mechanization and debugging of the program.

Computer hardware selections will sometimes impact model accuracy due to word length and double-precision considerations, among others. Thus, the same model implemented on two different computer systems may produce significantly different results. Related problems concern artifacts, which are false features that arise from some quirk of the computer, and which disappear when the software is written differently.

12.6.3 Validity

Validation and critical assessment are required of all theoretical conjectures, hypotheses, and models. One of the most difficult but important tasks in model construction is the specification of its limitations: what are its limits, and in what way is it an approximation? The consensus principle in science implies that the evaluation of models must be open, and cannot be accepted on the authority of the model developer alone. The model should not contain adjustable parameters or any hidden variables that have to be invoked to explain discrepancies between theory and experiment. The theoretical properties of the model should be sharply defined, and derived with sufficient mathematical rigor to be compared objectively with the observed phenomena. Elementary errors and misunderstandings will be detected by the independent repetition of experiments and by comparisons of calculations with experimental data, or by theoretical criticism (Ziman, 1978). Even when models are wrong, they can assist in structuring discussions.

The category of validity comprises three factors: theoretical validity, data validity, and operational validity. Theoretical validity entails review of the physics underlying the model and the major stated and implied assumptions. The applicability and restrictiveness of these assumptions must also be examined. In addition, the internal logic of the model should be reviewed.

For empirically based models, data validity is concerned with the accuracy and completeness of the original data and the manner in which the empirical model deals with the transformation of the original data.

Operational validity is concerned with assessing the impact of model errors (i.e., divergences between model predictions and reality) on decision processes. This aspect of model evaluation addresses accuracy. When evaluating model accuracy, it is important to examine the error budget. Errors in model predictions (e_p) are assumed to be the sum of two independent, random variables:

$$e_p = e_d + e_m \tag{12.3}$$

where e_d represents errors related to model input data and e_m represents model errors.

12.6.4 Maintainability

Maintainability considers the ease of incorporating new data and formulas as well as provisions for reviewing the accuracy of the model as more experimental data become available. A training program must be formalized to ensure that the operators understand how the model should be used and also to make revisions known to the computer-systems personnel.

12.6.5 Usability

Usability addresses the appropriateness of the model for the intended applications. In essence, the model should satisfy the user's specific requirements. This, of course, requires that users articulate (specify) their needs very precisely.

It is instructive to distinguish between research and operational models when discussing usability. Specifically, research models are intended to address a wide variety of often ill-posed scientific questions. In order to be responsive to such ambiguous issues, the research models are structured to allow (and often require) a high degree of operator intervention during execution. This allows the researcher to adjust parameters as the problem solution evolves. Alternatively, operational models are structured to minimize (if not eliminate) the need for operator intervention. Indeed, such intervention is viewed as a nuisance to the operator. The parallel between executive-versus-bundled system architectures and research-versus-operational models is valid (Section 11.4.1).

Model usability is heavily influenced by the model's inherent domains of applicability. Factors such as frequency coverage and problem geometry are implicit in these domains. For example, the use of normal mode models in the calculation of high-frequency bistatic reverberation is not practical at present due to excessive computation times. Model output options are also important. For example, wave-theoretical propagation models more easily generate transmission loss values in the range-depth plane while arrival structure information is more easily generated by ray-theoretical propagation models.

12.7 DOCUMENTATION STANDARDS

Documentation standards for computer models were reviewed by Gass (1979). Both the US Department of Defense (DOD) and the National Institute of Standards and Technology (formerly the National Bureau of Standards, or NBS) have issued formal guidelines:

- DOD-STD-7935A—Department of Defense standard: Automated data systems (ADS) documentation.
- DOD-STD-2167A—Military standard: Defense system software development.
- FIPS PUB 38—Guidelines for documentation of computer programs and ADS.
- DOD-STD-2168—Military standard: Defense system software quality program.

TABLE 12.4
Software Life Cycle and Documentation Types According to the Federal Information Processing Standards (FIPS)

	Development Phase				
Initiation Phase	**Definition Stage**	**Design Stage**	**Programming Stage**	**Test Stage**	**Operation Phase**
	Functional requirements document	System/ subsystem specification	User's manual		
		Program specification	Operations manual		
	Data requirements document	Database specification	Program maintenance manual		
			Test plan	Test analysis report	

Functional requirements document—provides a basis for the mutual understanding between users and designers of the initial definition of the software, including the requirements, operating environment, and development plan.
Data requirements document—provides data descriptions and technical information about the data collection requirements.
System/subsystem specification—specifies for analysts and programmers the requirements, operating environment, design characteristics, and program specifications for a system or subsystem.
Program specification—specifies for programmers the requirements, operating environment, and design characteristics of a computer program.
Database specification—specifies the identification, logical characteristics, and physical characteristics of a particular database.
User's manual—sufficiently describes the functions performed by the software in non-ADP terminology such that the user organization can determine its applicability, and when and how to use it; moreover, it should serve as a reference document for preparation of input data and parameters, and for interpretation of results.
Operations manual—provides computer operation personnel with a description of the software and of the operational environment so that the software can be run.
Program maintenance manual—provides the maintenance programmer with the information necessary to understand the programs, their operating environment, and their maintenance procedures.
Test plan—provides a plan for testing the software; provides detailed specifications, descriptions, and procedures for all tests; and provides test data reduction and evaluation criteria.
Test analysis report—documents the test analysis results and findings; presents the demonstrated capabilities and deficiencies for review; and provides a basis for preparing a statement of software readiness for implementation.

Source: Gass, S.I., Computer model documentation: A review and an approach. National Bureau of Standards Special Pub. 500–539, 1979.

According to the federal information processing standards (FIPS), model documentation is keyed to the various phases and stages of the software life cycle (Table 12.4). The major phases are:

Initiation phase—During this phase, the objectives and general definition of the requirements for the software are established. Feasibility studies, cost-benefit analyses, and the documentation prepared within this phase are determined by agency procedures and practices.
Development phase—During this phase, the requirements for the software are determined, and software is then defined, specified, programmed, and tested. Ten major documents are prepared in this phase to provide an adequate record of the technical information developed.
Operation phase—During this phase, the software is maintained, evaluated, and changed as additional requirements are identified. The documentation is maintained and updated accordingly.

The development phase of the software life cycle is further subdivided into four main stages as follows:

Definition stage—when the requirements for the software and documentation are determined.
Design stage—when the design alternatives, specific requirements and functions to be performed are analyzed and a design is specified.
Programming stage—when the software is coded and debugged.
Test stage—when the software is tested and related documentation is reviewed. The software and documentation are then evaluated in terms of readiness for implementation.

Formal documentation guidelines are frequently amended or even superseded by newer guidelines. Care should be taken to consult the latest governing instructions. For example, DOD-STD-7935A and DOD-STD-2167A (which were noted earlier) were later consolidated into MIL-STD-498 in an effort to implement governing ISO/IEC standards (DIS 12207—Software Life Cycle Processes). MIL-STD-498 (Software Development and Documentation) was issued on December 5, 1994. However, this standard was subsequently canceled on May 27, 1998 and replaced by IEEE/EIA 12207 (issued in three parts):

- IEEE/EIA 12207.0 (Industry implementation of international standard ISO/IEC 12207, Standard for Information Technology—Software Life Cycle Processes) contains concepts and guidelines to foster better understanding and application of the standard. This standard thus provides industry with a basis for software practices usable for both national and international business.
- IEEE/EIA 12207.1 (Guide for ISO/IEC 12207, Standard for Information Technology—Software Life Cycle Processes—Life Cycle Data) was adopted on May 27, 1998 for use by the US DOD. This document provides

guidance on life cycle data resulting from the processes of IEEE/EIA 12207.0 including relationships among content of life cycle data information items, references to documentation of life cycle data, and sources of detailed software product information.

- IEEE/EIA 12207.2 (Guide for ISO/IEC 12207, Standard for Information Technology—Software Life Cycle Processes—Implementation Considerations) was adopted on May 27, 1998 for use by the US DOD. This document provides implementation guidance based on software industry experience with the life cycle processes.

In recent developments, IEEE/EIA 12207-2008: "Standard for Information Technology –Software Life Cycle Processes" was issued on January 31, 2008 and replaced two earlier standards:

- IEEE/EIA 12207.0-1996: "Standard for Information Technology—Software Life Cycle Processes"
- IEEE/EIA 12207.2-1997: "Guide for ISO/IEC 12207, Standard for Information Technology—Software Life Cycle Processes—Implementation Considerations."

The remaining standard, IEEE/EIA 12207.1-1997 (IEEE Guide for Information Technology—Software Life Cycle Processes—Life Cycle Data), was withdrawn.

The transfer of modeling and simulation (M&S) technologies among members of the international community continues to stimulate new initiatives for improved international standards in simulation architecture. Such efforts seek to promote the large-scale interoperability of simulation software and hardware.

13 Simulation

13.1 BACKGROUND

This chapter discusses the structure and applications of simulation in underwater acoustics. Since simulation refers to a method for implementing a model over time, it is fitting that this topic is addressed after a firm foundation of modeling and evaluation has been established in the previous chapters. In the present context, the term "modeling and simulation" (M&S) refers to those techniques that can predict or diagnose the performance of complex acoustic systems operating in the dynamic undersea environment.

A widely used taxonomic scheme for classifying various types of simulation is based on the degree of human involvement and the realness of the system. This scheme distinguishes three categories of simulation: live, virtual, and constructive. Live simulation involves real people operating real systems. Virtual simulation involves real people operating simulated systems. Constructive simulation involves simulated people operating simulated systems. To complete the symmetry of this taxonomic scheme, a fourth category termed "smart systems" is added. In essence, smart systems involve simulated people operating real systems. The resulting taxonomic scheme comprising these four categories of simulation is summarized in Table 13.1.

Another term frequently encountered in discussions of simulation is "stimulation." Stimulation is the use of simulation to provide external stimuli to a system or subsystem. Stimulation often entails hardware-in-the-loop or software-in-the-loop configurations, which are commonly referred to as "constrained simulation" since the simulated time-advances have a specific relationship to wall clock time. McCammon (2004) documented a detailed survey of existing sonar stimulator models and facilities.

When simulating a complex assemblage of independent but interconnected systems, it is common to refer to such an assemblage as a "system-of-systems." Simulating the performance of such an assemblage is often referred to as "end-to-end" simulation. The "state" of a system is defined by the collection of variables necessary to describe that system at any given time.

Simulations are differentiated at three levels of system representations: static versus dynamic, deterministic versus stochastic, and continuous versus discrete. A static simulation represents a system state in which time is not a variable. Conversely, a dynamic simulation varies as a function of time. A deterministic simulation produces completely predictable values whereas a stochastic simulation produces values that must be represented by statistical variables (e.g., means and variances). A continuous simulation produces state variables that change continuously with changes in time while a discrete simulation produces values that change in a stepwise fashion as a function of time (Law and Kelton, 1991).

TABLE 13.1
Four Categories of Simulation Based on the Degree of Human Involvement

	Real Systems	Simulated Systems
Real people	Live Simulation	Virtual Simulation
Simulated people	Smart Systems	Constructive Simulation

Different mathematical approaches are used depending on the type of simulation employed. For example, continuous-system simulations are modeled using differential equations while time-stepped simulations are modeled using discrete-time approaches. Event-based simulations are modeled as discrete events.

Law and Kelton (1991: Chapter 12) provide a comprehensive introduction to the use of statistical experimental design and optimization techniques in simulation. Specifically, experimental design provides a way of deciding before any runs are made which particular configurations should be simulated so that the desired information can be obtained with the least amount of simulation. The term "design of experiment" (DOE) is sometimes used in the literature in reference to this process. DOE is particularly important when simulation is used to evaluate (or trade) alternative system configurations.

To structure subsequent discussions, the four hierarchical levels of simulation (engineering, engagement, mission, and theater) are first reviewed in Section 13.2. Next, simulation infrastructure is discussed in Section 13.3 using examples drawn from defense-related activities. Section 13.4 discusses the HLA, currently the highest priority effort within the defense M&S community. The role of testbeds is described in Section 13.5. Finally, a survey of current applications in Section 13.6 provides specific examples of generally accepted approaches and common practices.

13.2 HIERARCHICAL LEVELS

As previously mentioned in Chapter 1, simulation in support of naval applications can be decomposed into four fundamental levels: engineering, engagement, mission, and theater. Table 13.2 summarizes the outputs and applications associated with each of these four levels. Each level is discussed below in greater detail.

13.2.1 Engineering

Engineering-level simulation comprises the categories of environmental, propagation, noise, reverberation, and sonar performance models. This level of simulation generates measures of system performance that are used to design and evaluate systems and subsystems and also to support system testing. Representative measures of performance include probability of detection and median detection ranges. Sonar technologists and acoustical oceanographers routinely use this level of simulation for prognostic or diagnostic applications. These performance metrics are also useful in system design, cost, manufacturing, and supportability trade studies.

TABLE 13.2
Four Principal Levels of Simulation for Naval Applications

Level	Output	General Applications
Theater	Force dynamics	Evaluate force structures. Evaluate strategies.
Mission	Mission effectiveness	Evaluate force employment concepts. Evaluate system alternatives.
Engagement	System effectiveness	Train system operators. Evaluate tactics.
Engineering	System performance	Design and evaluate systems/subsystems. Support system testing.

Source: National Research Council. 1997. *Technology for the United States Navy and Marine Corps, 2000–2035. Becoming a 21st-Century Force*, Vol. 9, Modeling and Simulation. National Academy Press, Washington, DC.

13.2.2 Engagement

Engagement-level simulation executes engineering-level models to generate measures of system effectiveness in a particular spatial and temporal realization of an ocean environment when operating against (engaging) a particular target. This level of simulation is used to evaluate system alternatives, train system operators, and tactics. Engagement outputs can be used to estimate exchange ratios, which are useful in evaluating tactical effectiveness against known and postulated targets.

Tactical decision aids (TDAs) represent a form of engagement-level simulation products that blend environmental information with tactical rules garnered from higher-level, aggregate simulations. These decision aids guide system operators and scene commanders alike in planning missions and allocating resources by exploiting knowledge of the operating environment. While TDAs are usually associated with naval applications, the conceptual approach is valid in research and commercial applications as well.

Lam et al. (2009) established a complete chain of operational oceanographic and acoustic forecasting including METOC fields, *in situ* measurements, and data assimilation leading to the provision of tactical data tailored in the form of acoustic forecasts. This operational system had two components: an at-sea modeling capability for rapid coupled ocean-acoustic predictions, and onshore support for data quality control and management in addition to scientific and computational guidance. Acoustic modeling was performed using the sonar performance prediction package ALMOST, which is a range-dependent ray theoretical model.

Wathelet et al. (2008) developed a tactical planning aid for active monostatic and multistatic sonars. This aid incorporated a generic multistatic sonar performance model that could also be used to represent monostatic sonars. The performance model was combined with false-contact simulators for clutter and noise in addition to a tracker-fusion module. The output of the tracker was a series of simulated tracks resulting from targets, clutter objects, or random fluctuations in the background noise.

Li et al. (2014) analyzed submarine search-evasion-path planning using the artificial bee colony (ABC) algorithm. The ABC algorithm is a relatively new swarm-intelligence algorithm inspired by the foraging behavior of honey bees (e.g., Karaboga and Basturk, 2007). Various studies using different numerical benchmark tests have demonstrated that the ABC algorithm possesses competitive advantages compared to other swarm-intelligence and evolutionary algorithms. Specifically, several search-evasion cases were investigated in two-dimensional (2D) planes in which the anti-submarine vehicles were equipped with sensors having circular detection footprints of fixed radii. The invading submarine was assumed able to acquire the real-time locations of all anti-submarine assets in the search area. While the proposed method was shown to be a viable solution, future work may consider improved formulations of a model cast in three-dimensional (3D) space, and also replace straight-line search paths with smoothed segments to remove velocity discontinuities in the dynamic simulation.

13.2.3 Mission

Mission-level simulation aggregates multiple engagements to generate statistics useful in evaluating mission effectiveness. At this level, system concepts are evaluated within the context of well-defined mission scenarios. The outputs of this level of simulation are used to evaluate force employment concepts. The effectiveness of multiple platforms performing specific missions can be assessed using this level of simulation.

13.2.4 Theater

Theater-level simulation aggregates mission-level components to generate measures of force dynamics and analyze alternative system-employment strategies. This type of simulation is used in planning, budgeting, and operational analysis. Planning includes decisions regarding force structure, modernization, readiness, and sustainability. Budgeting includes decisions regarding specific line items in the defense budget. Operational analysis considers issues such as developing contingency plans, estimating logistics demands, and analyzing specific combat plans (Bracken et al., 1995). This level of simulation is useful in war-gaming with joint or combined forces.

13.3 SIMULATION INFRASTRUCTURE

The National Research Council (NRC) (1997, 2002) portrayed M&S as a foundation technology for many developments that will be central to the US Navy over the next several decades. Representative applications of simulation in the defense industry were summarized by Bracken et al. (1995), who edited a useful collection of papers coordinated by the Military Operations Research Society (MORS).

The Defense Modeling and Simulation Office (DMSO), now the Modeling and Simulation Coordination Office (M&SCO), was established in 1991 to provide a focal point for information concerning US Department of Defense (DOD) M&S activities. The M&SCO is the lead standardization activity for managing M&S standards and methodologies. Further details are available in Appendix C.

The US DOD (1994, 2009) officially adopted definitions for verification, validation, and accreditation (VV&A) that originated from the efforts of the MORS. These definitions are useful for applications in naval operations, offshore industries, and oceanographic research:

- **Verification**—The process of determining that a model implementation accurately represents the developer's conceptual description and specifications.
- **Validation—**The process of determining the degree to which a model is an accurate representation of the real world from the perspective of the intended uses of the model.
- **Accreditation—**The official certification that a model or simulation is acceptable for a specific purpose.

The US DOD (1996) assembled a very useful compendium of VV&A techniques from sources in government, industry, and academia. This evolving document provides practical guidelines for formulating VV&A procedures in a wide range of M&S environments.

Trends in system architectures are exemplified by the US Department of Defense architecture framework (DoDAF), which provides the rules, guidance, and product descriptions for developing and presenting architecture descriptions that ensure a common denominator for understanding, comparing, and integrating architectures.

Version 1.0 of the DoDAF (released in 2004) defined a common approach for DOD architecture description development, presentation, and integration of operations and processes. The framework, comprised of several views, was intended to ensure that architecture descriptions could be compared and related across organizational boundaries, including joint and multinational boundaries. Each view was composed of sets of architecture data elements that were depicted via graphic, tabular, or textual products. DoDAF Version 2.0 was released in 2009.

The unified modeling language (UML) representation was provided in Version 1.0 of the DoDAF to assist architects who chose to use object-oriented methodologies. The UML has emerged as the dominant and most prevalent language for object-oriented modeling, irrespective of the development process used. The UML has been characterized as a general-purpose modeling language for specifying, visualizing, constructing, and documenting the artifacts of software systems, business modeling, and other non-software systems.

The use of sonars in undersea warfare (USW) has been noted as a key topic in ocean-acoustic information engagement (Wang and Cai, 2006). Underwater warfare simulation involving multiple targets and multiple sonars is an important way to research sonar performance in the complicated ocean environment. For example, simulation software has been analyzed and designed by an object-oriented class model. The procedure of the design relied on UML, and the product of the design is both expansible and reusable. USW interoperability has been improved by using extensible markup language (XML) tagsets for system data interchange. A special USW-XML working group was formed to establish coherent battlespace visualization capabilities for network-centric USW. The tactical assessment markup language (TAML) created for anti-submarine warfare (ASW) tactical assessment system (A-TAS) was sponsored by the Fleet ASW Command (Childers et al., 2006).

13.4 HIGH-LEVEL ARCHITECTURE

The high-level architecture (Kuhl et al., 1999) is the highest priority effort within the DOD M&S community. The HLA has been adopted as IEEE Standard 1516 and has also been proposed for acceptance by the North Atlantic Treaty Organization (NATO) as the standard for simulations used within the alliance. The HLA is composed of three parts: the HLA rules, the HLA interface specification, and the object model template (OMT). The HLA rules describe the general principles defining the HLA and also delineate 10 basic rules that apply to HLA federations and their participating applications (called federates). The HLA interface specification defines the functional interface between federates and the runtime infrastructure (RTI). The OMT provides specifications for documenting key information about simulations and federations. Use of the OMT to describe simulation and federation object models (called SOMs and FOMs, respectively) is a key part of the HLA (Kuhl et al., 1999; NATO, 1998).

The NATO Alliance is generally dependent upon the M&S contributions of the member nations. These contributions include the cooperative development of technical capabilities such as the defense M&S technologies research program in the European Co-operation for the Long-term in Defence (EUCLID). In this environment, required simulations must either be specified and then developed anew, or else legacy simulations must be adapted to the meet the specified requirements.

The development of individual models and simulations has been occurring for decades and is therefore relatively well understood by NATO. The alliance, however, only has limited experience with the cooperative development of federations of diverse simulations. To be prudent, therefore, NATO has opted to demonstrate the viability of this innovative development approach by conducting a pathfinder development of an HLA-based federation of national simulations. This federation would be planned and centrally integrated and tested, but the individual national simulation developments would be executed by the involved nations. Ideally, such a pathfinder effort would be built on the experience base established during NATO's Distributed Multi-National Defence Simulation (DiMuNDS) project. Selected legacy simulations will have to fulfil the specified criteria and will, therefore, require some modifications (NATO, 1998, 2000).

13.5 TESTBEDS

Testbeds allow the simultaneous use of high-detail and low-detail system representations (i.e., variable resolution) in a single simulation. This flexibility enables an analyst to simulate a key system in high detail while simulating the less-critical contextual environment in lower detail.

In integrated hierarchical variable resolution (IHVR) simulations, high-level variables are expressed as functions of lower-level (but higher-resolution) variables. Here, hierarchies of variables can relate models at different resolutions (Davis, 1995). Complications may arise from so-called configural effects, which consider the influences of temporal and spatial correlations on simulated outcomes (NRC, 1997: 90 and 233). Specifically, the configuration of the simulated assets is inseparable from the outcome of the particular simulation. Consequently, if the assets were configured differently at the outset, the simulated outcome would likely be different.

An integrated testbed includes the processor on which the simulation software will run together with all other units that will interface with the processor. This arrangement affords the opportunity to perform early interface testing using the actual hardware. For example, the tactical oceanography simulation laboratory (TOSL) provides a testbed for the development, testing, and validation of high-fidelity underwater acoustic models and supporting databases (Ellis et al., 1996).

From a broader perspective, simulation testing can be accomplished either in laboratory-based testbeds or in at-sea tests. At-sea tests provide engineers the opportunity to validate sonar-system performance in real (versus synthetic) ocean environments. For example, the littoral warfare advanced development (LWAD) project provides at-sea tests (including platforms and coordination) to identify and resolve technical issues that arise from operating undersea systems in littoral environments (Spikes et al., 1997). Sea tests can range from simple focused technology experiments (FTE) to more complex system-concept validations (SCV). The penalty paid for testing in a real (versus synthetic) environment is the loss of experimental control and repeatability.

Situating testbeds at a fixed site in the field can create an interesting hybrid testbed configuration. Such arrangements, sometimes referred to as "natural laboratories," have attractive features over laboratory-based testbeds. For example, field sites permit sustained modeling and observing systems to be deployed so that models can be continually tested and refined in real (versus synthetic) environments. Moreover, operational training can be collocated with fielded testbeds so that realistic experience can be obtained. The loss of experimental control and repeatability is limited by selecting a fixed location in an ocean area that is well understood environmentally. An example of a fielded testbed is the northern Gulf of Mexico littoral initiative, or NGLI (Carroll and Szczechowski, 2001). The NGLI is a multi-agency program established through a partnership between the Commander, Naval Meteorology, and Oceanography Command and the Environmental Protection Agency's Gulf of Mexico Program Office. The goal of NGLI is to become a sustained comprehensive nowcasting/forecasting system for the coastal areas of Mississippi, Louisiana, and Alabama that will use model forecasts and observational data for training and coastal resource management. The program integrates a reliable and timely meteorological and oceanographic modeling scheme, combining 3D circulation, sediment transport, and atmosphere and wave models with *in situ* and remotely sensed observations via an extensive data distribution network that is available to a wide range of users in near-real time through an interactive website. The Naval Oceanographic Office, which manages the program, chose the Mississippi Bight as an ideal testbed to examine new modeling and observational technologies before they are applied to other littoral areas of interest. The NGLI directly addresses the US Navy's requirement to project oceanographic information from deep-water environments shoreward into littoral areas. Model nowcasts and forecasts are being applied to the ocean littoral environment by cascading information from large ocean basin models to shallow-water models. Lessons learned within this "natural laboratory" provide civil authorities with metrics by which to evaluate environmental stresses (e.g., sediment transport modifications and increased pollution) caused by growth in population and industry. For discussions of the complexity of littoral operations, see recent summaries by Bindi et al. (2005, 2008).

Arzelies et al. (2015) examined the possibility of using a novel acoustic test bench to predict the performance of a pair of modems operating at a center frequency of 60 kHz. The experimental setup employed two separate basins: a waterfront dock and an adjacent shore-based test facility. This approach avoided direct acoustic coupling between the transmitting and receiving modems and also reduced unwanted reflections. The signal emitted by the transmitting modem was acoustically recovered in the dock and then injected into the shore-based real-time propagation simulator using the RAYSON propagation model. This approach will also be used to test other pairs of modems to establish their suitability for operational use on French Research Institute for Exploitation of the Sea (IFREMER) underwater vehicles for tasks involving offshore-oil underwater intervention. This bench-testing approach offers the possibility of separately analyzing each underwater acoustic propagation parameter in isolation since the test environment is controlled.

13.6 APPLICATIONS

Simulations are used in diverse scientific and engineering disciplines. Although the following discussions will highlight many naval applications, the basic approaches and practices surveyed here are applicable to offshore industries and oceanographic research as well. Specific attention will be given to those activities relating to engineering-level and engagement-level simulations. It may be useful at this point to again refer to the discussions in Section 13.2 above (especially Table 13.2).

Discussions will start with engineering-level simulations that generate system-performance outputs. As indicated in Table 13.2, general applications include design and evaluation of systems (or subsystems) and system-testing support. As specific examples, Section 13.6.1 will discuss applications in systems engineering and Section 13.6.2 will discuss applications in simulation based acquisition.

The next level above engineering is engagement-level simulations, which generate system-effectiveness outputs. As indicated in Table 13.2, general applications include the evaluation of system alternatives, training of system operators, and evaluation of tactics. As specific examples, Section 13.6.3 will discuss applications in operations analysis and Section 13.6.4 will discuss applications in training.

These discussions emphasize *processes* as opposed to *specific implementations* of simulation packages. The intent is to familiarize the reader with generally accepted approaches to simulation as practiced in government (both civil and military) and in industry (both offshore and defense-related). Research applications may utilize variations of these approaches.

13.6.1 Systems Engineering

An objective of systems engineering is to gain visibility in the early stages of a system's development. The intent is to explore all feasible approaches to system design, identify and eliminate potential problems, select a preferred design configuration and thus reduce risks and costs (Blanchard, 1998). The use of simulation allows designers to investigate alternative design solutions prior to committing to any particular design. In a systems-engineering context, simulation is used to understand the behavior of a system or to evaluate alternative system-design considerations in trade-off studies. The use of simulation is most effective in the early stages of system development before any physical elements comprising the system are available for evaluation.

The systems engineering process is often represented by the so-called "V" diagram, as illustrated in Figure 13.1. The decomposition-and-definition process flows downward along the left leg of the "V" while the integration-and-verification process flows upward along the right leg of the "V." In the decomposition-and-definition process (on the left side of Figure 13.1), the utilization of simulation during the preliminary design phase would have the greatest impact on the final system configuration.

Boehm (1988) introduced the spiral model as one candidate for improving the systems engineering process. In essence, the spiral model creates a risk-driven approach in which each cycle of the spiral includes requirement identification, alternative generation, alternative evaluation, prototype development and testing,

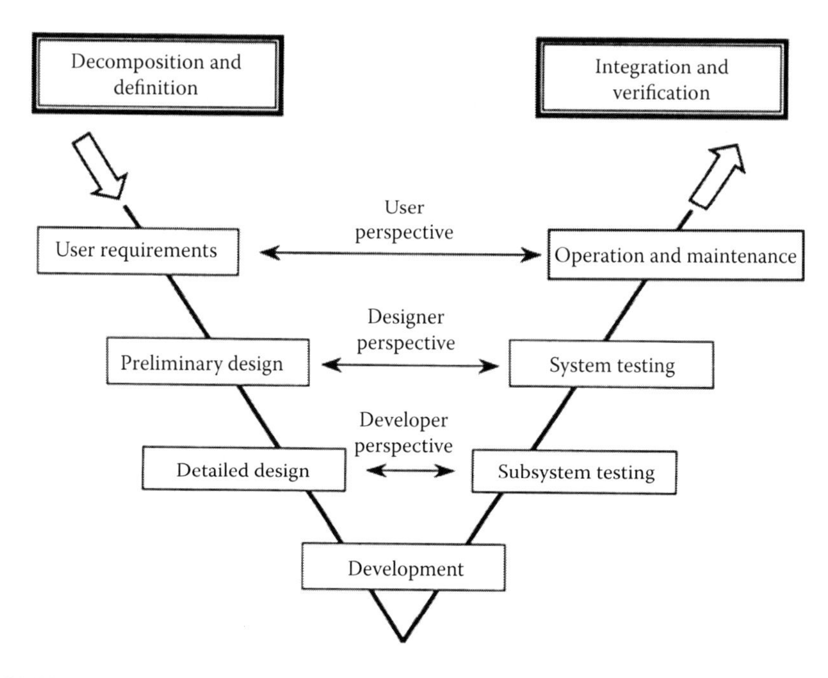

FIGURE 13.1 Systems engineering decomposition and definition process flow. (Reproduced from Blanchard, B.S., *System Engineering Management*. 2nd ed. 1998. Copyright Wiley-VCH Verlag GMBH & Co. KGaA. With permission.)

planning for the next cycle, cycle review outcome, risk evaluation, and abort points. Multiple prototypes are generated over the development phase, where each prototype reveals new information about the problem and requirements.

An important asset in the systems engineering process is the software engineering environment (SEE), which supports the software development process. The SEE comprises the facilities, integration methods, tools, procedures, and management information system (MIS) necessary to maintain productivity, simulation product quality, and software reuse.

13.6.2 Simulation-Based Acquisition

Simulation-based acquisition (SBA) is an acquisition process in which both the US DOD and industry collaborate in the use of simulation technologies that are integrated across acquisition phases and programs (US Department of the Navy, 2000a). Specifically, SBA entails the optimization of system performance versus total ownership cost (TOC) through exploration of the largest possible trade space. TOC comprises the cost to research, develop, acquire, own, operate, and dispose of primary and support systems, other equipment and real property; the costs to recruit, train, retain, separate, and otherwise support military and civilian personnel; and other costs of business operations in the DOD. Johnson et al. (1998) provided a very useful introduction to the concepts underpinning simulation based acquisition.

It is helpful to revisit the M&S hierarchy pyramid introduced previously in Chapter 1 (see Figure 1.3) in the context of system-design applications. In Figure 13.2 (US Department of the Navy, 2000a), the four categories of simulation (engineering, engagement, mission, and theater) are related to their principal outputs: engineering-level simulations output system or component performance;

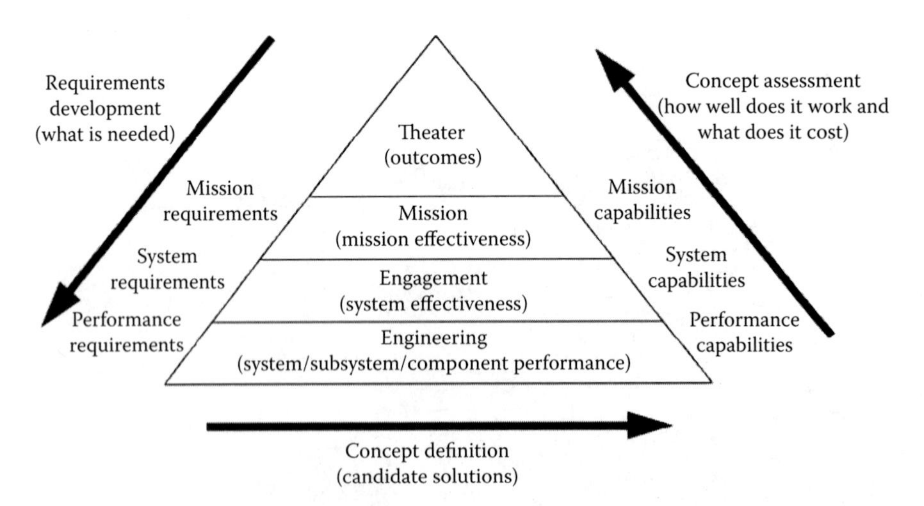

FIGURE 13.2 M&S in system design. (Adapted from US Department of the Navy. 2000a. SBA status and international implications. Prepared by the Department of the Navy acquisition reform office for the technical cooperation program [TTCP] panel on Systems Engineering for Defense Modernization.)

engagement-level simulations output system effectiveness; mission-level simulations output mission effectiveness; and theater-level simulations output the overall outcomes. Requirements are generally developed in a top-down approach as indicated by the downward-directed arrow (labeled "requirements development") on the left side of the pyramid in Figure 13.2. Specifically, mission requirements (derived from theater-level simulations) flow down to system requirements which, in turn, flow down to performance requirements. These performance requirements are then allocated among the various subsystems and components using engineering-level simulations. Alternatively, a bottoms-up approach may be used as indicated by the upward-directed arrow (labeled "concept assessment") on the right side of the pyramid. In this approach, performance capabilities (generated by engineering-level simulations) are translated into system capabilities which, in turn, are translated into mission capabilities whereafter the outcomes are evaluated in theater-level simulations. The arrow at the base of the pyramid in Figure 13.2 (labeled "concept definition") suggests a connection with the "V" diagram presented in Figure 13.1. Specifically, adhering to the discipline of the decomposition-and-definition process ensures that the perspectives of users, designers, and developers are reflected in the concept definition.

The US DOD divides the system acquisition process into three principal phases: concept and technology development, system development and demonstration, and production and deployment. A fourth phase (operations and support) covers life cycle sustainment (refer to Table 13.3). In each phase, it is possible to identify the generic functions performed by each of the four levels of simulation (engineering, engagement, mission, and theater). In Table 13.3, each phase of acquisition or sustainment is identified in the first column. In the second column, the relationship to notable requirements documents is indicated. The third through sixth columns identify the generic functions appropriate at each level of simulation to support that phase. The last column lists the metrics or tools associated with those simulations.

Further guidance on US DOD acquisition policy (sometimes referred to as "the 5000 model") is available in DOD Directive 5000.01 (Defense Acquisition System) along with DOD Instruction 5000.02 (Operation of the Defense Acquisition System). Together, these documents provide management principles and mandatory policies and procedures for managing all acquisition programs. (Directive 5000.1 dated October 23, 2000 is canceled.) DOD Regulation 5000.2-R (*Mandatory Procedures for Major Defense Acquisition Programs [MDAPs] and Major Automated Information System [MAIS] Acquisition Programs*) (US DOD, 2002) mandates that every defense acquisition program establish performance, schedule, and cost goals that can be defined by the least number of parameters necessary to characterize the program over its life cycle. Each parameter has a *threshold* value and an *objective* value.

For performance, the *threshold* is the minimum acceptable value necessary to satisfy the users' needs. For schedule and cost, the threshold is the maximum allowable value. The *objective* value is what the user desires and what the program manager tries to obtain. The objective value thus represents an incremental, operationally meaningful, time-critical, and cost-effective improvement to the threshold value of each program parameter.

TABLE 13.3
Functions of the Four Levels of Simulation Corresponding to Each Phase of the Acquisition Process

		Level of Simulation				
Phase	**Requirements**	**Engineering**	**Engagement**	**Mission**	**Theater**	**Metrics/Tools**
Concept and technology development	Mission needs statement (MNS)	Generate system/subsystem performance parameters Conduct performance trade-offs	Evaluate effectiveness Conduct cost and performance trade-offs	Evaluate mission effectiveness Evaluate consequences of different tactics	Examine outcomes of new system capabilities, technologies, and tactics	Key performance parameters (KPP) Cost as an independent variable (CAIV)
System development and demonstration	Operational requirements document (ORD)	Develop design specifications Plan prototype tests Reduce risk Support reliability and maintainability (R&M)	Evaluate mission effectiveness Define interoperability requirements		Examine how well proposed system meets identified needs	Measures of effectiveness (MOE) Measures of performance (MOP)
Production and deployment		Support pre-test planning and post-test analysis Verify compliance with specifications	Evaluate mission effectiveness		Evaluate consistency of measures of effectiveness (MOE)	Hardware-in-the-loop (HWIL)/ Software-in-the-loop (SWIL)
Operations and support		Evaluate and verify design changes	Conduct training on tactics and concepts of operation			Training systems

Performance threshold values represent true minima, with requirements stated in terms of capabilities rather than as technical solutions or specifications. If the performance-threshold values are not achieved, program performance may be seriously degraded and the utility of the system may become questionable. If schedule threshold values are not achieved, the program may no longer be timely. Cost threshold values represent true maxima. If cost threshold values are not achieved, the program may be too costly and the affordability of the system may become questionable.

In establishing realistic objectives, cost is now treated as a requirement. The cost-as-an-independent-variable (CAIV) process is therefore used to develop TOC, schedule, performance thresholds, and objectives. Cost is addressed in the operational requirements document (ORD). The CAIV trades consider the cost of delays as well as the potential for an early operational capability. The CAIV process relies heavily on M&S to perform trades in the multidimensional parameter space of threshold and objective values. Cost, schedule, and performance may be traded within the "trade space" between the objective and the threshold values. However, validated key performance parameters (KPPs) may not be traded off without proper authorization. The KPPs are a critical subset of the performance parameters found in the ORD. Each KPP has a threshold and an objective value. The significance of a KPP is that failure to meet the threshold value can cause the concept or system to be re-evaluated, or the program to be reassessed or terminated.

The best time to reduce TOC and shorten the program schedule is early in the acquisition process. Continuous cost-schedule-performance trade-off analyses can greatly facilitate cost and schedule reductions. When appropriate, commercial off-the-shelf (COTS) equipment may be considered. For example, Veenstra (1998) demonstrated the feasibility of using COTS hardware and software to build advanced sonar simulation-stimulation systems at costs that were significantly lower than traditional approaches.

13.6.3 Operations Analysis

Operations analysis is a scientific method for providing executives with a quantitative basis for making decisions regarding the operations under their control (Wagner et al., 1999). Operations research (or operational research, as it is sometimes called) is closely related to operations analysis. Modern operations analysis, which is a discipline that matured rapidly during World War II, has found application in a wide variety of military and civilian endeavors. The present focus is on naval operations analysis and those decisions that are useful to naval personnel in the conduct of naval operations. Wagner et al. (1999) presented instructive discussions concerning the role of M&S in sonar detection, search, and patrol.

The basis for decision-making entails predicting and describing the expected results of alternative courses of action. These results are presented in terms of appropriate measures of effectiveness (MOE). The MOE has four important attributes: (1) it is quantitative; (2) it is measurable from data or calculable from models; (3) it is highly correlated with an outcome gain or loss; and (4) it reflects the benefits and penalties of a given course of action (Wagner et al., 1999).

Many of the problems addressed in modern naval operations analysis have become so complex that simple analytical models can no longer be used to solve them. Recourse is therefore made to more sophisticated numerical models and Monte Carlo simulations. Typical problems to be solved in naval operations analysis include, for example, the determination of optimal deployment patterns for mobile or stationary sensor systems. Sonar performance models are used to generate appropriate MOE values such as median detection ranges for a stated probability of detection. Typically, Monte Carlo simulations are executed to create statistically meaningful data for subsequent analysis. Ultimately, sensor system performance may have to be traded against other metrics such as cost or asset availability (e.g., delivery platforms) to arrive at the best decision. Such approaches are commonly used in developing tactics for new sensor systems, or in comparing the merits of two or more candidate systems vying for the same acquisition funding. In the latter case, the process is often referred to formally as an analysis of alternatives (AOA) or a cost and operational effectiveness analysis (COEA).

Methods commonly associated with operations research—linear and dynamic programming, optimization techniques, queuing analysis and control theory—are also used as design tools in systems engineering (Blanchard, 1998).

Recent developments worthy of further exploration by serious students of naval operations analysis include effects-based (versus attrition-based) operations and capability-based (versus threat-based) analysis. Following Davis (2001), effects-based operations can be defined as those naval operations that are planned with full consideration of all possible direct, indirect, and cascading effects. The application of effects-based naval operations poses challenges to M&S since different degrees of probability must be assigned to each constituent (but interacting) effect. Capability-based analysis considers *how* (versus *where* and *with whom*) naval operations will be conducted. This approach leads to the development of generic (versus context-specific) systems and tactics. Here, the challenge to M&S is to develop analytical approaches that will provide usable products and guidance in specific (but as yet unrealized) contexts.

13.6.4 Training

Reductions in at-sea training opportunities have particularly encouraged (even necessitated) increased reliance on simulations for sonar-related training. For example, the use of computer-based training (CBT) has grown extensively, and sonar models have become common elements of simulations used in such learning environments. Two problems continue to plague advances in this area, however. First, the cost of developing quality courseware often becomes a limiting factor. Consequently, even the best-intentioned training products can quickly degenerate into page-turner programs that are useful for drill and practice among students of dissimilar educational backgrounds, but do not challenge students of more advanced topics. Second, the rapid evolution in computer technology often renders training systems prematurely obsolete by outpacing the financial capacity of educational centers to update their equipment. Moreover, installing upgraded software on aging equipment aggravates student-computer interactions by slowing computer response times, thereby frustrating students' experiences with CBT. An important consideration in the design of effective training environments is the creation of a standardized interface between the real-time host simulation devices and the image generator that provides realistic visualization of the simulated physics-based effects.

A notable development in underwater acoustic M&S is the interactive multisensor analysis training (IMAT) system. The IMAT system was originally developed to enhance the training of naval-aviation ASW operators (Ellis and Parchman, 1994; Wetzel-Smith et al., 1995; Wetzel-Smith and Czech, 1996), but has since expanded to include surface-ship and submarine ASW operators as well. Useful IMAT products include classroom multimedia systems and integrated curricula, PC-based learning systems, operator-console, and tactical simulations (US Department of the Navy, 1999). While primarily used to teach the physics of underwater acoustics, IMAT also has modules for optics, magnetics, and radar (Beatty, 1999).

The IMAT system couples scientific visualization with standard US Navy physics-based models and high-resolution databases to illustrate complex physical interactions. This combination of M&S creates a highly visual cause-and-effect-training tool. Variables characterizing targets, sensors, and environments can be manipulated to enable students to observe the overall impact of the changes. Many IMAT displays permit the creation of animations that illustrate how acoustic and oceanographic phenomena vary over a range of parameters. The US Navy air, surface, and submarine communities use the IMAT system to train officer and enlisted personnel in acoustics and oceanography (Beatty, 1999). In teaching acoustical oceanography (e.g., Foret et al., 1997), IMAT enables students to display temperature, salinity, and sound speed profiles worldwide during any month of the year. This information can be combined with tactical displays showing 3D representations of the ocean environment (Beatty, 1999).

Passive sonars detect the sounds emitted by surface or submerged targets. The IMAT system can display 3D graphics of various platforms to illustrate the source and operation of various sound-generating mechanisms aboard these vessels. It can turn the external hull transparent, enabling students to see the internal components of the platform. Sonar displays can be generated from either a real-time acoustic simulator or from actual recordings of *in situ* data. Additionally, an audio playback feature allows the student to listen to the time-series data represented by the display. A variety of processing options are available to emulate the various sonar systems and modes used by the naval air, surface, and submarine communities (Beatty, 1999).

The IMAT system contains standard US Navy range-dependent acoustic and electromagnetic propagation models. Passive acoustic propagation models include parabolic equation (PE), Gaussian ray bundle (GRAB), and ASTRAL. Passive sonar outputs can be combined with 3D tactical displays to visualize the environment along with the predicted propagation loss in that environment. Directional background noise can be incorporated into the display to illustrate the impact on signal excess. Active sonar predictions can be generated using ASPM (acoustic system performance model) or CASS (comprehensive acoustic system simulation). Active sonar outputs include echo, noise, and reverberation components (Beatty, 1999).

Instructors can also use IMAT to build tactical scenarios. The 3D tactical displays can be populated with objects such as sonobuoy fields, submarines, aircraft and surface ships. The instructor can then use the scenario to analyze a contact's motion and solve for its range, course, and speed by designating one animated object as the search platform and another as the target. It is also possible to match scenarios with specific sonar data and thus evaluate alternative tactics (Beatty, 1999).

Howard and Clark (2005) examined the impact of model diversity on interoperability in ASW training environments. Most legacy ASW stimulation trainers were developed without interoperability requirements. The choice of a propagation model is one such factor. Requirements peculiar to stimulation environments include computational speed and the ability to produce a representation of the acoustic field that can be utilized efficiently by the specific tactical equipment used in the trainers. Consistency in the context of stimulators means that similar sensors will see the same target signature components with the same temporal characteristics when using the same ASW scenario.

References

Abawi, A.T. and Porter, M.B. (2007) Propagation in an elastic wedge using the virtual source technique. *J. Acoust. Soc. Amer.*, **121**, 1374–82.

Abawi, A.T., Kuperman, W.A. and Collins, M.D. (1997) The coupled mode parabolic equation. *J. Acoust. Soc. Amer.*, **102**, 233–8.

Abkar, M. and Porté-Agel, F. (2015) A new wind-farm parameterization for large-scale atmospheric models. *J. Renew. Sustain. Energy*, **7**, 013121, doi: 10.1063/1.4907600.

Abraham, D.A. (2007) The effect of multipath on the envelope statistics of bottom clutter. *IEEE J. Oceanic Engr.*, **32**, 848–61.

Abraham, D.A. and Lyons, A.P. (2004) Simulation of non-Rayleigh reverberation and clutter. *IEEE J. Oceanic Engr.*, **29**, 347–62.

Abramowitz, M. and Stegun, I.A. (eds) (1964) *Handbook of Mathematical Functions with Formulas, Graphs and Mathematical Tables.* National Bureau of Standards Applied Mathematics Series 55. Washington, DC: US Government Printing Office.

Abrantes, A.A.M., Smith, K.B., and Larraza, A. (1999) Examination of time-reversal acoustics and applications to underwater communications. *J. Acoust. Soc. Amer.*, **105**, 1364 (abstract).

Ainslie, M.A. (1999) Interface waves in a thin sediment layer: Review and conditions for high loss. *Proc. Inst. Acoust.*, **21** (Part 9), 33–47.

Ainslie, M.A. (2000) Bistatic sonar performance assessment. Proc. Undersea Defence Technology Conf. UDT Europe, London, paper 2A.1.

Ainslie, M.A. (2004) The sonar equation and the definitions of propagation loss. *J. Acoust. Soc. Amer.*, **115**, 131–4.

Ainslie, M.A. (2010) *Principles of Sonar Performance Modeling*, New York, NY: Springer Praxis.

Ainslie, M.A., Hamson, R.M., Horsley, G.D., James, A.R., Laker, R.A., Lee, M.A. et al. (2000) Deductive multi-tone inversion of seabed parameters. *J. Comput. Acoust.*, **8**, 271–84.

Ainslie, M.A. and Harrison, C.H. (1990) Diagnostic tools for the ocean acoustic modeller. In *Computational Acoustics. Vol. 3: Seismo-Ocean Acoustics and Modeling*, eds D. Lee, A. Cakmak, and R. Vichnevetsky, Amsterdam, the Netherlands: North-Holland, pp. 107–30.

Ainslie, M.A., Harrison, C.H., and Burns, P.W. (1994) Reverberation modelling with INSIGHT. *Proc. Inst. Acoust.*, **16** (Part 6), 105–112.

Ainslie, M.A., Harrison, C.H., and Burns, P.W. (1996) Signal and reverberation prediction for active sonar by adding acoustic components. *IEE Proc. Radar Sonar Navig.*, **143**(3), 190–5.

Ainslie, M.A., Laker, R.A., and Hamson, R.M. (2001) Deductive geoacoustic inversion: Application to measurements in the Strait of Sicily. In *Acoustical Oceanography*, eds T.G. Leighton, G.J. Heald, H.D. Griffiths, and G. Griffiths. Proc. Inst. Acoust., **23** (Part 2), 66–73.

Ainslie, M.A. and McColm, J.G. (1998) A simplified formula for viscous and chemical absorption in sea water. *J. Acoust. Soc. Amer.*, **103**, 1671–2.

Ainslie, M.A. and Morfey, C.L. (2005) "Transmission loss" and "propagation loss" in undersea acoustics. *J. Acoust. Soc. Amer.*, **118**, 603–4.

Ainslie, M.A., Packman, M.N., and Harrison, C.H. (1998a) Fast and explicit Wentzel-Kramers-Brillouin mode sum for the bottom-interacting field, including leaky modes. *J. Acoust. Soc. Amer.*, **103**, 1804–12.

Ainslie, M.A. and Robins, A.J. (2003) Erratum: "Benchmark solutions of plane wave bottom reflection loss" (*J. Acoust. Soc. Am.*, **104**, 3305–3312 [1998]). *J. Acoust. Soc. Amer.*, **113**, 2180.

Ainslie, M.A., Robins, A.J., and Prior, M.K. (1998b) Benchmark solutions of plane wave reflection loss. *J. Acoust. Soc. Amer.*, **104**, 3305–12.

Ainslie, M.A., Robins, A.J., and Simons, D.G. (2004) Caustic envelopes and cusp coordinates due to the reflection of a spherical wave from a layered sediment. *J. Acoust. Soc. Amer.*, **115**, 1449–59.

Akal, T. and Berkson, J.M. (eds) (1986) *Ocean Seismo-Acoustics: Low-Frequency Underwater Acoustics*. NATO Conf. Series IV, Marine Sciences, Vol. 16. New York: Plenum.

Aksoy, S., Serim, H.A., Bölükbaş, D., and Akgün, S. (2007) Modeling and simulation of passive underwater acoustic wave propagation by normal mode method. Proc. USMOS 2007 Conf., Ankara, pp. 138–44 (in Turkish).

Akyildiz, I.F., Pompili, D., and Melodia, T. (2004) Challenges for efficient communication in underwater acoustic sensor networks. *ACM SIGBED Rev.*, **1**(2), 3–8.

Akyildiz, I.F., Pompili, D., and Melodia, T. (2005) Underwater acoustic sensor networks: Research challenges. *Ad. Hoc. Netw.*, **3**, 257–79.

Alapati, N.K., Kirklin, R.H., and Etter, P.C. (1993) Analysis of chaotic waveforms for application to active sonar systems. *Radix Systems, Inc.*, TR-93-081.

Albers, V.M. (1965) *Underwater Acoustics Handbook—II*, University Park, PA: The Pennsylvania State Univ. Press.

Alexander, P., Duncan, A., Bose, N., and Smith, D. (2013) Modelling acoustic transmission loss due to sea ice cover. *Acoust. Australia*, **41**(1), 79–87.

Alksne, R. (2000) Rapidly deployable systems (RDS) underwater acoustic telemetry trials report. Australia Department of Defence, Defence Science and Technology Organisation, Maritime Operations Division, Aeronautical and Maritime Research Laboratory, DSTO-TN-0259.

Almeida, R.J. Jr. and Medeiros, R.C. (1985) RAYMODE accuracy, limitations and transportability. Nav. Underwater Syst. Ctr, Tech. Memo. 841032.

Alvarez, A., Harrison, C., and Siderius, M. (2001) Predicting underwater ocean noise with genetic algorithms. *Phys. Lett. A*, **280**, 215–20.

Alves, J., Furfaro, T., LePage, K., Munafò, A., Pelekanakis, K., Petroccia, R. et al. (2016) Moving JANUS forward: A look into the future of underwater communications interoperability. Proc. MTS/IEEE Oceans 2016 Conf., Monterey, California, pp. 1–6.

Amato, I. (1993) A sub surveillance network becomes a window on whales. *Science*, **261**, 549–50.

Ammicht, E. and Stickler, D.C. (1984) Uniform asymptotic evaluation of the continuous spectrum contribution for a stratified ocean. *J. Acoust. Soc. Amer.*, **76**, 186–91.

Andersen, N.R. and Zahuranec, B.J. (eds) (1977). *Oceanic Sound Scattering Prediction*, New York, NY: Plenum.

Anderson, V.C. (1958) Arrays for the investigation of ambient noise in the ocean. *J. Acoust. Soc. Amer.*, **30**, 470–7.

Anderson, V.C. (1979) Variation of the vertical directionality of noise with depth in the North Pacific. *J. Acoust. Soc. Amer.*, **66**, 1446–52.

Anderson, A.L. and Hampton, L.D. (1980a) Acoustics of gas-bearing sediments I. Background. *J. Acoust. Soc. Amer.*, **67**, 1865–89.

Anderson, A.L. and Hampton, L.D. (1980b) Acoustics of gas-bearing sediments II. Measurements and models. *J. Acoust. Soc. Amer.*, **67**, 1890–903.

Andrew, R.K., Ganse, A., White, A.W., Mercer, J.A., Dzieciuch, M.A., Worcester, P.F. et al. (2016) Low-frequency pulse propagation over 510 km in the Philippine Sea: A comparison of observed and theoretical pulse spreading. *J. Acoust. Soc. Amer.*, **140**, 216–28.

Andrew, R.K., Howe, B.M., Mercer, J.A., and Dzieciuch, M.A. (2002) Ocean ambient sound: Comparing the 1960s with the 1990s for a receiver off the California coast. *Acoust. Res. Lett. Online*, **3**(2), 65–70.

Andrew, R.K., White, A.W., Mercer, J.A., Dzieciuch, M.A., Worcester, P.F., and Colosi, J.A. (2015) A test of deep water Rytov theory at 284 Hz and 107 km in the Philippine Sea. *J. Acoust. Soc. Amer.*, **138**, 2015–23.

Aparicio, J., Jiménez, A., Álvarez, F.J., Ureña, J., De Marziani, C., and Diego, C. (2013) Influence of different phenomena on the errors in distance measurement using underwater acoustics coded signals. Proc. MTS/IEEE Oceans 2013 Conf., Bergen, Norway, pp. 1–8.

Aparicio, J., Jiménez, A., Álvarez, F.J., Ureña, J., De Marziani, C., de Diego, D. et al. (2015) Accurate detection of spread-spectrum modulated signals in reverberant underwater environments. *Appl. Acoust.*, **88**, 57–65.

Apel, J.R. (1987) *Principles of Ocean Physics*. International Geophysics Series, Vol. 38. San Diego, CA: Academic Press.

Apel, J.R., Badiey, M., Chiu, C.-S., Finette, S., Headrick, R., Kemp, J. et al. (1997) An overview of the 1995 SWARM shallow-water internal wave acoustic scattering experiment. *IEEE J. Oceanic Engr.*, **22**, 465–500.

Apel, J.R., Ostrovsky, L.A., Stepanyants, Y.A., and Lynch, J.F. (2007) Internal solitons in the ocean and their effect on underwater sound. *J. Acoust. Soc. Amer.*, **121**, 695–722.

Applied Hydro-Acoustics Research, Inc. (2001) *ASPECT User's Guide*. Version 2.0.1 Build 664. Revised January 2, 2001.

Applied Physics Laboratory, University of Washington. (1994) APL-UW high-frequency ocean environmental acoustic models handbook. APL-UW TR 9407 (also, Office of Naval Research Tech. Rept AEAS 9501).

Applied Radar & Sonar Technologies GmbH. (2002) Sonar performance prediction system (SPPS). Product description literature.

Arakawa, A. and Lamb, V.R. (1977). Computational design of the basic dynamical processes of the UCLA general circulation model. In *Methods in Computational Physics: Advances in Research and Applications*, **17**, 173–265, doi: 10.1016/B978-0-12-460817-7.50009-4.

Ardhuin, F., Lavanant, T., Obrebski, M., Marié, L., Royer, J.-Y., d'Eu, J.-F. et al. (2013) A numerical model for ocean ultra-low frequency noise: Wave-generated acoustic-gravity and Rayleigh modes. *J. Acoust. Soc. Amer.*, **134**, 3242–59.

Aredov, A.A. and Furduev, A.V. (1994) Angular and frequency dependencies of the bottom reflection coefficient from the anisotropic characteristics of a noise field. *Acoust. Phys.*, **40**, 176–80.

Arnold, J.M. and Felsen, L.B. (1983) Rays and local modes in a wedge-shaped ocean. *J. Acoust. Soc. Amer.*, **73**, 1105–19.

Arrhenius, G. (1963) Pelagic sediments. In *The Sea. Vol. 3: The Earth Beneath the Sea, History*, ed. M.N. Hill, New York, NY: Interscience Publishers, pp. 655–727.

Arvelo, J. (1998) Array response advanced modal integrated simulator (ARAMIS) user's guide. Johns Hopkins APL, STX-98-010.

Arvelo, J.I. and Überall, H. (1990) Adiabatic normal-mode theory of sound propagation including shear waves in a range-dependent ocean floor. *J. Acoust. Soc. Amer.*, **88**, 2316–25.

Arzelies, P., Drogou, M., Opderbecke, J., Noel, C., Beauchene, S., Martin, C. et al. (2015) A novel acoustic test bench for evaluating underwater acoustic modems by numerical real-time simulation of the acoustic channel. Proc. IEEE Oceans 2015 Conf., Genova, doi: 10.1109/OCEANS-Genova.2015.7271337.

Atlas Elektronik UK. (2015) ODIN—The underwater warfare software simulation toolset. Product description literature.

Au, W.W.L. and Lammers, M.O. (eds) (2016) *Listening in the Ocean: New Discoveries and Insights on Marine Life from Autonomous Acoustic Recorders*, New York, NY: Springer.

Au, W.W.L., Penner, R.H., and Turl, C.W. (1987) Propagation of Beluga echolocation signals. *J. Acoust. Soc. Amer.*, **82**, 807–13.

Auer, S.J. (1987) Five-year climatological survey of the Gulf Stream system and its associated rings. *J. Geophys. Res.*, **92**, 11709–26.

Auner, L.C. (2015) Evaluation of acoustic seep detection methods and analysis of methane seep spatial distribution in the Gulf of Mexico. Submitted in partial fulfillment of the requirements for a Bachelor of Arts degree from Carleton College, Northfield, MN.

Austin, M.E. and Chapman, N.R. (2009) Computational grid design to improve three-dimensional parabolic equation modeling efficiency. *J. Acoust. Soc. Amer.*, **126**, 2305 (abstract).

Austin, M.E. and Chapman, N.R. (2011) The use of tessellation in three-dimensional parabolic equation modeling. *J. Comput. Acoust.*, **19**, 221–39.

Austin, M., Chorney, N., Ferguson, J., Leary, D., O'Neill, C., and Sneddon, H. (2009) Assessment of underwater noise generated by wave energy devices. Prepared by JASCO Applied Sciences on behalf of Oregon Wave Energy Trust. Tech Report, P001081-001, Version 1.0.

Avilov, K.V. (1995) Pseudodifferential parabolic equations of sound propagation in the slowly range-dependent ocean and their numerical solutions. *Acoust. Phys.*, **41**, 1–7.

Axelrod, E.H., Schoomer, B.A., and VonWinkle, W.A. (1965) Vertical directionality of ambient noise in the deep ocean at a site near Bermuda. *J. Acoust. Soc. Amer.*, **37**, 77–83.

Baer, R.N. (1981) Propagation through a three-dimensional eddy including effects on an array. *J. Acoust. Soc. Amer*, **69**, 70–5.

Baer, R.N., Berman, D.H., Perkins, J.S., and Wright, E.B. (1985). A three-dimensional model for acoustic scattering from rough ocean bathymetry. *Comp. Maths. Applic.*, **11**, 863–71.

BAeSEMA Ltd. (1998) Tactical support systems. Product description literature.

Baggeroer, A.B. (2012) An overview of acoustic communications from 2000–2012. Proc. Workshop on Underwater Communications (UComms): Channel Modelling and Validation, Sestri Levante, Italy, pp. 1–9.

Baggeroer, A.B., Kuperman, W.A., and Mikhalevsky, P.N. (1993) An overview of matched field methods in ocean acoustics. *IEEE J. Oceanic Engr.*, **18**, 401–24.

Baggeroer, A.B., Scheer, E.K., and the NPAL Group (Colosi, J.A., Cornuelle, B.D., Dushaw, B.D., Dzieciuch, M.A., Howe, B.M., Mercer, J.A. et al.). (2005) Statistics and vertical directionality of low-frequency ambient noise at the North Pacific acoustics laboratory site. *J. Acoust. Soc. Amer.*, **117**, 1643–65.

Bailey, H., Senior, B., Simmons, D., Rusin, J., Picken, G., and Thompson, P.M. (2010) Assessing underwater noise levels during pile-driving at an offshore windfarm and its potential effects on marine mammals. *Marine Poll. Bull.*, **60**, 888–97, doi: 10.1016/j.marpolbul.2010.01.003.

Bailey, J.A. (2013) Uniform and multi-grid modeling of acoustic wave propagation with cellular automaton techniques. M.S. thesis, Naval Postgraduate School, Monterey, CA.

Bailey, R.S., Beck, M.A., Fink, I.M., Garcia, M.L., Kaduchak, G., and Lawrence, M.Z. (1997) Installation and user's guide for Mineray 3 sonar performance prediction model, X-Windows version (XMineray). Appl. Res. Labs., Univ. Texas at Austin, USA.

Baker, C.L. (1976) FANIN preprocessor for the FANM-I ambient noise model. Ocean Data Syst., Inc.

Baker, C.L. and Spofford, C.W. (1974) The FACT model, Vol II. Acoustic Environmental Support Detachment, Off. Nav. Res., Tech. Note TN-74-04.

Baldacci, A. and Harrison, C.H. (2002) SUPREMO prototype version 01 user's guide. SACLANT Undersea Res. Ctr, M-141.

Baldo, N., Miozzo, M., Guerra, F., Rossi, M., and Zorzi, M. (2010) Miracle: The multi-interface cross-layer extension of ns2. *EURASIP J. Wirel. Commun. Netw.*, **2010**, 1–16, doi: 10.1155/2010/761792.

Ballard, M.S. (2012) Modeling three-dimensional propagation in a continental shelf environment. *J. Acoust. Soc. Amer.*, **131**, 1969–77.

Ballard, M.S., Goldsberry, B.M., and Isakson, M.J. (2015) Normal mode analysis of three-dimensional propagation over a small-slope cosine shaped hill. *J. Comput. Acoust.*, **23**, 1–19, doi: 10.1142/S0218396X15500058.

Bannister, R.W. (1986) Deep sound channel noise from high-latitude winds. *J. Acoust. Soc. Amer.*, **79**, 41–8.

Bannister, R.W., Kewley, D.J., and Burgess, A.S. (1989) Directional underwater noise estimates—The DUNES model. Australia Department of Defence, Defence Science and Technology Organisation, Maritime Systems Division, Weapons Systems Research Laboratory, WSRL-TN-34/89.

Barker, R. (2004) New subsea acoustics model from BMT. *Coast Map News*, Issue 7, p. 31. Centre for Environment, Fisheries and Aquaculture Science (CEFAS), Essex, England.

Barkhatov, A.N. (1968) Modeling of sound propagation in the sea. Gidrometeorologicheskoe Izdatel'stvo. (Translated from Russian by Wood J.S. (1971) *Consultants Bureau*, New York, NY: A Division of Plenum Publishing Corp.)

Bartberger, C.L. (1965) Lecture notes on underwater acoustics. Nav. Air Devel. Ctr, Rept NADC-WR-6509.

Bartberger, C.L. (1978a) PLRAY: A ray propagation loss program. Nav. Air Devel. Ctr, Rept NADC-77296-30.

Bartberger, C.L. (1978b) AP-2 normal mode program. Nav. Air Devel. Ctr, Rept.

Bartberger, C.L. (1985) The NADC bistatic active sonar model. Nav. Air Devel. Ctr, Rept.

Bartberger, C.L. (1991a) The physics of the bistatic acoustic model. Nav. Air Devel. Ctr, Rept NADC-91020-50.

Bartberger, C.L. (1991b) The NADC monostatic sonar model (version 1.0). Nav. Air Devel. Ctr, TN-5044-7-91.

Bartel, D.W. (2010) On some rigorous computational ocean-acoustic modelling tools. Proc. IEEE Oceans 2010 Conf., Sydney.

Bass, S.J. and Hay, A.E. (1997) Ambient noise in the natural surf zone: Wave-breaking frequencies. *IEEE J. Oceanic Engr.*, **22**, 411–24.

Bathen, K.H. (1972) On the seasonal changes in the depth of the mixed layer in the North Pacific Ocean. *J. Geophys. Res.*, **77**, 7138–50.

Battle, D.J., Gerstoft, P., Hodgkiss, W.S., Kuperman, W.A., and Nielsen, P.L. (2004) Bayesian model selection applied to self-noise geoacoustic inversion. *J. Acoust. Soc. Amer.*, **116**, 2043–56.

Beatty, W.F. (January–February 1999) *Interactive Multisensor Analysis Training. Wavelengths*, West Bethesda, MD: Published by Naval Surface Warfare Center, Carderock Division.

Becken, B.A. (1961) The directional distribution of ambient noise in the ocean. Scripps Inst. Oceanogr., Rept 61-4.

Beckmann, P. and Spizzichino, A. (1963) *The Scattering of Electromagnetic Waves from Rough Surfaces*. International Series of Monographs on Electromagnetic Waves, Vol. 4. New York, NY: Pergamon Press.

Bedard, A.J. Jr. and Georges, T.M. (2000) Atmospheric infrasound. *Phys. Today*, **53**(3), 32–7.

Beebe, J.H., McDaniel, S.T., and Rubano, L.A. (1982) Shallow-water transmission loss prediction using the Biot sediment model. *J. Acoust. Soc. Amer.*, **71**, 1417–26.

Benchekroun, N. and Mansour, A. (2006) Blind separation of underwater acoustic signals. Proc. 2nd International Symposium on Communications, Control and Signal Processing (ISCCSP), Marrakech, Morocco.

Benders, F.P.A., Beerens, S.P., and Verboom, W.C. (2004) SAKAMATA: The ideas and algorithms behind it. Proc. 7th European Conf. Underwater Acoustics, Delft, the Netherlands, pp. 1251–6.

Bennett, A.F. (1992) *Inverse Methods in Physical Oceanography*, New York, NY: Cambridge University Press.

Benoit-Bird, K.J. and Au, W.W.L. (2002) Energy: Converting from acoustic to biological resource units. *J. Acoust. Soc. Amer.,* **111**, 2070–5.

Berger, M.D., Boucher, C.E., Daley, E.M., Renner, W.W., Pastor, V.L., Haines, L.C. et al. (1994) Acoustic system performance model (ASPM) 4.0A: User's guide. Sci. Appl. Inter. Corp., SAIC-94/1000.

Bergman, D.R. (2005) Symmetry and Snell's law. *J. Acoust. Soc. Amer.,* **118**, 1278–82.

Berman, D.H., Wright, E.B., and Baer, R.N. (1989) An optimal PE-type wave equation. *J. Acoust. Soc. Amer.*, **86**, 228–33.

Bertuccelli, H.C. (1975) Digital computer programs for analyzing acoustic search performance in refractive waters, Vol. 3. Nav. Undersea Ctr, Tech. Pub.164.

Bialek, E.L. (1966) *Handbook of Oceanographic Tables*. Nav. Oceanogr. Off., Special Pub. SP-68. Washington DC: US Government Printing Office.

Bindi, V., Baker, J., Billington, R., Gallassero, T., Gueary, J., Harts, J. et al. (2005) Littoral undersea warfare in 2025. Naval Postgraduate School, NPS-97-06-001.

Bindi, V., Strunk, J., Baker, J., Bacon, R., Boensel, M.G., Shoup, F.E. et al. (2008) Littoral undersea warfare: A case study in process modeling for functionality and interoperability of complex systems. *Int. J. Syst. Syst. Engr.*, **1**, 18–58.

Bini-Verona, F., Nielsen, P.L., and Jensen, F.B. (2000) PROSIM broadband normal-mode model: A users' guide. SACLANT Undersea Res. Ctr, Memo. SM-358.

Biot, M.A. (1956a) Theory of propagation of elastic waves in a fluid-saturated porous solid. I. Low-frequency range. *J. Acoust. Soc. Amer.*, **28**, 168–78.

Biot, M.A. (1956b) Theory of propagation of elastic waves in a fluid-saturated porous solid. II. Higher frequency range. *J. Acoust. Soc. Amer.*, **28**, 179–91.

Bishop, G.C. (1987) A bistatic, high-frequency, under-ice, acoustic scattering model. *J. Acoust. Soc. Amer.*, **82** (Suppl. 1), S30 (abstract).

Bishop, G.C. (1989a) A bistatic, high-frequency, under-ice, acoustic scattering model. I: Theory. *J. Acoust. Soc. Amer.*, **85**, 1903–11.

Bishop, G.C. (1989b) A bistatic, high-frequency, under-ice, acoustic scattering model. II: Applications. *J. Acoust. Soc. Amer.*, **85**, 1912–24.

Bishop, G.C., Ellison, W.T., and Mellberg, L.E. (1987) A simulation model for high-frequency under-ice reverberation. *J. Acoust. Soc. Amer.*, **82**, 275–86.

Bishop, G.C., Mellberg, L.E., and Ellison, W.T. (1986) A simulation model for high-frequency, under-ice reverberation. Nav. Underwater Syst. Ctr, Tech. Rept 6268.

Bjørnø, L. (1998) Sources of ambient noise in littoral waters. *Arch. Acoust.,* **23**, 211–25.

Bjørnø, L. (2002) 40 years of nonlinear underwater acoustics. *Acta Acust. United Acust.*, **88**, 771–5.

Blanchard, B.S. (1998) *System Engineering Management*, 2nd edn. New York, NY: John Wiley & Sons.

Blatstein, I.M. (1974) Comparisons of normal mode theory, ray theory, and modified ray theory for arbitrary sound velocity profiles resulting in convergence zones. Nav. Ordnance Lab., NOLTR 74-95.

Blondel, P. and Pace, N.G. (2009) Bistatic sonars: Sea trials, laboratory experiments and future surveys. *Arch. Acoust.*, **34**, 95–109.

Blouin, S., Heard, G.J., and Pecknold, S. (2015) Autonomy and networking challenges of future underwater systems. Proc. IEEE 28th Canadian Conf. Electrical Computer Engineering (CCECE), Halifax, pp. 1514–19.

Blumen, L.S. and Spofford, C.W. (1979) The ASTRAL model, Vol. II: Software implementation. Sci. Appl., Inc., SAI-79-743-WA.

BMT Cordah. (2010) EIA and environmental services. Offshore renewables. Noise assessment and modelling. Product description literature.

Boehm, B.W. (1988) A spiral model of software development and enhancement. *Computer*, **21**(5), 61–72.

Bogart, C.W. and Yang, T.C. (1992) Comparative performance of matched-mode and matched-field localization in a range-dependent environment. *J. Acoust. Soc. Amer.*, **92**, 2051–68.

Bom, N. (1969) Effect of rain on underwater noise level. *J. Acoust. Soc. Amer.*, **45**, 150–6.

Bonomo, A.L., Chotiros, N.P., and Isakson, M.J. (2015) On the validity of the effective density fluid model as an approximation of a poroelastic sediment layer. *J. Acoust. Soc. Amer.*, **138**, 748–57.

Bonomo, A.L., Chotiros, N.P., and Isakson, M.J. (2016) Erratum: "On the validity of the effective density fluid model as an approximation of a poroelastic sediment layer" (*J. Acoust. Soc. Am.* 138, 748–757 [2015]). *J. Acoust. Soc. Amer.*, **139**, 1702.

Borejko, P. (2004) An exact representation of the image field in a perfect wedge. *Acta Mechanica*, **169**, 23–36.

Botseas, G., Lee, D., and Gilbert, K.E. (1983) IFD: Wide angle capability. Nav. Underwater Syst. Ctr, Tech. Rept 6905.

Botseas, G., Lee, D., and King, D. (1987) FOR3D: A computer model for solving the LSS three-dimensional, wide angle wave equation. Nav. Underwater Syst. Ctr, Tech. Rept 7943.

Botseas, G., Lee, D., and Siegmann, W.L. (1989) IFD: Interfaced with Harvard open ocean model forecasts. Nav. Underwater Syst. Ctr, Tech. Rept 8367.

Bouchage, G. and LePage, K.D. (2002) A shallow-water reverberation PE model. *Acta Acust. United Acust.*, **88**, 638–41.

Bowditch, N. (1977) *American Practical Navigator*, Vol. 1. Washington, DC: Defense Mapping Agency Hydrographic Center, Pub. No. 9. (Continuously maintained since first published in 1802.)

Bowlin, J.B., Spiesberger, J.L., Duda, T.F., and Freitag, L.F. (1992) Ocean acoustical ray-tracing software RAY. Woods Hole Oceanogr. Inst., Tech. Rept WHOI-93-10.

Boyles, C.A. (1984) *Acoustic Waveguides. Applications to Oceanic Science*. New York, NY: John Wiley & Sons.

Braca, P., Goldhahn, R., LePage, K.D., Marano, S., Matta, V., and Willett, P. (2014) Cognitive multistatic AUV networks. Proc. 17th International Conf. Information Fusion, Salamanca, Spain, pp. 1–7.

Bracken, J., Kress, M., and Rosenthal, R.E. (eds) (1995) *Warfare Modeling*, New York, NY: John Wiley & Sons.

Bradley, M.R. and Bradley, B.W. (1984) Computer program performance specification for the geophysics ambient noise model. Planning Syst., Inc., Tech. Rept TR-S310018.

Breeding, J.E., Pflug, L.A., Bradley, M., Hebert, M., and Wooten, M. (1994) RANDI 3.1 user's guide. Nav. Res. Lab., Rept NRL/MR/7176-94-7552.

Breeding, J.E. Jr. (1993) Description of a noise model for shallow water: RANDI-III. *J. Acoust. Soc. Amer.*, **94**, 1820 (abstract).

Breeding, J.E. Jr., Pflug, L.A., Bradley, M., Walrod, M.H., and McBride, W. (1996) Research ambient noise directionality (RANDI) 3.1 physics description. Nav. Res. Lab., Rept NRL/FR/7176-95-9628.

Breitzke, M. and Bohlen, T. (2010) Modelling sound propagation in the Southern Ocean to estimate the acoustic impact of seismic research surveys on marine mammals. *Geophys. J. Int.*, **181**, 818–46, doi: 10.1111/j.1365-246X.2010.04541.x.

Brekhovskikh, L.M. and Godin, O.A. (1999) *Acoustics of Layered Media II. Point Sources and Bounded Beams*, 2nd edn. New York, NY: Spring-Verlag.

Brekhovskikh, L. and Lysanov, Yu. (1982) *Fundamentals of Ocean Acoustics*. New York, NY: Springer-Verlag.

Brekhovskikh, L.M. and Lysanov, Yu.P. (2003) *Fundamentals of Ocean Acoustics*, 3rd edn. New York, NY: Springer-Verlag.

Brock, H.K. (1978) The AESD parabolic equation model. Nav. Ocean Res. Devel. Activity, Tech. Note 12.

Brock, H.K., Buchal, R.N., and Spofford, C.W. (1977) Modifying the sound-speed profile to improve the accuracy of the parabolic-equation technique. *J. Acoust. Soc. Amer.*, **62**, 543–52.

Broecker, W.S. (1997) Thermohaline circulation, the Achilles heel of our climate system: Will man-made CO_2 upset the current balance? *Science*, **278**, 1582–8.

Broecker, W. (2010) *The Great Ocean Conveyor: Discovering the Trigger for Abrupt Climate Change*. Princeton, NJ: Princeton University Press.

Brooke, G.H. and Thomson, D.J. (2000) Non-local boundary conditions for high-order parabolic equation algorithms. *Wave Motion,* **31**, 117–29.

Brooke, G.H., Thomson, D.J., and Ebbeson, G.R. (2001) PECAN: A Canadian parabolic equation model for underwater sound propagation. *J. Comput. Acoust.,* **9**, 69–100.

Brooks, L.A. (2008) Ocean acoustic interferometry. Doctoral thesis, The University of Adelaide, Australia.

Brown, J.R. (1964) Reverberation under Arctic ice. *J. Acoust. Soc. Amer.*, **36**, 601–3.

Brown, M.G. (1994) Global acoustic propagation modeling using MaCh1. *J. Acoust. Soc. Amer.*, **95**, 2880 (abstract).

Brown, M.G., Tappert, F.D., and Goni, G. (1991a) An investigation of sound ray dynamics in the ocean volume using an area preserving mapping. *Wave Motion*, **14**, 93–9.

Brown, M.G., Tappert, F.D., Goni, G.J., and Smith, K.B. (1991b) Chaos in underwater acoustics. In *Ocean Variability and Acoustic Propagation*, eds J. Potter and A. Warn-Varnas, Dordrecht, the Netherlands: Kluwer Academic Publishers, pp. 139–60.

Brown, M.G., Beron-Vera, F.J., Rypina, I., and Udovydchenkov, I.A. (2005) Rays, modes, wavefield structure, and wavefield stability. *J. Acoust. Soc. Amer.,* **117**, 1607–10.

Brown, N.R., Leighton, T.G., Richards, S.D., and Heathershaw, A.D. (1998) Measurement of viscous sound absorption at 50–150 kHz in a model turbid environment. *J. Acoust. Soc. Amer.*, **104**, 2114–20.

Brown, W.E. and Barlett, M.L. (2005) Midfrequency "through-the-sensor" scattering measurements: A new approach. *IEEE J. Oceanic Engr.*, **30**, 733–47.

Browning, D.G., Christian, R.J., and Petitpas, L.S. (1994) A global survey of the impact on acoustic propagation of deep water warm core ocean eddies. *J. Acoust. Soc. Amer.*, **95**, 2880 (abstract).

Browning, D.G., Scheifele, P.M., and Mellen, R.H. (1988) Attenuation of low frequency sound in ocean surface ducts: Implications for surface loss values. Proc. MTS/IEEE Oceans '88 Conf., Baltimore, pp. 318–22.

Brutzman, D.P., Kanayama, Y., and Zyda, M.J. (1992) Integrated simulation for rapid development of autonomous underwater vehicles. Proc. 1992 IEEE Symposium on Autonomous Underwater Vehicle Technology, pp. 3–10.

Bryan, G.M. (1967) Underwater sound in marine geology. In *Underwater Acoustics*, ed. V.M. Albers, Vol. 2. New York, NY: Plenum, pp. 351–61.

Buchanan, J.L., Gilbert, R.P., Wirgin, A., and Xu, Y.S. (2004) *Marine Acoustics: Direct and Inverse Problems*. Philadelphia, PA: Society for Industrial and Applied Mathematics.

Buck, B.M. (1968) Arctic acoustic transmission loss and ambient noise. In *Arctic Drifting Stations*, ed. J.E. Sater, Calgary, Canada: Arctic Institute of North America, pp. 427–38.

Buck, B.M. (1981) Preliminary underice propagation models based on synoptic ice roughness. Polar Res. Lab., TR-30.

Buck, B.M. (1985) Long term statistical measurements of environmental acoustics parameters in the Arctic (AEAS Report No. 2—Low frequency transmission loss measurements in the central Arctic Ocean). Polar Res. Lab., TR-55.

Bucker, H.P. (1971) Some comments on ray theory with examples from current NUC ray trace models. SACLANT ASW Res. Ctr, Conf. Proc. No. 5, Part I, pp. 32–6.

Bucker, H.P. (1976) Use of calculated sound fields and matched-field detection to locate sound sources in shallow water. *J. Acoust. Soc. Amer.*, **59**, 368–73.

Bucker, H.P. (1983) An equivalent bottom for use with the split-step algorithm. *J. Acoust. Soc. Amer.*, **73**, 486–91.

Bucker, H.P. (1986) Use of SALT tables for rapid calculation of sound angle, level and travel time. *J. Acoust. Soc. Amer.*, **80** (Suppl. 1), S63 (abstract).

Bucker, H.P. (1994) A simple 3-D Gaussian beam sound propagation model for shallow water. *J. Acoust. Soc. Amer.*, **95**, 2437–40.

Bucker, H.P. and Morris, H.E. (1968) Normal-mode reverberation in channels or ducts. *J. Acoust. Soc. Amer.*, **44**, 827–8.

Buckingham, M.J. (1991) On acoustic transmission in ocean-surface waveguides. *Phil. Trans. R. Soc. Lond.*, **335**, 513–55.

Buckingham, M.J. (1992) Ocean-acoustic propagation models. *J. Acoust.*, **3**, 223–87.

Buckingham, M.J. (1993) Theory of acoustic imaging in the ocean with ambient noise. *J. Comput. Acoust.*, **1**, 117–40.

Buckingham, M.J. (2005) Compressional and shear wave properties of marine sediments: Comparisons between theory and data. *J. Acoust. Soc. Amer.*, **117**, 137–52.

Buckingham, M.J., Berkhout, B.V., and Glegg, S.A.L. (1992) Imaging the ocean with ambient noise. *Nature*, **356**, 327–9.

Buckingham, M.J., Epifanio, C.L., and Readhead, M.L. (1996a) Passive imaging of targets with ambient noise: Experimental results. *J. Acoust. Soc. Amer.*, **100**, 2736 (abstract).

Buckingham, M.J. and Jones, S.A.S. (1987) A new shallow-ocean technique for determining the critical angle of the seabed from the vertical directionality of the ambient noise in the water column. *J. Acoust. Soc. Amer.*, **81**, 938–46.

Buckingham, M.J. and Potter, J.R. (eds) (1996) *Sea Surface Sound '94*, Singapore: World Scientific Publishing.

Buckingham, M.J., Potter, J.R., and Epifanio, C.L. (1996b) Seeing underwater with background noise. *Sci. Am.*, **274**(2), 86–90.

Buckingham, M.J. and Tolstoy, A. (1990) An analytical solution for benchmark problem 1: The "ideal" wedge. *J. Acoust. Soc. Amer.*, **87**, 1511–3.

Buffett, G.G., Biescas, B., Pelegrí, J.L., Machín, F., Sallarès, V., Carbonell, R. et al. (2009) Seismic reflection along the path of the Mediterranean Undercurrent. *Cont. Shelf Res.*, **29**, 1848–60.

Burdic, W.S. (1991) *Underwater Acoustic System Analysis*, 2nd edn. Englewood Cliffs, NJ: PTR Prentice Hall.

Bureau Veritas. (2014) Underwater Radiated Noise (URN), Rule Note NR 614 DT R00 E, Marine & Offshore Division, Neuilly sur Seine Cedex, France.

Burnett, W., Harding, J., and Heburn, G. (2002) Overview of operational ocean forecasting in the U.S. Navy: Past, present and future. *Oceanography*, **15**(1), 4–12.

Burnett, W., Harper, S., Preller, R., Jacobs, G., and LaCroix, K. (2014) Overview of operational ocean forecasting in the US Navy: Past, present, and future. *Oceanography*, **27**(3), 24–31, doi: 10.5670/oceanog.2014.65.

Burridge, R. and Weinberg, H. (1977) Horizontal rays and vertical modes. In *Wave Propagation and Underwater Acoustics*, eds J.B. Keller and J.S. Papadakis. Lecture Notes in Physics, Vol. 70. New York, NY: Springer-Verlag, pp. 86–152.

Burrowes, G.E., Brown, J., and Khan, J.Y. (2014) Impact of reverberation levels on short-range acoustic communication in an underwater swarm sensor network (USSN) and application to transmitter power control. Proc. MTS/IEEE Oceans 2014 Conf., St John's, Canada, doi: 10.1109/OCEANS.2014.7003039.

Cai, C., Ferrari, S., and Qian, M. (2007) Bayesian network modeling of acoustic sensor measurements. Proc. IEEE Sensors 2007 Conf., Atlant, pp. 345–8.

Cai, Z., Zheng, Z., and Yang, S. (2002) Chaos characteristic analysis of underwater reverberation. *Acta Acust.*, **27**, 497–501 (in Chinese).

Caiti, A., Chapman, N.R., Hermand, J.-P., and Jesus, S.M. (eds) (2006) *Acoustic Sensing Techniques for the Shallow Water Environment: Inversion Methods and Experiments*, New York, NY: Springer-Verlag.

Caiti, A., Hermand, J.-P., Jesus, S., and Porter, M. (eds) (2000) *Experimental Acoustic Inversion Methods for Exploration of the Shallow Water Environment*, Dordrecht, the Netherlands: Kluwer Academic Publishers.

Calnan, C. (2006) DMOS—Bellhop Extension. DRDC Atlantic CR 2006–005. Defence R&D Canada–Atlantic.

Candy, J.V. and Sullivan, E.J. (1992) Ocean acoustic signal processing: A model-based approach. *J. Acoust. Soc. Amer.*, **92**, 3185–201.

Candy, J.V. and Sullivan, E.J. (1993) Sound velocity profile estimation: A system theoretic approach. *IEEE J. Oceanic Engr.*, **18**, 240–52.

Canepa, E. and Irwin, J.S. (2005) Evaluation of Air Pollution Models. In *Air Quality Modeling—Theories, Methodologies, Computational Techniques, and Available Data Bases and Software, Vol. II—Advanced Topics*, ed. P. Zannetti, Fremont, CA: The EnvironComp Institute and Air & Waste Management Association, Chapter 17, pp. 503–56.

Canepa, G. and Berron, C. (2006) Characterization of seafloor geo-acoustic properties from multibeam data. Proc. MTS/IEEE Oceans 2006 Conf., Boston.

Carbone, N.M., Deane, G.B., and Buckingham, M.J. (1998) Estimating the compressional and shear wave speeds of a shallow water seabed from the vertical coherence of ambient noise in the water column. *J. Acoust. Soc. Amer.*, **103**, 801–13.

Carey, W.M. (1995) Standard definitions for sound levels in the ocean. *IEEE J. Oceanic Engr.,* **20**, 109–13.

Carey, W.M. (2006) Sound sources and levels in the ocean. *IEEE J. Oceanic Engr.*, **31**, 61–75.

Carey, W.M. and Evans, R.B. (2011) *Ocean Ambient Noise: Measurement and Theory.* New York, NY: Springer.

Carey, W.M., Evans, R.B., and Davis, J.A. (1987) Downslope propagation and vertical directionality of wind noise. *J. Acoust. Soc. Amer.*, **82** (Suppl. 1), S63 (abstract).

Carey, W.M., Evans, R.B., Davis, J.A., and Botseas, G. (1990) Deep-ocean vertical noise directionality. *IEEE J. Oceanic Engr.*, **15**, 324–34.

Carey, W.M. and Wagstaff, R.A. (1986) Low-frequency noise fields. *J. Acoust. Soc. Amer.*, **80**, 1523–6.

Carlson, J.T. (1994) IUSS community adopts collateral uses. *Sea Technol.*, **35**(5), 25–8.

Carnes, M.R. (2009) Description and evaluation of GDEM-V3.0. Nav. Res. Lab., Rept NRL/MR/7330-09-9165.

Carroll, S.N. and Szczechowski, C. (2001) The northern Gulf of Mexico littoral initiative. Proc. MTS/IEEE Oceans 2001 Conf., Honolulu, pp. 1311–17.

Caruthers, J.W. and Novarini, J.C. (1993) Modeling bistatic bottom scattering strength including a forward scatter lobe. *IEEE J. Oceanic Engr.*, **18**, 100–7.

Caruthers, J.W., Sandy, R.J., and Novarini, J.C. (1990) Modified bistatic scattering strength model (BISSM2). *Nav. Oceanogr. Atmos. Res. Lab.*, SP 023:200:90.

Caruthers, J.W. and Yoerger, E.J. (1993) ARSRP reconn results and BISSM modeling of direct path backscatter. In *Ocean Reverberation*, eds D.D. Ellis, J.R. Preston, and H.G. Urban. Dordrecht, the Netherlands: Kluwer Academic Publishers, pp. 203–8.

Castor, K., Gerstoft, P., Roux, P., Kuperman, W.A., and McDonald, B.E. (2004) Long-range propagation of finite-amplitude acoustic waves in an ocean waveguide. *J. Acoust. Soc. Amer.,* **116**, 2004–10.

Cato, D.H. and McCauley, R.D. (2002) Australian research in ambient sea noise. *Acoust. Australia*, **30**(1), 13–20.

Cato, D.H. and Tavener, S. (1997) Ambient sea noise dependence on local, regional and geostrophic wind speeds: Implications for forecasting noise. *Appl. Acoust.*, **51**, 317–38.

Cavanagh, R.C. (1974a) Fast ambient noise model I (FANM I). Acoustic Environmental Support Detachment, Off. Nav. Res.

Cavanagh, R.C. (1974b) Fast ambient noise model II (FANM II). Acoustic Environmental Support Detachment, Off. Nav. Res.

Cavanagh, R.C. (1978a) Acoustic fluctuation modeling and system performance estimation, Vol. I. Sci. Appl., Inc., SAI-79-737-WA.

Cavanagh, R.C. (1978b) Acoustic fluctuation modeling and system performance estimation, Vol. II. Sci. Appl., Inc., SAI-79-738-WA.

Cederberg, R.J. and Collins, M.D. (1997) Application of an improved self-starter to geoacoustic inversion. *IEEE J. Oceanic Engr.,* **22**, 102–9.

Chadwick, E. and Bettess, P. (1997) Modelling of progressive short waves using wave envelopes. *Int. J. Numer. Meth. Engng.*, **40**, 3229–45.

Chaika, E.D., Granum, R.M., and Marusich, R.B. (1979) Program for integrated sonar modeling (PRISM) user's guide. Nav. Ocean Syst. Ctr, Tech. Note 851.

Chamberlain, S.G. and Galli, J.C. (1983) A model for numerical simulation of nonstationary sonar reverberation using linear spectral prediction. *IEEE J. Oceanic Engr.*, **8**, 21–36.

Chandrasekhar, V., Seah, W.K.G., Choo, Y.S., and Ee, H.V. (2006) Localization in underwater sensor networks—Survey and challenges. Proc. WUWNet'06, The First ACM International Workshop on Underwater Networks, Los Angeles, CA.

Chapman, D.M.F. (2000) Decibels, SI units, and standards. *J. Acoust. Soc. Amer.,* **108**, 480.

Chapman, D.M.F., Ward, P.D., and Ellis, D.D. (1989) The effective depth of a Pekeris ocean waveguide, including shear wave effects. *J. Acoust. Soc. Amer.,* **85**, 648–53.

Chapman, N.R. and Ebbeson, G.R. (1983) Acoustic shadowing by an isolated seamount. *J. Acoust. Soc. Amer.*, **73**, 1979–84.

Chapman, R.P. and Harris, J.H. (1962) Surface backscattering strengths measured with explosive sound sources. *J. Acoust. Soc. Amer.*, **34**, 1592–7.

Chapman, R.P. and Scott, H.D. (1964) Surface backscattering strengths measured over an extended range of frequencies and grazing angles. *J. Acoust. Soc. Amer.*, **36**, 1735–7.

Chen, C., Beardsley, R.C., Cowles, G., Qi, J., Lai, Z., Gao, G. et al. (2011) *An Unstructured-Grid, Finite-Volume Community Ocean Model FVCOM User Manual (3rd edn).* Cambridge, MA: Sea Grant College Program Massachusetts Institute of Technology, MITSG 12-25.

Chen, C.-T. and Millero, F.J. (1977) Speed of sound in seawater at high pressures. *J. Acoust. Soc. Amer.*, **62**, 1129–35.

Chen, F., Shapiro, G.I., Bennett, K.A., Ingram, S.N., Thompson, D., Vincent, C. et al. (2017) Shipping noise in a dynamic sea: A case study of grey seals in the Celtic Sea. *Mar. Pollut. Bull.*, **114**, 372–83, doi: 10.1016/j.marpolbul.2016.09.054.

Cheney, R.E. and Winfrey, D.E. (1976) Distribution and classification of ocean fronts. Nav. Oceanogr. Off., Tech. Note 3700-56-76.

Cheng, E., Duan, L.-Z., and Yuan, F. (2013) Time-varying model of underwater acoustic channel with Doppler effect. *J. Convergent Info. Technol.*, **8**, 480–6, doi: 10.4156/jcit.vol8.issue10.59.

Childers, C., Brutzman, D., Blais, C., and Young, P. (2006) *A Case Study in Applying Semantic Web Technologies to the XML-Based Tactical Assessment Markup Language (TAML). AAAI 2006 Fall Symposium on Semantic Web for Collaborative Knowledge Acquisition*, Arlington, VA: AAAI Press.

Chin-Bing, S.A., Davis, J.A., and Evans, R.B. (1982) Nature of the lateral wave effect on bottom loss measurements. *J. Acoust. Soc. Amer.*, **71**, 1433–7.

Chin-Bing, S.A., Gilbert, K.E., Evans, R.B., Werby, M.F., and Tango, G.J. (1986) Research and development of acoustic models at NORDA. In *The NORDA Review* (31 March 1976–31 March 1986). Stennis Space Center, MS, pp. 39–44.

Chin-Bing, S.A., King, D.B., and Boyd, J.D. (1994) The effects of ocean environmental variability on underwater acoustic propagation forecasting. In *Oceanography and Acoustics: Prediction and Propagation Models*, eds A.R. Robinson and D. Lee, New York, NY: American Institute of Physics, Chapter 2, pp 7–49.

Chin-Bing, S.A., King, D.B., Davis, J.A., and Evans, R.B. (eds) (1993b) *PE Workshop II. Proceedings of the Second Parabolic Equation Workshop.* Nav. Res. Lab., Rept NRL/BE/7181-93-0001.

Chin-Bing, S.A., King, D.B., and Murphy, J.E. (1993a) Numerical simulations of lower-frequency acoustic propagation and backscatter from solitary internal waves in a shallow water environment. In *Ocean Reverberation*, eds D.D. Ellis, J.R. Preston, and H.G. Urban, Dordrecht, the Netherlands; Kluwer Academic Publishers, pp. 113–18.

Chin-Bing, S.A. and Murphy, J.E. (1987) Two numerical models—one simplistic, the other sophisticated—applied to under-ice acoustic propagation and scattering. *J. Acoust. Soc. Amer.*, **82** (Suppl. 1), S31 (abstract).

Chin-Bing, S.A. and Murphy, J.E. (1988) Long-range, range-dependent, acoustic propagation simulation using a full-wave, finite-element model coupled with a one-way parabolic equation model. *J. Acoust. Soc. Amer.*, **84** (Suppl. 1), S90 (abstract).

Chin-Bing, S.A., Warn-Varnas, A., King, D.B., Hawkins, J., and Lamb, K. (2009) Effects on acoustics caused by ocean solitons, Part B: Acoustics. *Nonlinear Anal. Theory Methods Appl.*, **71**, 2194–204.

Chin-Bing, S.A., Warn-Varnas, A., King, D.B., Lamb, K.G., Teixeira, M., and Hawkins, J.A. (2003) Analysis of coupled oceanographic and acoustic soliton simulations in the Yellow Sea: A search for soliton-induced resonances. *Math. Comput. Simul.*, **62**, 11–20.

Chitre, M. (2007) A high-frequency warm shallow water acoustic communications channel model and measurements. *J. Acoust. Soc. Amer.*, **122**, 2580–6.

Chitre, M., Shahabudeen, S., and Stojanovic, M. (2008) Underwater acoustic communications and networking: Recent advances and future challenges. *Mar. Tech. Soc. J.*, **42**(1), 103–16.

Chiu, C.-S. and Ehret, L.L. (1990) Computation of sound propagation in a three-dimensionally varying ocean: A coupled normal mode approach. In *Computational Acoustics. Vol. 1: Ocean-Acoustic Models and Supercomputing*, eds D. Lee, A. Cakmak, and R. Vichnevetsky, Amsterdam, the Netherlands: North-Holland, pp 187–202.

Chiu, C.-S. and Ehret, L.L. (1994) Three-dimensional acoustic normal mode propagation in the Gulf Stream. In *Oceanography and Acoustics: Prediction and Propagation Models*, eds A.R. Robinson and D. Lee, New York, NY: American Institute of Physics, Chapter 8, pp 179–97.

Chiu, L.Y.S., Lin, Y.-T., Chen, C.-F., Duda, T.F., and Calder, B. (2011) Focused sound from three-dimensional sound propagation effects over a submarine canyon. *J. Acoust. Soc. Amer.*, **129**(6), EL260–6, doi:10.1121/1.3579151.

Choi, J.W. and Dahl, P.H. (2005) On spectral and amplitude properties of first- and zeroth-order head waves, and implications for geoacoustic inversion. *J. Acoust. Soc. Amer.*, **118**, 1969 (abstract).

Choi, J.W. and Dahl, P.H. (2006) First-order and zeroth-order head waves, their sequence, and implications for geoacoustic inversion. *J. Acoust. Soc. Amer.*, **119**, 3660–8.

Choi, J.W., Yoon, K.-S., Na, J., Park, J.-S., and Na, Y.N. (2002) Shallow water high-frequency reverberation model. *J. Acoust. Soc. Korea*, **21**, 671–8 (in Korean).

Chotiros, N. P. (2017) *Acoustics of the Seabed as a Poroelastic Medium*. New York, NY: Springer. [Book not yet published.]

Chotiros, N.P. and Isakson, M.J. (2004) A broadband model of sandy ocean sediments: Biot-Stoll with contact squirt flow and shear drag. *J. Acoust. Soc. Amer.*, **116**, 2011–22.

Çinar, T. and Örencik, M.B. (2009) An underwater acoustic channel model using ray tracing in ns-2. Proc. Second IFIP Wireless Days 2009 Conf., Paris.

Claerbout, J.F. (1976) *Fundamentals of Geophysical Data Processing with Applications to Petroleum Prospecting*, International Series in the Earth and Planetary Sciences. New York, NY: McGraw-Hill.

Clancy, R.M. (1999) Operational systems, products, and applications today and tomorrow: The Navy perspective. In *Coastal Ocean Prediction,* ed. C.N.K. Mooers, Coastal and Estuarine Studies, Vol. 56. Washington, DC: American Geophysical Union, pp. 501–11.

Clancy, R.M., Cummings, J.A., and Ignaszewski, M.J. (1991a) New ocean thermal structure model operational at FNOC. *Nav. Oceanography Command News,* **11** (5 and 6), 1–5.

Clancy, R.M. and Johnson, A. (1997) An overview of naval operational ocean modeling. *Mar. Tech. Soc. J.*, **31**(1), 54–62.

Clancy, R.M. and Martin, P.J. (1979) The NORDA/FLENUMOCEANCEN thermodynamical ocean prediction system (TOPS): A technical description. Nav. Ocean Res. Devel. Activity, Tech. Note 54.

Clancy, R.M., Martin, P.J., Piacsek, S.A., and Pollak, K.D. (1981) Test and evaluation of an operationally capable synoptic upper-ocean forecast system. Nav. Ocean Res. Devel. Activity, Tech. Note 92.

Clancy, R.M. and Pollak, K.D. (1983) A real-time synoptic ocean thermal analysis/forecast system. *Prog. Oceanogr.*, **12**, 383–424.

Clancy, R.M., Price, J.F., and Hawkins, J.D. (1991b) The FNOC diurnal ocean surface layer (DOSL) model. *Nav. Oceanography Command News*, **11**(7), 8–10.

Clark, C.A. (2005) Acoustic wave propagation in horizontally variable media. *IEEE J. Oceanic Engr.*, **30**, 188–97.

Clay, C.S. (1999) Underwater sound transmission and SI units. *J. Acoust. Soc. Amer.,* **106**, 3047.

Clay, C.S. and Medwin, H. (1964) High-frequency acoustical reverberation from a rough-sea surface. *J. Acoust. Soc. Amer.*, **36,** 2131–4.

Clay, C.S. and Medwin, H. (1977) *Acoustical Oceanography: Principles and Applications.* New York, NY: Wiley-Interscience.

Climent, S., Sanchez, A., Capella, J.V., Meratnia, N., and Serrano, J.J. (2014) Underwater acoustic wireless sensor networks: Advances and future trends in physical, MAC and routing layers. *Sensors*, **14**, 795–833, doi: 10.3390/s140100795.

Coates, R., de Cogan, D., and Willison, P.A. (1990) Transmission line matrix modeling applied to problems in underwater acoustics. Proc. IEEE Oceans 90 Conf., Washington, DC, pp. 216–20.

Cockrell, K.L. and Schmidt, H. (2010) A relationship between the waveguide invariant and wavenumber integration. *J. Acoust. Soc. Amer.*, **128**(1), EL63–8, doi: 10.1121/1.3453768.

Coelho, E.F., Rixen, M., and Signell, R. (2005) NATO tactical ocean modelling system concept applicability. SACLANT Undersea Res. Ctr, Rept SR-411.

Cohen, J.S. and Einstein, L.T. (1970) Continuous gradient ray tracing system (CONGRATS) II: Eigenray processing programs. Navy Underwater Sound Lab., Rept 1069.

Cohen, J.S. and Weinberg, H. (1971) Continuous gradient ray-tracing system (CONGRATS) III: Boundary and volume reverberation. Nav. Underwater Syst. Ctr, Rept 4071.

Cole, B., Davis, J., Leen, W., Powers, W., and Hanrahan, J. (2004) Coherent bottom reverberation: Modeling and comparisons with at-sea measurements. *J. Acoust. Soc. Amer.,* **116**, 1985–94.

Colilla, R.A. (1970) Improvement in the efficiency of the present model and documentation of the program RP-70. Ocean Data Syst., Inc.

Collins, A.D. (1970) Equation comparison shows similarity in sonar/radar work. *UnderSea Technol.,* **11**(5), 47–8.

Collins, J., Foreman, T., and Speicher, D. (2002) A new design for a sonar environmental effects server. Simulation Interoperability Standards Organization, Proc. Simulation Interoperability Workshop, paper 02F-SIW-114.

Collins, J.B. and Scannell, C.G. (2005) Natural environment data services in distributed modeling and simulation. In *Net-Centric Approaches to Intelligence and National Security*, eds R. Ladner and F.E. Petry, New York, NY: Springer, pp. 149–74.

Collins, M.D. (1988a) FEPE user's guide. Nav. Ocean Res. and Devel. Activity, Tech. Note 365.

Collins, M.D. (1988b) The time-domain solution of the wide-angle parabolic equation including the effects of sediment dispersion. *J. Acoust. Soc. Amer.*, **84**, 2114–25.

Collins, M.D. (1990a) The rotated parabolic equation and sloping ocean bottoms. *J. Acoust. Soc. Amer.*, **87**, 1035–7.

Collins, M.D. (1990b) Benchmark calculations for higher-order parabolic equations. *J. Acoust. Soc. Amer.*, **87**, 1535–8.

Collins, M.D. (1991) Higher-order Padé approximations for accurate and stable elastic parabolic equations with application to interface wave propagation. *J. Acoust. Soc. Amer.*, **89**, 1050–7.

Collins, M.D. (1992) A self-starter for the parabolic equation method. *J. Acoust. Soc. Amer.*, **92**, 2069–74.

Collins, M.D. (1993a) A split-step Padé solution for the parabolic equation method. *J. Acoust. Soc. Amer.*, **93**, 1736–42.

Collins, M.D. (1993b) A two-way parabolic equation for elastic media. *J. Acoust. Soc. Amer.*, **93**, 1815–25.

Collins, M.D. (1993c) An energy-conserving parabolic equation for elastic media. *J. Acoust. Soc. Amer.*, **94**, 975–82.

Collins, M.D. (1993d) The adiabatic mode parabolic equation. *J. Acoust. Soc. Amer.*, **94**, 2269–78.

Collins, M.D. (1994) Generalization of the split-step Padé solution. *J. Acoust. Soc. Amer.*, **96**, 382–5.

Collins, M.D. (1998) New and improved parabolic equation models. *J. Acoust. Soc. Amer.*, **104**, 1808 (abstract).

Collins, M.D. (1999) The stabilized self-starter. *J. Acoust. Soc. Amer.*, **106**, 1724–6.

Collins, M.D. (2015) Treatment of ice cover and other thin elastic layers with the parabolic equation method. *J. Acoust. Soc. Amer.*, **137**, 1557–63.

Collins, M.D. and Chin-Bing, S.A. (1990) A three-dimensional parabolic equation model that includes the effects of rough boundaries. *J. Acoust. Soc. Amer.*, **87**, 1104–9.

Collins, M.D. and Dacol, D.K. (2000) A mapping approach for handling sloping interfaces. *J. Acoust. Soc. Amer.*, **107**, 1937–42.

Collins, M.D. and Evans, R.B. (1992) A two-way parabolic equation for acoustic backscattering in the ocean. *J. Acoust. Soc. Amer.*, **91**, 1357–68.

Collins, M.D. and Kuperman, W.A. (1994a) Overcoming ray chaos. *J. Acoust. Soc. Amer.*, **95**, 3167–70.

Collins, M.D. and Kuperman, W.A. (1994b) Inverse problems in ocean acoustics. *Inverse Problems*, **10**, 1023–40.

Collins, M.D. and Siegmann, W.L. (1999) A complete energy conservation correction for the elastic parabolic equation. *J. Acoust. Soc. Amer.*, **105**, 687–92.

Collis, J.M., Frank, S.D., Metzler, A.M., and Preston, K.S. (2016) Elastic parabolic equation and normal mode solutions for seismo-acoustic propagation in underwater environments with ice covers. *J. Acoust. Soc. Amer.*, **139**, 2672–82.

Collison, N. and Dosso, S. (2000) A comparison of modal decomposition algorithms for matched-mode processing. *Canadian Acoust.*, **28**(4), 15–27.

Condron, T.P., Onyx, P.M., and Dickson, K.R. (1955) Contours of propagation loss and plots of propagation loss vs range for standard conditions at 2, 5, and 8 kc. Navy Underwater Sound Lab., Tech. Memo. 1110-14-55.

Coppens, A.B. (1981) Simple equations for the speed of sound in Neptunian waters. *J. Acoust. Soc. Amer.*, **69**, 862–3.

Coppens, A.B. and Sanders, J.V. (1980) Propagation of sound from a fluid wedge into a fast fluid bottom. In *Bottom-Interacting Ocean Acoustics*, eds W.A. Kuperman and F.B. Jensen, New York, NY: Plenum, pp. 439–50.

Copping, A., Hanna, L., Whiting, J., Geerlofs, S., Grear, M., Blake, K. et al. (2013). Environmental effects of marine energy development around the world: Annex IV final report. Report by Ocean Energy Systems (OES) and Pacific Northwest National Laboratory (PNNL), p. 96.

Copping, A.E. and O'Toole, M.J. (2010) OES-IA annex IV: Environmental effects of marine and hydrokinetic devices. Report from the experts' workshop, September 27th—28th 2010, Clontarf Castle, Dublin, Ireland. Pacific Northwest National Laboratory, PNNL-20034. Prepared for the US Department of Energy under Contract DE-AC05-76RL01830, p. 64.

Cornuelle, B., Wunsch, C., Behringer, D., Birdsall, T., Brown, M., Heinmiller, R., Knox, R., Metzger, K., Munk, W., Spiesberger, J., Spindel, R., Webb, D., and Worcester, P. (1985) Tomographic maps of the ocean mesoscale, Part 1: Pure acoustics. *J. Phys. Oceanogr.*, **15**, 133–52.

Cornyn, J.J. (1973a) GRASS: A digital-computer ray-tracing and transmission-loss-prediction system. Vol. I—Overall description. Nav. Res. Lab., Rept 7621.

Cornyn, J.J. (1973b) GRASS: A digital-computer ray-tracing and transmission-loss-prediction system. Vol. II—User's manual. Nav. Res. Lab., Rept 7642.

Cornyn, J.J. (1980) A simple analytical directional ambient noise model. *J. Acoust. Soc. Amer.*, **67** (Suppl. 1), S97 (abstract).

Costa, E.D., and Medeiros, E.B. (2010) Numerical modeling and simulation of acoustic propagation in shallow water. In *Mecánica Computacional*, eds E. Dvorkin, M. Goldschmit, and M. Storti, Vol. 29. Buenos Aires, Argentina, pp. 2215–28.

Coulter, R.L. and Kallistratova, M.A. (1999) The role of acoustic sounding in a high-technology era. *Meteorol. Atmos. Phys.*, **71**, 3–13.

Cox, H. (1989) Fundamentals of bistatic active sonar. In *Underwater Acoustic Data Processing*, ed. Y.T. Chan, NATO Advanced Science Institutes Series E: Applied Sciences, Vol. 161. Dordrecht, the Netherlands: Kluwer Academic Publishers, pp. 3–24.

Crum, L.A., Roy, R.A., and Prosperetti, A. (1992) The underwater sounds of precipitation. *Nav. Res. Rev.*, **44**(2), 2–12.

Culkin, F. and Ridout, P. (1989) Salinity: Definitions, determinations, and standards. *Sea Tech.*, **30**(10), 47–9.

Culver, R.L. and Camin, H.J. (2008) Sonar signal processing using probabilistic signal and ocean environmental models. *J. Acoust. Soc. Amer.*, **124**, 3619–31.

Cummings, W.C. and Holliday, D.V. (1987) Sounds and source levels from Bowhead whales off Pt Barrow, Alaska. *J. Acoust. Soc. Amer.*, **82**, 814–21.

Dagnino, D., Sallarès, V., Biescas, B., and Ranero, C.R. (2016) Fine-scale thermohaline ocean structure retrieved with 2D prestack full-waveform inversion of multichannel seismic data: Application to the Gulf of Cadiz (SW Iberia), *J. Geophys. Res.*, **121**, 5454–69, doi: 10.1002/2016JC011844.

Dahal, P., Peng, D., Yang, Y., and Sharif, H. (2013) RSS based bridge scour measurement using underwater acoustic sensor networks. *Commun. Netw.*, **5**, 641–8.

Dahl, P.H., Miller, J.H., Cato, D.H., and Andrew, R.K. (2007) Underwater ambient noise. *Acoust. Today*, **3**(1), 23–33.

Dantzler, H.L. Jr., Sides, D.J., and Neal, J.C. (1993) An automated tactical oceanographic monitoring system. *Johns Hopkins APL Tech. Dig.*, **14**(3), 281–92.

Darehshoorzadeh, A. and Boukerche, A. (2015) Underwater sensor networks: A new challenge for opportunistic routing protocols. *IEEE Commun. Mag.*, **53**(11), 98–107, doi:10.1109/MCOM.2015.7321977.

Dashen, R. and Munk, W. (1984) Three models of global ocean noise. *J. Acoust. Soc. Amer.*, **76**, 540–54.

Davies, G.L. (2002) Project 03F—1: Phase I completion report. SACLANT Undersea Res. Ctr, Memo SM-399.

Davies, G.L. and Signell, E.P. (2006a) Espresso—scientific user guide. NATO Undersea Res. Ctr, Rept NURC-SP-2006-003.

Davies, G.L. and Signell, E.P. (2006b) Espresso(m)—user guide. NATO Undersea Res. Ctr, Rept NURC-SP-2006-002.

Davis, J.A., White, D., and Cavanagh, R.C. (1982) NORDA parabolic equation workshop, 31 March–3 April 1981. Nav. Ocean Res. Devel. Activity, Tech. Note 143.

Davis, P.K. (1995) An introduction to variable-resolution modeling. In *Warfare Modeling*, eds J. Bracken, M. Kress and R.E. Rosenthal, New York, NY: John Wiley & Sons, pp. 5–35.

Davis, P.K. (2001) *Effects-Based Operations (EBO): A Grand Challenge for the Analytical Community.* MR-1477-USJFCOM/AF. Santa Monica, CA: RAND.

Dawe, R.L. (1997) *Detection Threshold Modelling Explained.* Australia Department of Defence, Defence Science and Technology Organisation, Maritime Operations Division, Aeronautical and Maritime Research Laboratory, DSTO-TR-0586.

Dawoud, W.A., Negm, A.M., Saleh, N.M., and Bady, M.F. (2016) Impact assessment of off-shore pile driving noise on Red Sea marine mammals. *Int. J. Environ. Sci. Devel.*, **7**(1), 10–15, doi: 10.7763/IJESD.2016.V7.733.

Deane, G.B. (1997) Sound generation and air entrainment by breaking waves in the surf zone. *J. Acoust. Soc. Amer.,* **102**, 2671–89.

Deane, G.B. and Stokes, M.D. (2002) Scale dependence of bubble creation mechanisms in breaking waves. *Nature*, **418**, 839–44.

Deane, G.B. and Stokes, M.D. (2006) The acoustic signature of bubbles fragmenting in sheared flow. *J. Acoust. Soc. Amer.,* **120**(6), EL84–9, doi: 10.1121/1.2364466.

Deane, G.B. and Stokes, M.D. (2010) Model calculations of the underwater noise of breaking waves and comparison with experiment. *J. Acoust. Soc. Amer.*, **127**, 3394–410.

Deavenport, R.L. and DiNapoli, F.R. (1982) Evaluation of Arctic transmission loss models. Nav. Underwater Syst. Ctr, Tech. Memo. 82–1160.

DeFerrari, H., Williams, N., and Nguyen, H. (2003) Focused arrivals in shallow water propagation in the Straits of Florida. *Acoust. Res. Lett. Online*, **4**(3), 106–11.

Dekeling, R.P.A., Tasker, M.L., Van der Graaf, A.J., Ainslie, M.A, Andersson, M.H., André, M. et al. (2014a) *Monitoring Guidance for Underwater Noise in European Seas, Part I: Executive Summary. JRC Scientific and Policy Report EUR 26557 EN*, Luxembourg: Publications Office of the European Union, doi:10.2788/29293.

Dekeling, R.P.A., Tasker, M.L., Van der Graaf, A.J., Ainslie, M.A, Andersson, M.H., André, M. et al. (2014b) *Monitoring Guidance for Underwater Noise in European Seas, Part II: Monitoring Guidance Specifications. JRC Scientific and Policy Report EUR 26555 EN*, Luxembourg: Publications Office of the European Union, doi:10.2788/27158.

Dekeling, R.P.A., Tasker, M.L., Van der Graaf, A.J., Ainslie, M.A, Andersson, M.H., André, M. et al. (2014c) *Monitoring Guidance for Underwater Noise in European Seas, Part III: Background Information and Annexes. JRC Scientific and Policy Report EUR 26556 EN*, Luxembourg: Publications Office of the European Union, doi:10.2788/2808.

Del Balzo, D.R., Feuillade, C., and Rowe, M.M. (1988) Effects of water-depth mismatch on matched-field localization in shallow water. *J. Acoust. Soc. Amer.*, **83**, 2180–5.

Del Grosso, V.A. (1974) New equation for the speed of sound in natural waters (with comparisons to other equations). *J. Acoust. Soc. Amer.*, **56**, 1084–91.

Delory, E., André, M., Navarro Mesa, J.-L., and van der Schaar, M. (2007) On the possibility of detecting surfacing sperm whales at risk of collision using others' foraging clicks. *J. Mar. Biol. Ass. U.K.*, **87**, 47–58.

DeSanto, J.A. (1979) Derivation of the acoustic wave equation in the presence of gravitational and rotational effects. *J. Acoust. Soc. Amer.*, **66**, 827–30.

Desharnais, F. and Ellis, D.D. (1997) Data-model comparisons of reverberation at three shallow-water sites. *IEEE J. Oceanic Engr.*, **22**, 309–16.

Diachok, O.I. (1976) Effects of sea-ice ridges on sound propagation in the Arctic Ocean. *J. Acoust. Soc. Amer.*, **59**, 1110–20.

Diachok, O. (1980) Arctic hydroacoustics. *Cold Regions Sci. Tech.*, **2**, 185–201.

Diachok, O., Caiti, A., Gerstoft, P., and Schmidt, H. (eds) (1995) *Full Field Inversion Methods in Ocean and Seismo-Acoustics,* Modern Approaches in Geophysics, Vol. 12. Dordrecht, the Netherlands: Kluwer Academic Publishers.

Diachok, O., Liorzou, B., and Scalabrin, C. (2001) Estimation of the number density of fish from resonance absorptivity and echo sounder data. *ICES J. Mar. Sci.*, **58**, 137–53.

Diachok, O. and Wales, S. (2005) Concurrent inversion of geo- and bio-acoustic parameters from transmission loss measurements in the Yellow Sea. *J. Acoust. Soc. Amer.,* **117**, 1965–76.

Diachok, O.I. and Winokur, R.S. (1974) Spatial variability of underwater ambient noise at the Arctic ice-water boundary. *J. Acoust. Soc. Amer.*, **55**, 750–3.

DiNapoli, F.R. (1971) Fast field program for multi-layered media. Nav. Underwater Syst. Ctr, Rept 4103.

DiNapoli, F.R. and Deavenport, R.L. (1979) Numerical models of underwater acoustic propagation. In *Ocean Acoustics*, ed. J.A. DeSanto, Topics in Current Physics, Vol. 8, New York, NY: Springer-Verlag, pp. 79–157.

DiNapoli, F.R. and Deavenport, R.L. (1980) Theoretical and numerical Green's function field solution in a plane multilayered medium. *J. Acoust. Soc. Amer.*, **67**, 92–105.

Dol, H.S., Colin, M.E.G.D., Ainslie, M.A., van Walree, P.A., and Janmaat, J. (2013) Simulation of an underwater acoustic communication channel characterized by wind-generated surface waves and bubbles. *IEEE J. Oceanic Engr.*, **38**, 642–54.

Dombestein, E. (2017) *LYBIN 6.2 2200—User Manual.* Norway: Forsvarets forskningsinstitutt/Norwegian Defence Research Establishment, FFI-rapport 17/00412.

Dombestein, E. and Ektvedt, K.W. (2014) *LYBIN 6.2—User Manual.* Norway: Forsvarets forskningsinstitutt/Norwegian Defence Research Establishment, FFI-rapport 2014/00512.

Dombestein, E., Gjersøe, A., and Bosseng, M. (2010) *LybinCom 6.0—Description of the Binary Interface.* Norway: Forsvarets forskningsinstitutt/Norwegian Defence Research Establishment, FFI-rapport 2009/02267.

Domingo, M.C. (2008) Overview of channel models for underwater wireless communication networks. *Phy. Commun.*, **1**, 163–82.

Dong, Y., Dong, H., and Zhang, G. (2014) Study on denial of service against underwater acoustic networks. *J. Commun.*, **9**(2), 135–43.

Donn, W.L. and Rind, D. (1971) Natural infrasound as an atmospheric probe. *Geophys. J. Roy. Astron. Soc.*, **26**, 111–33.

Doolittle, R., Tolstoy, A., and Buckingham, M. (1988) Experimental confirmation of horizontal refraction of cw acoustic radiation from a point source in a wedge-shaped ocean environment. *J. Acoust. Soc. Amer.*, **83**, 2117–25.

Doran, S.L. and Fredricks, A.J. (2007) High frequency acoustic propagation using level set methods. Proc. MTS/IEEE Oceans 2007 Conf., Vancouver.

Dosso, S.E. (2002) Benchmarking range-dependent propagation modeling in matched-field inversion. *J. Comput. Acoust.*, **10**, 231–42.

Dosso, S.E., Yeremy, M.L., Ozard, J.M., and Chapman, N.R. (1993) Estimation of ocean-bottom properties by matched-field inversion of acoustic field data. *IEEE J. Oceanic Engr.*, **18**, 232–9.

Dourado, N. (2015) Sediment laden ice detection using broadband acoustic backscattering measurements from calibration targets in ice. M.S. thesis, Dalhousie University, Halifax, Nova Scotia, Canada.

Dozier, L.B. (1984) PERUSE: A numerical treatment of rough surface scattering for the parabolic wave equation. *J. Acoust. Soc. Amer.*, **75**, 1415–32.

Dozier, L.B. and Cavanagh, R.C. (1993) Overview of selected underwater acoustic propagation models. Office of Naval Research, Advanced Environmental Acoustic Support Program, AEAS Rept 93-101.

Dozier, L. and Lallement, P. (1995) Parallel implementation of a 3-D range-dependent ray model for replica field generation. In *Full Field Inversion Methods in Ocean and Seismo-Acoustics*, eds O. Diachok, A. Caiti, P. Gerstoft, and H. Schmidt, Dordrecht, the Netherlands: Kluwer Academic Publishers, pp. 45–50.

Dozier, L.B. and White, D. (1988) Software requirements specification for the ASTRAL model. Sci. Appl. Inter. Corp., OAML-SRS-23.

Dreini, G., Isoppo, C., and Jensen, F.B. (1995) PC-based propagation and sonar prediction models. SACLANT Undersea Res. Ctr, Rept SR-240.

Drob, D.P., Meier, R.R., Picone, J.M., and Garcés, M.M. (2010) Inversion of infrasound signals for passive atmospheric remote sensing, In *Infrasound Monitoring for Atmospheric Studies*, eds A. Le Pichon, E. Blanc and A. Hauchecorne, New York, NY: Springer, Chapter 24, pp. 701–32.

D'Spain, G.L., Murray, J.J., Hodgkiss, W.S., Booth, N.O., and Schey, P.W. (1999) Mirages in shallow water matched field processing. *J. Acoust. Soc. Amer.*, **105**, 3245–65.

Duda, T.F. (2006a) Temporal and cross-range coherence of sound traveling through shallow-water nonlinear internal wave packets. *J. Acoust. Soc. Amer.*, **119**, 3717–25.

Duda, T.F. (2006b) Initial results from a Cartesian three-dimensional parabolic equation acoustical propagation code. Woods Hole Oceanogr. Inst., Tech. Rept WHOI-2006-14.

Duda, T.F., Howe, B.M., and Miller, J.H. (2007) Acoustics in global process ocean observatories. *Sea Technol.*, **48**(2), 35–8.

Dumuid, P.M. (2012) The application of Tikhonov regularised inverse filtering to digital communication through multi-channel acoustic systems. PhD Dissertation, The University of Adelaide, Australia.

Duncan, A.J. and Maggi, A.L. (2006) A consistent, user friendly interface for running a variety of underwater acoustic propagation codes. Proc. Acoustics 2006 Conf., Christchurch, New Zealand, pp. 471–7.

Dunn, R.A. (2015) Ocean acoustic reverberation tomography. *J. Acoust. Soc. Amer.*, **138**, 3458–69.

Dushaw, B.D. (2014) Assessing the horizontal refraction of ocean acoustic tomography signals using high-resolution ocean state estimates. *J. Acoust. Soc. Amer.*, **136**, 122–9.

Dushaw, B.D. and Colosi, J.A. (1998) Ray tracing for ocean acoustic tomography. Appl. Phys. Lab., Univ. Washington, APL-UW TM 3-98.

Dushaw, B.D., Worcester, P.F., Cornuelle, B.D., and Howe, B.M. (1993) On equations for the speed of sound in seawater. *J. Acoust. Soc. Amer.*, **93**, 255–75.

Duston, M.D., Gilbert, R.P., and Wood, D.H. (1986) A wave propagation computation technique using function theoretic representation. In *Numerical Mathematics and Applications*, eds R. Vichnevetsky and J. Vignes, Amsterdam, the Netherlands: North-Holland, pp. 281–8.

Dyer, I. (1973) Statistics of distant shipping noise. *J. Acoust. Soc. Amer.*, **53**, 564–70.

Dyer, I. (1984) The song of sea ice and other Arctic melodies. In *Arctic Technology and Policy*, eds I. Dyer and C. Chryssostomidis, Philadelphia, PA: Hemisphere Publishing Corp, pp. 11–37.

Dyer, I. (1988) Arctic ambient noise: Ice source mechanisms. Physics news in 1987—Acoustics. *Phys. Today*, **41**(1), S5–6.

Dzieciuch, M., Munk, W., and Rudnick, D.L. (2004) Propagation of sound through a spicy ocean, the SOFAR overture. *J. Acoust. Soc. Amer.*, **116**, 1447–62.

Dzieciuch, M., Worcester, P., and Munk, W. (2001) Turning point filters: Analysis of sound propagation on a gyre-scale. *J. Acoust. Soc. Amer.*, **110**, 135–49.

Earle, M.D. and Bishop, J.M. (1984) *A Practical Guide to Ocean Wave Measurement and Analysis*, Marion, MA: ENDECO, Inc.

Ebbeson, G.R. and Turner, R.G. (1983) Sound propagation over Dickins Seamount in the northeast Pacific Ocean. *J. Acoust. Soc. Amer.*, **73**, 143–52.

Ebihara, T. and Mizutani, K. (2014) Underwater acoustic communication with anorthogonal signal division multiplexing scheme in doubly spread channels. *IEEE J. Oceanic Engr.*, **39**, 47–58.

Ebihara, T., Mizutani, K., and Wakatsuki, N. (2009) Chip-interleaved multiple access for underwater acoustic communication and its performance evaluation in propagation simulation. *Jpn. J. Appl. Phys.*, **48**, 1–6, doi:10.1143/JJAP.48.114505.

Ebihara, T., Mizutani, K., and Wakatsuki, N. (2010) Basic study of orthogonal signal division multiplexing for underwater acoustic communication: A comparative study. *Jpn. J. Appl. Phys.*, **49**, 1–8, doi:10.1143/JJAP.49.07HG09.

Ebor Computing. (2005) MODRAY. Product description literature, Mile End, South Australia.

Eckart, C. (1953) The scattering of sound from the sea surface. *J. Acoust. Soc. Amer.*, **25**, 566–70.

Edelmann, G.F. (2005) An overview of time-reversal acoustic communications. Proc. Turkish Inter. Conf. Acoustics, TICA '05, New Concepts for Harbour Protection, Littoral Security and Underwater Acoustic Communications, Istanbul.

Edelmann, G.F., Song, H.C., Kim, S., Hodgkiss, W.S., Kuperman, W.A., and Akal, T. (2005) Underwater acoustic communications using time reversal. *IEEE J. Oceanic Engr.*, **30**, 852–64.

Eggen, C., Howe, B., and Dushaw, B. (2002) A MATLAB GUI for ocean acoustic propagation. Proc. MTS/IEEE Oceans '02 Conf., Biloxi, MS, pp. 1415–21.

Einstein, P.A. (1975) Underwater sonic ray tracing in three dimensions. *J. Sound Vib.*, **42**, 503–8.

Eller, A.I. (1984a) Findings and recommendations of the surface loss model working group: Final report, 1984. Nav. Ocean Res. Devel. Activity, Tech. Note 279.

Eller, A.I. (1984b) Acoustics of shallow water: A status report. Nav. Res. Lab., Memo. Rept 5405.

Eller, A.I. (1985) Findings and recommendations of the under-ice scattering loss model working group. Nav. Ocean Res. Devel. Activity, Tech. Note 255.

Eller, A.I. (1986) Shallow water sonar performance prediction assessment. Nav. Ocean Res. Devel. Activity, Tech. Note 330.

Eller, A.I. (1990) Automated long range acoustic modeling. Proc. IEEE Oceans 90 Conf., Washington, DC, pp. 205–8.

Eller, A.I. and Haines, L. (1987) Identification and acoustic characterization of seamounts. Sci. Appl. Inter. Corp., SAIC-87/1722.

Eller, A.I. and Venne, H.J. Jr. (1981) Evaluation procedure for environmental acoustic models. IEEE Inter. Conf. Acoust. Speech Signal Process., Atlanta, pp. 1026–9.

Eller, A.I., Venne, H.J. Jr., and Hoffman, D.W. (1982) Evaluation of ocean acoustic reverberation models. Proc. MTS/IEEE Oceans 82 Conf., Washington, DC, pp. 206–10.

Ellis, D.D. (1980) NORM2L: An interactive computer program for acoustic normal mode calculations for the Pekeris model. Defence Research Establishment Atlantic (Canada), Tech. Memo. 80/K.

Ellis, D.D. (1985) A two-ended shooting technique for calculating normal modes in underwater acoustic propagation. Defence Research Establishment Atlantic (Canada), DREA Rept 85/105.

Ellis, D.D. (1993) Shallow water reverberation: Normal-mode model predictions compared with bistatic towed-array measurements. *IEEE J. Oceanic Engr.*, **18**, 474–82.

Ellis, D.D. (1995) A shallow-water normal-mode reverberation model. *J. Acoust. Soc. Amer.*, **97**, 2804–14.

Ellis, D.D. (2007) Measurements and analysis of reverberation and clutter data. DRDC Atlantic ECR 2007-065. Defence R&D Canada–Atlantic.

Ellis, D.D. (2008) Normal mode models OGOPOGO and NOGRP applied to the 2006 ONR reverberation modeling workshop problems. DRDC Atlantic TM 2006-289. Defence R&D Canada–Atlantic.
Ellis, D.D. and Chapman, D.M.F. (1985) A simple shallow water propagation model including shear wave effects. *J. Acoust. Soc. Amer.*, **78**, 2087–95.
Ellis, D.D. and Crowe, D.V. (1991) Bistatic reverberation calculations using a three-dimensional scattering function. *J. Acoust. Soc. Amer.*, **89**, 2207–14.
Ellis, D.D., Preston, J.R., and Urban, H.G. (eds) (1993) *Ocean Reverberation*, Dordrecht, the Netherlands: Kluwer Academic Publishers.
Ellis, J.A. and Parchman, S. (1994) The interactive multisensor analysis training (IMAT) system: A formative evaluation in the aviation antisubmarine warfare operator (AW) class "A" school. NPRDC TN-94-20. Navy Personnel Research and Development Center, San Diego, CA.
Ellis, J.W., Miller, J.E., and Fernandez, M.R. (1996) NRL's tactical oceanography simulation lab. *Sea Technol.*, **37**(5), 31–7.
Elston, G.R. and Bell, J.M. (2004) Pseudospectral time-domain modeling of non-Rayleigh reverberation: Synthesis and statistical analysis of a sidescan sonar image of sand ripples. *IEEE J. Oceanic Engr.*, **29**, 317–29.
Emery, W.J. and Meincke, J. (1986) Global water masses: Summary and review. *Oceanol. Acta.*, **9**, 383–91.
Epifanio, C.L., Potter, J.R., Deane, G.B., Readhead, M.L., and Buckingham, M.J. (1999) Imaging in the ocean with ambient noise: The ORB experiments. *J. Acoust. Soc. Amer.*, **106**, 3211–25.
Erbe, C., McCauley, R., McPherson, C., and Gavrilov, A. (2013) Underwater noise from offshore oil production vessels. *J. Acoust. Soc. Amer.*, **133**(6), EL465–70, doi:10.1121/1.4802183.
Errico, R.M. (1997) What is an adjoint model? *Bull. Amer. Meteor. Soc.*, **78**, 2577–91.
Estalote, E. (1984) NORDA acoustic models and data bases. Nav. Ocean Res. Devel. Activity, Tech. Note 293.
Estalote, E.A., Kerr, G.A., and King, D.B. (1986) Recent advances in application of acoustic models. In *The NORDA Review* (31 March 1976—31 March 1986). Stennis Space Center, MS, pp. 45–50.
Etter, P.C. (1981) Underwater acoustic modeling techniques. *Shock Vib. Dig.*, **13**(2), 11–20.
Etter, P.C. (1984) Underwater acoustic modeling techniques. *Shock Vib. Dig.*, **16**(1), 17–23.
Etter, P.C. (1987a) Underwater acoustic modeling techniques. *Shock Vib. Dig.*, **19**(2), 3–10.
Etter, P.C. (1987b) Numerical modeling techniques in underwater acoustics. *J. Acoust. Soc. Amer.*, **82** (Suppl. 1), S102 (abstract).
Etter, P.C. (1987c) Arctic oceanography and meteorology review. Nav. Training Syst. Ctr, NTSC-TR-87-032.
Etter, P.C. (1989) Underwater acoustic modeling for antisubmarine warfare. *Sea Technol.*, **30**(5), 35–7.
Etter, P.C. (1990) Underwater acoustic modeling techniques. *Shock Vib. Dig.* **22**(5), 3–12.
Etter, P.C. (1991) *Underwater Acoustic Modeling: Principles, Techniques and Applications*, London, England: Elsevier Applied Science.
Etter, P.C. (1993) Recent developments in computational ocean acoustics. Radix Systems, Inc., Tech. Rept TR-93-082.
Etter, P.C. (1995) The status of Naval underwater acoustic modeling. *J. Acoust. Soc. Amer.*, **97**, 3312 (abstract).
Etter, P.C. (1996) *Underwater Acoustic Modeling: Principles, Techniques and Applications*, 2nd edn. London, England: E & FN Spon.
Etter, P.C. (1999) Updated technology baseline in underwater acoustic modeling, simulation, and analysis. *J. Acoust. Soc. Amer.*, **106**, 2298 (abstract).
Etter, P.C. (2000) Challenges in environmental acoustics. Proc. MTS/IEEE Oceans 2000 Conf., Boston, pp. 877–85.

Etter, P.C. (2001a) Recent advances in underwater acoustic modelling and simulation. *J. Sound Vib.,* **240**, 351–83.

Etter, P.C. (2001b) Progress in underwater acoustic modeling. In *Acoustic Interactions with Submerged Elastic Structures. Part II: Propagation, Ocean Acoustics and Scattering,* eds A. Guran, G. Maugin, J. Engelbrecht, and M. Werby, Singapore: World Scientific Publishing, pp. 112–23.

Etter, P.C. (2001c) Models for the analysis and prediction of acoustic phenomena in the coastal ocean. Proc. Fourth Conf. Coastal Atmospheric Oceanic Prediction and Processes, American Meteorological Society, St. Petersburg, FL, pp. 76–83.

Etter, P.C. (2003a) *Underwater Acoustic Modeling and Simulation*, 3rd edn. London, England: Spon Press.

Etter, P.C. (2003b) Recent developments in underwater acoustic modeling. *J. Acoust. Soc. Amer.,* **114**, 2430 (abstract).

Etter, P.C. (2009) Review of ocean-acoustic models. Proc. MTS/IEEE Oceans 2009 Conf., Biloxi, MS.

Etter, P.C. (2011) A review of recent developments in underwater acoustic modeling *J. Acoust. Soc. Amer.,* **129**, 2631 (abstract).

Etter, P.C. (2012a) Advanced applications for underwater acoustic modeling. *Adv. Acoust. Vib.*, **2012**, 1–28, Article ID 214839, doi:10.1155/2012/214839.

Etter, P.C. (2012b) New frontiers for underwater acoustic modeling. Proc. MTS/IEEE Oceans 2012 Conf., Virginia Beach, VA.

Etter, P.C. (2013) *Underwater Acoustic Modeling and Simulation*, 4th edn. Boca Raton, FL: CRC Press.

Etter, P.C. (2014) Overview of sound propagation model types. Proc. Sound Exposure Modeling Workshop (Sponsored by NMFS and BOEM), Washington, DC.

Etter, P.C. (2017) Assessing the impacts of marine-hydrokinetic energy (MHK) device noise on marine systems by using underwater acoustic models as enabling tools. In *Marine Renewable Energy: Resource Characterization and Physical Effects*, eds Z. Yang and A. Copping, New York, NY: Springer Publishing, pp. 305–22.

Etter, P.C. and Cochrane, J.D. (1975) Water temperature on the Texas-Louisiana shelf. Texas A & M Univ., TAMU-SG-75-604.

Etter, P.C. and Flum, R.S. Sr. (1978) A survey of marine environmental/acoustic data banks and basic acoustic models with potential application to the acoustic performance prediction (APP) program. ASW Syst. Proj. Off., ASWR-78-117.

Etter, P.C. and Flum, R.S. Sr. (1979) An overview of the state-of-the-art in naval underwater acoustic modeling. *J. Acoust. Soc. Amer.*, **65** (Suppl. 1), S42 (abstract).

Etter, P.C. and Flum, R.S. Sr. (1980) A survey of underwater acoustic models and environmental-acoustic data banks. ASW Syst. Proj. Off., ASWR-80-115.

Etter, P.C., Deffenbaugh, R.M., and Flum, R.S. Sr. (1984) A survey of underwater acoustic models and environmental-acoustic data banks. ASW Syst. Prog. Off., ASWR-84-001.

Etter, P.C., Haas, C.H., and Ramani, D.V. (2014) Advanced concepts for underwater acoustic channel modeling. Proc. American Geophysical Union, Fall Meeting, Paper OS11B-1274, San Francisco, CA.

Etter, P.C., Haas, C.H., and Ramani, D.V. (2015) Evolving trends and challenges in applied underwater acoustic modeling, Proc. MTS/IEEE Oceans 2015 Conf., Washington, DC.

Etter, P.C., Howard, M.K., and Cochrane, J.D. (2004) Heat and freshwater budgets of the Texas-Louisiana shelf. *J. Geophys. Res.*, **109**, C02024, doi:10.1029/2003JC001820.

Etter, P.C., Lamb, P.J., and Portis, D.H. (1987) Heat and freshwater budgets of the Caribbean Sea with revised estimates for the Central American Seas. *J. Phys. Oceanogr.*, **17**, 1232–48.

Evans, R.B. (1983) A coupled mode solution for acoustic propagation in a waveguide with stepwise depth variations of a penetrable bottom. *J. Acoust. Soc. Amer.*, **74**, 188–95.

Evans, R.B. (1986) COUPLE: A user's manual. Nav. Ocean Res. Devel. Activity. Tech. Note 332.

Evans, R.B. (2006) Stepwise coupled mode scattering of ambient noise by a cylindrically symmetric seamount. *J. Acoust. Soc. Amer.*, **119**, 161–7.

Evans, R.B. and Gilbert, K.E. (1985) Acoustic propagation in a refracting ocean waveguide with an irregular interface. *Comp. Maths. Applic.*, **11**, 795–805.

Ewart, T.E. and Reynolds, S.A. (1984) The mid-ocean acoustic transmission experiment, MATE. *J. Acoust. Soc. Amer.*, **75**, 785–802.

Ewing, M. and Worzel, J.L. (1948) Long-range sound transmission. *Geol. Soc. Amer. Mem.*, **27** (Part 3), 1–35.

Fabre, J.P. and Wilson, J.H. (1997) Noise source level density due to surf—Part II: Duck, NC. *IEEE J. Oceanic Engr.*, **22**, 434–44.

Fabre, J.P., Wilson, J.H., Earle, M.D., and McDermid, J.G. (1997) An inversion method for determination of ambient noise due to surf. *J. Acoust. Soc. Amer.*, **102**, 3103–4 (abstract).

Farcas, A., Thompson, P.M., and Merchant, N.D. (2016) Underwater noise modelling for environmental impact assessment. *Environ. Impact Asses.*, **57**, 114–22, doi:10.1016/j.eiar.2015.11.012.

Farina, A. (2014) *Soundscape Ecology: Principles, Patterns, Methods and Applications,* Dordrecht, the Netherlands: Springer.

Farmer, D. and Li, M. (1995) Patterns of bubble clouds organized by Langmuir circulation. *J. Phys. Oceanogr.*, **25**, 1426–40.

Farquhar, G.B. (ed.) (1970) *Proceedings of an International Symposium on Biological Sound Scattering in the Ocean,* Washington, DC: US Government Printing Office.

Farrell, T. and LePage, K. (1996) Development of a comprehensive hydroacoustic coverage assessment model. BBN Systems and Technologies, Tech. Memo W1278. Formally released as Phillips Lab. Tech. Rept PL-TR-96-2248, Air Force Materiel Command, Hanscom Air Force Base, MA.

Fawcett, J.A., Westwood, E.K., and Tindle, C.T. (1995) A simple coupled wedge mode propagation method. *J. Acoust. Soc. Amer.*, **98**, 1673–81.

Felizardo, F.C. and Melville, W.K. (1995) Correlations between ambient noise and the ocean surface wave field. *J. Phys. Oceanogr.*, **25**, 513–32.

Felsen, L.B. (1986) Benchmarks: Are they helpful, diversionary, or irrelevant? *J. Acoust. Soc. Amer.*, **80** (Suppl. 1), S36 (abstract).

Felsen, L.B. (1987) Chairman's introduction to session on numerical solutions of two benchmark problems. *J. Acoust. Soc. Amer.*, **81** (Suppl. 1), S39 (abstract).

Felsen, L.B. (1990) Benchmarks: An option for quality assessment. *J. Acoust. Soc. Amer.*, **87**, 1497–8.

Ferla, C.M., Isoppo, C., Martinelli, G., and Jensen, F.B. (2001) Performance assessment of the LYBIN-2.0 propagation-loss model. SACLANT Undersea Res. Ctr, Memo SM-384.

Ferla, C.M., Porter, M.B., and Jensen, F.B. (1993) C-SNAP: Coupled SACLANTCEN normal mode propagation loss model. SACLANT Undersea Res. Ctr, Memo. SM-274.

Ferla, M.C., Jensen, F.B., and Kuperman, W.A. (1982) High-frequency normal-mode calculations in deep water. *J. Acoust. Soc. Amer.*, **72**, 505–9.

Ferris, R.H. (1972) Comparison of measured and calculated normal-mode amplitude functions for acoustic waves in shallow water. *J. Acoust. Soc. Amer.*, **52**, 981–8.

Fialkowski, L.T., Lingevitch, J.F., Perkins, J.S., Dacol, D.K., and Collins, M.D. (2003) Geoacoustic inversion using a rotated coordinate system and simulated annealing. *IEEE J. Oceanic Engr.*, **28**, 370–79.

Finette, S. (2005) Embedding uncertainty into ocean acoustic propagation models. *J. Acoust. Soc. Amer.*, **117**, 997–1000.

Finette, S. (2006) A stochastic representation of environmental uncertainty and its coupling to acoustic wave propagation in ocean waveguides. *J. Acoust. Soc. Amer.,* **120**, 2567–79.
Finneran, J.J. (2015) Noise-induced hearing loss in marine mammals: A review of temporary threshold shift studies from 1996 to 2015. *J. Acoust. Soc. Amer.,* **138**, 1702–26.
Fisher, A. Jr. and Pickett, R. (1973) Evaluation of methods for merging BT-derived and deep climatological sound speeds. Nav. Oceanogr. Off., Tech. Note 7700-12-73.
Fisher, F.H. and Simmons, V.P. (1977) Sound absorption in sea water. *J. Acoust. Soc. Amer.,* **62**, 558–64.
Fizell, R.G. (1987) Application of high-resolution processing to range and depth estimation using ambiguity function methods. *J. Acoust. Soc. Amer.*, **82**, 606–13.
Fizell, R.G. and Wales, S.C. (1985) Source localization in range and depth in an Arctic environment. *J. Acoust. Soc. Amer.*, **78** (Suppl. 1), S57–8 (abstract).
Flanagan, R.P., Weinberg, N.L., and Clark, J.G. (1974) Coherent analysis of ray propagation with moving source and fixed receiver. *J. Acoust. Soc. Amer.*, **56**, 1673–80.
Flatté, S.M. (ed.), Dashen, R., Munk, W.H., Watson, K.M., and Zachariasen, F. (1979) *Sound Transmission through a Fluctuating Ocean*, New York, NY: Cambridge Univ. Press.
Focke, K.C. (1984) Acoustic attenuation in ocean sediments found in shallow water regions. Appl. Res. Labs., Univ. Texas at Austin, USA, ARL-TR-84-6.
Fofonoff, N.P. (1985) Physical properties of seawater: A new salinity scale and equation of state for seawater. *J. Geophys. Res.*, **90**, 3332–42.
Forbes, A. (1994) Acoustic monitoring of global ocean climate. *Sea Techn.*, **35**(5), 65–7.
Foreman, T.L. (1977) Acoustic ray models based on eigenrays. Appl. Res. Labs., Univ. Texas at Austin, USA, ARL-TR-77-1.
Foreman, T.L. (1982) The ARL:UT range-dependent ray model MEDUSA. Appl. Res. Labs., Univ. Texas at Austin, USA, ARL-TR-82-64.
Foreman, T.L. (1983) Ray modeling methods for range dependent ocean environments. Appl. Res. Labs., Univ. Texas at Austin, USA, ARL-TR-83-41.
Foreman, T. and Speicher, D. (2003) Expanded sonar environmental acoustic effects server capabilities for fleet battle experiments. Simulation interoperability standards organization, Proc. Simulation Interoperability Workshop, paper 03F-SIW-062.
Foret, J.A., Korman, M.S., Schuler, J.W., and Holmes, E. (1997) Design and development of PC-IMAT: Teaching strategies for acoustical oceanography. *J. Acoust. Soc. Amer.,* **101**, 3097 (abstract).
Fortuin, L. (1970) Survey of literature on reflection and scattering of sound waves at the sea surface. *J. Acoust. Soc. Amer.*, **47**, 1209–28.
Franchi, E.R., Griffin, J.M., and King, B.J. (1984) NRL reverberation model: A computer program for the prediction and analysis of medium-to-long-range boundary reverberation. Nav. Res. Lab., Rept 8721.
Francois, R.E. and Garrison, G.R. (1982a) Sound absorption based on ocean measurements: Part I: Pure water and magnesium sulfate contributions. *J. Acoust. Soc. Amer.,* **72**, 896–907.
Francois, R.E. and Garrison, G.R. (1982b) Sound absorption based on ocean measurements. Part II: Boric acid contribution and equation for total absorption. *J. Acoust. Soc. Amer.,* **72**, 1879–90.
Frank, S.D., Badiey, M., Lynch, J.F., and Siegmann, W.L. (2005) Experimental evidence of three-dimensional acoustic propagation caused by nonlinear internal waves. *J. Acoust. Soc. Amer.,* **118**, 723–34.
Frankel, A.S., Ellison, W.T., and Buchanan, J. (2002) Application of the acoustic integration model (AIM) to predict and minimize environmental impacts. Proc. MTS/IEEE Oceans '02 Conf., Biloxi, MS, pp. 1438–43.
Franz, G.J. (1959) Splashes as sources of sound in liquids. *J. Acoust. Soc. Amer.*, **31**, 1080–96.

Fredricks, A., Colosi, J.A., Lynch, J.F., Gawarkiewicz, G., Chiu, C.-S., and Abbot, P. (2005) Analysis of multipath scintillations from long range acoustic transmissions on the New England continental slope and shelf. *J. Acoust. Soc. Amer.*, **117**, 1038–57.

Fricke, M.B. and Rolfes, R. (2015) Towards a complete physically based forecast model for underwater noise related to impact pile driving. *J. Acoust. Soc. Amer.*, **137**, 1564–75.

Frisk, G.V. (1994) *Ocean and Seabed Acoustics: A Theory of Wave Propagation,* Englewood Cliffs, NJ: PTR Prentice-Hall.

Frison, T.W., Abarbanel, H.D.I., Cembrola, J., and Neales, B. (1996) Chaos in ocean ambient "noise." *J. Acoust. Soc. Amer.*, **99**, 1527–39.

Frith, K. (2003) MODRAY underwater sound simulator user manual, Version 5.0. Australia Department of Defence, Defence Science and Technology Organisation, Maritime Operations Division, Aeronautical and Maritime Research Laboratory.

Fromm, D.M. (1999) Multistatic active system performance modeling in littoral waters. Proc. 28th Meeting of The Technical Cooperation Program, Technical Panel MAR TP-9, Nav. Res. Lab., Washington, DC.

Fromm, D.M. (2011) A computationally efficient multistatic reverberation algorithm. *J. Acoust. Soc. Amer.*, **129**, 2631 (abstract).

Fromm, D.M., Crockett, J.P., and Palmer, L.B. (1996) BiRASP—The bistatic range-dependent active system prediction model. Nav. Res. Lab., Rept NRL/FR/7140-95-9723.

Frye, D., Freitag, L., Detrick, R., Collins, J., Delaney, J., Kelley, D. et al. (2006) An acoustically linked moored-buoy ocean observatory. *EOS, Trans. Amer. Geophys. Union*, **87**, 213–18.

Frye, H.W. and Pugh, J.D. (1971) A new equation for the speed of sound in seawater. *J. Acoust. Soc. Amer.*, **50**, 384–6.

Fulford, J.K. (1991) Shallow water RASP upgrade. Nav. Oceanogr. Atmos. Res. Lab., Tech. Note 135.

Futa, K. and Kikuchi, T. (2001) Finite difference time domain analysis of bottom effect on sound propagation in shallow water. *Acoust. Sci. Technol.*, **22**, 303–5.

Gabrielson, T.B. (1982) Mathematical foundations for normal mode modelling in waveguides. Nav. Air Devel. Ctr, Rept NADC-81284-30.

Gaillard, F. (1985) Ocean acoustic tomography with moving sources or receivers. *J. Geophys. Res.*, **90**, 11891–8.

Galkin, O.P., Gostev, V.S., Popov, O.E., Shvachko, L.V., and Shvachko, R.F. (2006) Illumination of the shadow zone in a two-channel oceanic waveguide with a fine structure of sound speed inhomogeneities. *Acoust. Phys.*, **52**, 252–8.

Gammaitoni, L., Hänggi, P., Jung, P., and Marchesoni, F. (1998) Stochastic resonance. *Rev. Modern Phys.*, **70**, 223–87.

Gao, C.-X., Zhang, Y.-X., Cheng, E., and Yuan, F. (2013) Investigation of SNR estimation algorithms of FM signal for the underwater acoustic channel. *J. Comput.*, **8**, 2042–50, doi:10.4304/jcp.8.8.2042-2050.

Garon, H.M. (1975) FACTEX: FACT extended to range-dependent environments. Acoustic Environmental Support Detachment, Off. Nav. Res.

Garon, H.M. (1976) SHALFACT: A shallow water transmission loss model. Acoustic Environmental Support Detachment, Off. Nav. Res.

Garrett, C. and Munk, W. (1979) Internal waves in the ocean. *Ann. Rev. Fluid Mech.*, **11**, 339–69.

Garrett, C., Li, M., and Farmer, D. (2000) The connection between bubble size spectra and energy dissipation rates in the upper ocean. *J. Phys. Oceanogr.*, **30**, 2163–71.

Garwood, R.W. Jr. (1979) Air-sea interaction and dynamics of the surface mixed layer. *Rev. Geophys. Space Phys.*, **17**, 1507–24.

Gass, S.I. (1979) Computer model documentation: A review and an approach. National Bureau of Standards Special Pub. 500-39.

Gavrilov, A.N. and Mikhalevsky, P.N. (2006) Low-frequency acoustic propagation loss in the Arctic Ocean: Results of the Arctic climate observations using underwater sound experiment. *J. Acoust. Soc. Amer.,* **119**, 3694–706.

Gemmill, W. and Khedouri, E. (1974) A note on sound ray tracing through a Gulf Stream eddy in the Sargasso Sea. Nav. Oceanogr. Off., Tech. Note 6150-21-74.

Georges, T.M., Jones, R.M., and Lawrence, R.S. (1990) A PC version of the HARPO ocean acoustic ray-tracing program. Wave Propagation Lab., NOAA Tech. Memo. ERL WPL-180.

Gerstoft, P. (1994) Inversion of seismoacoustic data using genetic algorithms and *a posteriori* probability distributions. *J. Acoust. Soc. Amer.*, **95**, 770–82.

Gerstoft, P. (2004) *SAGA User Manual 5.1: An Inversion Software Package*, San Diego, CA: Marine Physical Lab., Scripps Inst. Oceanogr.

Gerstoft, P., Hodgkiss, W.S., Kuperman, W.A., and Song, H. (2003) Phenomenological and global optimization inversion. *IEEE J. Oceanic Engr.*, **28**, 342–54.

Gerstoft, P. and Schmidt, H. (1991) A boundary element approach to ocean seismoacoustic facet reverberation. *J. Acoust. Soc. Amer.*, **89**, 1629–42.

Gervaise, C., Aulanier, F., Simard, Y., and Roy, N. (2015) Mapping probability of shipping sound exposure level. *J. Acoust. Soc. Amer.*, **137**(6), EL429–35, doi:10.1121/1.4921673.

Ghadimi, P., Bolghasi, A., and Chekab, M.A.F. (2015a) Sea surface effects on sound scattering in the Persian Gulf region based on empirical relations. *J. Marine Sci. Appl.*, **14**, 113–25, doi:10.1007/s11804-015-1306-x.

Ghadimi, P., Bolghasi, A., and Chekab, M.A.F. (2016) Acoustic simulation of scattering sound from a more realistic sea surface: Consideration of two practical underwater sound sources. *J. Braz. Soc. Mech. Sci. Eng.*, **38,** 773–87, doi:10.1007/s40430-014-0285-1.

Ghadimi, P., Bolghasi, A., Chekab, M.A.F., and Zamanian, R. (2015b) Numerical investigation of transmission of low frequency sound through a smooth air-water interface. *J. Marine Sci. Appl.*, **14**, 334–42, doi:10.1007/s11804-015-1315-9.

Gilbert, K.E. and Evans, R.B. (1986) A Green's function method for one-way wave propagation in a range-dependent ocean environment In *Ocean Seismo-Acoustics*, eds T. Akal and J.M. Berkson, New York, NY: Plenum, pp. 21–8.

Gilbert, K.E., Evans, R.B., Chin-Bing, S.A., White, D., and Kuperman, W.A. (1983) Some new models for sound propagation in bottom-limited ocean environments. Conf. Proc., Acoustics and the Sea Bed, Bath Univ. Press, England, pp. 243–50.

Gilbert, R.P. and Wood, D.H. (1986) A transmutation approach to underwater sound propagation. *Wave Motion*, **8**, 383–97.

Gill, A.E. (1982) *Atmosphere-Ocean Dynamics*, International Geophysics Series, Vol. 30. San Diego, CA: Academic Press.

Glover, F. and Laguna, M. (1997) *Tabu Search,* Dordrecht, the Netherlands: Kluwer Academic Publishers.

Goddard, R.P. (1989) The sonar simulation toolset. Proc. IEEE Oceans 89 Conf., Seattle, DC, pp. 1217–22.

Goddard, R.P. (1993) Simulating ocean reverberation: A review of methods and issues. Appl. Phys. Lab., Univ. Washington, APL-UW TR 9313.

Goddard, R.P. (1994) Sonar simulation toolset software description. Release 2.5. Appl. Phys. Lab., Univ. Washington, APL-UW TR 9211.

Goddard, R.P. (2008) The sonar simulation toolset, Release 4.6: Science, mathematics, and algorithms. Appl. Phys. Lab., Univ. Washington, APL-UW TR 0702.

Godin, O.A. (1999) Reciprocity and energy conservation within the parabolic approximation. *Wave Motion*, **29**, 175–94.

Godin, O.A. (2002) Wide-angle parabolic equations for sound in a 3D inhomogeneous moving medium. *Dokl. Phys.*, **47**, 643–6.

Godin, O.A. (2006) Anomalous transparency of water-air interface for low-frequency sound. *Phys. Rev. Lett.*, **97**, 164301, doi:10.1103/PhysRevLett.97.164301.

Godin, O.A. (2011) An exact wave equation for sound in inhomogeneous, moving, and non-stationary fluids. Proc. MTS/IEEE Oceans 2011 Conf., Kona, HI.

Godin, O.A. (2012) On the possibility of using acoustic reverberation for remote sensing of ocean dynamics. *Acoust. Phys.*, **58**, 129–38.

Godin, O.A. and Palmer, D.R. (2008) *History of Russian Underwater Acoustics*, Singapore: World Scientific Publishing.

Godin, O.A., Zavorotny, V.U., Voronovich, A.G., and Goncharov, V.V. (2006) Refraction of sound in a horizontally inhomogeneous, time-dependent ocean. *IEEE J. Oceanic Engr.*, **31**, 384–401.

Goff, J.A. (1995) Quantitative analysis of sea ice draft: 1. Methods for stochastic modeling. *J. Geophys. Res.*, **100**(C4), 6993–7004, doi:10.1029/94JC03200.

Goh, J.T. and Schmidt, H. (1996) A hybrid coupled wave-number integration approach to range-dependent seismoacoustic modeling. *J. Acoust. Soc. Amer.*, **100**, 1409–20.

Goh, J.T., Schmidt, H., Gerstoft, P., and Seong, W. (1997) Benchmarks for validating range-dependent seismo-acoustic propagation codes. *IEEE J. Oceanic Engr.*, **22**, 226–36.

Gold, B.A. and Renshaw, W.E. (1978) Joint volume reverberation and biological measurements in the tropical Western Atlantic. *J. Acoust. Soc. Amer.*, **63**, 1809–19.

Goldberg, E.D., McCave, I.N., O'Brien, J.J., and Steele, J.H. (eds) (1977) *The Sea*, Vol. 6, Marine Modeling. New York, NY: Wiley Interscience.

Goldhahn, R., Braca, P., LePage, K.D., Willett, P, Marano, S., and Matta, V. (2014) Environmentally sensitive particle filter tracking in multistatic AUV networks with port-starboard ambiguity. Proc. IEEE ICASSP, Florence, Italy, pp. 1458–62.

Goldman, J. (1974) A model of broadband ambient noise fluctuations due to shipping. Bell Telephone Labs. Rept OSTP-31 JG.

Goldstein, C.I. (1984) The numerical solution of underwater acoustic propagation problems using finite difference and finite element methods. Nav. Res. Lab., Rept 8820.

Gong, Z., Jain, A.D., Tran, D., Yi, D.H., Wu, F., Zorn, A., Ratilal, P., and Makris, N.C. (2014) Ecosystem scale acoustic sensing reveals Humpback whale behavior synchronous with herring spawning processes and re-evaluation finds no effect of sonar on Humpback song occurrence in the Gulf of Maine in fall 2006. *PLOS ONE*, **9**(10), e104733, doi:10.1371/journal.pone.0104733.

Goni, G.J. and Johns, W.E. (2001) A census of North Brazil Current rings observed from TOPEX/POSEIDON altimetry: 1992–1998. *Geophys. Res. Lett.*, **28**, 1–4.

Gonzalez, R. and Hawker, K.E. (1980) The acoustic normal mode model NEMESIS. Appl. Res. Labs., Univ. Texas at Austin, USA, ARL-TR-80-13.

Gonzalez, R. and Payne, S.G. (1980) User's manual for NEMESIS and PLMODE. Appl. Res. Labs., Univ. Texas at Austin, USA, ARL-TM-80-6.

Goodman, R.R. and Farwell, R.W. (1979) A note on the derivation of the wave equation in an inhomogeneous ocean. *J. Acoust. Soc. Amer.*, **66**, 1895–6.

Gordon, D.F. (1964) Extension of the ray intensity procedure for underwater sound based on a profile consisting of curvilinear segments. Nav. Electron. Lab., Res. Rept 1217.

Gordon, D.F. (1979) Underwater sound propagation-loss program. Computation by normal modes for layered oceans and sediments. Nav. Ocean Syst. Ctr, Tech. Rept 393.

Gordon, D.F. and Bucker, H.P. (1984) Arctic acoustic propagation model with ice scattering. Nav. Ocean Syst. Ctr, Tech. Rept 985.

Gostev, V.S. and Shvachko, R.F. (2000) Caustics and volume prereverberation in a surface oceanic waveguide. *Acoust. Phys.*, **46**, 559–62.

Graff, J. (2004) e-Science potential in the ocean data environment. *Inter. Ocean Syst.*, **8**(6), 6–7.

Gragg, R.F. (1985) The broadband normal mode model PROTEUS. Appl. Res. Labs., Univ. Texas at Austin, USA, ARL-TM-85-6.

Grassle, J.F. (2000) The ocean biogeographic information system (OBIS): An on-line, worldwide atlas for accessing, modeling and mapping marine biological data in a multidimensional geographic context. *Oceanography*, **13**(3); 5–7.

Graver, J.G. (2005) Underwater gliders: Dynamics, control and design. PhD Dissertation, Princeton University, Princeton, NJ.

Greaves, R.J. and Stephen, R.A. (1997) Seafloor acoustic backscattering from different geological provinces in the Atlantic natural laboratory. *J. Acoust. Soc. Amer.,* **101**, 193–208.

Greaves, R.J. and Stephen, R.A. (2000) Low-grazing-angle monostatic acoustic reverberation from rough and heterogeneous seafloors. *J. Acoust. Soc. Amer.,* **108**, 1013–25.

Green, D.M. (1994) Sound's effects on marine mammals need investigation. *EOS, Trans. Amer. Geophys. Union*, **75**, 305–6.

Greene, C.H. and Wiebe, P.H. (1988) New developments in bioacoustical oceanography. *Sea Tech.*, **29**(8), 27–9.

Greene, R.R. (1984) The rational approximation to the acoustic wave equation with bottom interaction. *J. Acoust. Soc. Amer.*, **76**, 1764–73.

Grigorieva, N.S. and Fridman, G.M. (2008) Relationship between the axial wave and first normal modes for ducted propagation in a waveguide. *J. Comput. Acoust.*, **16**, 117–35.

Grilli, S., Pedersen, T., and Stepanishen, P. (1998) A hybrid boundary element method for shallow water acoustic propagation over an irregular bottom. *Eng. Anal. Bound. Elem.,* **21**, 131–45.

Guelton, S., Clorennec, D., Pardo, É., Brunet, P., and Folegot, T. (2013) Quonops©, la prévision opérationnelle en acoustique sous-marine sur grille de calcul. *Journées SUCCES,* **2013**, 1–4.

Guerra, F., Casari, P., and Zorzi, M. (2009) World ocean simulation system (WOSS): A simulation tool for underwater networks with realistic propagation modeling. *WUWNet'09,* the Fourth ACM International Workshop on Underwater Networks, Berkeley, California, USA (ACM 978-1-60558-821-6).

Guigné, J.Y. and Blondel, P. (2016) *Acoustic Investigation of Complex Seabeds,* New York, NY: Springer.

Guizzo, E. (2008) Defense contractors snap up submersible robot gliders. *IEEE Spectr.*, **45**(9) (North America), 11–12.

Haines, L.C., Renner, W.W., and Eller, A.I. (1988) Prediction system for acoustic returns from ocean bathymetry. Proc. MTS/IEEE Oceans 88 Conf., Baltimore, pp. 295–7.

Hall, H.R. and Watson, W.H. (1967) An empirical bottom reflection loss expression for use in sonar range prediction. Nav. Undersea Warfare Ctr, Tech. Note 10.

Hall, M.V. (1989) A comprehensive model of wind-generated bubbles in the ocean and predictions of the effects on sound propagation at frequencies up to 40 kHz. *J. Acoust. Soc. Amer.*, **86**, 1103–17.

Hall, M.V. (1995) Dimensions and units of underwater acoustic parameters. *J. Acoust. Soc. Amer.,* **97**, 3887–9.

Hall, M.V. (1996) Erratum: Dimensions and units of underwater acoustic parameters (J. Acoust. Soc. Am. 97, 3887–3889 [1995]). *J. Acoust. Soc. Amer.,* **100**, 673.

Hall, M.V. and Irving, M.A. (1989) Application of adiabatic mode theory to the calculation of horizontal refraction through a mesoscale eddy. *J. Acoust. Soc. Amer.*, **86**, 1465–77.

Hamilton, E.L. (1980) Geoacoustic modeling of the sea floor. *J. Acoust. Soc. Amer.*, **68**, 1313–40.

Hamilton, L.J., Anstee, S., Chapple, P.B., Hall, M.V., and Mulhearn, P.J. (2002) Scattering in the ocean. *Acoust. Australia*, **30**(2), 71–7.

Hamson, R.M. (1997) The modelling of ambient noise due to shipping and wind sources in complex environments. *Appl. Acoust.,* **51**, 251–87.

Hamson, R.M. and Ainslie, M.A. (1998) Broadband geoacoustic deduction. *J. Comput. Acoust.*, **6**, 45–59.

Hamson, R.M. and Wagstaff, R.A. (1983) An ambient-noise model that includes coherent hydrophone summation for sonar system performance in shallow water. SACLANT ASW Res. Ctr, Rept SR-70.

Han, Z., Sun, Y.L., and Shi, H. (2008) Cooperative transmission for underwater acoustic communications. IEEE Inter. Conf. Communications, ICC '08, Beijing, China, pp. 2028–32.

Hanna, J.S. (1976) Example of acoustic model evaluation and data interpretation. *J. Acoust. Soc. Amer.*, **60**, 1024–31.

Hanna, J.S. and Rost, P.V. (1981) Parabolic equation calculations versus North Pacific measurement data. *J. Acoust. Soc. Amer.*, **70**, 504–15.

Hanna, S.R. (1989) Plume dispersion and concentration fluctuations in the atmosphere. In *Encyclopedia of Environmental Control Technology*, ed. P.N. Cheremisinoff, Air Pollution Control, Vol. 2. Houston, TX: Gulf Publishing Company, pp. 547–82.

Hannay, D.E. and Racca, R.G. (2005) Acoustic Model Validation. Prepared by JASCO Research Ltd for Sakhalin Energy Investment Company Ltd. Doc. No. 0000-S-90-04-T-7006-00-E, Rev. 02.

Haralabus, G. and Baldacci, A. (2006) Impact of seafloor and water column parameter variations on signal-to-background ratio estimates—A sensitivity analysis based on sonar performance modeling. Proc. 8th European Conf. Underwater Acoustics, Carvoeiro, Portugal.

Hardin, R.H. and Tappert, F.D. (1973) Applications of the split-step Fourier method to the numerical solution of nonlinear and variable coefficient wave equations. *SIAM Rev.*, **15**, 423.

Harding, E.T. (1970) Geometry of wave-front propagation by integration of rays. Meteorology International, Inc.

Harlan, J.A., Georges, T.M., and Jones, R.M. (1991a) PROFILE—A program to generate profiles from HARPO/HARPA environmental models. Wave Propagation Lab., NOAA Tech. Memo. ERL WPL-198.

Harlan, J.A., Jones, R.M., and Georges, T.M. (1991b) PSGRAPH—A plotting program for PC-HARPO, PROFILE, CONPLT, and EIGEN. Wave Propagation Lab., NOAA Tech. Memo. ERL WPL-203.

Harrison, B.F., Vaccaro, R.J., and Tufts, D.W. (1998) Robust matched-field localization in uncertain ocean environments. *J. Acoust. Soc. Amer.*, **103**, 3721–4.

Harrison, C.H. (1989) Ocean propagation models. *Appl. Acoust.*, **27**, 163–201.

Harrison, C.H. (1996) Formulas for ambient noise level and coherence. *J. Acoust. Soc. Amer.*, **99**, 2055–66.

Harrison, C.H. (1997a) CANARY: A simple model of ambient noise and coherence. *Appl. Acoust.*, **51**, 289–315.

Harrison, C.H. (1997b) Noise directionality for surface sources in range-dependent environments. *J. Acoust. Soc. Amer.*, **102**, 2655–62.

Harrison, C.H. (1998) DINAMO: A noise directionality model suitable for operational use. Proc. Undersea Defence Technology Conf. UDT Europe, London, pp. 11–15.

Harrison, C.H. (2002a) Scattering strength uncertainty. *J. Acoust. Soc. Amer.*, **112**, 2253 (abstract).

Harrison, C.H. (2002b) SUPREMO: A multistatic sonar performance model. SACLANT Undersea Res. Ctr, Memo SM-396.

Harrison, C.H. (2003) Closed-form expressions for ocean reverberation and signal excess with mode stripping and Lambert's law. *J. Acoust. Soc. Amer.*, **114**, 2744–56.

Harrison, C.H. (2004a) Sub-bottom profiling using ocean ambient noise. *J. Acoust. Soc. Amer.*, **115**, 1505–15.

Harrison, C.H. (2004b) Geoacoustic inversion and subbottom profiling with ambient noise. *J. Acoust. Soc. Amer.*, **116**, 2558 (abstract).

Harrison, C.H. (2005) Closed form bistatic reverberation and target echoes with variable bathymetry and sound speed. *IEEE J. Oceanic Engr.*, **30**, 660–75.

Harrison, C. (2008) ARTEMIS: A fast general environment reverberation model. In *Proceedings of the International Symposium on Underwater Reverberation and Clutter*, eds P.L. Nielsen, C.H. Harrison, and J.-C. LeGac, Lerici, Italy.

Harrison, C.H. (2013) Ray convergence in a flux-like propagation formulation. *J. Acoust. Soc. Amer.,* **133,** 3777–89.

Harrison, C.H. (2015) Efficient modeling of range-dependent ray convergence effects in propagation and reverberation. *J. Acoust. Soc. Amer.,* **137,** 2982–5.

Harrison, C.H., Ainslie, M.A., and Packman, M.N. (1990) INSIGHT: A fast, robust propagation loss model providing clear understanding. Proc. Undersea Defence Technology Conf. UDT Europe, London, pp. 317–22.

Harrison, C.H., Brind, R., and Cowley, A. (1999) Computation of noise directionality and array response in range dependent media with CANARY. In *Theoretical and Computational Acoustics '97*, eds Y.-C. Teng, E.-C. Shang, Y.-H. Pao, M.H. Schultz, and A.D. Pierce, Singapore: World Scientific Publishing, pp. 527–51.

Harrison, C.H., Brind, R., and Cowley, A. (2001) Computation of noise directionality, coherence and array response in range dependent media with CANARY. *J. Comput. Acoust.*, **9**, 327–45.

Harrison, C.H. and Nielsen, P.L. (2007) Multipath pulse shapes in shallow water: Theory and simulation. *J. Acoust. Soc. Amer.,* **121,** 1362–73.

Harrison, C.H. and Siderius, M. (2008) Bottom profiling by correlating beam-steered noise sequences. *J. Acoust. Soc. Amer.,* **123,** 1282–96.

Harrison, C.H. and Simons, D.G. (2001) Geoacoustic inversion of ambient noise: A simple method. In *Acoustical Oceanography*, eds T.G. Leighton, G.J. Heald, H.D. Griffiths, and G. Griffiths. *Proc. Inst. Acoust.*, **23** (Part 2), 91–8.

Harrison, C.H. and Simons, D.G. (2002) Geoacoustic inversion of ambient noise: A simple method. *J. Acoust. Soc. Amer.*, **112**, 1377–89.

Hassab, J.C. (1989) *Underwater Signal and Data Processing*, Boca Raton, FL: CRC Press.

Hauge, L.H. and Hetland, F.I. (2015) Hydroacoustic channel emulator—HACE. Norwegian University of Science and Technology, NTNU—Trondheim.

Hawker, K.E. (1980) The use and evaluation of acoustic transmission loss models. Appl. Res. Labs., Univ. Texas at Austin, USA, ARL-TR-80-1.

Hawkins, J.D. (1992) TOTS: Three dimensional ocean thermal structure analysis. *Sea Technol.*, **33**(1), 62–3.

Hawkins, J.D., Clancy, R.M., and Price, J.F. (1993) Use of AVHRR data to verify a system for forecasting diurnal sea surface temperature variability. *Int. J. Remote Sens.*, **14**, 1347–57.

Hazelwood, R.A. and Connelly, J. (2005) Estimation of underwater noise—a simplified method. *Int. J. Soc. Underwater Technol.*, **26**(3), 97–103.

Heaney, K.D. and Campbell, R.L. (2013) Effective ice model for under-ice propagation using the fluid-fluid parabolic equation. *Proc. Meet. Acoust.*, **19**, 070052, doi:10.1121/1.4801397.

Heaney, K.D. and Campbell, R.L. (2016) Three-dimensional parabolic equation modeling of mesoscale eddy deflection. *J. Acoust. Soc. Amer.*, **139**, 918–26.

Heaney, K.D., Kuperman, W.A., and McDonald, B.E. (1991) Perth-Bermuda sound propagation (1960): Adiabatic mode interpretation. *J. Acoust. Soc. Amer.*, **90**, 2586–94.

Heathershaw, A.D., Stretch, C.E., and Maskell, S.J. (1991) Coupled ocean-acoustic model studies of sound propagation through a front. *J. Acoust. Soc. Amer.*, **89**, 145–55.

Heathershaw, A.D., Ward, P.D., and David, A.M. (2001) The environmental impact of underwater sound. *Proc. Inst. Acoust.*, **23**(4), 1–12.

Heindsmann, T.E., Smith, R.H., and Arneson, A.D. (1955) Effect of rain upon underwater noise levels. *J. Acoust. Soc. Amer.*, **27**, 378–9.

Heinemann, M., Larraza, A., and Smith, K.B. (2003) Experimental studies of applications of time-reversal acoustics to noncoherent underwater communications. *J. Acoust. Soc. Amer.,* **113**, 3111–6.

Heinis, F., de Jong, C.A.F., Ainslie M.A., Borst W., and Vellinga T. (2013) Monitoring programme for the Maasvlakte 2, Part III—The effects of underwater sound. *Terra. Et. Aqua.,* **132** (September), 21–32 (Quarterly publication of International Association of Dredging Companies).

Heitmeyer, R.M. (2006) A probabilistic model for the noise generated by breaking waves. *J. Acoust. Soc. Amer.,* **119**, 3676–93.

Helber, R.W., Barron, C.N., Carnes, M.R., and Zingarelli, R.A. (2008) Evaluating the sonic layer depth relative to the mixed layer depth. *J. Geophys. Res.*, **113**, C07033, doi:10.1029/2007JC004595.

Helble, T.A., D'Spain, G.L., Hildebrand, J.A., Campbell, G.S., Campbell, R.L., and Heaney, K.D. (2013) Site specific probability of passive acoustic detection of humpback whale calls from single fixed hydrophones. *J. Acoust. Soc. Amer.,* **134**, 2556–70.

Henrick, R.F. (1983) A cautionary note on the use of range-dependent propagation models in underwater acoustics. *J. Acoust. Soc. Amer.*, **73**, 810–12.

Hermand, J.-P., Meyer, M., Asch, M., and Berrada. M. (2006) Adjoint-based acoustic inversion for the physical characterization of a shallow water environment. *J. Acoust. Soc. Amer.,* **119**, 3860–71.

Hester, K.C., Peltzer, E.T., Kirkwood, W.J., and Brewer, P.G. (2008) Unanticipated consequences of ocean acidification: A noisier ocean at lower pH. *Geophys. Res. Lett.*, **35**, L19601, doi:10.1029/2008GL034913.

Hickling, R. (1999) Noise control and SI units. *J. Acoust. Soc. Amer.,* **106**, 3048.

Higham, C.J. and Tindle, C.T. (2003) Coupled perturbed modes over a sloping penetrable bottom. *J. Acoust. Soc. Amer.,* **114,** 3119–24.

Hildebrand, J.A. (2009) Anthropogenic and natural sources of ambient noise in the ocean. *Mar. Ecol. Prog. Ser.*, **395**, 5–20.

Hjelmervik, K.T. and Sandsmark, G.H. (2008) In ocean evaluation of low frequency active sonar systems. Proc. Acoustics '08 Conf., Paris, pp. 2839–43.

Hodges, P. (1987) Three decades by the numbers. *Datamation*, **33**(18), 77–87.

Hodgkiss, W.S. (1980) Reverberation model: I. Technical description and user's guide. Marine Physical Lab., Scripps Inst. Oceanogr., Tech. Memo. 319.

Hodgkiss, W.S. Jr. (1984) An oceanic reverberation model. *IEEE J. Oceanic Engr.*, **9**, 63–72.

Hodgkiss, W.S. Jr and Alexandrou, D. (1985) Under-ice reverberation rejection. *IEEE J. Oceanic Engr.*, **10**, 285–9.

Hodgkiss, W.S. Jr and Fisher, F.H. (1990) Vertical directionality of ambient noise at 32°N as a function of longitude and wind speed. *IEEE J. Oceanic Engr.*, **15**, 335–9.

Hoffman, D.W. (1976) LORA: A model for predicting the performance of long-range active sonar systems. Nav. Undersea Ctr, Tech. Pub. 541.

Hoffman, D.W. (1979) LIRA: A model for predicting the performance of low-frequency active-sonar systems for intermediate surveillance ranges. Nav. Ocean Syst. Ctr, Tech. Doc. 259.

Holbrook, W.S. (2013) Ten years of seismic oceanography: Accomplishments and challenges. *J. Acoust. Soc. Amer.*, **134**, 4088 (abstract).

Holbrook, W.S. and Fer, I. (2005) Ocean internal wave spectra inferred from seismic reflection transects. *Geophys. Res. Lett.*, **32**: L15604, doi:10.1029/2005GL023733.

Holbrook, W.S., Páramo, P., Pearse, S. and Schmitt, R.W. (2003). Thermohaline fine structure in an oceanographic front from seismic reflection profiling. *Science*, **301**, 821–4.

Holland, C.W. (2002) Geoacoustic inversion for fine-grained sediments. *J. Acoust. Soc. Amer.,* **111,** 1560–4.

Holland, C.W. (2006) Constrained comparison of ocean waveguide reverberation theory and observations. *J. Acoust. Soc. Amer.,* **120**, 1922–31.

Holland, C.W. (2007) Erratum: "Constrained comparison of ocean waveguide reverberation theory and observations" (*J. Acoust. Soc. Am.* 120(4), 1922–1931 [2006]). *J. Acoust. Soc. Amer.,* **121**, 1802.

Holland, C.W. (2010) Propagation in a waveguide with range-dependent seabed properties. *J. Acoust. Soc. Amer.,* **128**, 2596–609.

Holland, C.W. and Brunson, B.A. (1988) The Biot-Stoll sediment model: An experimental assessment. *J. Acoust. Soc. Amer.*, **84**, 1437–43.

Holland, C.W. and Neumann, P. (1998) Sub-bottom scattering: A modeling approach. *J. Acoust. Soc. Amer.,* **104**, 1363–73.

Holmes, E.S. (1988) Software requirements specification for the parabolic equation model. Sci. Appl. Inter. Corp., OAML-SRS-22.

Holmes, E.S. and Gainey, L.A. (1992a) Software requirements specification for the parabolic equation model version 3.4. Anal. & Technol., Inc. and Sci. Appl. Inter. Corp., OAML-SRS-22.

Holmes, E.S. and Gainey, L.A. (1992b) Software test description for the parabolic equation model version 3.4. Anal. & Technol., Inc. and Sci. Appl. Inter. Corp., OAML-STD-22.

Holmes, E.S. and Gainey, L.A. (1992c) Software design document for the parabolic equation model version 3.4. Anal. & Technol., Inc. and Sci. Appl. Inter. Corp., OAML-SDD-22.

Holmes, E.S., Miller, E.C., and Stephens, R.H. (1990) A PC-based acoustic model operating system. Proc. IEEE Oceans 90 Conf., Washington, DC, pp. 227–31.

Holt, R.M. (1985) FACT-10B version description document. ODSI Defense Syst., Inc., DSIR-PU-85-0124.

Houser, D.S. (2006) A method for modeling marine mammal movement and behavior for environmental impact assessment. *IEEE J. Oceanic Engr.*, **31**, 76–81.

Hovem, J.M. (1993) Mechanisms of bottom loss in underwater acoustics. In *Acoustic Signal Processing for Ocean Exploration*, eds J.M.F. Moura and I.M.G. Lourtie, Dordrecht, the Netherlands: Kluwer Academic Publishers, pp. 21–40.

Hovem, J.M. (2008) PlaneRay: An acoustic underwater propagation model based on ray tracing and plane-wave reflection coefficients. Forsvarets forskningsinstitutt/Norwegian Defence Research Establishment (FFI), FFI-rapport 2008/00610.

Hovem, J.M. and Dong, H. (2006) PlaneRay: An underwater acoustic propagation model using ray tracing and plane wave reflection coefficients. *J. Acoust. Soc. Amer.,* **120**, 3221 (abstract).

Hovem, J.M. and Knobles, D.P. (2002) A range-dependent propagation model based on a combination of ray theory and plane-wave reflection coefficients. *J. Acoust. Soc. Amer.,* **112**, 2393 (abstract).

Hovem, J.M. and Knobles, D.P. (2003) A range dependent propagation model based on a combination of ray theory and plane wave reflection coefficients. Proc. Tenth International Congress on Sound and Vibration, pp. 2593–2600.

Hovem, J.M., Richardson, M.D., and Stoll, R.D. (eds) (1991) *Shear Waves in Marine Sediments*, Dordrecht, the Netherlands: Kluwer Academic Publishers.

Hovem, J.M., Yan, S., Bao, X., and Dong, H. (2008) Modeling underwater communication links. Proc. Second Inter. Conf. Sensor Technologies and Applications (SENSORCOMM '08), IEEE Computer Society, Cap Esterel, France, pp 679–86. Sponsored by IARIA.

Hovland, M. (1988) Organisms: The only cause of scattering layers? *EOS, Trans. Amer. Geophys. Union*, **69**, 760.

Howard, R.J. and Clark, D. (2005) Propagation models and anti-submarine warfare (ASW) trainers. Proc. Interservice/Industry Training, Simulation and Education Conf., Orlando, FL.

Howard, R.J., Foreman, T., and Clark, D. (2000) Architectural and design considerations in propagation model selection and design. Simulation Interoperability Standards Organization, Proc. Simulation Interoperability Workshop, paper 00S-SIW-054.

HR Wallingford Ltd. (2016) HAMMER—Hydro-Acoustic model for mitigation and ecological response. Product description literature.

Huang, C.-F., Gerstoft, P., and Hodgkiss, W.S. (2006) Validation of statistical estimation of transmission loss in the presence of geoacoustic inversion uncertainty. *J. Acoust. Soc. Amer.*, **120**, 1932–41.

Huang, C.-F., Gerstoft, P., and Hodgkiss, W.S. (2007) On the effect of error correlation on matched-field geoacoustic inversion. *J. Acoust. Soc. Amer.*, **121**(2), EL64–9, doi:10.1121/1.2424267.

Hursky, P., Porter, M.B., Cornuelle, B.D., Hodgkiss, W.S., and Kuperman, W.A. (2004) Adjoint modeling for acoustic inversion. *J. Acoust. Soc. Amer.*, **115**, 607–19.

Ikpekha, O.W., Soberon, F., and Daniels, S. (2014) Modelling the propagation of underwater acoustic signals of a marine energy device using finite element method. Inter. Conf. Renewable Energies and Power Quality (ICREPQ'14), Cordoba, Spain.

Incze, B.I. and Dasinger, S.B. (2006) A Bayesian method for managing uncertainties relating to distributed multistatic sensor search. Proc. 9th International Conf. Information Fusion, ICIF '06, Florence, Italy.

Ingenito, F., Ferris, R.H., Kuperman, W.A., and Wolf, S.N. (1978) Shallow water acoustics summary report (first phase). Nav. Res. Lab., Rept 8179.

Institute of Electrical and Electronics Engineers, Inc. (1989) IEEE standard glossary of modeling and simulation terminology. IEEE Std 610.3-1989.

International Energy Agency. (1996) *Global Offshore Oil Prospects to 2000*, Paris, France: Organisation for Economic Co-operation and Development.

Isakson, M.J., Goldsberry, B., and Chotiros, N.P. (2014) A three-dimensional, longitudinally-invariant finite element model for acoustic propagation in shallow water waveguides. *J. Acoust. Soc. Amer.*, **136**(3), EL206–11, doi:10.1121/1.4890195.

Ivansson, S. (1994) Shear-wave induced transmission loss in a fluid-solid medium. *J. Acoust. Soc. Amer.*, **96**, 2870–5.

Ivansson, S. and Bishop, J. (2003) Travelling-wave representations of diffraction using leaky-mode Green function expansions. *J. Sound Vib.*, **262**, 1223–34.

Ivansson, S. and Karasalo, I. (1992) A high-order adaptive integration method for wave propagation in range-independent fluid-solid media. *J. Acoust. Soc. Amer.*, **92**, 1569–77.

Ivansson, S. and Karlsson, P.A. (2016) Sea-bed scattering and reflection contributions to the short-range acoustic impulse response: Measurements and modelling. *Hydroacoustics*, **19**, 153–64.

Ivansson, S., Abrahamsson, L., Karasalo, I., and Ainslie, M.A. (2009) Aspects on reverberation modelling and inversion with physical scattering kernels.Proc. Sixteenth International Congress on Sound and Vibration, ICSV16, Kraków, Poland.

Ivansson, S.M. (2006) Stochastic ray-trace computations of transmission loss and reverberation in 3-D range-dependent environments. Proc. 8th European Conf. Underwater Acoustics, Carvoeiro, Portugal, pp. 131–6.

Jackson, D.R. and Richardson, M.D. (2007) *High-Frequency Seafloor Acoustics*, New York: Springer-Verlag.

Jackson, D.R., Winebrenner, D.P., and Ishimaru, A. (1986) Application of the composite roughness model to high-frequency bottom backscattering. *J. Acoust. Soc. Amer.*, **79**, 1410–22.

Jacobs, G. (1974) Multiple profile restructuring and supplemental plot programs. Ocean Data Syst., Inc.

Jacobs, G. (1982) FACT 9-H: Version description document. Nav. Ocean Res. Devel. Activity, Tech. Note 133T.

Jagannathan, S., Bertsatos, I., Symonds, D., Chen, T., Nia, H.T., Jain, A.D. et al. (2009) Ocean acoustic waveguide remote sensing (OAWRS) of marine ecosystems. *Mar. Ecol. Prog. Ser.*, **395**, 137–60, doi:10.3354/meps08266.

Jain, S. and Ali, M.M. (2006) Estimation of sound speed profiles using artificial neural networks. *IEEE Geosci. Remote Sens. Lett.*, **3**, 467–70.

Jasny, M. (1999) *Sounding the Depths: Supertankers, Sonar, and the Rise of Undersea Noise*, New York, NY: Natural Resources Defense Council.

Jasny, M., Reynolds, J., Horowitz, C., and Wetzler, A. (2005) *Sounding the Depths II: The Rising Toll of Sonar, Shipping and Industrial Ocean Noise on Marine Life*, New York, NY: Natural Resources Defense Council.

Jaster, C.E. and Boehme, H. (1984) The MINERAY sonar performance prediction computer model baseline description. Appl. Res. Labs., Univ. Texas at Austin, USA, ARL-TR-84-9.

Jennette, R.L., Sander, E.L., and Pitts, L.E. (1978) The USI array noise model, Version I documentation. Underwater Syst., Inc. USI-APL-R-8.

Jensen, F. (1993) CW and pulse propagation modeling in ocean acoustics. In *Acoustic Signal Processing for Ocean Exploration*, eds J.M.F. Moura and I.M.G. Lourtie, Dordrecht, the Netherlands: Kluwer Academic Publishers, pp. 3–20.

Jensen, F.B. (1982) Numerical models of sound propagation in real oceans. Proc. MTS/IEEE Oceans 82 Conf., Washington, DC, pp. 147–54.

Jensen, F.B. (1984) Numerical models in underwater acoustics. In *Hybrid Formulation of Wave Propagation and Scattering*, ed. L.B. Felsen, Dordrecht, the Netherlands: Martinus Nijhoff, pp. 295–335.

Jensen, F.B. (1986) The art of generating meaningful results with numerical codes. *J. Acoust. Soc. Amer.*, **80** (Suppl. 1), S20–1 (abstract).

Jensen, F.B. (1988) Wave theory modeling: A convenient approach to CW and pulse propagation modeling in low-frequency acoustics. *IEEE J. Oceanic Engr.*, **13**, 186–97.

Jensen, F.B. (1998) On the use of stair steps to approximate bathymetry changes in ocean acoustic models. *J. Acoust. Soc. Amer.*, **104**, 1310–15.

Jensen, F.B. (2001) Acoustics, Shallow Water. In *Encyclopedia of Ocean Sciences*, eds J.H. Steele, S.A. Thorpe, and K.K. Turekian, San Diego, CA: Academic Press, pp. 89–96.

Jensen, F.B. (2008) Propagation and signal modeling. In *Handbook of Signal Processing in Acoustics*, eds D. Havelock, S. Kuwano, and M. Vorländer. New York, NY: Springer, pp. 1669–93.

Jensen, F.B. (2009) Acoustics, shallow water. In *Measurement Techniques, Sensors and Platforms. A Derivative of Encyclopedia of Ocean Sciences*, 2nd edn, eds J.H. Steele, S.A. Thorpe, and K.K. Turekian, San Diego, CA: Academic Press, pp. 417–24.

Jensen, F.B. and Ferla, M.C. (1979) SNAP: The SACLANTCEN normal-mode acoustic propagation model. SACLANT ASW Res. Ctr, Memo. SM-121.

Jensen, F.B. and Ferla, C.M. (1988) Numerical solutions of range-dependent benchmark problems in ocean acoustics. SACLANT Undersea Res. Ctr, Rept SR-141.

Jensen, F.B. and Ferla, C.M. (1990) Numerical solutions of range-dependent benchmark problems in ocean acoustics. *J. Acoust. Soc. Amer.*, **87**, 1499–1510.

Jensen, F.B., Ferla, C.M., LePage, K.D., and Nielsen, P.L. (2001) Acoustic models at SACLANTCEN: An update. SACLANT Undersea Res. Ctr, Rept SR-354.

Jensen, F.B., Ferla, C.M., Nielsen, P.L., and Martinelli, G. (2003) Broadband signal simulation in shallow water. *J. Comput. Acoust.*, **11**, 577–91.

Jensen, F.B. and Krol, H. (1975) The use of the parabolic equation method in sound propagation modelling. SACLANT ASW Res. Ctr, Memo. SM-72.

Jensen, F.B. and Kuperman, W.A. (1979) Environmental acoustical modelling at SACLANTCEN. SACLANT ASW Res. Ctr, Rept SR-34.

Jensen, F.B. and Kuperman, W.A. (1983) Optimum frequency of propagation in shallow water environments. *J. Acoust. Soc. Amer.*, **73**, 813–19.

Jensen, F.B., Kuperman, W.A., Porter, M.B., and Schmidt, H. (1994) *Computational Ocean Acoustics*, New York, NY: American Institute of Physics (Reprinted 1997 by Springer-Verlag).

Jensen, F.B., Kuperman, W.A., Porter, M.B., and Schmidt, H. (2011) *Computational Ocean Acoustics*, 2nd edn. New York, NY: Springer-Verlag.

Jensen, F.B. and Martinelli, G. (1985) Accurate PE calculations for range-dependent environments based on c_0 updates. *J. Acoust. Soc. Amer.*, **78** (Suppl. 1), S23 (abstract).

Jensen, F.B. and Schmidt, H. (1984) Review of numerical models in underwater acoustics, including recently developed fast-field program. SACLANT ASW Res. Ctr, Rept SR-83.

Jensen, F.B. and Schmidt, H. (1987) Subcritical penetration of narrow Gaussian beams into sediments. *J. Acoust. Soc. Amer.*, **82**, 574–9.

Jenserud, T. and Knudsen, T. (2004) *Rumble Final Report (DE 19).* Kjeller, Norway: Forsvarets forskningsinstitutt/Norwegian Defence Research Establishment, FFI/Rapport-2004/03327.

Jerzak, W., Siegmann, W.L., and Collins, M.D. (2005) Modeling Rayleigh and Stoneley waves and other interface and boundary effects with the parabolic equation. *J. Acoust. Soc. Amer.,* **117**, 3497–503.

Jiang, W., Guan, J., Zhou, H., and Wang, J. (2012) To construct an underwater acoustic communication software with the interface technology between Vc++6.0 and Matlab2007a. *Adv. Mat. Res.*, **403–8**, 2115–18.

Jiang, Y., Chapman, N.R., and DeFerrari, H.A. (2006) Geoacoustic inversion of broadband data by matched beam processing. *J. Acoust. Soc. Amer.,* **119**, 3707–16.

Jochens, A., Biggs, D., Engelhaupt, D., Gordon, J., Jaquet, N., Johnson, M. et al. (2006) Sperm whale seismic study in the Gulf of Mexico; Summary Report, 2002–2004. U.S. Dept. of the Interior, Minerals Management Service, Gulf of Mexico OCS Region, New Orleans, USA. OCS Study MMS 2006-034.

Johnson, H.R., Backus, R.H., Hersey, J.B., and Owen, D.M. (1956) Suspended echo-sounder and camera studies of midwater sound scatterers. *Deep-Sea Res.*, **3**, 266–72.

Johnson, M.V.R. Sr, McKeon, M.F., and Szanto, T.R. (1998) *Simulation based acquisition: A new approach. Report of the Military Research Fellows DSMC 1997–1998*, Fort Belvoir, VA: Defense Systems Management College Press. (Available from the US Government Printing Office [GPO 008-020-01461-3].)

Johnson, O.G. (1984) Three-dimensional wave equation computations on vector computers. *Proc. IEEE*, **72**, 90–5.

Jones, A.D., Duncan, A.J., and Maggi, A. (2013) An initial assessment of effects of seafloor roughness on coherent sound reflection from the seafloor. Proc. Acoustics 2013 Conf., Victor Harbor, Australia, pp. 1–7.

Jones, A.D., Duncan, A.J., Maggi, A., Clarke, P.A., and Sendt, J. (2010) Aspects of practical models of acoustic reflection loss at the ocean surface. Proc. 20th Inter. Congress on Acoustics, Sydney, Australia.

Jones, A.D., Maggi, A.L., Clarke, P.A., and Duncan, A.J. (2006) Analysis and simulation of an extended data set of waveforms received from small explosions in shallow oceans. Proc. Acoustics 2006 Conf., Christchurch, New Zealand, pp. 485–92.

Jones, A.D., Sendt, J.S., Clarke, P.A., and Exelby, J.R. (2002) Seafloor data for operational predictions of transmission loss in shallow ocean areas. *Acoust. Australia*, **30**(1), 27–31.

Jones, A.D., Sendt, J., Duncan, A.J., Clarke, P.A., and Maggi, A. (2009) Modelling the acoustic reflection loss at the rough ocean surface. Proc. Acoustics 2009 Conf., Adelaide, Australia.

Jones, D. and Marten, K. (2016) Dredging sound levels, numerical modelling and EIA. *Terra. Et. Aqua.*, **144** (September), 21–9 (Quarterly publication of International Association of Dredging Companies).

Jones, R.M. (1982) Algorithms for reflecting rays from general topographic surfaces in a ray tracing program. Wave Propagation Lab., NOAA Tech. Memo. ERL WPL-98.

Jones, R.M. (1983) A survey of underwater-acoustic ray tracing techniques. Wave Propagation Lab., NOAA Tech. Memo. ERL WPL-111.

Jones, R.M. and Georges, T.M. (1991) HARPX—A program to extend HARPO or HARPA ray paths in horizontally uniform media. Wave Propagation Lab., NOAA Tech. Memo. ERL WPL-201.

Jones, R.M., Georges, T.M., Nesbitt, L., and Weickmann, A. (1991) Ocean acoustic tomography inversion in the adiabatic-invariant approximation. Wave Propagation Lab., NOAA Tech. Memo. ERL WPL-217.

Jones, R.M., Riley, J.P., and Georges, T.M. (1982) A versatile three-dimensional Hamiltonian ray-tracing computer program for acoustic waves in the atmosphere. Wave Propagation Lab., NOAA Tech. Memo. ERL WPL-103.

Jones, R.M., Riley, J.P., and Georges, T.M. (1986) HARPO: A versatile three-dimensional Hamiltonian ray-tracing program for acoustic waves in an ocean with irregular bottom. Wave Propagation Lab., NOAA Tech. Rept.

Kalinowski, A.J. (1979) A survey of finite element-related techniques as applied to acoustic propagation in the ocean. Part I: Finite element method and related techniques. *Shock Vib. Dig.*, **11**(3), 9–16.

Kalogerakis, M.A., Skarsoulis, E.K., Papadakis, J.S., Flouri, E., Piperakis, G., Fountoulakis, R. et al. (2004) An integrated computer system for underwater acoustic detection analysis. Proc. Undersea Defence Technology Conf. UDT Europe, Nice, France.

Kampanis, N.A., Mitsoudis, D.A., and Dracopoulos, M.C. (2007) Benchmarking two simulation models for underwater and atmospheric sound propagation. *Environ. Model. Softw.*, **22**, 308–14.

Kanabis, W.G. (1975) A shallow water acoustic model for an ocean stratified in range and depth, Vol. I. Nav. Underwater Syst. Ctr, Tech. Rept 4887-I.

Kanabis, W.G. (1976) A shallow water acoustic model for an ocean stratified in range and depth, Vol. II. Nav. Underwater Syst. Ctr, Tech. Rept 4887-II.

Kapoor, T.K. and Schmidt, H. (1997) Acoustic scattering from a three-dimensional protuberance on a thin, infinite, submerged elastic plate. *J. Acoust. Soc. Amer.*, **102**, 256–65.

Karaboga, D. and Basturk, B. (2007) A powerful and efficient algorithm for numerical function optimization: Artificial bee colony (ABC) algorithm. *J. Glob. Optim.*, **39**, 459–71.

Katsnelson, B.G. and Petnikov, V.G. (2002) *Shallow-Water Acoustics*, New York, NY: Springer Praxis.

Katsnelson, B., Petnikov, V., and Lynch, J. (2012) *Fundamentals of Shallow Water Acoustics*, New York, NY: Springer.

Keenan, R.E. (2000) An introduction to GRAB eigenrays and CASS reverberation and signal excess. Proc. MTS/IEEE Oceans 2000 Conf., Providence, Rhode Island, pp. 1065–70.

Keenan, R.E. and Weinberg, H. (2001) Gaussian ray bundle (GRAB) model shallow water acoustic workshop implementation. *J. Comput. Acoust.*, **9**, 133–48.

Keenan, R.E., Weinberg, H., and Aidala, F.E. Jr. (1998) GRAB: Gaussian ray bundle (GRAB) eigenray propagation model version 2.0. Software Test Description. Naval Oceanographic Office, Stennis Space Center, MS.

Keiffer, R.S. and Novarini, J.C. (1990) A wedge assemblage method for 3-D acoustic scattering from sea surfaces: Comparison with a Helmholtz-Kirchhoff method. In *Computational Acoustics. Vol. 1: Ocean-Acoustic Models and Supercomputing*, eds D. Lee, A. Cakmak and R. Vichnevetsky, Amsterdam, the Netherlands: North-Holland, pp. 67–81.

Keller, J.B. and Papadakis, J.S. (eds) (1977) *Wave Propagation and Underwater Acoustics*, Lecture Notes in Physics, Vol. 70. New York, NY: Springer-Verlag.

Kelly, K.M. (2002) Variability effects due to shallow sediment gas in acoustic propagation: A case study from the Malta Plateau. In *Impact of Littoral Environmental Variability on Acoustic Predictions and Sonar Performance*, eds N.G. Pace and F.B. Jensen, Dordrecht, the Netherlands: Kluwer Academic Publishers, pp. 263–70.

Kennedy, M. C. and O'Hagan, A. (2001) Bayesian calibration of computer models. *J. Roy. Statist. Soc. Ser. B*, **63**, 425–64, doi:10.1111/1467-9868.00294.

Ker, S., Le Gonidec, Y., and Marié, L. (2016) Multifrequency seismic detectability of seasonal thermoclines assessed from ARGO data. *J. Geophys. Res. Oceans*, **121**, 6035–6060, doi:10.1002/2016JC011793.

Kerman, B.R. (ed.) (1988) *Sea Surface Sound. Natural Mechanisms of Surface Generated Noise in the Ocean*, NATO Advanced Science Institutes Series C: Mathematical and Physical Sciences 238, Dordrecht, the Netherlands: Kluwer Academic Publishers.

Kerman, B.R. (ed.) (1993) *Natural Physical Sources of Underwater Sound, Sea Surface Sound (2)*, Dordrecht, the Netherlands: Kluwer Academic Publishers.

Kerr, D.E. (ed.) (1951) *Propagation of Short Radio Waves*, MIT Radiation Laboratory Series, Vol. 13. New York, NY: McGraw-Hill.

Kerr, R.A. (1977) Oceanography: A closer look at Gulf Stream rings. *Science*, **198**, 387–430.

Kewley, D.J. and Bucker, H.P. (1987) A fast bistatic reverberation and systems model. *J. Acoust. Soc. Amer.*, **82** (Suppl. 1), S75 (abstract).

Khan, R., Gang, Q., and Ismail, A. (2013) Climatical characterization of northern Arabian Sea for OFDM based underwater acoustic communication. *Res. J. Appl. Sci. Engr. Technol.*, **6**, 1252–61.

Kibblewhite, A.C. and Ewans, K.C. (1985) Wave-wave interactions, microseisms and infrasonic ambient noise in the ocean. *J. Acoust. Soc. Amer.*, **78**, 981–94.

Kibblewhite, A.C., Shooter, J.A., and Watkins, S.L. (1976) Examination of attenuation at very low frequencies using the deep-water ambient noise field. *J. Acoust. Soc. Amer.*, **60**, 1040–7.

Kim, S., Kuperman, W.A., Hodgkiss, W.S., Song, H.C., Edelmann, G., and Akal, T. (2004) Echo-to-reverberation enhancement using a time reversal mirror. *J. Acoust. Soc. Amer.*, **115**, 1525–31.

King, D. and White, D. (1986a) FACT version 10B. Volume 1: Physics description. Nav. Ocean Res. Devel. Activity, Tech. Note 319.

King, D. and White, D. (1986b) FACT version 10B. Volume 2: Software description. Nav. Ocean Res. Devel. Activity, Tech. Note 320.

Kinsler, L.E., Frey, A.R., Coppens, A.B., and Sanders, J.V. (1982) *Fundamentals of Acoustics*, 3rd edn. New York, NY: John Wiley & Sons.

Klamper, A. (2007) Future net. *Sea Power*, **50** (2), 16–19.

Knight, W.C., Pridham, R.G., and Kay, S.M. (1981) Digital signal processing for sonar. *Proc. IEEE*, **69**, 1451–506.

Knobles, D.P. and Vidmar, P.J. (1986) Simulation of bottom interacting waveforms. *J. Acoust. Soc. Amer.*, **79**, 1760–6.

Knudsen, V.O., Alford, R.S., and Emling, J.W. (1948) Underwater ambient noise. *J. Mar. Res.*, **7**, 410–29.

Koch, R.A. and Knobles, D.P. (1995) A practical application of the Galerkin method to the broadband calculation of normal modes for underwater acoustics. *J. Acoust. Soc. Amer.*, **98**, 1682–98.

Koch, R.A. and Knobles, D.P. (2005) Geoacoustic inversion with ships as sources. *J. Acoust. Soc. Amer.*, **117**, 626–37.

Koch, R.A. and LeMond, J.E. (2001a) Software design description (SDD) for the NAUTILUS model, version 1.0. Appl. Res. Labs., Univ. Texas at Austin, USA, ARL- TL-EV-01-17.

Koch, R.A. and LeMond, J.E. (2001b) Software requirements specification for the NAUTILUS model (version 1.0). Appl. Res. Labs., Univ. Texas at Austin, USA, ARL- TL-EV-01-18.

Koch, R.A. and LeMond, J.E. (2001c) Software test description for the NAUTILUS model (version 1.0). Appl. Res. Labs., Univ. Texas at Austin, USA, ARL- TL-EV-01-19.

Koch, R.A., Rutherford, S.R., and Payne, S.G. (1983) Slope propagation: Mechanisms and parameter sensitivities. *J. Acoust. Soc. Amer.*, **74**, 210–18.

Korakas, A. and Hovem, J.M. (2013) Comparison of modeling approaches to low-frequency noise propagation in the ocean. Proc. MTS/IEEE Oceans 2013 Conf., Bergen, Norway, pp. 1–7.

Kormann, J., Cobo, P., Biescas, B., Sallarés, V., Papenberg, C., Recuero, M. et al. (2010) Synthetic modelling of acoustical propagation applied to seismic oceanography experiments. *Geophys. Res. Lett.*, **37**, L00D90, doi:10.1029/2009GL041763.

Koyunbakan, H. (2009) The transmutation method and Schrödinger equation with perturbed exactly solvable potential. *J. Comput. Acoust.*, **17**, 1–10.

Kraus, E.B. (ed.) (1977) *Modelling and Prediction of the Upper Layers of the Ocean,* New York, NY: Pergamon Press.

Kristiansen, U. (2010) Sound propagation in regions of variable oceanography—A summary of student work at NTNU. Norwegian Defence Research Establishment, FFI-rapport 2010/00812.

Krupin, V.D. (2005) Application of the WKB method to calculating the group velocities and attenuation coefficients of normal waves in the Arctic underwater waveguide. *Acoust. Phys.*, **51**, 313–20.

Kuhl, F., Weatherly, R., and Dahmann, J. (1999) *Creating Computer Simulation Systems. An Introduction to the High Level Architecture*, Upper Saddle River, NJ: Prentice Hall PTR.

Kumagai, J. (2006) Drowning in sound. *IEEE Spectrum*, **43**(4) (North America), 54–60.

Kumar, P.V.H., Kumar, T.P., Sunil, T., and Gopakumar, M. (2005) Observations on the relationship between scattering layer and mixed layer. *Curr. Sci.*, **88**, 1799–802.

Kuo, E.Y.T. (1988) Sea surface scattering and propagation loss: Review, update, and new predictions. *IEEE J. Oceanic Engr.,* **13**, 229–34.

Kuperman, W.A. (1985) Models of sound propagation in the ocean. *Nav. Res. Rev.*, **37**(3), 32–41.

Kuperman, W.A., Hodgkiss, W.S., Song, H.C., Akal, T., Ferla, C., and Jackson, D.R. (1998) Phase conjugation in the ocean: Experimental demonstration of an acoustic time-reversal mirror. *J. Acoust. Soc. Amer.,* **103**, 25–40.

Kuperman, W.A. and Ingenito, F. (1980) Spatial correlation of surface generated noise in a stratified ocean. *J. Acoust. Soc. Amer.*, **67**, 1988–96.

Kuperman, W.A. and Lynch, J.F. (2004) Shallow-water acoustics. *Physics Today*, **57**(10), 55–61.

Kuperman, W.A., Porter, M.B., and Perkins, J.S. (1987) Three-dimensional oceanographic acoustic modeling of complex environments. *J. Acoust. Soc. Amer.*, **82** (Suppl. 1), S42 (abstract).

Kuperman, W.A., Porter, M.B., Perkins, J.S., and Piacsek, A.A. (1988) Rapid three-dimensional ocean acoustic modeling of complex environments. In *Proceedings of the IMACS 12th World Congress on Scientific Computation*, eds R. Vichnevetsky, P. Borne, and J. Vignes, Villeneuve d'Ascq, France: GERFIDN—Cite Scientifique, pp. 231–3.

Kuperman, W.A., Porter, M.B., Perkins, J.S., and Evans, R.B. (1991) Rapid computation of acoustic fields in three-dimensional ocean environments. *J. Acoust. Soc. Amer.*, **89**, 125–33.

Küsel, E.T., Siegmann, W.L., and Collins, M.D. (2007) A single-scattering correction for large contrasts in elastic layers. *J. Acoust. Soc. Amer.,* **121**, 808–13.

Kutschale, H.W. (1973) Rapid computation by wave theory of propagation loss in the Arctic Ocean. Lamont-Doherty Geol. Obs., CU-8-73.

Kutschale, H.W. (1984) Arctic marine acoustics. Lamont-Doherty Geol. Obs., Final Rept, ONR Contr, N00014-80-C-0021.

Kutschale, H.W. and DiNapoli, F.R. (1977) Pulse propagation in the ocean. II: The fast field program method. *J. Acoust. Soc. Amer.*, **62** (Suppl. 1), S18 (abstract).

Kutschale, H.W. and Lee, T. (1983) Bottom-interacting acoustic signals in the Arctic channel: Long-range propagation. *J. Acoust. Soc. Amer.*, **74** (Suppl. 1), S1 (abstract).

Kwon, Y.W. (2016) *Multiphysics and Multiscale Modeling: Techniques and Applications*, Boca Raton, FL: CRC Press.

Labianca, F.M. (1973) Normal modes, virtual modes, and alternative representations in the theory of surface-duct sound propagation. *J. Acoust. Soc. Amer.*, **53**, 1137–47.

Lacaze, B. (2007) A stochastic model for acoustic attenuation. *Wave. Random Complex Media*, **17**, 343–56.

LaFond, E.C. (1962) Internal waves. Part I. In *The Sea*, ed. M.N. Hill, Physical Oceanography, Vol. 1. New York, NY: Interscience Publishers, pp. 731–51.

Lai, C.-C.A., Qian, W., and Glenn, S.M. (1994) Data assimilation and model evaluation experiment datasets. *Bull. Amer. Meteor. Soc.*, **75**, 793–809.

Lai, D.Y. and Richardson, P.L. (1977) Distribution and movement of Gulf Stream rings. *J. Phys. Oceanogr.*, **7**, 670–83.

Lam, F.-P.A., Konijnendijk, N.J., Groen, J., and Simons, D.G. (2006) Non-Rayleigh wideband sonar reverberation modeling including hybrid multi-paths. Proc. MTS/IEEE Oceans 2006 Conf., Boston.

Lam, F.-P.A., Haley, P.J. Jr., Janmaat, J., Lermusiaux, P.F.J., Leslie, W.G., Schouten et al. (2009). At-sea real-time coupled four-dimensional oceanographic and acoustic forecasts during Battlespace Preparation 2007. *J. Marine Syst.*, **78** (Suppl. 1), S306–20. doi:10.1016/j.jmarsys.2009.01.029.

Lamb, P.J. (1984) On the mixed-layer climatology of the north and tropical Atlantic. *Tellus*, **36A**, 292–305.

Land, P.E., Shutler, J.D., Findlay, H.S., Girard-Ardhuin, F., Sabia, R., Reul, N. et al. (2015) Salinity from space unlocks satellite-based assessment of ocean acidification. *Environ. Sci. Technol.*, **49**, 1987–94, doi:10.1021/es504849s.

Langmuir, I. (1938) Surface motion of water induced by wind. *Science*, **87**, 119–23.

Lapinski, A.-L.S. and Chapman, D.M.F. (2005) The effects of ignored seabed variability in geoacoustic inversion. *J. Acoust. Soc. Amer.*, **117**, 3524–38.

Lasky, M. and Colilla, R. (1974) FANM-I fast ambient noise model. Program documentation and user's guide. Ocean Data Syst., Inc.

Lau, R.L., Lee, D., and Robinson, A.R. (eds) (1993) *Computational Acoustics. Vol. 1: Scattering, Supercomputing and Propagation*. Amsterdam, the Netherlands: North-Holland.

Lauer, R.B. (1979) Acoustic model evaluation: Issues and recommendations incorporating the experience of the panel on sonar system models (POSSM). Nav. Underwater Syst. Ctr, Tech. Rept 6025.

Lauer, R.B. and Sussman, B. (1976) A methodology for the comparison of models for sonar system applications, Vol. I. Nav. Sea Syst. Command, SEA 06H1/036-EVA/MOST-10.

Lauer, R.B. and Sussman, B. (1979) A methodology for the comparison of models for sonar system applications—Results for low frequency propagation loss in the Mediterranean Sea, Vol. II. Nav. Sea Syst. Command, SEA 06H1/036-EVA/MOST-11.

Laurinolli, M.H., Tollefsen, C.D.S., Carr, S.A., and Turner, S.P. (2005) Assessment of the effects of underwater noise from the proposed Neptune LNG project. Part (3): Noise sources of the Neptune project and propagation modeling of underwater noise. Prepared by JASCO Research Ltd for LGL Ltd, LGL Rept TA4200-3.

Laville, F., Abbott, G.D., and Miller, M.J. (1991) Underwater sound generation by rainfall. *J. Acoust. Soc. Amer.*, **89**, 715–21.

Law, A.M. and Kelton, W.D. (1991) *Simulation Modeling and Analysis*, 2nd edn. New York, NY: McGraw-Hill.

Lawson, J.W. (2009) The use of sound propagation models to determine safe distances from a seismic sound energy source. Dept. Fisheries and Oceans, Canadian Science Advisory Secretariat, Res. Doc. 2009/060.

Lee, D. (1983) Effective methods for predicting underwater acoustic wave propagation. Proc. AIAA 8th Aeroacoustics Conf., Atlanta, Paper No. 83-0683.

Lee, D. and Botseas, G. (1982) IFD: An implicit finite-difference computer model for solving the parabolic equation. Nav. Underwater Syst. Ctr, Tech. Rept 6659.

Lee, D., Botseas, G., and Papadakis, J.S. (1981) Finite-difference solution to the parabolic wave equation. *J. Acoust. Soc. Amer.*, **70**, 795–800.

Lee, D., Botseas, G., and Siegmann, W.L. (1992) Examination of three-dimensional effects using a propagation model with azimuth-coupling capability (FOR3D). *J. Acoust. Soc. Amer.*, **91**, 3192–202.

Lee, D., Cakmak, A., and Vichnevetsky, R. (eds) (1990a) *Computational Acoustics. Vol. 1: Ocean-Acoustic Models and Supercomputing*, Amsterdam, the Netherlands: North-Holland.

Lee, D., Cakmak, A., and Vichnevetsky, R. (eds) (1990b) *Computational Acoustics. Vol. 2: Scattering, Gaussian Beams, and Aeroacoustics*, Amsterdam, the Netherlands: North-Holland.

Lee, D., Cakmak, A., and Vichnevetsky, R. (eds) (1990c) *Computational Acoustics. Vol. 3: Seismo-Ocean Acoustics and Modeling*, Amsterdam, the Netherlands: North-Holland.

Lee, D. and McDaniel, S.T. (1983) Wave field computations on the interface: An ocean acoustic model. *Math. Model.*, **4**, 473–88.

Lee, D. and McDaniel, S.T. (1987) Ocean acoustic propagation by finite difference methods. *Comp. Maths. Applic.*, **14**(5), 305–423.

Lee, D. and Papadakis, J.S. (1980) Numerical solutions of the parabolic wave equation: An ordinary-differential-equation approach. *J. Acoust. Soc. Amer.*, **68**, 1482–8.

Lee, D. and Pierce, A.D. (1995) Parabolic equation development in recent decade. *J. Comput. Acoust.*, **3**, 95–173.

Lee, D., Pierce, A.D., and Shang, E.-C. (2000) Parabolic equation development in the twentieth century. *J. Comput. Acoust.*, **8**, 527–637.

Lee, D., Saad, Y., and Schultz, M.H. (1988) An efficient method for solving the three-dimensional wide angle wave equation. In *Computational Acoustics, Vol. 1, Wave Propagation*, eds D. Lee, R.L. Sternberg, and M.H. Schultz, Amsterdam, the Netherlands: North-Holland, pp. 75–89.

Lee, D. and Schultz, M.H. (1995) *Numerical Ocean Acoustic Propagation in Three Dimensions*, Singapore: World Scientific Publishing.

Lee, D. and Siegmann, W.L. (1986) A mathematical model for the 3-dimensional ocean sound propagation. *Math. Model.*, **7**, 143–62.

Lee, D., Vichnevetsky, R., and Robinson, A.R. (eds) (1993) *Computational Acoustics. Vol. 2: Acoustic Propagation*, Amsterdam, the Netherlands: North-Holland.

Lee, K., Chu, Y., and Seong, W. (2013) Geometrical ray-bundle reverberation modeling. *J. Comput. Acoust.*, **21**(3), 1–17, doi:10.1142/S0218396X13500112.

Lee, P.W.Q. and Seah, W.K.G. (2007) A comparison of two data delivery schemes for underwater sensor networks. Proc. Oceans '07 Conf., Aberdeen, Scotland.

Le Gall, Y. (2015) Problèmes inverses en acoustique sous-marine: Prédiction de performances et localisation de source en environnement incertain. Thèse de Doctorat, Université de Bretagne Occidentale, Français.

Le Gall, Y. and Bonnel, J. (2013) Passive estimation of the waveguide invariant per pair of modes. *J. Acoust. Soc. Amer.*, **134**(2), EL230–6, doi:10.1121/1.4813846.

Lei, B., Yang, K., and Ma, Y. (2014) Forward scattering detection of a submerged object by a vertical hydrophone array. *J. Acoust. Soc. Amer.*, **136**, 2998–3007.

Leibiger, G.A. (1968) Wave propagation in an inhomogeneous medium with slow spatial variation. Ph.D. Dissertation, Stevens Institute of Technology, Hoboken, NJ.

Leibiger, G.A. (1977) Criteria for propagation loss model assessment for APP application. Nav. Underwater Syst. Ctr, Tech. Memo. 771245.

Leibovich, S. (1983) The form and dynamics of Langmuir circulations. *Ann. Rev. Fluid Mech.*, **15**, 391–427.

Leighton, T.G. (1994) *The Acoustic Bubble*, San Diego, CA: Academic Press.

Leighton, T.G. (2012) The use of extra-terrestrial oceans to test ocean acoustics students. *J. Acoust. Soc. Amer.*, **131**, 2551–5.

Lemon, D.D. and Duddridge, G. (1987) A numerical model for the calibration of CASP WOTAN wind data. Proc. MTS/IEEE Oceans 87 Conf., Halifax, Nova Scotia, pp. 167–71.

LeMond, J.E. and Koch, R.A. (1997) Finite correlation and coherent propagation effects in the normal-mode description of bottom reverberation. *J. Acoust. Soc. Amer.*, **102**, 266–77.

LePage, K. (1999) Bottom reverberation in shallow water: Coherent properties as a function of bandwidth, waveguide characteristics, and scatterer distributions. *J. Acoust. Soc. Amer.*, **106**, 3240–54.

LePage, K.D. (2002) Bistatic reverberation modeling for range-dependent waveguides. *J. Acoust. Soc. Amer.*, **112**, 2253–4 (abstract).

LePage, K. (2003) Monostatic reverberation in range dependent waveguides: The R-SNAP model. SACLANT Undersea Res. Ctr, Rept SR-363.

LePage, K.D. (2004) Statistics of broad-band bottom reverberation predictions in shallow-water waveguides. *IEEE J. Oceanic Engr.*, **29**, 330–46.

LePage, K.D. (2010) Higher moment estimation for shallow-water reverberation prediction. *IEEE J. Oceanic Engr.*, **35**, 185–98.

LePage, K.D. and Harrison, C.H. (2003a) Bistatic reverberation benchmarking exercise: BiStaR versus analytic formulas. *J. Acoust. Soc. Amer.*, **113**, 2333–4 (abstract).

LePage, K.D. and Harrison, C. (2003b) Effects of refraction on the prediction of bistatic reverberation in range dependent shallow water waveguides. *J. Acoust. Soc. Amer.*, **114**, 2302 (abstract).

LePage, K. and Schmidt, H. (1994) Modeling of low-frequency transmission loss in the central Arctic. *J. Acoust. Soc. Amer.*, **96**, 1783–95.

LePage, K.D., Neumann, P., and Holland, C.W. (2006) Broad-band time domain modeling of sonar clutter in range dependent waveguides. Proc. MTS/IEEE Oceans 2006 Conf., Boston.

Leroy, C.C. (1969) Development of simple equations for accurate and more realistic calculation of the speed of sound in seawater. *J. Acoust. Soc. Amer.*, **46**, 216–26.

Leroy, C.C. (2007) Erratum: "Depth-pressure relationships in the oceans and seas" (*J. Acoust. Soc. Am.* 103(3), 1346–1352 [1998]). *J. Acoust. Soc. Amer.*, **121**, 2447.

Leroy, C.C. and Parthiot, F. (1998) Depth-pressure relationships in the oceans and seas. *J. Acoust. Soc. Amer.*, **103**, 1346–52.

Leroy, C.C., Robinson, S.P., and Goldsmith, M.J. (2008) A new equation for the accurate calculation of sound speed in all oceans. *J. Acoust. Soc. Amer.*, **124**, 2774–82.

Leroy, C.C., Robinson, S.P., and Goldsmith, M.J. (2009) Erratum: "A new equation for the accurate calculation of sound speed in all oceans" (*J. Acoust. Soc. Am.* 124, 2774–2783 [2008]). *J. Acoust. Soc. Amer.*, **126**, 2117.

Levinson, S.J., Westwood, E.K., Koch, R.A., Mitchell, S.K., and Sheppard, C.V. (1995) An efficient and robust method for underwater acoustic normal-mode computations. *J. Acoust. Soc. Amer.*, **97**, 1576–85.

Levitus, S. (1982) Climatological atlas of the world ocean. NOAA Professional Paper 13.

Lewis, E.L. (1980) The practical salinity scale 1978 and its antecedents. *IEEE J. Oceanic Engr.*, **OE-5**, 3–8.

Lewis, J.K. and Denner, W.W. (1988) Higher frequency ambient noise in the Arctic Ocean. *J. Acoust. Soc. Amer.*, **84**, 1444–55.

Li, B., Chiong, R., and Gong, L.-G. (2014) Search-evasion path planning for submarines using the artificial bee colony algorithm. Proc. 2014 IEEE Congress on Evolutionary Computation (CEC), Beijing, China, pp. 528–35.

Li, F.-H. and Liu, J.-J. (2002) Bistatic reverberation in shallow water: Modelling and data comparison. *Chinese Phys. Lett.*, **19**, 1128–30.

Li, M. and Garrett, C. (1997) Mixed layer deepening due to Langmuir circulation. *J. Phys. Oceanogr.*, **27**, 121–32.

Li, M., Zahariev, K., and Garrett, C. (1995) Role of Langmuir circulation in the deepening of the ocean surface mixed layer. *Science*, **270**, 1955–7.

Li, Y. and Çalişal, S.M. (2010) Numerical analysis of the characteristics of vertical axis tidal current turbines. *Renew. Energ.*, **35**, 435–42, doi:10.1016/j.renene. 2009. 05.024.

Lin, Y.-T., Duda, T.F., and Newhall, A.E. (2013) Three-dimensional sound propagation models using the parabolic-equation approximation and the split-step Fourier method. *J. Comput. Acoust.*, **21**(1), 1–24, doi:10.1142/S0218396X1250018X.

Lingevitch, J.F. (2008) A parabolic equation method for modeling rough interface reverberation. In *Proceedings of the International Symposium on Underwater Reverberation and Clutter*, eds P.L. Nielsen, C.H. Harrison, and J.-C. LeGac, Lerici, Italy.

Lingevitch, J.F. and Collins, M.D. (1998) Wave propagation in range-dependent poro-acoustic waveguides. *J. Acoust. Soc. Amer.*, **104**, 783–90.

Lingevitch, J.F., Collins, M.D., Mills, M.J., and Evans, R.B. (2002a) A two-way parabolic equation that accounts for multiple scattering. *J. Acoust. Soc. Amer.*, **112**, 476–80.

Lingevitch, J.F., Song, H.C., and Kuperman, W.A. (2002b) Time reversed reverberation focusing in a waveguide. *J. Acoust. Soc. Amer.*, **111**, 2609–14.

Linton, L. (2016) Iterative and adaptive processing for multiuser communication systems. PhD thesis, Victoria University, Melbourne, Australia.

Linton, L., Conder, P., and Faulkner, M. (2008) Multiuser communications for underwater acoustic networks using MIMO-OFDM-IDMA. Proc. Second Inter. Conf. Signal Processing and Communication Systems, ICSPCS 2008, Gold Coast, Australia.

Linton, L., Conder, P., and Faulkner, M. (2009) Multiple-access communications for underwater acoustic sensor networks using OFDM-IDMA. Proc. MTS/IEEE Oceans 2009 Conf., Biloxi, MS.

Lippert, S., Nijhof, M., Lippert, T., Wilkes, D., Gavrilov, A., Heitmann, K. et al. (2016) COMPILE—A generic benchmark case for predictions of marine pile-driving noise. *IEEE J. Oceanic Engr.*, **41**, 1061–71.

Lippert, T., Heitmann, K., Ruhnau, M., Lippert, S., and von Estorff, O. (2014) On the estimation of prediction accuracy in numerical offshore pile driving noise modelling. Proc. INTER.NOISE 2014 Conf., 43rd International Congress on Noise Control Engineering, Melbourne, Australia, pp. 1–7.

Lippert, T. and von Estorff, O. (2014) The significance of parameter uncertainties for the prediction of offshore pile driving noise. *J. Acoust. Soc. Amer.*, 136, 2463–71.

Liu, J., Huang, Y., Sun, H., and Xiao, M. (2013) High-order numerical methods for wave equations with van der Pol type boundary conditions. Proc. SIAM Conf. Control and Applications (CT13), San Diego, pp. 144–51.

Liu, J., Huang, Y., Sun, H., and Xiao, M. (2016) Numerical methods for weak solution of wave equation with van der Pol type nonlinear boundary conditions. *Numer. Methods Partial Differ. Eq.*, **32**, 373–98, doi:10.1002/num.21997.

Liu, J.-J., Li, F.-H., and Peng, Z.-H. (2003) Stochastic inversion of seabottom scattering coefficients from shallow-water reverberation. *Chinese Phys. Lett.*, **20**, 2188–91.

Liu, Q., Li, F., Guo, L., Gong, Z., and Li, X. (2001) Applications of BDRM theory to numerical predictions of acoustic transmission losses in shallow water. *Acta Acust.* **26**, 410–16 (In Chinese).

Lloyd, T.P., Turnock, S.R., and Humphrey, V.F. (2011) Modelling techniques for underwater noise generated by tidal turbines in shallow waters. Proc. 30th Inter. Conf. Ocean, Offshore and Arctic Engr. (OMAE2011), Rotterdam, the Netherlands, pp. 1–9.

Locklin, J.H. and Etter, P.C. (1988) Sonar model operating system functional requirement. SYNTEK Engr. Comp. Syst., Inc.

Locklin, J. and Webster, J. (1980) NORDA model operating system functional description and FY 80 implementation plan. Ocean Data Syst., Inc.

Long, D. (1979) FANM/SIAM noise plot program. Ocean Data Syst., Inc.

Long, W., Yang, Z., Copping, A., Jung, K.W., and Deng, Z.D. (2015) The development of a finite volume method for modeling sound in coastal ocean environment. Proc. MTS/IEEE Oceans 2015 Conf., Washington, DC.

Love, R.H. (1975) Predictions of volume scattering strengths from biological trawl data. *J. Acoust. Soc. Amer.*, **57**, 300–6.

Love, R.H. (1978) Resonant acoustic scattering by swimbladder-bearing fish. *J. Acoust. Soc. Amer.*, **64**, 571–80.

Love, R.H. (1993) A comparison of volume scattering strength data with model calculations based on quasisynoptically collected fishery data. *J. Acoust. Soc. Amer.*, **94**, 2255–68.

Love, R.H., Thompson, C.H., and Nero, R.W. (1996) Volume reverberation in littoral waters. *J. Acoust. Soc. Amer.*, **100**, 2799 (abstract).

Lovett, J.R. (1978) Merged seawater sound-speed equations. *J. Acoust. Soc. Amer.*, **63**, 1713–18.

Lu, Y.Y. and Zhu, J. (2007) Perfectly matched layer for acoustic waveguide modeling—benchmark calculations and perturbation analysis. *CMES*, **22**, 235–47.

Luby, J.C. and Lytle, D.W. (1987) Autoregressive modeling of nonstationary multibeam sonar reverberation. *IEEE J. Oceanic Engr.*, **12**, 116–29.

Lucke, K., Lepper, P.A., Blanchet, M.-A., and Siebert, U. (2011) The use of an air bubble curtain to reduce the received sound levels for harbor porpoises (Phocoena phocoena). *J. Acoust. Soc. Amer.*, **130**, 3406–12.

Ludwig, D. (1966) Uniform asymptotic expansions at a caustic. *Commun. Pure Appl. Math.*, **19**, 215–50.

Lukas, I.J., Hess, C.A., and Osborne, K.R. (1980a) ASERT/ASEPS version 4.1 FNOC user's manual. Ocean Data Syst., Inc.

Lukas, I.J., Hess, C.A., and Osborne, K.R. (1980b) DANES/ASEPS version 4.1 FNOC user's manual. Ocean Data Syst., Inc.

Lurton, X. (1992) The range-averaged intensity model: A tool for underwater acoustic field analysis. *IEEE J. Oceanic Engr.*, **17**, 138–49.

Lurton, X. (2002) *An Introduction to Underwater Acoustics: Principles and Applications*, New York, NY: Springer-Verlag.

Lurton, X. (2016) Modelling of the sound field radiated by multibeam echosounders for acoustical impact assessment. *Appl. Acoust.*, **101**, 201–21.

Lynch, J. and Tang, D. (2008) Overview of Shallow Water 2006 *JASA EL* special issue papers. *J. Acoust. Soc. Amer.*, **124**(3), EL63–5, doi:10.1121/1.2972156.

Lynch, J.F., Newhall, A.E., Chiu, C.-S., and Miller, J.H. (1994) Three-dimensional ray acoustics in a realistic ocean. In *Oceanography and Acoustics: Prediction and Propagation Models*, eds A.R. Robinson and D. Lee, New York, NY: American Institute of Physics, Chapter 9, pp 198–232.

Lyons, A.P., Anderson, A.L., and Dwan, F.S. (1994) Acoustic scattering from the seafloor: Modeling and data comparison. *J. Acoust. Soc. Amer.*, **95**, 2441–51.

Ma, B. B. and Nystuen, J.A. (2005) Passive acoustic detection and measurement of rainfall at sea. *J. Atmos. Oceanic Technol.*, **22**, 1225–48.

Ma, B.B., Nystuen, J.A., and Lien, R.-C. (2005) Prediction of underwater sound levels from rain and wind. *J. Acoust. Soc. Amer.*, **117**, 3555–65.

Macaskill, C. and Ewart, T.E. (1996) Numerical solution of the fourth moment equation for acoustic intensity correlations and comparison with the mid-ocean acoustic transmission experiment. *J. Acoust. Soc. Amer.*, **99**, 1419–29.

MacGillivray, A., Warner, G., Racca, R., and O'Neill, C. (2011) Tappan Zee Bridge construction hydroacoustic noise modeling. Prepared by JASCO Applied Sciences for AECOM, New York. Final Report, P001116-001, Version 1.0.

MacGillivray, A.O. (2006) An acoustic modelling study of seismic airgun noise in Queen Charlotte Basin. M.S. thesis, University of Victoria, British Columbia, Canada.

Mackenzie, K.V. (1961) Bottom reverberation for 530- and 1030-cps sound in deep water. *J. Acoust. Soc. Amer.*, **33**, 1498–504.

Mackenzie, K.V. (1981) Nine-term equation for sound speed in the oceans. *J. Acoust. Soc. Amer.*, **70**, 807–12.
Macpherson, J.D. and Daintith, M.J. (1967) Practical model of shallow-water acoustic propagation. *J. Acoust. Soc. Amer.*, **41**, 850–4.
Mahdy, A. (2008a) A perspective on marine wireless sensor networks. *J. Comput. Sci. Colleges*, **23**, 89–96.
Mahdy, A.M. (2008b) Marine wireless sensor networks: Challenges and applications. Proc. Seventh Inter. Conf. Networking (ICN 2008), Cancun, Mexico, pp. 530–5.
Mahler, J.I., Sullivan, F.J.M., and Moll, M. (1975) Statistical methodology for the estimation of noise due to shipping in small sectors and narrow bands. Bolt, Beranek and Newman, Inc., Tech. Memo. W273.
Makarov, D., Prants, S., Virovlyansky, A., and Zaslavsky, G. (2010) *Ray and Wave Chaos in Ocean Acoustics: Chaos in Waveguides*. Singapore: World Scientific Publishing.
Makarov, D.V., Uleysky, M. Yu., and Prants, S.V. (2004) Ray chaos and ray clustering in an ocean waveguide. *Chaos*, **14**, 79–95.
Makhija, D., Kumaraswamy, P., and Roy, R. (2006) Challenges and design of MAC protocol for underwater acoustic sensor networks. Proc. Fourth IEEE International Symposium on Modeling and Optimization in Mobile, Ad Hoc and Wireless Networks.
Makrakis, G.N. (2014) Transmutation of non-local boundary conditions in ocean acoustics. *Appl. Anal.*, **93**, 1319–26, doi:10.1080/00036811.2013.829566.
Makrakis, G.N. and Skarsoulis, E.K. (2004) Asymptotic approximation of ocean-acoustic pulse propagation in the time domain. *J. Comput. Acoust.*, **12**, 197–215.
Makris, N.C. (1993) Imaging ocean-basin reverberation via inversion. *J. Acoust. Soc. Amer.*, **94**, 983–93.
Makris, N.C. and Dyer, I. (1986) Environmental correlates of pack ice noise. *J. Acoust. Soc. Amer.*, **79**, 1434–40.
Makris, N.C. and Dyer, I. (1991) Environmental correlates of Arctic ice-edge noise. *J. Acoust. Soc. Amer.*, **90**, 3288–98.
Makris, N.C., Ingenito, F., and Kuperman, W.A. (1994) Detection of a submerged object insonified by surface noise in an ocean waveguide. *J. Acoust. Soc. Amer.*, **96**, 1703–24.
Mani, T.K. and Pillai, P.R.S. (2004) Drop parameter estimation from underwater noise produced by raindrop impact. *Acoust. Res. Lett. Online*, **5**(3), 118–24.
Mansour, A. (2012) Enhancement of acoustic tomography using spatial and frequency diversities. *EURASIP J. Adv. Signal Process.*, **2012**, 1–20, doi:10.1186/1687-6180-2012-225.
Mansour, A. and Leblond, I. (2013) Ecosystem monitoring and port surveillance systems. *Adv. Appl. Acoust.*, **2**, 91–110.
Marconi, V., Nielsen, P.L., and Holland, C. (2004) S-SCARAB: SACLANTCEN-Scattering reverberation and backscatter user guide and reference manual. NATO Undersea Res. Ctr, NURC/SM-418.
Marin, F. de Oliveira. (2015) Ocean acoustic tomography based on modal travel time in shallow water. Doctoral thesis, Alberto Luiz Coimbra Institute for Graduate Studies and Research in Engineering, Federal University of Rio de Janeiro (in Portugese).
Marsh, H.W. (1950) Theory of the anomalous propagation of acoustic waves in the ocean. Navy Underwater Sound Lab., Rept 111.
Marsh, H.W. (1963) Sound reflection and scattering from the sea surface. *J. Acoust. Soc. Amer.*, **35**, 240–4.
Marsh, H.W. and Mellen, R.H. (1963) Underwater sound propagation in the Arctic Ocean. *J. Acoust. Soc. Amer.*, **35**, 552–63.
Marsh, H.W. and Schulkin, M. (1962a) Underwater sound transmission. AVCO Corp. Marine Electronics Office Rept.
Marsh, H.W. and Schulkin, M. (1962b) Shallow-water transmission. *J. Acoust. Soc. Amer.*, **34**, 863–4.

Marsh, H.W., Schulkin, M., and Kneale, S.G. (1961) Scattering of underwater sound by the sea surface. *J. Acoust. Soc. Amer.*, **33**, 334–40.

Marsh, H.W. Jr. and Schulkin, M. (1955) Report on the status of project AMOS (Acoustic, Meteorological, and Oceanographic Survey) (1 January 1953–31 December 1954). Navy Underwater Sound Lab., Res. Rept 255.

Marsh, P. (1976) A computer program for studying the Doppler content of reverberation. Nav. Sea Syst. Command, OD 52258.

Marsh, P. and Poynter, A.B. (1969) Digital computer programs for analyzing acoustic search performance in refractive waters, Vols 1 and 2. Nav. Undersea Ctr, Tech. Pub. 164.

Marshalls, S.W. and Cornyn, J.J. (1974a) Ambient-noise prediction. Vol. 1—Model of low-frequency ambient sea noise. Nav. Res. Lab., Rept 7755.

Marshall, S.W. and Cornyn, J.J. (1974b) Ambient-noise prediction. Vol. 2—Model evaluation with IOMEDEX data. Nav. Res. Lab., Rept 7756.

Mason, T. (2013) Modelling of subsea noise during the proposed piling operations at the Dudgeon Wind Farm. Subacoustech Environmental Ltd Rept E438R0106, prepared for Royal Haskoning DHV.

Martin, A. (2012) U.S. expands use of underwater unmanned vehicles. *National Defense*, **96**(701), 34–6.

Martin, P.J. (1993) Sensitivity of acoustic transmission loss prediction to mixed-layer hindcasts calculated with data from Ocean Weather Station Papa. Nav. Res. Lab., Rept NRL/FR/7322-93-9426.

Martin, R.L. (1995) Acoustic beacon navigation. Proc. MTS/IEEE Oceans '95 Conf., San Diego, pp. 1614–9.

Matthews, M.-N.R. and Zykov, M. (2012) Underwater acoustic modeling of construction activities. Marine Commerce South Terminal in New Bedford, MA. JASCO Document 00420, Version 3.0. Prepared by JASCO Applied Sciences for Apex Companies, LCC.

Mazur, M.A. and Gilbert, K.E. (1997a) Direct optimization methods, ray propagation, and chaos. I. Continuous media. *J. Acoust. Soc. Amer.*, **101**, 174–83.

Mazur, M.A. and Gilbert, K.E. (1997b) Direct optimization methods, ray propagation, and chaos. II. Propagation with discrete transitions. *J. Acoust. Soc. Amer.*, **101**, 184–92.

McCabe, B.J. (1976) Ambient noise effects in the modeling of detection by a field of sensors. Report to the Office of Naval Research. Daniel H. Wagner Assoc., Paoli, PA.

McCall, P.L. and Tevesz, M.J.S. (eds) (2013) *Animal-Sediment Relations: The Biogenic Alteration of Sediments*. Topics in Geobiology, Vol. 2. New York, NY: Springer Science+Business Media (softcover reprint of original 1982 edition).

McCammon, D. (2004) Active sonar modelling with emphasis on sonar stimulators. Defence Research and Development Canada–Atlantic, Rept CR 2004-130.

McCammon, D. (2010) A literature survey of reverberation modeling with emphasis on Bellhop compatibility for operational applications. Defence Research and Development Canada–Atlantic, Rept CR 2010-119.

McCammon, D.F. (1988) Fundamental relationships between geoacoustic parameters and predicted bottom loss using a thin layer model. *J. Geophys. Res.*, **93**, 2363–9.

McCammon, D.F. (1991) Underwater acoustic modeling. *Sea Technol.*, **32**(8), 53–5.

McCammon, D.F. and Crowder, D.C. (1981) NEPBR—numerable energy paths by RAYMODE computer program. Appl. Res. Lab., Pennsylvania State Univ., Tech. Memo. TM 81-50.

McCammon, D.F. and McDaniel, S.T. (1985) The influence of the physical properties of ice on reflectivity. *J. Acoust. Soc. Amer.*, **77**, 499–507.

McCarthy, E. (2004) *International Regulation of Underwater Sound: Establishing Rules and Standards to Address Ocean Noise Pollution*, Dordrecht, the Netherlands: Kluwer Academic Publishers.

McCarthy, E.M. and Sabol, B. (2000) Acoustic characterization of submerged aquatic vegetation: Military and environmental monitoring applications. Proc. MTS/IEEE Oceans 2000 Conf., Providence, RI, pp. 1957–61.

McConnell, S.O. (1983) Remote sensing of the air-sea interface using microwave acoustics. Proc. MTS/IEEE Oceans 83 Conf., San Francisco, pp. 85–92.

McDaniel, S.T. (1993) Sea surface reverberation: A review. *J. Acoust. Soc. Amer.*, **94**, 1905–22.

McDaniel, S.T. (2003) Coupled-mode prediction of backscatter. *J. Comput. Acoust.*, **11**, 551–61.

McDaniel, S.T. (2008) Underwater acoustic sensing of sea surface waves. *J. Comput. Acoust.*, **16**, 55–70.

McDonald, B.E. and Kuperman, W.A. (1987) Time domain formulation for pulse propagation including nonlinear behavior at a caustic. *J. Acoust. Soc. Amer.*, **81**, 1406–17.

McDonald, M.A., Hildebrand, J.A., and Wiggins, S.M. (2006) Increases in deep ocean ambient noise in the Northeast Pacific west of San Nicolas Island, California. *J. Acoust. Soc. Amer.*, **120**, 711–18.

McDonald, M.A., Hildebrand, J.A., Wiggins, S.M., and Ross, D. (2008) A 50 year comparison of ambient ocean noise near San Clemente Island: A bathymetrically complex coastal region off Southern California. *J. Acoust. Soc. Amer.*, **124**, 1985–92.

McGirr, R., White, D., and Bartberger, C. (1984) Technical evaluation of FACT 10A. Nav. Ocean Res. Devel. Activity, Rept 70.

McGirr, R.W. (1979) Acoustic model evaluation procedures: A review. Nav. Ocean Syst. Ctr, Tech. Doc. 287.

McGirr, R.W. (1980) Applications of statistical distributions in acoustic model evaluation. Nav. Ocean Syst. Ctr, Tech. Doc. 387.

McGirr, R.W. and Hall, J.C. (1974) FLIRT: A fast linear intermediate-range transmission loss model. Nav. Undersea Ctr, Tech. Note 1282.

McHugh, R., McLaren, D., Wilson, M., and Dunbar, R. (2005) The underwater environment—A fluctuating acoustic medium rich in marine life. Implications for active military sonar. *Acta. Acust. United ACUST.*, **91**, 51–60.

McKinney, C.M. and Anderson, C.D. (1964) Measurements of backscattering of sound from the ocean bottom. *J. Acoust. Soc. Amer.*, **36**, 158–63.

Means, S.L. (2004) Low-frequency sound generation from breaking surf. *Acoust. Res. Lett. Online*, **5**(2), 13–18.

Medeiros, R.C. (1982a) RAYMODE passive propagation loss program performance specification. New England Tech. Services, Doc. 8205.

Medeiros, R.C. (1982b) Active acoustic performance prediction using RAYMODE integration. New England Tech. Services, Doc. 8203.

Medeiros, R.C. (1985a) 1985 baseline RAYMODE passive propagation loss program performance specification. New England Tech. Services, Doc. 8501 Rev. 1.

Medeiros, R.C. (1985b) 1985 baseline active RAYMODE propagation loss prediction computer program performance specification. New England Tech. Services, Doc. 8502 Rev. 1.

Medwin, H. (1975) Speed of sound in water: A simple equation for realistic parameters. *J. Acoust. Soc. Amer.*, **58**, 1318–19.

Medwin, H. (ed.) (2005) *Sounds in the Sea: From ocean acoustics to acoustical oceanography*, New York, NY: Cambridge University Press.

Medwin, H., Browne, M.J., Johnson, K.R., and Denny, P.L. (1988) Low-frequency backscatter from Arctic leads. *J. Acoust. Soc. Amer.*, **83**, 1794–803.

Medwin, H., Childs, E., Jordan, E.A., and Spaulding, R.A. Jr. (1984) Sound scatter and shadowing at a seamount: Hybrid physical solution in two and three dimensions. *J. Acoust. Soc. Amer.*, **75**, 1478–90.

Medwin, H. and Clay, C.S. (1998) *Fundamentals of Acoustical Oceanography*, San Diego, CA: Academic Press.

Medwin, H., Nystuen, J.A., Jacobus, P.W., Ostwald, L.H., and Snyder, D.E. (1992) The anatomy of underwater rain noise. *J. Acoust. Soc. Amer.*, **92**, 1613–23.

Mellen, R.H. and Marsh, H.W. (1965) Underwater sound in the Arctic Ocean. AVCO Marine Electronics Office, Rept MED-65-1002.

Mellen, R.H., Scheifele, P.M., and Browning, D.G. (1987a) Global model for sound absorption in sea water. Nav. Underwater Syst. Ctr, Tech. Rept 7923.

Mellen, R.H., Scheifele, P.M., and Browning, D.G. (1987b) Global model for sound absorption in sea water. Part II: GEOSECS pH data analysis. Nav. Underwater Syst. Ctr, Tech. Rept 7925.

Mellen, R.H., Scheifele, P.M., and Browning, D.G. (1987c) Global model for sound absorption in sea water. Part III: Arctic regions. Nav. Underwater Syst. Ctr, Tech. Rept 7969.

Mellor, G.L. and Yamada, T. (1974) A hierarchy of turbulence closure models for planetary boundary layers. *J. Atmos. Sci.*, **31**, 1791–806.

Melville, W.K. and Matusov, P. (2002) Distribution of breaking waves at the ocean surface. *Nature*, **417**, 58–63.

Menemenlis, D., Campin, J.-M., Heimbach, P., Hill, C., Lee, T., Nguyen, A., Schodlok, M., and Zhang, H. (2008) ECCO2: High Resolution Global Ocean and Sea Ice Data Synthesis. *Mercator Ocean Q. Newsl.*, **31**, 13–21.

Menke, W. (1989) *Geophysical Data Analysis: Discrete Inverse Theory* (rev. edn), International Geophysics Series, Vol. 45. San Diego, CA: Academic Press.

Merchant, N.D. (2013) Measuring underwater noise exposure from shipping. PhD thesis, University of Bath, Claverton Down, Bath, UK.

Merklinger, H.M. (ed.) (1987) *Progress in Underwater Acoustics*. Proc. 12th International Congress on Acoustics Associated Symposium on Underwater Acoustics, 16–18 July 1986, Halifax, Nova Scotia, Canada. Plenum, New York.

Metzger, E.J., Wallcraft, A.J., Posey, P.G., Smedstad, O.M., and Franklin, D.S. (2013) The switchover from NOGAPS to NAVGEM 1.1 atmospheric forcing in GOFS and ACNFS. Nav. Res. Lab., Rept NRL/MR/7320--13-9486.

Metzler, A.M., Moran, D., Collis, J.M., Martin, P.A., and Siegmann, W.L. (2014) A scaled mapping parabolic equation for sloping range-dependent environments. *J. Acoust. Soc. Amer.*, **135**(3), EL172–8, doi:10.1121/1.4865265.

Meyer, M. and Davies, G.L. (2002) A beam-based high-frequency reverberation model. SACLANT Undersea Res. Ctr, Rept SR-365.

Meyer, M. and Hermand, J.-P. (2006) Backpropagation techniques in ocean acoustic inversion: Time reversal, retrogation and adjoint modeling—A review. In *Acoustic Sensing Techniques for the Shallow Water Environment: Inversion methods and experiments*, eds A. Caiti, N.R Chapman, J.-P. Hermand, and S.M. Jesus, New York, NY: Springer, pp. 29–46.

Michalopoulou, Z.-H. and Ghosh-Dastidar, U. (2004) Tabu for matched-field source localization and geoacoustic inversion. *J. Acoust. Soc. Amer.*, **115**, 135–45.

Mignerey, P.C. (1995) Horizontal refraction of 3-D curvilinear wedge modes. *J. Acoust. Soc. Amer.*, **98**, 2912 (abstract).

Mikhalevsky, P.N., Sagen, H., Worcester, P.F., Baggeroer, A.B., Orcutt, J., Moore, S.E. et al. (2015) Multipurpose acoustic networks in the integrated Arctic ocean observing system. *Arctic*, **68** (Suppl. 1), pp. 11–27, doi:10.14430/arctic4449.

Mikhin, D. (2004) Exact discrete nonlocal boundary conditions for high-order Padé parabolic equations. *J. Acoust. Soc. Amer.*, **116**, 2864–75.

Milder, D.M. (1969) Ray and wave invariants for SOFAR channel propagation. *J. Acoust. Soc. Amer.*, **46**, 1259–63.

Milinazzo, F.A., Zala, C.A., and Brooke, G.H. (1997) Rational square-root approximations for parabolic equation algorithms. *J. Acoust. Soc. Amer.*, **101**, 760–6.

Miller, J.F. (1982) Range-dependent ocean acoustic transmission loss calculations in a real-time framework. 4th Interservice/Industry Training Equip. Conf., Kissimmee, FL, pp. 287–99.

Miller, J.F. (1983) Acoustic propagation loss calculations in a complex ocean environment for simulator-based training. Proc. IEE Inter. Conf. Simulators. Conf. Pub. 226, Brighton, England, pp. 215–21.

Miller, J.F. and Ingenito, F. (1975) Normal mode FORTRAN programs for calculating sound propagation in the ocean. Nav. Res. Lab., Memo. Rept 3071.

Miller, J.F., Nagl, A., and Überall, H. (1986) Upslope sound propagation through the bottom of a wedge-shaped ocean beyond cutoff. *J. Acoust. Soc. Amer.*, **79**, 562–5.

Miller, J.F. and Wolf, S.N. (1980) Modal acoustic transmission loss (MOATL): A transmission-loss computer program using a normal-mode model of the acoustic field in the ocean. Nav. Res. Lab., Rept 8429.

Miller, J.H., Lynch, J.F., and Chiu, C.-S. (1989) Estimation of sea surface spectra using acoustic tomography. *J. Acoust. Soc. Amer.*, **86**, 326–45.

Millero, F.J. and Li, X. (1994) Comments on "On equations for the speed of sound in seawater". *J. Acoust. Soc. Amer.*, **95**, 2757–9.

Milne, A.R. (1964) Underwater backscattering strengths of Arctic pack ice. *J. Acoust. Soc. Amer.*, **36**, 1551–6.

Milne, A.R. (1967) Sound propagation and ambient noise under sea ice. In *Underwater Acoustics*, ed. V.M. Albers, Vol. 2. New York, NY: Plenum, pp. 103–38.

Mjølsnes, S. (2000) *LYBIN SGP-180(C)—Model Description*, Bergen, Norway: The Royal Norwegian Navy Materiel Command.

Moler, C.B. and Solomon, L.P. (1970) Use of splines and numerical integration in geometrical acoustics. *J. Acoust. Soc. Amer.*, **48**, 739–44.

Moll, M., Zeskind, R.M., and Scott, W.L. (1979) An algorithm for beam noise prediction. Bolt, Beranek and Newman, Inc., Rept 3653.

Moll, M., Zeskind, R.M., and Sullivan, F.J.M. (1977) Statistical measures of ambient noise: Algorithms, program, and predictions. Bolt, Beranek and Newman, Inc., Rept 3390.

Monjo, C.L. and DeFerrari, H.A. (1994) Analysis of pulse propagation in a bottom-limited sound channel with a surface duct. *J. Acoust. Soc. Amer.*, **95**, 3129–48.

Mooers, C.N.K., Piacsek, S.A., and Robinson, A.R. (eds) (1982) Ocean prediction: The scientific basis and the Navy's needs. A status and prospectus report. Proc. Ocean Prediction Workshop, Monterey, CA, May 1981.

Moore-Head, M.E., Jobst, W., and Showalter, J.A. (1989) The calculation-frequency method for fast parabolic-equation modeling at high frequencies. *J. Acoust. Soc. Amer.*, **85**, 1527–30.

Morfey, C.L. (ed.) (2000) *Dictionary of Acoustics*, San Diego, CA: Academic Press.

Morgan, J.G. Jr. (1998) Networking ASW systems: Anti-submarine warfare dominance. *Sea Technol.*, **39**(11), 19–22.

Moritz, E. (1982) Reverberation research overview. Nav. Coastal Syst. Ctr, Tech. Memo. NCSC TM 362-82.

Morley, M.G., Dosso, S.E., and Chapman, N.R. (2008) Parameter estimate biases in geoacoustic inversion from neglected range dependence. *IEEE J. Oceanic Engr.*, **33**, 255–65.

Morris, G.B. (1978) Depth dependence of ambient noise in the northeastern Pacific Ocean. *J. Acoust. Soc. Amer.*, **64**, 581–90.

Moskowitz, L. (1964) Estimates of the power spectrums for fully developed seas for wind speeds of 20 to 40 knots. *J. Geophys. Res.*, **69**, 5161–79.

Mountain, J.A.R., Ainslie, M.A., Martin, P.L.R., Hughes, M.R., Seto, L.Y., Laker, R.A. et al. (2001) The effectiveness of monostatic and bistatic deployment of low frequency active sonar. Proc. IEE seminar, Multifunction Radar and Sonar Sensor Management Techniques, London, UK, pp. 1–8.

Munk, W. (1994) Heard Island and beyond. *J. Acoust. Soc. Amer.*, **95**, 567.

Munk, W. and Baggeroer, A. (1994) The Heard Island papers: A contribution to global acoustics. *J. Acoust. Soc. Amer.*, **96**, 2327–9.

Munk, W., Worcester, P., and Wunsch, C. (1995) *Ocean Acoustic Tomography*, New York, NY: Cambridge University Press.

Munk, W. and Wunsch, C. (1979) Ocean acoustic tomography: A scheme for large scale monitoring. *Deep-Sea Res.*, **26**, 123–61.

Munk, W.H. (1974) Sound channel in an exponentially stratified ocean, with application to SOFAR. *J. Acoust. Soc. Amer.*, **55**, 220–6.

Munk, W.H. (1980) Horizontal deflection of acoustic paths by mesoscale eddies. *J. Phys. Oceanogr.*, **10**, 596–604.

Munk, W.H. (1991) Bermuda shadow. *Atmos.-Ocean*, **29**, 183–96.

Munk, W.H. and Forbes, A.M.G. (1989) Global ocean warming: An acoustic measure? *J. Phys. Oceanogr.*, **19**, 1765–78.

Munk, W.H. and Zachariasen, F. (1991) Refraction of sound by islands and seamounts. *J. Atmos. Ocean. Technol.*, **8**, 554–74.

Murphy, J.E. and Chin-Bing, S.A. (1988) A finite element model for ocean acoustic propagation. *Math. Comput. Model.*, **11**, 70–4.

Murphy, J.E. and Chin-Bing, S.A. (1989) A finite-element model for ocean acoustic propagation and scattering. *J. Acoust. Soc. Amer.*, **86**, 1478–83.

Murphy, J.E. and Chin-Bing, S.A. (1991) A seismo-acoustic finite element model for underwater acoustic propagation. In *Shear Waves in Marine Sediments*, eds J.M. Hovem, M.D. Richardson, and R.D. Stoll, Dordrecht, the Netherlands: Kluwer Academic Publishers, pp. 463–70.

Murphy, J.E. and Chin-Bing, S.A. (2002) Computational ocean-seismoacoustic modeling using finite elements. In *Acoustic Interactions with Submerged Elastic Structures. Part IV: Nondestructive Testing, Acoustic Wave Propagation and Scattering*, eds A. Guran, A. Boström, O. Leroy, and G. Maze, Singapore: World Scientific Publishing, pp. 163–213.

Myagotin, A.V. and Burdinsky, I.N. (2010) A framework of an acoustic navigation network servicing multiple autonomous underwater vehicles. Proc. IASTED Int. Conf. Automation, Control and Information Technology (ACIT-ICT 2010), Novosibirsk, Russia, pp. 261–6.

National Defense Research Committee. (1946) *Physics of Sound in the Sea*, Summary Technical Report of Division 6, NDRC, Vol. 8. Reprinted by Dept of the Navy, Naval Material Command, 1969, NAVMAT P-9675, Washington, DC.

National Oceanographic Data Center. (1992) *Users Guide*. Key to Oceanographic Records Documentation No. 14, 2nd edn. Washington, DC.

National Research Council. (1994) *Low-Frequency Sound and Marine Mammals: Current Knowledge and Research Needs*, Washington, DC: National Academy Press.

National Research Council. (1996) *Undersea Vehicles and National Needs*, Washington, DC: National Academy Press.

National Research Council. (1997) *Technology for the United States Navy and Marine Corps, 2000–2035. Becoming a 21st-Century Force. Vol. 9: Modeling and Simulation*, Washington, DC: National Academy Press.

National Research Council. (2000) *Marine Mammals and Low-Frequency Sound: Progress Since 1994*, Washington, DC: National Academy Press.

National Research Council. (2002) *Modeling and Simulation in Manufacturing and Defense Systems Acquisition: Pathways to Success*, Washington, DC: National Academy Press.

National Research Council. (2003a) *Ocean Noise and Marine Mammals*. Washington, DC: The National Academies Press.

National Research Council. (2003b) *Environmental Information for Naval Warfare*, Washington, DC: The National Academies Press.

National Research Council. (2005) *Marine Mammal Populations and Ocean Noise: Determining When Noise Causes Biologically Significant Effects.* Washington, DC: The National Academies Press.

Naval Oceanographic Office. (1967) *Oceanographic Atlas of the North Atlantic Ocean. Section II—Physical Properties.* Pub. No. 700, Washington, DC: US Dept of the Navy.

Naval Oceanographic Office. (1972) *Environmental-Acoustics Atlas of the Caribbean Sea and Gulf of Mexico. Volume II—Marine Environment.* SP-189II, Washington, DC: US Dept of the Navy.

Naval Oceanographic Office. (1991a) Software design document for the 1985 baseline passive RAYMODE computer program. OAML-SDD-01A, US Dept of the Navy, Stennis Space Center, MS.

Naval Oceanographic Office. (1991b) Software design document for the 1985 baseline active RAYMODE computer program. OAML-SDD-02A, US Dept of the Navy, Stennis Space Center, MS.

Naval Oceanographic Office. (1991c) Test specification for the 1985 baseline passive RAYMODE computer program. OAML-TS-01G, US Dept of the Navy, Stennis Space Center, MS.

Naval Oceanographic Office. (1991d) Test specification for the 1985 baseline active RAYMODE computer program. OAML-TS-02F, US Dept of the Navy, Stennis Space Center, MS.

Naval Oceanographic Office. (1999) Oceanographic and atmospheric master library summary. OAML-SUM-21G, US Dept of the Navy, Stennis Space Center, MS.

Naval Oceanographic Office. (2003) Database description for Bottom Sediment Type. OAML-DBD-86, US Dept of the Navy, Stennis Space Center, MS.

Naval Oceanographic Office. (2004) Oceanographic and atmospheric master library summary. OAML-SUM-21L, US Dept of the Navy, Stennis Space Center, MS.

Navy Modeling and Simulation Management Office. (1999) *Navy M&S Catalog*, Washington, DC: NAVMSMO.

Nedwell, J. and Howell, D. (2004) A review of offshore windfarm related underwater noise sources. Subacoustech Ltd Rept 544R0308, commissioned by COWRIE.

Nedwell, J.R., Lovell, J., and Turnpenny, A.W.H. (2005) Experimental validation of a species-specific behavioral impact metric for underwater noise. *J. Acoust. Soc. Amer.*, **118**, 2019 (abstract).

Nedwell, J.R., Turnpenny, A.W.H., Lovell, J., Parvin, S.J., Workman, R., Spinks, J.A.L. et al. (2007) A validation of the dB_{ht} as a measure of the behavioural and auditory effects of underwater noise. Subacoustech Ltd Rept 534R1231, prepared for UK Department of Business, Enterprise and Regulatory Reform under Project No. RDCZ/011/0004.

Nesbitt, L. and Jones, R.M. (1994a) A C program for generating publication-quality plots. NOAA Tech. Memo. ERL ETL-242.

Nesbitt, L. and Jones, R.M. (1994b) A FORTRAN program for performing nonperturbative ocean acoustic tomography inversion. NOAA Tech. Memo. ERL ETL-243.

Nesbitt, L. and Jones, R.M. (1994c) A FORTRAN program to generate comparison data to test tomography inversion programs. NOAA Tech. Memo. ERL ETL-245.

Neumann, G. and Pierson, W.J. Jr. (1966) *Principles of Physical Oceanography.* Englewood Cliffs, NJ: Prentice-Hall.

Newhall, A.E., Lynch, J.F., Chiu, C.-S., and Daugherty, J.R. (1990) Improvements in three-dimensional raytracing codes for underwater acoustics. In *Computational Acoustics. Vol. 1: Ocean-Acoustic Models and Supercomputing*, eds D. Lee, A. Cakmak, and R. Vichnevetsky, Amsterdam, the Netherlands: North-Holland, pp. 169–85.

Newman, A.V. and Ingenito, F. (1972) A normal mode computer program for calculating sound propagation in shallow water with an arbitrary velocity profile. Nav. Res. Lab., Memo. Rept 2381.

Newman, F.C., Biondo, A.C., Mandelberg, M.D., Matthews, C.C., and Rottier, J.R. (2002) Enhancing realism in computer simulations: Environmental effects. *Johns Hopkins APL Tech. Dig.*, **23**(4), 443–53.

Newton, J. and Galindo, M. (2001) Hydroacoustic monitoring network. *Sea Technol.*, **42** (9), 41–7.

Nghiem-Phu, L., Daubin, S.C., and Tappert, F. (1984) A high-speed, compact, and interactive parabolic equation solution generator (PESOGEN) system. *J. Acoust. Soc. Amer.*, **75** (Suppl. 1), S26 (abstract).

Nicholas, M., Ogden, P.M., and Erskine, F.T. (1998) Improved empirical descriptions for acoustic surface backscatter in the ocean. *IEEE J. Oceanic Engr.*, **23**, 81–95.

Nichols, S.M. and Bradley, D.L. (2016) Global examination of the wind-dependence of very low frequency underwater ambient noise. *J. Acoust. Soc. Amer.*, **139**, 1110–23.

Nielsen, R.O. (1991) *Sonar Signal Processing*, Boston, MA: Artech House.

Nijhof, M.J.J., Binnerts, B., de Jong, C.A.F., and Ainslie, M.A. (2014) An efficient model for prediction of underwater noise due to pile driving at large ranges. Proc. INTER.NOISE 2014 Conf., 43rd International Congress on Noise Control Engineering, Melbourne, Australia, pp. 5505–14.

North Atlantic Treaty Organization. (1998) *NATO Modelling and Simulation Master Plan.* Document AC/323 (SGMS)D/2, Version 1.0. Brussels, Belgium: North Atlantic Council.

North Atlantic Treaty Organization. (2000) *Follow-On NIAG Prefeasibility Studies on Common Technical Framework and Multi-National Force Rehearsal. PATHFINDER Experiment Definition. Final Report.* NIAG-D(99)6 AC/323 (NMSG).

Norton, G.V. and Novarini, J.C. (1996) The effect of sea-surface roughness on shallow water waveguide propagation: A coherent approach. *J. Acoust. Soc. Amer.*, **99**, 2013–21.

Norton, G.V., Novarini, J.C., and Keiffer, R.S. (1995) Coupling scattering from the sea surface to a one-way marching propagation model via conformal mapping: Validation. *J. Acoust. Soc. Amer.*, **97**, 2173–80.

Norton, G.V., Novarini, J.C., and Keiffer, R.S. (1998) Modeling the propagation from a horizontally directed high-frequency source in shallow water in the presence of bubble clouds and sea surface roughness. *J. Acoust. Soc. Amer.*, **103**, 3256–67.

Noutary, E. and Plaisant, A. (1996) 2D and 3D propagation modelling with coupled modes. In *Theoretical and Computational Acoustics '95*, eds D. Lee, Y.-H. Pao, M.H Schultz, and Y.-C. Teng, Singapore: World Scientific Publishing, pp. 61–74.

Nystuen, J.A. (1986) Rainfall measurements using underwater ambient noise. *J. Acoust. Soc. Amer.*, **79**, 972–82.

Nystuen, J.A. (1994) Acoustical rainfall analysis. *J. Acoust. Soc. Amer.*, **95**, 2882–3 (abstract).

Nystuen, J.A. (2001) Listening to raindrops from underwater: An acoustic disdrometer. *J. Atmos. Oceanic Technol.*, **18**, 1640–57.

Nystuen, J.A., McPhaden, M.J., and Freitag, H.P. (2000) Surface measurements of precipitation from an ocean mooring: The underwater acoustic log from the South China Sea. *J. Appl. Meteorol.*, **39**, 2182–97.

Nystuen, J.A. and Medwin, H. (1995) Underwater sound produced by rainfall: Secondary splashes of aerosols. *J. Acoust. Soc. Amer.*, **97**, 1606–13.

Oba, R. and Finette, S. (2002) Acoustic propagation through anisotropic internal wave fields: Transmission loss, cross-range coherence, and horizontal refraction. *J. Acoust. Soc. Amer.*, **111**, 769–84.

Ocean Acoustic Developments Ltd. (1999a) WADER global ocean information system. Product description literature.

Ocean Acoustic Developments Ltd. (1999b) Global sonar assessment and range prediction system. Product description literature.

Ocean Acoustic Developments Ltd. (2013) HARCAM. Product description literature.

Officer, C.B. (1958) *Introduction to the Theory of Sound Transmission with Application to the Ocean*, New York, NY: McGraw-Hill.

Ogden, P.M. and Erskine, F.T. (1994a) Surface scattering measurements using broadband explosive charges in the Critical Sea Test experiments. *J. Acoust. Soc. Amer.*, **95**, 746–61.

Ogden, P.M. and Erskine, F.T. (1994b) Surface and volume scattering measurements using broadband explosive charges in the Critical Sea Test 7 experiment. *J. Acoust. Soc. Amer.*, **96**, 2908–20.

Ogden, P.M. and Erskine, F.T. (1997) Bottom scattering strengths measured using explosive sources in the Critical Sea Test Program. Nav. Res. Lab., Rept NRL/FR/7140-97-9822.

Oğuz, H.N. (1994) A theoretical study of low-frequency oceanic ambient noise. *J. Acoust. Soc. Amer.*, **95**, 1895–912.

Ohta, K., Okabe, K., Morishita, I., Ozaki, S., and Frisk, G.V. (2005) Inversion for seabed geoacoustic properties in shallow water experiments. *Acoust. Sci. Technol.*, **26**, 326–37.

Oikawa, M., Ohta, K., Okabe, K., Morishita, I., and Kanda, H. (2007) Estimation of sea surface and bottom scattering strengths from measured and modeled reverberation levels. *Acoust. Sci. Technol.*, **28**, 375–84.

Olsen, H.S. (2008) Lydutbredelse i havområder med avstandsavhengig oseanografi. Norges teknisk-naturvitenskapelige universitet, Institutt for elektronikk og telekommunikasjon, Trondheim, Norway.

Orchard, B.J., Siegmann, W.L., and Jacobson, M.J. (1992) Three-dimensional time-domain paraxial approximations for ocean acoustic wave propagation. *J. Acoust. Soc. Amer.*, **91**, 788–801.

Orcutt, J.A. (1988) Ultralow- and very-low-frequency seismic and acoustic noise in the Pacific. *J. Acoust. Soc. Amer.*, **84** (Suppl. 1), S194 (abstract).

Oreskes, N., Shrader-Frechette, K., and Belitz, K. (1994) Verification, validation, and confirmation of numerical models in the Earth sciences. *Science*, **263**, 641–6.

Orme, E.A., Johns, P.B., and Arnold, J.M. (1988) A hybrid modelling technique for underwater acoustic scattering. *Int. J. Numer. Model.: Electron. Netw., Devices Fields*, **1**, 189–206.

Orris, G.J. and Collins, M.D. (1994) The spectral parabolic equation and three-dimensional backscattering. *J. Acoust. Soc. Amer.*, **96**, 1725–31.

Osborne, A.R. (2010) *Nonlinear Ocean Waves and the Inverse Scattering Transform*. International Geophysics Series, Vol. 97. San Diego, CA: Academic Press.

Osborne, K.R. (1979) DANES—a directional ambient noise prediction model for FLENUMOCEANCEN. Ocean Data Syst., Inc.

Osher, S. and Sethian, J.A. (1988) Fronts propagating with curvature dependent speed: Algorithms based on Hamilton-Jacobi formulations. *J. Comput. Phys.*, **79**, 12–49.

Otnes, R., Asterjadhi, A., Casari, P., Goetz, M., Husøy, T., Nissen, I. et al. (2012) *Underwater Acoustic Networking Techniques*, New York, NY: Springer.

Otnes, R., van Walree, P.A., and Jenserud, T. (2013a) Validation of replay-based underwater acoustic communication channel simulation. *IEEE J. Oceanic Engr.*, **38**, 689–700.

Otnes, R., van Walree, P.A., and Jenserud, T. (2013b) Erratum to "Validation of Replay-Based Underwater Acoustic Communication Channel Simulation" (R. Otnes, P. A. van Walree, T. Jenserud, IEEE J. Ocean. Eng., doi:10.1109/JOE.2013.2262743). *IEEE J. Oceanic Engr.*, **38**, 809.

Outing, D.A., Siegmann, W.L., Collins, M.D., and Westwood, E.K. (2006) Generalization of the rotated parabolic equation to variable slopes. *J. Acoust. Soc. Amer.*, **120**, 3534–8.

Pace, N.G. and Jensen, F.B. (eds) (2002) *Impact of Littoral Environmental Variability on Acoustic Predictions and Sonar Performance*, Dordrecht, the Netherlands: Kluwer Academic Publishers.

Pace, N.G., Pouliquen, E., Bergem, O., and Lyons, A.P. (eds) (1997) High Frequency Acoustics in Shallow Water. SACLANT Undersea Res. Ctr, Conf. Proc. CP-45.

Pack, P.M.W. (1986) The finite element method in underwater acoustics. Ph.D. thesis, Institute of Sound and Vibration Research, University of Southampton, Southampton, England.

Packman, M.N., Harrison, C.H., and Ainslie, M.A. (1992) INSIGHT—A fast, robust, propagation loss model with physical intuition. *Acoust. Bull.*, **17**(4), 21–4.

Packman, M.N., Harrison, C.H., and Ainslie, M.A. (1995) INSTANT calculations of range dependent ocean acoustic propagation loss. In *Transactions on the Built Environment, Vol. 11—Computational Acoustics and its Environmental Applications*. ed. C.A. Brebbia, Southampton, England: WIT Press, pp. 127–34.

Packman, M.N., Harrison, C.H., and Ainslie, M.A. (1996) Rapid calculations of acoustic propagation loss in range dependent oceans. *IEE Proc.Radar Sonar Nav.*, **143**(3), 184–9.

Palmer, D.R. (2002) A parabolic approximation method with application to global wave propagation. *J. Math. Phys.*, **43**, 1875–905.

Palmer, D.R., Brown, M.G., Tappert, F.D., and Bezdek, H.F. (1988a) Classical chaos in nonseparable wave propagation problems. *Geophys. Res. Lett.*, **15**, 569–72.

Palmer, D.R., Lawson, L.M., Daneshzadeh, Y.-H., and Behringer, D.W. (1988b) Computational studies of the effect of an El Niño/southern oscillation event on underwater sound propagation. In *Computational Acoustics, Vol. 2, Algorithms and Applications*, eds D. Lee, R.L. Sternberg, and M.H. Schultz, Amsterdam, the Netherlands: North-Holland, pp. 335–56.

Palmer, L.B. and Fromm, D.M. (1992) The range-dependent active system performance prediction model (RASP). Nav. Res. Lab., Rept NRL/FR/5160-92-9383.

Papadakis, J.S., Dougalis, V.A., Kampanis, N.A., Flouri, E.T., Pelloni, B., Plaisant, A. et al. (1998) Ocean acoustic models for low frequency propagation in 2D and 3D environments. *ACUST. Acta Acust.*, **84**, 1031–41.

Papadakis, J.S. and Flouri, E.T. (2007) Geoacoustic inversions based on an adjoint parabolic equation with a Neumann to Dirichlet map boundary condition. *Acta Acust. United ACUST.*, **93**, 924–33.

Papadakis, J.S. and Flouri, E.T. (2008) A Neumann to Dirichlet map for the bottom boundary of a stratified sub-bottom region in parabolic approximation. *J. Comput. Acoust.*, **16**, 409–25.

Parsons, M.J.G. and Duncan, A.J. (2011) The effect of seabed properties on the receive beam pattern of a hydrophone located on the sea floor. *Acoust. Australia*, **39**(3), 106–12.

Partan, J. (2006) A survey of practical issues in underwater networks. Computer Science Department Faculty Publication Series, Paper 133, University of Massachusetts http://scholarworks.umass.edu/cs_faculty_pubs/133.

Patterson, T.A., Thomas, L., Wilcox, C., Ovaskainen, O., and Matthiopoulos, J. (2008) State–space models of individual animal movement. *Trends Ecol. Evol.*, **23**, 87–94, doi:10.1016/j.tree.2007.10.009.

Paul, M., Lefebvre, A., Manca, E., and Amos, C.L. (2011) An acoustic method for the remote measurement of seagrass metrics. *Estuar. Coast. Shelf Sci.*, **93**, 68–79.

Pedersen, M.A. (1961) Acoustic intensity anomalies introduced by constant velocity gradients. *J. Acoust. Soc. Amer.*, **33**, 465–74.

Pedersen, M.A. and Gordon, D.F. (1965) Normal-mode theory applied to short-range propagation in an underwater acoustic surface duct. *J. Acoust. Soc. Amer.*, **37**, 105–18.

Pedersen, M.A. and Gordon, D.F. (1967) Comparison of curvilinear and linear profile approximation in the calculation of underwater sound intensities by ray theory. *J. Acoust. Soc. Amer.*, **41**, 419–38.

Pedersen, M.A., Gordon, D.F., and Keith, A.J. (1962) A new ray intensity procedure for underwater sound based on a profile consisting of curvilinear segments. Nav. Electron. Lab., Res. Rept 1105.

Pedersen, M.A. and McGirr, R.W. (1982) Use of theoretical controls in underwater acoustic model evaluation. Nav. Ocean Syst. Ctr, Tech. Rept 758.

Pedersen, M.A., Gordon, D.F., and Hosmer, R.F. (1984) Comparison of experimental detailed convergence-zone structure with results of uniform asymptotic ray theory. Nav. Ocean Syst. Ctr, Tech. Rept 997.

Peixoto, J.P. and Oort, A.H. (1992) *Physics of Climate*, New York, NY: American Institute of Physics.

Pekeris, C.L. (1946) Theory of propagation of sound in a half-space of variable sound velocity under conditions of formation of a shadow zone. *J. Acoust. Soc. Amer.*, **18**, 295–315.

Pekeris, C.L. (1948) Theory of propagation of explosive sound in shallow water. *Geol. Soc. Amer. Mem.*, **27**, 1–117.

Peloquin, R.A. (1992) The Navy ocean modeling and prediction program. *Oceanography*, **5**(1), 4–8.

Penrose, J.D., Pauly, T.J., Arcus, W.R., Duncan, A.J., and Bush, G. (1993) Remote sensing with underwater acoustics. *Acoust. Aust.*, **21**(1), 19–21.

Perkin, R.G. and Lewis, E.L. (1980) The practical salinity scale 1978: Fitting the data. *IEEE J. Oceanic Engr.*, **5**, 9–16.

Perkins, J.S., Adams, B.B., and McCoy, J.J. (1984) Vertical coherence along a macroray path in an inhomogeneous anisotropic ocean. Nav. Res. Lab., Rept 8792.

Perkins, J.S. and Baer, R.N. (1978) A corrected parabolic-equation program package for acoustic propagation. Nav. Res. Lab., Memo. Rept 3688.

Perkins, J.S. and Baer, R.N. (1982) An approximation to the three-dimensional parabolic-equation method for acoustic propagation. *J. Acoust. Soc. Amer.*, **72**, 515–22.

Perkins, J.S., Baer, R.N., Roche, L.F., and Palmer, L.B. (1983) Three-dimensional parabolic-equation-based estimation of the ocean acoustic field. Nav. Res. Lab., Rept 8685.

Perkins, J.S., Baer, R.N., Wright, E.B., and Roche, L.F. (1982) Solving the parabolic equation for underwater acoustic propagation by the split-step algorithm. Nav. Res. Lab., Rept 8607.

Perkins, J.S., Kuperman, W.A., Ingenito, F., Fialkowski, L.T., and Glattetre, J. (1993) Modeling ambient noise in three-dimensional ocean environments. *J. Acoust. Soc. Amer.*, **93**, 739–52.

Perkins, J.S. and Thorsos, E.I. (2007) Overview of the reverberation modeling workshops. *J. Acoust. Soc. Amer.*, **122**, 3074 (abstract).

Perkins, J.S., Williamson, M., Kuperman, W.A., and Evans, R.B. (1990) Sound propagation through the Gulf Stream: Current status of two three-dimensional models. In *Computational Acoustics. Vol. 1: Ocean-Acoustic Models and Supercomputing*, eds D. Lee, A. Cakmak, and R. Vichnevetsky, Amsterdam, the Netherlands: North-Holland, pp. 203–15.

Pershing, A.J., Wiebe, P.H., Manning, J.P., and Copley, N.J. (2001) Evidence for vertical circulation cells in the well-mixed area of Georges Bank and their biological implications. *Deep-Sea Res. II*, **48**, 283–310.

Peters, H. and Gregg, M.C. (1987) Equatorial turbulence: Mixed layer and thermocline. In *Dynamics of the Oceanic Surface Mixed Layer*, eds P. Muller and D. Henderson, Honolulu, HI: Hawaii Inst. Geophys. Special Pub., pp. 25–45.

Petrov, P.S. and Sturm, F. (2016) An explicit analytical solution for sound propagation in a three-dimensional penetrable wedge with small apex angle. *J. Acoust. Soc. Amer.*, **139**, 1343–52.

Pflug, L.A. (1996) Modification of the RANDI model for shallow-water applications. Nav. Res. Lab., Rept NRL/MR/7176-96-8003.

Piao, S., Tian, L., and Yang, J. (2008) Underwater low frequency sound field simulation with the digital waveguide mesh method. Proc. Acoustics '08 Conf., Paris, pp. 1895–1900.

Pickard, G.L. (1963) *Descriptive Physical Oceanography: An Introduction.* New York, NY: Pergamon Press.

Pickard, G.L. and Emery, W.J. (1990) *Descriptive Physical Oceanography: An Introduction*, 5th edn. New York, NY: Pergamon Press.

Pierce, A.D. (1965) Extension of the method of normal modes to sound propagation in an almost-stratified medium. *J. Acoust. Soc. Amer.*, **37**, 19–27.

Pierce, A.D. (1982) Guided mode disappearance during upslope propagation in variable depth shallow water overlying a fluid bottom. *J. Acoust. Soc. Amer.*, **72**, 523–31.

Pierce, A.D. (1989) *Acoustics: An introduction to its physical principles and applications*, New York, NY: Acoustical Society of America.

Pierce, A.D. and Thurston, R.N. (eds) (1993) *Underwater Scattering and Radiation, Physical Acoustics XXII*, San Diego, CA: Academic Press.

Piercy, J.J.B. (2015) The relevance of coral reef soundscapes to larval fish responses. PhD thesis, University of Essex, Colchester, England.

Pierson, W.J. Jr. (1964) The interpretation of wave spectrums in terms of the wind profile instead of the wind measured at a constant height. *J. Geophys. Res.*, **69**, 5191–203.

Pierson, W.J., Jr. (1991) Comment on "Effects of Sea Maturity on Satellite Altimeter Measurements" by Roman E. Glazman and Stuart H. Pilorz. *J. Geophys. Res.*, **96**, 4973–7.

Pierson, W.J., Jr. and Moskowitz, L. (1964) A proposed spectral form for fully developed wind seas based on the similarity theory of S.A. Kitaigorodskii. *J. Geophys. Res.*, **69**, 5181–90.

Pignot, P. and Chapman, N.R. (2001) Tomographic inversion of geoacoustic properties in a range-dependent shallow-water environment. *J. Acoust. Soc. Amer.*, **110**, 1338–48.

Piskarev, A.L. (1992) Soviet research in underwater acoustic propagation modeling. EG&G, Tech. Rept 6K1-011A.

Plotkin, A.M. (1996) LYCH 2-D ray program functional description. ISDCO unpublished manuscript.

Podeszwa, E.M. (1969) Computer program for Colossus II shallow-water transmission loss equations. Navy Underwater Sound Lab., Tech. Memo. 2040-85-69.

Poikonen, A. (2012) Measurements, analysis and modeling of wind-driven ambient noise in shallow brackish water. PhD dissertation, Aalto University, Helsinki, Finland.

Pomerenk, K. and Novick, A. (1987) *BISAPP User's Manual.* Mission Sci. Corp., pp. 1–13.

Popper, A.N. and Hawkins, A. (eds) (2012) *The Effects of Noise on Aquatic Life.* Advances in Experimental Medicine and Biology 730. New York, NY: Springer.

Popper, A.N. and Hawkins, A. (eds) (2016) *The Effects of Noise on Aquatic Life II.* Advances in Experimental Medicine and Biology, Volume 875. New York, NY: Springer.

Porter, M. (1991) The KRAKEN normal mode program. SACLANT Undersea Res. Ctr, Memo. SM-245.

Porter, M. and Reiss, E.L. (1984) A numerical method for ocean-acoustic normal modes. *J. Acoust. Soc. Amer.*, **76**, 244–52.

Porter, M.B. (1990) The time-marched fast-field program (FFP) for modeling acoustic pulse propagation. *J. Acoust. Soc. Amer.*, **87**, 2013–23.

Porter, M.B. (1993) Acoustic models and sonar systems. *IEEE J. Oceanic Engr.*, **18**, 425–37.

Porter, M.B. (2011) *The BELLHOP Manual and User's Guide: Preliminary Draft*, La Jolla, CA: Heat, Light, and Sound Research, Inc., pp. 1–57.

Porter, M.B. (2015) Out-of-plane effects in three-dimensional oceans. *J. Acoust. Soc. Amer.*, **137**, 2419 (abstract).

Porter, M.B. and Bucker, H.P. (1987) Gaussian beam tracing for computing ocean acoustic fields. *J. Acoust. Soc. Amer.*, **82**, 1349–59.

Porter, M.B. and Jensen, F.B. (1993) Anomalous parabolic equation results for propagation in leaky surface ducts. *J. Acoust. Soc. Amer.*, **94**, 1510–16.

Porter, M.B. and Reiss, E.L. (1985) A numerical method for bottom interacting ocean acoustic normal modes. *J. Acoust. Soc. Amer.*, **77**, 1760–7.

Porter, M.B., Siderius, M., and Kuperman, W.A. (eds) (2004) *High Frequency Ocean Acoustics*. American Institute of Physics Conference Proceedings 728. New York, NY: Springer-Verlag.

Potter, J.R. (1994) Acoustic imaging using ambient noise: Some theory and simulation results. *J. Acoust. Soc. Amer.*, **95**, 21–33.

Potter, J.R. and Chitre, M. (1999) Ambient noise imaging in warm shallow seas; second-order moment and model-based imaging algorithms. *J. Acoust. Soc. Amer.*, **106**, 3201–10.

Pouliquen, E., Lyons, A.P., and Pace, N.G. (1998) The Helmholtz-Kirchhoff approach to modeling penetration of acoustic waves into rough seabeds. *J. Acoust. Soc. Amer.*, **104**, 1762 (abstract).

Powers, W.J. (1987) Bistatic active signal excess model—An extension of the Generic Sonar Model. Nav. Underwater Syst. Ctr, Tech. Memo. 871025.

Preisig, J.C. and Deane, G.B. (2004) Surface wave focusing and acoustic communications in the surf zone. *J. Acoust. Soc. Amer.*, **116**, 2067–80.

Preston, J.R. (1999) Report on the 1999 ONR shallow-water reverberation focus workshop. Appl. Res. Lab., Pennsylvania State Univ., Tech. Memo. TM 99-155.

Prideaux, G. (ed.) (2016) *CMS Family Guidelines on Environmental Impact Assessment for Marine Noise-generating Activities*, Bonn, Germany: Convention on Migratory Species of Wild Animals, pp. 96.

Prideaux, G. and Prideaux, M. (2015) Environmental impact assessment guidelines for offshore petroleum exploration seismic surveys. *Impact Assess. Proj. Apprais.*, **34**, 33–43, doi:10.1080/14615517.2015.1096038.

Primack, H. and Gilbert, K.E. (1991) A two-dimensional downslope propagation model based on coupled wedge modes. *J. Acoust. Soc. Amer.*, **90**, 3254–62.

Princehouse, D.W. (1977) REVGEN, a real-time reverberation generator. Proc. IEEE Inter. Conf. Acoust. Speech Signal Process., pp. 827–35.

Prior, M.K. (2007a) SUPREMO v.2.0 user guide. NATO Undersea Res. Ctr, Rept NURC-FR-2007-009.

Prior, M.K. (2007b) SUPREMO v.2.0 program description. NATO Undersea Res. Ctr, Rept NURC-FR-2007-011.

Prior, M.K. and Baldacci, A. (2005) Comparison between predictions made by the SUPREMO sonar performance model and measured data. SACLANT Undersea Res. Ctr, Rept SR-429.

Prior, M.K., Harrison, C.H., and LePage, K.D. (2002) Reverberation comparisons between RSNAP, SUPREMO and analytical solutions. SACLANT Undersea Res. Ctr, Rept SR-361.

Pritchard, R.S. (1990) Sea ice noise-generating processes. *J. Acoust. Soc. Amer.*, **88**, 2830–42.

Prosperetti, A. and Oǧuz, H.N. (1993) The impact of drops on liquid surfaces and the underwater noise of rain. *Ann. Rev. Fluid Mech.*, **25**, 577–602.

Pumphrey, H.C. (1994) Underwater rain noise—The initial impact component. *Acoust. Bull.*, **19**(2), 19–28.

Pumphrey, H.C. and Crum, L.A. (1990) Free oscillations of near-surface bubbles as a source of the underwater noise of rain. *J. Acoust. Soc. Amer.*, **87**, 142–8.

Qarabaqi, P. and Stojanovic, M. (2013) Statistical characterization and computationally efficient modeling of a class of underwater acoustic communication channels. *IEEE J. Oceanic Engr.*, **38**, 701–17.

Radford, C.A., Tindle, C.T., Montgomery, J.C., and Jeffs, A.G. (2011) Modelling a reef as an extended sound source increases the predicted range at which reef noise may be heard by fish larvae. *Mar. Ecol. Prog. Ser.*, **438**, 167–74.

Rajan, S.D. (1992) Determination of geoacoustic parameters of the ocean bottom—data requirements. *J. Acoust. Soc. Amer.*, **92**, 2126–40.

Ramsdale, D.J. and Posey, J.W. (1987) Understanding underwater acoustics under the Arctic ice canopy. *Sea Technol.*, **28**(7), 22–8.

Rankin, S. and Barlow, J. (2005) Source of the North Pacific "boing" sound attributed to minke whales. *J. Acoust. Soc. Amer.*, **118**, 3346–51.

Rasband, S.N. (1990) *Chaotic Dynamics of Nonlinear Systems*, New York, NY: John Wiley & Sons.

Raveau, M. and Feuillade, C. (2016) Resonance scattering by fish schools: A comparison of two models. *J. Acoust. Soc. Amer.*, **139**, 163–75.

Rayborn, G.H. (2006) Littoral acoustic demonstration center. USM Grant Portion Final Report (ONR Contract N00014-01-1-0914). Univ. Southern Mississippi, USA, Rept 64-6000818.

Rayleigh, J.W.S. (1945) *The Theory of Sound*, Vols 1 and 2. New York, NY: Dover Publications. (Republication of the 1894 and 1896 editions of Vols 1 and 2, respectively.)

Reeder, D.B. (2016) Field observation of low-to-mid-frequency acoustic propagation characteristics of an estuarine salt wedge. *J. Acoust. Soc. Amer.*, **139**, 21–9.

Reilly, S.M., Potty, G.R., and Goodrich, M. (2016) Computing acoustic transmission loss using 3D Gaussian ray bundles in geodetic coordinates. *J. Comput. Acoust.*, **24**(1), 1–24, doi:10.1142/S0218396X16500077.

Reise, B. and Etter, P.C. (1997) Performance assessment of active sonar configuration options. Proc. Undersea Defence Technology Conf. UDT Europe, Hamburg, Germany, pp. 408–13.

Renner, W. (1995a) Software requirements specification for the ANDES model (Version 4.2). Sci. Appl. Inter. Corp. Tech. Rept.

Renner, W. (1995b) User's guide for the ANDES model (Version 4.2). Sci. Appl. Inter. Corp. Tech. Rept.

Renner, W.W. (1986a) Ambient noise directionality estimation system (ANDES) user's guide (HP 9000 installations). Sci. Appl. Inter. Corp., SAIC-86/1705.

Renner, W.W. (1986b) Ambient noise directionality estimation system (ANDES) technical description. Sci. Appl. Inter. Corp., SAIC-86/1645.

Renner, W.W. (1988) Ambient noise directionality estimation system (ANDES) II user's guide (VAX 11/78X installations). Sci. Appl. Inter. Corp., SAIC-88/1567.

Richards, S.D. (1998) The effect of temperature, pressure, and salinity on sound attenuation in turbid seawater. *J. Acoust. Soc. Amer.*, **103**, 205–11.

Richards, S.D. and Leighton, T.G. (2001a) Acoustic sensor performance in coastal waters: Solid suspensions and bubbles. In *Acoustical Oceanography*, eds T.G. Leighton, G.J. Heald, H.D. Griffiths, and G. Griffiths. *Proc. Inst. Acoust.*, **23** (Part 2), 399–406.

Richards, S.D. and Leighton, T.G. (2001b) Sonar performance in coastal environments: Suspended sediments and microbubbles. *Acoust. Bull.*, **26**(1), 10–17.

Richards, S.D., Heathershaw, A.D., and Thorne, P.D. (1996) The effect of suspended particulate matter on sound attenuation in seawater. *J. Acoust. Soc. Amer.*, **100**, 1447–50.

Richardson, W.J., Greene, C.R. Jr, Malme, C.I., and Thomson, D.H. (1995) *Marine Mammals and Noise*. Academic Press, San Diego.

Riordan, J., Toal, D., and Flanagan, C. (2005) Towards real-time simulation of the sides-can sonar imaging process. Proc. 4th WSEAS International Conf. Signal Processing, Robotics and Automation, Salzburg, Austria.

Risch, D., Corkeron, P.J., Ellison, W.T., and van Parijs, S.M. (2012) Changes in Humpback whale song occurrence in response to an acoustic source 200 km away. *PLOS ONE*, **7**(1), e29741, doi:10.1371/journal.pone.0029741.

Roberts, B.G. Jr. (1974) Horizontal-gradient acoustical ray-trace program TRIMAIN. Nav. Res. Lab., Rept 7827.

Robertson, J.S. (1989) A classification scheme for computational ocean acoustic benchmark problems. *Appl. Acoust.*, **27**, 65–8.

Robertson, J.S., Arney, D.C., Jacobson, M.J., and Siegmann, W.L. (1989) An efficient enhancement of finite-difference implementations for solving parabolic equations. *J. Acoust. Soc. Amer.*, **86**, 252–60.

Robertson, J.S., Siegmann, W.L., and Jacobson, M.J. (1991) OS2IFD: A microcomputer implementation of the parabolic equation for predicting underwater sound propagation. *Comput. Geosci.*, **17**, 731–57.

Robertsson, J.O.A., Levander, A., and Holliger, K. (1996) A hybrid wave propagation simulation technique for ocean acoustic problems. *J. Geophys. Res.*, **101**, 11225–41.

Robins, A.J. (1991) Reflection of a plane wave from a fluid layer with continuously varying density and sound speed. *J. Acoust. Soc. Amer.*, **89**, 1686–96.

Robinson, A.R. (1987) Predicting open ocean currents, fronts and eddies. In *Three-Dimensional Models of Marine and Estuarine Dynamics*, eds J.C.J. Nihoul and B.M. Jamart, Elsevier Oceanography Series 45, Amsterdam, the Netherlands: Elsevier Science Publishers, pp. 89–111.

Robinson, A.R. (1992) Shipboard prediction with a regional forecast model. *Oceanography*, **5**(1), 42–8.

Robinson, A.R. (1999) Realtime forecasting of the multidisciplinary coastal ocean with the littoral ocean observing and predicting system (LOOPS). Proc. 3rd Conf. Coastal Atmospheric Oceanic Prediction and Processes, American Meteorological Society, New Orleans, LA.

Robinson, A.R., Carton, J.A., Mooers, C.N.K., Walstad, L.J., Carter, E.F., Rienecker, M.M. et al. (1984) A real-time dynamical forecast of ocean synoptic/mesoscale eddies. *Nature*, **309**, 781–3.

Robinson, A.R., Carman, J.C., and Glenn, S.M. (1994) A dynamical system for acoustic applications. In *Oceanography and Acoustics: Prediction and Propagation Models*, eds A.R. Robinson and D. Lee, New York, NY: American Institute of Physics, Chapter 4, pp 80–117.

Robinson, E.R. and McConnell, S.O. (1983) Sensitivity of high frequency surface-generated noise to sonar and environmental parameters. Proc. MTS/IEEE Oceans 83 Conf., San Francisco, pp. 11–15.

Robinson, S.P., Lepper, P. A., and Hazelwood, R.A. (2014) Good Practice Guide for Underwater Noise Measurement. National Measurement Office, Marine Scotland, The Crown Estate, National Physical Laboratory Good Practice Guide No. 133, ISSN: 1368-6550.

Robinson, T. (2001) ODIN—An underwater warfare simulation environment. Proc. 2001 Winter Simulation Conf., eds B.A. Peters, J.S. Smith, D.J. Medeiros, and M.W. Rohrer. Arlington, VA, pp. 672–9.

Rodríguez, O.C., Jesus, S., Stéphan, Y., Demoulin, X., Porter, M., and Coelho, E. (2000) Nonlinear soliton interaction with acoustic signals: Focusing effects. *J. Comput. Acoust.*, **8**, 347–63.

Rogers, P.H. (1981) Onboard prediction of propagation loss in shallow water. Nav. Res. Lab., Rept 8500.

Rohr, J. and Detsch, R. (1992) A low sea-state study of the quieting effect of monomolecular films on the underlying ambient-noise field. *J. Acoust. Soc. Amer.*, **92**, 365–83.

Roll, H.U. (1965) *Physics of the Marine Atmosphere.* International Geophysics Series, Vol. 7. New York, NY: Academic Press.

Rosenberg, A.P. (1999) A new rough surface parabolic equation program for computing low-frequency acoustic forward scattering from the ocean surface. *J. Acoust. Soc. Amer.*, **105**, 144–53.

Ross, D. (1976) *Mechanics of Underwater Noise*, New York, NY: Pergamon Press.

Ross, D. (1993) On ocean underwater ambient noise. *Acoust. Bull.*, **18**(1), 5–8.

Ross, T. and Lavery, A. (2010) Acoustic detection of oceanic double-diffusive convection: A feasibility study. *J. Atmos. Oceanic Technol.*, **27**, 580–93.

Rouseff, D. and Tang, D. (2006) Internal wave effects on the ambient noise notch in the East China Sea: Model/data comparison. *J. Acoust. Soc. Amer.*, **120**, 1284–94.

Rouseff, D. and Tang, D. (2010) Internal waves as a proposed mechanism for increasing ambient noise in an increasingly acidic ocean. *J. Acoust. Soc. Amer.*, **127**(6), EL235–9, doi:10.1121/1.3425741.

Rouseff, D., Turgut, A., Wolf, S.N., Finette, S., Orr, M.H., Pasewark, B.H. et al. (2002) Coherence of acoustic modes propagating through shallow water internal waves. *J. Acoust. Soc. Amer.*, **111**, 1655–66.

Roux, P. and Kuperman, W.A. (2005) Time reversal of ocean noise. *J. Acoust. Soc. Amer.*, **117**, 131–6.

Roux, P., Kuperman, W.A., and Sabra, K.G. (2011) Ocean acoustic noise and passive coherent array processing. *C. R. Geosci.*, **343**, 533–47.

Roux, P., Kuperman, W.A., and the NPAL Group (Colosi, J.A., Cornuelle, B.D., Dushaw, B.D., Dzieciuch, M.A., Howe, B.M., Mercer, J.A. et al.). (2004) Extracting coherent wave fronts from acoustic ambient noise in the ocean. *J. Acoust. Soc. Amer.*, **116**, 1995–2003.

Rubenstein, D. and Greene, R. (1991) Modeling the acoustic scattering by under-ice-ridge keels. *J. Acoust. Soc. Amer.*, **89**, 666–72.

Rudnick, D.L. and Munk, W. (2006) Scattering from the mixed layer base into the sound shadow. *J. Acoust. Soc. Amer.*, **120**, 2580–94.

Runyan, L. (1991) 40 years on the frontier. *Datamation*, **37**(6), 34–57.

Sachs, D.A. and Silbiger, A. (1971) Focusing and refraction of harmonic sound and transient pulses in stratified media. *J. Acoust. Soc. Amer.*, **49**, 824–40.

Sadowski, W., Katz, R., and McFadden, K. (1984) Ambient noise standards for acoustic modeling and analysis. Nav. Underwater Syst. Ctr, Tech. Doc. 7265.

Saenger, R.A. (1984) Volume scattering strength algorithm: A first generation model. Nav. Underwater Syst. Ctr, Tech. Memo. 841193.

Sagen, H., Johannessen, O.M., and Sandven, S. (1990) The influence of sea ice on ocean ambient sound. In *Ice Technology for Polar Operations*, eds T.K.S. Murthy, J.G. Paren, W.M. Sackinger, and P. Wadhams, Second Inter. Conf. Ice Technology, Cambridge, England: Computational Mechanics Publications, pp. 415–26.

Sammelmann, G.S. (1998) SWAT acoustic modeling. Coastal Systems Station, Dahlgren Division, Naval Surface Warfare Center.

Santiago, J.A.F. and Wrobel, L.C. (2000) A boundary element model for underwater acoustics in shallow water. *Comput. Model. Eng. Sci.*, **1**,73–80.

Schecklman, S., Houser, D., Cross, M., Hernandez, D., and Siderius, M. (2011) Comparison of methods used for computing the impact of sound on the marine environment. *Marine Environ. Res.*, **71**, 342–50.

Schilling, R. (1998) Innovations in remote intervention: Pressures for change from the offshore oil industry. *UnderWater Mag.*, **10**(2), 21–4.

Schippers, P. (1995) The ALMOST PC-model for propagation and reverberation in range dependent environments. Proc. Undersea Defence Technology Conf. UDT Europe, Cannes, France, pp. 430–5.

Schippers, P. (1996) Modelling of sediment bottom propagation using ray theory. Proc. Undersea Defence Technology Conf. UDT Europe, London, pp. 43–47.

Schippers, P. (1998) First results of a newly developed ambient noise time signal model. Proc. Undersea Defence Technology Conf. UDT Europe, Hamburg, Germany, pp. 6–10.

Schmidt, H. (1988) SAFARI: Seismo-acoustic fast field algorithm for range-independent environments. User's guide. SACLANT Undersea Res. Ctr, Rept SR-113.

Schmidt, H. (1991) Numerical modeling in ocean seismo-acoustics. Proc. IEEE Oceans 91 Conf., Honolulu, pp 84–92.

Schmidt, H. (1999) *OASES, Version 2.2 User Guide and Reference Manual*. Cambridge, MA: Department of Ocean Engineering, Massachusetts Institute of Technology.

Schmidt, H. (2001) Bistatic scattering from buried targets in shallow water. In *Autonomous Underwater Vehicle and Ocean Modelling Networks: GOATS 2000 Conference Proceedings*, eds E. Bovio, R. Tyce, and H. Schmidt. SACLANT Undersea Res. Ctr, Conf. Proc. CP-46.

Schmidt, H. (2004) *OASES Version 3.1 User Guide and Reference Manual*. Cambridge, MA: Department of Ocean Engineering, Massachusetts Institute of Technology.

Schmidt, H. and Glattetre, J. (1985) A fast field model for three-dimensional wave propagation in stratified environments based on the global matrix method. *J. Acoust. Soc. Amer.*, **78**, 2105–14.

Schmidt, H. and Kuperman, W.A. (1995) Spectral representations of rough interface reverberation in stratified ocean waveguides. *J. Acoust. Soc. Amer.*, **97**, 2199–209.

Schmidt, H., Maguer, A., Bovio, E., Fox, W.L.J., LePage, K., Pace, N.G. et al. (1998) Generic oceanographic array technologies (GOATS) '98—Bistatic seabed scattering measurements using autonomous underwater vehicles. SACLANT Undersea Res. Ctr, Rept SR-302.

Schmidt, H., Seong, W., and Goh, J.T. (1995) Spectral super-element approach to range-dependent ocean acoustic modeling. *J. Acoust. Soc. Amer.*, **98**, 465–72.

Schmidt, H., Veljkovic, I., and Zampolli, M. (2004) Bistatic scattering from buried targets in shallow water—Experiment and modeling. Proc. 7th European Conf. Underwater Acoustics, Delft, the Netherlands, pp. 475–82.

Schmitt, R.W. (1987) The Caribbean sheets and layers transects (C-SALT) program. *EOS, Trans. Amer. Geophys. Union*, **68**, 57–60.

Schmitt, R.W., Nandi, P., Ross, T., Lavery, A., and Holbrook, W.S. (2005) Acoustic detection of thermocline staircases in the ocean and laboratory. Proc. MTS/IEEE Oceans 2005 Conf., Washington, DC, pp. 1052–5.

Schmitt, R.W., Perkins, H., Boyd, J.D., and Stalcup, M.C. (1987) C-SALT: An investigation of the thermohaline staircase in the western tropical North Atlantic. *Deep-Sea Res.*, **34**, 1655–65.

Schneider, H.G. (1990) MOCASSIN: Sound propagation and sonar range prediction model for shallow water environments. User's guide. Tech. Rept 1990-9, Forschungsanstalt der Bundeswehr für Wasserschall- und Geophysik, Kiel, Germany.

Schneider, H.G. (1993) Surface loss, scattering, and reverberation with the split-step parabolic wave equation model. *J. Acoust. Soc. Amer.*, **93**, 770–81.

Schramm, W.G. (1993) NEONS: A database management system for environmental data. *Earth Syst. Monit.*, **3**(4), 7–8. (Published by Dept of Commerce, NOAA, Silver Spring, Maryland, USA.)

Schreiner, H.F., Jr. (1990) The RANDI-PE noise model. Proc. IEEE Oceans 90 Conf., Washington, DC, pp. 576–7.

Schulkin, M. and Marsh, H.W. (1962) Sound absorption in sea water. *J. Acoust. Soc. Amer.*, **34**, 864–5.

Schulkin, M. and Marsh, H.W. (1963) Errata: Sound absorption in sea water (*J. Acoust. Soc. Am.* 34, 864 [1962]). *J. Acoust. Soc. Amer.*, **35**, 739.

Schulkin, M. and Mercer, J.A. (1985) Colossus revisited: A review and extension of the Marsh-Schulkin shallow water transmission loss model. (1962). Appl. Phys. Lab., Univ. Washington, APL-UW 8508.

Schurman, I.W., Siegmann, W.L., and Jacobson, M.J. (1991) An energy-conserving parabolic equation incorporating range refraction. *J. Acoust. Soc. Amer.*, **89**, 134–44.

Science Applications, Inc. (1977) Review of models of beam-noise statistics. Sci. Appl., Inc., SAI-78-696-WA.

Scott, I.J.G. and de Cogan, D. (2008) An improved transmission line matrix model for the 2D ideal wedge benchmark problem. *J. Sound Vib.*, **311**, 1213–27.

Scrimger, J.A., Evans, D.J., McBean, G.A., Farmer, D.M., and Kerman, B.R. (1987) Underwater noise due to rain, hail, and snow. *J. Acoust. Soc. Amer.*, **81**, 79–86.

Scully-Power, P.D. and Lee, D. (eds) (1984) Recent progress in the development and application of the parabolic equation. Nav. Underwater Syst. Ctr, Tech. Doc. 7145.

Seah, W.K.G. and Tan, H.-X. (2006) Multipath virtual sink architecture for underwater sensor networks. Proc. IEEE Oceans '06 Asia Pacific Conf., Singapore.

Sehgal, A., Tumar, I., and Schönwälder, J. (2010) AquaTools: An underwater acoustic networking simulation toolkit. Proc. IEEE Oceans 2010 Conf., Sydney.

Selsor, H.D. (1993) Data from the sea: Navy drifting buoy program. *Sea Technol.*, **34**(12), 53–8.

Semantic T.S. (2002) RAYSON: Underwater acoustic rays software. Product description literature.

Seong, W. (1990) Hybrid Galerkin boundary element—wavenumber integration method for acoustic propagation in laterally inhomogeneous media. Ph.D. thesis, Massachusetts Institute of Technology, Cambridge, MA.

Seong, W. and Choi, B. (2001) Multiplicative Padé PE formulation applied to SWAM'99 test cases. *J. Comput. Acoust.*, **9**, 227–41.

Sertlek, H. Ö. and Ainslie, M.A. (2014) A depth-dependent formula for shallow water propagation. *J. Acoust. Soc. Amer.*, **136**, 573–82.

Sertlek, H. Ö. and Aksoy, S. (2013) Analytical time domain normal mode solution of an acoustic waveguide with perfectly reflecting walls. *J. Comput. Acoust.*, **21**(2), 1–12, doi:10.1142/S0218396X12500269.

Shang, E.-C., Wang, Y.Y., and Gao, T.F. (2001) On the adiabaticity of acoustic propagation through nongradual ocean structures. *J. Comput. Acoust.*, **9**, 359–65.

Shapiro, G., Chen, F., and Thain, R. (2014) The effect of ocean fronts on acoustic wave propagation in the Celtic Sea. *J. Marine Syst.*, **139**, 217–26.

Shaw, P.T., Watts, D.R., and Rossby, H.T. (1978) On the estimation of oceanic wind speed and stress from ambient noise measurements. *Deep-Sea Res.*, **25**, 1225–33.

Shenderov, E.L. (1998) Some physical models for estimating scattering of underwater sound by algae. *J. Acoust. Soc. Amer.*, **104**, 791–800.

Shyu, H.-J. and Hillson, R. (2006) A software workbench for estimating the effects of cumulative sound exposure in marine mammals. *IEEE J. Oceanic Engr.*, **31**, 8–21.

Siderius, M., Harrison, C.H., and Porter, M.B. (2006) A passive fathometer technique for imaging seabed layering using ambient noise. *J. Acoust. Soc. Amer.*, **120**, 1315–23.

Siderius, M., Nielsen, P.L., Sellschopp, J., Snellen, M., and Simons, D. (2001) Experimental study of geo-acoustic inversion uncertainty due to ocean sound-speed fluctuations. *J. Acoust. Soc. Amer.*, **110**, 769–81.

Siderius, M. and Porter, M.B. (2006) Modeling techniques for marine-mammal risk assessment. *IEEE J. Oceanic Engr.*, **31**, 49–60.

Siderius, M. and Porter, M.B. (2008) Modeling broadband ocean acoustic transmissions with time-varying sea surfaces. *J. Acoust. Soc. Amer.,* **124**, 137–50.

Siderius, M., Snellen, M., Simons, D.G., and Onken, R. (2000) An environmental assessment in the Strait of Sicily: Measurement and analysis techniques for determining bottom and oceanographic properties. *IEEE J. Oceanic Engr.*, **25**, 364–86.

Siderius, M., Song, H., Gerstoft, P., Hodgkiss, W.S., Hursky, P., and Harrison, C.H. (2010) Adaptive passive fathometer processing. *J. Acoust. Soc. Amer.,* **127**, 2193–220.

Sidorovskaia, N.A. (2003) A new normal mode program SWAMP as a tool for modeling scattering effects in oceanic waveguides. In *Modelling and Experimental Measurements in Acoustics III*, eds D. Almorza, C.A. Brebbia, and R. Hernandez, Southampton, England: WIT Press, pp. 267–75.

Sidorovskaia, N.A. (2004) Systematic studies of pulse propagation in ducted oceanic waveguides in normal mode representation. *Eur. Phys. J. Appl. Phys.*, **25**, 113–31.

Sidorovskaia, N., Ackleh, A.S., Tiemann, C.O., Ioup, J.W., and Ioup, G.E. (2013) Prospects for short-term and long-term passive acoustic monitoring of environmental change impact on marine mammals. *J. Acoust. Soc. Amer.*, **134**, 4176 (abstract).

Siegmann, W.L., Jacobson, M.J., and Law, L.D. (1987) Effects of bottom attenuation on acoustic propagation with a modified ray theory. *J. Acoust. Soc. Amer.*, **81**, 1741–51.

Siegmann, W.L., Kriegsmann, G.A., and Lee, D. (1985) A wide-angle three-dimensional parabolic wave equation. *J. Acoust. Soc. Amer.*, **78**, 659–64.

Siemes, K., Snellen, M., Simons, D.G., Hermand, J.-P., Meyer, M., and LeGac, J.-C. (2008) High-frequency multibeam echosounder classification for rapid environmental assessment. Proc. Acoustics '08 Conf., Paris, pp. 4261–6.

Sienkiewicz, C.G., Boyd, M.L., Rosenzweig, J.R., and Pelton, R.M. (1975) Computer-programmed models used at the University of Washington's Applied Physics Laboratory for evaluating torpedo sonar performance. Appl. Phys. Lab., Univ. Washington, APL-UW 7611.

Silva, A., Rodriguez, O., Zabel, F., Huilery, J., and Jesus, S.M. (2010) Underwater acoustic simulations with a time variable acoustic propagation model. Proc. 10th European Conf. Underwater Acoustics, Istanbul, Turkey, pp. 989–96.

Simmen, J., Flatté, S.M., and Wang, G.-Y. (1997) Wavefront folding, chaos, and diffraction for sound propagation through ocean internal waves. *J. Acoust. Soc. Amer.*, **102**, 239–55.

Simons, D.G. and Laterveer, R. (1995) Normal-mode sound propagation modelling for matched-field processing. TNO Physics Electronics Lab., Rept FEL-95-A084.

Simons, D.G., McHugh, R., Snellen, M., McCormick, N.H., and Lawson, E.A. (2001) Analysis of shallow-water experimental acoustic data including a comparison with a broad-band normal-mode-propagation model. *IEEE J. Oceanic Engr.,* **26**, 308–23.

Simons, D.G. and Snellen, M. (1998) Multi-frequency matched-field inversion of benchmark data using a genetic algorithm. *J. Comput. Acoust.*, **6**, 135–50.

Simpson, S. D., Munday, P. L., Wittenrich, M. L., Manassa, R., Dixson, D. L., Gagliano, M. et al. (2011) Ocean acidification erodes crucial auditory behaviour in a marine fish. *Biol. Lett.*, **7**, 917–20, doi:10.1098/rsbl.2011.0293.

Simpson, S.D., Piercy, J.J.B., King, J., and Codling, E.A. (2013) Modelling larval dispersal and behaviour of coral reef fishes. *Ecol. Complex.*, **16**, 68–76.

Skretting, A. and Leroy, C.C. (1971) Sound attenuation between 200 Hz and 10 kHz. *J. Acoust. Soc. Amer.*, **49**, 276–82.

Smith, E.M. (1982) Active RAYMODE program user's guide. Nav. Underwater Syst. Ctr, Tech. Memo. 821081.

Smith, G.B. (1997) "Through the sensor" environmental estimation. *J. Acoust. Soc. Amer.*, **101**, 3046 (abstract).

Smith, K.B., Abrantes, A.A.M., and Larraza, A. (2003) Examination of time-reversal acoustics in shallow water and applications to noncoherent underwater communications. *J. Acoust. Soc. Amer.,* **113**, 3095–110.

Smith, K.B., Brown, M.G., and Tappert, F.D. (1992a) Ray chaos in underwater acoustics. *J. Acoust. Soc. Amer.*, **91**, 1939–49.

Smith, K.B., Brown, M.G., and Tappert, F.D. (1992b) Acoustic ray chaos induced by mesoscale ocean structure. *J. Acoust. Soc. Amer.*, **91**, 1950–9.

Smith, K.B., Hodgkiss, W.S., and Tappert, F.D. (1996) Propagation and analysis issues in the prediction of long-range reverberation. *J. Acoust. Soc. Amer.*, **99**, 1387–404.

Smith, K.B. and Tappert, F.D. (1993) UMPE: The University of Miami parabolic equation model, version 1.0. Marine Physical Lab., Scripps Inst. Oceanogr., Tech. Memo. 432.

Smith, K.B., Tappert, F.D., and Hodgkiss, W.S. (1993) Comparison of UMPE/PEREV bistatic reverberation predictions with observations in the ARSRP natural lab. *J. Acoust. Soc. Amer.*, **94**, 1766 (abstract).

Smith, P.W. Jr. (1974) Averaged sound transmission in range-dependent channels. *J. Acoust. Soc. Amer.,* **55**, 1197–204.

Snellen, M. and Simons, D.G. (2008) An assessment of the performance of global optimization methods for geo-acoustic inversion. *J. Comput. Acoust.*, **16**, 199–223.

Snellen, M., Simons, D.G., Siderius, M., Sellschopp, J., and Nielsen, P.L. (2001) An evaluation of the accuracy of shallow water matched field inversion results. *J. Acoust. Soc. Amer.,* **109**, 514–27.

Socheleau, F.-X., Laot, C., and Passerieux, J.-M. (2010) A maximum entropy framework for statistical modeling of underwater acoustic communication channels. Proc. IEEE Oceans 2010 Conf., Sydney.

Solomon, L.P., Ai, D.K.Y., and Haven, G. (1968) Acoustic propagation in a continuously refracted medium. *J. Acoust. Soc. Amer.*, **44**, 1121–9.

Solomon, L.P., Barnes, A.E., and Lunsford, C.R. (1977) Ocean route envelopes. Planning Syst., Inc., Tech. Rept TR-036049.

Song, H., Roux, P., Akal, T., Edelmann, G., Higley, W., Hodgkiss, W.S. et al. (2004) Time reversal ocean acoustic experiments at 3.5 kHz: Applications to active sonar and undersea communications. In *High Frequency Ocean Acoustics*, eds M.B. Porter, M. Siderius and W.A. Kuperman. American Institute of Physics Conference Proceedings 728, New York, NY: Springer-Verlag, pp. 522–9.

Song, H.C., Hodgkiss, W.S., Kuperman, W.A., Roux, P., Akal, T., and Stevenson, M. (2005) Experimental demonstration of adaptive reverberation nulling using time reversal. *J. Acoust. Soc. Amer.,* **118**, 1381–7.

Song, H.C., Kuperman, W.A., and Hodgkiss, W.S. (1998) A time-reversal mirror with variable range focusing. *J. Acoust. Soc. Amer.,* **103**, 3234–40.

Song, J. and Peng, Z. (2006) On far-field approximation of parabolic equation. *Acta Acust.*, **31**, 85–90 (in Chinese).

Soukup, R.J., Canepa, G., Simpson, H.J., Summers, J.E., and Gragg, R.F. (2007) Small-slope simulation of acoustic backscatter from a physical model of an elastic ocean bottom. *J. Acoust. Soc. Amer.*, **122**, 2551–9.

Southall, B.L., Bowles, A.E., Ellison, W.T., Finneran, J.J., Gentry, R.L., Greene, C.R. Jr. et al. (2007) Marine mammal noise exposure criteria: Initial scientific recommendations. *Aquatic Mammals*, **33**(4), 411–521.

Sozer, E.M., Stojanovic, M., and Proakis, J.G. (2000) Underwater acoustic networks. *IEEE J. Oceanic Engr.*, **25**, 72–83.

Spiga, I. (2015a) Review: Ocean acoustics modelling for environmental impact assessments. Natural environment research council (NERC), marine renewable energy knowledge exchange programme, Newcastle University, UK, pp. 1–50.

Spiga, I. (2015b) Developing and testing models of fish behaviour around tidal turbines. NERC MRE internship report. Natural environment research council (NERC), marine renewable energy knowledge exchange programme. In partnership with Sustainable Marine Energy Ltd, HR Wallingford Ltd, and University of Exeter, UK, pp. 1–32.

Spiesberger, J.L. (1989) Remote sensing of western boundary currents using acoustic tomography. *J. Acoust. Soc. Amer.*, **86**, 346–51.

Spiesberger, J.L. and Metzger, K. (1991a) Basin-scale tomography: A new tool for studying weather and climate. *J. Geophys. Res.*, **96**, 4869–89.

Spiesberger, J.L. and Metzger, K. (1991b) A new algorithm for sound speed in seawater. *J. Acoust. Soc. Amer.*, **89**, 2677–88.

Spiesberger, J.L., Metzger, K., and Furgerson, J.A. (1992) Listening for climatic temperature change in the northeast Pacific: 1983–1989. *J. Acoust. Soc. Amer.*, **92**, 384–96.

Spikes, C.H., Erskine, F.T., McEachern, J.F., and Backes, D.A. (1997) Littoral warfare technology testing. *Sea Technol.*, **38**(6), 71–7.

Spindel, R.C. (1985) Sound transmission in the ocean. *Ann. Rev. Fluid Mech.*, **17**, 217–37.

Spindel, R.C. and Worcester, P.F. (1990a) Ocean acoustic tomography programs: Accomplishments and plans. Proc. IEEE Oceans 90 Conf., Washington, DC, pp. 1–10.

Spindel, R.C. and Worcester, P.F. (1990b) Ocean acoustic tomography. *Sci. Am.*, **263**(4), 94–9.

Spindel, R.C. and Worcester, P.F. (1991) Ocean acoustic tomography: A decade of development. *Sea Technol.*, **32**(7), 47–52.

Spofford, C.W. (1973a) The Bell Laboratories multiple-profile ray-tracing program. Bell Telephone Labs.

Spofford, C.W. (1973b) A synopsis of the AESD workshop on acoustic-propagation modeling by non-ray-tracing techniques, 22–25 May 1973, Washington, DC. Acoustic Environmental Support Detachment, Off. Nav. Res., Tech. Note TN-73-05.

Spofford, C.W. (1974) The FACT model, Vol. I. Acoustic Environmental Support Detachment, Off. Nav. Res., Maury Ctr Rept 109.

Spofford, C.W. (1979) The ASTRAL model, Vol. I: Technical description. Sci. Appl. Inter., Inc., SAI-79-742-WA.

Spofford, C.W. and Haynes, J.M. (eds) (1983) Stochastic modeling workshop. Proc. workshop, Nav. Res. Lab., Oct. 26–28, 1982, Washington, DC.

Staelens, P. (2009) Defining and modeling the limits of high-resolution underwater acoustic imaging. Doctoral dissertation, Universiteit Gent, Belgium.

Stagpoole, V., Schenke, H.W., and Ohara, Y. (2016) A name directory for the ocean floor. *EOS, Trans. Amer. Geophys. Union*, **97**, 1–7, doi:10.1029/2016EO063177.

Stephen, R.A. (1990) Solutions to range-dependent benchmark problems by the finite-difference method. *J. Acoust. Soc. Amer.*, **87**, 1527–34.

Stewart, G.A. and James, S.D. (1992) Computer simulation of underwater sound propagation as a teaching aid. *Eur. J. Phys.*, **13**, 264–7.

Stickler, D.C. (1975) Normal-mode program with both the discrete and branch line contributions. *J. Acoust. Soc. Amer.*, **57**, 856–61.

Stickler, D.C. and Ammicht, E. (1980) Uniform asymptotic evaluation of the continuous spectrum contribution for the Pekeris model. *J. Acoust. Soc. Amer.*, **67**, 2018–24.

Stojanovic, M. and Preisig, J. (2009) Underwater acoustic communication channels: Propagation models and statistical characterization. *IEEE Commun. Mag.*, **47**(1), 84–9.

Stoll, R.D. (1974) Acoustic waves in saturated sediments. In *Physics of Sound in Marine Sediments*, ed. L. Hampton, New York, NY: Plenum, pp. 19–39.

Stoll, R.D. (1980) Theoretical aspects of sound transmission in sediments. *J. Acoust. Soc. Amer.*, **68**, 1341–50.

Stoll, R.D. (1989) *Sediment Acoustics*. Lecture Notes in Earth Sciences, Vol. 26. New York, NY: Springer-Verlag.

Stotts, S.A. (2002) Coupled-mode solutions in generalized ocean environments. *J. Acoust. Soc. Amer.*, **111**, 1623–43.

Stotts, S.A. and Bedford, N.R. (1991) Development of an Arctic ray model. *J. Acoust. Soc. Amer.*, **90**, 2299 (abstract).

Stotts, S.A., Knobles, D.P., and Koch, R.A. (2011) Scattering in a Pekeris waveguide from a rough bottom using a two-way coupled mode approach. *J. Acoust. Soc. Amer.*, **129**(5), EL172–8, doi:10.1121/1.3554724.

Stotts, S.A., Knobles, D.P., Koch, R.A., Grant, D.E., Focke, K.C., and Cook, A.J. (2004) Geoacoustic inversion in range-dependent ocean environments using a plane wave reflection coefficient approach. *J. Acoust. Soc. Amer.*, **115,** 1078–1102.

Stotts, S.A. and Koch, R.A. (2015) A two-way coupled mode formalism that satisfies energy conservation for impedance boundaries in underwater acoustics. *J. Acoust. Soc. Amer.*, **138**, 3383–96.

Stotts, S.A., Koch, R.A., and Bedford, N.R. (1994) Development of an Arctic ray model. *J. Acoust. Soc. Amer.*, **95**, 1281–98.

Strode, C., Mourre, B., and Rixen, M. (2012) Decision support using the multistatic tactical planning aid (MSTPA). *Ocean Dynam.*, **62**, 161–75, doi:10.1007/s10236-011-0483-7.

Sturm, F. (2002) Examination of signal dispersion in a 3-D wedge-shaped waveguide using 3DWAPE. *Acta Acust. United ACUST.*, **88**, 714–17.

Sturm, F. (2005) Numerical study of broadband sound pulse propagation in three-dimensional oceanic waveguides. *J. Acoust. Soc. Amer.*, **117**, 1058–79.

Sturm, F. and Fawcett, J.A. (2003) On the use of higher-order azimuthal schemes in 3-D PE modeling. *J. Acoust. Soc. Amer.,* **113**, 3134–45.

Sturm, F., Ivansson, S., Jiang, Y.-M., and Chapman, N.R. (2008) Numerical investigation of out-of-plane sound propagation in a shallow water experiment. *J. Acoust. Soc. Amer.,* **124**(6), EL341–6, doi:10.1121/1.3008068.

Su, R., Venkatesan, R., and Li, C. (2010) A review of channel modeling techniques for underwater acoustic communications. Eighteenth Annual Newfoundland Electrical and Computer Engineering Conference, St. John's, Newfoundland, Canada.

Sun, P. and Seah, W.K. (2007) Adaptive data delivery for underwater sensor networks. Proc. MTS/IEEE Oceans 2007 Conf., Vancouver, Canada.

Sundaram, B. and Zaslavsky, G.M. (1999) Wave analysis of ray chaos in underwater acoustics. *Chaos*, **9**, 483–92.

Svensson, E., Karasalo, I., and Hermand, J.-P. (2004) Hybrid raytrace modelling of an underwater acoustics communication channel. Proc. 7th European Conf. Underwater Acoustics, Delft, the Netherlands, pp. 1211–16.

Sverdrup, H.U., Johnson, M.W., and Fleming, R.H. (1942) *The Oceans. Their Physics, Chemistry, and General Biology*, Englewood Cliffs, NJ: Prentice-Hall.

Swift, S.A. and Stephen, R.A. (1994) The scattering of a low-angle pulse beam from seafloor volume heterogeneities. *J. Acoust. Soc. Amer.*, **96**, 991–1001.

Tactical Oceanography Wide Area Network (TOWAN). (1999) http://www7180.nrlssc.navy .mil/homepages/TOWAN/towanpgs/TOWANDataIndex.html

Tadeu, A., Costa, E.G.A., António, J., and Amado-Mendes, P. (2013) 2.5D and 3D Green's functions for acoustic wedges: Image source technique versus a normal mode approach. *J. Comput. Acoust.* **21**, 1–26, doi:10.1142/S0218396X12500257.

Talham, R.J. (1964) Ambient-sea-noise model. *J. Acoust. Soc. Amer.*, **36**, 1541–4.

Tamendarov, I.M. and Sidorovskaia, N.A. (2004) Unified acoustic model for simulating propagation and scattering effects in oceanic waveguides. *J. Acoust. Soc. Amer.,* **116**, 2527 (abstract).

Tan, H.-P., Diamant, R., Seah, W.K.G., and Waldmeyer, M. (2011) A survey of techniques and challenges in underwater localization. *Ocean Eng.*, **38**, 1663–76.

Tan, H.-X. and Seah, W.K.G. (2007) Distributed CDMA-based MAC protocol for underwater sensor networks. Proc. 32nd IEEE Conf. Local Computer Networks, LCN 2007, Dublin, Ireland, pp. 26–36.

Tang, D., Moum, J.N., Lynch, J.F., Abbot, P., Chapman, R., Dahl, P.H. et al. (2007). Shallow Water '06: A joint acoustic propagation/nonlinear internal wave physics experiment. *Oceanography*, **20**(4): 156–67.

Tang, X. and Tappert, F.D. (1997) Effects of internal waves on sound pulse propagation in the Straits of Florida. *IEEE J. Oceanic Engr.*, **22**, 245–55.

Tango, G.J. (1988) Numerical models for VLF seismic-acoustic propagation prediction: A review. *IEEE J. Oceanic Engr.*, **13**, 198–214.

Tao, Y. and Xu, X. (2007) Simulation study of multi-path characteristics of acoustic propagation in shallow water wireless channel. Proc. Inter. Conf. Wireless Communications, Networking and Mobile Computing, WiCom 2007, Shanghai, China, pp. 1068–70.

Tappert, F., Lee, D., and Weinberg, H. (1984) High-frequency propagation modeling using HYPER. *J. Acoust. Soc. Amer.*, **75** (Suppl. 1), S30 (abstract).

Tappert, F.D. (1977) The parabolic approximation method. In *Wave Propagation and Underwater Acoustics*, eds J.B. Keller and J.S. Papadakis. Lecture Notes in Physics, Vol. 70. New York, NY: Springer-Verlag, pp. 224–87.

Tappert, F.D. (1998) Parabolic equation modeling with the split-step Fourier algorithm in four dimensions. *J. Acoust. Soc. Amer.*, **103**, 2990 (abstract).

Tappert, F.D., Brown, M.G., Palmer, D.R., and Bezdek, H. (1988) Chaos in underwater acoustics. *J. Acoust. Soc. Amer.*, **83** (Suppl. 1), S36 (abstract).

Tappert, F.D., Brown, M.G., and Goni, G. (1991) Weak chaos in an area-preserving mapping for sound ray propagation. *Phys. Lett. A*, **153**, 181–5.

Tappert, F.D. and Lee, D. (1984) A range refraction parabolic equation. *J. Acoust. Soc. Amer.*, **76**, 1797–803.

Tappert, F.D., Spiesberger, J.L., and Wolfson, M.A. (2002) Study of a novel range-dependent propagation effect with application to the axial injection of signals from the Kaneohe source. *J. Acoust. Soc. Amer.*, **111**, 757–62.

Tappert, F.D. and Tang, X. (1996) Ray chaos and eigenrays. *J. Acoust. Soc. Amer.*, **99**, 185–95.

Taroudakis, M.I. and Makrakis, G.N. (eds) (2001) *Inverse Problems in Underwater Acoustics*, New York, NY: Springer-Verlag.

Taylor, K.E. (2000) Summarizing multiple aspects of model performance in a single diagram. Program for Climate Model Diagnosis and Intercomparison (PCMDI), University of California, Lawrence Livermore National Laboratory, Report No. 55 (UCRL-JC-138644).

Taylor, K.E. (2001) Summarizing multiple aspects of model performance in a single diagram. *J. Geophys. Res.*, **106**, 7183–92, doi:10.1029/2000JD900719.

Teague, W.J., Carron, M.J., and Hogan, P.J. (1990) A comparison between the generalized digital environmental model and Levitus climatologies. *J. Geophys. Res.*, **95**, 7167–83.

Tennessen, J.B. and Parks, S.E. (2016) Acoustic propagation modeling indicates vocal compensation in noise improves communication range for North Atlantic right whales. *Endang. Species. Res.*, **30**, 225–37, doi:10.3354/esr00738.

The Acoustic Mid-Ocean Dynamics Experiment Group. (1994) Moving ship tomography in the North Atlantic. *EOS, Trans. Amer. Geophys. Union*, **75**, 17–23.

Theriault, J.A. and Ellis, D.D. (1997) Shallow-water low-frequency active sonar modelling issues. Proc. MTS/IEEE Oceans '97 Conf., Halifax, Nova Scotia, pp. 672–8.

Theriault, J.A., Pecknold, S., Collison, N., and Calnan, C. (2006) Modelling multibeam reverberation with an N × 2D model. Proc. 8th European Conf. Underwater Acoustics, Carvoeiro, Portugal.

The Ring Group. (1981) Gulf Stream cold-core rings: Their physics, chemistry, and biology. *Science*, **212**, 1091–100.

Thode, A., Mellinger, D.K., Stienessen, S., Martinez, A., and Mullin, K. (2002) Depth-dependent acoustic features of diving sperm whales (*Physeter macrocephalus*) in the Gulf of Mexico. *J. Acoust. Soc. Amer.*, **112**, 308–21.

Thompson, B.B., Marks, R.J., II, El-Sharkawi, M.A., Fox, W.J., and Miyamoto, R.T. (2003) Inversion of neural network underwater acoustic model for estimation of bottom parameters using modified particle swarm optimizers. Proc. IEEE International Joint Conf. Neural Networks, Toulouse, France, pp. 1301–6.

Thompson, P.M., Hastie, G.D., Nedwell, J., Barham, R., Brookes, K.L., Cordes, L.S. et al. (2013) Framework for assessing impacts of pile-driving noise from offshore wind farm construction on a harbour seal population. *Environ. Impact Assess. Rev.*, **43**, 73–85.

Thomson, D.J. (1990) Wide-angle parabolic equation solutions to two range-dependent benchmark problems. *J. Acoust. Soc. Amer.*, **87**, 1514–20.

Thomson, D.J., Brooke, G.H., and DeSanto, J.A. (1990) Numerical implementation of a modal solution to a range-dependent benchmark problem. *J. Acoust. Soc. Amer.*, **87**, 1521–6.

Thomson, D.J. and Chapman, N.R. (1983) A wide-angle split-step algorithm for the parabolic equation. *J. Acoust. Soc. Amer.*, **74**, 1848–54.

Thomson, D.J. and Mayfield, M.E. (1994) An exact radiation condition for use with the *a posteriori* PE method. *J. Comput. Acoust.*, **2**, 113–32.

Thomson, D.J. and Wood, D.H. (1987) *A posteriori* phase corrections to the parabolic equation. In *Progress in Underwater Acoustics*, ed. H.M. Merklinger, New York, NY: Plenum, pp. 425–31.

Thorp, W.H. (1967) Analytic description of the low-frequency attenuation coefficient. *J. Acoust. Soc. Amer.*, **42**, 270.

Thorpe, S.A. (1992) The breakup of Langmuir circulation and the instability of an array of vortices. *J. Phys. Oceanogr.*, **22**, 350–60.

Thorpe, S.A. (2004) Langmuir circulation. *Ann. Rev. Fluid Mech.*, **36**, 55–79.

Thorpe, S.A., Cure, M.S., Graham, A., and Hall, A.J. (1994) Sonar observations of Langmuir circulation and estimation of dispersion of floating particles. *J. Atmos. Ocean. Technol.*, **11**, 1273–94.

Thorsos, E.I. (1990) Acoustic scattering from a "Pierson-Moskowitz" sea surface. *J. Acoust. Soc. Amer.*, **88**, 335–49.

Thorsos, E.I. and Perkins, J.S. (2007) Reverberation modeling issues highlighted by the first reverberation modeling workshop. *J. Acoust. Soc. Amer.*, **122**, 3091 (abstract).

Thorsos, E.I. and Perkins, J.S. (2008) Overview of the reverberation modeling workshops. In *Proceedings of the International Symposium on Underwater Reverberation and Clutter*, eds P.L. Nielsen, C.H. Harrison, and J.-C. LeGac, Lerici, Italy, pp. 3–14.

Tielbürger, D., Finette, S., and Wolf, S. (1997) Acoustic propagation through an internal wave field in a shallow water waveguide. *J. Acoust. Soc. Amer.*, **101**, 789–808.

Tindle, C.T. (2002) Wavefronts and waveforms in deep-water sound propagation. *J. Acoust. Soc. Amer.*, **112**, 464–75.

Tindle, C.T. and Bold, G.E.J. (1981) Improved ray calculations in shallow water. *J. Acoust. Soc. Amer.*, **70**, 813–9.

Tindle, C.T. and Deane, G.B. (2005) Shallow water sound propagation with surface waves. *J. Acoust. Soc. Amer.*, **117**, 2783–94.

Tindle, C.T., Deane, G.B., and Preisig, J.C. (2009) Reflection of underwater sound from surface waves. *J. Acoust. Soc. Amer.*, **125**, 66–72.

Tindle, C.T., Hobaek, H., and Muir, T.G. (1987) Downslope propagation of normal modes in a shallow water wedge. *J. Acoust. Soc. Amer.*, **81**, 275–86.

Tindle, C.T., O'Driscoll, L.M., and Higham, C.J. (2000) Coupled mode perturbation theory of range dependence. *J. Acoust. Soc. Amer.*, **108**, 76–83.

Tindle, C.T. and Zhang, Z.Y. (1992) An equivalent fluid approximation for a low shear speed ocean bottom. *J. Acoust. Soc. Amer.*, **91**, 3248–56.

Tindle, C.T. and Zhang, Z.Y. (1997) An adiabatic normal mode solution for the benchmark wedge. *J. Acoust. Soc. Amer.*, **101**, 606–9.

Todd, V.L.G., Todd, I.B., Gardiner, J.C., and Morrin, E.C.N. (2015) *Marine Mammal Observer and Passive Acoustic Monitoring Handbook*, Exeter, England: Pelagic Publishing.

Tolstoy, A. (1992) Review of matched field processing for environmental inverse problems. *Inter. J. Modern Phys. C*, **3**, 691–708.

Tolstoy, A. (1993) *Matched Field Processing for Underwater Acoustics*, Singapore: World Scientific Publishing.

Tolstoy, A. (1996) 3-D propagation issues and models. *J. Comput. Acoust.*, **4**, 243–71.

Tolstoy, A. (2000) Applications of matched-field processing to inverse problems in underwater acoustics. *Inverse Probl.*, **16**, 1655–66.

Tolstoy, A. (2001) What about adiabatic normal modes? *J. Comput. Acoust.*, **9**, 287–309.

Tolstoy, A. (2008) Volumetric (tomographic) three-dimensional geoacoustic inversion in shallow water. *J. Acoust. Soc. Amer.*, **124**, 2793–804.

Tolstoy, A., Berman, D.H., and Franchi, E.R. (1985) Ray theory versus the parabolic equation in a long-range ducted environment. *J. Acoust. Soc. Amer.*, **78**, 176–89.

Tolstoy, A., Smith, K., and Maltsev, N. (2001) The SWAM'99 Workshop—An overview. *J. Comput. Acoust.*, **9**, 1–16.

Tolstoy, I. and Clay, C.S. (1966) *Ocean Acoustics*, New York, McGraw-Hill (reprinted 1987 by the Acoustical Society of America).

Tozzi, J.J., Slaughter, S., and Levinson, B. (2006) Testing, validation and transparency of marine mammal acoustic effects models. Proc. Ninth International Marine Environmental Modelling Seminar, Rio de Janeiro.

Train, K.E. (2009) *Discrete Choice Methods with Simulation*, 2nd edn. New York, NY: Cambridge University Press.

Tseng, C.-H. and Cole, M. (2007) A hybrid algorithm based on neuron-fuzz and wavelet transforms for wideband sonar detection in a reverberation-limited environment. Proc. 15th European Signal Processing Conf., EUSIPCO 2007, Poznań, Poland.

Tseng, C.-H. and Cole, M. (2009) Optimum multi-target detection using an ANC neuro-fuzzy scheme and wideband replica correlator. Proc. IEEE ICASSP 2009, Taipei, Taiwan, pp. 1369–72.

Tsouvalas, A. and Metrikine, A.V. (2016) Noise reduction by the application of an air-bubble curtain in offshore pile driving. *J. Sound Vib.*, **371**, 150–70, doi:10.1016/j.jsv.2016.02.025.

Tsuchiya, T., Okuyama, T., Endoh, N., and Anada, T. (1999) Numerical analysis of acoustical propagation in ocean with warm and cold water mass used by the three-dimensional wide-angle parabolic equation method. *Jpn. J. Appl. Phys.*, **38**, 3351–5.

Tsuru, H. and Iwatsu, R. (2010) Accurate numerical prediction of acoustic wave propagation. *Int. J. Adapt. Control Signal Process.*, **24**, 128–41, doi:10.1002/acs.1118.

Tuovila, S.M. (1989) SEARAY sonar simulation model. Nav. Coastal Syst. Ctr., Tech. Note NCSC TN 946-88.

Twersky, V. (1957) On scattering and reflection of sound by rough surfaces. *J. Acoust. Soc. Amer.*, **29**, 209–25.

UK Ministry of Defence, Director Naval Surveying, Oceanography and Meteorology. (1995) User Guide. Hodgson propagation loss model version 4.0.

Untersteiner, N. (1966) Sea ice. In *The Encyclopedia of Oceanography*, ed. R.W. Fairbridge, New York, NY: Van Nostrand Reinhold Co., pp. 777–81.

Upton, Z.M., Pulli, J.J., Myhre, B., and Blau, D. (2006) A reflected energy prediction model for long-range hydroacoustic reflection in the oceans. *J. Acoust. Soc. Amer.*, **119**, 153–60.

Urick, R.J. (1979) *Sound Propagation in the Sea*, Washington, DC: US Government Printing Office.

Urick, R.J. (1982) *Sound Propagation in the Sea*, Los Altos, CA: Peninsula Publishing.

Urick, R.J. (1983) *Principles of Underwater Sound*, 3rd edn. New York, NY: McGraw-Hill.

Urick, R.J. (1984) *Ambient Noise in the Sea*, Washington, DC: Naval Sea Systems Command.

Uscinski, B.J. and Nicholson, J.R.S. (2008) Horizontal structure of acoustic intensity fluctuations in the ocean. *J. Acoust. Soc. Amer.*, **124**, 1963–73.

Uscinski, B.J. and Stanek, C.J. (2002) Acoustic scattering from a rough sea surface: The mean field by the integral equation method. *Waves Random Media*, **12**, 247–63.

US Department of Defense. (1994) DoD modeling and simulation (M&S) management. DoD Directive 5000.59 (updated with Change 1 in 1998).

US Department of Defense. (1996) Verification, validation and accreditation (VV&A) recommended practices guide (RPG). Office of the Director of Defense Research and Engineering, Defense Modeling and Simulation Office, Washington, DC.

US Department of Defense. (2002) Mandatory procedures for major defense acquisition programs (MDAPS) and major automated information system (MAIS) acquisition programs. DOD Regulation 5000.2-R. Office of the Secretary of Defense, Washington, DC.

US Department of Defense. (2009) DoD Modeling and Simulation (M&S) Verification, Validation, and Accreditation (VV&A). DoD Instruction 5000.61. Office of the Secretary of Defense, Washington, DC.

US Department of Energy. (2009) Report to Congress on the Potential Environmental Effects of Marine and Hydrokinetic Energy Technologies. Office of Energy Efficiency & Renewable Energy, Wind and Hydropower Technologies Program, DOE/GO-102009-2955, Washington, DC.

US Department of Labor, Bureau of Labor Statistics. (2016) Occupational Outlook Handbook, 2016–17 Edition., Wind Turbine Technicians http://www.bls.gov/ooh/installation-maintenance-and-repair/wind-turbine-technicians.htm.

US Department of the Navy. (1999) Interactive multisensor analysis training. Joint publication of the SPAWAR Systems Center and the Naval Surface Warfare Center, Carderock Division.

US Department of the Navy. (2000a) Simulation based acquisition (SBA) status and international implications. Prepared by the Department of the Navy Acquisition Reform Office for The Technical Cooperation Program (TTCP) panel on Systems Engineering for Defence Modernisation.

US Department of the Navy. (2000b) Strategy for research & development: A roadmap to a vision of operational oceanography. Office of the Chief of Naval Operations, Oceanographer of the Navy (CNO N096), Washington, DC.

US Department of the Navy. (2007) Science and Technology, Ocean Battlespace Sensing. Office of Naval Research, Ocean Acoustics (Code 321), Arlington, Virginia, USA.

US General Accounting Office. (1979) Guidelines for model evaluation. PAD-79-17.

US Navy Air ASW Project Office. (1996a) User's guide for the environmental acoustics server software package. Program Executive Office, Air ASW, Assault, and Special Mission Programs (PMA-264).

US Navy Air ASW Project Office. (1996b) Operating system interface description for the environmental acoustics server software package. Program Executive Office, Air ASW, Assault, and Special Mission Programs (PMA-264).

Vaccaro, R.J. (ed.) (1998) The past, present, and future of underwater acoustic signal processing. *IEEE Sig. Process. Mag.*, **15**(4), 21–51.

Vadov, R.A. (2001) Long-range sound propagation in the central region of the Baltic Sea. *Acoust. Phys.*, **47**, 150–9.

Vadov, R.A. (2005) On the predictability of the positions of the convergence zones in the ocean. *Acoust. Phys.*, **51**, 265–70.

Vagle, S., Large, W.G., and Farmer, D.M. (1990) An evaluation of the WOTAN technique of inferring oceanic winds from underwater ambient sound. J. *Atmos. Oceanic Technol.*, **7**, 576–95.

van Aken, H.M. (2006) *The Oceanic Thermohaline Circulation: An Introduction*, New York, NY: Springer.

van Leijen, A.V. (2010) The sound of sediments: Acoustic sensing in uncertain environments. PhD Dissertation, Universiteit van Amsterdam, Amsterdam, the Netherlands.

van Leijen, A.V., Rothkrantz, L., and Groen, F. (2011) Metaheuristic optimization of acoustic inverse problems. *J. Comput. Acoust.*, **19**, 407–31.

van Leijen, A.V., van Norden, W.L., and Bolderheij, F. (2009) Unification of radar and sonar coverage modeling. Proc. 12th International Conf. Information Fusion, Seattle, pp. 1673–8.

van Moll, C.A.M., Ainslie, M.A., and van Vossen, R. (2009) A simple and accurate formula for the absorption of sound in seawater. *IEEE J. Oceanic Engr.*, **34**, 610–16.

van Parijs, S.M., Clark, C.W., Sousa-Lima, R.S., Parks, S.E., Rankin, S., Risch, D. et al. (2009) Management and research applications of real-time and archival passive acoustic sensors over varying temporal and spatial scales. *Mar. Ecol. Prog. Ser.*, **395**, 21–36, doi:10.3354/meps08123.

van Uffelen, L.J., Worcester, P.F., Dzieciuch, M.A., and Rudnick, D.L. (2009) The vertical structure of shadow-zone arrivals at long range in the ocean. *J. Acoust. Soc. Amer.*, **125**, 3569–88.

van Valkenburg-Haarst, T.Y.C., van Norden, W.L., van der Meiden, H.A., and ten Holter, K.P.A. (2010) Deployment optimization of electro-optical sensor systems for naval missions. In *Proc. SPIE 7834, Electro-Optical and Infrared Systems: Technology and Applications VII*, eds D.A. Huckridge and R.R. Ebert, Toulouse, France, doi:10.1117/12.864829.

van Walree, P.A. (2013) Propagation and scattering effects in underwater acoustic communication channels. *IEEE J. Oceanic Engr.*, **38**, 614–31.

van Walree, P.A., Jenserud, T., and Smedsrud, M. (2008) A discrete-time channel simulator driven by measured scattering functions. *IEEE J. Sel. Areas Commun.*, **26**, 1628–37.

van Walree, P., Otnes, R., and Jenserud, T. (2016) WATERMARK: A realistic benchmark for underwater acoustic modems. Proc. IEEE Third Underwater Communications and Networking Conf. (UComms), pp. 1–4, doi:10.1109/UComms.2016.7583423.

Vastano, A.C. and Owens, G.E. (1973) On the acoustic characteristics of a Gulf Stream cyclonic ring. *J. Phys. Oceanogr.*, **3**, 470–8.

Veenstra, D. (1998) A COTS approach to sonar simulation/stimulation. *RTC*, **7**(12), 135–7.

Vendetti, A., Zeidler, E., and Bartberger, C. (1993a) The NAWC bistatic acoustic model (version 3.0) software modifications and user's guide. Nav. Air Warfare Ctr Rept.

Vendetti, A., Zeidler, E., and Bartberger, C. (1993b) The NAWC monostatic acoustic model (version 1.1) software modifications and user's guide. Nav. Air Warfare Ctr Rept.

Vent, R.J. (1972) Acoustic volume-scattering measurements at 3.5, 5.0, and 12.0 kHz in the eastern Pacific Ocean: Diurnal and season variations. *J. Acoust. Soc. Amer.*, **52**, 373–82.

Viala, C., Noël, C., and Lapierre, G. (2004) RAYSON: A real time underwater communication simulator and performance estimator. Proc. 7th French Workshop on Underwater Acoustics, Brest, France.

Viala, C., Noël, C., Stéphan, Y., and Asch, M. (2005) Real-time geoacoustic inversion of large band signals. Proc. IEEE Oceans 2005 Europe Conf., Brest, France, pp. 1392–5.

Viala, C., Noël, C., Stéphan, Y., and Asch, M. (2006) *Real-Time Geoacoustic Inversion of Broad Band Signals in Deep Water*, Brest, France: CMM'06: Caractérisation du Milieu Marin.

Virovlyansky, A.L. (2005a) Statistical description of ray chaos in an underwater acoustic waveguide. *Acoust. Phys.*, **51**, 71–80.

Virovlyansky, A.L. (2005b) Signal travel times along chaotic rays in long-range sound propagation in the ocean. *Acoust. Phys.*, **51**, 271–81.

Virovlyansky, A.L. (2017) Matched shadow processing. *J. Acoust. Soc. Amer.*, **142**(1), EL136–42, doi:10.1121/1.4994684.

Virovlyansky, A.L. and Zaslavsky, G.M. (2007) Ray and wave chaos in problems of sound propagation in the ocean. *Acoust. Phys.*, **53**, 282–97.

von Benda-Beckmann, A.M., Wensveen, P.J., Kvadsheim, P.H., Lam, F.-P.A., Miller, P.J.O., Tyack, P.L. et al. (2014) Modeling effectiveness of gradual increases in source level to mitigate effects of sonar on marine mammals. *Conserv. Biol.*, **28**(1), 119–28. doi:10.1111/cobi.12162.

Von Winkle, W.A. (1963) Vertical directionality of deep ocean noise. Navy Underwater Sound Lab., Rept 600.

Vosbein, H.T. (2002) Towards better sonar performance predictions. Proc. MTS/IEEE Oceans '02 Conf., Biloxi, MS, pp. 352–7.

Wagner, D.H., Mylander, W.C., and Sanders, T.J. (eds) (1999) *Naval Operations Analysis*, 3rd edn. Annapolis, MD: Naval Institute Press.

Wagstaff, M.D., Meredith, R.W., and Terrill-Stolper, H.A. (1997) Multi-beam high frequency imaging in a range-dependent environment. Proc. MTS/IEEE Oceans '97 Conf., Halifax, Nova Scotia, pp. 927–31.

Wagstaff, R.A. (1973) RANDI: Research ambient noise directionality model. Nav. Undersea Ctr, Tech. Pub. 349.

Wagstaff, R.A. (1981) Low-frequency ambient noise in the deep sound channel—The missing component. *J. Acoust. Soc. Amer.*, **69**, 1009–14.

Wagstaff, R.A. (1982) Noise field calculation or measurement simulation? Some comments on ambient noise modeling. Proc. MTS/IEEE Oceans 82 Conf., Washington, DC, pp. 187–91.

Wainman, C.K. (2012) Estimating the upper ocean vertical temperature structure from surface temperature as applied to the southern Benguela. PhD thesis, University of Cape Town, Cape Town, South Africa.

Wales, S.C. and Heitmeyer, R.M. (2002) An ensemble source spectra model for merchant ship-radiated noise. *J. Acoust. Soc. Amer.*, **111**, 1211–31.

Wallcraft, A.J., Hurlburt, H.E., Metzger, E.J., Rhodes, R.C., Shriver, J.F., and Smedstad, O.M. (2002) Real-time ocean modeling systems. *Comput. Sci. Engr*, **4**(2), 50–7.

Walker, S.C., Roux, P., and Kuperman, W.A. (2005) Focal depth shifting of a time reversal mirror in a range-independent waveguide. *J. Acoust. Soc. Amer.*, **118**, 1341–7.

Wall, G. R., Nystrom, E. A., and Litten, S. (2006) Use of an ADCP to Compute Suspended-Sediment Discharge in the Tidal Hudson River, New York. US Geological Survey Scientific Investigations Report 2006-5055, 26 pp.

Walsh, J., Bashir, I., Thies, P.R., Johanning, L., and Blondel, P. (2016) Modelling the propagation of underwater acoustic emissions for condition monitoring of marine renewable energy. *In Progress in Renewable Energies Offshore*. ed. C.G. Soares, Proc. 2nd International Conf. on Renewable Energies Offshore (RENEW2016), Lisbon, Portugal, 24–26 October 2016, CRC Press/Balkema, Leiden, the Netherlands.

Wan, L., Badiey, M., and Knobles, D.P. (2016) Geoacoustic inversion using low frequency broadband acoustic measurements from L-shaped arrays in the Shallow Water 2006 Experiment. *J. Acoust. Soc. Amer.*, **140**, 2358–73.

Wang, J., Cai, P., and Yuan, D. (2010) An underwater acoustic channel simulator for UUV communication performance testing. Proc. 2010 IEEE Inter. Conf. Information and Automation, Harbin, China, pp. 2286–90.

Wang, J., Stewart, R., Dyaur, N., and Bell, M.L. (2014) Marine guided-waves: Applications and filtering using physical modeling data. Proc. 2014 SEG Annual Meeting, Denver, pp. 4804–9.

Wang, J., Stewart, R.R., Dyaur, N.I., and Bell, M.L. (2016) Marine guided waves: Subbottom property estimation and filtering using physical modeling data. *Geophysics*, **81**, 303–15 doi:10.1190/geo2015-0401.1.

Wang, L.S., Heaney, K., Pangerc, T., Theobald, P., Robinson, S.P., and Ainslie, M. (2014) Review of underwater acoustic propagation models. NPL Rept AC 12, National Physical Laboratory, Teddington, UK.

Wang, N. and Liu, J.-Z. (2002) New approach to the normal mode method in underwater acoustics. *Chin. Phys.*, **11**, 456–60.

Wang, P., Zhang, L., and Li, V.O.K. (2013) A stratified acoustic model accounting for phase shifts for underwater acoustic networks. *Sensors*, **13**, 6183–203, doi:10.3390/s130506183.

Wang, Q. and Gong, X. (2007) Shallow-water bottom target detection based on time, frequency dispersive channel and adaptive beamforming algorithm. In *Bio-Inspired Computational Intelligence and Applications: International Conference on Life System Modeling and Simulation, LSMS 2007, Shanghai, China, September 2007, Proceedings (Lecture Notes in Computer Science, LNCS 4688)*, eds K. Li, M. Fei, G.W. Irwin, and S. Ma, New York, NY: Springer-Verlag, pp. 561–70.

Wang, X. and Cai, Z. (2006) Modeling simulation system for oceanic acoustic information engagement with UML. *J. Syst. Simulat.* (China), **18**, 547–69.

Ward, P.D. (1989) A novel method of finding normal modes in shallow water. *J. Sound Vib.*, **128**, 349–54.

Ward, P.D., Varley, P., and Clarke, T. (2004) Detecting marine mammals using active sonar—A theoretical consideration. *Proc. Inst. Acoust.*, **26** (Part 6), Symposium on Bio-Sonar Systems and Bioacoustics.

Warn-Varnas, A.C., Chin-Bing, S.A., King, D.B., Hallock, Z., and Hawkins, J.A. (2003) Ocean-acoustic solitary wave studies and predictions. *Surveys in Geophys.*, **24**, 39–79.

Warn-Varnas, A., Chin-Bing, S.A., King, D.B., Hawkins, J., and Lamb, K. (2009) Effects on acoustics caused by ocean solitons, Part A: Oceanography. *Nonlinear Anal. Theory Methods Appl.*, **71**, 1807–17.

Wathelet, A., Strode, C., Vermeij, A., and Been, R. (2008) Optimisation in the multistatic tactical planning aid (MSTPA). NATO Undersea Res. Ctr, Rept NURC-FR-2008-013.

Watkins, W.A. and Schevill, W.E. (1977) Sperm whale codas. *J. Acoust. Soc. Amer.*, **62**, 1485–90.

Watkins, W.A., Tyack, P., Moore, K.E., and Bird, J.E. (1987) The 20-Hz signals of finback whales (Balaenoptera physalus). *J. Acoust. Soc. Amer.*, **82**, 1901–12.

Watson, A.G.D. (1958) The effect of the Earth's sphericity in the propagation of sound in the sea. Admiralty Res. Lab., ARL/N29/L.

Watson, W.H. and McGirr, R. (1975) RAYWAVE II: A propagation loss model for the analysis of complex ocean environments. Nav. Undersea Ctr, Tech. Note 1516.

Weickmann, A. and Jones, R.M. (1994) A FORTRAN program for performing Abel transforms. NOAA Tech. Memo. ERL ETL-244.

Weickmann, A.M., Riley, J.P., Georges, T.M., and Jones, R.M. (1989) EIGEN—A Program to compute eigenrays from HARPA/HARPO raysets. Wave Propagation Lab., NOAA Tech. Memo. ERL WPL-160.

Weinberg, H. (1969) CONGRATS I: Ray plotting and eigenray generation. Navy Underwater Sound Lab., Rept 1052.

Weinberg, H. (1971) A continuous-gradient curve-fitting technique for acoustic-ray analysis. *J. Acoust. Soc. Amer.*, **50**, 975–84.

Weinberg, H. (1973) Navy interim surface ship model (NISSM) II. Nav. Underwater Syst. Ctr, Tech. Rept 4527 (also, Nav. Undersea Ctr, Tech. Pub. 372).

Weinberg, H. (1975) Application of ray theory to acoustic propagation in horizontally stratified oceans. *J. Acoust. Soc. Amer.*, **58**, 97–109.

Weinberg, H. (1981) Effective range derivative for acoustic propagation loss in a horizontally stratified ocean. *J. Acoust. Soc. Amer.*, **70**, 1736–42.

Weinberg, H. (1982) Generic sonar model. Proc. MTS/IEEE Oceans 82 Conf., Washington, DC, pp. 201–5.

Weinberg, H. (1985a) Multilayer expansion for computing acoustic pressure in a horizontally stratified ocean. Proc. 11th IMACS World Congress, System Simulation and Scientific Computation, Oslo, Norway, pp. 135–8.

Weinberg, H. (1985b) Generic sonar model. Nav. Underwater Syst. Ctr, Tech. Doc. 5971D.

Weinberg, H. (2000) CASS roots. Proc. MTS/IEEE Oceans 2000 Conf., Providence, Rhode Island, pp. 1071–6.

Weinberg, H. and Burridge, R. (1974) Horizontal ray theory for ocean acoustics. *J. Acoust. Soc. Amer.*, **55**, 63–79.

Weinberg, H. and Keenan, R.E. (1996) Gaussian ray bundles for modeling high-frequency propagation loss under shallow-water conditions. *J. Acoust. Soc. Amer.*, **100**, 1421–31.

Weinberg, H., Keenan, R.E., and Aidala, F.E. Jr. (1997). Uniqueness problems in extracting environmental parameters from high-frequency, shallow-water reverberation measurements. In *High Frequency Acoustics in Shallow Water*, eds N.G. Pace, E. Pouliquen, O. Bergem, and A.P. Lyons, SACLANT Undersea Res. Ctr, Conf. Proc. CP-45, pp. 587–92.

Weinberg, N.L. and Dunderdale, T. (1972) Shallow water ray tracing with nonlinear velocity profiles. *J. Acoust. Soc. Amer.*, **52**, 1000–10.

Weinberg, N.L. and Zabalgogeazcoa, X. (1977) Coherent ray propagation through a Gulf Stream ring. *J. Acoust. Soc. Amer.*, **62**, 888–94.

Wells, D.K. and Wargelin, R.M (1985) Programs and products of the Naval Oceanographic Office. *Mar. Tech. Soc. J.*, **19** (3), 18–25.

Wenz, G.M. (1962) Acoustic ambient noise in the ocean: Spectra and sources. *J. Acoust. Soc. Amer.*, **34**, 1936–56.

Weston, D.E. (1960) A Moiré fringe analog of sound propagation in shallow water. *J. Acoust. Soc. Amer.*, **32**, 647–54.

Weston, D.E. (1971) Intensity-range relations in oceanographic acoustics. *J. Sound Vib.*, **18**, 271–87.

Weston, D.E. (1976) Propagation in water with uniform sound velocity but variable-depth lossy bottom. *J. Sound Vib.*, **47**, 473–83.

Weston, D.E. (1980a) Acoustic flux formulas for range-dependent ocean ducts. *J. Acoust. Soc. Amer.*, **68**, 269–81.

Weston, D.E. (1980b) Acoustic flux methods for oceanic guided waves. *J. Acoust. Soc. Amer.*, **68**, 287–96.

Weston, D.E. (1998) Ray acoustics for fluids. In *Handbook of Acoustics*, ed. M.J. Crocker, New York, NY: Wiley-Interscience, pp. 39–45.

Weston, D.E. and Rowlands, P.B. (1979) Guided acoustic waves in the ocean. *Rep. Prog. Phys.*, **42**, 347–87.

Westwood, E.K. (1989a) Complex ray methods for acoustic interaction at a fluid-fluid interface. *J. Acoust. Soc. Amer.*, **85**, 1872–84.

Westwood, E.K. (1989b) Ray methods for flat and sloping shallow-water waveguides. *J. Acoust. Soc. Amer.*, **85**, 1885–94.

Westwood, E.K. (1989c) Acoustic propagation modeling in shallow water using ray theory. Appl. Res. Labs., Univ. Texas at Austin, USA, ARL-TR-89-6.

Westwood, E.K. (1990) Ray model solutions to the benchmark wedge problems. *J. Acoust. Soc. Amer.*, **87**, 1539–45.

Westwood, E.K. (1992) Broadband matched-field source localization. *J. Acoust. Soc. Amer.*, **91**, 2777–89.

Westwood, E.K. and Koch, R.A. (1999) Elimination of branch cuts from the normal-mode solution using gradient half spaces. *J. Acoust. Soc. Amer.*, **106**, 2513–23.

Westwood, E.K. and Tindle, C.T. (1987) Shallow water time-series simulation using ray theory. *J. Acoust. Soc. Amer.*, **81**, 1752–61.

Westwood, E.K., Tindle, C.T., and Chapman, N.R. (1996) A normal mode model for acousto-elastic ocean environments. *J. Acoust. Soc. Amer.*, **100**, 3631–45.

Westwood, E.K. and Vidmar, P.J. (1987) Eigenray finding and time series simulation in a layered-bottom ocean. *J. Acoust. Soc. Amer.*, **81**, 912–24.

Wetzel-Smith, S.K. and Czech, C. (1996) The interactive multisensor analysis training system: Using scientific visualization to teach complex cognitive skills. NPRDC TR-96-9. Navy Personnel Research and Development Center, San Diego, California, USA.

Wetzel-Smith, S.K., Ellis, J.A., Reynolds, A.M., and Wulfeck, W.H. (1995) The interactive multisensor analysis training (IMAT) system: An evaluation in operator and tactician training. NPRDC TR-96-3. Navy Personnel Research and Development Center, San Diego, California, USA.

White, A.W., Andrew, R.K., Mercer, J.A., Worcester, P.F., Dzieciuch, M.A., and Colosi, J.A. (2013) Wavefront intensity statistics for 284-Hz broadband transmissions to 107-km range in the Philippine Sea: Observations and modeling. *J. Acoust. Soc. Amer.*, **134**, 3347–58.

White, D. (1992) Software requirements specification for the ASTRAL model (version 4.1). Sci. Appl. Inter. Corp., OAML-SRS-23.

White, D. and Corley, M. (1992a) Software test description for the ASTRAL model (version 4.1). Sci. Appl. Inter. Corp., OAML-STD-23.

White, D. and Corley, M. (1992b) Software design document for the ASTRAL model (version 4.1). Sci. Appl. Inter. Corp., OAML-SDD-23.

White, D., Dozier, L.B., and Pearson, C. (1988) Software product specification for the ASTRAL 2.2 model, Vols 1 and 2. Sci. Appl. Inter. Corp., OAML-SPS-23.

Whitehouse, B.G., Hines, P., Ellis, D., and Barron, C.N. (2004) Rapid environmental assessment within NATO. *Sea Technol.*, **45** (11), 10–14.

Whitman, E.C. (1994) Defense conversion in marine technology. *Sea Technol.*, **35** (11), 21–5.

Wiebe, P.H., Prada, K.E., Austin, T.C., Stanton, T.K., and Dawson, J.J. (1995) New tool for bioacoustical oceanography. *Sea Technol.*, **36** (2), 10–14.

Wilcock, W.S.D, Stafford, K.M., Andrew, R.K., and Odom, R.I. (2014) Sounds in the ocean at 1–100 Hz. In *Annual Review of Marine Science*, eds C.A. Carlson and S.J. Giovannoni, Annual Reviews, Vol. 6. Palo Alto, CA, pp. 117–40, doi:10.1146/annurev-marine-121211-172423.

Wille, P.C. (2005) *Sound Images of the Ocean in Research and Monitoring*, Berlin, Germany: Springer-Verlag.

Williams, A.O. Jr. (1976) Hidden depths: Acceptable ignorance about ocean bottoms. *J. Acoust. Soc. Amer.*, **59**, 1175–9.

Williams, K.L. (2001) An effective density fluid model for acoustic propagation in sediments derived from Biot theory. *J. Acoust. Soc. Amer.*, **110**, 2276–81.

Williams, K.L. (2013) Adding thermal and granularity effects to the effective density fluid model. *J. Acoust. Soc. Amer.*, **133** (5), EL431–7, doi:10.1121/1.4799761.

Williams, K.L., Thorsos, E.I., and Elam, W.T. (2004) Examination of coherent surface reflection coefficient (CSRC) approximations in shallow water propagation. *J. Acoust. Soc. Amer.*, **116**, 1975–84.

Willis, C.L. (1992) Oceanographic and atmospheric master library. *Nav. Oceanography Command News*, **12** (2), 1–5.

Wilson, C.J., Wilson, P.S., and Dunton, K.H. (2013) Assessing the low frequency acoustic characteristics of *Macrocystis pyrifera, Egregia menziessi*, and *Laminaria solidungula*. *J. Acoust. Soc. Amer.*, **133**, 3819–26.

Wilson, D.K., Frisk, G.V., Lindstrom, T.E., and Sellers, C.J. (2003) Measurement and prediction of ultralow frequency ocean ambient noise off the eastern U.S. coast. *J. Acoust. Soc. Amer.*, **113**, 3117–33.

Wilson, J.D. and Makris, N.C. (2006) Ocean acoustic hurricane classification. *J. Acoust. Soc. Amer.*, **119**, 168–81.

Wilson, O.B., Stewart, M.S., Wilson, J.H., and Bourke, R.H. (1997) Noise source level density due to surf—Part I: Monterey Bay, CA. *IEEE J. Oceanic Engr.*, **22**, 425–33.

Wilson, O.B. Jr., Wolf, S.N., and Ingenito, F. (1985) Measurements of acoustic ambient noise in shallow water due to breaking surf. *J. Acoust. Soc. Amer.*, **78**, 190–5.

Wilson, W.D. (1960) Equation for the speed of sound in sea water. *J. Acoust. Soc. Amer.*, **32**, 1357.

Wood, A.B. (1930) *A Textbook of Sound: Being an Account of the Physics of Vibrations with Special Reference to Recent Theoretical and Technical Developments.* New York, NY: Macmillan.

Wood, D.H. and Papadakis, J.S. (1985) An overview of stochastic signal modeling. Nav. Underwater Syst. Ctr, Tech. Rept 7267.

Worcester, P.F. and Spindel, R.C. (2005) North Pacific acoustic laboratory. *J. Acoust. Soc. Amer.*, **117**, 1499–1510.

World Meteorological Organization. (1964) *Annual Report of the World Meteorological Organization*, Geneva: WMO No. 163.RP.60.

Wright, E.B., Berman, D.H., Baer, R.N., and Perkins, J.S. (1988) The ocean-refraction bathymetric-scattering (ORBS) model. Nav. Res. Lab., Rept 9123.

Würsig, B., Greene, C.R, Jr., and Jefferson, T.A. (2000) Development of an air bubble curtain to reduce underwater noise of percussive piling. *Mar. Environ. Res.*, **49**, 79–93.

Xu, B., Zeng, L., and Li, J. (2007) Application of stochastic resonance in target detection in shallow-water reverberation. *J. Sound Vib.*, **303**, 255–63.

Xu, C., Zhang, X., and Xu, Z. (2010) Detection in present of reverberation combined with blind source separation and beamforming. Proc. Second Inter. Conf. Advanced Computer Control, Shenyang, China, pp. 158–62, doi:10.1109/ICACC.2010.5486980.

Xu, G., Jackson, D.R., and Bemis, K.G. (2017) The relative effect of particles and turbulence on acoustic scattering from deep sea hydrothermal vent plumes revisited. *J. Acoust. Soc. Amer.*, **141**, 1446–58.

Xu, J., Li, K., and Min, G. (2013) Asymmetric multi-path division communications in underwater acoustic networks with fading channels. *J. Comput. Syst. Sci.*, **79**, 269–78, doi:10.1016/j.jcss.2012.05.007.

Xu, X., Tong, F., Qing, L., and Tao, Y. (2006) Characterization of wireless shallow-water communication channel based on Gaussian beam tracing. Proc. Inter. Conf. Wireless Communications, Networking and Mobile Computing, WiCom 2006, Wuhan, China, doi:10.1109/WiCOM.2006.192.

Yan, J. (1999) Effects of earth curvature on two-dimensional ray tracing in underwater acoustics. *App. Acoust.*, **57**, 163–77.

Yang, K., Ma, Y., Sun, C., Miller, J.H., and Potty, G.R. (2004) Multistep matched-field inversion for broad-band data from ASIAEX2001. *IEEE J. Oceanic Engr.*, **29**, 964–72.

Yang, X. (ed.) (2010) *Underwater Acoustic Sensor Networks*, Boca Raton, FL: CRC Press.

Yaremchuk, M.I. and Yaremchuk, A.I. (2001) Variational inversion of the ocean acoustic tomography data using quadratic approximation to travel times. *Geophys. Res. Lett.*, **28**, 1767–70.

Yarger, D.F. (1976) The user's guide for the RAYMODE propagation loss program. Nav. Underwater Syst. Ctr, Tech. Memo. 222-10-76.

Yarger, D.F. (1982) The user's guide for the passive RAYMODE propagation loss program. Nav. Underwater Syst. Ctr, Tech. Memo. 821061.

Yesson, C., Clark, M.R., Taylor, M.L., and Rogers, A.D. (2011) The global distribution of seamounts based on 30 arc seconds bathymetry data. *Deep-Sea Res. I*, **58**, 442–53.

Yevick, D. and Thomson, D.J. (1994) Split-step/finite-difference and split-step/Lanczos algorithms for solving alternative higher-order parabolic equations. *J. Acoust. Soc. Amer.*, **96**, 396–405.

Yudichak, T.W., Royal, G.S., Knobles, D.P., Gray, M., Koch, R.A., and Stotts, S.A. (2006) Broadband modeling of downslope propagation in a penetrable wedge. *J. Acoust. Soc. Amer.,* **119**, 143–52.

Zabal, X., Brill, M.H., and Collins, J.L. (1986) Frequency and angle spreads of acoustic signals reflecting from a fixed rough boundary. *J. Acoust. Soc. Amer.*, **79**, 673–80.

Zanolin, M., Ingram, I., Thode, A., and Makris, N.C. (2004) Asymptotic accuracy of geoacoustic inversions. *J. Acoust. Soc. Amer.,* **116**, 2031–42.

Zărnescu, G. (2014) On the variability of optimal transmission frequency for underwater acoustic communication in the north-western part of the Black Sea. Proc. IEEE COMM 2014 International Conf. on Communications, Bucharest, Romania, pp. 1–6.

Zedel, L., Gordon, L., and Osterhus, S. (1999) Ocean ambient sound instrument system: Acoustic estimation of wind speed and direction from a subsurface package. *J. Atmos. Oceanic Technol.,* **16**, 1118–26.

Zhang, H., Xu, B., Jiang, Z.-P., and Wu, X. (2008) Target detection in shallow-water reverberation based on parameter-induced stochastic resonance. *J. Phys. A: Math. Theor.*, **41**, 1–13.

Zhang, L., Da, L., and Zhou, Y. (2009) A new ocean acoustic model in computational oceanographic physics. Proc. Inter. Conf. Computational Intelligence and Software Engineering Conf., CiSE 2009, Wuhan, China.

Zhang, R., He, Y., Liu, H., and Akulichev, V.A. (1995) Applications of the WKBZ adiabatic mode approach to sound propagation in the Philippine Sea. *J. Sound Vib.,* **184**, 439–51.

Zhang, R. and Jin, G. (1987) Normal-mode theory of average reverberation intensity in shallow water. *J. Sound Vib.*, **119**, 215–23.

Zhang, R. and Li, F. (1999) Beam-displacement ray-mode theory of sound propagation in shallow water. *Sci. Chin. (Series A)*, **42**, 739–49.

Zhang, X., Zhang, A., Fang, J., and Yang, S. (2000) Study on blind separation of underwater acoustic signals. Proc. Fifth Inter. Conf. Signal Processing (ICSP), Beijing, China, pp. 1802–5, doi:10.1109/ICOSP.2000.893451.

Zhang, Z.Y. and Tindle, C.T. (1993) Complex effective depth of the ocean bottom. *J. Acoust. Soc. Amer.*, **93**, 205–13.

Zhang, Z.Y. and Tindle, C.T. (1995) Improved equivalent fluid approximations for a low shear speed ocean bottom. *J. Acoust. Soc. Amer.,* **98**, 3391–6.

Zhou, J.-X., Zhang, X.-Z., and Rogers, P.H. (1991) Resonant interaction of sound wave with internal solitons in the coastal zone. *J. Acoust. Soc. Amer.*, **90**, 2042–54.

Zhou, J.-X. and Zhang, X.-Z. (2013) Integrating the energy flux method for reverberation with physics-based seabed scattering models: Modeling and inversion. *J. Acoust. Soc. Amer.*, **134**, 55–66.

Zhu, D. and Bjørnø, L. (1999) A hybrid 3-D, two-way IFD PE model for 3-D acoustic backscattering. *J. Comput. Acoust.*, **7**, 133–45.

Zielinski, A. and Geng, X. (1996) Traveling wavefront ray tracing and its applications. *Marine Geodesy*, **19**, 165–79.

Ziman, J. (1978) *Reliable Knowledge*, New York, NY: Cambridge Univ. Press.

Ziomek, L.J. (1985) *Underwater Acoustics. A Linear Systems Theory Approach*, San Diego, CA: Academic Press.

Ziomek, L.J. (1989) Three-dimensional ray acoustics: New expressions for the amplitude, eikonal, and phase functions. *IEEE J. Oceanic Engr.*, **14**, 396–99.

Ziomek, L.J. (1994) Sound-pressure level calculations using the RRA algorithm for depth-dependent speeds of sound valid at turning points and focal points. *IEEE J. Oceanic Engr.*, **19**, 242–48.

Ziomek, L.J. (1995) *Fundamentals of Acoustic Field Theory and Space-Time Signal Processing*, Boca Raton, FL: CRC Press.

Ziomek, L.J. and Polnicky, F.W. (1993) The RRA algorithm: Recursive ray acoustics for three-dimensional speeds of sound. *IEEE J. Oceanic Engr.*, **18**, 25–30.

Zornig, J.G. (1979) Physical modeling of underwater acoustics. In *Ocean Acoustics*, ed. J.A. DeSanto, New York, NY: Springer-Verlag, pp. 159–186.

Zoubir, A.M., Koivunen, V., Chakhchoukh, Y., and Muma, M. (2012) Robust estimation in signal processing: A tutorial-style treatment of fundamental concepts. *IEEE Signal Process. Mag.*, **29** (4), 61–80.

Zuba, M., Jiang, Z., Yang, T.C., Su, Y., and Cui, J.-H. (2013) An advanced channel framework for improved underwater acoustic network simulations. Proc. MTS/IEEE Oceans 2013 Conf., San Diego, pp. 1–8.

Zurk, L.M. and Tracey, B.H. (2005) Depth-shifting of shallow water guide source observations. *J. Acoust. Soc. Amer.,* **118**, 2224–33.

Zykov, M. (2013) Underwater sound modeling of low energy geophysical equipment operations. JASCO Document 00600, Version 2.0. Prepared by JASCO Applied Sciences for CSA Ocean Sciences Inc.

Appendix A: Abbreviations and Acronyms

1D	One Dimensional
2D	Two Dimensional
2½D	Quasi-3D Modeling Approach
3D	Three Dimensional
3DTDPA	3D Time-Domain Parabolic Approximation
3DVAR	3D Variational Data Assimilation
3DWAPE	3D Wide-Angle Parabolic Equation
3D Ocean	3D Normal Mode Model
3DPE	3D Parabolic Equation
3MB	Marine Mammal Movement and Behavior
4DVAR	4D Variational Data Assimilation
AAAI	American Association for Artificial Intelligence
AAIW	Antarctic Intermediate Water
AASM	Airgun Array Source Model
ABC	Artificial Bee Colony
ABL	Atmospheric Boundary Layer
ACCOBAMS	Agreement on the Conservation of Cetaceans in the Black Sea Mediterranean Sea and Contiguous Atlantic Area
ACCURAY	Range-Dependent Shallow-Water Ray Propagation Model
ACIT	Automation, Control, and Information Technology
ACM	Association for Computing Machinery
ACNFS	Arctic Cap Nowcast/Forecast System
ACO	Ant Colony Optimization
Acomms	Acoustic Communications
ACOUS	Arctic Climate Observations Using Underwater Sound
AcTUP	Acoustic Toolbox User Interface and Post Processor
ADCIRC	Advanced Circulation Model (US Army)
ADCP	Acoustic Doppler Current Profiler
ADIAB	Adiabatic Normal Mode
ADONIS	Acoustic Daylight Ocean Noise Imaging System
ADP	Automated Data Processing
ADS	Automated Data Systems
AE	Acoustic Emission
AEAS	Advanced Environmental Acoustic Support
AEP	Auditory Evoked Potential
AESD	Acoustic Environmental Support Detachment
AESS	Allied Environmental Support System

AHA	Applied Hydro-Acoustics Research, Inc. (now Adaptive Methods)
AIAA	American Institute of Aeronautics and Astronautics
AIM	Acoustic Integration Model
AIP	American Institute of Physics
AIS	Automatic Identification System
ALMOST	Acoustic Loss Model for Operational Studies and Tasks
ALS	Area Localization Scheme
AMBENT	Ambient Noise Model
AMDC	Asymmetric Multipath Division Communications
AMLD	Asymmetric Multiple Layer Division
AMEC	Acoustic Model Evaluation Committee
AMFP	Adaptive MFP
AMODE	Acoustic Mid-Ocean Dynamics Experiment
AMOS	Acoustic, Meteorological, and Oceanographic Survey
AMPE	Adiabatic-Mode Parabolic Equation
AN	Ambient Noise
ANC	Adaptive Noise Cancellation
ANDES	Ambient Noise Directionality Estimation System
ANI	Ambient-Noise Imaging
ANSI	American National Standards Institute
AoA	Analysis of Alternatives; Angle of Arrival
AODV	Ad hoc On-demand Distance Vector
AP	Arbitrary Profile Model
APIT	Approximate Point-in-Triangle
APL	Applied Physics Laboratory
APM	Acoustic Propagation Model
APP	Acoustic Performance Prediction
AQUO	Achieve Quieter Ocean by Shipping Noise Footprint Reduction
ARA	Acoustic Rainfall Analysis
ARAMIS	Array Response Advanced Modal Integrated Simulator
ARG	Acoustic Rain Gauge
ARGO	Global network of temperature-salinity profiling floats
ARL	Applied Research Laboratory
ARLO	Acoustics Research Letters Online
ARSRP	Acoustic Reverberation Special Research Program
ARTEMIS	Adiabatic Reverberation and Target Echo Mode Incoherent Sum
ARU	Acoustic Recording Unit
ASA	Acoustical Society of America
ASCII	American Standard Code for Information Interchange
ASCOBANS	Agreement on the Conservation of Small Cetaceans of the Baltic, North East Atlantic, Irish and North Seas
ASEPS	Automated Signal Excess Prediction System
ASERT	ASTRAL System for Estimation of Radial Transmission
ASIAEX	Asian Sea International Acoustic Experiment

ASME	American Society of Mechanical Engineers
ASPECT	Active System Performance Estimate Computer Tool
ASPM	Acoustic System Performance Model
ASTM	American Society for Testing and Materials
ASTRAL	ASEPS Transmission Loss
ASW	Anti-Submarine Warfare
A-TAS	ASW Tactical Assessment System
ATLoS	Acoustic Transmission Loss Server
ATOC	Acoustic Thermometry of Ocean Climate
AUAMP	Advanced Underwater Acoustic Modeling Program
AUTEC	Atlantic Undersea Test and Evaluation Center
AUTO OCEAN	Automated Oceanographic Database
AUTO SHIPS	Automated Shipping Density Database
AUV	Autonomous Underwater Vehicle
AVHRR	Advanced Very High Resolution Radiometer
AVO	Amplitude-versus-Offset (Seismic)
AW	Acoustic Mode Generation Program Using Chebyshev Polynomials as Basis Functions
BAM	Bistatic Acoustic Model
BB	Bottom Bounce
BBN	Bolt, Beranek, and Newman, Inc.
BDRM	Beam-Displacement Ray-Mode Propagation Model
BEAMPL	Beam Program Library
BELLHOP	Gaussian-Beam, Finite-Element, Range-Dependent Propagation Model
BELLHOP3D	BELLHOP Extended to 3D
BEM	Boundary Element Method
BER	Bit Error Rate
BIAS	Baltic Sea Information on Acoustic Soundscape
BICSQS	BISQ plus Grain Contact Squirt Flow and Viscous Drag
BiKR	Bistatic Shallow-Water Reverberation Model Based on KRAKEN
BIMGS	Biot Modified Gap Stiffness Model
BIO-C-SNAP	C-SNAP Modified to Simulate Effects of Bio-Acoustic Absorption
BiRASP	Bistatic Range-Dependent Active System Prediction Model
BISAPP	Bistatic Acoustic Performance Prediction
BISQ	Biot plus Squirt Flow
BISSM	Bistatic Scattering Strength Model
BISTAR	Incoherent Bistatic Reverberation for Range-Dependent Environments
BLM	Bureau of Land Management (now BOEM)
BLUG	Bottom Loss Upgrade
BMT	British Maritime Technology Ltd
BN	Beam Noise
BOEM	Bureau of Ocean Energy Management

BOEMRE	Bureau of Ocean Energy Management, Regulation, and Enforcement (now BOEM)
BORIS-SSA	Bottom Reverberation from Inhomogeneities and Surfaces—Small-Slope Approximation
BP	Beam Pattern; Breakdown Point
BPSK	Binary Phase Shift Keying
BS	Beam Signal
BSS	Blind Separation of Sources
BST	Bottom Sediment Type
BT	Bathythermograph
BTL	Bell Telephone Laboratories
CA	Cellular Automaton
CAAM	Composite Area Analysis Model
CAIV	Cost as an Independent Variable
CAFI	Computation of Acoustic Fluctuations from Internal Waves
CALYPSO	Model Operating System
CAM	Cumulative Accuracy Measure
CAMOS	Coastal and Arctic Maritime Operations and Surveillance
CANARD	CANARY (Version 6)
CANARY	Coherence and Ambient Noise for Arrays
CAPARAY	Broadband Eigenray Model
Cartesian 3DPE	3DPE Employing Cartesian Coordinates in the Numerical Scheme
CASP	Canadian Atlantic Storms Program
CASS	Comprehensive Acoustic System Simulation
CASTAR	Computer Aided Sonar Tactical Recommendations
CBT	Computer-Based Training
CCECE	Canadian Conference on Electrical and Computer Engineering
CCUB	Finite Element PE Model
CDMA	Code Division Multiple Access
CDPE	Complex Density Parabolic Equation
CEC	Congress on Evolutionary Computation
CED	Complex Effective Depth
CEDA	Central Dredging Association
CEE	Controlled Exposure Experiments
CENTRO	Normal Mode Model with Shear Wave Effects
CetMap	Cetacean Density and Distribution Mapping
CFM	Calculation Frequency Method
CICE	Los Alamos Community Ice Model/Community Ice Code
CIMA	Chip-Interleaved Multiple-Access (communication method)
CiSE	Computational Intelligence and Software Engineering
C-MAN	Coastal-Marine Automated Network
CMES	Computer Modeling in Engineering and Sciences
CMM	Caractérisation du Milieu Marin
CMMS	Conceptual Models of the Mission Space
CMM3D	Three-Dimensional Coupled Normal Mode Model

CMPE	Coupled Mode PE
CMRE	Centre for Maritime Research and Experimentation (formerly NURC)
CMS	Convention on Migratory Species
CMST	Centre for Marine Science and Technology (Curtin University)
CNOISE	Noise Model
CNP1	Finite Element PE Model
COADS	Comprehensive Ocean-Atmosphere Data Set
COAMPS	Coupled Ocean-Atmosphere Mesoscale Prediction System/ Coupled Ocean Atmosphere Prediction System
COAMPS-TC	COAMPS-Tropical Cyclone
COEA	Cost and Operational Effectiveness Analysis
Coherent DELTA	3D Range-Dependent Ray Model
COI	Community of Interest
COMODE	Normal Mode Model
COMPILE	Generic Benchmark Case for Predictions of Marine Pile-Driving Noise
CONGRATS	Continuous Gradient Ray-Tracing System
CORDA	Centre for Operational Research and Defence Analysis
CORE	Coupled OASES for Range-Dependent Environments
Corrected PE	PE Model Simulating a Tilted Acoustic Array
COTS	Commercial Off-the-Shelf
COUPLE	Coupled Mode Model
COVIS	Cabled Observatory Vent Imaging Sonar
COWRIE	Collaborative Offshore Wind Energy Research Into the Environment
CPMS	Coupled Perturbed Mode Solution
CRAM	ANSI-standard C version of RAM
CRE	Center for Regulatory Effectiveness
CRLB	Cramer-Rao Lower Bound
C-SALT	Caribbean Sheets and Layers Transects Program
CSI	Channel-State Information
CSMA	Carrier-Sense Multiple Access
C-SNAP	Coupled SNAP
C-SNAP-REV	Reverberation Using the C-SNAP Normal-Mode Model
CSRC	Coherent Surface Reflection Coefficient
CST	Critical Sea Test
CTBT	Comprehensive Test-Ban Treaty
CTD	Conductivity, Temperature, Depth Sensor
CTF	Common Technical Framework
CTS	Clear to Send
CW	Continuous Wave
CZ	Convergence Zone
CZMA	Coastal Zone Management Act
DANES	Directional Ambient Noise Estimation System

DANM	Dynamic Ambient Noise Model
DAU	Defense Acquisition University
dB	Decibel
DBDB	Digital Bathymetric Database
DCDB	Data Center for Digital Bathymetry (IHO)
DELFT3D	Delft Hydraulics 3D Modeling Suite for Coastal/River/Lake/ Estuarine Areas
DFE	Decision Feedback Equalizer
DHL	Density-Aware Hop-Count Localization
DI	Directivity Index
DIC	Dissolved Inorganic Carbon
DiMuNDS	Distributed Multi-National Defence Simulation
DINAMO	Directional Noise Array Model
DIS	Draft International Standard
DMA	Defense Mapping Agency (now NIMA)
DMOS	DRDC Atlantic Model Operating System
DMSO	Defense Modeling and Simulation Office
DMSP	Defense Meteorological Satellite Program
DOD	Department of Defense
DoDAF	Department of Defense Architecture Framework
DODGE	Normal Mode Model
DOE	Design of Experiment
DOI	Digital Object Identifier
DOP	Doppler Content of Reverberation Model
DoS	Denial of Service
DOSL	Diurnal Ocean Surface Layer Model
DP	Direct Path
DRDC	Defence Research and Development Canada
DREA	Defence Research Establishment Atlantic (Canada)
DREAM	Deductive Rapid Environmental Assessment Model
DREP	Defence Research Establishment Pacific (Canada)
DSBN	Discrete Shipping Beam Noise Model
DSC	Deep Sound Channel
DSL	Deep Scattering Layer
DSMC	Defense Systems Management College
DSR	Dynamic Source Routing
DSTO	Defence Science and Technology Organisation (Australia)
DT	Detection Threshold
D-tag	Digital-Recording Acoustic Tags
DtN	Dirichlet to Neumann Map
DTN	Disruption-Tolerant Networking
DUNES	Directional Underwater Noise Estimates Model
DV	Distance Vector
DWG	Digital Waveguide Mesh
EAST	Environmentally Adaptive Sonar Technology
EBCM	Extended Boundary Condition Method

ECCO2	Estimating the Circulation and Climate of the Ocean
EDA	European Defence Agency
EDFM	Effective Density Fluid Model
EEZ	Exclusive Economic Zone
EFEPE	Exponential FEPE (superseded by RAM)
EIA	Electronic Industries Association, Environmental Impact Assessment
EIGEN/REVERB	Eigenray/Reverberation Model
EIS	Environmental Impact Statement
EOD	Estimated Ocean Detector
EOF	Empirical Orthogonal Function
E&P	Exploration and Production
EPSI	Equivalent Plane Wave Intensity
EPWI	Equivalent Plane Wave Intensity
ERL	Environmental Research Laboratory
ERSEM	European Regional Seas Ecosystem Model
ESA	Endangered Species Act
ESL	Echo Signal Level, Energy Source Level
ESME	Effects of Sound on the Marine Environment
ESPRESSO	Extensible Performance and Evaluation Suite for Sonar
ETOP05	Earth Topography Five-Minute Grid
EUCA	European Conference on Underwater Acoustics
EUCLID	European Co-operation for the Long-Term in Defence
EURASAP	European Association for the Science of Air Pollution
EURASIP	European Association for Signal Processing
EUSIPCO	European Signal Processing Conference
EVA	Environmental Acoustics
FACT	Fast Asymptotic Coherent Transmission
FACTEX	FACT Extended to Range-Dependent Environments
FAMA	Floor Acquisition Multiple Access
FAME	Fast Multipath Expansion Model
FANM	Fast Ambient Noise Model
FDHB3D	Hybrid 3D, Two-Way IFD PE Model for Computing 3D Backscattering
FDM	Finite-Difference Method
FDMA	Frequency-Division Multiple Access
FDTD	Finite-Difference Time-Domain
FE	Finite Element
FEM	Finite-Element Model
FELMODETNO	Fysisch en Elektronisch Laboratorium (FEL) Normal Mode Model
FEMODE	Module in FEPE that generates eigenvalues and modes
FENL	Finite-Element PE Model
FEPE	Finite-Element PE Model
FEPE-CM	FEPE with Conformal Mapping
FEPES	Finite-Element PE Model with Shear Wave Effects
FESTA	Finite-Element Structural Acoustics (3D)

FeyRay	Broadband, Range-Dependent Gaussian-Beam Propagation Model
FFI	Forsvarets forskningsinstitutt (Norwegian Defence Research Establishment)
FFP	Fast Field Program
FFT	Fast Fourier Transform
FIPS	Federal Information Processing Standards
FIR	Finite Impulse Response
FLIP	Floating Instrument Platform
FLIRT	Fast Linear Intermediate Range Transmission Model
FM	Frequency Modulation
FNMOC	Fleet Numerical Meteorology and Oceanography Center (formerly FNOC)
FNMSS	Fast Normal Mode with Surface Scattering Integrals Model
FNOC	Fleet Numerical Oceanography Center (now FNMOC)
FOAM	Finite Element Ocean Acoustic Model
FOARAM	Federal Ocean Acidification Research and Monitoring Act
FOM	Figure of Merit; Federation Object Model
FOR3D	Finite Difference Methods, Ordinary Differential Equations, and Rational Function Approximations to Solve the LSS 3D Wave Equation
FORTRAN	Formula Translation
FPSO	Floating Production, Storage and Off-Loading
FTE	Focused Technology Experiments
FVCOM	Finite Volume Community Ocean Model
FWRAM	Far-field Waveform Synthesis Model
GA	Genetic Algorithm
GAIN	Geoacoustic Inversion of Noise
GAMARAY	Broadband Ray Propagation Model
GAMUT	Generic Acoustic Modeling Universal Toolkit
GASS	Generic Acoustic Stimulator System
GDEM	Generalized Digital Environmental Model
GEBCO	General Bathymetric Chart of the Oceans
GEOSECS	Geochemical Ocean Section Study
GFDN	Geophysical Fluid Dynamics Laboratory—Navy Model
GFMPL	Geophysics Fleet Mission Program Library
GIS	Geographic Information System
GmbH	Gesellschaft mit beschränkter Haftung (company with limited liability)
GOATS	Generic Ocean Array Technology Sonar
GODAE	Global Ocean Data Assimilation Experiment
GOFS	Global Ocean Forecast System
GOTM	General Ocean Turbulence Model
GRAB	Gaussian Ray Bundles (Propagation Model in CASS)
GRASS	Germinating Ray-Acoustics Simulation
G-S	Grain-Shearing Model

GSM	Generic Sonar Model
GUI	Graphical User Interface
GUNDALF	Gun Design and Linear Filtering
HACE	Hydroacoustic Channel Emulator
HAMMER	Hydro-Acoustic Model for Mitigation and Ecological Response
HAPE	High-Angle PE Model
HARCAM	HODGSON and RAM Composite Acoustic Model
HARORAY	(Haro Strait) 2D Broadband Propagation Model Based on Ray Theory
HARPO	Hamiltonian Acoustic Raytracing Program—Ocean
HARVEST	Hybrid Adaptive Regime Visco-Elastic Simulation Technique
HBEM	Hybrid BEM
HFBL	High-Frequency Bottom Loss
HIE	Historical Ice Edge
HIFT	Heard Island Feasibility Test
HITS	Historical Temporal Shipping
HLA	High-Level Architecture
HODGSON	Range-Dependent Ray Theoretical Propagation Model
HOMER	Hierarchical ODIN Modeling Environment
HOOM	Harvard Open Ocean Model
HOS	High-Order Statistics
HPC	High Performance Computing
HVMS	HITS Vessel Motion Simulation
HWIL	Hardware in the Loop
HWS	Historical Wind Speed
HWT 3D	Huygens Wavefront Tracing in 3D (for moving media)
HYCOM	Hybrid Coordinate Ocean Model
HydroCAM	Hydroacoustic Coverage Assessment Model
HYDROMAP™	Generates Current/Water Level Predictions for Coastal Waters
HYPER	Hybrid PE-Ray Model
HYREV	HanYang University Reverberation Model
IA	Implementing Agreement
IACM-FORTH	Institute of Applied and Computational Mathematics—Foundation for Research and Technology—Hellas
IADC	International Association of Dredging Companies
IAGC	International Association of Geophysical Contractors
IAPSO	International Association for the Physical Sciences of the Ocean
IARIA	International Academy, Research, and Industry Association
IASTED	International Association of Science and Technology for Development
ICA	Independent Component Analysis
ICACC	International Conference on Advanced Computer Control
ICAPS	Integrated Command ASW Prediction System
ICASSP	International Conference on Acoustics, Speech, and Signal Processing

ICC	International Conference on Communications
ICECAP	Interactive Computer Environment for Comprehensive Arctic Predictions
ICERAY	Under-Ice Ray Propagation Model
ICEX	Ice Exercise
ICIA	International Conference on Information and Automation
ICIF	International Conference on Information Fusion
ICREPQ	International Conference on Renewable Energies and Power Quality
ICSP	International Conference on Signal Processing
ICSPCS	International Conference on Signal Processing and Communication Systems
ICT	Information and Communication Technology
IDMA	Interleave-Division Multiple Access
IEC	International Electrotechnical Commission
IECM	Integral Equation Coupled-Mode
IEE	Institution of Electrical Engineers
IEEE	Institute of Electrical and Electronics Engineers
IEER	Improved Extended Echo Ranging
IF	Influence Function
IFD	Implicit Finite Difference
IFD Wide Angle	IFD Model with Arbitrary and Irregular Boundaries
IFIP	International Federation for Information Processing
Ifremer	Institut français de recherche pour l'exploitation de la mer (French Research Institute for Exploitation of the Sea)
IHO	International Hydrographic Organization
IHVR	Integrated Hierarchical Variable Resolution
Im	Imaginary
IMACS	International Association for Mathematics and Computers in Simulation
IMAT	Interactive Multisensor Analysis Training
IMEMS	International Marine Environmental Modelling Seminar
IMP3D	Finite Difference PE Model with Elastic Impedance Bottom Boundary
IMR	Inspection, Maintenance, and Repair
IMREC	International Marine Renewable Energy Conference
IMS	International Monitoring System
INSACT	Analytical Active/Passive Sonar Model
INSIGHT	Analytical Range-Independent Propagation Model
INSPIRE	Impulse Noise Sound Propagation and Impact Range Estimator
INSTANT	Analytical Range-Dependent Propagation Model
Integrated Mode	Multipath Expansion Method Extended to Range-Dependent Environments
INTIMATE	Internal Tide Measurements with Acoustic Tomography Experiments
IOA	Institute of Acoustics

IOC	Intergovernmental Oceanographic Commission
IODA	Integrated Ocean Dynamics and Acoustics
IOMEDEX	Ionian Sea/Mediterranean Sea Exercise (part of LRAPP)
IRFC	Industry Research Funders Coalition
IRSIM	Impulse Response Simulator
ISAAC	Impact on Species from Anthropogenic Acoustic Channels
ISCCSP	International Symposium on Communications, Control, and Signal Processing
ISI	Inter-Symbol Interference
ISO	International Organization for Standardization
ISVRFEM	Institute of Sound and Vibration Research Finite Element Model
ITCZ	Intertropical Convergence Zone
IUSS	Integrated Undersea Surveillance System
JANUS	Standard modulation and coding scheme to support interoperability in digital underwater communications
JASA	Journal of the Acoustical Society of America
Kanabis	Range-Dependent Normal Mode Model
KdV	Korteweg-de Vries
KPP	Key Performance Parameters
KRAKEN	Adiabatic/Coupled Normal Mode Model
KRAKENC	Version of KRAKEN for the Complex Plane
Kutschale FFP	Fast Field Program for Ice-Covered Arctic Ocean
LABL	Littoral Atmospheric Boundary Layer
LADC	Littoral Acoustic Demonstration Center
LAN	Local Area Network
LBSF&I	Littoral Battlespace Sensing, Fusion, and Integration
LFBL	Low-Frequency Bottom Loss (formerly BLUG)
LIDO	Listening to the Deep-Ocean Environment
LIRA	Low-Frequency Intermediate-Surveillance-Range Active Model
LNCS	Lecture Notes in Computer Science
LOFAR	Low-Frequency Analysis and Recording
LOGPE	PE Model Using Logarithmic Expression for Index of Refraction
LOOPS	Littoral Ocean Observing and Predicting System
LORA	Long Range Active Model
LRAPP	Long Range Acoustic Propagation Project
LSMS	Life System Modeling and Simulation
LSS	Lee-Saad-Schultz Method
LWAD	Littoral Warfare Advanced Development
LYBIN	Range-Dependent Ray-Theoretical Propagation Model (LYdBane og Intensitetsprogram)
LYCH	Range-Dependent, Ray-Theoretical Propagation Model
M2M4	Second Order and Fourth Order Moment Method
MAC	Medium Access Control, Multi-Static Active Coherent

MACA	Multiple Access with Collision Avoidance
MaCh1	Broadband, Range-Dependent Propagation Model
MAIS	Major Automated Information System
MAM	Monostatic Acoustic Model
MAP	Maximum *a posteriori* Probability
MAST	Marine Science and Technology (European Commission Program)
MATE	Mid-Ocean Acoustic Transmission Experiment
MATLAB	Matrix Laboratory (Mathworks, Inc.)
MBC	Maximum Bias Curve
MBT	Mechanical Bathythermograph
MCM	Mine Countermeasures
MCMC	Monte Carlo Markov Chain
MCPE	Monte Carlo Parabolic Equation
MCS	Multichannel Seismic
MDAP	Major Defense Acquisition Program
MEDUSA	Propagation Model
MEL	Master Environmental Library
MEP	Model Evaluation Program
METOC	Meteorology and Oceanography
METS	Marine Energy Technology Symposium
MFP	Matched-Field Processing
MGS	Marine Geophysical Survey
MHK	Marine Hydro-Kinetic
MIL	Military
MIME	Channel Simulator
MIMIC	Low-Frequency, Range-Dependent, Ray-Theoretical Propagation Model
MIMO	Multiple-Input/Multiple-Output
MINDIS	Graphics Library
MINEOS	MSASM Interactive Execution and Optimization System
MINERAY	Active Sonar Model Used in Mine-Hunting Scenarios
MINEX	Mine Deployment Exercise
MIPE	(University of) Miami PE Model
MIRACLE	Multi-Interface Cross-Layer Extension
MIS	Management Information System
MIW	Mediterranean Intermediate Water
MIZ	Marginal Ice Zone
MLD	Mixed Layer Depth
MMPA	Marine Mammal Protection Act
MMPE	Monterey-Miami PE (formerly UMPE, now PE-SSF)
MMS	Minerals Management Service
MNS	Mission Needs Statement
MOATL	Modal Acoustic Transmission Loss Model
MOC3D	3D Version of MOCASSIN
MOCASSIN	Monte Carlo Schall-Strahlen Intensitäten (Monte Carlo Sound Ray Intensities)

MOCC	Mean Oceanic Communication Channel
MOCTESUMA	Coupled Normal Mode Model
MOD	Ministry of Defence
MODAS	Modular Ocean Data Assimilation System
MODELAB	Normal Mode Model
MODRAY	Maritime Operations Division Ray-Tracer
MOE	Measures of Effectiveness
MONM	Marine Operations Noise Model
MONM3D	3D PE Model for MONM
MONOGO	Reverberation Module in SWAMI
MOODS	Master Oceanographic Observation Data Set
MOP	Measures of Performance
MOREPE	Modified Refraction PE Model
MORS	Military Operations Research Society
MOS	Model Operating System
MOSPEF	Modal Spectrum of the PE Field (method for computing modal travel times under strong mode-coupling environments)
MOST	Mobile Sonar Technology
MOVES	Modeling, Virtual Environments and Simulation Institute
MPC	Multiple Profile Configuration Model
MPE	Mode Parabolic Equation
MPI	Message Passing Interface
MPIRAM	Parallel Version of RAM that uses either OPENMP or MPI
MPP	Multiple Profile Program
MRE	Marine Renewable Energy
M&S	Modeling and Simulation
MSASM	Multistatic Active System Model, Multistatic Anti-Submarine Model
MSEAS	Multidisciplinary Simulation, Estimation, and Assimilation Systems
MSFD	Marine Strategy Framework Directive
MSP	Mean Square Pressure/Matched Shadow Processing
MSPFFP	Multiple Scattering Pulse FFP Model
MST	Moving Ship Tomography
MSTPA	Multistatic Tactical Planning Aid
MTS	Marine Technology Society
MULE	Multilayer Expansion Model
MURAL	Multistatic Reverberation Algorithm
MURI	Multidisciplinary University Research Initiative
MVOI	Multi-Variate Optimal Interpolation System
N	Noise Sources
N × 2D	Quasi-3D Modeling Approach
NABTAM	Narrow Beam Towed Array Model
NADC	Naval Air Development Center (now NAWC)
NAEMO	Navy Acoustic Effects Model
NATO	North Atlantic Treaty Organization

NAUTILUS	Broadband Adiabatic Normal-Mode Model for Shallow-Water Areas
NAVDAB	Navy Ocean Environmental Acoustic Data Bank
NAVGEM	Navy Global Environmental Model (replaces NOGAPS)
NAVMSMO	Navy Modeling and Simulation Management Office (now NMSO)
NAVOCEANO	Naval Oceanographic Office
NAWC	Naval Air Warfare Center
NBS	National Bureau of Standards (now NIST)
NCCOSC	Naval Command Control and Ocean Surveillance Center
NCDC	National Climatic Data Center (now NCEI)
NCDDC	National Coastal Data Development Center (now NCEI)
NCEI	National Centers for Environmental Information
NCODA	Navy Coupled Ocean Data Assimilation
NCOM	NRL Coastal Ocean Model/Navy Coastal Ocean Model
NDRC	National Defense Research Committee
NECTA	Naval Environmental Command Tactical Aid
NEMESIS	Propagation Model
NEMO	NUWC Exposure Model
NEONS	Naval Environmental Operational Nowcasting System
NEPA	National Environmental Policy Act
NEPBR	Numerable Energy Paths by RAYMODE
NERC	Natural Environment Research Council (UK)
NetCDF	Network Common Data Form
NGA	National Geospatial-Intelligence Agency (formerly NIMA)
NGDC	National Geophysical Data Center (now NCEI)
NGLI	Northern Gulf of Mexico Littoral Initiative
NHA	National Hydropower Association
NIAG	NATO Industrial Advisory Group
NIMA	National Imagery and Mapping Agency (formerly DMA, now NGA)
NISSM	Navy Interim Surface Ship Model
NIST	National Institute of Standards and Technology (formerly NBS)
NL	Noise Level
NLAYER	N-Layer Normal Mode Model
NLBC	Non-Local Boundary Condition
NLBCPE	Nonlocal Boundary Condition PE
NLNM	N-Layer Normal Mode Model
NLOM	NRL Layered Ocean Model
NMCI	Navy Marine Corps Intranet
NMFS	National Marine Fisheries Service
nmi	Nautical Mile
NML	Noise-Masking Level
NMPQ	Normal Mode P-Q (description of the bottom reflection coefficient using reflection phase-shift parameter P and attenuation parameter Q)

NMSG	NATO Modeling and Simulation Group
NMSIS	Navy Modeling and Simulation Information Service
NMSO	Navy Modeling and Simulation Office (formerly NAVMSMO)
NOAA	National Oceanic and Atmospheric Administration
NOARL	Naval Oceanographic and Atmospheric Research Laboratory (now part of NRL)
NODC	National Oceanographic Data Center (now NCEI)
NODDS	Navy/NOAA Oceanographic Data Distribution System
NOGAPS	Navy Operational Global Atmospheric Prediction System (replaced by NAVGEM)
NOGRP	Fast Normal Mode (Monostatic) Reverberation Model
NORAPS	Navy Operational Regional Atmospheric Prediction System
NORDA	Naval Ocean Research and Development Activity (now part of NRL)
NORM2L	Normal Mode 2-Layer Model
NORMOD3	Normal Mode Model
NOSC	Naval Ocean Systems Center
NPAL	North Pacific Acoustic Laboratory
NPE	Nonlinear Progressive-Wave Equation
NRDA	Natural Resource Damage Assessment
NRDC	Natural Resources Defense Council
NRL	Naval Research Laboratory
NS2	Network Simulator
NSF	National Science Foundation
NSIDC	National Snow and Ice Data Center
NSODB	NATO Standard Oceanographic Data Base
NSPE	Navy Standard Parabolic Equation
NSSM	Navy Standard Surf Model
NtD	Neumann to Dirichlet Map
NTNU	Norwegian University of Science and Technology
NTRM	Noise-Based Time Reversal Mirror
NURC	NATO Undersea Research Centre (formerly SACLANTCEN; now CMRE)
NUSC	Naval Underwater Systems Center (now NUWC)
NUWC	Naval Undersea Warfare Center (formerly NUSC)
OA	Ocean Acidification
OALIB	Ocean Acoustics Library
OAML	Oceanographic and Atmospheric Master Library
OASAN	OASES Ambient Noise
OASES	Ocean Acoustics and Seismic Exploration Synthesis
OASES-3D	Version of OASES Incorporating 3D Scattering Effects
OASI	OASES Module for Environmental Inversion
OASIS	Ocean Ambient Sound Instrument System
OASM	OASES Match Field Processing Module
OASN	OASES Module for Noise, Covariance Matrices and Signal Replicas

OASP	OASES Module for 2D Wideband Transfer Functions
OASP3D	OASES Module for 3D Wideband Transfer Functions
OASR	OASES Reflection Coefficient Module
OASS	OASES Scattering and Reverberation Module
OASSP	OASES Module for 2D Waveguide Reverberation Realizations
OAST	OASES Transmission Loss Module
OAT	Ocean Acoustic Tomography
OBIS	Ocean Biogeographic Information System
OCEAN MVOI 3D	Ocean Multi-Variate Optimal Interpolation System
OCS	Outer Continental Shelf
ODE	Ordinary Differential Equation
ODIN	Underwater Warfare Software Simulation Toolset
OE	Ocean Environment
OES	Ocean Energy Systems (International Energy Association)
OFDM	Orthogonal Frequency Division Multiplexing
OGOPOGO	Normal-Mode Reverberation Model
OI	Optimum Interpolation
O&M	Operations and Maintenance
OMAE	Offshore Mechanics and Arctic Engineering
OMG	Object Management Group
OMS	Optimum Mode Selection
OMT	Object Model Template
ONR	Office of Naval Research
OOAE	Ocean, Offshore, and Arctic Engineering
OOI	Ocean Observatories Initiative
OPEC	Organization of Petroleum Exporting Countries
OpenMP	Open Multi-Processing
ORB	Research Platform
ORBS	Ocean Refraction and Bathymetric Scattering Model
ORCA	Normal Mode Model for Acousto-Elastic Ocean Environments
ORD	Operational Requirements Document
ORION	Ocean Research Interactive Observatory Networks
OS2IFD	Implicit Finite Difference PE Model
OSDM	Orthogonal Signal-Division Multiplexing
OSI	Open Systems Interconnection
OTIS	Optimum Thermal Interpolation System
OWARS	Ocean Acoustic Waveguide Remote Sensing
OWWE	One-Way Wave Equation
PAM	Passive Acoustic Monitoring
PAREQ	Parabolic Equation Model
PAREQ-REV	Reverberation Using the PAREQ Parabolic Equation Model
PARSIFAL	Plane Wave Acoustic Reflection from a Sediment of Inhomogeneous Fluid
PAT	Passive Acoustic Tomography
PCMDI	Program for Climate Model Diagnosis and Intercomparison
PC TIDES	Relocateable Tide/Surge Prediction System

PDF	Probability Density (or Distribution) Function
PDPE	Pseudo-Differential PE
PE	Parabolic Equation
PECan	Canadian Parabolic Equation
Pedersen	Range-Dependent Ray Model
PE-FFRAME	Finite-Element, Full-Wave Range-Dependent, Acoustic Marching Element Model
PER	Packet Error Rate
PEREGRINE	PE Model
PEREV	PE Reverberation Model
PERM-2D	Reverberation Model
PERUSE	PE Rough Surface
PESOGEN	PE Solution Generator
PE-SSF	PE—Split-Step Fourier
PGS	Petroleum Geo-Services
PIPS	Polar Ice Prediction System
PLAN	Protocol for Long-Latency Access Networks
PlaneRay	Eigenray Model for Range-Dependent Environments
PLoS	Public Library of Science
PLRAY	Ray Propagation Loss Model
PML	Perfectly Matched Layer
PMODES	Range-independent normal-mode transmission-loss module in SWAMI
PODgrams	Probability-of-Detection Diagrams
POLCOMS	Proudman Oceanographic Laboratory Coastal Ocean Modelling System
POM	Princeton Ocean Model
POPP	Range-Independent Version of PROLOS
POSSM	Panel on Sonar System Models
PP	Post-Processor
ppt	Parts per Thousand (‰)
PRISM	Program for Integrated Sonar Modeling
PROLOSS	Propagation Loss Model
PROPLOSS	Transmission Loss Module in ALMOST
PROSIM	Propagation Channel Simulator (Broadband Adiabatic Normal-Mode Propagation Model)
PROTEUS	Propagation Model
PSC	Propagation of Surfaces under Curvature
PSI	Planning Systems, Inc.
PSSEL	Probabilistic Shipping Sound Exposure Level
PSTD	Pseudospectral Time-Domain
psu	Practical Salinity Unit
PTS	Permanent Threshold Shift
Pty Ltd	Private Company (South Africa)
Pulse FFP	Time-Marched FFP Model
PWE	Progressive Wave Equation

PWRC	Plane Wave Reflection Coefficient
QUONOPS©	Ocean Noise Anthropogenic Forecasting Platform
RACUN	Robust Acoustic Communications in Underwater Networks
RAIM	Range-Averaged Intensity Model
RAM	Range-Dependent Acoustic Model
RAMGEO	RAM Modified for Range-Dependent Sediment Layers
RAMS	RAM for Acousto-Elastic Problems
RAMSURF	RAM Rough Surface
RANDI	Research Ambient Noise Directionality Model
RANGER	Range-Invariant Ray Model
RAP	Reliable Acoustic Path
RASP	Range-Dependent Active System Performance Model
RAY	Range-Dependent Raytracing Program
RAY2C	2D Ray Model
Ray5	Range-Dependent Ray Model
RAYMODE	Ray/Normal Mode Model
RAYSON	Range-Dependent Ray-Theoretic Model for High Frequencies
RAYWAVE	Ray/Wave Propagation Model
RBR	Refracted-Bottom-Reflected
RD	Recognition Differential; Range-Dependent
R&D	Research and Development
RDFFP	Range-Dependent FFP Model
RD-OASES	Range-Dependent OASES
RDOASP	OASES Module for 2D Range-Dependent Transfer Functions
RDOAST	OASES Range-Dependent Transmission Loss Module
RDS	Rapidly Deployable System
Re	Real
re	Relative to
REA	Rapid Environmental Assessment
REACT	Active Sonar Range Prediction Module in ALMOST
REFMS	Reflection and Refraction Multilayered Ocean/Ocean Bottoms with Shear Wave Effects (range-independent shock wave model)
ReloPOM	Relocateable Princeton Ocean Model
REP	Reflected Energy Prediction Model
REPAS	Passive Sonar Range Prediction Module in ALMOST
REV3D	Deterministic version of MOC3D reverberation model
REVGEN	Reverberation Generator
REVMOD	Reverberation Spectrum Model
REVPA	Parabolic-Equation Reverberation Model
REVSIM	Reverberation Simulation
RF	Radio Frequency
RI	Range-Independent
rimLG	Legendre-Galerkin Code for Calculating Normal Modes
RITSHPA	Reverberation Including the Statistics of Hybrid Path Arrivals
RL	Reverberation Level; Received Level

R&M	Reliability and Maintainability
RML	Reverberation-Masking Level
RMPE	Ray-Mode Parabolic Equation Model
rms	Root Mean Square
ROMS	Regional Ocean Modeling System
ROSELLA	Extension of NOGRP to Handle Beam Patterns
ROV	Remotely Operated Vehicle
RP-70	Ray-Tracing Program—1970
RPG	Recommended Practices Guide
RPRESS	Model for Computing Seismoacoustic Wavefields
RR	Refracted-Refracted
RRA	Recursive Ray Acoustics
RRC	Rayleigh Reflection Coefficient
Rrsfc	RAM Rough Surface Fine Coarse
R-SNAP	Coherent Monostatic Reverberation Model
RSR	Refracted-Surface-Reflected
RSRBR	Refracted-Surface-Reflected-Bottom-Reflected
RSS	Received Signal Strength
RSSI	Received Signal Strength Indicator
RTI	Runtime Infrastructure
RTS	Request to Send
RUMBLE	Reverberation Underwater Model, Bistatic Level Estimation; Reverberation Undersea Model for Bottom-Limited Environments
SABLE	Sonar Active Boundary Loss Estimation
SACLANT	Supreme Allied Commander, Atlantic
SACLANTCEN	SACLANT Undersea Research Centre
SAFARI	Seismo-Acoustic Fast-Field Algorithm for Range-Independent Environments
SAFE	Seismo-Acoustic Finite Element Model
SAFESIMM	Statistical Algorithms for Estimating the Sonar Influence on Marine Megafauna
SAFRAN	Hybrid BEM and WI Model
SAGA	Seismo-Acoustic Inversion Using Genetic Algorithms
SAKAMATA	Sea Animal Kind Area-Dependent Mitigated Active Transmission Aid
SALT	Sound Angle, Level, and Travel Time
SAM	Stratified Acoustic Model
SBA	Simulation Based Acquisition
SC	System Characteristics
SCOOTER	FFP, Finite-Element, Range-Independent Propagation Model
SCUFN	GEBCO Subcommittee on Undersea Feature Names
SCV	System Concept Validation
SDD	Software Design Document, Software Development and Documentation
SDMA	Space-Division Multiple Access

SEARAY	Mine-Hunting Sonar Performance Model
SEDRIS	Synthetic Environment Data Representation and Interchange Specification
SEE	Software Engineering Environment
SEG	Society of Exploration Geophysicists
SEL	Sound Exposure Level
SHALFACT	Shallow-Water FACT
SHAZAM	Shallow Water Model
SHEAR2	Normal Mode Model with Shear Wave Effects
SI	Système International (d'Unités) (International System [of Units])
SIAM	Simulated Ambient Noise Model
SIGBED	Special Interest Group on Embedded Systems (ACM)
SIMAS	Sonar In Situ Mode Assessment System
SIO	Scripps Institution of Oceanography
SL	Source Level, Signal Level
SLD	Sonic Layer Depth
SMGC	Surface Marine Gridded Climatology (derived from COADS)
SMOD	Sound Modeling and Ranging Software
SN	Self-Noise, Shipping Noise
SNAP	SACLANTCEN Normal-Mode Acoustic Propagation Model
SNIR	Signal to Noise-plus-Interference Ratio
SNR	Signal-to-Noise Ratio
SNUPE	Seoul National University Parabolic Equation
SOFAR	Sound Fixing and Ranging
SOM	Simulation Object Model, Self-Organizing Map
SONAR	Sound Navigation and Ranging
SONIC	Suppression of Underwater Noise Induced by Cavitation
SP	Sonar Performance, Source Probe
SPARC	SACLANTCEN Pulse Acoustic Research Code
SPAWAR	US Navy Space and Naval Warfare Systems Command
SPD	Spectral Probability Density
Spectral PE	PE Model with 3D Backscattered Energy
SPL	Sound Pressure Level
SPLN	Finite Element PE Model
SPM	Small-Perturbation Method
SPPS	Sonar Performance Prediction System
SPUR	Synthetic Pulse Reception Model
SRA	Source-Receiver Array
SRBR	Surface-Reflected-Bottom-Reflected
SRS	Software Requirements Specification
SSAS	Sea-Surface Acoustic Simulator
SSC	Suspended Sediment Concentration
S-SCARAB	SACLANTCEN-Scattering Reverberation and Backscatter
SSF	Split-Step Fourier
SSFEPE	SSF PE Model

SSM/I	Special Sensor Microwave/Imager
SSPE	Split-Step Parabolic Equation
SSPPE	Split-Step Padé PE Model
SSS	Sea-Surface Salinity
SST	Sonar Simulation Toolset, Sea-Surface Temperature
S-tag	Satellite-Tracked Radio Telemetry Tags
STB	System Test Bed
STD	Salinity, Temperature, Depth Sensor; Software Test Description; Standard
Stickler	Normal Mode Model with Discrete and Continuous Modes
STRIVE	Science, Technology, Research, and Innovation for the Environment
STWAVE	Steady State Spectral Wave Model
SuperSNAP	Enhanced SNAP
SUPRA-GAIN	Subbottom Profiling Using Ambient Noise
SUPREMO	Multistatic Sonar Model
SWAFS	Shallow Water Analysis and Forecasting System, Shallow Water Assimilation Forecast System
SWAM	Shallow Water Acoustic Model Workshop
SWAMI	Shallow Water Active-Sonar Modelling Initiative
SWAMP	Shallow Water Acoustic Modal Propagation
SWAPS	Spectral Wave Prediction System
SWARM	Shallow Water Acoustics in a Random Medium
SWAT	Shallow Water Acoustics Toolset, Shallow Water Acoustic Technology
SWIL	Software in the Loop
SWSS	Sperm Whale Seismic Study
SYNBAPS	Synthetic Bathymetric Profiling System
TA	Transmission Anomaly, Total Alkalinity
TAMAR	Towed Array Model of Acoustic Reverberation
TAMDA	Tactical Acoustic Measurement and Decision Aid
TAML	Tactical Assessment Markup Language
TAMU	Texas A&M University
TDA	Tactical Decision Aid
TDGF	Time-Domain Green's Function
TDMA	Time-Division Multiple Access
TDoA	Time Difference of Arrival
TDPA	Time-Domain Parabolic Approximation Model
TDPE	Time-Domain Parabolic Equation
TENAR	Target Echo, Noise, and Reverberation
TESS	Tactical Environmental Support System
TIAPS	Towed Integrated Active-Passive Sonar
TICA	Turkish International Conference on Acoustics
TKE	Turbulent Kinetic Energy
TL	Transmission Loss
TLM	Transmission Line Matrix Modeling

TNO	Netherlands Organization for Applied Scientific Research
ToA	Time of Arrival
TOC	Total Ownership Cost
ToF	Time of Flight
TOPAS	Topographic Parametric Sonar (used in GOATS'98 experiment)
TOPEX	Topography Experiment for Ocean Circulation (unflown NASA precursor mission to TOPEX / Poseidon)
TOPEX/Poseidon	Joint US-French orbital mission launched in 1992 to track changes in sea-level height using radar altimeters
TOPS	Thermodynamical Ocean Prediction System
TOSL	Tactical Oceanography Simulation Laboratory
TOTS	Tactical Ocean Thermal Structure
TOWAN	Tactical Oceanography Wide Area Network
TPEM	Tropospheric Parabolic Equation Model
TRACEO3D	Beam Tracing Code for 3D
TREX	Target and Reverberation Experiment
TRIMAIN	Range-Dependent Acoustic Propagation Model Based on Triangular Segmentation of the Range-Depth Plane
TRM	Time-Reversal Mirror
TS	Target Strength
TSC	Tactical Support Center
TSPS	Tide and Surge Prediction System
TTCP	The Technical Cooperation Program
TTS	Temporary Threshold Shift
TUHH	Hamburg University of Technology
TUS	Thales Underwater Systems
TV-APM	Time-Variable Acoustic Propagation Model
TVIR	Time-Varying Impulse Response
TWF	Traveling Wavefront
Two-Way PE	PE Model with Backscattered Energy
TWTT	Two-Way Travel Time
UAIM	Underwater Acoustic Imaging Model
UAN	Underwater Acoustic Network
UComms	Underwater Communications
UDT	Undersea Defence Technology
UK	United Kingdom
ULETA	Propagation Model
ULF	Ultra-Low Frequency
UML	Unified Modeling Language (Object Management Group–OMG)
UMPE	University of Miami PE (now MMPE)
UMS	Unmanned Maritime Systems
UNIMOD	Propagation Model
URN	Underwater Radiated Noise
US	United States
USGS	United States Geological Survey

USI	Underwater Systems, Inc.
USM	University of Southern Mississippi
USML	Under Sea Modeling Library
USMOS	Turkish National Modeling and Simulation Conference
USSN	Underwater Swarm Sensor Network
USW	Undersea Warfare
USWTR	Undersea Warfare Training Range
UUV	Unmanned Undersea Vehicle
UWA	Underwater Acoustic
UWSN	Underwater Wireless Sensor Network
VirTEX	Virtual Time Series Experiment
VISA	Virtual Source Algorithm
VLA	Vertical Line Array
VLF	Very Low Frequency
VSS	Volume Scattering Strength
VSTACK	Near-Field Wavenumber Integration Model
VV&A	Verification, Validation, and Accreditation
WADER	Global Ocean Information System
WAM	Wave Action Model
WATERMARK	Underwater Acoustic Channel Replay Benchmark
WaveQ3D	Wavefront Queue 3D
WDC	World Data Center
WEDGE	Range-Dependent Shallow-Water Normal Mode Model
WHOI	Woods Hole Oceanographic Institution
WI	Wavenumber Integration
WIT	Wessex Institute of Technology
WKB	Wentzel, Kramers, and Brillouin
WKBJ	Wentzel, Kramers, Brillouin, and Jeffreys
WKBZ	Adiabatic Normal Mode Model
WMO	World Meteorological Organization
WOSS	World Ocean Simulation System
WOTAN	Weather Observation Through Ambient Noise
WRAP	Wide-Area Rapid Acoustic Prediction
WRN	Wind and Residual Noise
WSEAS	World Scientific and Engineering Academy and Society
WUWNet	Workshop on Underwater Networks
WW3	Wave Watch (third generation model)
XBT	Expendable Bathythermograph
XML	Extensible Markup Language
XRAY	Range-Dependent Ray Theoretical Model Combined with Full-Field Modeling of Seabed Interactions
XSV	Expendable Sound Velocimeter

Appendix B: Glossary of Terms

Absorption: loss of acoustic energy due to conversion to heat.

Accreditation: the official certification that a model or simulation is acceptable for a specific purpose.

Acoustic daylight: imaging of underwater objects using the ambient noise field.

Acoustic impedance: characteristic acoustic impedance is quantitatively equal to the product of the density and sound speed of the medium.

Acoustic tomography: inverse technique that uses acoustic signals to sample the interior of a water body.

Acoustical oceanography: describes the role of the ocean as an acoustic medium by relating oceanic properties to the behavior of underwater acoustic propagation, noise, and reverberation; acoustical oceanography includes both the study of acoustics in the ocean and the use of acoustics to study the ocean.

Adiabatic process: thermodynamic change in which there is no transfer of heat or mass across the system boundaries.

Advection: movement of oceanic properties by currents.

Ambient noise: background level of unwanted sound in the sea, exclusive of occasional (transient) noise sources.

Ambient noise models: mathematical models that predict the levels and directionality of noise in the ocean due to surface weather and shipping sources.

Analog models: controlled acoustic experiments in water tanks using appropriate scaling factors.

Analytical models: same as physical (physics-based) models.

Anticyclonic: gyral pattern of motion, induced by Earth's rotation, that is clockwise in the northern hemisphere but counter-clockwise in the southern hemisphere.

Archipelagic apron: a gentle slope with a smooth surface on the seafloor.

Attenuation: loss of acoustic energy due to the combined effects of absorption and scattering.

Backscattering: scattering of sound in the direction of the source.

Bank: an elevation of the seafloor located on a shelf.

Baroclinic: state of stratification in the ocean in which surfaces of constant pressure intersect surfaces of constant density.

Barotropic: state of stratification in the ocean in which surfaces of constant pressure and surfaces of constant density are parallel.

Basic acoustic models: category of models containing underwater acoustic propagation, noise, and reverberation models.

Basin: a depression of variable extent, generally in a circular or oval form.

Bathythermograph: instrument used to measure water temperature versus depth.

Beam displacement: lateral displacement of an acoustic beam, of finite width, undergoing reflection at a water-sediment interface.

Beam noise statistics models: mathematical models that predict the levels and directionality of low-frequency shipping noise for application to large-aperture, narrow-beam passive sonar systems.

Bistatic: geometry in which the acoustic source and receiver are not at the same position.

Borderland: a region adjacent to a continent that is highly irregular with depths in excess of those typical of a shelf.

Bottom bounce: ray paths that have reflected off the seafloor.

Bottom limited: ocean environment characterized by a water depth and sound speed profile that will not support long-range refracted paths.

Bottom loss: reflection loss at the sea floor.

Boundary conditions: constraints imposed on possible solutions of the wave equation by adjacent surfaces.

Buoyancy frequency: frequency at which a water parcel displaced from equilibrium will oscillate.

Canonical sound speed profile: standard sound speed profile applicable to ocean areas having a deep sound channel; the profile normally assumes an exponential form in the region of the sound channel axis.

Canyon: a narrow, deep depression with steep slopes.

Caustics: envelopes formed by sets of tangential rays.

Cell scattering models: mathematical models of reverberation based on the assumption that the scatterers are uniformly distributed throughout the ocean.

Cellular automata: a mathematical model composed of units that have simple rules governing their replication and destruction; they can be used to model complex systems comprising simpler units.

Chaotic: a deterministic process for which certain ranges of parameters are unpredictable.

Composite roughness model: model that partitions treatment of boundary scattering into two regimes according to large-scale and small-scale surface roughness.

Conjugate depth: depth below the sound channel axis at which the value of sound speed equals that at a shallower depth above the axis.

Constructive simulation: models and simulations involving simulated people operating simulated systems.

Continental margin: a zone separating the continent from the deeper sea bottom, generally consisting of the rise, slope, and shelf.

Continental rise: a gentle slope rising toward the foot of the continental slope.

Continental shelf: zone adjacent to a continent or island from the waterline to the depth at which there is usually a marked increase of slope to greater depth (shelf break).

Continental slope: zone between the continental shelf and the continental rise.

Convergence zone: ray paths formed by long-range refraction in deep-water environments having both a critical depth and sufficient depth excess.

Cordillera: an entire underwater mountain system including all the subordinate ranges, interior plateaus, and basins.

Coriolis force: an apparent force acting on moving particles resulting from Earth's rotation; the force is proportional to the speed and latitude of the particle, and results in a deflection to the right of motion in the northern hemisphere, or to the left of motion in the southern hemisphere.

Crank-Nicolson method: method for numerically solving partial differential equations of the parabolic type (can be considered integration by the trapezoidal rule).

Critical angle: angle in the vertical plane separating the angular region of total reflection from that of partial reflection and partial transmission at a boundary.

Critical depth: depth at which the value of sound speed equals that of the near-surface maximum sound speed.

Cusped caustic: intersection of two (smooth) caustics.

Cyclonic: gyral pattern of motion, induced by Earth's rotation, that is counterclockwise in the northern hemisphere but clockwise in the southern hemisphere.

Cylindrical spreading: form of geometrical spreading that is confined between two parallel planes.

Decibel: unit of measure of acoustic intensity based on a logarithmic scale.

Deep scattering layer: layer of biological organisms associated with an increased scattering of sound.

Deep sound channel axis: location of the deep sound speed minimum in the ocean.

Deterministic: state of being completely predictable (as distinguished from stochastic or chaotic).

Diagnostic information: information used to analyze the past or present state of a system, particularly with reference to identifying sonar system pathologies.

Diel: characterized by a 24-hour period.

Diffraction: frequency-dependent interference effects.

Direct path: ray paths directly connecting source and receiver over short ranges without boundary interaction.

Dispersion: separation of sound into component frequencies; condition in which the phase velocity is frequency dependent.

Diurnal: having a daily cycle.

Domains of applicability: the spatial, temporal, or spectral ranges over which a model's output can be considered valid; in general, such limitations are imposed by the model's physical or mathematical basis.

Doppler: frequency shift resulting from relative motion of source and receiver.

Downwelling: downward-directed water motion in the ocean.

Ducted precursors: acoustic energy that leaks out of a surface duct, travels via a convergence zone or bottom bounce path, or both, before coupling back into the surface duct down range.

Earth curvature correction: correction applied to the sound speed profile to adjust for propagation of sound over the curved Earth surface rather than over a flat surface.

Eddies: isolated patterns of gyral motion in the ocean.

Eigenray: ray that connects the source and receiver.

Eigenvalue: solution to the normal mode equation.

Eikonal: acoustic path length as a function of the endpoints.

Ekman drift current: surface current produced by the wind, but directed perpendicular to the wind direction.

Empirical models: mathematical models based on observations.

Environmental models: empirical algorithms used to quantify the boundary conditions and volumetric effects of the ocean environment.

Escarpment: an elongated and comparatively steep slope of the seafloor separating flat or gently sloping areas.

Fan: a gently sloping, fan-shaped feature normally located near the lower termination of a canyon.

Fast field models: mathematical models of acoustic propagation based on a variation of the normal mode technique.

Fathometer returns: multiply reflected echoes arriving from the sea surface directly above or from the seafloor directly below a monostatic sonar.

Feature model: statistical representation of a common synoptic structure in the ocean such as a front or an eddy.

Fidelity: the accuracy of a simulation representation when compared to the real world.

Figure of merit (FOM): quantitative measure of sonar performance; value equals allowable one-way transmission loss in passive sonars, or two-way transmission loss in noise-limited active sonars.

Finite difference: mathematical technique for solving differential equations.

Finite element: mathematical technique for solving complex problems; in underwater acoustics, this method is implemented by dividing the range-depth plane into a gridded pattern of small elements.

Fracture zone: an extensive linear zone of unusually irregular topography of the sea floor characterized by large seamounts, steep-sided ridges, troughs, or escarpments.

Front: boundary between two different water masses.

Galerkin's method: a member of the class of so-called weighted-residual methods and a variational formulation that has been the generally accepted basis of finite-element discretizations.

Gap: a depression cutting transversely across a ridge or rise.

Gaussian beam tracing: method that associates each acoustic ray with a beam having a Gaussian intensity profile normal to the ray.

Geodesic path: shortest distance between two points on the surface of an elliptical Earth; the shortest distance between two points on the surface of a spherical Earth is referred to as a great-circle path.

Grazing angle: angle in the range-depth plane measured from the horizontal.

Half channel: ducted environment defined by the equivalent sound speed profile either above or below the deep-sound channel axis (the deep sound channel is considered to be a full channel).

Harmonic solution: solution to the wave equation based on a single frequency.

Hidden depths: concept that regards as unimportant the deep ocean sediment structure below the ray turning point.

Hill: a small elevation rising less than about 200 m above the seafloor.

Hindcast: estimate of a previous state of a system.

Hole: a small depression on the seafloor.

Hybrid models: models based on multiple physical or mathematical approaches in order to broaden the domains of applicability.

Incidence angle: angle in the range-depth plane measured from the vertical.

Inertial motion: periodic, quasi-circular oceanic motions influenced by Earth's rotation.

Isotherm: line connecting values of equal temperature.

Knoll: an elevation rising less than about 1 km above the sea floor and of limited extent across the summit.

Levee: an embankment bordering either one of both sides of a sea channel, or the low-gradient seaward part of a canyon or valley.

Limiting angle: angle of ray measured relative to a horizontal surface that leaves the source tangent to the horizontal plane.

Littoral zone: region between the shore and water depths of approximately 200 meters. (The usage and interpretation of this term can vary widely.)

Live simulation: simulation involving real people operating real systems.

Lloyd mirror effect: near-field interference patterns associated with a shallow acoustic source and receiver.

Marginal ice zone (MIZ): area of ice cover separating pack ice from the open ocean.

Mathematical models: category of models containing empirical models and numerical models.

Mesoscale features: dynamic features of the ocean with characteristic length scales on the order of 100 m to 100 km; examples include eddies and internal waves.

Microbaroms: class of atmospheric infrasonic waves generated in marine storms by nonlinear interactions of ocean surface waves with the atmosphere in the frequency range 0.1–1 Hz.

Mixed layer: surface layer of uniform temperature that is well mixed by wind or wave action or by thermohaline convection.

Moat: a depression located at the base of many seamounts or islands.

Mode capture: for downslope propagation in a wedge geometry, the modes that initially propagate in the sediment (continuous modes) in the shallower regions of the wedge waveguide are progressively converted (captured) into modes trapped in the water column (trapped or discrete modes) in the deeper regions of the wedge waveguide.

Modified data banks: data banks containing modified or extrapolated data derived from a primary data bank.

Monostatic: geometry in which the acoustic source and receiver are at the same position.

Mountains: a well-delineated subdivision of a large and complex feature, generally part of a cordillera.

Multipath: propagation conditions characterized by a combination of different types of paths.

Multipath expansion models: mathematical models of acoustic propagation based on a variation of the normal mode technique.

Noise models: mathematical models of noise levels and directionality in the ocean consisting of two categories: ambient noise and beam-noise statistics models.

Normal mode models: mathematical models of acoustic propagation based on the normal mode technique.

Normal modes: natural frequencies at which a medium vibrates in response to an excitation.

Nowcast: estimate of the present state of a system.

Numerical models: mathematical models based on the governing physics.

Ocean acidification: a process that occurs when CO_2 in the atmosphere reacts with sea water to create carbonic acid (H_2CO_3); this process may be accelerated by the emission of carbon into the atmosphere through the effects of fossil-fuel combustion.

Ocean impulse response function: time-variant and space-variant Green's function that describes the response of the ocean medium to a unit impulse.

Parabolic equation models: mathematical models of acoustic propagation based on a solution of the parabolic versus the elliptic-reduced wave equation.

Pathological test case: specification of inputs to a model that prove to be particularly troublesome; a test case that produces disorders in an otherwise well-functioning model.

Perfectly matched layer: technique for truncating unbounded domains in numerical simulations of wave propagation problems.

Physical models: theoretical or conceptual (physics-based) representations of the physical and acoustical processes occurring within the ocean.

Plain: a flat, gently sloping or nearly level region of the seafloor.

Plateau: a comparatively flat-topped elevation of the seafloor of considerable extent across the summit and usually rising more than 200 m above the sea floor.

Point scattering models: mathematical models of reverberation based on the assumption that the scatterers are randomly distributed throughout the ocean.

Primary data banks: data banks containing original or modestly processed environmental acoustic data.

Prognostic information: information used to forecast the future state of a system, particularly with reference to predicting sonar system performance.

Propagation: transmission of acoustic energy through the ocean medium.

Propagation models: mathematical models of underwater acoustic transmission loss consisting of five categories: ray theory, normal mode, multipath expansion, fast field, and parabolic equation models.

Province: a region composed of a group of similar bathymetric features whose characteristics are markedly in contrast with the surrounding areas.

Range: a series of generally parallel ridges or seamounts.

Rayleigh parameter: measure of the acoustic roughness of a surface.

Ray theory models: mathematical models of acoustic propagation based on ray-tracing techniques.

Reef: an offshore consolidation with a depth below the sea surface of less than about 20 meters.

Reflection: return of a portion of the incident energy in the forward direction after an encounter with a boundary.

Refraction: directional changes in acoustic propagation caused by density discontinuities.

Reliable acoustic path: ray paths formed when a source or receiver is located at the critical depth.

Reverberation: backscattered sound.

Reverberation models: mathematical models of reverberation based on boundary and volumetric scattering processes consisting of two categories: cell-scattering and point-scattering models.

Ridge: a long, narrow elevation of the seafloor with steep sides.

Saddle: a low part on a ridge, or between seamounts.

Salinity: a measure of the quantity of dissolved salts in sea water.

Scattering: random dispersal of sound after encounters with boundaries or with volumetric inhomogeneities.

Seachannel: a long, narrow, shallow depression of the seafloor, usually occurring on a gently sloping plain or fan, with either a U-shaped or V-shaped cross section.

Seamount: an elevation rising 1 kilometer or more above the seafloor, and of limited extent across the summit.

Seismic oceanography: a technique that employs low-frequency marine seismic reflection data to image thermohaline fine-structure in the oceanic water column.

Self noise: noise background originating from own ship or sonar structure.

Shadow zone: region of the range-depth plane into which little, if any, acoustic energy penetrates owing to the refractive properties of the water column.

Shoal: an offshore hazard to navigation with a depth below the sea surface of 20 meters or less, usually composed of unconsolidated material.

Signal processing models: mathematical models of signal detection processes.

Significant wave height: the average height of the one-third highest waves of a given wave group.

Sill: the low part of a ridge or rise separating ocean basins from one another of from the adjacent seafloor.

Simulation: dynamic execution of models to generate prognostic or diagnostic information.

Skip distance: distance traveled by a ray that just grazes the sea surface, the seafloor, or some intermediate layer in the ocean such as the sonic layer depth.

Smart systems: simulated people operating real systems.

Soliton: internal solitary waves in the ocean, often generated by the nonlinear deformation of internal tides.

Sonar performance models: mathematical models organized to solve specific sonar applications problems; contain environmental models, basic acoustic models, and signal processing models.

Sonic layer: acoustic equivalent of mixed layer; surface layer characterized by near isovelocity conditions.

Spherical spreading: form of geometrical spreading that is unbounded.

Spur: a subordinate elevation, ridge, or rise projecting from a larger feature.

Stochastic: state of being random; describable in probabilistic terms.

Surface duct: propagation channel bounded by the sea surface and the base of the sonic layer.

Surface loss: reflection loss at the sea surface.

Tablemount: a seamount having a comparatively smooth, flat top (also called a guyot).

Terrace: a bench-like feature bordering an undersea feature (also called a bench).

Thermocline: region of the water column characterized by a strong negative temperature gradient.

Thermohaline: of or relating to the effects of water temperature (thermo) and salinity (haline).

Transmission anomaly: difference between the observed transmission loss and the loss expected from spherical spreading alone.

Transparent resonances: modes for which coupling that occurred while propagating over (or through) the first half of a symmetric feature (on the bottom or in the water column) were reversed, or uncoupled, while propagating over (or through) the second half of the feature.

Trench: a long, narrow and deep depression of the seafloor, with relatively steep sides.

Trough: a long depression of the sea floor, normally wider and shallower than a trench.

Turning point: position in the range-depth plane at which an upgoing or a downgoing ray reverses direction.

Upwelling: upward-directed motion in the ocean.

Validation: the process of determining the degree to which a model is an accurate representation of the real world from the perspective of the intended uses of the model.

Valley: a relatively shallow, wide depression with gentle slopes, the bottom of which grades continually downward (as opposed to a canyon).

Verification: the process of determining that a model implementation accurately represents the developer's conceptual description and specifications.

Virtual mode: mode associated with acoustic propagation in the surface duct.

Virtual simulation: simulation involving real people operating simulated systems (human-in-the-loop).

Water mass: body of water characterized by a unique temperature-salinity relationship.

Wave equation: equation describing motion of acoustic waves in the ocean.

Whispering gallery: phenomenon relating to the propagation of sound along the curved walls of a gallery, especially those of a hemispherical dome.

Appendix C: Websites

The Internet has become a data-access "bus" for work in modeling and simulation (M&S). This appendix contains references to selected sites of practical use to sonar technologists and acoustical oceanographers. Addresses for these websites sometimes change or disappear entirely. In such cases, web searches using appropriate key words may help in locating new addresses or alternate sites. Many of these websites themselves contain helpful directories that refer readers to related websites.

OALIB http://oalib.hlsresearch.com/

The Ocean Acoustics Library provides access to some of the stand-alone propagation models reviewed in this book. This access is provided directly to downloadable software or indirectly by reference to other authoritative websites. The Ocean Acoustics Library contains acoustic modeling software and data. It is supported by the US Office of Naval Research (Ocean Acoustics Program) as a means of publishing software of general use to the international ocean acoustics community. Table C.1 summarizes the propagation models and other information currently available from the Ocean Acoustics Library. The contents of this site are subject to change.

- Also see the Acoustic Toolbox User Interface and Post processor (AcTUP V2.2L) developed by Alec Duncan at the Centre for Marine Science and Technology at Curtin University in Perth, Australia. The AcTUP, which runs under MATLAB, provides a consistent, menu-driven user interface to some of the underwater acoustic propagation codes found in OALIB http://cmst.curtin.edu.au/products/underwater/.

TABLE C.1
Contents of the Ocean Acoustics Library (OALIB) Website

Category	Contents
Rays	Acoustic Channel Simulator, BELLHOP, EIGENRAY, HARPO, HWT 3D, TRACEO3D, TV-APM, TRIMAIN, USML, VirTEX
Normal Modes	AW, AMODES, COUPLE, KRAKEN, MOATL, NLAYER, NMPQ, Numerov Mode Code, rimLG, WKBZ
Wavenumber Integration	OASES, RPRESS, SCOOTER, SPARC
Parabolic Equation	FOR3D, MMPE, NLBCPE, PDPE, PECan, RAM/RAMS, RAMsurf, MPIRAM, MATLAB RAM, UMPE
Other	Related modeling software and data sets to support oceanographic and acoustic analyses.

Note: Abbreviations and acronyms are defined in Appendix A.

DoDM&SCO https://www.msco.mil/

The Department of Defense Modeling and Simulation Coordination Office (DoDM&SCO) performs those key corporate-level coordination functions necessary to encourage cooperation, synergism, and cost-effectiveness among the M&S activities of the DOD Components. (The *DOD Modeling and Simulation [M&S] Glossary* [DOD 5000.59-M] prescribes a uniform M&S terminology).

NCEI https://www.ncei.noaa.gov/

NOAA's former three data centers have merged into the National Centers for Environmental Information (NCEI): the National Climatic Data Center (NCDC), the National Geophysical Data Center (NGDC), and the National Oceanographic Data Center (NODC), which includes the National Coastal Data Development Center (NCDDC).

- The NCDC archives surface marine data from around the world. Surface marine observations contain various meteorological elements, which, over time, can describe the nature of the climate of a location or region. These elements include temperature, dew point, relative humidity, precipitation, snowfall, snow depth on ground, wind speed, wind direction, cloudiness, visibility, atmospheric pressure, evaporation, soil temperatures, and various types of weather occurrences such as hail, fog, thunder, and glaze. NCDC receives and archives meteorological data from ships at sea as well as from buoys, both fixed and free floating. The weather observations are normally hourly (although they can be more frequent) and there are daily and monthly summaries. Many of the summaries are published on paper, electronically and via CD-ROM.
- The NGDC Marine Geology & Geophysics Division (and the collocated World Data Center for Marine Geology & Geophysics) compiles and maintains extensive databases in both coastal and open ocean areas. Key data types include bathymetry and gridded relief, trackline geophysics (gravity, magnetics, and seismic reflection), sediment thickness, data from ocean drilling and seafloor sediment and rock samples, digital coastlines, and data from the Great Lakes. The NGDC also operates the International Hydrographic Organization Data Center for Digital Bathymetry (IHO DCDB), and is an active participant in international ocean mapping projects sponsored by the IHO and the Intergovernmental Oceanographic Commission (IOC). The National Snow and Ice Data Center (NSIDC) (http://nsidc.org/) monitors sea ice concentrations and snow extent using near-real-time satellite imagery. The NSIDC is part of the University of Colorado Cooperative Institute for Research in Environmental Sciences, and is affiliated with the National Oceanic and Atmospheric Administration National Geophysical Data Center through a cooperative agreement.
- The NODC archives physical, chemical, and biological oceanographic data collected by US federal agencies (including the DOD—primarily the US

Navy), state, and local government agencies, universities and research institutions, and private industry. The NODC serves as a repository and dissemination facility for data collected by other organizations.

SEDRIS http://www.sedris.org

Synthetic Environment Data Representation and Interchange Specification (SEDRIS) is concerned with the representation of environmental data and the interchange of data sets for terrain, atmosphere, ocean, or space. It is an infrastructure technology that facilitates the expression, comprehension, sharing, and reuse of environmental data. SEDRIS comprises a data representation model, an environmental data coding specification, and a spatial reference model that enables users to articulate their own environmental data clearly and also to comprehend other users' data unambiguously.

World Ocean Simulation System (WOSS) http://telecom.dei.unipd.it/ns/woss/doxygen/

This document provides a short technical description of the World Ocean Simulation System (WOSS) library and of its integration into Multi InteRfAce Cross Layer Extension (NS-Miracle). WOSS is a multi-threaded framework that permits the integration of any existing underwater channel simulator that expects environmental data as input and provides as output a channel realization represented using the channel profiles just mentioned. The WOSS library comes in two forms:

1. A simplified version uses empirical equations for the calculation of delay, attenuation, and noise power.
2. A more powerful configuration models channel-power delay or frequency-attenuation profiles using more accurate representations of propagation phenomena. Currently, WOSS integrates the BELLHOP ray-tracing program.

Appendix D: Problem Sets

This appendix contains both *discussion questions* and *analytical questions*, organized by chapter.

- *Discussion questions* are intended to encourage students to interpret the technical content and thus assess their comprehension.
- *Analytical questions* are intended to challenge mastery of the technical content by solving basic quantitative problems. Note that some of these problems may require access to computing devices.

In advanced classes, students could be further challenged by assigning outside research projects that require access to the current scientific and technical literature contained in academic journals. Such assignments could entail individual study of an assigned journal article followed by a presentation to the assembled class for group discussion.

Leighton (2012) noted that the existence of extraterrestrial oceans offers opportunities to pose examination questions for which students in underwater acoustics do not already know the answers. Specifically, the limited set of scenarios in Earth's oceans that can be presented to students as tractable examination questions means that students often rely on knowledge from previous examples in formulating their solutions. A wide range of acoustical phenomenon in gas, liquid, and solid materials can be tested without a priori knowledge of the answer for extraterrestrial applications. The current lack of data on other worlds allows scenarios to be sufficiently simplified to make them tractable for the student, with appropriate wording to cover any future mismatch between prevailing conditions and those assumed by the examiner.

D.1 INTRODUCTION

a. Characterize the challenges associated with naval, offshore, and research applications of underwater sound. Make note of common as well as unique challenges in these three areas of application.
b. What is the difference in dB of between a reference sound pressure of 20 μPa versus 1 μPa?
c. Can simulation replace the need for direct observations and measurements? Provide examples to defend your position.
d. Discuss the utility of inverse methods versus direct-sensing methods. Are there classes of ocean-acoustic sensing that might be better suited to one method over the other?

D.2 ACOUSTICAL OCEANOGRAPHY

a. Use Equation 2.2 to compute the speed of sound in sea water at a depth of 1000 m, a temperature of 6°C and a salinity of 35 ‰. Repeat your computations using Equation 2.3 and compare the results.
b. Estimate $\Delta c/\Delta D$ by evaluating the derivative $\partial c/\partial D$ using the expression for c found in Equation 2.2, where c is sound speed (m s^{-1}) and D is depth (m). Assuming isothermal and isohaline seawater, what does this tell us about sound speed changes due to depth (or pressure) variations?
c. Estimate the first-order dependencies of c on T and S by evaluating $\partial c/\partial T$ and $\partial c/\partial S$, respectively, using Equation 2.2.
d. How are ox-bow lakes and Gulf Stream rings related?
e. Distinguish *sea* from *swell.*
f. Distinguish between *water mass* and *water type.*
g. What are the colligative properties of sea water? Why are they important in underwater acoustics?
h. Can salinity be considered to be a conservative tracer of water movements in the ocean? Provide examples to defend your position.

D.3 PROPAGATION I

a. For a fixed wave height ($2a$), show how R varies with frequency using Equation 3.6. How sensitive is R to the wave height type used (i.e., rms, one-third significant, average)?
b. Program a crude ray-tracing routine based on Snell's law. Use this program to generate ray traces for each of the channels and ducts discussed in the text.
c. Compute the Rayleigh reflection coefficient (RRC) from a homogeneous fluid half-space assuming (1) silty-clay bottoms and (2) sand bottoms using Equation D.1. In Figure D.1, subscript 1 refers to the water column while subscript 2 refers to the bottom (refer to Frisk, 1994 for additional details). Also compute the angle of intromission and critical angle, as appropriate.

$$\mathrm{RRC} = \frac{m\cos\phi_1 - n\cos\phi_2}{m\cos\phi_1 + n\cos\phi_2} = \frac{m\cos\phi_1 - \sqrt{n^2 - \sin^2\phi_1}}{m\cos\phi_1 + \sqrt{n^2 - \sin^2\phi_1}} \tag{D.1}$$

where

n = index of refraction = c_1/c_2
$m = \rho_2/\rho_1$
ϕ = incidence angle (not grazing angle)

1. For silty-clay bottoms:
$m > n > 1$
$\rho_2 > \rho_1$
$c_2 < c_1$

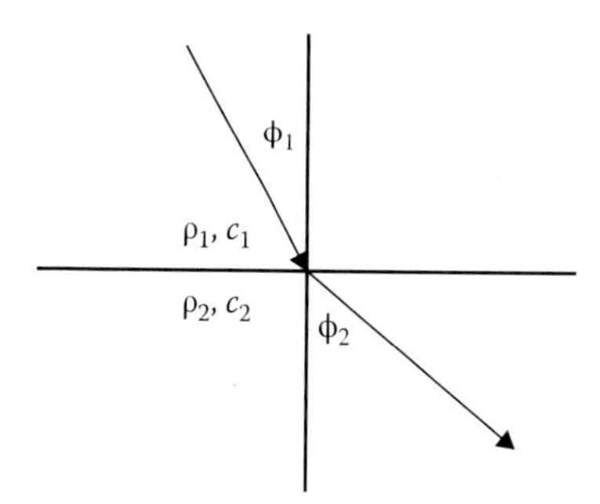

FIGURE D.1 Geometry for problem D.3c

Angle of intromission (ϕ_B) as defined in Equation D.2:

$$\phi_B = \sin^{-1}\sqrt{\frac{m^2 - n^2}{m^2 - 1}} \tag{D.2}$$

2. For sand bottoms: $n < 1$, $m > 1$
   $\rho_2 > \rho_1$
   $c_2 > c_1$
   Critical angle (ϕ_C) as defined in Equation D.3:

$$\phi_C = \sin^{-1} n \tag{D.3}$$

d. Calculate the effects of Lloyd's mirror using Equation D.4:

$$\text{TL} = -10\ \log_{10} I_{\text{LM}} = -10\ \log_{10}\left[4\sin^2\left(\frac{kz_s z_r}{r}\right)\right] + 20\log_{10}(r) \tag{D.4}$$

where

k = wavenumber = $\omega/c = 2\pi f/c = 2\pi/\lambda$
c = sound speed
r = range
z_s = source depth
z_r = receiver depth
f = acoustic frequency
λ = wavelength
$\text{TL}_{\text{LM}} = -10\ \log_{10}\ \text{LM (dB)}$
$\text{LM} = \dfrac{4}{r^2}\sin^2\left(\dfrac{kz_s z_r}{r}\right)$

Assuming a calm sea surface, plot TL_{LM} for selected values of f, z_s, and z_r versus r (refer to Figure 3.8 for the problem geometry). Calculate ranges r_1 and r_2 in Figure 3.9.

e. Is it possible for a Lloyd's mirror interference pattern to be encountered in the vicinity of the seafloor? Explain the necessary boundary conditions for such an interference pattern to occur at the seafloor.

f. How would Thorp's (1967) formula for absorption be modified to handle low frequencies? Is there a minimum value for absorption as the acoustic frequency decreases?

g. Use Figure 3.22 (or Figure 3.23) to compare the transmission losses associated with different source-receiver geometries relative to the mixed layer: (1) in-layer geometry, (2) cross-layer geometry, and (3) below-layer geometry. Discuss the implications for source-receiver geometries that are in-layer, below-layer, or cross-layer. If a submarine were trying to avoid sonar detection by a surface ship, what is the best position in depth for the submarine? If the surface ship wanted to improve detection against submarines at unknown depths, what type of sonar arrangement would be best? (You don't have to be restricted by monostatic sonar geometries.)

D.4 PROPAGATION II (PART 1)

a. Derive the wave Equation 4.1 from first principles. You will first need to postulate the equations of state, motion, and continuity.

b. Under what conditions would it be appropriate to use a range-independent propagation model? Consider the effects of water depth, acoustic frequency, and prediction range in your response.

c. By imposing a harmonic solution on the wave equation, how has its application been limited? For what sonar applications would the harmonic solution be best suited? Where would it not be appropriate?

d. Construct a table summarizing the assumptions necessary to arrive at the five basic modeling approaches discussed in the text. Also, summarize the resulting restrictions on applications. Is it possible for one propagation model to satisfy all possible combinations of acoustic frequency, water depth, and prediction range? Explain your answer.

e. Compute the beam displacements (Δ) for Figure 4.9 assuming fixed values for c_1, c_2, ρ_1, ρ_2, and θ. Use ω as the free variable. Plot the variation of Δ versus ω.

f. Discuss methods by which to account for uncertainty in transmission loss predictions generated by deterministic propagation models. What are the likely sources of uncertainty in predicting transmission loss? How could you construct a meaningful envelope around the mean transmission loss to account for possible uncertainties?

g. Plot Equation 4.8 against the cut-off frequency data presented in Figure 3.24. How do these two formulas agree near the commonly assumed 200-m maximum limit for shallow water?

D.5 PROPAGATION II (PART 2)

a. Plot the canonical sound speed profile according to Equation 5.14 using the values cited in the adjacent paragraph.
b. Evaluate the derivative $\partial c/\partial z$ using the expression for c found in Equation 5.14 where c is sound speed (m s^{-1}) and z is depth (m). How does this compare with the value of $\Delta c/\Delta D$ from Equation 2.2?
c. Modify the profile obtained in part (a) for Earth curvature corrections using Equations 5.15 and 5.16. Compare this modified profile with the original one. Comment on your results.

D.6 SPECIAL APPLICATIONS AND INVERSE TECHNIQUES

a. Outline the steps involved in the scientific and logistical planning of an inverse-acoustic field experiment to measure biomass on the continental shelf using only fixed or moored assets. As part of your plan, you must also account for protection of marine mammals.
b. Expand on the discussion of question D.6a by including an underwater acoustic network (UAN) as part of an experimental design that includes autonomous underwater vehicles (AUVs). Comment on how the AUVs could be used to improve your experimental design.
c. Reconsider question D.4f in view of the discussion of stochastic modeling in Section 6.2.

D.7 NOISE I

a. Why are noises of nearby ships and marine organisms, or of passing rain showers, not considered as ambient noise? How would such noises be factored into an assessment of overall noise levels at a given place and time?
b. For what types of sound-speed profiles would a noise notch not exist? Provide a graphical representation to illustrate your answer.
c. Discuss possible mechanisms for propagating surf noise into the deeper ocean. Consider the effects of waves.
d. Explain possible causes for seasonal variations in the levels of bioacoustic noise.
e. Use Figure 7.2 and Equation 7.1 to estimate the ambient noise level in the open ocean for sea state 3 and heavy shipping at a frequency of 200 Hz.

D.8 NOISE II

a. Discuss how the shipping routes in Figure 8.8 might be altered during the hurricane season in the North Atlantic Ocean.
b. Use Equations 8.5 and 8.6 to calculate the omnidirectional noise level (N) in Equation 8.7a. Sketch the coordinate system used in this system of equations.
c. Define steradian.

D.9 REVERBERATION I

a. Compare the Chapman-Harris formula for surface scattering with the critical sea test (CST) formula. Plot the results and discuss your findings. What are the implications for sonar performance assessments?
b. In Equation 9.6, what does the term μ represent physically?
c. What is meant by "the traveling acoustician problem" discussed in Section 9.4?

D.10 REVERBERATION II

a. Develop a mathematical proof to show that the extension in range of the scattering volume is $c\tau/2$ and not $c\tau$.
b. In Figure 10.10, discuss what types of differences between measured and modeled results might be due to significant scatterers present in the under-ice surface but not present in either the measured or modeled data.
c. How might under-ice scattering change under the influence of global warming? Speculate on possible impacts on sonar performance in Arctic waters.

D.11 SONAR PERFORMANCE MODELS

a. Describe how Equation 11.16 behaves as a function of frequency. Specifically, examine the pressure in the shadow zone as frequency decreases. Compare these results with those presented in Figure 3.24.
b. In Table 11.2, the figure of merit (FOM) for active sonars must be restricted to noise-limited (versus reverberation-limited) conditions. Explain why the FOM is not a valid metric for reverberation-limited sonar conditions.
c. Construct a simplified version of Figure 11.12 appropriate for a hull-mounted echo sounder intended for deep-water applications. How would this diagram aid in the design, fabrication, and testing of the echo sounder?
d. Explain why the Doppler shift for a passive sonar is half that for an active sonar (refer to Equation 11.25).

D.12 MODEL EVALUATION

a. Under what conditions would the methodology suggested in Figure 12.3 be appropriate for comparing and evaluating a model against field data? Would this methodology be appropriate for comparing two different models?
b. In order to grade the performance of a new sonar system, the program manager has decided that the results of the sonar acceptance tests will be compared against predictions generated by a sonar performance model that was initialized for the specified test location and sonar system parameters. Discuss the possible merits or complications arising from this decision if an important buy/no-buy decision will be based on these results.
c. You suspect that your model output might be corrupted by a pathological computation, but no field measurements are available for comparison. How could you confirm your suspicion?

D.13 SIMULATION

a. Provide real-life examples for each of the four elements in Table 13.1 and include brief descriptions.
b. Discuss the importance of visualization in presenting the outputs of a simulation.
c. Reconsider question D.12b in light of simulation testbeds. How might the program manager benefit from this approach?

Author Index

Page numbers appearing in *italic* indicate pages on which references are given in full.

C

D

E

F

G

H

M

N

O

Q

R

S

T

U

V

W

X

Y

Z

Subject Index

A

B

C

E

F

G

K

L

M

S

T

X